Electronic Structure and Reactivity of Metal Surfaces

NATO ADVANCED STUDY INSTITUTES SERIES

A series of edited volumes comprising multifaceted studies of contemporary scientific issues by some of the best scientific minds in the world, assembled in cooperation with NATO Scientific Affairs Division.

Series B: Physics

RECENT VOLUMES IN THIS SERIES

Volume 7 – Low-Dimensional Cooperative Phenomena
edited by H.J. Keller

Volume 8 – Optical Properties of Ions in Solids
edited by Baldassare Di Bartolo

Volume 9 – Electronic Structure of Polymers and Molecular Crystals
edited by Jean-Marie André and János Ladik

Volume 10 – Progress in Electro-Optics
edited by Ezio Camatini

Volume 11 – Fluctuations, Instabilities, and Phase Transitions
edited by Tormod Riste

Volume 12 – Spectroscopy of the Excited State
edited by Baldassare Di Bartolo

Volume 13 – Weak and Electromagnetic Interactions at High Energies (Parts A and B)
edited by Maurice Lévy, Jean-Louis Basdevant, David Speiser, and Raymond Gastmans

Volume 14 – Physics of Nonmetallic Thin Films
edited by C.H.S. Dupuy and A. Cachard

Volume 15 – Nuclear and Particle Physics at Intermediate Energies
edited by J. B. Warren

Volume 16 – Electronic Structure and Reactivity of Metal Surfaces
edited by E. G. Derouane and A. A. Lucas

Volume 17 – Linear and Nonlinear Electron Transport in Solids
edited by J.T. Devreese and V. van Doren

The series is published by an international board of publishers in conjunction with NATO Scientific Affairs Division

A	**Life Sciences**	Plenum Publishing Corporation
B	**Physics**	New York and London
C	**Mathematical and Physical Sciences**	D. Reidel Publishing Company Dordrecht and Boston
D	**Behavioral and Social Sciences**	Sijthoff International Publishing Company Leiden
E	**Applied Sciences**	Noordhoff International Publishing Leiden

Electronic Structure and Reactivity of Metal Surfaces

Edited by
E. G. Derouane and A. A. Lucas
Facultés Universitaires N.-D. de la Paix
Namur, Belgium

PLENUM PRESS • NEW YORK AND LONDON
Published in cooperation with NATO Scientific Affairs Division

Library of Congress Cataloging in Publication Data

Nato Advanced Study Institute on Electronic Structure and Structure and Reactivity of Metal Surfaces, Facultés universitaires Notre-Dame de la Paix, 1975.
Electronic structure and reactivity of metal surfaces.

(NATO advanced study institutes series: Series B, Physics; v. 16)
Includes index.
1. Surface chemistry–Congresses. 2. Metallic surfaces–Congresses. 3. Chemisorption–Congresses. 4. Catalysis–Congresses. 5. Reactivity (Chemistry)–Congresses. I. Derouane, E. G. II. Lucas, A. A. III. Title. IV. Series.
QD506.AlN37 1975 546'.3 76-10692
ISBN 0-306-35716-X

Lectures presented at the NATO Advanced Study Institute on Electronic Structure and Reactivity of Metal Surfaces held at The Facultés Universitaires Notre-Dame de la Paix, Namur, Belgium, August 31-September 13, 1975

A Division of Plenum Publishing Corporation
227 West 17th Street, New York, N.Y. 10011

United Kingdom edition published by Plenum Press, London
A Division of Plenum Publishing Company, Ltd.
Davis House (4th Floor), 8 Scrubs Lane, Harlesden, London, NW10 6SE, England

Printed in the United States of America

Foreword

What's new since Langmuir ?

Imagine that a young physicist would approach a granting agency and propose to contribute to heterogeneous catalysis by studying the heat conductivity of gases in contact with a hot filament. How would he be received now ? How would he have been treated sixty years ago ?

Yet, more than sixty years ago, Irving Langmuir, through his study of heat transfer from a tungsten filament, uncovered most of the fundamental ideas which are used to-day by the scientific community in pure and applied heterogeneous catalysis. Through his work with what were for the first time "clean" metal surfaces, Langmuir formulated during a period of a little over ten years until the early thirties, the concepts of chemisorption, monolayer, adsorption sites, adsorption isotherm, sticking probability, catalytic mechanisms by way of the interaction between chemisorbed species, behavior of non-uniform surfaces and repulsion between adsorbed dipoles.

It is fair to say that many of these ideas constituting the first revolution in surface chemistry have since been refined through thousands of investigations. Countless papers have been published on the subject of the Langmuir adsorption isotherm, the Langmuir catalytic kinetics and the Langmuir site-exclusion adsorption kinetics. The refinements have been significant. The original concepts in their primitive or amended form are used everyday by catalytic chemists and chemical engineers all over the world in their treatment of experimental data, design of reactors or invention of new processes.

On the other hand, during the past ten years, a second revolution has taken place in surface chemistry, especially that concerned with metallic surfaces. Technological advances have taken place in ultra-high vacuum technique, physical instrumentation affecting old methods such as low energy electron diffraction. New ways to look at photoelectron spectroscopy have evolved. Progress in solid state theory and computing has been made. As a result, new concepts have emerged or are emerging which are essentially non-Langmuirian. Their effect on the science of chemical reactivity of surfaces will be profound. The new surface chemistry or physics or chemical physics is particularly vital in the countries disposing of an aggressive industry which equips the scientific worker with new instruments and ever faster computers.

It seems most timely to survey such advances in a didactic manner during an international meeting grouping the countries where this new revolution has taken place and in a leisurely way through the physics and chemistry, the theory and practice of metallic surfaces. The limitation to metals, wisely decided by the organizers of the NATO Institute whose proccedings follow, is not fundamental. It is just a reflection of reality as much of the new work deals at the moment with metallic surfaces.

What are the new concepts ? The list depends very much on what part of the surface science he is considering. I am suggesting here a few which are of considerable importance in heterogeneous catalysis where most of the current important applications of surface science can be found. First, from the large number of ordered structures on single crystal metal surfaces following chemisorption of atoms and molecules, the idea of island or patch chemisorption has emerged with attractive interactions between chemisorbed species. Second, the Langmuirian picture of site-exclusion kinetics in chemisorption has been modified significantly by the frequent mechanism of chemisorption through precursor states. Third, the idealized concept of adsorption sites has been relaxed in two important ways through surface reconstruction and through adsorption of compressed layers forming coincidence overlayer lattices. Fourth, the rigid band theory of alloys which dominated alloy catalysis in the fifties and sixties has been largely abandoned as a result of new findings in photoelectron spectroscopy. Fifth, many binding states of a given species exist even on a single crystal face.

The challenge to-day is that many of the simple ideas have been replaced by a wealth of new observations without consistent pattern or new guiding principles. If surface reconstruction does occur, when is it expected, both in the absence or in the presence of what impurities ? If the Pauling theory of the metallic bond with its convenient percentage d bond character is inadequate, what should we use instead to explain patterns of catalytic activity ? If attractive as well as repulsive forces between chemisorbed species are important, when do we guess that attraction will overwhelm repulsion ? If many binding states exist for a given species on a certain metal, which ones are important in a given process ?

But there is another challenge which faces the contemporay catalytic chemist as well as the theoretical and experimental physicist. Many catalytic metals and alloys are used in form of very small particles, called clusters if most of the atoms in them are exposed to the surface. What are the properties of these clusters ? Their phase diagram ? Their superparamagnetic by opposition to ferromagnetic beavior ? Their surface composition in the case of alloys ? Their interaction with the support or carrier which preserves their existence ? Their electronic structure insofar as it differs from that of larger aggregates ? Their anomalous and mobile atomic structures ?

It is my guess that the third revolution in surface chemistry will deal with these small particles. They may well be easier to tackle theoretically than large crystals but they are certainly elusive objects for the experimental scientist. Yet many of the emerging tools of surface physics can be used in their investigation e.g. Mössbauer Spectroscopy, Extended X-ray Absorption Fine Structure and Small Angle X-ray Scattering. I have summarized elsewhere some of the recent catalytic results involving such object, as illuminated by recent results in surface science (1,2,3). Readers of this book who look for applications of their brand of surface science to heterogeneous catalysis may find in these references some facts to bolster their justification, if they need one, for studying the electronic structure of metals.

M. Boudart

Stanford, November 1975

(1) Chemisorption During Catalytic Reaction on Metal Surfaces. J. Vac. Sci. Technol., Vol. 12, No. 1 (1975) (M. Boudart)

(2) "Concepts in Heteregeneous Catalysis", in "Interactions on Metal Surfaces", R. Gomer, Ed., Chapter 7, Springer Verlag, New York 1975 (M. Boudart)

(3) "Heterogeneous Catalysis", Chapter 7 in Volume 7 of Physical Chemistry : An Advanced Treatise (Eds.: H. Eyring, W. Jost, and D. Henderson), Academic Press 1975 (M. Boudart)

Preface

The lectures collected in this volume were presented at the NATO Advanced Study Institute on "Electronic Structure and Reactivity of Metal Surfaces" which was held at the Facultés Universitaires Notre-Dame de la Paix in Namur from August 31st to September 13th, 1975.

As indicated by its title, the initial purpose of the Institute was to provide a progressive and comprehensive course on fundamental problems in the physics and chemistry of metal surfaces with emphasis on their electronic structure, adsorptive, catalytic and reactivity behaviours.

The possibility for the attendence to familiarize itself with such basic properties of metal surfaces was made particularly timely by the all too persistent divorce between the languages used by surface physicists and surface chemists in this field and also by the highly relevant nature of the subject for our present day energy problems. It is hoped that the school and the present proceedings will help in bridging the two point of views by preparing some younger scientists to work and contribute in this most important field of surface physical chemistry.

The ordering of the lectures in this book closely follows the actual timetable of the Institute. The material is divided into four part.

First, a general introduction provides a critical overview of the field.

Second, the various theoretical physics and quantum chemistry approaches to the electronic structure of bare metal surfaces and chemisorption systems are presented.

The third part describes theoretical and experimental aspects of physical techniques for the investigation of the metal-gas interface.

The last part is devoted to more specific problems of metal surface reactivity and other catalytic properties. The present proceedings also list the short communications which were presented in addition to the main lecture programme.

All lecturers are to be complimented and thanked for the clarity of both their oral and written contributions.

We wish to express our deepest gratitude to the Scientific Affairs Division of NATO, the main sponsor of this Institute, and to the Facultés Universitaires de Namur and their Academic Authorities who gave us a generous financial help as well as all accomodation supports for the School.

We are particularly indebted to Prof. J.-M. André, our Scientific Secretary, who provided us with his invaluable experience in setting up the Study Institute.

A meeting of this size and length does not succeed without offering a lively social program to the participants and accompanying people. We are much obliged to Mr. G. Kelner of our public relation office who fulfilled this responsibility with inexhaustible imagination and we acknowledge the gracious help of Mrs Derouane, Mrs Lucas and Mrs J.-M. André for entertaining the ladies.

The secretarial burden fell on Miss P. Lonnoy who, throughout the Institute up to the final preparation of these proceedings, worked expertly and smilingly. We wish to thank her most heartedly.

Finally we gratefully acknowledge the further help to all the other people, hostesses, members of the Chemistry Department, students, etc. who took an active and usefull part in arranging many practical details during the Institute.

E.G. DEROUANE

A.A. LUCAS

Namur, November 1975

Contents

PART III - THEORETICAL AND EXPERIMENTAL ASPECTS OF PHYSICAL TECHNIQUES FOR THE INVESTIGATION OF THE METAL-GAS INTERFACE

PART IV - REACTIVITY AND CATALYTIC ACTIVITY OF METAL SURFACES

INTRODUCTION TO PHENOMENOLOGICAL MODELS AND ATOMISTIC CONCEPTS OF CLEAN AND CHEMISORBED SURFACES

Thor RHODIN
School of Applied and Engineering Physics
Cornell University
Ithaca, N.Y. 14853
and
David ADAMS
Xerox Research Lab.
Webster, N.Y. 14644

1. INTRODUCTION

There has been a steady acceleration during the last twenty-five years in efforts to observe and to study the properties unique to the physical and chemical boundary separating a metal from its environment. The study of the metal-vacuum interface itself, one of the most important subsets of this effort was greatly implemented by the early recognition by Taylor(1), Langmuir(2), Hinshlelwood(3), Laidler(4) and others(5) of the specific features of chemical processes on metals, by the development of ultrahigh vacuum techniques pioneered by Alpert(6) and others for controlling and measuring the gaseous environment and by the ability of metallurgists to prepare single metal and semiconductor crystals of highy purity. The current emphasis on the measurement and interpretation of surface variables in which atomistic and microscopic considerations are emphasized is a more recent development in contrast to the earlier work based mainly on indirect interpretations derived from kinetic studies. The availability of new experimental methods to measure atomistic and microscopic variables in a detailed and reliable manner has made possible a current and more recent effort to extend the principles of quantum chemistry and of Bloch wave mechanics to the description of the behavior of atoms and electrons at or near metallic surfaces. It is the large expansion in this investigatory effort with which this institute will principally be concerned. The corresponding efforts to expand the base of data relating to phenomenological models and atomistic concepts of metal surfaces is the main objective of this specific lecture.

It is the availability of detailed and reliable data which stimulates the subsequent development of theoretical interpretation and analysis. Hence, we are concerned here mainly with the experimental approach to the subject. Furthermore, to make the subject manageable and finite we will deal chiefly with reactions of simple gases on single component single crystal surfaces. We will limit consideration to a set of typical models and approaches chosen because of their atomistic importance and because they represent active and fruitful areas of inquiry. Neglect of some vital aspects of this area of surface research is unavoidable. Some arbitrary choices are inevitable.

The seven subject areas chosen for discussion are :

1) atomic composition of the surface
2) atomic geometry and surface crystallography
3) work function and surface charge transfer
4) surface overlayer coverage
5) energetics and kinetics of adsorption
6) collective interactions in adsorption
7) vibrational properties at surfaces and of adsorbed layers.

It should be noted that none of these topics can be treated in more than a cursory manner in an introductory lecture. The structural presentation of this material is based to a large degree on a monograph of the subject by D. Adams and the author. Reference is recommended to that review[7] for a detailed and systematic presentation of the subject and of the published literature upon which it is derived. The rapid expansion of work on metal surfaces since 1950 is also documented in a number of conference proceedings and texts[8-17].

2. ATOMIC COMPOSITION OF SURFACES AND SURFACE LAYERS[7,8,20-23,26-35]

The determination of surface composition on an atomic scale is a primary prerequisite for a well defined base of surface characterization. It is essential not only to begin with well defined atomic composition but to follow associated compositional changes ocurring from adsorption and reaction. In this sense the ideal objective is to combine with the composition, the spatial distribution of the surface atoms to the same limit of precision. Under ideal conditions these determinations are done *in situ* with negligible perturbation of the surface during the observation. These determinations can be very important in studying the dynamics of surface transport as well as segregation effects associated with thermodynamic redistribution between the surface and the bulk. The largest application in effort is in the preparation and monitoring of clean surfaces or of monitoring the surface coverage of an adsorbate under favorable circumstances. Clean surface preparation may include *in situ* treatments by thermal, chemical, ion bombardment, crystal cleavage or evaporated film deposition techniques[18]. With the introduction of Auger spectroscopy in 1967-68[19-22] this has become the dominant technique for this purpose. The use of this method to obtain qualitative analysis and order-of-magnitude estimates of the surface composition has greatly extended the range of cleaning procedures and materials for surface studies.

Actually, there are several so-called core-level excitation methods all of which can provide compositional information on the surface. Elemental identification is made by relating spectral peak positions to known energy transitions. Composition is associated with the intensity of a specific transition. Although electron spectroscopy for chemical analysis of solids (ESCA) using x-ray photoemission (XPS) was pioneered by Siegbahn and co-workers[23], it has never become widely used for characterization of metals under ultrahigh vacuum conditions. Electron beam excitation techniques such as Auger electron spectroscopy (AES)[20-22] and appearance potential (APS)[24,25] spectroscopy have been widely used in uhv surface studies. The theory and practice of AES[26-29], XPS[30-32] and APS[33-35] in metal surface studies have been extensively reviewed[26-35]. Their application to metal-gas interfaces will be discussed in detail later in this meeting. We consider here briefly their physical distinctions and particularly the features of AES, the most common technique of the three.

The electronic transitions involved in these three spectroscopies[7] are indicated schematically in Figure 1, where the structure of an element in the first transition metal series is indicated using the conventional x-ray level notation.

In all cases a core-level electron is excited by either a high energy electron or a photon. Subsequent deexcitation by recombination of the core hole with an electron from a higher lying energy level is accompanied by either an Auger electron (AES) or an x-ray photon (APS). In XPS the energy of the incident photon, $h\nu$ is spent removing a core electron of energy, E_1. The measured kinetic energy of the photoelectron is given by the photon energy minus the work function and the energy difference between the core level and the Fermi level. In APS the value of the incident electron energy for the excitation threshold of a core electron to a vacant site just above the Fermi level is followed by detection of an abrupt change in the x-ray fluorescent yield.The core binding energy in this case is given by the corresponding threshold energy plus the work function. In AES the measured kinetic energy of the emitted electron depends upon differences among energy levels and gives a spectra which becomes increasingly more complicated with atomic number of the metal. A more detailed analysis to be covered in later lectures of all three spectra must consider perturbation of the electronic structure

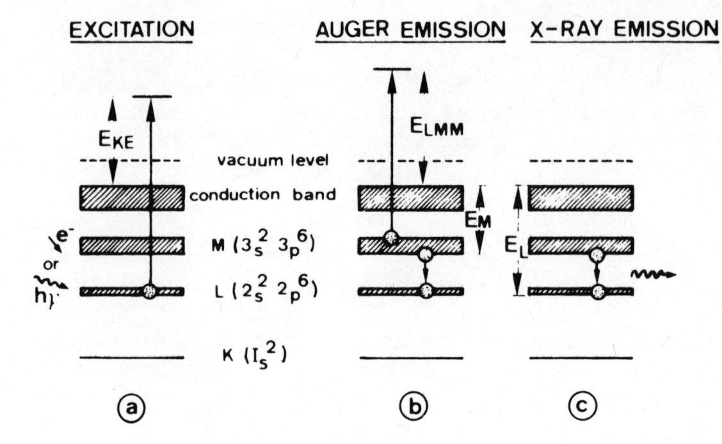

Figure 1. : Schematic of the electronic transitions involved in appearance potential spectroscopy, Auger electron spectroscopy and X-ray photoemission spectroscopy.
a) Excitation of a core level electron by incident electrons or X-ray photons
b) Core-hole deexcitation with Auger electron emission
c) Core-hole deexcitation with X-ray emission (For a more detailed discussion see Rhodin and Adams, ref.7).

of the atom caused by the ionization process and to shifts in the one-electron levels due to relaxation of the remaining close shell electrons in the potential field of the ion. The surface sensitivity of all three spectroscopic techniques is due to the large cross-sections for inelastic collision of electrons giving small free paths[27] which increase slowly with energy of about 50 eV from about 5 to 30 Å as indicated in Figure 2. It is the shallow escape

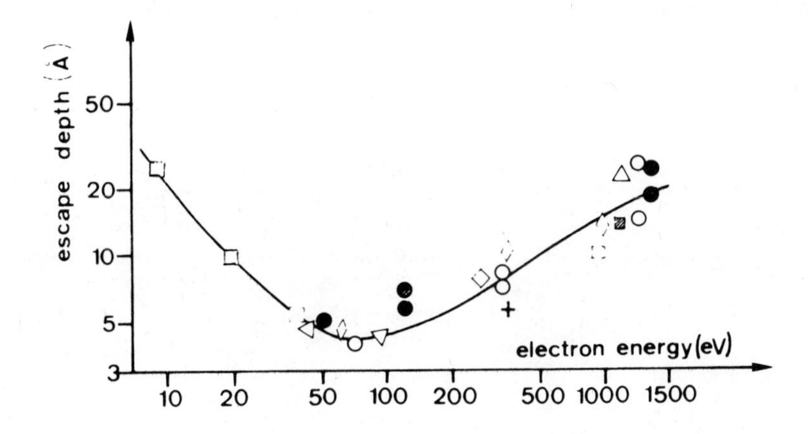

Figure 2. : Experimental measurements of electron escape depth as a function of electron energy (From Tracy, ref. 27)

depth for XPS and AES and the shallow penetration depth for APS which makes all three methods so surface-sensitive. It should be noted that the excitation processes need not be understood to make good use of the methods as semi-quantitative tools since they are generally calibrated on an empirical basis using standards.

Since AES is so widely used it is helpful to make some specific remarks on its usefulness to define the use of this parameter for surface composition. A typical Auger spectrum(28) obtained from real-time differentiation of the emitted intensity as well as the original intensity is illustrated in Figure 3. Despite ambiguities in comparison of experimental and calculated Auger spectral energies, qualitative analysis is widely achieved by comparison with spectra for known materials. Tabulated cross-sections of Auger intensities from which absolute concentrations can be calculated for quantitative applications without standards are not yet available. There are also other limitations of AES at present. One such problem originates from the additional production of Auger electrons due to incident excitation-electrons which are scattered back into the escape-depth region. Independent and accurate measurement of coverage is another problem. The response of the electron spectrometer and possible modifications of the spectral line shape must be accounted for. It is important to note that whereas radiation damage effects to the clean metal surface are rather unlikely, serious perturba-

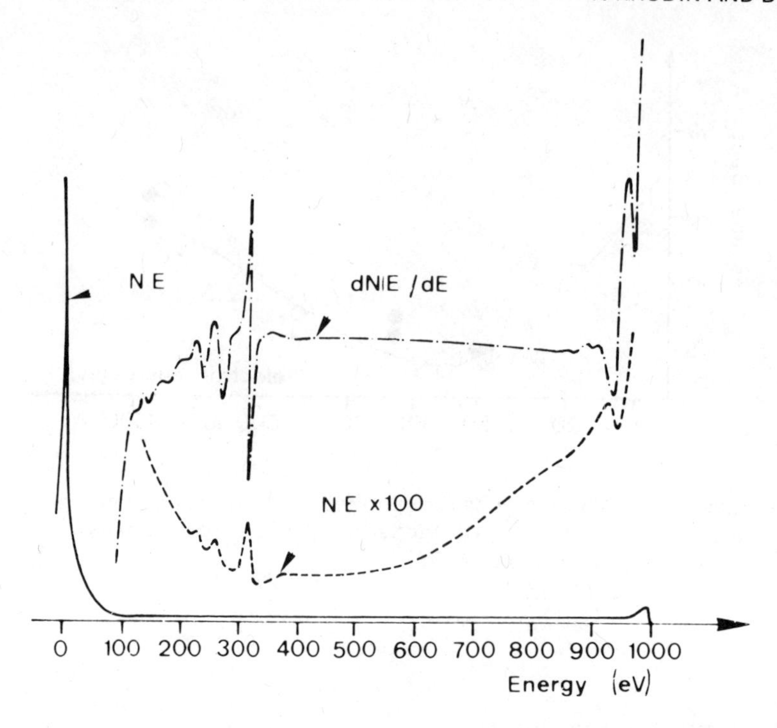

Figure 3. : Auger electron spectrum observed for clean Pt(100). See text for discussion (From Palmberg, ref. 28).

tions of adsorbed layers can be caused by the incident electron beam. In summary, whereas these and other factors must be provided for in special applications, the AES method nevertheless finds wide usage for monitoring the composition of metal surfaces.

3. ATOMIC GEOMETRY AND SURFACE CRYSTALLOGRAPHY[36-59]

The parameter of atomic structure for the clean metal substrate and that of adsorbed layers provides critical information on the bonding forces and configurations characteristic of the surface region. Two very important aspects of this approach are the use of low-energy electron-diffraction (LEED) to study the qualitative aspects of surface crystal structure through determination of the two-dimensional order and periodicity of adsorbed layers and the use of LEED intensity analysis to provide quantitative information on the coordination, lengths and orientation characteristic of chemical bonding at surfaces. We are all familiar with the classical experiment of Davisson and Germer[36] who measured the energy and angular distribution of backscattered electrons from a Ni(111) surface in the 1920's, thereby confirming the de Broglie hypothesis concerning the wave nature of electrons. The substantial expansion in the use of LEED for determining surface structure parameters since then was greatly stimulated by the pioneering research applications of L.H. Germer[37], H.E. Farnsworth[38] and others and by the commercial availability of post-acceleration LEED systems[39,43]. This major application of LEED to determine the parameter of surface structure is thoroughly covered in the review literature[41-52]. Details on the techniques and applications of LEED will also be covered here in later lectures. It is significant to point out its essential usefulness in terms of both atomic geometry and surface crystallography as an approach to the evaluation of models and concepts of surfaces.

The diffraction pattern results from the constructive interference between electrons scattered from large numbers of atoms in the plane of the surface and other sub-planes parallel to it within the escape distance of the electron (Figure 2).For many simple surface structures the effect of multiple scattering is to modulate the intensities of the diffracted beams but not the angular distribution, i.e. the diffraction pattern. In this case the diffraction pattern corresponds to a superposition of the reciprocal lattices of the nonequivalent layers. The great value of LEED pattern analysis is that the order of the clean metal surface as well as the adsorbent-adsorbate periodicities can be determined with little ambiguity for a great number of overlayer systems. There are many complications when the reciprocal lattice relationships are not simple but produce superlattice structures, when the surface is but partially ordered, when large differences in diffraction intensities exist between thesubstrate and the overlayer and when diffraction complications arise from the presence of point defects and line imperfections. Several types of disorder are found, for example, in adsorption systems. Their general characteristic is the occurrence of structurally equivalent domains which may scatter out-of-phase by virtue of their mutual orientation on the substrate. This occurs if application of a translational, rotational or reflection symmetry operation characteristic of the domain structure does not superpose the individual domains. Examples of translational and reflection domains degeneracy are shown in Figures 4 and 5 respectively.

A great number of 2-D simple surface lattices on metals and semiconductors containing domain structures and other disordered features have been observed and are discussed in detail[41,42]. A p(2x1) diffraction pattern formed by thermal annealing after CO-adsorption on W(210) to half a monolayer coverage is shown for example in Figure 6 together with a possible surface structure consistent with this pattern. A large amount of useful information is obtainable from LEED-pattern analysis of this kind. It is important

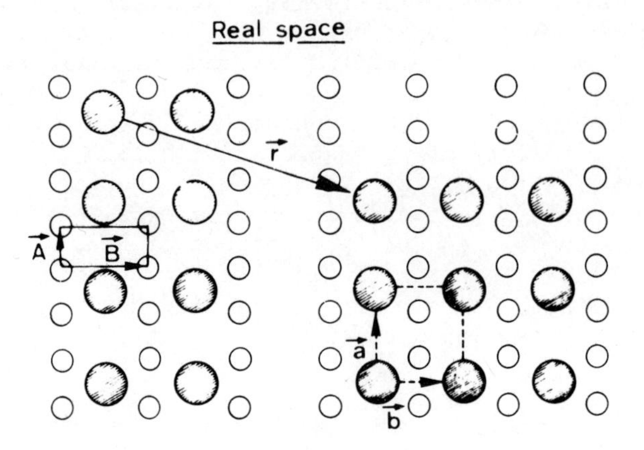

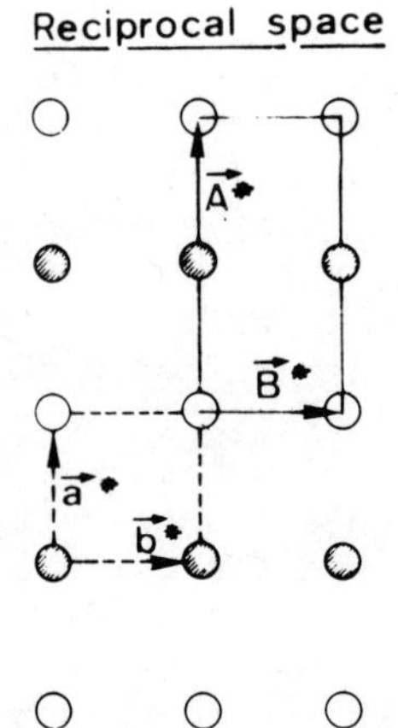

Figure 4 : Schematic of domain structures exhibiting translational degeneracy. The domains scatter in-phase if the translation vector r is formed by integral multiples of the domain unit mesh vectors a and b. Open circles in the real-space structure represent absorbent surface atoms and closed circles represent adsorbate atoms. The corresponding reciprocal lattice is also shown in the figure (From Rhodin and Adams, Ref. 7)

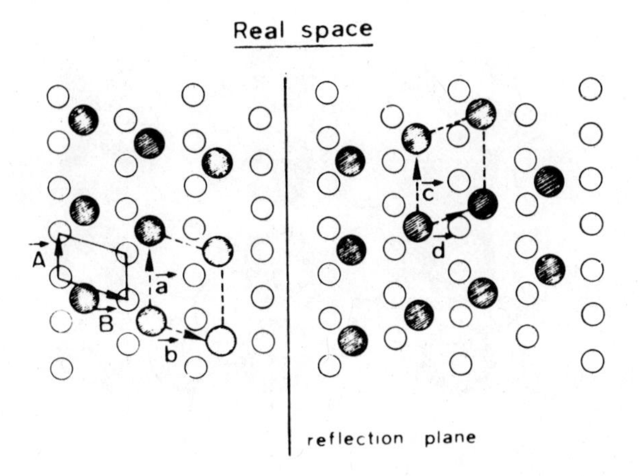

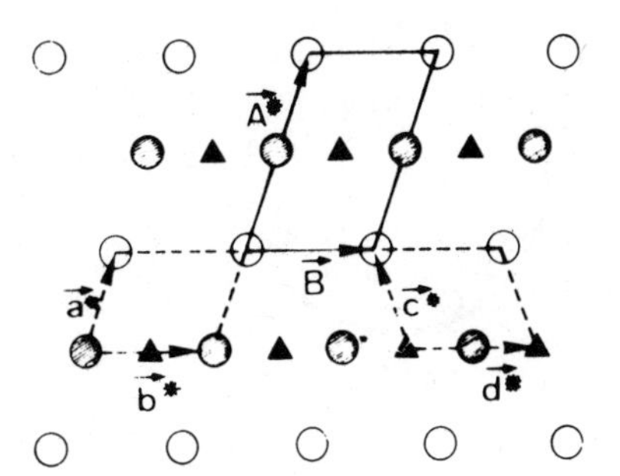

Figure 5. : Schematic of domain structure exhibiting reflection degeneracy. The corresponding reciprocal lattice is also shown in the figure. Open and closed circles represent adsorbent and adsorbate atoms respectively in the real-space structure. Superposition of the reciprocal lattices of the two domain orientations leads to sets of diffracted beams characteristic of each orientation. (From Rhodin and Adams, ref. 7)

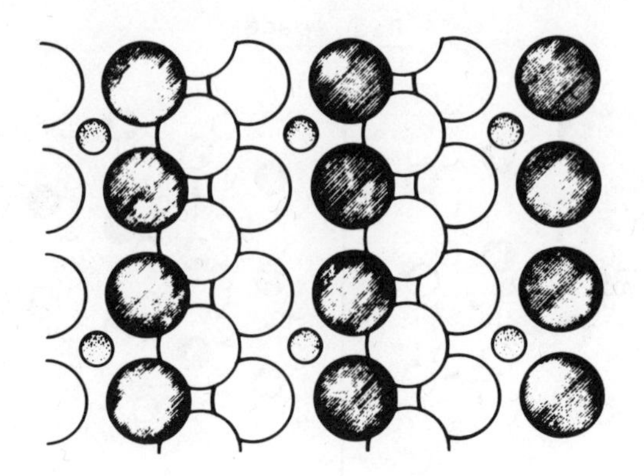

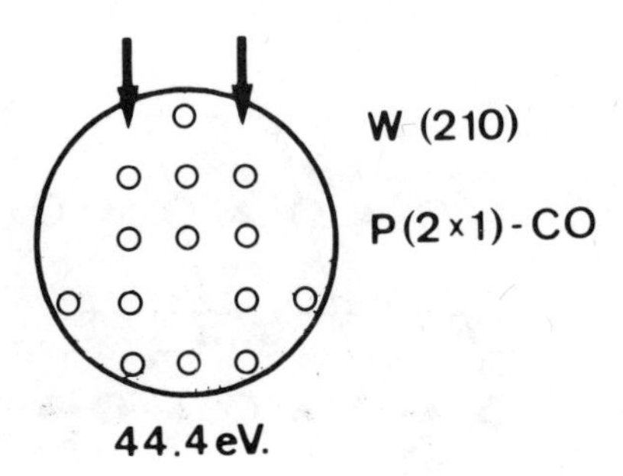

Figure 6. : LEED pattern obtained at half monolayer coverage of carbon monoxide on W(210) after thermal annealing to 1100°K. Vertical rows of new diffraction spots due to carbon monoxide adsorption are marked by arrows. The geometrical arrangement of an ideal W(210) surface is shown in the sketch. Topmost tungsten atoms are cross-hatched. The filled circles represent possible locations of CO molecules on this surface consistent with the diffraction pattern (From Rhodin and Adams, ref. 7)

however to emphasize that the inferences drawn from such studies on periodicity are rather reliable whereas those relating to location are inherently ambiguous.

A complete surface structure crystallographic determination requires an analysis of intensity spectra for a number of different incident beam angles and sample azimuthal orientations over a range of incident energies from about 30-200 eV. Under ideal circumstances only in those cases where comparison between theory and experiment show unique and good agreement in the position, shape and relative intensities of the spectra over the whole energy range can the bond distances and coordination assumed in the trial geometry be considered established. Many individuals and research groups have contributed significantly to establishing suitable models for describing the scattering of low-energy electrons from solid surfaces and to subsequent applications of these models to surface structure analysis[44-59]. The selective sensitivity of LEED intensity spectra to the geometric positions of atoms over surface dynamical quantities is responsible for the capability of this technique in determining the surface crystallography. This applies particularly to relatively simple overlayer systems[52]. Its successful extension to more complicated materials such as layered compounds and heavy transition metals is also likely to be achieved[52,53].

It should be noted that there are a variety of calculational routes to the evaluation of the intensity spectra[52]. The dynamical methods where all the pertinent multiple scattering contributions are summed in detail have been most widely applied and have so far proved the most successful. The "exact" approaches in this category include all orders of scattering whereas the convergent "iteration-perturbation" approaches sum up only a critical number of multiple scattering steps. The perturbation-approach of van Hove and Tong[53] has been fruitfully applied in selective cases as indicated in Table 1 for the c(2x2) and p(2x2) overlayer structure of the chalcogen adsorbates on Ni(001) where comparison is made with the calculations of Demuth, Marcus and Jepsen[54] using the "exact" KKR-layer method of Jepsen, Marcus and Jona[55].

A useful example of a simple crystallographic analysis in which both methods were more directly compared is that of the c(2x2) sodium overlayer on Al(001) by Hutchins et al[59]. This is an interesting system because it involves a relatively simple alkali metal atom sitting on a simple free-electron-like aluminium surface. Useful comparison can also be made to similar results for sodium adsorption on a transition metal (nickel) the crystallography of which has also been established independently by two different research groups[54,62]. Comparison between theory and experiment for the fractional order beam determined by the KKR-layer method[54,55] is shown in Figure 7. Similar good agreement was obtained using the convergent iteration-perturbation approach of van Hove and Tong[53].

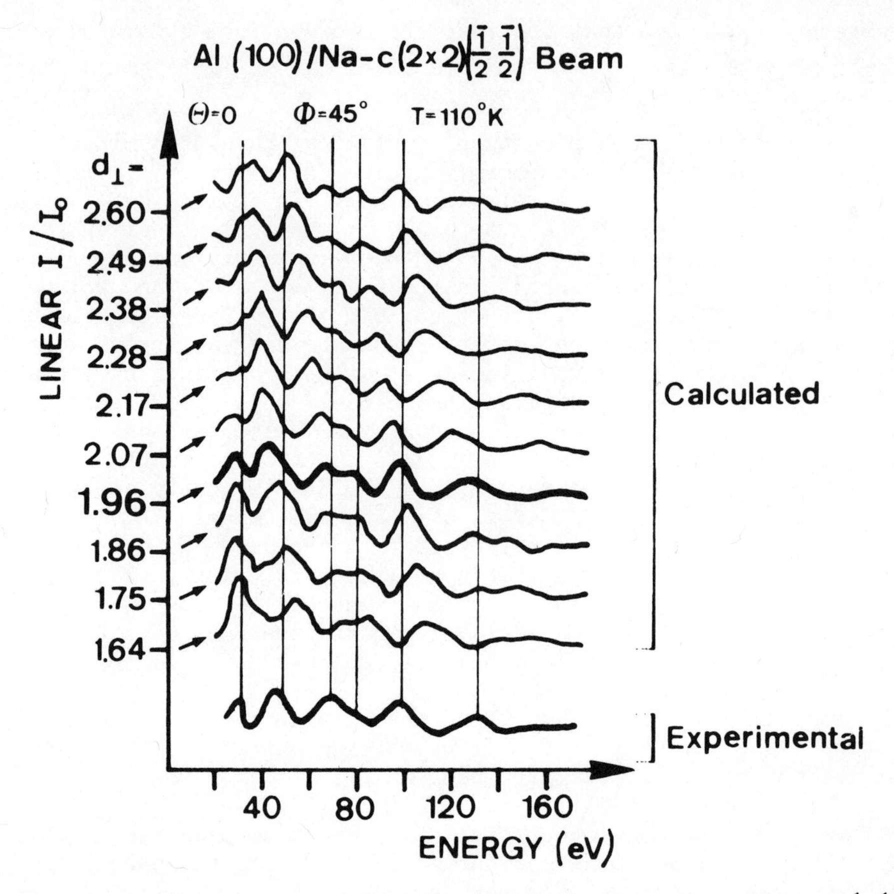

Figure 7 : Structure analysis of c(2x2). Na on (001)Al, the $(\frac{1}{2}\ \frac{1}{2})$ beam at $\theta = 0^o$. The interlayer spacing (d in Å) of best fit with experiment for this beam is determined to be 1.96 ± 0.1 Å. (From Hutchins, Rhodin and Demuth, ref. 59)

A comparison of the crystal structure deduced from analysis(56) of a large set of such data for c(2x2)Na(001)Al is compared to that obtained(54) for c(2x2)Na(001)Ni in Figure 8. It is possible to obtain reasonably good agreement for the determined bond lengths of both the Al/Na and the Ni/Na systems by using single bond covalent radii for both the metal and the overlayer.

In summary it should be noted that the dynamical methods referred to above require rather involved calculational analysis as well as the assumption of a trial geometry. It would be desireable to avoid both of these complications. Two data-reduction systems which promise to achieve this objective are the constant momentum transfer averaging scheme proposed by Webb, Gnoc and Lagally(57) and the Fourier-transform-deconvolution method being developed by Landman and Adams58. Neither have been asfully tested and developed as the corresponding dynamical approaches but both appear to be fruitful directions for further development. It can be concluded that when accurate and well-defined dynamical calculations of LEED are carried out and compared in depth to accurate experimental intensity spectra, they have the ability to extract reliable structural information for simple overlayer systems on a number of important metal-systems. It should be stressed that present results are still confined to relatively simple overlayer systems and to overlayer displacement distances of not less than about 0.5 A. It is possible that quantitative chemisorption bond-lengths and bonding sites will be achieved for more complex systems with the use of streamlined dynamical methods or alternatively, with the successful development of data-reduction methods. Thus, after 50 years since Davisson and Germer(36) first demonstrated the coherent scattering of electrons by a crystalline solid, the goal of studying surface structure with this technique is now becoming a reality.

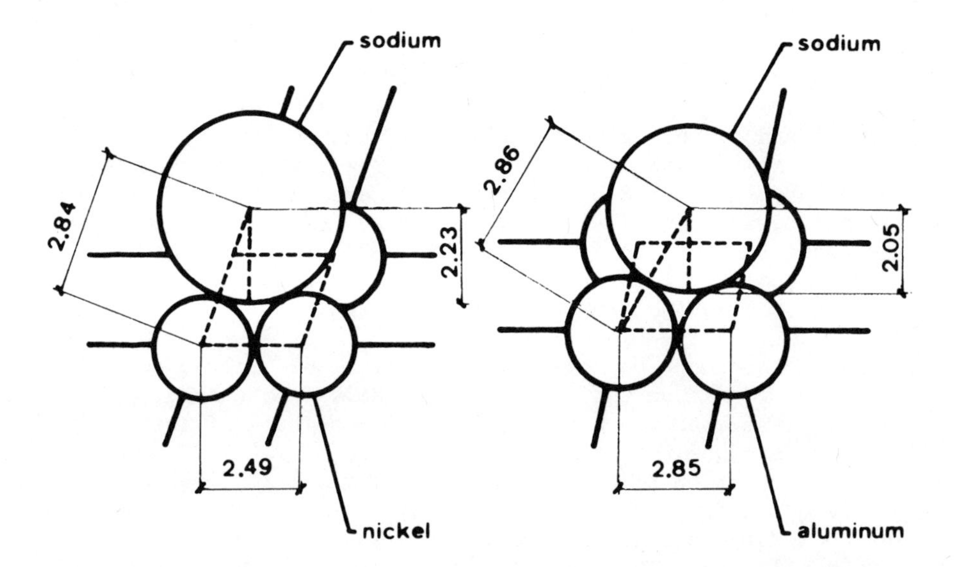

Figure 8. : Hard sphere model showing the local geometry and dimension for Na or Ni(001) (Left - from Demuth, Marcus and Jepsen, ref. 54) and Na on Al(001) (Right - from Hutchins, Rhodin and Demuth, ref. 59)

4. WORK FUNCTION AND SURFACE CHARGE TRANSFER[60-70]

The distribution of electrons both parallel and normal to the metal surface and the modification of this distribution by surface adsorption or reaction are fundamental to the electronic properties of the metal and are reflected, of course, in all the chemical bonding or physical polarization effects local to the surface. The charge transfer or electron redistribution which occurs at a local point with the magnitude of the physical displacement can be expressed in terms of the surface dipole. The collective effect of surface dipoles produces modifications in the magnitude of the work function which in turn can be conveniently and accurately measured. It is unfortunately not feasible to work backwards from experimental data on work function changes to the original distribution of surface charge. Nevertheless, measurement of the work function and the inferences derivable from its variation with the nature of the metal, its crystallography and the effect of overlayers are extremely useful in characterizing the parameters associated with dipole moment and charge transfer at the surface of a metal. In terms of thermodynamic quantities as discussed by Herring and Nichols[135], and by Lang and Kohn[60], the work function is the work done against the chemical potential of the electron in the bulk, μ and the difference, $\Delta\Phi$ of the electrostatic potential across the surface of the metal i.e.

$$e\phi = e\Delta\Phi - \mu \tag{1}$$

The importance of the surface contribution to the work function is shown empirically by the variation in ϕ with crystal plane of the same metal and by the large changes in ϕ that can result from adsorption. For example, the work function of the most closepacked (110) plane of tungsten is 5.3 eV, whereas the work function of the fourth most dense plane, the (310) face, is 4.3 eV[61]. Adsorption of cesium reduces the work function of W(001) 3.1 eV from 4.7 to 1.6 eV[13]. This dramatically illustrates the importance of adsorption and polarization at the surface in transfer of charge.

Distinction must be made between surface and volume contributions to the work function[63,64]. The chemical potential, μ is identified with the electron binding energy in the bulk of the crystal whereas the surface potential, $\Delta\Phi$ is associated with a dipole layer caused by separation of charge at the surface. Electrons from the surface spread out into the vacuum to smooth the potential discontinuity and to lower their kinetic energy and thus produces a negative outward dipole layer. A dipole layer of opposite sign is produced by the lateral flow of negative charge to smooth the variation in potential along the surface. This smoothening effect is greater for less close-packed plances thus accounting for their lower work functions. These effects have been discussed in detail by Smolushowski[65] and by Herring[66].

The effects of adsorption on work function are very interesting and even more complicated. Where a large fraction of electronic charge transfer occurs as for alkali metal adsorption on metal, the discrete dipole model discussed in detail by Schmidt and Gomer[67] is reasonable. For cases where surface binding is essentially covalent or metallic, a simple physical interpretation is rendered more difficult by the smaller (and largely unknown) bonding distances and by the more complex spatial redistribution of charge associated with electron sharing (covalency) compared to electron transfer (ionic bonding). It is clear that the discrete dipole picture breaks down if the adatom is located very close to or even below the image plane as pointed out by Boudart[68] for proton adsorption on metals. A further difficulty ensues for heteronuclear adsorbates like carbon monoxide where consideration of applications associated with the charge rearrangement upon adsorption must take into account those occurring in the admolecule as well. Nevertheless in general, work functions correlate reasonably well with differences in electronegativity between the adsorbate and adsorbent. They also often appear to vary continuously, although not in a simple fashion, with the coverage. An interesting exception to the generally observed simple relationship between work function and electronegativity is for nitrogen chemisorbed on single crystal tungsten surfaces where the dipole moment apparently varies in sign from one plane to another[69]. Although the basic mechanism is not understood, a simple empirical relationship between the work function change and the atomic configuration of the crystal substrate is valid. This effect is illustrated in Figure 9 from the work of Adams and Germer[70].

In conclusion, it is clear that relatively little is understood of the microscopic nature of charge transfer for any direct measurements on either the clean or chemisorbed metal surface. On the other hand, work function changes are a sensitive indirect indication of such changes and continue to provide valuable information on the effect of adsorption on surface charge effects. It is significant that surface electron charge has an important effect on the character of electron emission spectra (as observed in photo and field emission spectroscopy). Further development of theoretical understanding of these spectra may lead to additional microscopic information on the details of the surface potential.

5. SURFACE-OVERLAYER COVERAGE[71-81]

Surface monolayer coverage is probably one of the most fundamental and generally useful surface parameters. It is also the one associated with the greatest error of measurement especially for relatively small area well-defined single crystal surfaces. The most simple and direct way to determine the amount of gas adsorbed on a solid surface is to measure the change in mass of the sample. This can be done with reasonable sensitivity using vacuum microba-

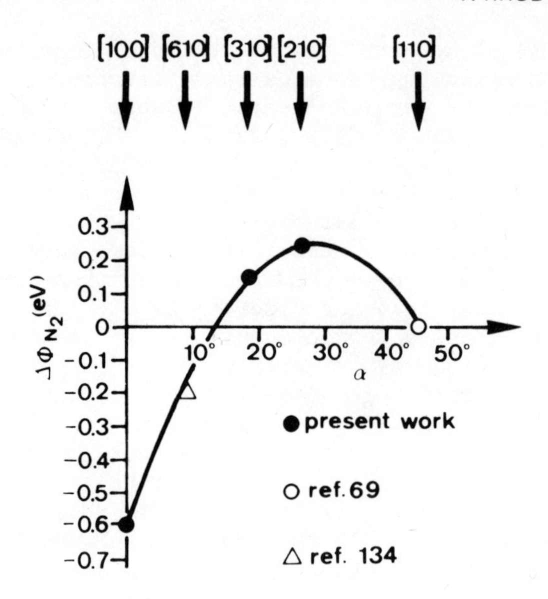

Figure 9. : Δϕ values for saturation nitrogen coverage on planes (h,k,0) of the (001) zone plotted against the angle α between (kk0) and (100). Values are all reported for ordered structures. (After Adams and Germer, ref. 70)

lances[71-73] or better, the quartz crystal oscillator[74]. Serious experimental operational limitations of these techniques preclude, however their wide application. The most widely used method for measurement of coverages and rate of adsorption and desorption is the thermal desorption technique in which, following adsorption and evacuation of the system, the adsorbate is desorbed by raising the temperature of the sample. A desorption spectra obtained by Tamm and Schmidt[75] after saturation adsorption of hydrogen on tungsten (111) is shown in Figure 10. Analysis of the desorption kinetics as well as the spectral line shape can give valuable information on the sticking coefficient and surface coverage. On the other hand errors do arise in analysis if quantitative information is sought due to various experimental approximations and inaccuracies. However, the widespread use of the thermal desorption thechnique, despite difficulties in achieving quantitative measurements reflects its primary application to the identification of adsorbate binding states and measurements of binding energies. These will be discussed in more detail in section 6 of this lecture on atomistic parameters associated with the energetics and kinetics of adsorption on metals.

There are many other measurements of surface physical properties that can be related to surface coverage. One of the most useful and yet rather underdeveloped is the radiotracer technique. Use of this technique has been very effectively developed by Oudar and co-workers[76] together with LEED to interpret atomistic mechanisms of sulfur chemisorbed on copper and other metals. Ultrahigh vacuum electron detectors for β-decay have been constructed and used in adsorption studies by Dillon and Farnsworth[77], by Crowell[78,79] and by Klier[80,81]. The very fruitful area for measurement of atomistic parameters by combining this approach with LEED and other forms of electron spectroscopy such as AES have hardly been adequately explored. Other measurements can also be effectively used to monitor surface coverage such as optical properties, LEED intensities and work function changes. In almost all cases however, non-linear relationship result between the physical measurement and the surface coverage. In addition, ambiguities resulting from uncorrected variations of other inter related atomistic parameters must be considered.

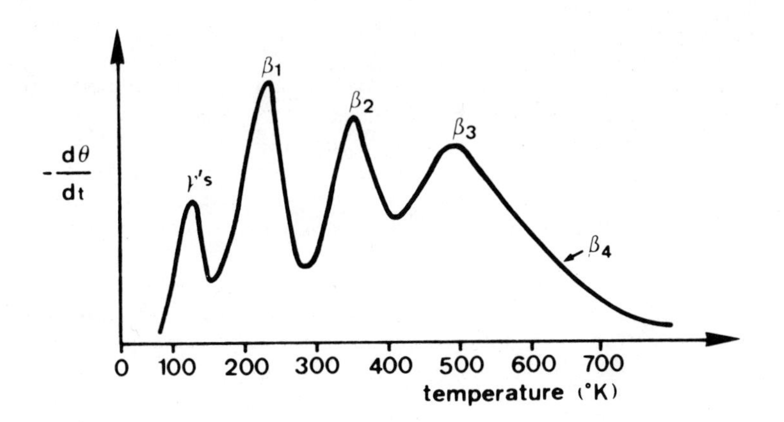

Figure 10 : Thermal desorption spectrum obtained after saturation adsorption of hydrogen on W(111). (From Tamm and Schmidt, ref. 75)

6. ENERGETICS AND KINETICS OF ADSORPTION[82-104]

The use of relatively homogeneous and uniform surfaces has led to an improved understanding of the influence of surface structure on the energetics and kinetics of adsorption. This has brought in turn the important question of the relative importance of induced versus intrinsic heterogeneity more sharply into focus. The most significant new observations concernthe occurrence of multiple adsorbate binding states on individual crystal planes of a metal adsorbent. These are most often investigated using thermal-[82-84] and electron-[85-86] induced desorption techniques and measurements of heat of adsorption versus coverage based on the Clausius-Clapeyron equation[87-89]. Correlation of these kinetic and thermodynamic studies with information concerning surface arrangement and electronic structure obtained from more recently developed spectroscopic techniques represents a major fraction of current work in the field. Here we are concerned mainly with measurements of adsorbate binding energies using thermal desorption spectra and equilibrium heats of adsorption.

In the common case of nonactivated adsorption,the heat of adsorption is given by the activation energy for desorption to a good approximation. The adsorbate binding energy is equal to the heat of adsorption for nondissociative adsorption and to the heat of adsorption plus the heat of dissociation in the case of dissociative adsorption as illustrated in Figure 11.

The thermal desorption technique is the most widely used method for estimation of activation energies for desorption. A plot of partial pressure of the desorbing species versus sample temperature is usually referred to as a thermal desorption spectrum. Under appropriate experimental conditions the increase in partial pressure is proportional to the rate of desorption. We should caution that determination of activation energies form desorption spectra requires a detailed model of the desorption mechanism. Lack of independent knowledge of the mechanism in general gives rise to ambiguities which limit the quantitative application of the method. This point has been emphasized recently in a useful review by Petermann[84].

The commonly used rate equation is,

$$-d\theta/dt = \nu f(\theta) \exp(-E/RT) \qquad (2)$$

where $f(\theta)$ is a function of surface coverage and the preexponential constant ν can be identified as

$$\nu = \tau(kT/h)(g^{*}/g) \qquad (3)$$

where g^{*} and g are the partition functions of the activated complex and adsorbed species respectively and ν is a function of both temperature and coverage. The incomplete description of the temperature and coverage-dependence of the rate of desorption given by Eqn.(2) when ν and the activation energy, E are taken to be independent of θ and T is a common source of uncertainty. We believe that it accounts in part for observations of correlation[90] between ν and E. This appears to be the cause of conclusions that either ν or E must be allowed to vary with coverage to fit experimental data[91-93].

Application of Eqn. (2) is frequently based on the analysis

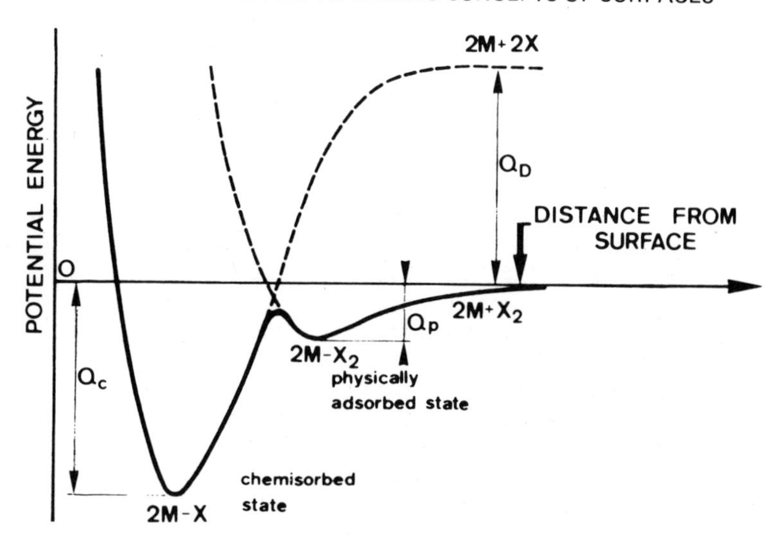

Figure 11. : Schematic of the variation in potential energy with distance from the surface associated with dissociative adsorption of a diatomic molecule X_2. Intersection of the potential energy curves for the molecule and atom below the energy zero results in non-activated adsorption. Q_P is the heat of physical adsorption of the molecule, Q_C is the heat of chemisorption, Q_D is the heat of dissociation of the free molecule. The binding energy per atom is $(Q_C + Q_D)/2$. (From Rhodin and Adams, ref. 7)

of desorption peak temperatures as given by Redhead[83]. For a linear heating rate, $T = T_o + \alpha t$ and assuming that ν and E are independent of θ and T, it can be readily shown that for first and second order desorption respectively

$$E_1/RT_p^2 = (\nu_1/\alpha)\exp(-E_1/RT_p) \quad (4)$$

and

$$E_2/RT_p^2 = (\nu_2\theta_o/\alpha)\exp(-E_2/RT_p) \quad (5)$$

T_p is the temperature of a peak in the desorption spectrum and θ_o in Eqn. (5) is the initial coverage. Equations (4) and (5) are commonly used in attempts to determine activation energies and preexponential factors and to distinguish between first and second order mechanisms from measurements of the dependence of peak temperature upon initial coverage and heating rate. Even in the fairly rare cases of single peak or well resolved multiple peak spectra a dependence of E or ν upon θ often occurs[91-93] with a corresponding uncertainty in the desorption mechanism.

The familiar analysis of multiple peak spectra in terms of desorption from independent binding states each associated with a characteristic activation energy, preexponential factor and desorption order has essentially been carried over from earlier studies on polycrystalline samples. A large number of recent studies on single-crystal adsorbents, however have removed the possibility that multiple desorption peaks observed after adsorption on polycrystalline samples could be solely explained by the existence of binding states of different energy on different crystal planes. A very important question of current interest is the extent to which multiple peak spectra observed on single crystal planes can be understood in terms of independent binding states. These can result from the site-heterogeneity that exists even on such surfaces, <u>or</u> from the coexistence of different adsorbate binding configurations, molecules and atoms, for example, <u>or</u> from a reflection at least in part of the activation energy and preexponential factor with coverage due to adsorbate-interactions.

For a number of systems[94-97] correlation between desorption spectra and LEED structural measurements have actually been observed. This is discussed in more detail in the next section. Prompted by these observations attempts[98-100] have been made to include the role of lateral interactions in desorption rate expressions. Some success in the interpretation of thermal desorption spectra has been achieved. It is evident however that reliable discrimination between the two concepts of independent binding states and of lateral interactions cannot be achieved solely on the basis of the current analysis of desorption spectra. This stems from the fact that agreement between experimental and calculated spectra is a necessary but not sufficient condition for the validity of a proposed desorption mechanism.

With the exception of some early work on oriented metal films (101), direct calorimetric measurements of heats of adsorption have not been carried out on single-crystal adsorbents. Equilibrium measurements of the isosteric heat of adsorption, q have been made however, using methods based on application of the Clausius-Clapeyron equation

$$(d(\ln P)/dT)_{\theta} = q/RT^2 \qquad (6)$$

Tracy and Palmberg[87-89,102] have studied the coverage dependence of the isosteric heat of adsorption of carbon monoxide on the (100) planes of Pd, Ni and Cu. At coverages less than half monolayer, adsorption isobars (temperature versus coverage at constant pressure) can be constructed using measurements of change in work function, $\Delta\phi$ as an indicator of surface coverage. Adsorption isosteres (p versus T at constant θ) can be derived from the set of adsorption isobars and the heat of adsorption determined for different θ using Eqn. (6). Although nonlinearity in $\Delta\phi$ versus θ may cause distortion of the θ-axis in plots of q versus θ obtained by this method it does not affect the measurement of q itself.

At coverages greater than half monolayer, the work function-based measurements can be supplemented by direct construction of adsorption isosteres using LEED observations of the two-dimensional surface structures formed by CO. Compression of the surface structure with increasing coverage above half monolayer causes monotonic changes in the position of diffracted beams characteristic of the adsorbate structure. The CO-coverage can be determined from the diffraction pattern based upon a reasonable model of the surface structure. Adsorption isosteres are obtained by increasing the sample temperature at constant CO pressure until the diffraction pattern reaches a predetermined state characteristic of a particular coverage. The variation in heat of adsorption with coverage for CO adsorption on Pd(100), Ni(100) and Cu(100) is shown in Fig.(12). Abrupt changes in heat of adsorption can be correlated with changes in surface structure observed in the LEED measurements and interpreted in terms of repulsive lateral interactions between the adsorbed molecules. (See the discussion which follows on collective interactions). Isosteric heats of adsorption for CO on Cu(100) have also been measured by Alexander and Pritchard[103,104].

Finally, in the interpretation of thermal desorption spectra, variation in heat of adsorption with coverage may be attributed alternatively to the effect of adsorbate-adsorbate interactions or to changes in the mechanism of adsorption. For the systems referred to above, correlation with LEED measurements indicates that the former explanation is probably applicable.

7. COLLECTIVE INTERACTIONS IN ADSORPTION[105-117]

The understanding of the collective properties of molecules on metal surfaces is of considerable importance from several aspects. First of all, the effect of collective surface oscillations of free electrons in a clean metal surface is basic to the surface electron gas of the metal itself. A second collective effect also peculiar to the clean metal surface is that of the components of the electron wave functions which are characteristic of the surface. These are indeed most sensitive to the electron nature and crystallography of the clean surface as well as to the influence of adsorbed atoms. Indirect overlap of the wave functions on two adsorbed atoms via the conduction electrons at the surface have been discussed from a quantum mechanical viewpoint by Koutecky[105], by Grim-

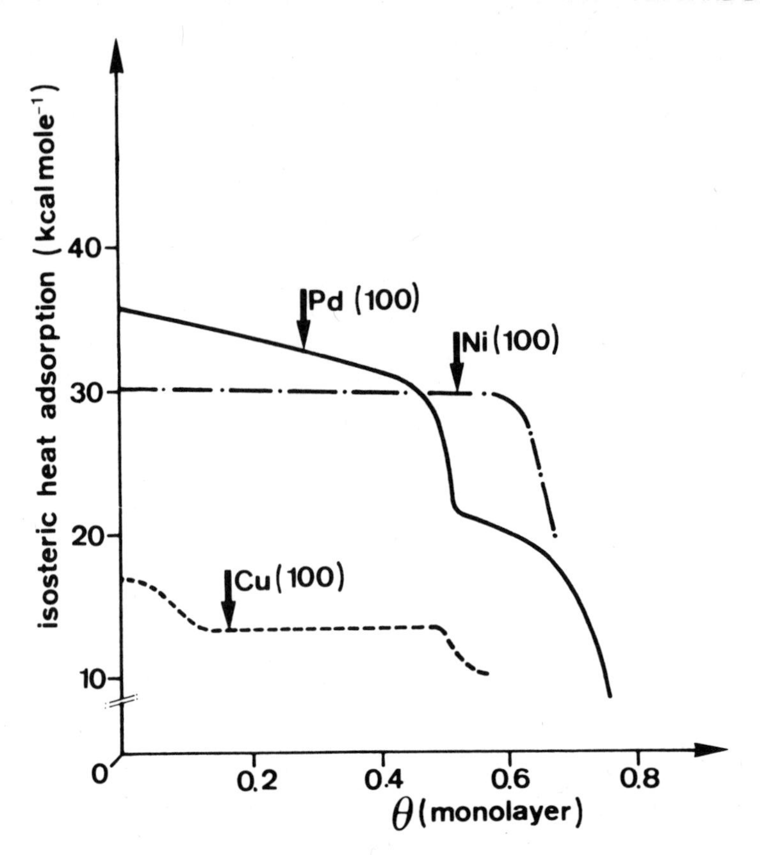

Figure 12. : Variation in isosteric heat of adsorption for carbon monoxide on the (100) planes of Pd, Ni and Cu. Abrupt decreases in heat of adsorption at high coverage have been attributed to strong repulsive interactions between the adsorbed molecules. (From Tracy and Palmberg, refs. 87 and 102 and Tracy, refs. 88 and 89)

ley[106] and by Einstein and Schrieffer[107].

It is evident from these studies that the importance of indirect interactions is related to the degree of delocalization of the adsorbate-adsorbent bond. In this vein, the statistical thermodynamic discussion of adsorbate interactions in terms of heats of adsorption and energies of activation also have a strong historical development in the work of Peierls[108], of Wang[113] and of Roberts[114]. It is significant from an experimental viewpoint that with the exception of carefully executed field ion microscope and field emission microscope studies developed from the original inventions of Müller[111,112] almost all experimental measurements tend to observe surface phenomena strongly influenced by collective interactions. The present useful procedure of extrapolating the latter measurements back to zero coverage to facilitate comparisons with simple models stressing localized interactions is susceptible to uncertainty.

The most graphic evidence for the importance of lateral interactions is found in LEED pattern studies of ordered overlayers. In this case the rather frequent occurrence of the same two-dimensional periodicity for quite different adsorbates on a particular metal crystal face suggests strongly the influence of indirect interactions with directional characteristics influenced by the adsorbent geometry[40-42]. Note for example, the prevalence of the c(2x2) structure for a variety of reactive molecular gases on the (001) crystal face of tungsten and nickel as well as other transition metals[41].

The existence of definite correlations between structure and binding is observed for a number of gas-metal adsorption systems where coverage beyond 0.5 monolayer is accompanied by an abrupt decrease in the heat of adsorption or the occurrence of additional peaks in the lower temperature range of a thermal desorption spectra. Following Wang[113] and Roberts[114], correlation between surface order and binding energy can be usefully considered in terms of the model shown in Figure 13a. This is based on a mobile monolayer with a coverage-independent pairwise repulsive interaction, V between admolecules in nearest-neighbor sites. Neglecting thermal vibration the minimum energy configuration for $\theta \leqslant 0.5$ has unoccupied nearest-neighbor sites as shown in Figure 13a. The differential heat of adsorption remains constant at its zero-coverage value, H_o in this coverage range (see Figure 13b-curve i). There is an abrupt decrease in H due to a repulsive interaction of zV, where z = number of nearest-neighbor sites. With increasing coverage beyond $\theta = 0.5$ H remains constant. That is, $H = H_o - zV$ as the vacant sites are randomly filled. Now at higher temperatures, where thermal migration provides surface mobility, for the adsorbed layer to adjust its configuration to a state of minimum energy, the tendency to adopt the configuration of lowest energy is opposed by the thermal migration of the adsorbed molecules. This leads to the sigmoidal variation of H with θ (see Figure 13b-curve ii).

It is interesting now to compare these predictions with the experimental results of Tracy and Palmberg[87,102] illustrated in Figure 12 for the heat of adsorption of CO on Pd[87,102], on Ni[88], and on Cu[89]. The experimental measurements show a region of nearly constant heat of adsorption followed by an abrupt decrease at about $\theta = 0.5$. Similar observations have been reported by Ertl _et al._[115,116] on the various planes of palladium. They suggest that formation of double-spaced structures occur on the (001) planes at coverages close to half monolayer in all cases as indicated by LEED pattern

a

$\theta < 0.5$

$\theta = 0.5$

$\theta > 0.5$

$\theta = 1.0$

b

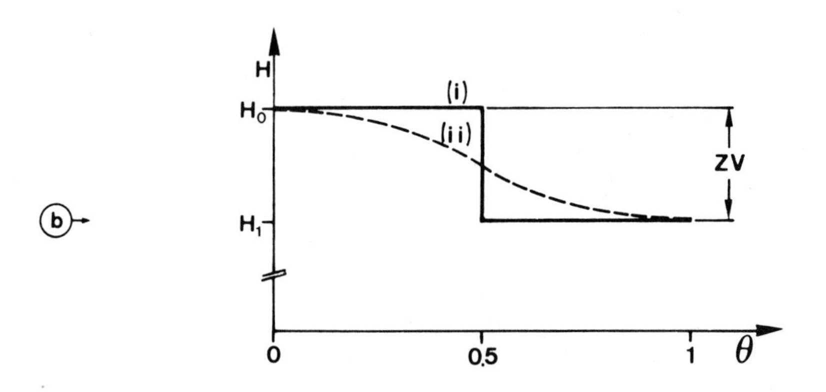

Figure 13 : a) Schematic adsorbate configurations resulting from nearest-neighbor repulsive interactions.
b) Associated change in differential heat of adsorption, H with coverage, θ : (i) for a perfectly ordered structure, (ii) for a structure with imperfect order due to thermal migration (From Adams, ref. 99)

analysis. It should be noted that for other surfaces with stronger adsorbate interactions, such as CO on (001) Ni, the lattice gas model breaks down and a more complex surface structure results.

The influence of lateral interactions upon thermal desorption spectra has also been considered within approximate treatments of the lattice-gas model by Toya[98], by Adams[99], and by Goymour and King[100]. It has been established by these studies that multiple peak thermal desorption spectra can result from the variation of activation energy for desorption with coverage due to lateral interactions. Quantitative agreement between calculated and experimental multiple peak spectra have been observed for the case of CO on W(210) shown in Figure 14. Here good agreement was achieved only by allowing the repulsive interaction potential to increase in a simple manner with coverage. It should be noted that such agreement for this rather simple model is not generally achieved and that further refinement in the depiction of the interaction potential is required for a more accurate and general description of most such adsorption systems.

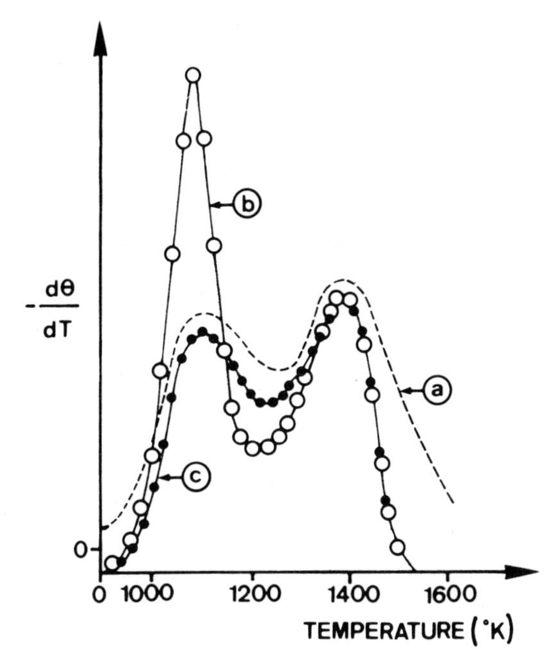

Figure 14: Calculated and experimental thermal desorption spectra for carbon nonoxide adsorbed on W(210) to saturation coverage. a) Experimental spectrum with origin displaced for clarity. b) Calculated spectrum for coverage-independent repulsive interaction. c) Calculated spectrum for coverage-dependent repulsive interaction. (From Adams, ref. 99)

Mention should also be made that useful estimations have also been made of the magnitude of the adsorbate-adsorbate interactions from LEED observations of the temperature-dependence of the intensity of diffracted beams characteristic of the adsorbent structures. An adsorbed layer behaving as a two-dimensional lattice gas will show critical behavior. In this case the critical temperature for half monolayer coverage for a square lattice is given by

$$\tanh(V/4rT_c) = \sqrt{2} - 1 \tag{7}$$

Experimental studies by Estrup[117] for a number of adsorbates on W(001) show a critical temperature of 550 K:(for hydrogen) which corresponds to a value of the pairwise interaction energy, V of 1.9 kcal/mole. This compares well with the value of 1.8 kcal/mole calculated by Adams[99] for the same system using the previously described model (see Figure 14).

The general conclusion based on both thermal desorption and LEED observations is that the lattice gas model although too simple to give precise results does provide useful estimates of the interaction energy. It is evident that the understanding of collective interactions on metal surfaces and their contribution to surface reactions is essential to understanding the mechanisms of monolayer adsorption and molecular interconversion on surfaces.

8. VIBRATIONAL PROPERTIES OF ADSORBED LAYERS[118-133]

The features of surface vibrational properties on metals are related directly to bonding modes of surface adsorbent and adsorbate atoms and hence are of great interest from both the experimental and theoretical viewpoints. This aspect of surface atomistics will be considered more in subsequent lectures on models for atom interactions at surface and on experimental interpretation of infrared as well as electron-loss spectroscopy measurements. Here we introduce briefly some of the concepts and models essential to this important approach to the theory of atomistic behavior of metal surfaces and to its connection with experiment. It is useful to start with consideration of analogies between the vibrational and electronic properties associated with the adsorbed atom. Two extreme limits may be usefully considered as follows,

a) The vibrational and electronic configurations in the vicinity of the adatom can be considered on the one hand to be essentially determined by the solid with the adatom providing a small perturbing effect.

b) The configurations can be considered in contrast to be dominated by a small cluster defined by local interactions of the adatom with a limited number of closely located metal surface atoms; the cluster being indented into the solid and the total effect being determined primarily by the cluster with small perturbation coming from the indentation.

In some sense, the former limit has been associated with the solid state physical approach whereas the latter has been identified to some extent with that of the quantum chemist. It should be pointed out clearly that evaluation of the relative value of various model calculations of both clean surface and adsorbed atom vibrational characteristics has been greatly limited by the lack of reliable microscopic data on metal surfaces with well defined atomistic properties. As a typical example of the first approach, Grimley(118) has discussed the case of an atom and a diatomic molecule adsorbed on the (001) face of a simple cubic metal within the context of a force-constant, lattice dynamical model. For the chemisorbed atom the familiar results are obtained that a localized mode is formed if the vibrational frequency of the metal-atom bond is greater than the highest frequency phonon state of the metal. Assuming that vibrational modes of the surface-complex are localized, then the interpretation of the experimental spectra is greatly simplified in principle with the modes treated as simple harmonic oscillators. In fact, such local modes are often difficult to measure because of inadequate instrumental sensitivity. It has not yet been possible to develop a substantial body of experimental knowledge based on the infrared spectroscopy of small crystals with well defined atomic and electronic surfaces largely for this reason(109,119). In contrast, the work of Eischens and Pliskin(120) on the infrared spectroscopy of dispersed materials where the sensitivity and surface characterization problems are not as critical, has been widely extended to measure the vibrational frequencies associated with adsorbed molecules. This approach has been applied by a limited number of workers to chemisorption on single crystal surfaces among whom that of Pritchard and co-workers(103),(104),(110),(121)-(124) and Greenler(125)-(127) are significant but rather isolated examples of this approach.

Microscopic techniques have also been used in special cases. Propst and Piper(128) observed characteristic energy losses due to adsorption of a number of simple gases on W(001) using high energy resolution of back-scattered low energy electrons. More recently Ibach <u>et al</u>.(129-132) have also observed inelastic loss peaks due to surface vibrational modes in electron scattering experiments for clean ZnO($1\bar{1}00$) and Si(111) as well as vibrational modes of adsorbed oxygen on Si(111). Energy losses due to vibrational excitations have also been observed in electron tunneling experiments by Plummer and Bell(133) as reflected in the spectral fine structure of the energy spectra of electrons field-emitted from a tungsten tip containing hydrogen. All of these methods involve significant experimental effort.

In summary, we may conclude that with reference to adsorption studies on metals, the infrared technique provides the better resolution but the electron scattering techniques are more sensitive especially in the low energy region. Vibrational spectra measured with suitable sensitivity and resolution can provide a wealth of information on the mutual interaction states of adsorbed molecules as well as the phonon states of the metal surface itself. Comparisons of both the vibrational and electronic properties of both the clean and chemisorbed surfaces can provide directly, information on the geometry and dynamics of surface bonding. Once the experimental obstacles associated with sensitivity and resolution are resolved this advance should come rapidly.

SUMMARY AND CONCLUSIONS

Substantial improvement has been made in our experimental capability to approach the problem of obtaining atomistic and microscopic information on metal surfaces. This applies first, to the preparation and characterisation of well defined reproducible single crystal samples, second, to the reliable measurement of critical properties such as atomic composition and geometry and third, to extension of spectral data relating to electron structure and excitations. On the cautionary site it should be noted that most of the work has been done on pure metal substrates corresponding to static conditions. The difficult connection existing in many cases in the interpretation of the measured quantity and its relationship to fundamental atomistic or microscopic parameters required to evaluate proposed mechanisms is far from achieved. Some examples include, the complex connection of LEED intensity measurements to the definition of the atomistic structure of a real surface, the precise resolution of compositional data in terms of position and distribution in the surface, and finally the detailed interpretation of electron spectral data in terms of established energy levels and transitions. Other measured quantities of great interest (with some spectacular exceptions) have been measured only within crude limits or almost not at all, such as surface coverage by adsorbates, differential and integral heats of adsorption and reaction, vibrational features on well defined surfaces and overlayers, and electron charge distribution at interfaces.

The greatest challenge in the immediate future is to extend these measurements of microscopic and atomistic parameters to multicomponent single crystal as well as dispersed samples characteristic of dynamic or non-steady state constraints with the same reliability with which they can now be applied to static features of simple systems. Many of the same limitations may be applied to the present state of theoretical interpretation of basic parameters relating to both clean and chemically modified metals. Some very powerful general approaches based on bulk solid state or atomistic or microscopic concepts have been brought to bear on surface phenomena. However, much remains to be done to develop theoretical models which deal directly and realistically with the intrinsic nature of metallic interfaces themselves. The achievements as well as the limitations in current efforts to develop a comprehensive theoretical description of the nature of the metallic surface will become more clearly developed in terms of the lectures at this meeting.

It is evident that the questions far outnumber the answers in both the experimental as well as the theoretical areas. Nevertheless, the subject of the elucidation of the basic parameters defining metallic properties at surfaces and interfaces is a fascinating and fast growing area for research effort. This state has evolved through no small degree from the major advances made in our abilities to make reliable measurements of atomistic and microscopic properties. It is also stimulated by the growing recognition that useful advances of a practical nature relating to the utilization of metals as engineering materials depend on possessing a much broader understanding than is yet available of the physical and chemical principles unique to the metal-vacuum boundary. Subsequent lectures dealing with the atomic and electronic structure of clean and chemisorbed metal surfaces, with the physical techniques for the study of the metal-gas interface and with the mechanisms and models by which molecules interact with metals will provide a detailed background against which some of these concepts and models may be more thoroughly understood.

REFERENCES

(1) H.S. Taylor, American Scientist, 34 (1946) 53
(2) I. Langmuir, J.A.C.S. 40 (1918) 1361
(3) C.N. Hinshelwood, The Kinetics of Chemical Change, (Oxford University Press, 1940)
(4) K.J. Laidler, Catalysis 1 (1954) 75
(5) D.O. Hayward and B.M.W. Trapnell, Chemisorption, Second Edition (Butterworths, 1964)
(6) D. Alpert, Proc. of 1st Congr. on Vac. Tech., 1958, Pergamon, p. 31
(7) Thor Rhodin and David Adams, "Adsorption of Gases on Solids", Chapter 5, Vol. 6, Treatise on Solid State Chemistry, editor, N. Hannay, Plenum Press, N.Y. (1975)
(8) E. Drauglis, R.D. Gretz and R.I. Jafee, editors, Molecular Processes on Solid Surfaces (McGraw-Hill, 1969)
(9) G.A. Somorjai, editor, Structure and Chemistry of Solid Surfaces, Proc. of the 4th International Materials Symposium, Berkeley, 1968 (Wiley, 1969)
(10) R. Park, editor, Solid Surfaces, Pro. of the 2nd International Conference on Solid Surfaces, New York, 1971 (American Vacuum Society, 1972)
(11) S. Kawaji, editor, Solid Surfaces, Proc. of the 3rd Interna-

tional Conference on Solid Surfaces, Kyoto, 1974 (Japanese Society of Applied Physics, 1974)
(12) E.A. Flood, editor, The Solid-Gas Interface, (Dekker, 1967)
(13) L.W. Swanson, A.E. Bell, C.M. Hinrichs, L.C. Crouser and B.E. Evans, Literature Review of Adsorption on Metal Surfaces, Vols. 1 and 2 (NASA CR-72402, 1967)
(14) A. Clark, The Chemisorptive Bond : Basic Concepts (Academic Press, 1974)
(15) J.M. Thomas, and W.J. Thomas, Introduction to the Priniples of Heterogeneous Catalysis (Academic Press, 1967)
(16) J.R. Anderson, editor, Chemisorption and Reactions on Metallic Films (Academic Press, 1971)
(17) R. Gomer, Chemisorption, Solid State Phys., to be published
(18) P.A. Redhead, J.B. Hobson and E.V. Kornelsen, The Physical Basis of Ultrahigh Vacuum (Chapman and Hall, 1968)
(19) T.L. Einstein and J.R. Schrieffer, Phys. Rev. 87 (1973) 3629
(20) R.E. Weber and W.T. Peria, J. Appl. Phys. 38 (1967) 4355
(21) P.W. Palmberg and T.N. Rhodin, J. Appl. Phys. 39 (1968) 2425
(22) C.C. Chang, Surf. Sci. 48 (1975) 9
(23) K. Siegbahn, C. Nordling, A. Fahlman, R. Nordberg, K. Hamrin, J. Hedman, G. Johansson, T. Bergmark, S. Karlson I. Lindgren, B. Lindberg, ESCA : Atomic, Molecular, and Solid State Structure Studied by Means of Electron Spectroscopy (Almqvist and Wiksells, 1967)
(24) R.L. Park, J.E. Houston and D.G. Schreiner, Rev. Sci. Inst. 41 (1970) 1810
(25) R.L. Park and J.E. Houston, Surface Science 26 (1971) 664
(26) N.J. Taylor, in : Techniques of Metals Research, Vol. 7, edited by R.F. Bunshah (Interscience, 1972)
(27) J.C. Tracy, in : Electron Emission Spectroscopy, edited by W. DeKeyser, L. Fiermans, G. Vanderkelen and J. Vennik (Reidel 1973)
(28) P.W. Palmberg, in : Electron Spectroscopy, edited by D. Shirley (North-Holland, 1972)
(29) E.N. Sickafus, J. Vac. Sci. and Tech. 11 (1974) 229 and references therein
(30) C.R. Brundle, in : Surface and Defect Properties of Solids, Vol. 1, Chapter 6 (The Chemical Society, 1972)
(31) D.T. Clark, in reference 27
(32) W.N. Delgass, T.R. Hughes and C.S. Fadley, Catalysis Rev. 4 (1971) 179
(33) R.L. Park and J.E. Houston, J. Vac. Sci. and Tech. 11 (1974) 1
(34) A.M. Bradshaw, in : Surface and Defect Properties of Solids, Vol. 3, Chapter 5 (The Chemical Society, 1974)
(35) R.L. Park and J.E. Houston, Advances in X-ray Analysis 15 (1972) 462
(36) C.J. Davisson and L.H. Germer, Nature 119 (1927) 558;Phys. Rev. 30 (1927) 705; Proc. Nat. Acad. Sci. 14 (1928) 317, 619
(37) L.H. Germer, Physics Today (1964) 19; Physik 54 (1929) 408
(38) H.E. Farnsworth, Advances in Catalysis 15 (1964) 31

(39) Varian Associates, Palo Alto, California
(40) E. Bauer, in : Techniques of Metals Research, Vol.2, edited by R.F. Bunshah (Willey, 1969)
(41) J.W. May, Ind. Eng. Chem. 57 (1965); Advances in Catalysis 21 (1970) 244
(42) P.J. Estrup and E.G. MacRae, Surface Science 25 (1971) 1
(43) Physical Electronics Industries, Edwina, Minnesota
(44) J.B. Pendry, Low Energy Electron Diffraction, Academic Press, London, New York (1974)
(45) C.B. Duke, Proc. of the International School of Physics "Enrico Fermi", Course LVIII, Academic Press, New York (1974); see also Adv. in Chem. Physics 27 (1974) 1
(46) M. Laznicka, editor, LEED Surface Structure of Solids, Union of Czechoslovak Mathematicians and Physicists, Prague (1972) Vol.2, pp. 7-291, 361-387
(47) M.B. Webb and M.G. LaGally, Solid State Physics 28 (1973) 301
(48) S.Y. Tong, Progress in Surface Science, editor S.G. Davison, Pergamon Press, New York (1975)
(49) J.A. Strozier, D.W. Jepsen and F.Jona, Surface Physics of Crystalline Solids, editor J. Blakely, Academic Press, New York
(50) G. Ertl and J. Küppers, Low Energy Electrons and Surface Chemistry, Verlag Chemie (1974), Weinheim (1973)
(51) E. Bauer, Topics in Applied Physics, Vol. 4, Chapter 6, Verlag Physik (1975)
(52) T.N. Rhodin and S.Y. Tong, Physics Today, October 1975
(53) S.Y. Tong, Solid State Comm. 16 (1975)91; S.Y. Tong and M. van Hove (to be published); M. van Hove, S.Y. Tong, and N. Stoner, Bull. APS 20 (1975) 388
(54) J.E. Demuth, D.W. Jepsen and P.M. Marcus, J. Phys. C Solid State Phys. 8 (1975) 8; Phys. Rev. Lett. 31 (1973) 540; Phys. Rev. Lett. 32 (1974) 1182; Surf. Sci. 45 (1974) 733; Jour. Vac. Sci. Tech. 11 (1974) 190
(55) D.W. Jepsen, P.M. Marcus and F. Jona, Phys. Rev. B5 (1972) 3933; Phys. Rev. B6 (1973) 3684
(56) S. Andersson and J.B. Pendry, Jour. Phys. Solid State Phys. 5 (1972) 141; S. Andersson, B. Kasemo, J.B. Pendry and M. van Hove, Phys. Rev. Lett. 31 (1973) 595; S. Andersson and J.B. Pendry, Jour. Phys. C. Solid. State Phys. 6 (1973) 601; S. Andersson and J.B. Pendry, Solid State Comm. 1975 (to be published)
(57) M.B. Webb and M.G. LaGally, see ref. 47, T.C. Ngoc, M.G. LaGallay and M.B. Webb, Surf. Sci. 25 (1973) 237; M.G. LaGally T.C. Gnoc and M.B. Webb, Phys. Rev. Lett. 26 (1971) 1557; T.C. Gnoc, M.G. LaGally and M.B. Webb, Jour. Vac. Sci. Tech. 9 (1971) 645
(58) D. Adams and U. Landman, Phys. Rev. Lett. 33 (1974) 585; U. Landman and D. Adams, Jour. Vac. Sci. Tech. 11 (1974) 195; D. Adams, U. Landman and J.C. Hamilton, Jour. Vac. Sci. Tech. 15 (1975) XXX; U. Landman, Discussions of the Faraday Soc. 60 (1975), Vancouver, Canada, July (1975)

(59) B. Hutchins, T.N. Rhodin and J.E. Demuth, Surf. Sci. (to be published) 1975
(60) N.D. Lang and W. Kohn, Phys. Rev. B3 (1971) 1215
(61) J.C. Rivière, in : Solid State Surface Science, edited by M. Green, Vol. 1 (Dekker, 1969)
(62) S. Andersson and J.B. Pendry, Solid State Comm. 16 (1975) 563
(63) E. Wigner and J. Bardeen, Phys. Rev. 48 (1935) 84
(64) J. Bardeen, Phys. Rev. 49 (1936) 653
(65) R. Smoluchowski, Phys. Rev. 60 (1941) 661
(66) C. Herring, Metal Interfaces, Am. Soc. for Metals, p. 1, Cleveland, Ohio (1951)
(67) L.D. Schmidt and R. Gomer, J. Chem. Phys. 42 (1965) 3573; 45 (1966) 1605
(68) M. Boudart, J.A.C.S. 74 (1952) 3556
(69) T.A. Delchar and G. Ehrlich, J. Chem. Phys. 42 (1965) 2686
(70) D.L. Adams and L.H. Germer, Surface Science 27 (1971) 21
(71) T.N. Rhodin, Advances in Catalysis 5 (1953) 39
(72) A.W. Czanderna, editor, Vacuum Microbalance Techniques, Vol.8 (Plenum, 1971)
(73) S.P. Wolsky and E.J. Zdanuk, editors, Ultra Micro Weight Determination (Wiley, 1969)
(74) See for example : C.E. Bryson,V. Cazcarra and L.L. Levenson, J. Vac. Sci. and Tech. 11 (1974) 411; M.P. Seah, Surface Science 32 (1972) 703
(75) P.W. Tamm and L.D. Schmidt, J. Chem. Phys. 54 (1971) 4775
(76) J.L. Domange, J. Oudar and J. Bernard, Molecular Processes on Solid Surfaces, p. 353. Ed. Drauglis, Gretz and Jaffee, McGraw-Hill, New York (1968)
(77) J.A. Dillon and H.E. Farnsworth, Rev.Sci. Instr. 25 (1954) 96; J. Chem. Phys. 22 (1954) 1601
(78) A.D. Crowell, J. Chem. Phys. 32 (1960) 1576
(79) A.D. Crowell and L.D. Mathews, Surface Science 7 (1967) 79
(80) K. Klier, Rev. Sci. Instr. 40 (1969) 15525
(81) K. Klier, A.C. Zettlemoyer and H. Leidheiser, J. Chem. Phys. 52 (1970) 589
(82) G. Ehrlich, Advances in Catalysis 14 (1963) 256
(83) P.A. Redhead, Vaccum 12 (1962) 203
(84) L.A. Pétermann, Progresses in Surface Science, Vol.3, edited by S.G. Davison (Pergamon Press, 1972); D.A. Kling, Surface Science 47 (1975) 384
(85) T.E. Maday and J.T. Yates, J. Vac. Sci. and Tech. 8 (1971) 525
(86) D. Menzel, Surface Science 47 (1975) 370
(87) J.C. Tracy and P.W. Palmberg, Surface Science 14 (1969) 274
(88) J.C. Tracy, J. Chem. Phys. 56 (1972) 2736
(89) J.C. Tracy, J. Chem. Phys. 56 (1972) 2748
(90) C. Pisani, G. Rabino and F. Ricca, Surface Science 41 (1974)277
(91) P.W. Tamm and L.D. Schmidt, J. Chem. Phys. 22 (1970) 365
(92) P.W. Tamm and L.D. Schmidt, J. Chem. Phys. 52 (1970) 1150
(93) T.E. Madey and J.T. Yates, in : Structure et Propriétés des Surfaces des Solides (Editions du C.N.R.S., Paris, 1970)

No. 187, p. 155
(94) D.L. Adams and L.H. Germer, Surface Science 23 (1970) 419
(95) K. Yonehara and L.D. Schmidt, Surface Science 25 (1971) 238
(96) D.L. Adams and L.H. Germer, Surface Science 32 (1972) 205
(97) T.E. Madey and D. Menzel, Japan. J. Appli. Phys. Suppl.2 (1974)
(98) T. Toya, J. Vac. Sci. and Tech. 9 (1972) 890
(99) D.L. Adams, Surface Science 42 (1974) 12
(100) C.G. Goymour and D.A. King, J. Chem. Soc. Faraday I 69 (1973) 749
(101) O. Beeck, Advances in Catalysis 2 (1950) 151
(102) J.T. Tracy and P.W. Palmberg, J. Chem. Phys. 51 (1969) 4852
(103) C.S. Alexander, Ph.D. Thesis, London, 1966; C.S. Alexander and J. Pritchard, J. Chem. Soc. Faraday Trans. I 68 (1972)202
(104) J. Pritchard, J. Vac. Sci. and Tech. 9 (1972) 895
(105) J. Koutécky, Trans. Farad. Soc. 54 (1958) 1038
(106) T.B. Grimley and M. Torrini, J. Phys. C 6 (1973) 868
(107) T.L. Einstein and J.R. Schrieffer, Phys. Rev. 87 (1973) 3629
(108) R.E. Peierls, Proc. Camb. Phil. Soc. 32 (1936) 471
(109) M.L. Hair , Infrared Spectroscopy in Surface Chemistry (Dekker, 1967)
(110) D.A. King, Surface Science 47 (1975) 384
(111) E.W. Müller, Z. Physik 106 (1937) 541
(112) E.W. Müller and T.T. Tsong, Field Ion Microscopy (Elsevier, 1969
(113) J.S. Wang, Proc. Roy. Soc. (London) A161 (1937) 127
(114) J.K. Roberts, Some Problems in Adsorption, Cambridge 1939
(115) G. Ertl and J. Koch, in : Adsorption - Desorption Phenomena, edited by F. Ricca (Academic Press, 1972) p. 345
(116) H. Conrad, G. Ertl, J. Koch and E.E. Latta, Surface Science 43 (1974) 462
(117) P.J. Estrup, in ref. 9 (19-1)
(118) T.B. Grimley, Proc. Phys. Soc. (London) 79 (1962) 1203
(119) G. Blyholder, in : Experimental Methods in Catalytic Research, edited by R.B. Anderson (Academic Press, 1968) p. 323
(120) R.P. Eischens and W.A. Plisken, Advances in Catalysis 10 (1958) 1
(121) M.A. Chesters, J. Pritchard and M.L. Sims, Chem. Comm. (1970) 1454
(122) J. Pritchard and M.L. Sims, Trans. Farad. Soc. 66 (1972) 427
(123) M.A. Chesters, J. Pritchard and M.L. Sims, in : Adsorption - Desorption Phenomena, edited by F. Ricca (Academic Press, 1972)
(124) M.A. Chesters and J. Pritchard, Surface Science 28 (1971) 460
(125) R.G. Greenler,J. Chem. Phys. 44 (1966) 310
(126) R.G. Greenler,J. Chem. Phys. 50 (1969) 1963
(127) R.G. Greenler, Japan.J.Appl. Phys. Suppl. 2 (1974) 265
(128) F.M. Propst and T.C. Piper, J. Vac. Sci. and Tech. 4 (1967)53
(129) H. Ibach, Phys. Rev. Lett. 24 (1970) 1416
(130) H. Ibach, Phys. Rev. Lett. 27 (1971) 253
(131) H. Ibach, J. Vac. Sci. and Tech. 9 (1972) 713
(132) H. Ibach, K. Horn, R. Dorn and H. Lüth, Surface Science 38

(1973) 433
(133) E.W. Plummer and A.E. Bell, J. Vac. Sci. and Tech. 9 (1972) 583
(134) T. Oguiri, J. Phys. Soc. Japan 19 (1964) 83
(135) C. Herring and M.H. Nichols, Phys. Rev. 21 (1949) 185

INTRODUCTORY LECTURE : SURVEY OF CURRENT IDEAS IN THE THEORY OF CHEMISORPTION BY METALS

T.B. GRIMLEY

Donnan Laboratories

University of Liverpool

1. INTRODUCTION

One of the most challenging problems in physical chemistry is the study of molecular processes on solid surfaces and chemisorption theory is part of this study. It is sometimes convenient to distinguish three situations in chemisorption :

(a) the admolecule coverage is self-limiting so that well-defined overlayers are formed,
(b) it is not, and there is continued attack on the metal to build up a reaction product as a new phase at the gas/metal interface,
(c) the admolecule is decomposed at the metal surface.

Often the distinction between (a) and (b) will be a matter of temperature only ; the chalcogens form well-defined overlayers on Ni, but they corrode the metal at high temperatures.

Theoretical studies tend to begin with single atoms or molecules on single crystal planes, and such studies lead naturally to the consideration of catalytic decomposition of molecules. Single crystals are not of course good catalysts, but even so, by studying molecular processes on such simple surfaces, we build up our fundamental knowledge of the factors determining the reactivity of metals. Once we can deal with single crystal planes, we can go on to study the effects of point defects, steps and so on. This is one approach. The other regards chemisorption as essentially a localized phenomenon so that it is useful to start by considering the adsorbate, and only a few metal atoms. This would seem to be the better approach to chemisorption by the highly disordered supported metal catalysts, not because these catalysts are in a sense small particles, they are not (they contain 100-100000 metal atoms), but because particular local geometries are best studied initially with a local model. The possibility of combining a good local model with a more approximate treatment of the rest of the semi-infinite metal has also been studied. I refer to these three approaches in more detail later, and particular examples of them will also be discussed by Professor Allan and Dr. Lang.

The aim of chemisorption theory must be complete quantum predictions of the equilibrium positions of all the nuclei, the ground state potential energy surface, the elementary excitations, and the responses to external probes. Such a goal is still a long way off, and at present we must accept a good deal less than this for systems of chemical interest. Simplified models can however be explored in some depth now.

2. THE ELECTRONIC STRUCTURES OF METAL SURFACES

Experimental workers in catalysis believe (1) that, if the electronic structure of a clean metal surface could be described quantitatively in molecular orbital language, they would be able to understand both the chemisorption properties, and the catalytic activity of the metal. But how is the information on electronic structure to be presented ? For small molecules one simply tabulates the energies, and wavefunctions, of the occupied, and low lying vacant molecular orbitals. The directional properties of the wavefunctions round a particular atom in the molecule tell us something about the directional properties of the molecule's free valence at this atom, its susceptibility to chemical attack, and so on. But even for a small piece of metal 20 nm across, there will be some 10^5 molecular orbitals for the valency electrons, all having some amplitude on the surface atoms, and it is now quite impracticable to embark on a straightforward tabulation ; for a semi-infinite metal it is impossible. What we do instead is to calculate the <u>local density of states</u> $\rho(r,\varepsilon)$ varying both with position throughout the whole of space (but particularly outside the metal), and with energy. $\rho(r,\varepsilon)drd\varepsilon$ is the number of states with energy in the range $d\varepsilon$ at ε in the volume element dr at r. It is defined by

$$\rho(r,\varepsilon) = \sum_\mu |\psi_\mu(r)|^2 \delta(\varepsilon-\varepsilon_\mu) \qquad (1)$$

where ψ_μ and ε_μ are the molecular orbital wavefunctions and energies. For an ordinary small molecule, this function simply describes the original tabulation of the molecular orbital energies ε_μ, and their wavefunction probability densities $|\psi_\mu(r)|^2$. For a semi-infinite metal we cannot calculate $\rho(r,\varepsilon)$ from this definition ; there is no question of solving the secular equation for the wavefunctions and energies. Instead we try to calculate $\rho(r,\varepsilon)$ directly from the Hamiltonian. This is how we first become involved with Green functions in chemisorption theory, because if $\hat{F}$ is the self-consistent one-electron operator for electrons in the metal, the Green operator $\hat{G}(\varepsilon)$ is

$$\hat{G}(\varepsilon) = (\zeta - \hat{F})^{-1} \quad , \quad \zeta = \varepsilon + i0$$

and since

$$\hat{G}(z) = \sum_{\mu} \frac{|\mu\rangle\langle\mu|}{z-\varepsilon_{\mu}} \qquad \text{(resolvent operator)}$$

we see that

$$\rho(r,\varepsilon) = -\frac{1}{\pi}\,\mathrm{Im}G(\zeta)$$

Suppose we now introduce the usual set $\{\phi\}$ of s, p, d,... atomic orbitals on every metal atom, then we can express $\rho(r,\varepsilon)$ in the form

$$\rho(r,\varepsilon) = \sum_{i} \phi_i(r)\phi_j(r)\tilde{\rho}_{ji}(\varepsilon) \qquad (2)$$

where the matrix $\tilde{\rho}$ is the local-and-overlap density of states matrix leading to the charge-and-bond order matrix P (the density matrix of MO theory), with elements

$$P_{ij} = \int_{-\infty}^{\varepsilon_F} d\varepsilon\, \tilde{\rho}_{ij}(\varepsilon)$$

The diagonal elements of $\tilde{\rho}$ are the net densities of states contributed by the atomic orbitals, the non-diagonal elements are the overlap densities. The gross densities of states $\rho_{ii}(\varepsilon)$, leading

to the gross populations (occupancies) are

$$n_i = \int_{-\infty}^{\varepsilon_F} d\varepsilon\, \rho_{ii}(\varepsilon)$$

are

$$\rho_{ii} = \sum_{j} S_{ij}\, \tilde{\rho}_{ji}(\varepsilon)$$

where S is the overlap matrix : $S_{ij} = \langle i|j\rangle$

For tight binding models where S = 1, the net/gross distinction disappears, and $\rho = \tilde{\rho}$, and powerful and direct methods have been developed to calculate $\tilde{\rho}$ by Heine and co-workers in England (2), and Cyrot-Lackmann and co-workers in France (3). Most attention has been paid to the diagonal elements, but it is now possible to obtain through (2), fair approximations to $\rho(r,\varepsilon)$ at the surface of a transition metal, and to analyze, if desired, how each d orbital $|xz\rangle$,

$|yz\rangle$, $|zx\rangle$, $|x^2-y^2\rangle$, $|3z^2-r^2\rangle$, on a chosen atom contributes states, either filled or empty at every energy. Professor Allan will discuss these developments in detail.

In the density-functional approach to the electronic structure (Dr. Lang's lectures), a local density of occupied states, $\rho(r,\varepsilon)$ for $\varepsilon < \varepsilon_F$, is computed in the course of achieving self-consistent solutions of the one-particle equations, although the connection between this density, and that of the single hole excitations needs to be considered. For the simple plane boundary uniform positive background model, ρ depends only on z, the distance from the surface plane, and so has no interesting directional properties. One needs to go at least to a corrugated boundary model (4) to find directional properties. The calculation of $\rho(r,\varepsilon)$ for $\varepsilon < \varepsilon_F$ should also be undertaken.

Finally, I note that computer calculations on clusters of metal atoms, which are becoming commonplace by the Xα technique, being ordinary molecular calculations, give $\rho(r,\varepsilon)$ directly through (1). The only question is whether the clusters are large enough to provide any sort of approximation to the actual situation at the surface of even a 10 nm particule.

3. ELECTRONICALLY ADIABATIC POTENTIAL ENERGY CURVES FOR CHEMISORPTION

Although the situation is not entirely clear (16), it is expected that electronically adiabatic potential energy curves for the interactions of molecules with metal surfaces will occupy as central a rôle in our thinking about reactions on metal surfaces as they do in homogeneous gas phase reactions. The calculations of these potential energy curves is a primary aim of chemisorption theory (another is to provide a basis for theoretical predictions in surface sensitive spectroscopies), because given these curves we could begin to study the molecular dynamics of some simple heterogeneous processes.

3.1. Cluster calculations

The idea that chemisorption by metals is a localized phenomenon, in the sense that only a few metal atoms near the adsorbate are strongly involved, is a common one in the literature. The possibility of treating the problem theoretically as just another problem in computational quantum chemistry should therefore be considered. For a chalcogen on (100)Ni for example, the cluster model of Figure 1 might be investigated.

Although it has well-known deficiencies, a good deal of quantum chemistry is built on the single configuration Hartree-Fock model, and practical computations on small molecules by the LCAO-MO method are routine (see ref.5 for a compilation of molecules treated recently). Moreover, Hartree-Fock electron densities are good starting points to estimate correlation energies. But the bottleneck with the LCAO method is the very large number of two-electron integrals to be evaluated and stored (the number increases like the fourth power of the number of electrons). For the cluster $X-Ni_5$ there are at least 5×10^8 such integrals, and although larger calculations than this have been performed, such a calculation is expensive. Because of this, the familiar semi-empirical molecular orbital schemes of quantum chemistry (extended Hückel, CNDO) have been used instead. These schemes are generally regarded by theoretical chemists as quantitatively unreliable (although there is a minority view to the contrary), and it would be unwise therefore to attach more significance to the results of such calculations on cluster models of chemisorption. In the last seven years we have seen the development of the $X\alpha$ scheme, and more recently its application to cluster models of chemisorption. Its great attraction is its ability to treat quite large molecular systems with realistic computing times. It also gives quantitatively good electron excitation energies, but total energies and binding energies are not reliable.

Accepting that improvements in the $X\alpha$ binding energy, and in the computing time of the LCAO approach will be made, we must ask whether cluster models like that of Figure 1 are useful. There are two points to be made :

(a) No Ni atom in the cluster $X-Ni_5$ has the correct number of nearest neighbours so that the wavefunctions over the Ni atoms may bear little resemblence to those with the same energy for a chalcogen on the semi-infinite metal.

(b) The cluster $X-Ni_5$ is really part of a much larger system, and therefore the mean number of electrons in it can be non-integral. This is because the computed cluster states are not stationary states of the extended system, but persist in it as virtual states having a certain width. Such states are filled to the Fermi level, and therefore have fractional occupancies. In the limit of "zero width" virtual levels, only a cluster level at the Fermi level can have fractional occupancy.

(a) is a basic defect of all cluster calculations, and, excepts perhaps for Li, we can never get enough metal atoms into the cluster to be able to neglect it. Consequently we have to treat it (Section 3.2)

(b) can, and should be investigated in cluster calculations. It has been allowed for in some simple model calculations (6,7), and it seems fairly straightforward to include it in the $X\alpha$ scheme.

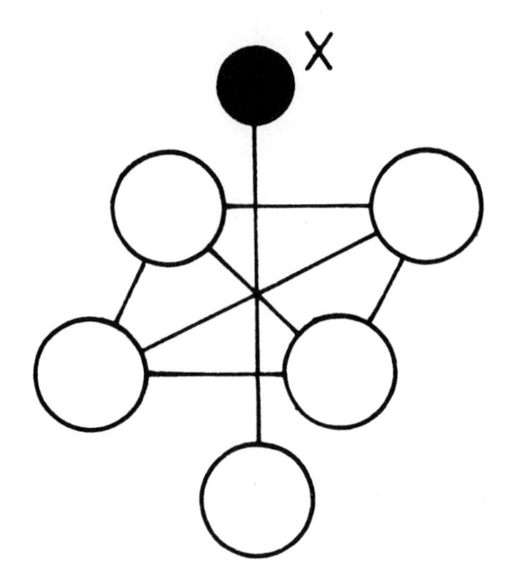

Fig. 1 : A six atom cluster model for the chemisorption of chalcogens by (100)Ni

3.2. Embedded clusters

If we attach the metal atoms in the cluster to an exactly solvable model of the metal, both deficiencies (a) and (b) above are mitigated ; the metal atoms are given their full complement of neighbours, and the cluster becomes part of a semi-infinite system. This modification is particularly easy to make, and to treat, if the cluster is describe by the SCF-LCAO-MO scheme, and attached to a tight binding model of the metal (8). Attaching the cluster to the rest of the metal destroys the self-consistency already existing in the isolated cluster, but Green function techniques (Dyson's equation) are used to establish the new self-consistency, and as a result a self-consistent cluster calculation is embedded in a simple non-self-consistent model for the rest of the metal. At the same time, the response of the rest of the metal to the self-consistent embedding of the cluster is also calculated, since this affects the total energy.

The cluster molecular orbitals persist in the semi-infinite system either as virtual states having certain widths in the metal bands, or as real discrete states with energies outside the metal bands. The cluster calculation will be valid if the virtual level widths are small compared to the widths of the metal bands. For simple models of chemisorption, Anderson's (9) for example, this criterian can be transferred to a similar statement about the level widths of certain group orbitals of the semi-infinite metal alone (10) and when this is done, it generally turns out (11) that clusters like that in Figure 1 would be expected to be very inadequate representations of chemisorption by semi-infinite metals. However simple models of chemisorption can be very misleading. Grimley and Pisani (8) calculated the widths of the virtual bonding and antibonding molecular orbitals of an embedded diatomic molecule H-M representing hydrogen chemisorption by M, and found them to be small when the simpler calculation outlined above would lead us to expect widths comparable to the metal band width.

Much more work is needed on isolated, and embedded clusters, and in view of the large number of $X\alpha$ calculations now being made, it is particularly important to consider the self-consistent embedding of an $X\alpha$ cluster calculation (17).

3.3. The density-functional method

In this approach we start from the beginning with a semi-infinite metal, and aim at a unified theory of bulk, surface, and chemisorption phenomena. Dr. Lang will give the details, I shall only refer briefly here to a recent paper (12) on hydrogen chemisorption by tungsten so as to see some of the problems involved.

The metal is represented by the plane surface uniform positive background model (a jellium) with $r_s = 1.5$, the density functional is taken as the first two terms of the gradient expansion

$$G[n] = \int g_0(n(r))dr + \int g_2(n(r))|\nabla n(r)|^2 dr \qquad (3)$$

(g_0 is the widely used local density term), and the perturbation of the semi-infinite metal due to the interacting species (the H^+ ion, see below) is treated in a linear approximation. The interaction energy of the proton with the jellium is calculated as the electrostatic potential of the bare jellium, plus the potential of the induced screening charge evaluated at the proton position. The calculation has electrostatic self-consistency ; the potential and the charge density satisfy Poisson's equation. The most important point to note in connection with this calculation is that, as the proton is moved away from the metal surface, the screening charge stays on the metal, i.e., H^+ is the desorbing species. This is contrary to experience. The neutral atom is desorbed, and this requires the screening charge to stay with the proton. This spurious result is presumably connected with the inability of the linear response approximation to provide any bound states for an electron in the hydrogen atom when it is far from the surface. Without bound states, the desorbing species can only be H^+. The exact calculations of Lang and Williams (18) are free from this defect. We shall hear more about this later.

3.4. Semi-empirical methods

Many phenomena in chemisorption have received their first investigation by analyzing the consequences of Anderson's (9) Hamiltonian when applied to chemisorption. The model is a simple one - non-interacting electrons in the energy bands of the metal, interacting electrons in localized states on the adsorbate - and because of this, qualitative investigations of complex phenomena can be contemplated. Effects of alloying, the through-bond interaction between adsorbates, the relation between cluster calculations, and the correct calculation for a semi-infinite metal, and so on, have all been examined using Anderson's model. At the present time we are using the model to investigate the role of surface states in chemisorption and this, I think, is the only application which will be discussed in the lectures. The model is also an important testing ground for calculations which go beyond the Hartree-Fock approximation (10,13-15). Such calculations are important ; in molecular quantum theory the single configuration Hartree-Fock approximation often leads to some very unfamiliar chemistry, because correlation effects associated with chemical bond formation are of crucial importance. The situation in chemisorption theory is expected to be similar.

4. DIRECT CALCULATION OF THE WAVEFUNCTIONS

It has become more widely accepted in the last five years that the technique of expanding wavefunctions in terms of a basis set may not be so useful in computational work on large systems as it has been for small ones, and that instead, a direct numerical integration of Schrödinger's equation should be undertaken. The widespread use of the $X\alpha$ computational scheme for large molecular systems is evidence to support this view.

In calculations on a semi-infinite metal (19), or on a solid with a chemisorbed overlayer (20), the first step is to choose a plane parallel to the surface, and a few atomic layers inside the solid, to be the dividing plane between a surface region, and an unperturbed bulk region. The position of this plane, and the consequent size of the surface region depends on the accuracy required. In the surface region the self-consistent potential seen by an electron is written as the sum of the Hartree potential, a local exchange-correlation potential, and an electron-ion pseudopotential. In the bulk, an empirical pseudopotential is used. Schrödinger's equation is integrated numerically in a Laue representation in the surface region, and the wavefunction matched on the dividing plane to a linear combination of bulk states of the same energy, and 2-dimensional translational symmetry. Thus, wavefunctions are found in the surface region which join onto bulk states. The calculation is made self-consistently ; the potential in the surface region is only known when the occupied wavefunctions are known.

No results for overlayers on metals have been published, but the method is evidently one of great importance in the study of semi-infinite systems.

REFERENCES

(1) G.C. Bond, Battelle Colloquium (1974), The Physical Basis for Heterogeneous Catalysis, to be published
(2) M.J. Kelly, Surface Sci., 43 (1974) 587 and references therein
(3) J.P. Gaspard and F. Cyrot-Lackmann, J. Phys. C., Solid State Phys., 6 (1973) 3077 and references therein
(4) R. Smoluchowski, Phys. Rev., 60 (1941) 661
(5) W.G. Richards, T.E.H. Walker, L. Farnell and P.R. Scott, Bibliography of Ab Initio Molecular Wavefunctions: Supplement for 1970-1973, Clarendon Press, Oxford (1974)
(6) T.B. Grimley, Molecular Processes on Solid Surfaces, E. Drauglis R.D. Gretz and R.I. Jaffee, eds., p. 181, McGraw-Hill, New York (1969)
(7) E. Mola, Surface Sci., to be published, treats the N_2-Fe cluster on iron as an open system
(8) T.B. Grimley and C. Pisani, J. Phys. C., Solid State Phys., 7 (1974) 2831
(9) P.W. Anderson, Phys. Rev., 124 (1961) 41
(10) T.B. Grimley, Prog. Surface Mem. Sci., 9 (1975) 71
(11) M.J. Kelly, Surface Sci., 43 (1974) 587
(12) S.C. Ying, J.R. Smith and W. Kohn, Phys. Rev., B11 (1975) 1483
(13) T.B. Grimley, Proc. Int. School of Physics 'Enrico Fermi', Course LVIII, to be published
(14) W. Brenig and K. Schönhammer, Z. Phys., 267 (1974) 201
(15) K. Schönhammer, to be published, has tested an approximate configuration interaction technique
(16) T.B. Grimley and J.R. Smith, Battelle Colloquium (1974), The Physical Basis for Heterogeneous Catalysis, to be published
(17) R.H. Paulson (Cornell) is working on this problem
(18) N.D. Lang and A.R. Williams, Phys. Rev. Lett., 34 (1975) 531
(19) J.A. Appelbaum and D.R. Hamann, Phys. Rev., B6 (1972) 2166
(20) J.A. Appelbaum and D.R. Hamann, Phys. Rev. Lett., 34 (1975) 806

ELECTRONIC STRUCTURE OF TRANSITION METAL SURFACES

G. ALLAN

Laboratoire de Physique des Solides

E.R.A. au CNRS ISEN, Lille, France

1. INTRODUCTION

In recent years a great deal of interest has been focused on the properties of surfaces. Many studies of transition metal have been made. It is not necessary to recall here the important part taken in the catalytic reactions by transition metals. It must be also noticed that this sudden interest has been allowed by the development of experimental techniques and by a quite good knowledge of bulk properties.

We shall first recall the main features of the bulk band structure of transition metals, emphasizing its description in the tight-binding approximation which seems to be the most natural way to study the properties of a narrow d valence band.

In spite of the simplicity of the model, the lack of periodicity perpendicularly to the surface plane compelled to further approximations. The earliest one was a description of the d band in a non degenerate s orbital scheme. It allows a complete and exact calculation of the surface states by the Green function method treating the surface as a perturbation of a bulk infinite crystal. However, it leads to heavy numerical calculations if one wants to have a detailed knowledge of the surface band structure. In such a case, one may use the moment method which is well suitable to a tight-binding approximation.

Moreover in section III, the combination of both these approximations allows a self-consistent description of the surface band structure taking into account the charge oscillations near the surface.

In section IV we display the application of the Green function method to surface of simple cubic lattices.

Finally in the last section we report some quantitative results concerning the d surface states.

2. BAND STRUCTURE OF TRANSITION METALS

The main features of the band structure of the transition metal are well known since a long time (1). In the bulk, a transition metal atom keeps strong d atomic characteristics. This is due to the following points (2) :

- the radial extension of the d atomic function is rather small (Fig.1) compared to the s one. This small extension must also be compared to the interatomic distances in transition metals which are close to 2.5.A. This shows that a d function on an atom will not be very much affected by the potentials of its neighbours.
- the parabolic variation of the d atomic function near the origin gives a very bad screening of the nucleus charge. So the d states are successively filled while the number of s or p valence electrons is almost constant along a transition series. This also increases the stability as the charge of the nucleus increases along a series.

The first point suggests that the width of the bulk d band will be rather small (5 to 10 eV). There are some experimental evidences of such a small width as X-ray spectra (3) or optical properties (4). Furthermore there are ten states in this narrow d band and only two in the broad s band which is often described in a free electron scheme. This gives a rather large density of d states (Fig. 2). Such a description of the d band is also consistent with experimental results related to the density of states. The transition metals have at low temperatures a large electronic specific heat and a large paramagnetic susceptibility. Let us remind that these properties are at low temperatures proportional to the density of states at the Fermi level. Many other properties (like cohesive energies or electrical conductivity) are also explained by large d densities of states.

A. The tight-binding approximation

Several techniques can be used to find the wave function of a valence-electron in a transition metal (5). The s and the d bands are often studied separatly. It is possible to treat afterwards the s-d interaction in perturbation theory. Keeping in mind the atomic characteristic of the d wave function, the one electron wave function is written as a linear combination of d atomic orbitals centered on each lattice site $\underset{\sim}{R}_i$:

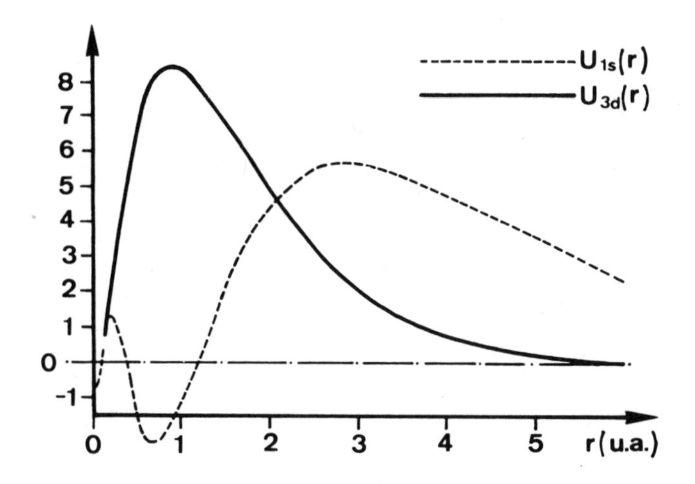

Fig. 1 : s and d radical function of a free transition atom

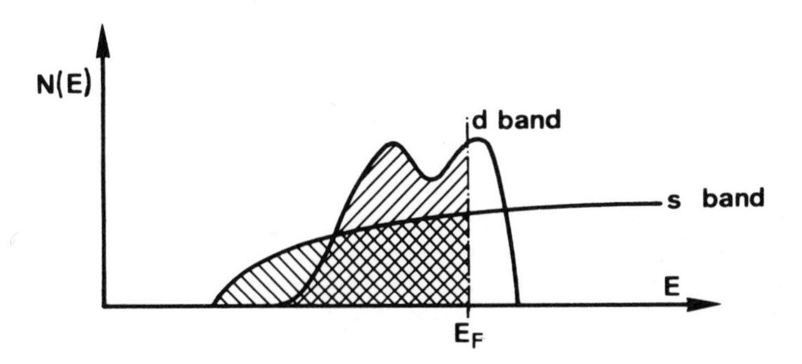

Fig. 2 : Density of states of transition metal valence bands (schematic)

$$\phi(\underset{\sim}{r}) = \sum_{i,j} a_{ij} \, |\psi_j(\underset{\sim}{r} - \underset{\sim}{R}_i)> \qquad (2.1)$$

where the summation over j is extended to the five d functions. We neglect the overlap over two sites of the d functions

$$<\psi_j(\underset{\sim}{r} - \underset{\sim}{R}_i) \mid \psi_k(\underset{\sim}{r} - \underset{\sim}{R}_\ell)> = \delta_{jk}\,\delta_{i\ell} \qquad (2.2)$$

The lattice potential is equal to the sum of atomic potentials $V(\underset{\sim}{r}-\underset{\sim}{R}_i)$

$$U(\underset{\sim}{r}) = \sum_i V(\underset{\sim}{r} - \underset{\sim}{R}_i) \qquad (2.3)$$

Solving the Schrödinger equation leads to a set of linear equations satisfied by the coefficients a_{ij} :

$$(E_o + \alpha_{im} - E)a_{im} + \sum_{j \neq i, m'} \beta_{im}^{jm'} \, a_{jm'} = 0 \qquad (2.4)$$

where E_o is the atomic level, α_{im} the crystal field integral defined by :

$$\alpha_{im} = < \psi_m(\underset{\sim}{r}-\underset{\sim}{R}_i) \mid \sum_{j \neq i} V(\underset{\sim}{r}-\underset{\sim}{R}_j) \mid \psi_m(\underset{\sim}{r}-\underset{\sim}{R}_i)> \qquad (2.5)$$

and $\beta_{im}^{jm'}$ a hopping integral :

$$\beta_{im}^{jm'} = < \psi_m(\underset{\sim}{r}-\underset{\sim}{R}_i) \mid \sum_{k \neq j} V(\underset{\sim}{r}-\underset{\sim}{R}_k) \mid \psi_{m'}(\underset{\sim}{r}-\underset{\sim}{R}_j)> \qquad (2.6)$$

Among the β integrals, we retain only the two center integrals with $\underset{\sim}{R}_i$ and $\underset{\sim}{R}_j$ being nearest neighbours or next nearest neighbours. Owing to the difficulty of the choice of the potential $V(\underset{\sim}{r})$, these integrals are often adjusted to reproduce more sophisticated band calculations or fitted to experimental results like the Fermi surface for example.

In a perfect infinite lattice, the wave function can be taken as a Bloch function built with atomic functions :

$$\phi_{\underset{\sim}{k}}(\underset{\sim}{r}) = \frac{1}{\sqrt{N}} \sum_j A_j \sum_i e^{i\underset{\sim}{k}\underset{\sim}{R}_i} \, |\psi_j(\underset{\sim}{r}-\underset{\sim}{R}_j)> \qquad (2.7)$$

The set of 5N equation (2.4) reduces to a set of 5 equations which must be solved for each value of the wave vector $\underset{\sim}{k}$.

At this state of the theory, it still remains difficult to study an extended defect like a surface and we simplify the band structure treating the d band as a non degenerate band with one orbital per atom. However the generalization to a five-fold degenerate d will be given in the last section.

Using (2.7), (2.4) can be reduced for example in the case of a simple cubic lattice to

$$E(\underset{\sim}{k}) = E_o - \alpha - 2\beta(\cos k_x a + \cos k_y a + \cos k_z a) \tag{2.8}$$

where a is the lattice parameter.

Even within such a simple analytic bulk band structure, it is rather difficult to get a complete self-consistent surface band structure. Some surface properties or tendancies do not depend on band structure details. A rough description of the density of states is enough.

B. The moment method

The moment method seems very well suited to roughly describe the density of states. The calculation of the moments of the state density fits very well the tight-binding approximation.

First of all let us remind ourselves the principles of the moment method. The moments of the density of states are defined by :

$$\mu_p = \frac{1}{N}\int E^p\, n(E)\, dE = \frac{1}{N}\,\mathrm{Tr}\, H^p \tag{2.9}$$

where Tr means the trace of the p-th power of the hamiltonian H. In the basis of the atomic functions $|\psi(\underset{\sim}{r}-\underset{\sim}{R}_i)>$ we can write :

$$\mu_p = \frac{1}{N}\sum_{i_1} <\psi(\underset{\sim}{r}-\underset{\sim}{R}_{i_1})\,|H^p|\,\psi(\underset{\sim}{r}-\underset{\sim}{R}_{i_1})>$$

$$= \frac{1}{N}\sum_{i_1 \ldots i_p}\sum <\psi(\underset{\sim}{r}-\underset{\sim}{R}_{i_1})|H|\psi(\underset{\sim}{r}-\underset{\sim}{R}_{i_2})><\psi(\underset{\sim}{r}-\underset{\sim}{R}_{i_2})|H|\ldots$$

$$\ldots <\psi(\underset{\sim}{r}-\underset{\sim}{R}_{i_p})|H|\,\psi(\underset{\sim}{r}-\underset{\sim}{R}_{i_1})> \tag{2.10}$$

If we take the origin of the energies at $(E_o - \alpha)$, with our approximations all the terms in (2.10) are equal to zero except that involving atomic functions centered on nearest neighbours. Then we get :

$$\mu_p = \beta^p P_p \tag{2.11}$$

where P_p is the number of walk from an atom to itself through (p-1) atoms nearest neighbours and only depends on the lattice geometry.

Then the problem is how to reconstruct the density of states n(E) when one knows its moments μ_p or only a finite number of its moments.

With only a few moments (μ_o, μ_1 and μ_2 for example) one can use a gaussian fitted to these moments :

$$n(E) = \frac{10}{\sqrt{2\pi\mu'_2}} \exp\left(-\frac{(E-\mu_1)^2}{2\mu'_2}\right) \tag{2.12}$$

$$\mu_2 = \mu'_2 + \mu_1^2 = N_V \beta^2 + \mu_1^2 \tag{2.13}$$

where N_V is the coordination number for a bulk atom. The value of α and β may be fitted to the cohesive energy. We define E_A equal to the attractive part of the cohesive energy :

$$E_A = \int^{E_F} E\, n(E)\, dE - N_d\alpha = -10\sqrt{\frac{\mu'_2}{2\pi}} \exp\left(-\frac{E_F^2}{2\mu'_2}\right) - N_d\alpha$$

$$N_d = \int^{E_F} n(E)\, dE \tag{2.14}$$

where N_d is the number of d electrons in the valence band. A repulsive term of the Born-Mayer type is then added to ensure the crystal stability (6). It reduces by a factor 2/3 the attractive energy. This leads to a band width (7) equal to 7.5, 9 and 10.5 eV for the three series. The crystal field integral α is small ($\sim$ 0.5 eV) and it still will be reduced if one takes into account the s-d interaction. Such a density of states (2.13) can only give a rough estimation of the band structure. It is necessary to go beyond a small number of moments to get an accurate density of states. Several methods have been used. It seems that the best way to introduce them is to display the recursive method developped by Haydock, Heine and Kelly (9).

C. The recursive method

Let us define the Green operator G^+ corresponding to the Hamiltonian H of a crystal and its intraatomic matrix elements :

$$G^{+} = \frac{1}{E-H+i\eta} \qquad (2.15)$$

$$\langle\psi(\underset{\sim}{r}-\underset{\sim}{R}_i)|G^{+}|\ \psi(\underset{\sim}{r}-\underset{\sim}{R}_i)\rangle = \phi\int\frac{n_i(E')}{E-E'}\,dE' - i\pi\, n_i(E)$$

ϕ is for the principal part and $n_i(E)$ is the local density of states on atom i

$$n_i(E) = \sum_{\substack{\underset{\sim}{k} \\ E(\underset{\sim}{k})=E}} |a_i(k)|^2 \qquad (2.16)$$

where the coefficients a_i are defined by (2.1) omitting here the orbital index j. In a perfect infinite lattice, $n_i(E)$ does not depend on index i and is equal to n(E). It is clear that this situation is changed when the crystal is perturbed by a defect.

Let us define the problem of a semi-infinite linear chain plus a free atom A and the corresponding hamiltonian H_o such as :

$$H_o = H_c + H_A \qquad (2.17)$$

where H_c and H_A are the hamiltonians of the linear chain and of the one of the free atom (Fig.3a). A coupling between the atom A and the atom R at the end of the chain may be represented by the potential V and let us put H the total hamiltonian

$$H = H_o + V \qquad (2.18)$$

If we define :

$$G_o = \frac{1}{E-H_o+i\eta}$$

$$G = \frac{1}{E-H+i\eta} \qquad (2.19)$$

straightforward manipulation leads to the well known Dyson equation

$$G = G_o + G_o VG$$

$$\text{or } G = \frac{G_o}{I - G_o V} \qquad (2.20)$$

We also define :

$$G_A^o = \langle A|G_o|A\rangle = \frac{1}{E+i\eta} \quad \text{and} \quad G_A = \langle A|G|A\rangle \qquad (2.21)$$

as E_o the free atom level is taken equal to zero and here α is neglected and :

$$G_R^o = \langle R|G_o|R\rangle \qquad (2.22)$$

The coupling between atoms R and A is equal to (Fig.3b)

$$\langle R|V|A\rangle = \langle A|V|R\rangle = -\beta \qquad (2.23)$$

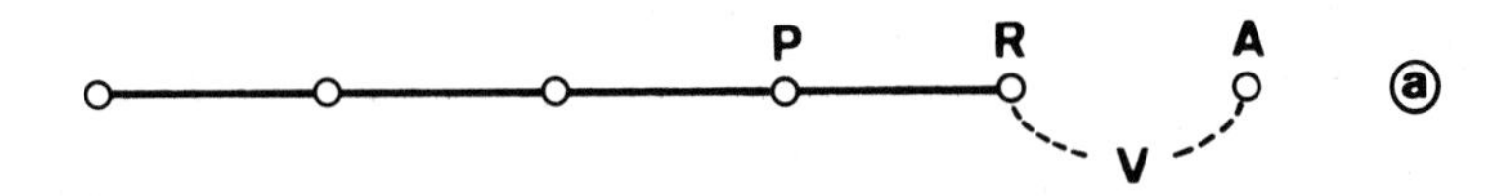

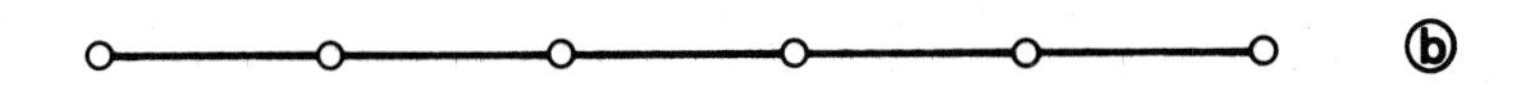

Fig.3 : Adsorption of an atom on a linear chain

Equation (2.20) leads to :

$$G_A = \frac{G_A^o}{1-\beta^2 G_A^o G_R^o} = \frac{1}{E-\beta^2 G_R^o} \qquad (2.24)$$

but the atom A is now at the same position as atom R before the adsorption so

$$G_A = G_R^o = \frac{1}{E-\beta^2 G_R^o} \qquad (2.25)$$

which gives :

$$G_R^o = \frac{E \pm (E^2-4\beta^2)^{1/2}}{2\beta^2}$$

We also have :

$$G_R = \frac{G_R^o}{1-\beta^2 G_R^o G_A^o} = G_P^o$$

and R is now the second atom of the linear chain P before adsorption. By degrees, it is possible to obtain all the intra-atomic matrix elements of G_o.

Then the idea is to apply the same method with a tri-dimensional crystal. We have to choose a basis of orthogonal functions which tri-diagonalizes the corresponding hamiltonian to get a similar form as the linear chain hamiltonian.

Taking for the first state $|1\}$ the atomic orbital $|0\rangle$ on the atom which we consider, the state $|2\}$ is defined by :

$$|2\} = H\,|1\} - a_1\,|1\} \qquad (2.26)$$

where the coefficient a_1 is chosen to orthogonalize $|2\}$ and $|1\}$. Similarly by repeated operations with H we get :

$$|n+1\} = H|n\} - a_n|n\} - b_{n-1}|n-1\}$$

$$H_{n\,n} = \frac{\{n|H|n\}}{\{n|n\}} = a_n \qquad (2.27)$$

$$H_{n-1,n} = H_{n,n-1} = (b_{n-1})^{1/2}$$

$$H_{n,m} = 0 \quad \text{for } |n-m| > 1$$

Using now repeated adsorption of an atom to this "linear chain" leads to :

$$G_{oo} = \langle 0|G\,|0\rangle = \frac{1}{E-a_1-b_1\,g_1(E)}$$

$$g_1(E) = \frac{1}{E-a_2-b_2\,g_2(E)} \qquad (2.28)$$

$$g_2(E) = \frac{}{E-a_3-b_3\,g_3(E)} \quad \ldots.$$

The coefficients a_n and b_n converge to constant values a and b related to the band limits E_T and E_B

$$a = \frac{E_T + E_B}{2} \qquad b = \frac{(E_T - E_B)^2}{16} \tag{2.29}$$

So one can truncate the process at the n-th stage putting a_m and b_m $(m \geqslant n)$ equal to the constant values a and b. There we have :

$$g(E) = \frac{1}{E-a-bg(E)} \text{ , so } g(E) = \frac{E-a}{2b}\left(1 - \left(1 - \frac{4b}{(E-a)^2}\right)^{1/2}\right) \tag{2.30}$$

The substitution of this equation into one another immediately generates a continued fraction which provides the density of states. It is also clear that :

$$G_{oo} = \langle 0|\frac{1}{E-H}|0\rangle = \langle 0|\frac{1}{E} + \frac{H}{E^2} + \dots \frac{H^n}{E^{n+1}} \tag{2.31}$$

$$= \frac{1}{E} + \frac{\mu_1}{E^2} + \dots \frac{\mu_n}{E^{n+1}} + \dots$$

for E outside the band.

So the coefficients a_n and b_n are related to the moments of the density of states.

Such a procedure may be generalized to a five orbital band and to a crystal with defects (see last section).

3. SELF-CONSISTENT APPROACH OF THE ELECTRONIC STRUCTURE OF TRANSITION METAL SURFACES

We shall first display the problem of a self-consistent approach of the surface band structure in a very simple band scheme. We have already said that the local density of states is perturbed by a defect. These charge oscillations change the potential seen by an electron and a self-consistent calculation must be performed. If one actually wants to do such a calculation, he has to make several approximations either on the band structure or on the self-consistent potential. A good approach seems to treat exactly the Coulomb terms and to use a very simple band structure such as (2.13). This will suggest some approximations very useful if one wants to describe accurately the surface band structure in a true d band scheme.

The first test of a surface model is certainly the calculation of the surface tension and if the model is self-consistent then one can calculate the work function.

A. Work function

Let us recall that the work function is the energy required to transfer an electron at the Fermi level inside the bulk to a small distance far away from the surface (Fig.4). This is equivalent to determine the surface dipole layer χ and the chemical potential μ which is a bulk quantity. In a band structure scheme, one can separate the problem into the following two steps :

- Firstly one has to calculate the distance between the Fermi level E_F and the average d level in the solid E_d (Fig.4) which was above taken equal to zero. This quantity may be easily determined if one knows the integral of n(E). Only a few moments are necessary to determine (E_F-E_d).
- Then remains the problem of finding the average d level E_d in the bulk. This has been done by several methods (10-11). The main effect seems the increasing of the s electron density which repells the d level upwards. This shift may be roughly estimated taking a Wigner-Seitz approximation to evaluate the s density in a crystal. It is equal to almost 4 eV for each transition metal (Fig.5).

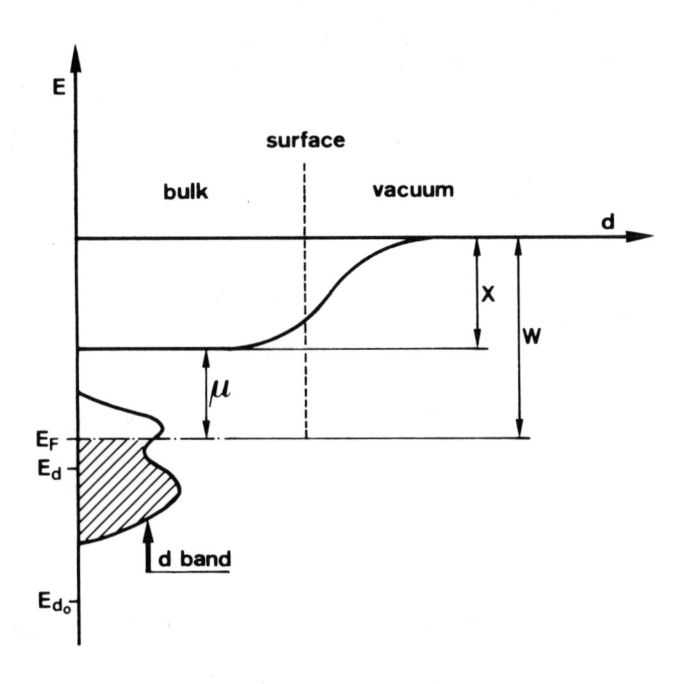

Fig.4 : Work function definition

Using for the free atom d level E_{do} the results of self-consistent Hartree-Fock calculations (12), this leads to a quite good overall agreement with the experimental work functions. So it seems that the corrections due to the surface dipole layer are small and would give rise to a work function anisotropy.

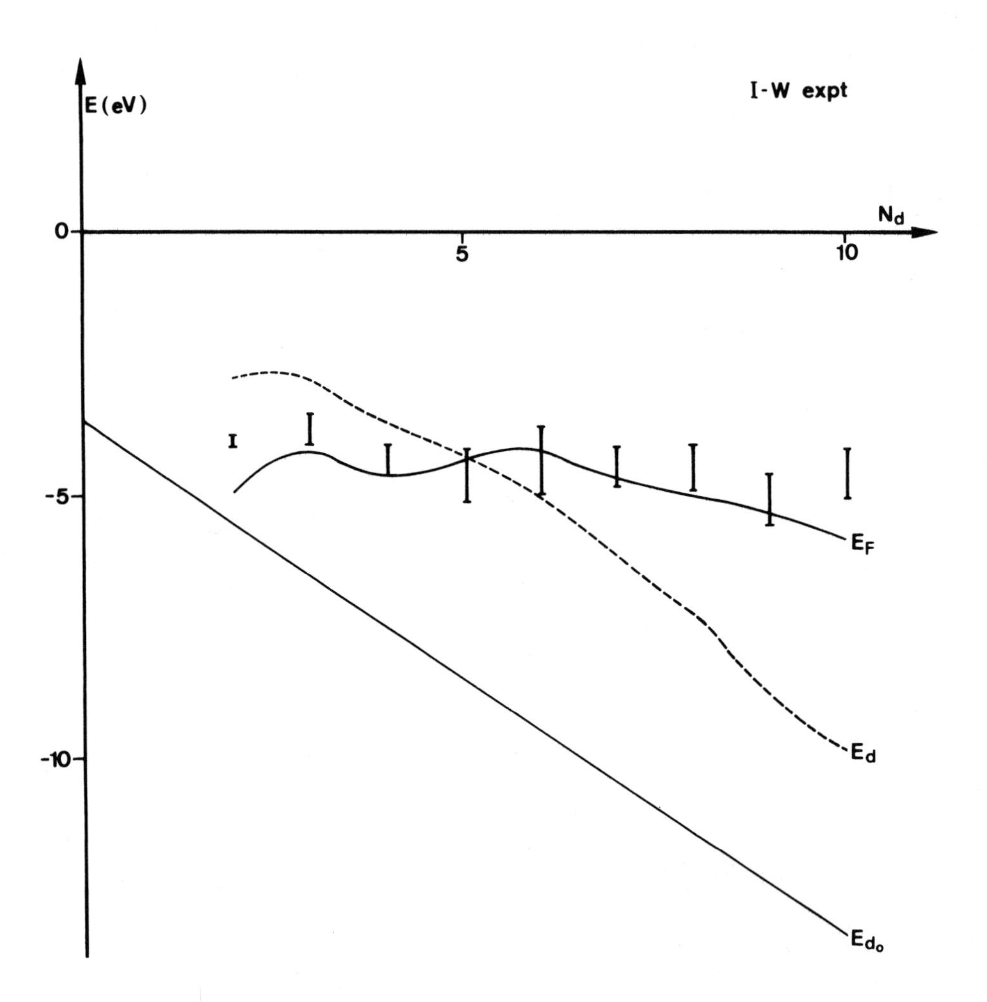

Fig. 5 : Comparison of the values of the Fermi level and the work function for a transition metal.
The experimental value is a mean value for the three series.

B. The d electron dipole layer

We shall only estimate the correction due to the d electrons ; many terms are involved in the s dipole layer notably surface relaxations which are not very well defined.

For simplicity we assume that the resonance integral β is independent of eventual charge oscillations. These charge oscillations create a potential on each atom i and we write its intraatomic matrix element as U_i :

$$U_i = \sum_j \gamma_{ij} \, \Delta N_j(E_F) \tag{3.1}$$

$\Delta N_j(E_F)$ is the net change of charge on j-th atom and γ_{ij} is nearly equal to the inverse interatomic distances except for γ_{ii} in which case it is equal to the intraatomic Coulomb repulsive term. On the other hand, $\Delta N_j(E_F)$ must be calculated considering the perturbation due to the surface and the potentials U_i. In a simplified band structure (2.13) the perturbation leads to a variation of the moments of the local density of states on atom j :

$$\begin{aligned} \mu_o &= 1 \\ \mu_1 &= U_j \\ \mu_2 &= N_j\beta^2 + U_j^2 \end{aligned} \quad . \tag{3.2}$$

where N_j is the coordination number of atom j. Hence we get :

$$n_j(E) = \frac{10}{(2\pi N_j)^{1/2}\beta} \exp\left[-(E-U_j)^2/2N_j\,\beta^2\right] \tag{3.3}$$

The charge neutrality must be ensured

$$\sum_j \Delta N_j(E_F) = 0$$

and the perturbation due to the surface is completely screened at large distances, hence the chemical potential μ is constant. These properties are formulated in the well known Friedel sum rule (13).

For a perfect surface, ΔN_j and U_j take the same values for all the atoms of a given atomic plane parallel to the surface. This reduces the numerical solution of equations (3.1) and (3.3). The explicit introduction of higher moments does not change appreciably the results :

- the charge oscillations are small. They are less than a few hundredth of an electron and are localized near the surface plane.

- the potentials U_j are almost constant inside the bulk except for the surface term.

These results lead to the approximation of figure 6 and using (3.3) to :

$$\Delta N_i(E_F) = \frac{10}{\sqrt{2\pi}} \int_{E_F/\mu^{1/2}}^{(E_F-U_i)/\mu_i^{1/2}} \exp - \frac{u^2}{2} \, du \sim 0 \quad (3.5)$$

where μ is the bulk second moment of the density of states and μ_i is defined by (3.2). Equation (3.5) leads to the following value of the self-consistent potential

$$U_i = E_F \left[1 - \left(\frac{\mu_i}{\mu}\right)^{1/2}\right] \sim E_F \times \frac{M}{2N_V} \quad (3.6)$$

where M is the number of dangling bonds. As U_i is nearly equal to the opposite of the dipole layer potential (Fig.6) and E_F is less than 4 eV (half the band width), (3.6) shows that U_i is always less than 1 eV as M is at most equal to $N_V/2$.

Let us summarize here the main results of concerning the surface band structure :

- the width of the d local surface density of states is reduced as μ_2(3.3)
- this surface density of states is shifted by a quantity which is linear in E_F and of the order of 1 eV (3.6) for a nearly empty or full d band.
- the perturbation due to the surface on the local d densities of states is localized near the surface plane as the potential and charge oscillations.

Such a band structure only gives a rough idea of the surface band structure and it is necessary to go beyond the second moment approximation to get localized surface states for example.

C. Surface energy and surface defects

However some properties like the surface energy which depends only on integrated density of states can be very well estimated with the band structure described above.

Within our approximations, the change in the electronic energy may conveniently be expressed as a sum of each atomic contribution

$$\Delta E_{Ai} = \int^{E_F} E \Delta n_i(E)\, dE - N_d U_i - \frac{1}{2} \Delta N_i(E_F) U_i \qquad (3.7)$$

The first term in (3.7) is equal to the difference between the one electron energies of the semi-infinite crystal and that of a perfect lattice. Both the last terms in (3.7) are equal to the Coulomb energy variation counted twice in the one-electron energies (14). If we make the approximation (3.5), simple manipulation of (3.7) leads to :

$$\Delta E_{Ai} = E_A \left(\sqrt{\frac{\mu_i}{\mu}} - 1 \right) \approx - E_A \frac{M}{2N_V}$$

where E_A is defined by (2.14) and M is the number of dangling bonds. The variation of the repulsive term (6) is equal to :

$$\Delta E_{Ri} = + \frac{E_A}{3} \times \frac{M}{N_V}$$

such as the total variation of the energy is equal to :

$$\Delta E_i = \Delta E_{Ai} + \Delta E_{Ri}$$

$$= - E_A \frac{1}{6} \times \frac{M}{N_V} = - \frac{M}{4N_V} E_c$$

instead of $- \frac{M}{N} E_c$ in a pairwise potential approximation.

Such an estimation may be also extended to surface vacancies or surface defects (steps for example). It seems that no calculations of surface states have been performed around surface defects even in very simple band structures. However it may be noticed that the self-consistent potential is certainly not negligible in such a calculation.

4. THE GREEN FUNCTION METHOD APPLIED TO THE SURFACE STATES

A. The Green function (or resolvent) method

We now want to improve the band structure and use a more accurate description of the density of states. However we keep one orbital per atom to display some results concerning a five fold degenerate d band in the last section. We shall see at once that any improvement in the treatment of the band structure increases rapidly the amount of calculation and restricts the field of application.

Some other approximations are then necessary. It is the case of the self-consistent determination of the potential near a surface. In fact, we shall take the approximation already used before. We assume that all the intraatomic matrix elements of the self-consistent potential are equal except on surface atoms (Fig.6). The value of U_o must satisfied the Friedel sum rule :

$$\Delta N(E_F, U_o) = 0 \qquad (4.1)$$

where $\Delta N(E_F, U_o)$ is the total variation of the number of states below the Fermi level.

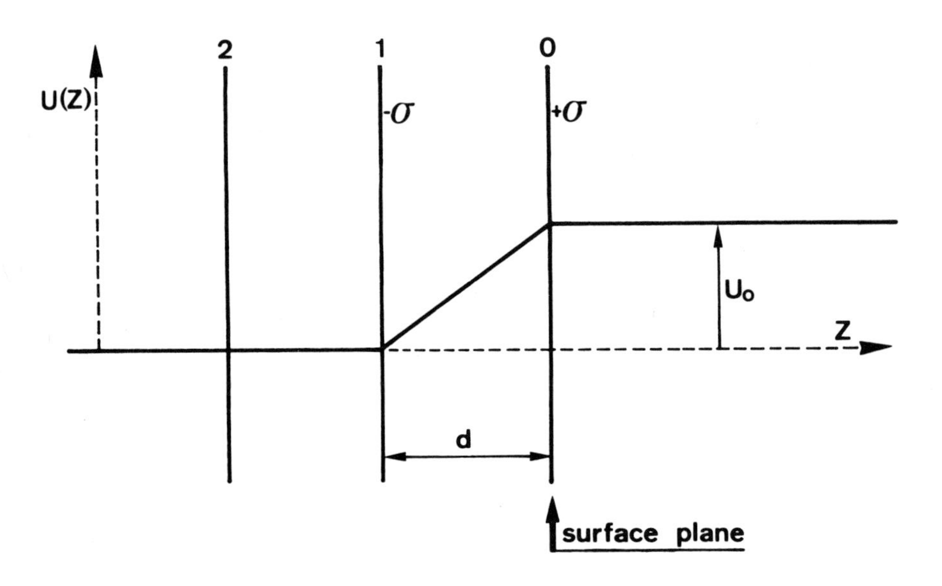

Fig. 6 : Approximation of the surface dipole layer and the corresponding potential energy

In an infinite perfect crystal, the creation of a surface may be represented by a potential V which cancels any interaction between atoms lying on both sides of the surface plane (Fig.7).

We denote the one electron hamiltonian of a perfect infinite crystal by H_o and H_S the hamiltonian of the two semi-infinite crystals

$$H_S = H_o + V \qquad (4.2)$$

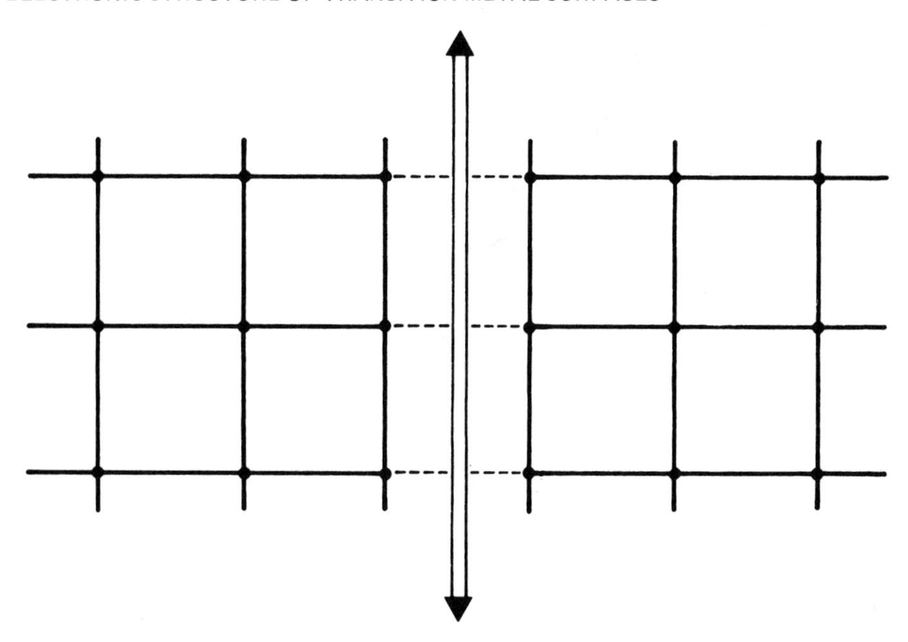

Fig. 7 : Creation of two surfaces by cutting an infinite crystal

Let us denote by G_o^+ and G_S^+ the corresponding Green operators :

$$G_o^+ = \frac{1}{E-H_o+i\eta} \quad ; \quad G_S^+ = \frac{1}{E-H_S+i\eta} = \frac{1}{I-G_o^+ \times V} \; G_o^+ \qquad (4.3)$$

We also put :

$$H_o|\phi_{\underset{\sim}{k}}> = E(\underset{\sim}{k})|\phi_{\underset{\sim}{k}}> \qquad (4.4a)$$

$$H \; |\psi> = E|\psi> \qquad (4.4b)$$

which are the equations to be solved for the perfect crystal before and after the perturbation.

Equation (4.4b) may be written if E is different of any $E(\underset{\sim}{k})$

$$(E - H_o)|\psi> = V|\psi> \qquad (4.5)$$

or

$$|\psi\rangle = \frac{1}{E - H_o} V|\psi\rangle$$

In the basis of atomic functions for example, (3.5) is equivalent to a linear set of homogeneous equations which admits a solution non equal to zero if

$$\det (1 - G_o^+ V) = 0 \qquad (4.6)$$

This equation determines the energies of bound states as a function of the characteristics of the perfect crystal band structure. Such states which cannot exist in a perfect infinite crystal are localized near the surface.

On the other hand, equation (4.4b) may be written if E lies in the bulk band

$$|\psi_{\underset{\sim}{k}}\rangle = |\phi_{\underset{\sim}{k}}\rangle + \frac{1}{E-H_o+i\eta} V|\psi_{\underset{\sim}{k}}\rangle \qquad (4.7)$$

The projection of this equation on the basis of atomic function for example allows to calculate the local density of states $n_i(E)$ on atom i and may also be deduced from G_S^+ using (2.15) and (2.16).

Moreover the total variation of the number of states $\Delta N(E)$ is simply equal to (15-16) :

$$\Delta N(E) = \sum_i \int \Delta n_i(E)\, dE = \frac{-1}{\pi} \operatorname{Arg} \left(\det(I - G_o^+ V)\right) \qquad (4.8)$$

So all the properties of the perturbed crystal may be deduced from those of the perfect one.

B. Application to surface band structure (17-20)

Since H is invariant under translations parallel to the surface, we shall use the formalism of Brown (21) who defines Wannier functions $|R_Z, r, \underset{\sim}{k}_{\parallel}\rangle$ localized in planes parallel to the surface. R_Z is the distance between the plane where the Wannier function is localized and the surface plane, $\underset{\sim}{k}_{\parallel}$ is the component of the wave vector parallel to the surface and k_Z its normal component :

$$|R_Z, r, \underset{\sim}{k}_{\parallel}\rangle = (N)^{1/2} \sum_{k_Z} e^{-ik_Z R_Z} \phi(\underset{\sim}{k}, \underset{\sim}{r}) \qquad (4.9)$$

where N is the number of planes parallel to the surface, each plane including N^2 atoms.

In this basis, the matrix elements of the potential V are :

$$V(\underset{\sim}{k}_{/\!/},R_Z,R'_Z) = \langle R'_Z,\underset{\sim}{r},\underset{\sim}{k}'_{/\!/}|V|R_Z,\underset{\sim}{r},\underset{\sim}{k}_{/\!/}\rangle \qquad (4.10)$$

$$= \delta_{\underset{\sim}{k}_{/\!/},\underset{\sim}{k}'_{/\!/}} \frac{1}{N^2} \sum_{\underset{\sim}{R}_{/\!/},\underset{\sim}{R}'_{/\!/}} e^{i\underset{\sim}{k}_{/\!/}(\underset{\sim}{R}_{/\!/}-\underset{\sim}{R}'_{/\!/})} \langle\psi(\underset{\sim}{r}-\underset{\sim}{R}_Z-\underset{\sim}{R}_{/\!/})|V| |\psi(\underset{\sim}{r}-\underset{\sim}{R}'_Z-\underset{\sim}{R}'_{/\!/})\rangle$$

Let us remind here our simplified d band model : we study a non degenerate band with only nearest neighbours interactions. If an atom has its nearest neighbors in two or three adjacent planes parallel to the surface, then it is possible to write :

$$E(\underset{\sim}{k}) = W(\underset{\sim}{k}_{/\!/}) + 2T(\underset{\sim}{k}_{/\!/}) \cos [k_Z d + \theta(\underset{\sim}{k}_{/\!/})]$$

$$W(\underset{\sim}{k}_{/\!/}) = \langle R_Z,\underset{\sim}{r},\underset{\sim}{k}_{/\!/}|H_o|R_Z,\underset{\sim}{r},\underset{\sim}{k}_{/\!/}\rangle \qquad (4.11)$$

$$T(\underset{\sim}{k}_{/\!/}) \exp [-i\theta(\underset{\sim}{k}_{/\!/})] = \langle R_Z,\underset{\sim}{r},\underset{\sim}{k}_{/\!/}|H_o|R_Z+d,\underset{\sim}{r},\underset{\sim}{k}_{/\!/}\rangle$$

where d is the distance between two adjacent planes parallel to the surface. The expressions of T and W are given in ref. 18 and 20. We have now reduced the problem of surface states to N^2 one-dimensional problem, one for each value of $\underset{\sim}{k}_{/\!/}$.

For $\underset{\sim}{k}_{/\!/}$ fixed, the limits of the one-dimensional bulk band are given by W ± 2T. For a fixed value of $\underset{\sim}{k}_{/\!/}$, if a surface bound state exists, its energy is necessarily outside this one-dimensional band. But as that band is always included in the bulk band and is smaller, a surface bound state may occur in the bulk band. However it remains localized near the surface and must not be confused with a resonant state.

In the basis of the Wannier functions defined above, the matrix of the perturbing potential V is 2 x 2 matrix. The non zero elements occur for two fonctions localized on the two surfaces if we neglect the relaxation of the atomic planes near the surfaces.

If we neglect the crystal field parameter α:

$$V = \left\| \begin{matrix} U_o & -T(\underset{\sim}{k}_{/\!/}) \exp [-i\theta(\underset{\sim}{k}_{/\!/})] \\ -T(\underset{\sim}{k}_{/\!/}) \exp + [i\theta(\underset{\sim}{k}_{/\!/})] & U_o \end{matrix} \right\| \qquad (4.12)$$

The off-diagonal elements just cancel the corresponding matrix elements of H_o. The diagonal terms are the intraatomic matrix elements of the self-consistent potential.

The calculation of the corresponding matrix elements of G_o^+ are analytic (18-19) and depends only on the distance between planes R_Z and R_Z'

$$G_o^+(n,\underset{\sim}{k}_*) = i(4T^2-\omega^2)^{-1/2} \left[\frac{\omega+i(4T^2-\omega^2)^{1/2}}{2T} \right]^n \exp(-in\theta)$$

$$\omega = E - W(\underset{\sim}{k}_*)$$

$$|R_Z-R_Z'| = n\,d$$

Then one can calculate $\Delta N(E_F,U_o)$ which gives the value of U_o.

The bound states energies are equal to :

$$E_L(\underset{\sim}{k}_*) = W(\underset{\sim}{k}_*) + U_o + T^2(\underset{\sim}{k}_*)/U_o$$

with the following condition :

$$|U_o| > |T|$$

Let us point here the importance of the potential U_o. One can notice that if U_o equal to zero as in non self-consistent model, there is no bound state. It is certainly not a general rule in the case of a degenerate d band, however this point shows the importance of self-consistency for the existence of surface bound states or their position.

Then the local density of states may be calculated as a sum of the extended states :

$$n_E(E,R_Z) = (\pi N^2)^{-1} \; \mathrm{Jm} \sum_{\underset{\sim}{k}_*} i\mu^{-1} \left[1+\left(\frac{\omega+\mu}{2T}\right)^{2m} \frac{\mu-i(\omega-2U_o)}{\mu+i(\omega-2U_o)} \right] u(4T^2-\omega^2)$$

and of the bound states

$$n_L(E,R_Z) = N^{-2} \sum_{\underset{\sim}{k}_*} u(|U_o|-|T|)\left(\frac{T}{U_o}\right)^{2m} (1-\frac{T^2}{U_o^2})\; \delta(E-W-U_o-T^2/U_o)$$

with : $R_Z = ma$

$\mu = (4T^2-\omega^2)^{1/2}$

$\omega = E - W$

$$u(x) = 1 \quad \text{if } x \geq 0$$
$$= 0 \quad \text{otherwise}$$

The quantity $N_L(R_Z)$ which is equal to the number of bound states at a given distance from the surface is equal to (7)

$$N_L(R_Z) = \int n_L(E,R_Z)\, dE \approx \left[1 - \left(\frac{T}{U_o}\right)^2\right] \left(\frac{T}{U_o}\right)^{2m}$$

where here T is now a mean value of $T(k_{*})$. This expression shows notably the experimental decrease of the number of band states as one goes inside the bulk. About 70 per cent of the bound states are localized in the surface plane (20).

We give on figures 8 and 9 the results concerning the (100) planes of the b.c.c. lattice near the middle of the band and are close to the top of the band for a f.c.c. lattice. In spite of the crude approximations, it is interesting to note here that such bound states will also be found with more sophisticated band structure although in this case they do not depend so much of the selfconsistent potential. In the case of tungsten ($N_d \sim 5$; $E_F \sim 0$) it seems that there is a bound state band or a resonant state band just below the Fermi level (22-24). The same surface states band is also observed for molybdenum (25) which is also a b.c.c. lattice with an almost half filled d band. In the case of the f.c.c. lattice, surface bound states near the top of the d band have been used to explain the magnetic susceptibility of small Pd particles (26) but these experimental results have not been confirmed.

5. d BAND SURFACE STATES

A. Local density of states

The models developed in the preceeding sections study the perturbation induced by the surface on a non degenerate s band. If some experimental results are not very sensitive to the true shape of n(E) as they depend only upon the integral of n(E), there are also many properties strongly related to $n(E_F)$ or more generally to n(E). Let us quote photoemission results for example. A non degenerate s band cannot reproduce the true density of d states and it seems necessary to go beyond this approximation. Nevertheless the investigation of surface effects on a degenerate d band needs some approximations notably for the determination of the self-consistent potential. It has been tested for an s band for which it is possible to perform an exact calculation of this term and then we now applied it to the d band.

We mainly report here the work of M.C. Desjonqueres and F. Cyrot-

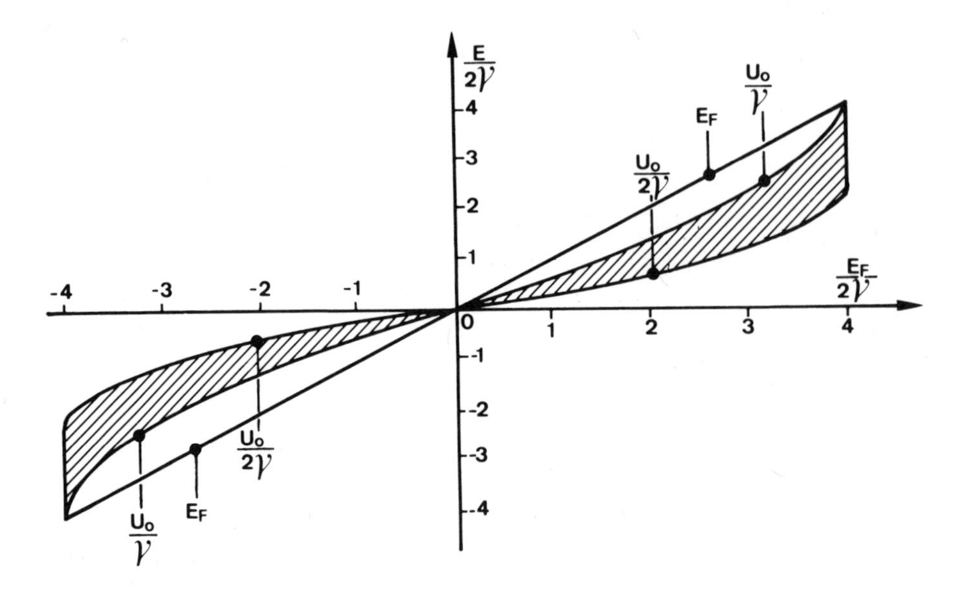

Fig.8 : Surface bound states band for a (100) surface of a b.c.c. lattice

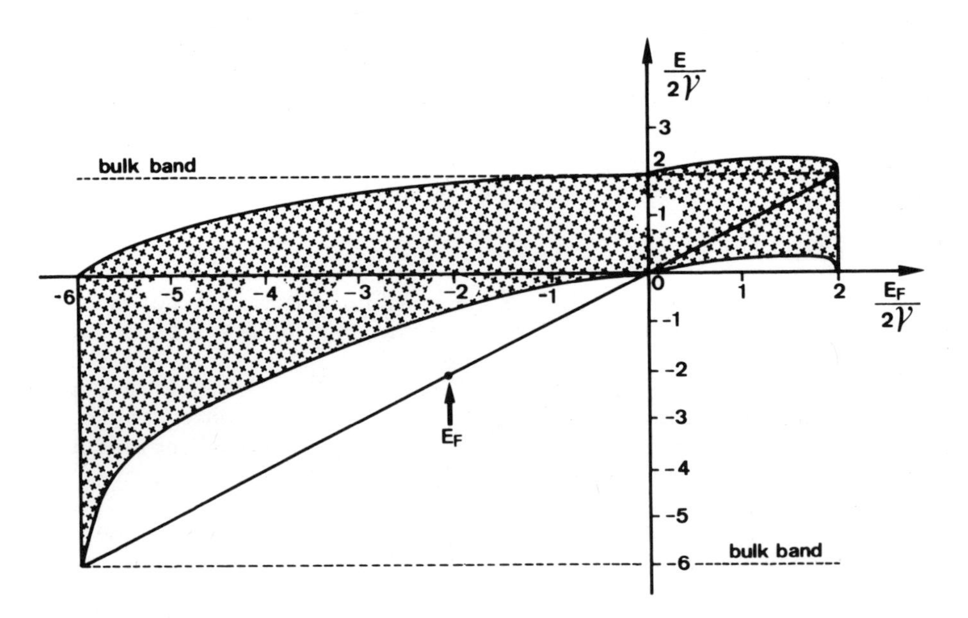

Fig.9 : Surface bound states band for a (100) surface of a f.c.c. lattice

Lackmann (27-29). They investigate the effects of charge transfers and of the self-consistent potential on transition metal surface electronic structure. The Friedel sum rule is satisfied fitting one matrix element of the self-consistent potential in the surface plane.

The figure 10 shows the value of the self-consistent potential (27) for f.c.c. and b.c.c. lattice surface. The potential U_o is roughly proportional to E_F and to the number of dangling bonds M as in equation (3.6). The discrepancies between equation (3.6) and fig.10 are due to the peaks in the density of states. In a rigid band model, to get very small charge oscillations ($\Delta N_i(E_F) \approx 0$), it is obvious that U_o decreases where $n_i(E_F)$ increases. The agreement of equation (3.6) and of the curves of figure 10 is better for a b.c.c. lattice than for a f.c.c. lattice. This point has also been raised for an s band by the Green function technique (20). The second moment approximation is obviously better for an b.c.c. lattice than for a f.c.c. one due to the third moment of the density of states. Let us also remark that U_o is roughly equal to the surface d dipole layer. The small values of U_o give some confidence in the evaluation of transition metal work functions developed in the second section.

Let us now look at the corresponding surface densities of states in the case of b.c.c. iron and f.c.c. nickel. We compare in both cases for the (100), (110) and (111) surfaces the bulk density of states with the surface densities of states for a self-consistent ($U_o \neq 0$) and a non self-consistent potential ($U_o = 0$).

For iron (fig.11-13) the self-consistent potential only shifts the d band but does not alter too much the overall shape. The large peak close to the Fermi level is kept in the surface densities. The peak is important in the discussion of surface magnetism. One also remarks the presence of a sharp peak near the middle of the d band for the (100) and (111) surfaces. This peak is assigned to surface bound or resonant states. It is also correlated to the observed surface states on the (100) face of W and M_o (22-25). Such a peak does not exist for the (110) surface whose local surface density of states is similar to the bulk one. Let us also remark that for such a plane, the number of dangling bonds is equal to two instead of four for a (100) and a (111) plane. This can also be seen on the width of the d band which is strongly reduced for the (100) and (111) surfaces according to the second moment of the density of states which is proportional to the number of nearest neighbours.

The following results (fig.14-16) concern nickel which has a nearly full d band. A fairly sharp peak appears near the middle of the d band for the (100) and the (110) surface but not for the (111) one. The main contribution to this peak arises from $(3Z^2-r^2)$, XZ and YZ orbitals which stick out from a (001) surface. These orbitals are the most affected by the creation of the surface (30-32). Bound states occur also near the top of the d band. This peak is mainly due

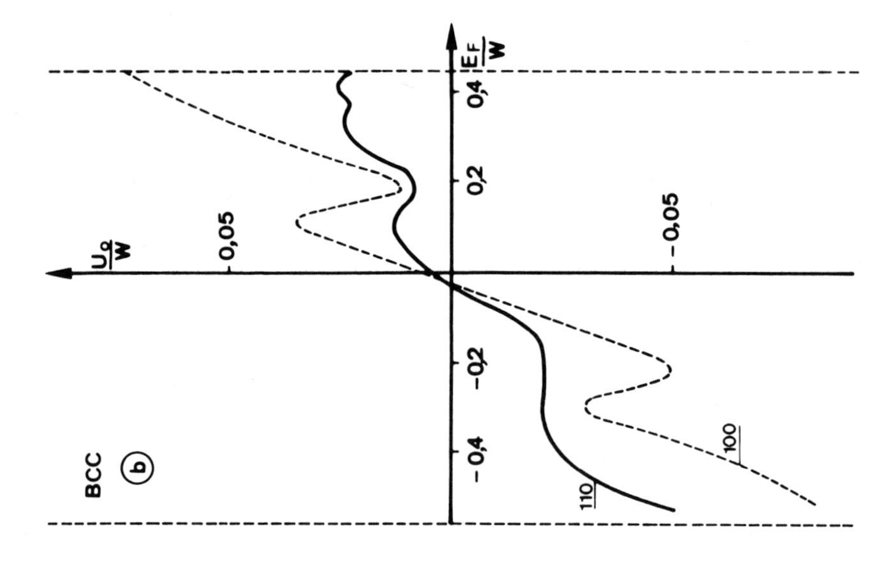

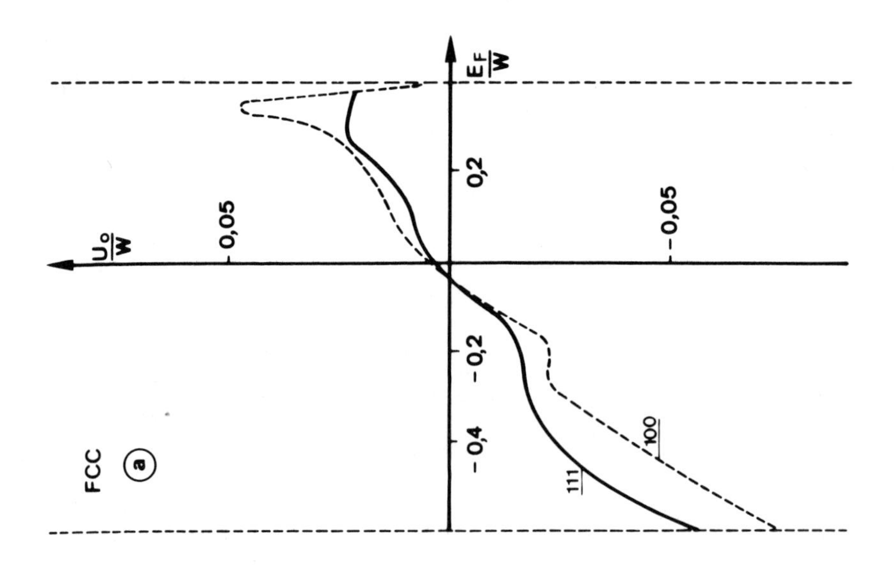

Fig.10 : Value of the surface self-consistent potential for a degenerate d band for b.c.c. and f.c.c. lattices

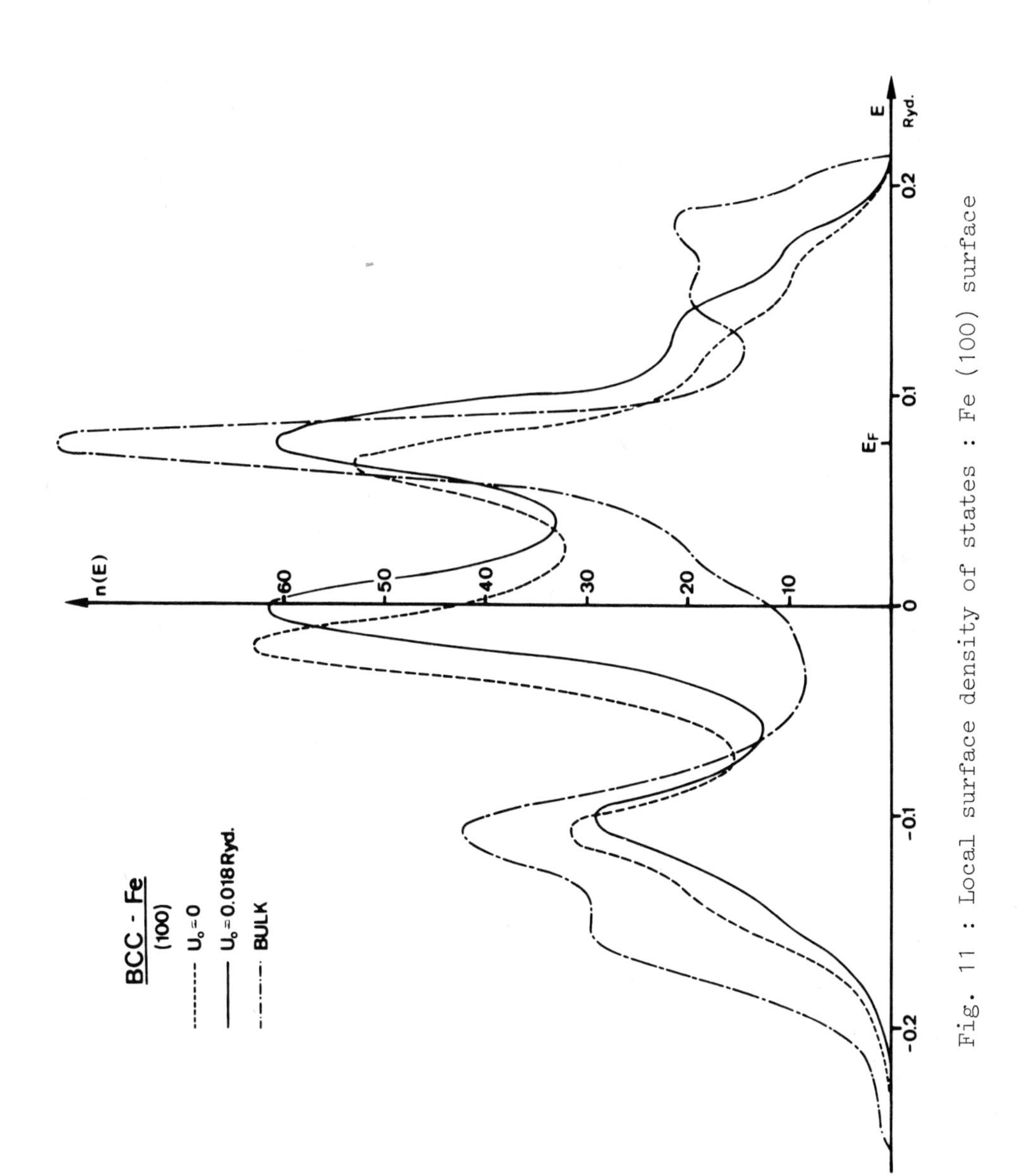

Fig. 11 : Local surface density of states : Fe (100) surface

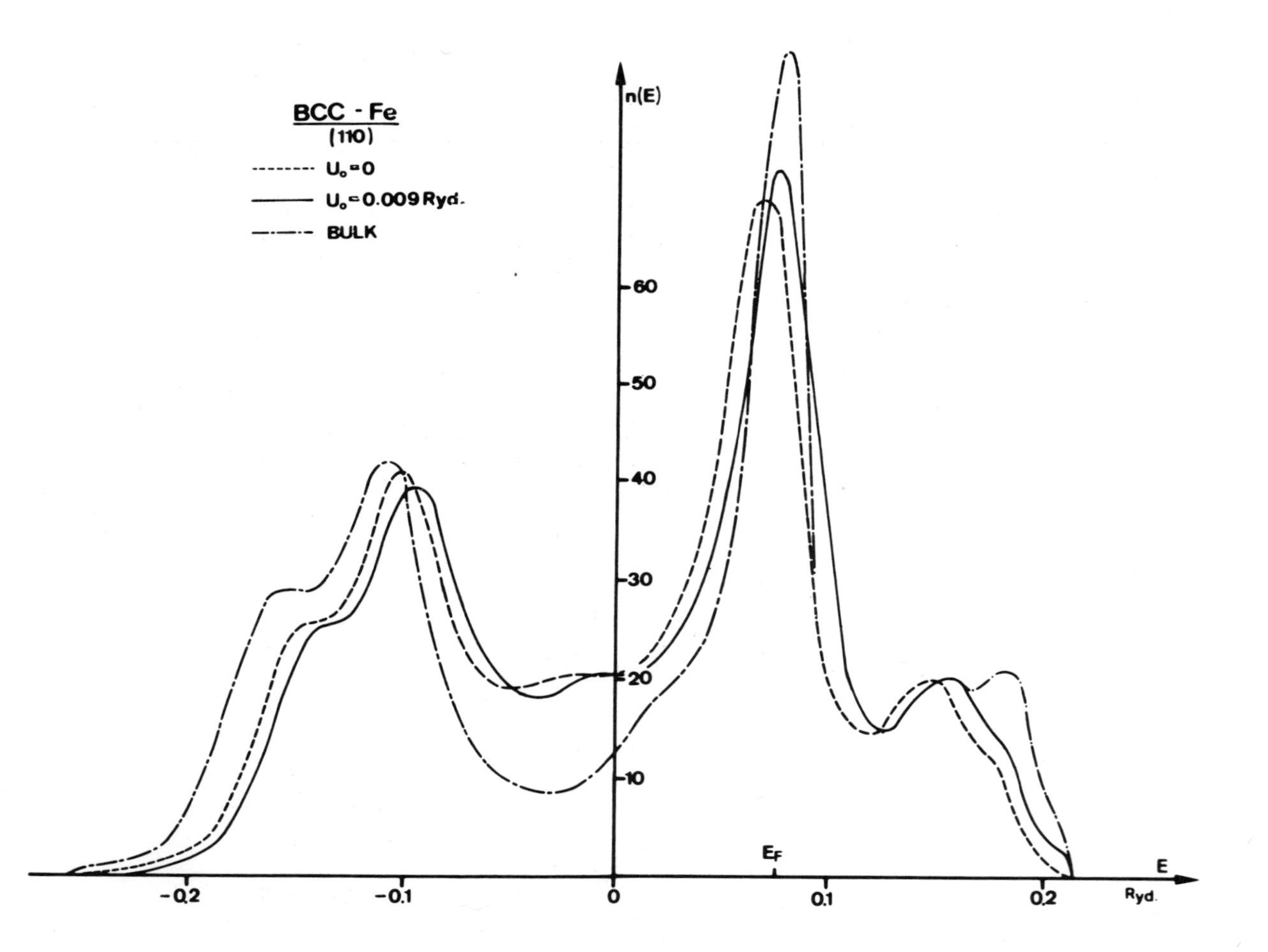

Fig.12 : Local surface density of states : Fe (110) surface

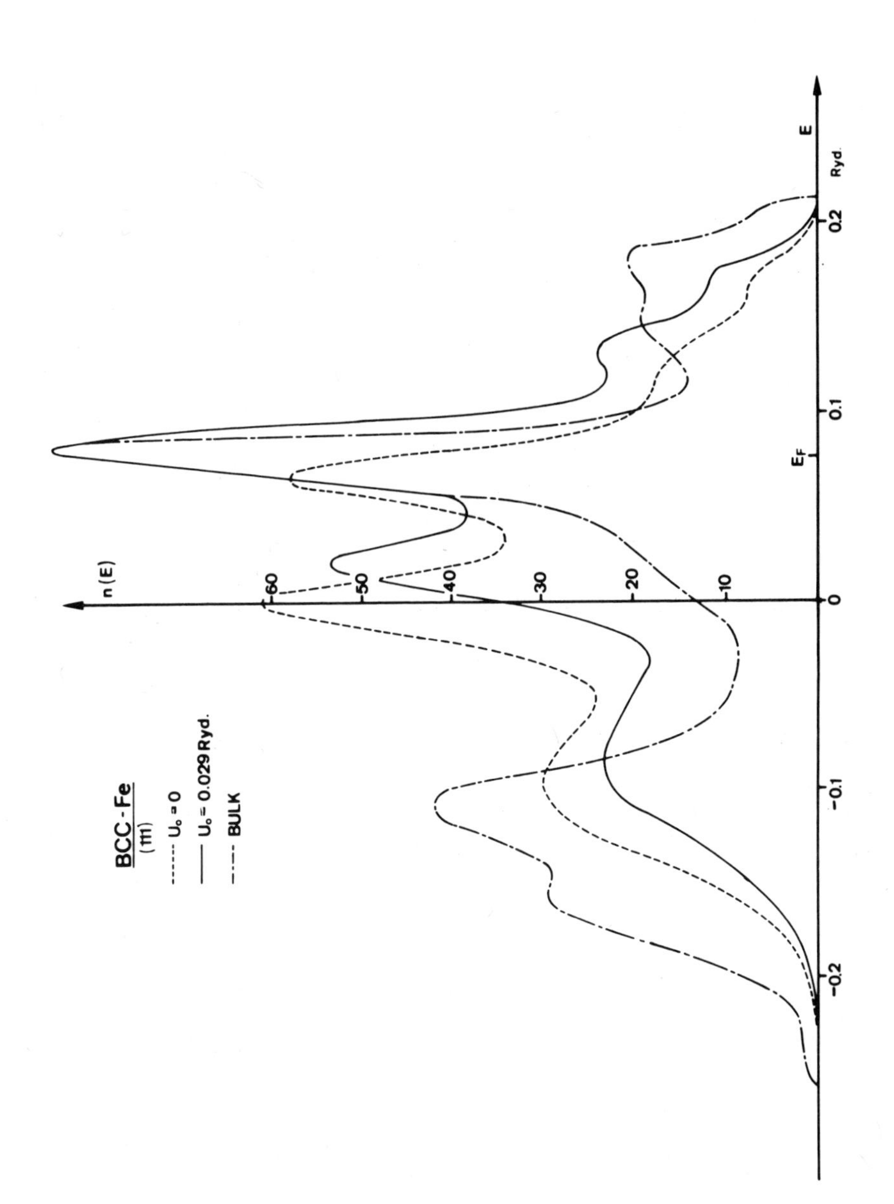

Fig.13 : Local surface density of states : Fe (111) surface

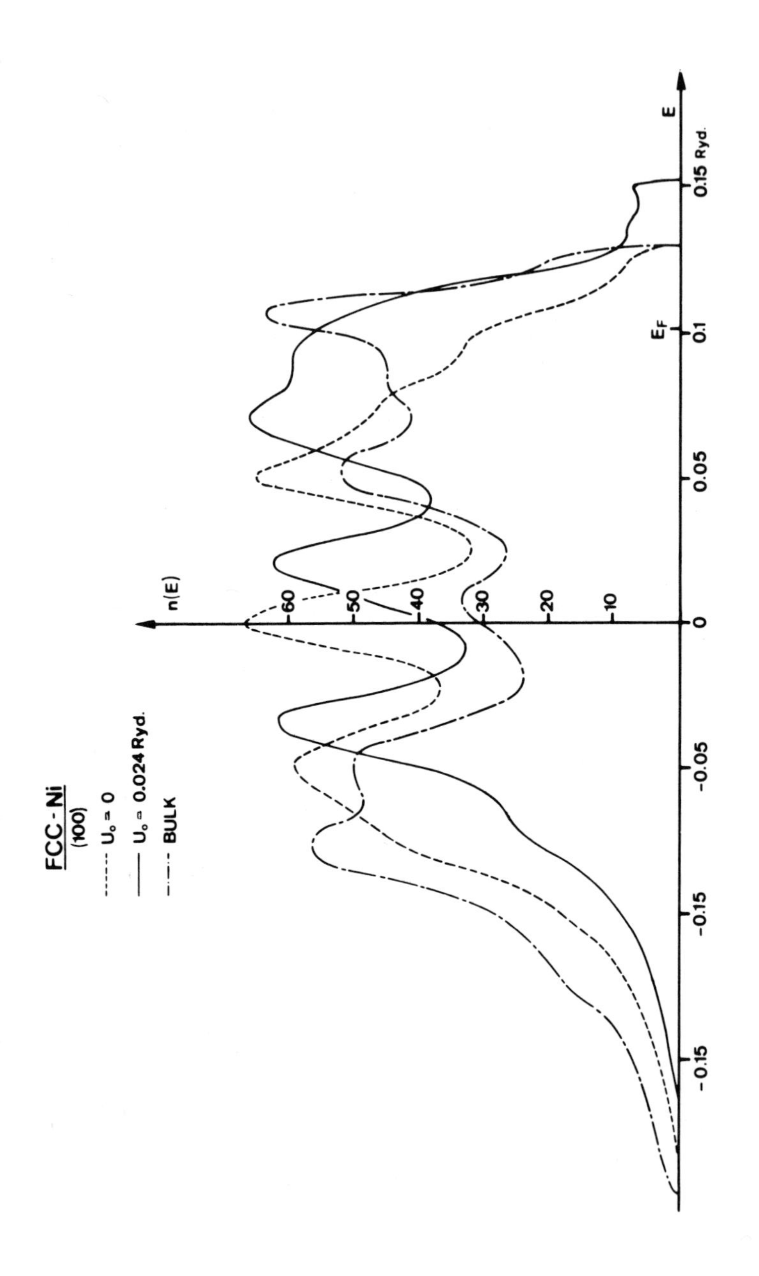

Fig.14 : Local surface density of states of Ni : (100) surface

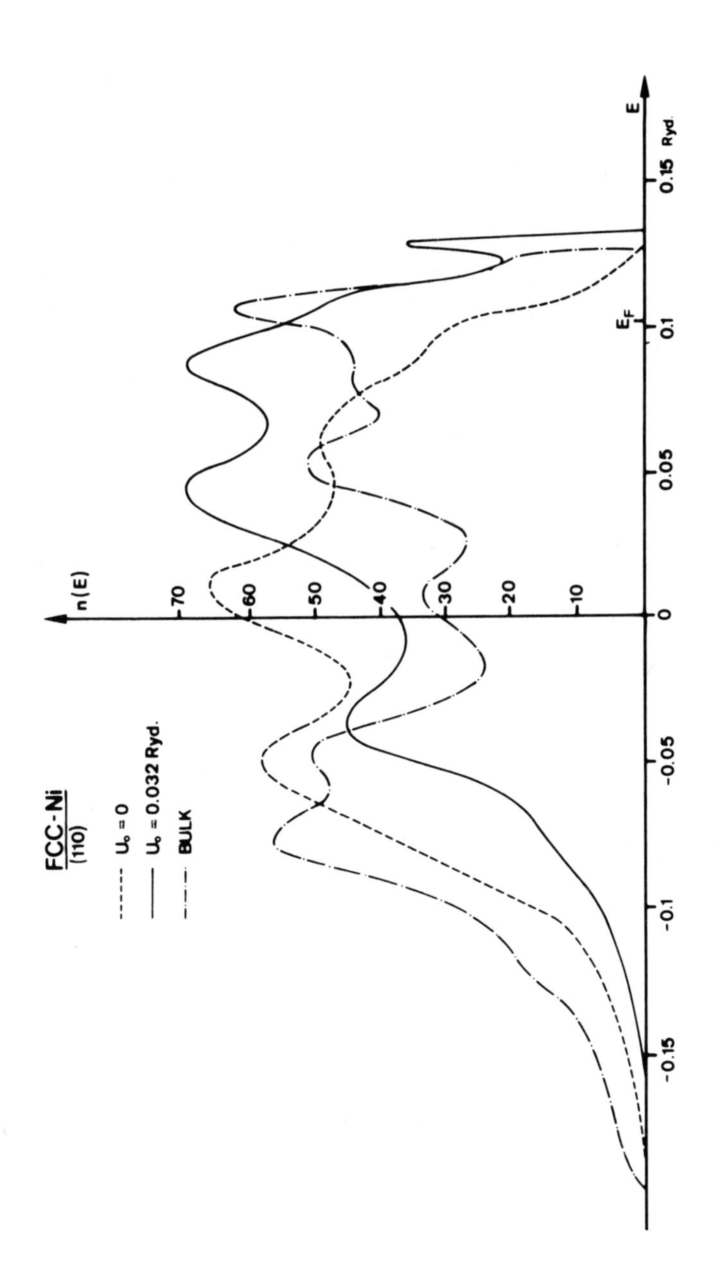

Fig. 15 : Local surface density of states of Ni : (110) surface

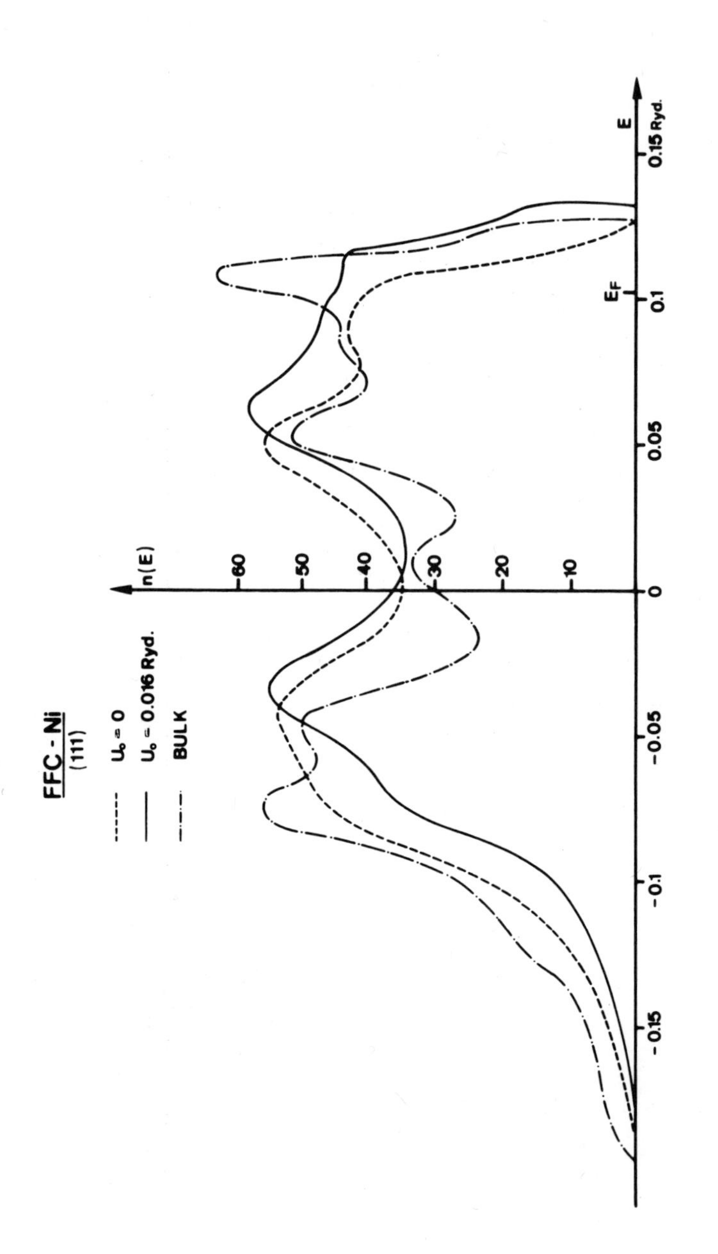

Fig. 16 : Local surface density of states of Ni : (111) surface

to one orbital XY lying in the (100) surface plane (27).

The influence on $n(E_F)$ of the self-consistent potential is here quite important. If one neglects U_o, we get a surface band narrower than the bulk one according to the value of the surface second moment. This would notably leads to a charge transfer of 0.4 electron (32) instead of the few per cent of electron we have above. In fact due to the large values of $n(E_F)$ this effect is compensated only by a small shift of the band due to U_o (less than 0.5 eV).

In both cases surface bound states occur for non dense surface planes. These bound states are well localized near the surface. They disappear in the local densities on the first plane under the surface. A study of these states for a set of values of $k_{\parallel}$ is needed to settle between resonant or surface bound states when they arise inside the bulk band.

B. d Orbital occupation (28)

The d orbital occupation on a surface atom has been calculated for Ni (001) and (110) surfaces (Table 1). One can see that the orbital occupations are not very different excepted for the xy orbital. These occupation numbers could change with adsorption. However it seems possible to predict an "on site adsorption" (the interaction being with $3Z^2-r^2$) or a "centered adsorption" (interaction being with yZ, Zx and x^2-y^2) for a (100) Ni surface and H, O, N, F .. adatoms. These atoms are expected to attract electrons. The same arguments lead to a bridge adsorption of this kind of atoms on a (110) plane (interaction being with x^2-y^2). Other factors (like the atomic size) must be involved to explain preferential adsorption sites because the orbital occupations are not very different.

It must be noticed that in the bulk transition metals the different orbitals also are almost equally occupied. This also means that the spatial distribution of the d electrons is relatively spherical with no strongly protubin lobe.

C. Influence of s-d interaction and spin-orbit coupling

The figure 17 (34) gives the values of E(k) in the case of Ni in the tight-binding approximation (fig.17a) and from a more sophisticated band structures including s and d electrons (fig.17b). The s band is not representated on the figure 17. However one can see (fig.17b) that the effect of the s-d interaction is small. It may be obtained in perturbation theory, the perturbation being of the order of 1 eV. This does not affect very much the bulk band structure. However the s-d interaction opens gaps in the dispersion curves. Localized states (35) may appear in such a gap. They would arise in the same energy region as the peak of the surface density of states calculated in the tight-binding approximation.

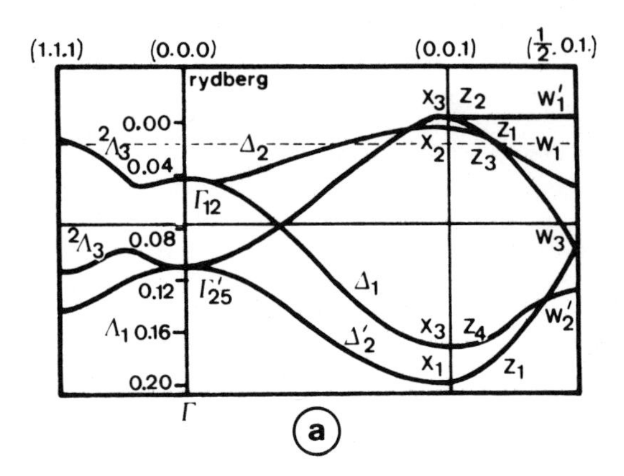

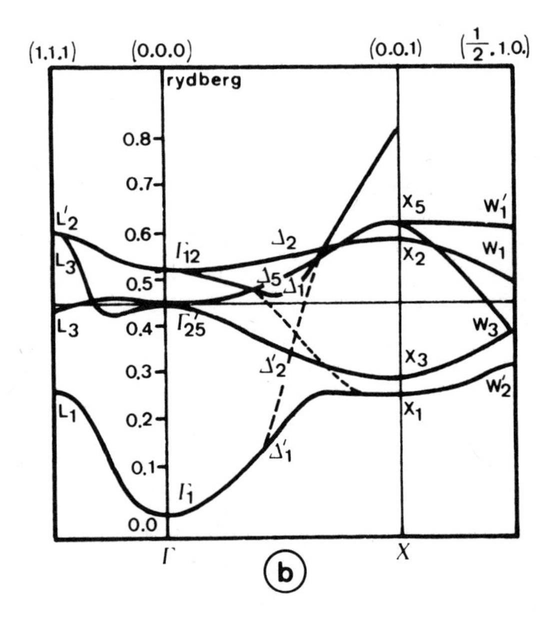

Fig. 17 : Band structure of Ni
a) Tight-binding approximation
b) Augmented plane waves calculation

TABLE 1 : d Orbital occupations for Ni surfaces (28) and bulk Ni (33)

	xy	yZ	Zx	x^2-y^2	$3Z^2-r^2$
BULK	0.870	0.870	0.870	0.946	0.946
100	0.774	0.931	0.931	0.945	0.923
110	0.800	0.904	0.904	0.968	0.922

As in the bulk the effect of s-d interaction may be included in the band structure using an A.P.W. scheme but this increases considerably the amount of numerical calculation (38). However in some effects like photoemission, the transition d → s may be strongly affected by the s-d interaction.

The spin-orbit coupling can also open gaps in the band structure (36). Such an effect is being investigated (37). As the spin orbit coupling increases as one goes from a series to an other, this effect will be certainly more important for elements of the third series (W for example) than for those of the first series (like Ni).

However these effects are only sensitive in small parts of the Brillouin zones and one may hope that they will not too much perturbed the density of states obtained in the simple tight-binding scheme.

6. CONCLUSION

We have tried in these lectures to show the evolution of the theory of surface electronic structure of transition metals. Using some very simple models, we have shown the effects of the surface in the density of states or on the appearance of localized states. We have also tried to report here the latter works on transition metals. It seems that we now are in position to get good quantitative agreement with experimental results.

REFERENCES

(1) See for example N.F. Mott, The Theory of the properties of metals and alloys, Dover Publications, N.Y. (1936)
(2) J. Friedel, The Physics of Metals (J.M. Ziman, Ed.) Vol.1, Cambridge University Press (1969), pp. 340-408
(3) Y. Cauchois, C. Bonnelle, Optical properties and electronic structure of metals (F. Abeles, Ed.) North Holland (1966), pp. 89-92
(4) H. Ehrenreich, Optical properties and electronic structure of metals (F. Abeles, Ed.) North Holland (1966), p. 109
(5) J. Callaway, Energy bands in Solids, Academic Press, N.Y. (1964)
(6) F. Ducastelle, J. Phys. (Paris) 31 (1970) 1055
(7) G. Allan, Electronic Structure of transition metal surfaces. To be published in "Surface properties. Surface states of materials" (L. Dobrzynski, Ed., M. Dekker, N.Y.)
(8) F. Cyrot-Lackmann, Thesis Orsay (1968) ; J. Phys. Chem. Sol., 29 (1968) 1235
(9) R. Haydock, V. Heine, M.J. Kelly, J. Phys. C. 5 (1972) 2845 and to be published
(10) L. Hodges, R.E. Watson, H. Ehrenreich, Phys. Rev. B5 (1972) 3953
(11) B.N. Cox, M.A. Conlthard, P. Lloyd, J. Phys. F, 4 (1974) 807
(12) F. Herman, S. Skillman, Atomic Structure Calculations, Prentice Hall, Engleword Cliffs, N.J. (1963)
(13) J. Friedel, Il Nuovo Cimento, Suppl. Vol. 7 (1958) 287
(14) M. Lannoo, G. Allan, J. Phys. Chem. Sol., 32 (1971) 637
(15) G. Toulouse, Solid State Comm., 4 (1966) 593
(16) B.S. Dewitt, Phys. Rev., 103 (1956) 1565
(17) G. Allan, P. Lenglart, Surface Sci., 15 (1969) 101
(18) G. Allan, Thesis Orsay (1970) ; Ann. Phys. (Paris), 5 (1970) 169
(19) D. Kalkstein, P. Soven, Surface Sci., 26, 85 (1971)
(20) G. Allan, P. Lenglart, Surface Sci., 30 (1972) 641
(21) R.A. Brown, Phys. Rev., 156 (1967) 889
(22) E.W. Plummer, J.W. Gadzuk, Phys. Rev. Lett., 25 (1970) 1493
(23) B.J. Waclawski, E.W. Plummer, Phys. Rev. Lett., 29 (1972) 783
(24) B. Feuerbacher, B. Fitton, Phys. Rev. Lett., 29 (1972) 786 ; Phys. Rev. Lett., 30 (1973) 923

(25) E.A.K. Nemeh, R.C. Cinti, J.B. Hudson, J. Phys. (Paris), 35 (1974) L179
(26) R.W. Zuehlke, J. Chem. Phys., 45 (1966) 411
(27) M.C. Desjonqueres, F. Cyrot-Lackmann, J. Phys. (Paris), 36 (1975) L45
(28) M.C. Desjonqueres, F. Cyrot-Lackmann, 9nth annual Conference on Surfaces, Coventry (March 1975). To be published in Surf. Sci.
(29) M.C. Desjonqueres, F. Cyrot-Lackmann (To be published)
(30) K. Terakura, J. Phys. Soc. Japan, 34 (1973) 1420
(31) K. Terakura, J. Kanamori (to be published)
(32) J.W. Davenport, T.L. Einstein, J.R. Schrieffer, Proc. 2nd Internl. Conf. on Solid Surfaces (Kyoto 1974) Japan ; J. Appl. Phys., Suppl. 2, Pt 2, 691 (1974)
(33) F. Cyrot-Lackmann, M.C. Desjonqueres, J.P. Gaspard, J. Phys. C, 7 (1974) 925
(34) F. Gautier, Propriétés électroniques des métaux et alliages, Masson, Ed. (Paris), pp. 255-330
(35) V. Heine, Proc. Phys. Soc., 81 (1962) 300
(36) K. Sturm, R. Feder, Sol. State Comm., 14 (1974) 1317
(37) M.C. Desjonqueres, F. Cyrot-Lackmann (to be published)
(38) D. Spanjaard (to be published)

DENSITY-FUNCTIONAL APPROACH TO THE ELECTRONIC STRUCTURE OF METAL SURFACES AND METAL-ADATOM SYSTEMS

N.D. LANG

IBM Thomas J. Watson Research Center

Yorktown Heights, New York 10598

1. DENSITY-FUNCTIONAL FORMALISM

A. Introduction

The density-functional formalism of Hohenberg, Kohn and Sham[1,2] provides a convenient framework for the study of the electronic structure of metal surfaces and of metal-adatom systems.

These authors consider a system of electrons in its ground state (assumed non-degenerate) moving in a static external potential v(r) (usually the potential due to the atomic nuclei). By showing that v(r) is a functional solely of the electron number density n(r), they demonstrate that the ground-state energy of the system is a universal functional of this density. The energy (excluding the self-energy of the array of nuclei) is written[2a]

$$E_v[n] = \int v(r)n(r)dr + \frac{1}{2}\int \frac{n(r)n(r')}{|r - r'|} drdr' + G[n] \quad (1.1)$$

where G[n] is the non-electrostatic energy of the system (kinetic, exchange and correlation energy).

It is then shown that $E_v[n]$ is a minimum for the correct density (assuming all densities considered to correspond to the same number of electrons N). This implies that

$$\phi(r) + \delta G[n]/\delta n(r) = \mu \quad (1.2)$$

where

$$\phi(r) = v(r) + \int \frac{n(r')}{|r - r'|} dr' \qquad (1.3)$$

and μ, a Lagrange multiplier, is the chemical potential in a large system.

B. Thomas-Fermi Method and Extensions

When n(r) varies sufficiently slowly in space, G[n] can be expanded in density gradients[1]:

$$G[n] = \int dr [g_0(n(r)) + g_2(n(r))|\nabla n(r)|^2 + \ldots] \qquad (1.4)$$

The coefficient $g_0(n)$ is just the non-electrostatic energy density of a uniform electron gas of density n. Often it is written

$$g_0(n) = [t(n) + \varepsilon_x(n) + \varepsilon_c(n)]n, \qquad (1.5)$$

where t and ε_x are respectively the average kinetic and exchange energies per particle of the uniform gas treated in the Hartree-Fock approximation, and ε_c is then the average correlation energy per particle of the uniform gas. Substitution of the above form for G[n] into Eq.(1.2) gives

$$\phi(r) = \left.\frac{dg_0(n)}{dn}\right|_{n = n(r)} - \left.\frac{dg_2(n)}{dn}\right|_{n = n(r)} |\nabla n(r)|^2$$

$$- 2g_2(n(r)) \nabla^2 n(r) + \ldots = \mu \qquad (1.6)$$

The Thomas-Fermi approximation consists of omitting all gradient terms in Eq. (1.6) and retaining only the kinetic energy term in $g_0(n)$. Since $t(n) = (3/10)(3\pi^2 n)^{2/3}$, this yields

$$n(r) = (2^{3/2}/3\pi^2)[\mu - \phi(r)]^{3/2} \qquad (1.7)$$

Combining this with Poisson's equation gives the Thomas-Fermi equation. Extensions of this can be obtained by including the other terms in Eqs. (1.5) and (1.6).

C. Hartree-like Forms

Kohn and Sham[2] consider a system of N fermions, each with the mass of an electron, that moves in a static external potential $v_s(r)$, but in which the particles do not interact with one another. The ground-state density distribution is denoted n(r). ($v_s(r)$ will in general be different from the external potential v(r) which would be required to produce this same n(r) in the presence of the inter-particle Coulomb interaction). The kinetic energy of this non-interacting system, a functional only of n(r), is denoted $T_s[n]$; and the analogue of Eq. (1.2) for this system is

$$v_s(r) + \delta T_s[n] / \delta n(r) = \mu_s \tag{1.8}$$

If the general form[3] of $T_s[n]$ were known, this equation could be used to find n(r), but we may observe instead that n(r) is simply the sum of the squares of the N lowest-lying orthonormal eigenfunctions of a one-particle Schrödinger equation with potential $v_s(r)$. This simple fact provides us with a means of solving equations that have the form of Eq. (1.8) (which is simply to solve the appropriate one-particle equations).

We return to the interacting system and define an exchange-correlation energy $E_{xc}[n]$ by using the functional $T_s[n]$ just defined :

$$E_{xc}[n] \equiv G[n] - T_s[n] \tag{1.9}$$

Equation (1.2) can then be written

$$\phi(r) + \frac{\delta E_{xc}[n]}{\delta n(r)} + \frac{\delta T_s[n]}{\delta n(r)} = \mu \tag{1.10}$$

which has the form of Eq. (1.8) if we take the function

$$v_{eff}[n;r] \equiv \phi(r) + \delta E_{xc}[n]/\delta n(r) \tag{1.11}$$

to play the role of the single-particle potential $v_s(r)$. We see therefore that we can obtain the n(r) determined by Eq.(1.2) by solving self-consistently the equations

$$\{-\frac{1}{2}\nabla^2 + v_{eff}[n;r]\}\psi_i(r) = \varepsilon_i\psi_i(r) \tag{1.12a}$$

$$n(r) = \sum_{i=1}^{N} |\psi_i(r)|^2 \tag{1.12b}$$

The functional $E_{xc}[n]$ can be expanded in density gradients just as G[n] was expanded in Eq.(1.4); a common approximation (the "local-density approximation") is to retain only the first (non-gradient) term in the series :

$$E_{xc}[n] \approx \int \varepsilon_{xc}(n(r))n(r)dr \tag{1.13}$$

In this case,

$$v_{eff}[n;r] \approx \phi(r) + \left.\frac{dn\varepsilon_{xc}(n)}{dn}\right|_{n=n(r)} \quad (1.14)$$

Sometimes $\varepsilon_c(n)$ is omitted in Eq.(1.14), yielding

$$v_{eff}[n;r] \approx \phi(r) - (3/\pi)^{1/3} n(r)^{1/3} \quad (1.15)$$

A quantity that is sometimes of interest is the local density of quasiparticle states, given by

$$n(\varepsilon,r) = \sum_i |\bar{\psi}_i(r)|^2 \delta(\varepsilon - \bar{\varepsilon}_i) \quad (1.16)$$

The eigenvalues $\bar{\varepsilon}_i$ and wave functions $\bar{\psi}_i(r)$ here are obtained from equations like (1.12a), in which v_{eff} is replaced by an appropriate approximation to the self-energy operator (details are given by Sham and Kohn[3a] and Hedin and Lundqvist[3b]). For the case in which the electron density varies sufficiently slowly, this self-energy operator can be replaced by an energy-dependent local potential, which, in the case of elementary excitations at the Fermi level, is simply equal to approximation (1.14) for v_{eff}[3a].

2. TWO MEASURABLE GROUND-STATE SURFACE PROPERTIES

A. Work Function

Consider a macroscopic crystal with all faces of comparable size. The work function Φ_i of a face i is the minimum work that must be done to remove an electron from the crystal to a region R_i, outside the central portion of face i, all of whose points are a distance O(l) away from the face, where l is large compared with the lattice spacing but small compared with the crystal dimensions (see Fig.1). Note that there are electric fields extending well outside the crystal such that moving an electron from region R_i to region R_j requires an amount of work $\Phi_j - \Phi_i$.

We can write Φ_i as

$$\Phi_i = \phi(r_i) + E_{N-1} - E_N \quad (2.1)$$

Here r_i is a point in R_i and E_N and E_{N-1} are the ground-state energies of the crystal with N and N-1 electrons respectively (but with N units of nuclear charge in both cases). From the definition of the chemical potential μ,

$$\Phi_i = \phi(r_i) - \mu \quad (2.2)$$

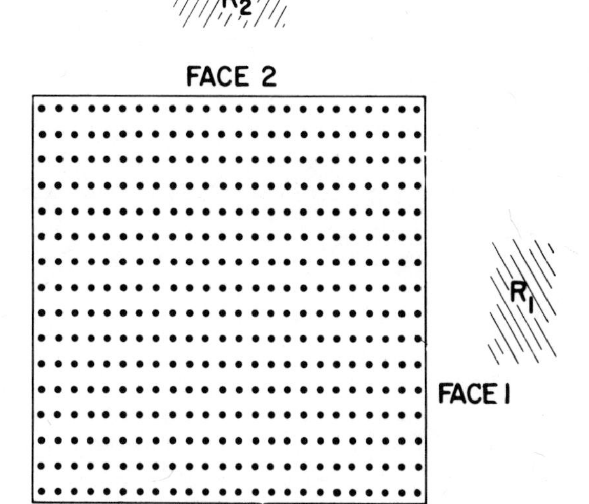

Fig.1 : The work function of face i of a crystal is measured by removing an electron to a region R_i, all of whose points are a distance from the face large compared with the lattice spacing but small compared with the face dimensions.

Let us now average Eq.(1.2) over the metal :

$$\mu = \bar{\phi} + \bar{\mu} \tag{2.3}$$

where

$$\bar{\phi} \equiv \Omega^{-1} \int_{metal} dr \phi(r) \tag{2.4a}$$

is the mean electrostatic potential in the metal and

$$\bar{\mu} \equiv \Omega^{-1} \int_{metal} dr \delta G[n]/\delta n(r) \tag{2.4b}$$

We can use Eq.(2.3) to divide the work function into what can be thought of as its surface and bulk components :(4)

$$\Phi_i = \Delta\phi_i - \bar{\mu} \tag{2.5}$$

where $\Delta\phi_i = \phi(r_i) - \bar{\phi}$. It is also sometimes convenient to write the work function as

$$\Phi_i = \phi(r_i) - \varepsilon_F \tag{2.6}$$

where ε_F is the eigenvalue of the highest occupied level in Eq.(1.12a).

This form has been derived by Schulte[5].

B. Surface Energy

The surface energy σ associated with a given crystal plane i is the work required, per unit area of new surface formed, to split the crystal along this plane. Using the notation introduced in Fig.2,

$$\sigma_i = \frac{1}{2A} \ (2E_i - E') \tag{2.7}$$

It is often convenient to write the surface energy as

$$\sigma = \sigma_s + \sigma_{xc} + \sigma_{es} \tag{2.8}$$

corresponding to the partitioning of the total energies E into $T_s[n]$, $E_{xc}[n]$, and $E_{es}[n]$ (the total electrostatic energy of the crystal, including the lattice self-energy).

3. PLANAR UNIFORM BACKGROUND MODEL OF A METAL SURFACE

A. Introduction

One of the simplest models of a metal surface is that in which a semi-infinite lattice of positive ions is replaced by a semi-infinite uniform positive background which terminates abruptly along a plane. We will take this background to fill the $x < 0$ half-space :

$$n_+(x) = \bar{n}\theta(-x) \tag{3.1}$$

where $\bar{n}$ is the number density of the background charge (equal to the mean interior electron number density). It is also useful to define a parameter r_s by

$$(4/3)\pi r_s^3 \equiv \bar{n}^{-1} \tag{3.2}$$

This model (sometimes called the "jellium" model) is appropriate for the surfaces of simple metals ; it is not particularly intended to described transition metals. The parameter r_s, for simple metals, ranges from about 2 to 6 au.

B. Thomas-Fermi Analysis

Solving the Thomas-Fermi equation for the uniform-background model[6] of a surface yields an electron density n(x) that shows the following behavior :

$$n(x) \propto x^{-6} \qquad (x \to \infty) \tag{3.3a}$$

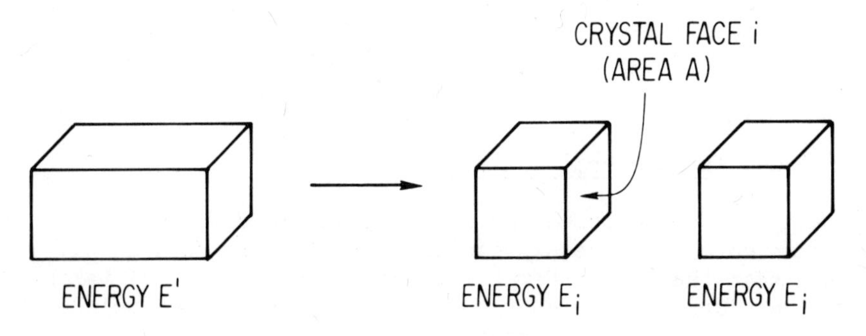

Fig. 2 : Quantities associated with the definition of the surface energy [Eq.(2.7)]

$$\bar{n} - n(x) \propto \exp(x/\lambda_{TF}) \qquad (x \to -\infty) \tag{3.3b}$$

where $\lambda_{TF} = (\pi/(4k_F))^{1/2}$ ($k_F^3 = 3\pi^2\bar{n}$). It may be seen directly from Eq.(1.2) that the surface electrostatic potential barrier $\Delta\phi$ is here equal to $(1/2)k_F^2$ and since $\bar{\mu}$ is also equal to $(1/2)k_F^2$ for the Thomas-Fermi case, Eq.(2.5) implies that

$$\Phi = 0 \tag{3.4}$$

independent of r_s. The surface energy can be calculated from the electron density distribution(6) and is given by

$$\sigma = -0.0763 \; r_s^{-9/2} \tag{3.5}$$

The results for the work function and surface energy are clearly not correct for real metals. We will see later that the zero result for Φ is simply an artifact of the Thomas-Fermi approximation, while the negative result for σ represents (at least for smaller r_s values) an intrinsic deficiency of the uniform-background model, which is remedied only when the discrete lattice is reintroduced.

We know on general quantum-mechanical grounds that the electron density outside a metal surface must decay exponentially, and that the density inside, in the vicinity of the surface, must exhibit quantum density oscillations (Friedel oscillations). The failure of Eqs. (3.3) to exhibit these features is remedied when the uniform-background model is treated wave mechanically (the treatment of Sec. IC). Eq. (3.3b) does however give the dominant behavior of the interior density when r_s is small, as will be seen below.

C. Extended Thomas-Fermi Analysis

Many of the deficiencies of the Thomas-Fermi method are remedied when more complete expressions for G[n] are used in Eq.(1.2). (Recall that the Thomas-Fermi approximation consists of taking $G[n]=\int t(n(r))n(r)dr$). When exchange-correlation terms are included in $g_o(n)$, for example, the computed work-function value becomes positive, but remains r_s-independent[7]. Smith[8] has shown that when, in addition to using the complete expression for $g_o(n)$, a simple form for the first gradient term in Eq.(1.4) is used (only kinetic-energy contributions to g_2 are included), quite satisfactory results for Φ and, at larger r_s, for σ are obtained. The negativity of σ for small r_s persists; this is due, as noted above, to the absence of lattice effects, and not to a deficiency in the treatment of G[n]. The reader is referred to Refs.7 and 8 for the details of this work, which showed the utility of the density functional formalism in the study of metal surfaces.

D. Wave Mechanical Analysis

Bardeen[9], Bennett and Duke[10] and Lang and Kohn[11-13] have given wave-mechanical (see Sec. IC) treatments of the uniform-background surface model[13a]. The treatments of refs.10-13 use the "local-density approximation" for exchange-correlation effects, as given in Eqs.(1.13)-(1.14) above. The electron density obtained in these treatments decays exponentially in the vacuum, while in the metal interior it exhibits Friedel oscillations (a characteristic feature of wave-mechanical analyses) :

$$\bar{n} - n(x) \propto \frac{\cos(2k_F x + \alpha)}{x^2} \qquad (x \to -\infty) \qquad (3.6)$$

Two of these density distributions are shown in Fig.3. Results for the work function and its bulk and surface components are given in Fig.4; and the calculated Φ is compared with experimental results for simple metals in Fig.5. The agreement is quite reasonable, and is improved when lattice effects are re-introduced into the model using perturbation theory[13].

Figure 6 gives results for the surface energy and its components. The reintroduction of lattice effects, which makes σ positive for all r_s (as it should be), is discussed in Sec. IV . Schmit and Lucas[14], Peuckert[15] and Craig[16] have discussed a contribution to the surface energy associated with the zero-point energy of the surface plasmons. The degree to which these effects are included in the local-

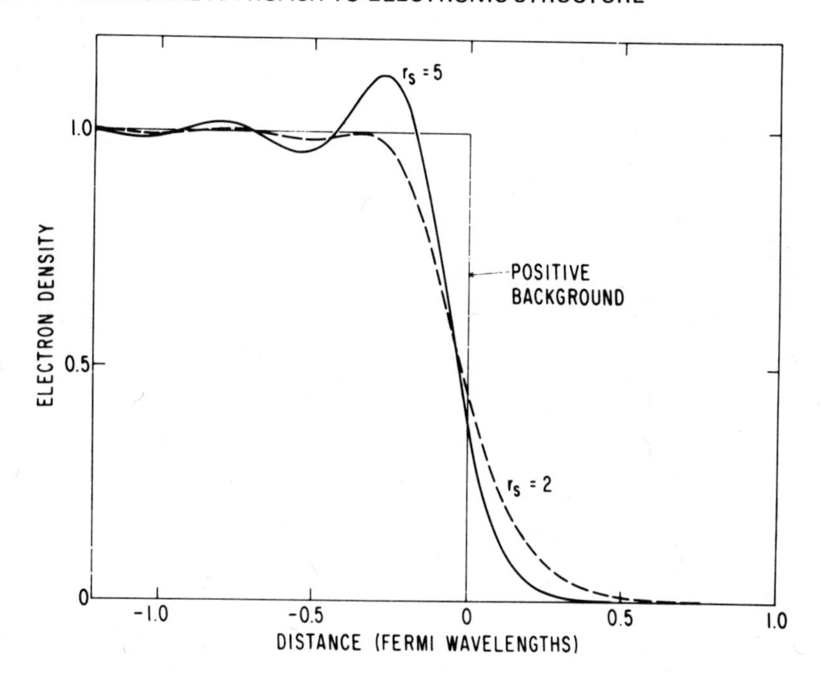

Fig. 3 : Electron density in surface region computed wave mechanically, with exchange-correlation effects included in the local-density approximation (Ref.11). One Fermi wavelength is equal to $2\pi/k_F$.

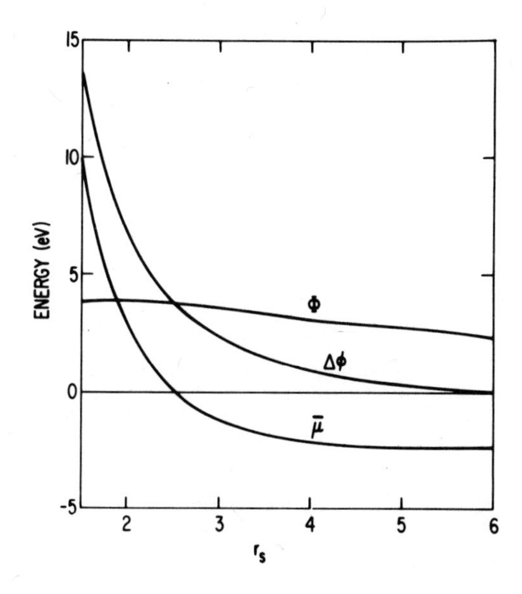

Fig.4 : Components of the work function (Ref.13)

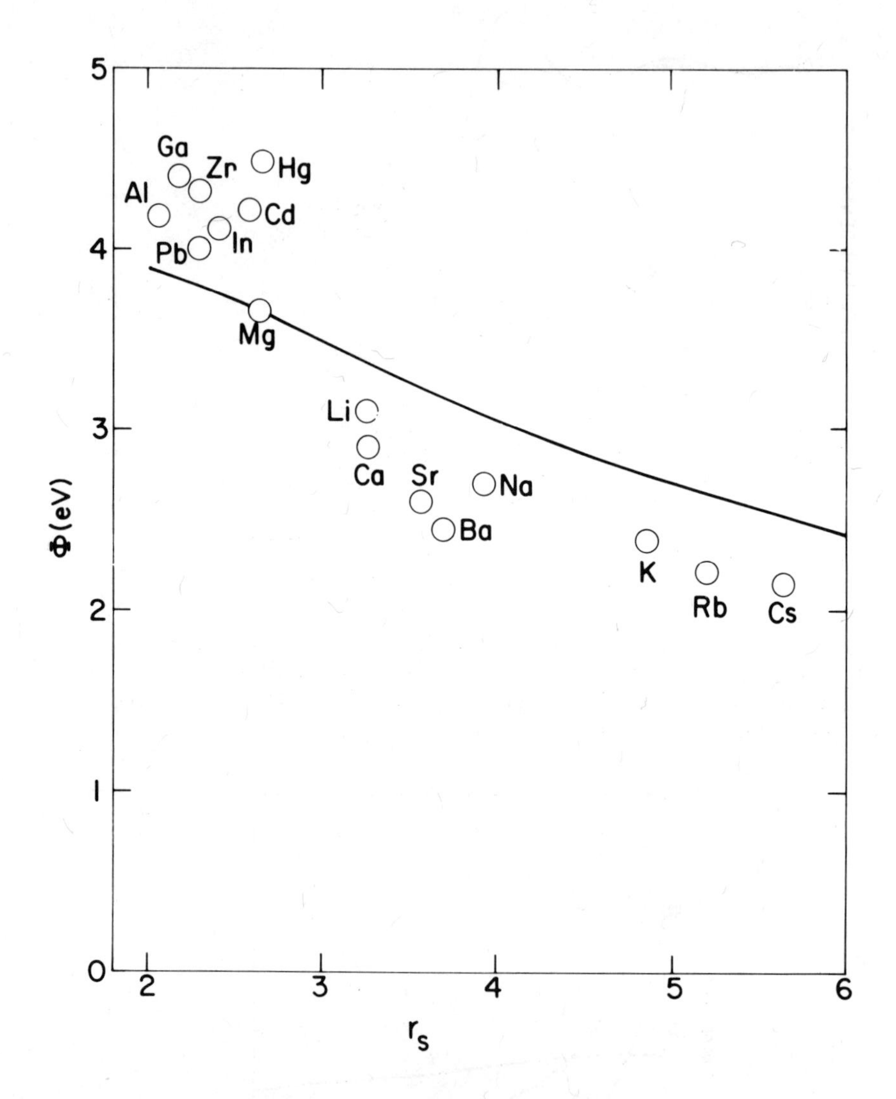

Fig. 5 : Solid line : work function computed for uniform-background model in Ref. 13.

Circles : measured work functions for polycrystalline samples (see Ref.13). The inclusion of lattice effects improves the agreement between theory and experiment (see Sec.IV B).

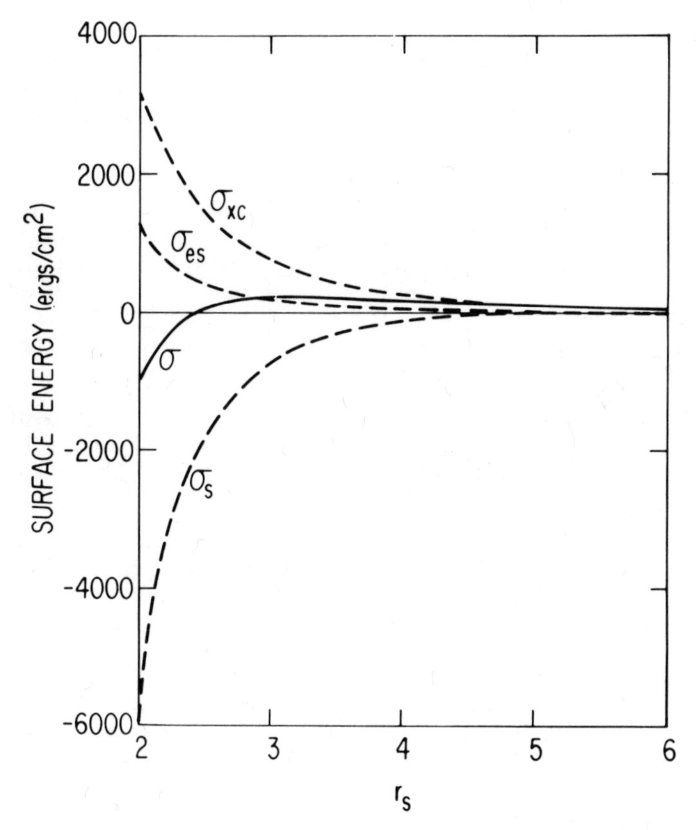

Fig. 6 : Components of the surface energy (Ref.12)

density approximation to σ_{xc} is difficult to assess a priori. The relative magnitude of this contribution to σ_{xc} continues to be a subject of active discussion(17).

Werner, Schulte and Bross(18) have calculated the local density of quasiparticle states [Eq.(1.16)] for the uniform-background model of the surface ; some of their results are given in Fig.7. Note in this figure the way the width of the filled portion of the local energy band decreases moving from the metal interior toward the surface. Such "band narrowing" is a characteristic feature of the local density of states at a surface. For energies ε sufficiently close to the band edge ε_o, these authors find that the local density of states in the surface region is proportional to $(\varepsilon-\varepsilon_o)^{\beta}$, with β changing from 1.6 for r_s=2 to 1.1 for r_s=6.

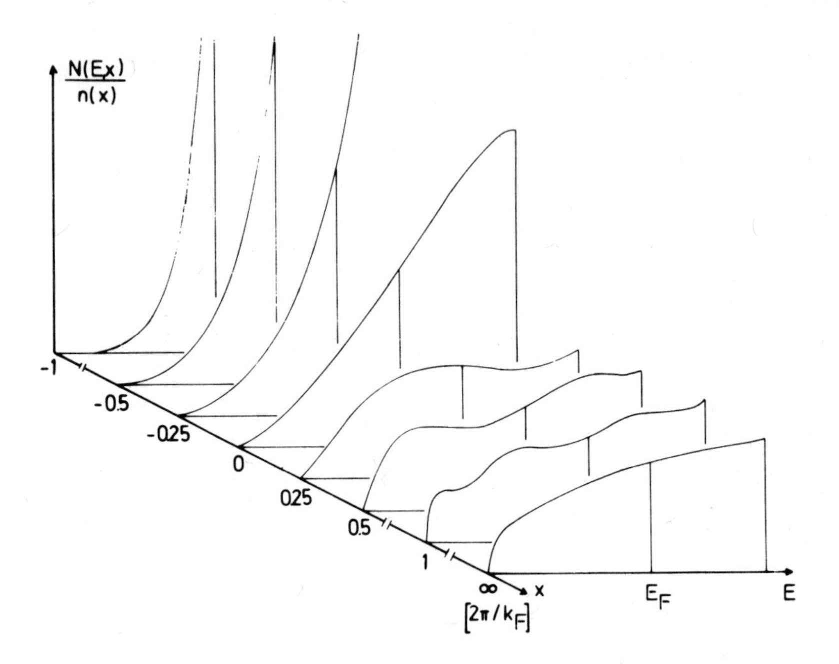

Fig. 7 : Local density of quasiparticle states as a function of energy for several distances x from the background edge, computed for $r_s = 4$ by Werner, Schulte and Bross (Ref.18). The x direction in this figure is opposite to that employed in the present article ($x = +\infty$ is inside the metal in the figure).

E. Sum Rules for Static Properties

Budd and Vannimenus[19] have derived two exact "sum rules" for the uniform-background model[20]. One of these relates the electrostatic potential at the surface to a derivative of the bulk energy density ; the other gives the derivative of the surface energy with respect to background density as an integral over the electrostatic potential.

Using the density functional formalism, these authors consider the change in total energy of a uniform-background model of a metal slab when the background is stretched along the normal to the large surfaces. In terms of the semi-infinite geometry defined by Eq.(3.1) they show that

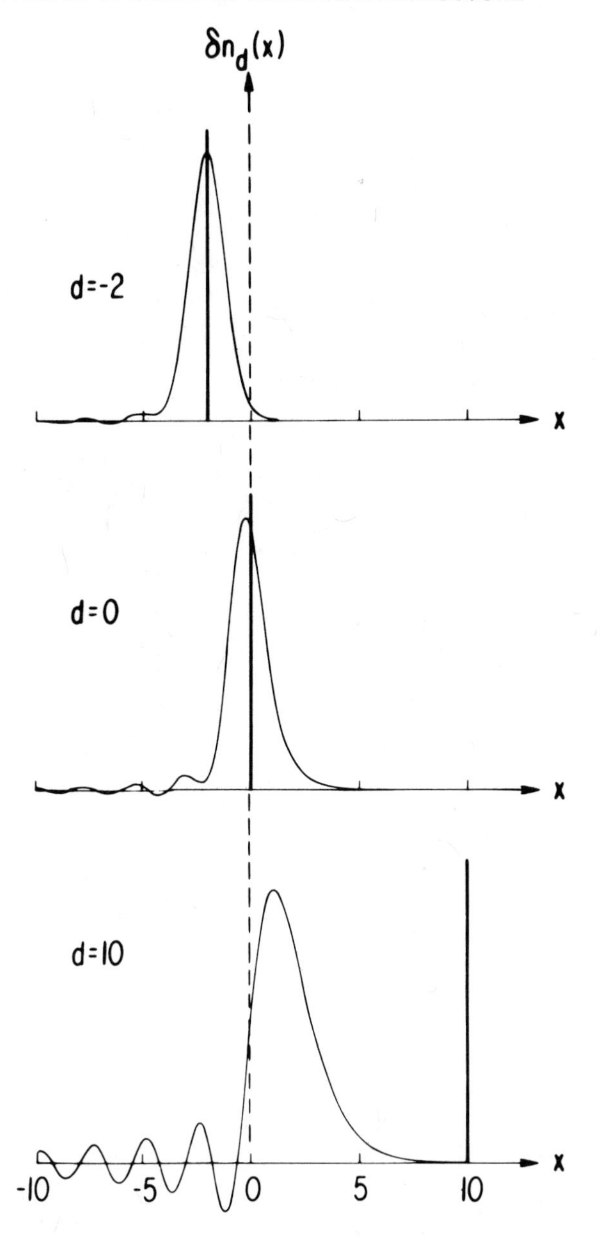

Fig. 8 : Changes in electron density induced by charged sheet (with small surface charge density) at position d (marked by heavy line). The x-coordinate is defined with reference to Eq.(3.1). All distances are in atomic units. The calculation, done at $r_s = 1.5$, is similar to that presented for the $d \to \infty$ case in Ref. 13.

$$\phi(0) - \phi(-\infty) = \bar{n}\,\frac{d(g_o(\bar{n})/\bar{n})}{d\bar{n}} \tag{3.7}$$

and that

$$\frac{d\sigma}{d\bar{n}} = \int_{-\infty}^{0} [\phi(-\infty) - \phi(x)]\, dx \tag{3.8}$$

The proof of these relations does not depend on the specific form of G[n], and so they permit testing the accuracy of approximate calculations. Taken together they also provide a useful way to discuss general features of the surface energy[19].

F. Static Response

We consider briefly here the response of the uniform-background model of the surface to a static perturbing charge distribution. We discuss first the response to a very simple configuration - a sheet of charge parallel to the surface. It will be recognized in particular that for a sheet well outside the surface, this is simply the problem of the response to a uniform electric field normal to the surface[13,21].

The change in electron number density $\delta n(x)$ induced by the sheet can be found simply by repeating the calculation described at the beginning of Sec. IIID with the sheet potential included in the total electrostatic potential $\phi(x)$ [Eqs.(1.11) and (1.12)], and subtracting from the resulting electron density distribution that obtained for the unperturbed case. Figure 8 gives distributions $\delta n_d(x)$ for several positions d of the sheet along the x-axis. (The results presented here are for sheets with small surface charge densities, i.e., they are in the linear-response regime).

Let us denote the centers of gravity of these distributions by $x_o(d)$:

$$x_o(d) \equiv \int_{-\infty}^{\infty} x\, \delta n_d(x)dx \Big/ \int_{-\infty}^{\infty} \delta n_d(x)dx \tag{3.9}$$

Within the linear-response regime, it is easily seen that $x_o(d)$ will also give the center of gravity of charge induced by a point charge at position d. Then the dipole moment μ induced by a small point charge q at d is simply equal to $q[d-x_o(d)]$. Figure 9 gives a graph of μ vs.d ; this is of interest in discussions of chemisorption[21-23]. For a point charge far outside the surface, it can be shown[21] that the distance-dependent part of the energy of interaction between the charge and the surface is given by

$$U = -\frac{q^2}{4[d-x_o(\infty)]} + O((d-x_o(\infty))^{-3}); \tag{3.10}$$

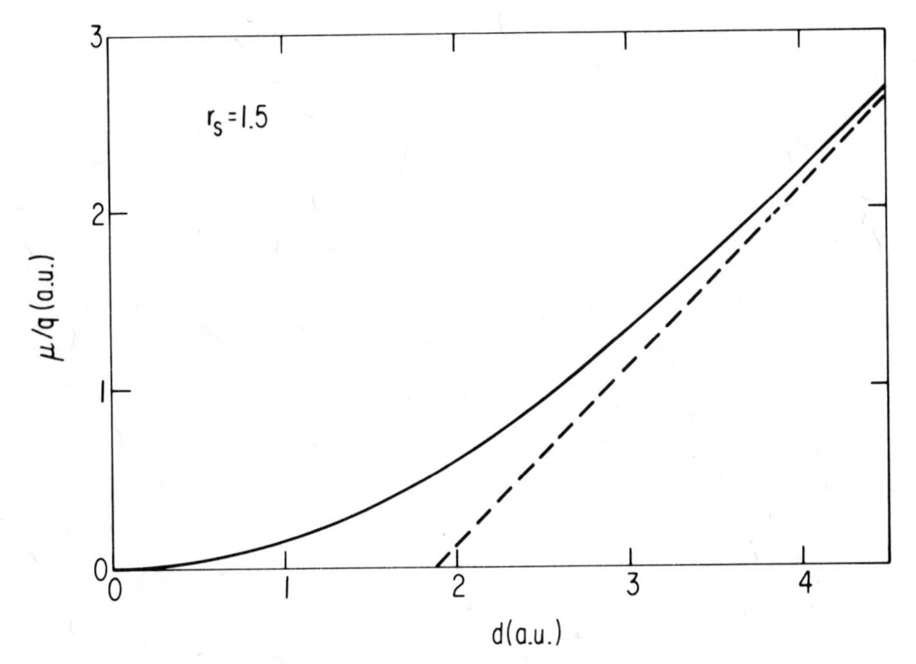

Fig.9 : Dipole moment μ induced by a point charge of strength q (taken to be small) at distance d from the positive background edge (d = +∞ is in the vacuum). The results are for r_s = 1.5, and are obtained from the centers of gravity of distributions some of which are presented in Fig.8. The curve is asymptotic to the dashed line, whose intercept with the d axis gives the position of the image plane (see text).

i.e. the position of the image plane for a static point charge is just given by the center of gravity of the charge induced by a uniform normal electric field.

4. LATTICE MODELS OF A METAL SURFACE

A. Introduction

Two major qualitative effects associated with the presence of a lattice in a surface model are anisotropy of the various ground-state properties and the presence of surface states. Surface states have wave functions localized in the surface region, and lie in general in band gaps of the bulk energy spectrum. These states are discussed elsewhere in these proceedings by Dr. Allan, and so we confine our attention to the effect of the discrete lattice on properties such as the charge density, surface energy and work function. We discuss first the latter two properties in a qualitative way using the classical neutralized lattice.

Consider a semi-infinite crystal consisting of a cubic lattice (for convenience) of point positive charges (in place of ions) and a rigid uniform distribution of negative charge that terminates abruptly along a plane at the surface (in place of conduction electrons). Denote the spacing of the lattice planes that are parallel to the surface by d. The boundary of the negative charge must be a distance $(1/2)d$ in front of the outermost lattice plane (in order that the electric field averaged over a macroscopic region vanish both inside and outside the crystal). This classical model crystal is shown schematically on the left-hand side of Fig. 10.

The potential difference $\Delta\phi$ in Eq. (2.5) depends only on the charge averaged parallel to the surface. To discuss this quantity, therefore, we can imagine that each lattice plane of point charges in the model crystal has been smeared out into a thin positive sheet, as in Fig. 10. A simple electrostatic calculation shows that $\Delta\phi$ is positive and proportional to d^2. Since the more closely packed a crystal face is the larger is the associated interplanar spacing, we see that the more closely packed crystal faces have higher work functions. This is in fact observed for most metals for which experiments have been performed[24].

A discussion of the surface energy for various faces of the classical neutralized lattice involves a more complicated electrostatic calculation[12]. When the model crystal is cleaved and the two halves are separated slightly there are electric fields in the gap that cause the halves to attract each other (Fig. 11). These fields would not be present if the lattice were smeared into a homogeneous positive distribution ; this is one of the chief reasons why the uniform-background model of the surface is not a satisfactory model for the computation of surface energies.

It is found in fact that for the model discussed here, the more closely packed principal lattice planes have higher surface energies[12]. This is generally observed experimentally[25].

B. Perturbative Treatments

For simple metals, the effect of the ionic lattice on the electrons can be described more accurately by $v_{pseudo}(r)$, a superposition of ionic pseudopotentials, than by $v_+(r)$, the potential due to a uniform positive background. In order to take account of this fact in a simple manner,

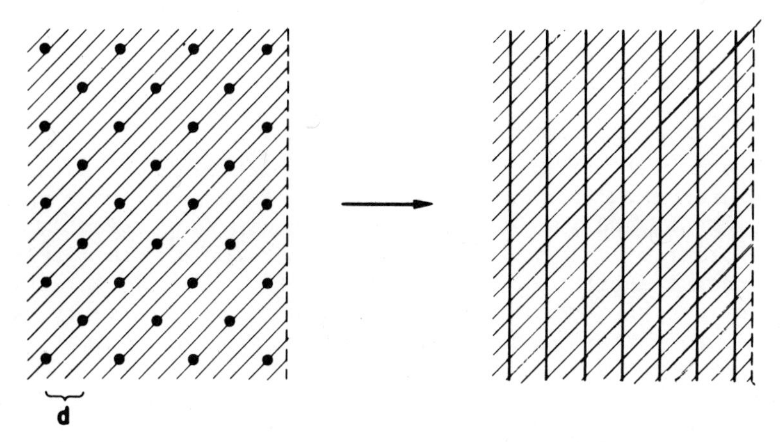

Fig.10 : Simple model for discussion of work-function anisotropy : Shaded regions represent negative charge distributions which terminate abruptly at a plane (dashed lines); dots represent point positive charges ; solid lines represent thin sheets of positive charge.

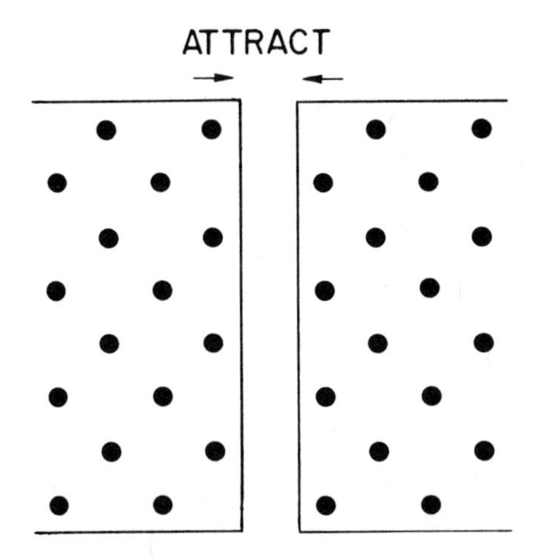

Fig.11 : A bcc classical neutralized lattice split along the (110) plane. Dots represent point charges lying on a selected plane through the crystal. Electric fields in the gap cause the two halves to attract. [The halves would repel for the simple cubic lattice split along the (100) plane.]

$$\delta v(r) = v_{pseudo}(r) - v_{+}(r) \tag{4.1}$$

can be treated as a perturbation, and the results of calculations for the uniform-background model can be corrected using ordinary perturbation theory.

A first-order perturbation treatment of the work function brings the calculated results shown in Fig.5 into somewhat closer agreement with experiment, and produces anisotropies of 1/4 - 1/2eV[13], which is of the order of the measured anisotropies for simple metals[26]. A first-order treatment of the surface energy, with the classical electrostatic contribution described above included exactly, yields a surface energy that is always positive, and in approximate agreement with measured values[12] (see Fig.12).

C. Non-Perturbative Treatments

We consider here only fully self-consistent treatments of the metal surface within the density-functional framework. We do not discuss tight-binding studies[27] or muffin-tin calculations for small clusters which are intended to model metal surfaces[28], or studies involving LEED-type wave-function matching across a non-self-consistent potential barrier (which usually focus on the analysis of surface states)[29].

Self-consistent calculations have been performed for Na(100) by Appelbaum and Hamann[30] and for Li(100) by Alldredge and Kleinman.[31] Both calculations represent the effect of the ionic lattice on the electrons by a pseudopotential, local in the Appelbaum-Hamann calculation for Na and non-local in the Alldredge-Kleinman calculation for Li. In both cases, Eqs.(1.12) are used, with the local-density approximation [Eq.(1.14)]. Advantage is taken of the lattice periodicity parallel to the surface by expanding the eigenfunctions of Eq.(1.12a) in the appropriate set of plane waves.

The Appelbaum-Hamann calculation considers a semi-infinite crystal geometry, and involves a step-by-step numerical integration of the set of coupled one-dimensional wave equations obtained from Eq. (1.12a) (after the plane-wave expansion) along the surface normal. The Alldredge-Kleinman calculation considers a film geometry, and involves the expression of the variation of the eigenfunctions along the surface normal in a set of sines and cosines that vanish at a distance outside the two film surfaces at which the charge density is negligible. A matrix eigenvalue equation is then solved.

Fig.13 shows a plot of the self-consistent potential in the Appelbaum-Hamann calculation. There is a fair variation, moving parallel to the surface, of the rate at which the vacuum barrier rises;

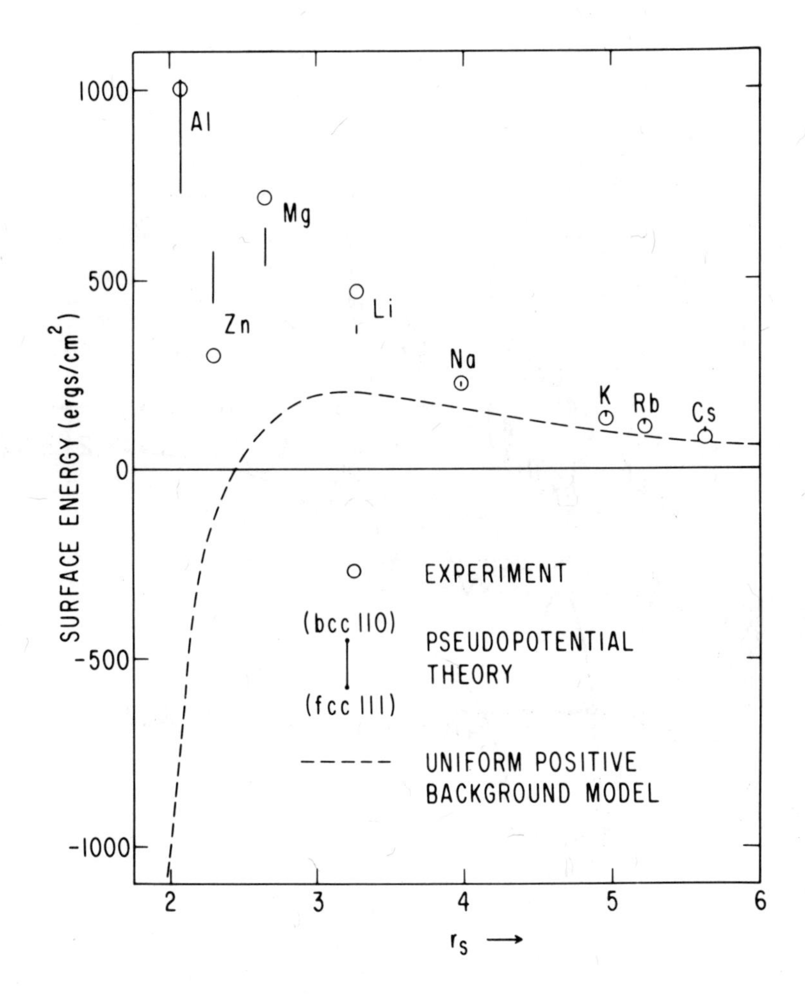

Fig. 12 : Comparison between theoretical values of the surface energy and extrapolations to zero temperature of measured liquid-metal surface tensions (Ref. 12). Dashed curve gives results for the planar uniform-background model. Vertical lines give computed values that include lattice effects : the lower endpoint represents the value appropriate to an fcc lattice, the upper endpoint that appropriate to a bcc lattice. The surface plane in both cases is taken to be the most closely packed plane of the lattice. (For the alkali metals of lower density, the lines are contracted almost to points.)

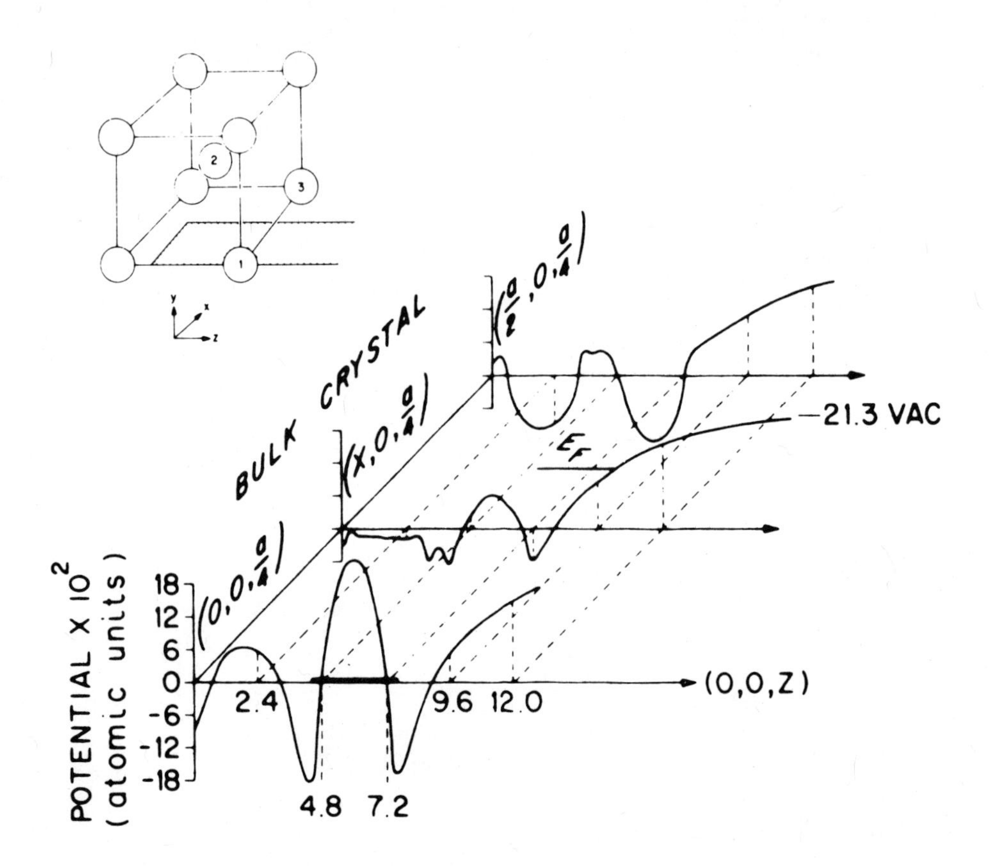

Fig.13 : Potential v_{eff} computed by Appelbaum and Hamann (Ref.30) for the Na(100) surface. The small figure at the upper left shows the unit cell containing the surface layer (atoms 1 and 3) and the next layer (atom 2). The z axis is along the surface normal in the figure. The potential is plotted vertically along three lines in the cross-hatched plane shown in the figure of the unit cell, whose proportions are distorted for clarity. The heavy line along the foreground abscissa indicates the location of the core of atom 1. Distances are given in atomic units.

the effective electric field in a plane 1Å outside the pseudopotential cores shows a variation of roughly a factor of 2. The average of the calculated charge density parallel to the surface is similar in many respects to that found for the uniform-background model of the surface (cf.Fig.3). The work-function values computed in this study and in that of Alldredge and Kleinman are similar to those obtained for the uniform-background model once lattice effects are included using first-order perturbation theory[13].

The Alldredge-Kleinman calculation gave particular attention to the surface states, which for Li lie above the Fermi level. Using the same method, but without making the potential self-consistent, Caruthers, Kleinman and Alldredge[32] have studied surface states on Al, some of which lie below the Fermi level.

5. CHEMISORPTION IN THE DENSITY-FUNCTIONAL FRAMEWORK

The properties of a chemical bond between an atom or molecule and a metal surface can be studied with the density functional formalism by including in the external potential v(r) (see Sec.I) the electrostatic potential due not only to the atomic nuclei of the substrate but to the nuclei of the chemisorbed species as well. The analysis can of course proceed via either Eqs.(1.12) (wave-mechanical) or Eq.(1.6) (Thomas-Fermi and extensions). We will not discuss here the simulation of a chemisorption system by a small cluster of atoms[28,33]; Prof. Grimley will consider this topic briefly in his article in this volume. So far, these cluster treatments (which are generally done wave-mechanically) have employed a number of central approximations such as muffin-tinning, the adequacy of which have been examined by Gunnarsson and Johansson[34] in a recent density-functional study of diatomic molecules and by Williams and Morgan.[34a]

Except for the cluster treatments, the metal substrate model used so far in density-functional studies of chemisorption has been the uniform positive background model (see Sec.III)[35]. In the simplest one of these studies[36], the sheet of positive ions of a layer of alkali atoms adsorbed on a metal surface is simulated by a uniform slab of positive charge (just as the lattice planes of substrate ions have been replaced by such slabs in creating the uniform substrate background). The charge configuration in this (wave-mechanical) calculation is given in Fig.14. The slab thickness was kept constant (at the interplanar spacing between closest-packed planes in the bulk alkali) and the slab density varied to model changes in number of adatoms per unit area. This model is sufficient to reproduce the experimentally observed minimum in the work function-vs-coverage curve, and to give the measured value at the minimum (Figs. 15 and 16). It can also be used to model the maximum in the curve that is sometimes observed when the first alkali layer is filled and second-layer formation commences[36].

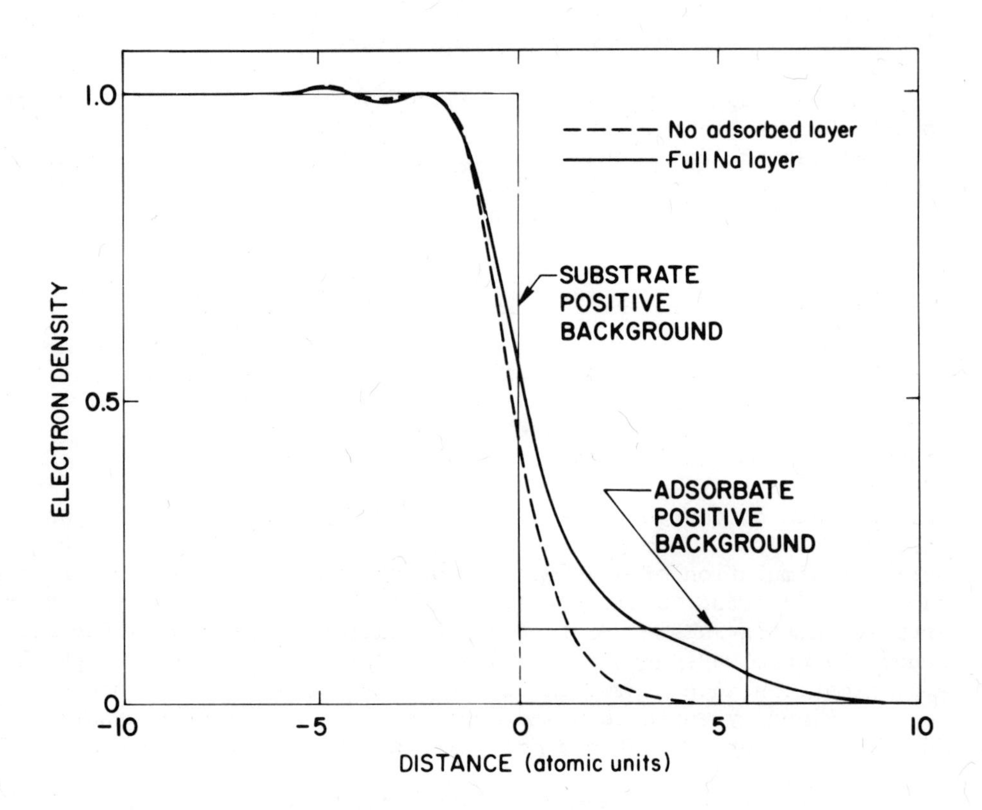

Fig. 14 : Electron density distribution at metal surface ($r_s = 2$) with (solid lines) and without (dashed lines) full adsorbed Na layer, as computed in Ref.36 using a uniform-background model. The adsorbed layer is deemed to be full when it has a packing density equal to that of the most closely packed lattice plane of the bulk alkali.

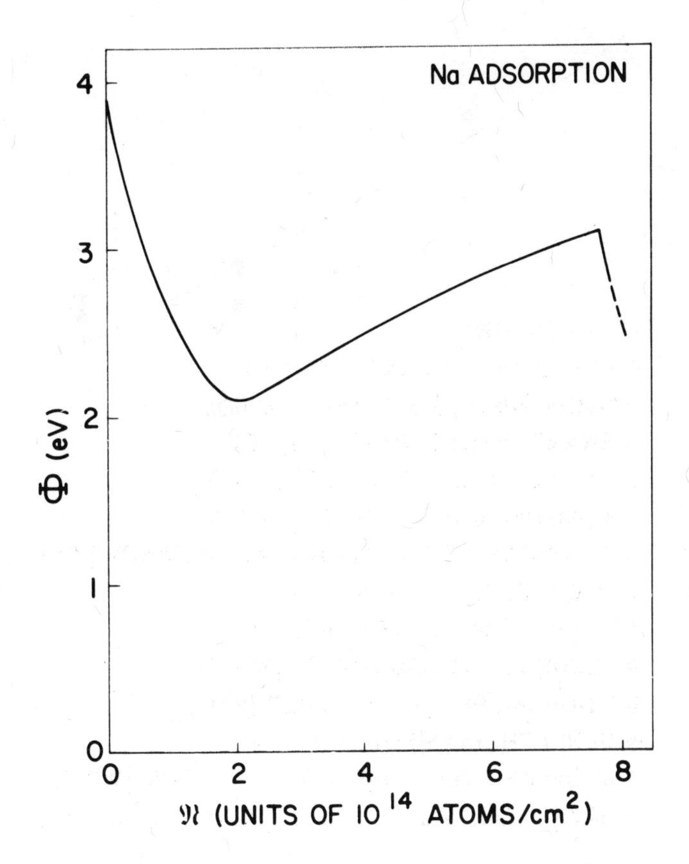

Fig.15 : Work function versus coverage for Na adsorption, as computed in Ref.36 using a uniform-background model of the substrate-adsorbate system. The curve exhibits a minimum, and a maximum which occurs at the commencement of second-layer formation.

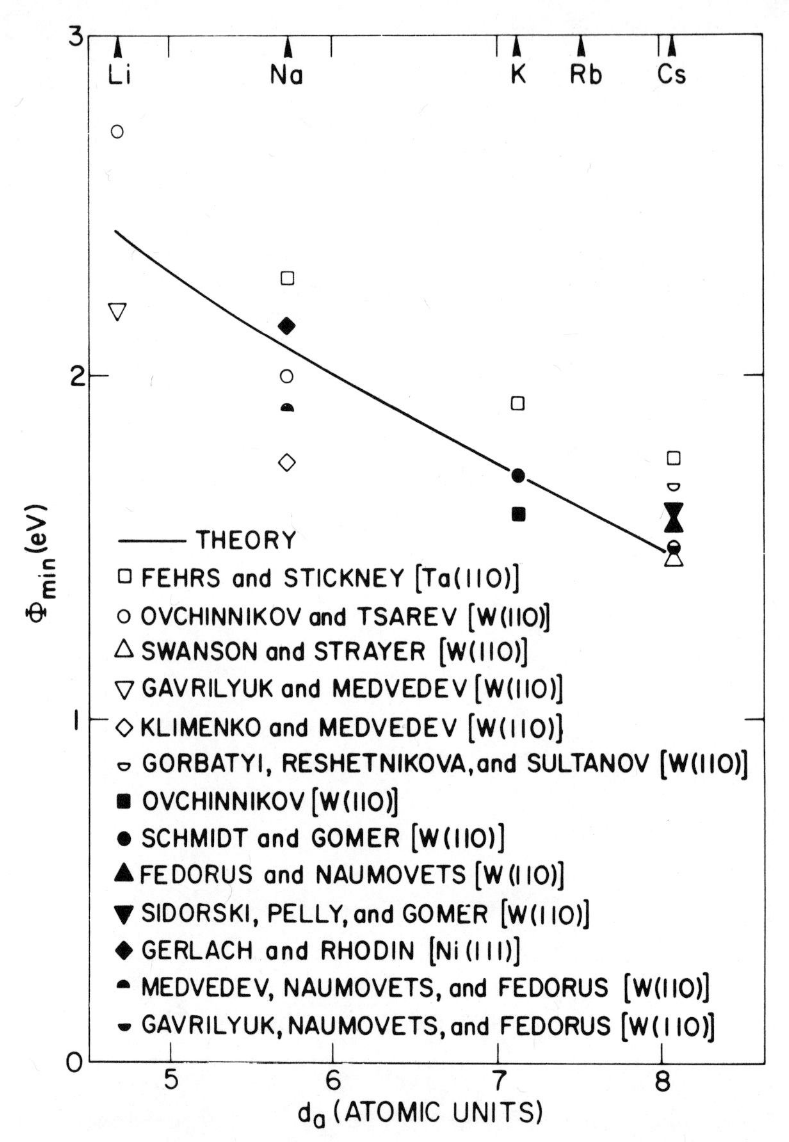

Fig. 16 : Symbols : measured values of the minimum work function for alkali adsorption on close-packed metal substrates of initially high work function (see Ref.36). Solid line : computed value (Ref. 36). A recent measurement has been made for Rb/W(100) by T.W. Hall and C.H.B. Mee (Φ_{min} = 1.50 eV), and discussed in relation to the theory on which this figure is based (Japan J. Appl. Phys. Suppl.2, Part 2, 741(1974)).

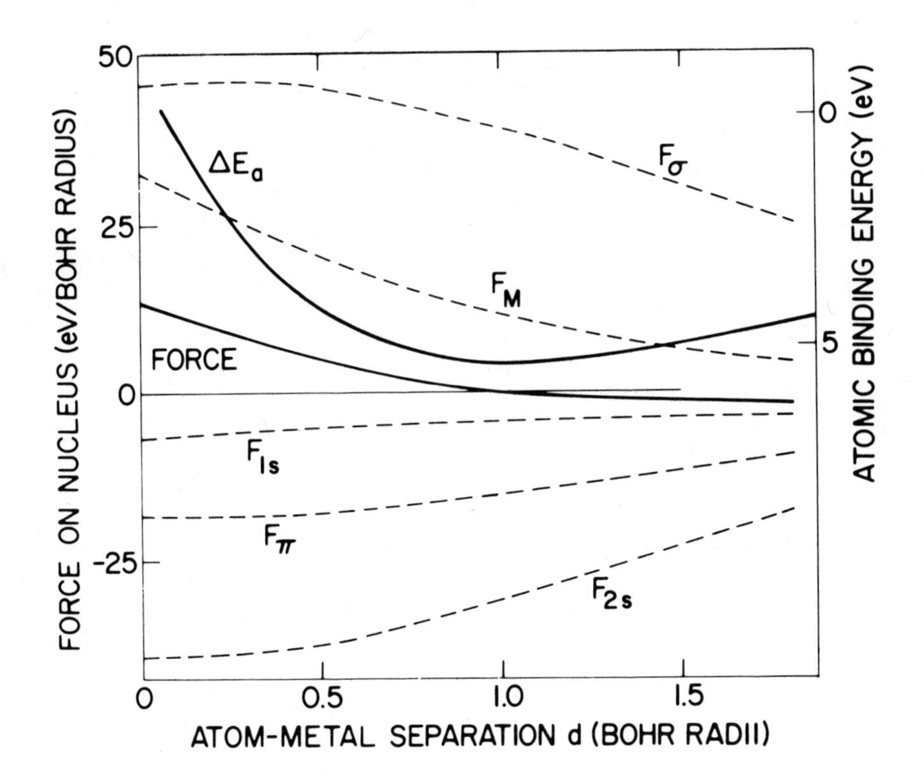

Fig.17 : Atomic binding energy ΔE_a and forces as function of distance d from the ($r_s = 2$) background edge for oxygen chemisorption, as computed in Ref.38. The "force" curve gives the total electrostatic force on the adatom nucleus. The different contributions to this force are shown as dashed curves. F_M is the force due to the bare metal. The remaining contributions are due to the various components of the additional electron density associated with the presence of the adatom : F_{1s} and F_{2s} arise from electrons in the 1s and 2s discrete states ; F_σ and F_π arise from electrons in continuum states characterized by azimuthal quantum numbers $m = 0$ and $m \geqslant 1$ (mostly $m = 1$), respectively.

TABLE 1 : Values of the dipole moment μ and atomic binding energy ΔE_a calculated in Ref.38 for an $r_s = 2$ substrate, compared with low-adatom-coverage experimental results. d in the second column is the calculated equilibrium distance from the background edge. The available experimental data are for transition-metal substrates (see Ref.38). The < sign in the table refers to the fact that, even at low coverages, H and O atoms cluster into islands, making the true zero-coverage limit of μ more negative than the measured low-coverage value.

	Theory			Experiment	
	d (a.u.)	μ (D)	ΔE_a (eV)	μ (D)	ΔE_a (eV)
H	1.1	-0.5	1.5	<-0.15	3
Li	2.5	2.6	1.3	∿1. 5-3	∿2.5-3
O	1.1	-1.7	5.4	<-0.4	5-6

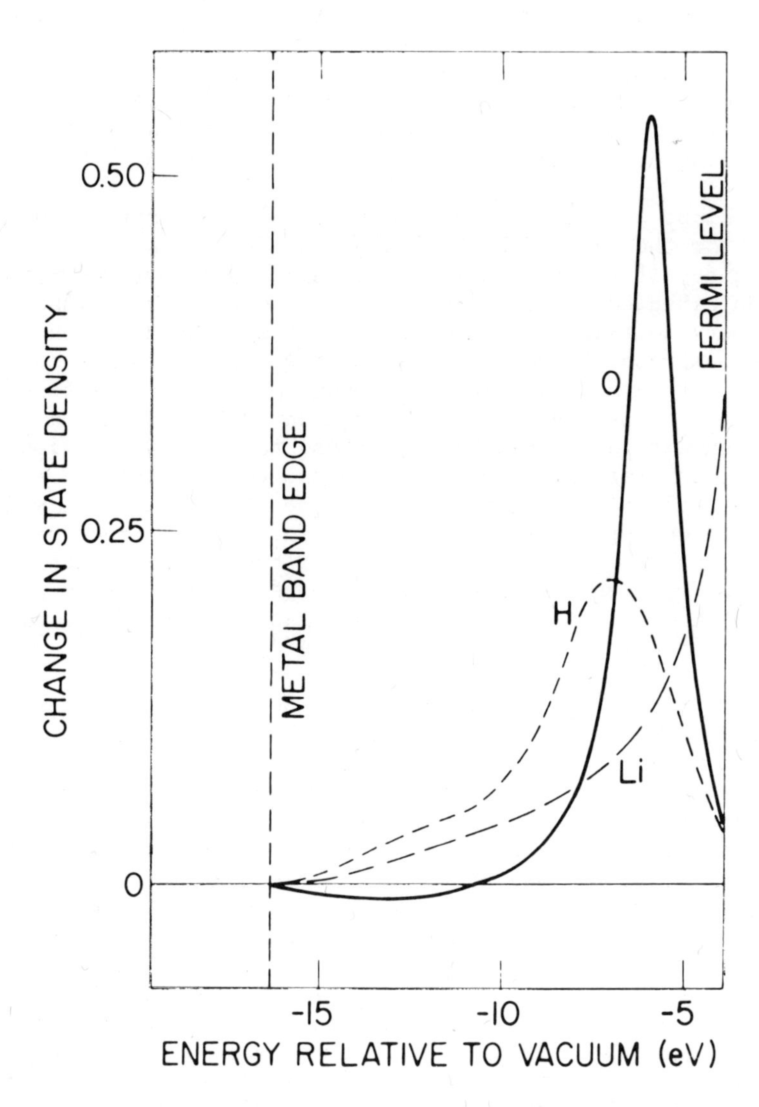

Fig.18 : Differences in continuum state density (see text) between chemisorbed system and bare substrate (r_s=2), as computed in Ref.38. Curves are each normalized to unity. Note that, in the case of hydrogen and oxygen, the valence-state resonances are below the Fermi level (and are therefore filled), reflecting the fact that these atoms, in contrast to Li, chemisorb electronegatively.

The simple 1-dimensional symmetry of the above calculation is lost once the discrete character of the adatom nuclei (or ions) is considered, which is of course necessary to study a case such as oxygen chemisorption. Density-functional studies of this type have so far dealt only with a single adatom on the surface, which has the advantage (when coupled with the uniform-background substrate model) of possessing cylindrical symmetry[35].

The first of these calculations was done by Smith, Ying and Kohn[37] for the case of H chemisorption. Two major approximations (the adequacy of which are discussed in Ref.38) were employed : the use of an extended Thomas-Fermi method, and the use of linear-response theory to treat the interaction between the H nucleus (proton) and the substrate. There are no parameters (other than the substrate r_s) in this treatment ; the equilibrium metal-adatom distance is found by minimizing the total energy. Huntington, Turk and White[39] employed an extended Thomas-Fermi method to study Na chemisorption, with no linear-response approximation but with the use of a small set of variational functions to model the charge density, and a pseudopotential to represent the Na ion potential. The procedure of Smith, Ying and Kohn was also applied to all of the alkalies, using pseudopotentials, by Kahn and Ying[40]. Quite reasonable agreement was obtained in this last study between calculated and measured dipole moments and binding energies for Na, K and Cs.

A completely self-consistent, wave-mechanical calculation has been done (using Eqs. (1.12)-(1.14)) by Lang and Williams[38] for the single atom chemisorbed on a uniform-background substrate. The cases of H, Li and O were considered. Figure 17 gives the curves of atomic binding energy and of the various components of the force on the nucleus for oxygen, as a function of metal-adatom separation. This figure shows the way in which the electrons in both core and continuum states polarize in the non-spherical self-consistent potential. Table 1 compares calculated dipole moments and atomic binding energies with the only available experimental data (on transition-metal substrates, to which the uniform-background model is not particularly meant to apply). The differences in state density between the chemisorbed system and the bare substrate were calculated using Eq.(1.12a)[41], and are shown in Fig.18. It should be emphasized that the only input to these calculations is the adatom atomic number Z and the substrate density (specified by r_s).

REFERENCES

(1) P. Hohenberg and W. Kohn, Phys. Rev. 136, B864 (1964)
(2) W. Kohn and L.J. Sham, Phys. Rev. 140, A1133 (1965)
(2a)Atomic units are used.
(3) Cf. J.C. Stoddart and N.H. March, Proc. Roy. Soc.A299, 279(1967)

(3a) L.J. Sham and W. Kohn, Phys. Rev. 145, 561 (1966)
(3b) L. Hedin and B.I. Lundqvist, J. Phys. C 4, 2064 (1971)
(4) Other partitionings into "surface" and "bulk" terms can be found in the literature. Confusion sometimes arises when surface components defined in different ways are compared (see discussion by N.D. Lang and W. Kohn, Phys. Rev. B8, 6010 (1973)). A partioning of the work function into bulk and surface terms was recognized by E. Wigner and J. Bardeen, Phys. Rev. 48, 84 (1935).
(5) F.K. Schulte, J. Phys. C8, L001 (1974)
(6) For discussion and references, see N.D. Lang, "The Density-Functional Formalism and the Electronic Structure of Metal Surfaces", Solid State Physics, ed. F. Seitz, D. Turnbull and H. Ehrenreich (Academic Press, New York, 1973) Vol.28, p.225
(7) J.R. Smith, Ph.D. Thesis, Ohio State University, Columbus (1968)
(8) J.R. Smith, Phys. Rev. 181, 522 (1969)
(9) J. Bardeen, Phys. Rev. 49, 653 (1936)
(10) A. J. Bennett and C.B. Duke, in Structure and Properties of Solid Surfaces, ed.G.A. Somorjai (Wiley, New York, 1969) Ch.25
(11) N.D. Lang, Solid State Commun. 7, 1047 (1969)
(12) N.D. Lang and W. Kohn, Phys. Rev. B1, 4555 (1970)
(13) N.D. Lang and W. Kohn, Phys. Rev. B3, 1215 (1971)
(13a) Cf. also F.K. Schulte, Ph.D. Thesis, Ludwig-Maximilians-Universität München (1973) (thin films)
(14) J. Schmit and A.A. Lucas, Solid State Commun. 11, 415 (1972)
(15) V. Peuckert, Z. Phys. 241, 191 (1971)
(16) R.A. Craig, Phys. Rev. B 6, 1134 (1972)
(17) N.D. Lang and L.J. Sham, Solid State Commun. 17 (to be published); J. Harris and R.O. Jones, J. Phys. F4, 1170 (1974); E. Wikborg and J.E. Inglesfield, Solid State Commun. 16, 335 (1975) ; M. Jonson and G. Srinivasan, Physica Scripta 10, 262 (1974) ; J. Vannimenus and H.F. Budd, Solid State Commun. 15, 1739 (1974) ; A. Griffin, H. Kranz and J. Harris, J. Phys. F 4, 1744 (1974) ; G. Paasch, Phys. Stat. Sol. (b) 65, 221 (1974); J. Heinrichs, Phys. Rev. B 11, 3637 (1975) ; P.J. Feibelman, Solid State Commun. 13, 319 (1973) ; W. Kohn, Solid State Commun. 13, 323 (1973).
(18) C. Werner, F.K. Schulte and H. Bross, to be published
(19) H.F. Budd and J. Vannimenus, Phys. Rev. Lett. 31, 1218 (1973) and erratum ; and Ref.17 above. Cf. also G.D. Mahan and W.L. Schaich, Phys. Rev. B 10, 2647 (1974)
(20) Some other sum rules for static and dynamic properties that we do not discuss here include A. Sugiyama, J. Phys. Soc. Japan 15, 965 (1960) ; D.C. Langreth, Phys. Rev. B 5, 2842 (1972) and 11, 2155 (1975) ; P. Feibelman, Phys. Rev. B3, 220 (1971); J.A. Appelbaum and E.I. Blount, Phys. Rev. B 8, 483 (1973); J.E. Inglesfield and E. Wikborg, Solid State Commun. 15, 1727 (1974); D. Wagner, to be published; J. Heinrichs and N. Kumar Phys. Rev. B (to be published). Cf. also V. Peuckert, J. Phys.

C7, 2221 (1974) (high-density limit of the work function)
(21) N.D. Lang and W. Kohn, Phys. Rev. B 7, 3541 (1973). See also V.E. Kenner, R.E. Allen, and W.M. Saslow, Phys. Rev. B 8, 576 (1973) ; A.K. Theophilou and A. Modinos, Phys. Rev. B 6, 801 (1972) ; A.K. Theophilou, J. Phys. F 2, 1124 (1972).
(22) Cf. N.D. Lang and A.R. Williams, Phys. Rev. Lett. 34, 531 (1975)
(23) H.F. Budd and J. Vannimenus, Phys. Rev. (to be published), have been able to obtain curves such as that of Fig.9 by using only $\delta n_d(x)$ for $d \to \infty$
(24) Eg., A.W. Dweydari and C.H.B. Mee, Phys. Stat.Sol.(a) 27, 223 (1975) ; P.O. Gartland, S. Berge and B.J. Slagsvold, Phys. Rev. Lett. 28, 738 (1972)
(25) E.g., H.O.K. Kirchner and G.A. Chadwick, Phil.Mag.[8] 20,405 (1969); M. McLean and B. Gale, ibid., p. 1033. But note, for example, the results of U. Jeschkowski and E. Menzel, Surface Sci. 15, 333 (1968)
(26) E.g.,J.K.Grepstad, P.O.Gartland,B.J. Slagsvold, to be published.
(27) An excellent review of this work has been given G. Allan, in Surface Properties and Surface States of Materials, ed. L. Dobrzynski (Dekker, New York, to be published)
(28) Cf., e.g., K.H. Johnson and R.P. Messmer, J. Vac. Sci. Technol. 11, 236 (1974)
(29) E.g., F. Forstmann, Z. Phys. 235, 69 (1970) ; D.S. Boudreaux, Surf. Sci. 28, 344 (1971); V. Hoffstein, Solid State Commun. 10, 605 (1972) ; J.B. Pendry and S.J. Gurman, Surf. Sci. 49, 87 (1975) ; D. Spanjaard, D.W. Jepsen and P.M. Marcus, to be published.
(30) J.A. Appelbaum and D.R. Hamann, Phys. Rev. B 6, 2166 (1972)
(31) G.P. Alldredge and L.Kleinman, Phys. Rev. B 10, 559 (1974)
(32) E. Caruthers, L. Kleinman and G.P. Alldredge, Phys. Rev. B 8, 4570 (1973); 9, 3325 (1974); 9,3330 (1974)
(33) E.g., I.P. Batra and O. Robaux, J. Vac. Sci. Technol. 12, 242 (1975)
(34) O. Gunnarsson and P. Johansson, Int. J. Quant. Chem. (to be published)
(34a) A.R. Williams and J. van W. Morgan, J. Phys. C 7, 37 (1974)
(35) But cf. J.A. Appelbaum and D.R. Hamann, Phys. Rev. Lett. 34, 806 (1975) (H/Si)
(36) N.D. Lang, Solid State Commun. 9, 1015 (1971) ; Phys. Rev. B4, 4234 (1971). See also C. Warner, "Thermionic Conversion Specialists Conference, San Diego, 1971" (Institute of Electrical and Electronics Engineers, New York, 1972) (extended Thomas-Fermi analysis)
(37) J.R. Smith, S.C. Ying and W. Kohn, Phys. Rev. Lett. 30, 610 (1973) ; S.C. Ying, J.R. Smith and W. Kohn, Phys. Rev. B 11, 1483 (1975)
(38) N.D. Lang and A.R. Williams, Ref. 22
(39) H.B. Huntington, L.A. Turk and W.W. White, III, Surf. Sci. 48, 187 (1975)

(40) L.M. Kahn and S.C. Ying, Solid State Commun. 16, 799 (1975)
(41) The formal relationship between the eigenstate density associated with Eq.(1.12a) and the quasiparticle energy spectrum is discussed above in connection with Eq.(1.16), and in references 3a and 3b.

CHEMISORPTION THEORY, ELECTRONIC STRUCTURE, AND REACTIVITY OF METAL SURFACES

T.B. GRIMLEY

Donnan Laboratories

University of Liverpool

1. WAVEFUNCTION EXPANSIONS

1.1. General

The technique of expanding a wavefunction in terms of a complete set of functions (a basis set) is familiar in formal quantum theory. It is encountered in Rayleigh-Schrödinger perturbation theory, and in the proof of the Variation Theorem for example. The use of an incomplete, and non-orthogonal basis set is familiar to all students of quantum chemistry in the linear combination of atomic orbitals molecular orbital (LCAOMO) scheme. The wavefunction being represented in this way may be either a one-electron state (an orbital), or a many-electron function representing the true states (ground or excited) of a complicated many-electron system. In the former case one has of course ultimately to build up many-electron states by allocating electrons to orbitals. In chemisorption theory we encounter both approaches, but the approach which begins by searching for the one-electron states of the interacting system adsorbate + substrate, i.e., the molecular orbital (MO) scheme, is certainly the more familiar, as indeed it is in molecular quantum mechanics. It is however interesting to observe that the other approach of expressing the many-electron ground state of the interacting system in terms of the many-electron states of the non-interacting adsorbate and substrate, i.e., the Heitler-London (HL) scheme, was used in the first serious work in chemisorption theory by Toya[1,2] just as it had been used in the early work on molecular binding by Heitler and London[3], Ireland[4], and others. We refer to Toya's work later (section 1.10), but first, because its concepts are generally familiar, and easily visualized, we consider the MO theory.

We are interested in chemisorption by a semi-infinite metal occupying the half-space $z > 0$ say (Fig.1). A self-consistent potential $V(r)$, the same for all valency electrons is assumed to exist so that the one-electron states ψ satisfy the Schrödinger equation

$$-(\hbar^2/2m)\nabla^2\psi + V(r)\psi = \varepsilon\psi \qquad (1)$$

In the interior of the metal, far from the admolecule, we know that ψ must resemble very closely one of the states ϕ_k of the semi-infinite metal, and we also know that there will be some solutions of (1) which, over the admolecule, resemble the admolecule states ϕ_A. Consequently it is natural to assume that ψ can be expanded in terms of the basis set $\{\phi_A, \phi_k\}$:

$$\psi = \sum_A a_A\phi_A + \sum_k b_k\phi_k \qquad (2)$$

This basis set is <u>overcomplete</u> ; the eigenstates of either the admolecule, or the semi-infinite metal alone are complete sets. We avoid the problem on the admolecule by limiting the set $\{\phi_A\}$ to only those orbitals which our chemical knowledge and intuition tell us are important in bonding, but we cannot easily limit the set $\{\phi_k\}$ because they form a quasi-continuous spectrum going up through the vacuum level. Overcompleteness is therefore an essential feature of the expansion (2).

It is also important to realize that the states ϕ_k, which satisfy Schrödinger's equation for the semi-infinite metal

$$(-(\hbar^2/2m)\nabla^2 + V_{metal})\phi_k = \varepsilon_k\phi_k \qquad (3)$$

are not the usual Bloch states of solid state theory. The semi-infinite crystal has no translational symmetry in the z-direction, and the states ϕ_k are decaying with z in the half-space $z < 0$ for energies below the vacuum level. In addition, some members of $\{\phi_k\}$ will be surface states decaying in the half-space $z > 0$ as well (Section 1.5). It is clear therefore that the determination of the semi-infinite metal states ϕ_k is a major problem (see Dr. Lang's lectures). But any disturbance in a metal, an impurity in the interior, or an admolecule on the surface, is very efficiently screened (the Thomas-Fermi screening length in Al is 0.07 nm), so that chemisorption is, in a certain sense, a localized phenomenon, and we can argue that a good solution to the local problem of the admolecule, and a few metal atoms near it in the surface, is the most important part of the chemisorption problem. If this is so, then chemisorption by metals can be treated in exactly the same way as quantum chemists treat molecular binding, and instead of introducing the set $\{\phi_k\}$, we introduce a set $\{\phi_m\}$ of localized atomic orbitals on the metal atoms, and write

$$\psi = \sum_A a_A \phi_A + \sum_m b_m \phi_m \tag{4}$$

This basis set $\{\phi_A, \phi_m\}$ will usually be limited to those orbitals which our chemical knowledge tells us are important in bonding, and it will therefore be incomplete as well as being non-orthogonal. If the set $\{\phi_m\}$ is limited to a few metal atoms near the admolecule, we are aiming at a local cluster calculation, and our chemisorption theory no longer involves any special problems. If $\{\phi_m\}$ goes over all metal atoms, it might seem that we have made no advance on the expansion (2) because the only change is to express the states$\{\phi_k\}$ in terms of the set $\{\phi_m\}$. But if chemisorption is a localized phenomenon, (4) is much more convenient than (2) because we can divide the localized basis set $\{\phi_m\}$ into orbitals on a few metal atoms near the admolecule ($m\varepsilon B$ say), and those beyond ($m\varepsilon D$ say), and calculate the wavefunction coefficients b_m in the domain B as accurately as possible, but calculate in domain D using a much simpler model (section 2.5. Also Professor Allan's lectures). However, we begin our study of chemisorption with the wavefunction expansion (2). This approach is best suited to the simple metals in Groups I and II whose electronic structures including surface state structures are reasonably well understood.

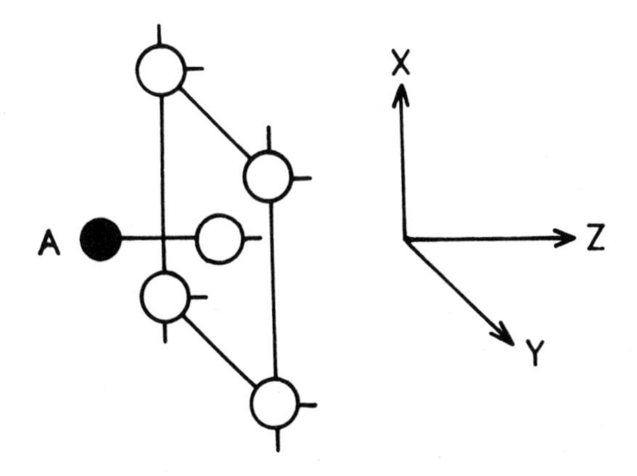

Fig. 1 : An admolecule A on a semi-infinite metal

1.2. Overcompleteness

Because the basis {A,k} is overcomplete, the expansion (2) is not unique ; the coefficients a_A and b_k can be chosen in any number of different ways, all equally valid. This is illustrated in Figure 2 where we have set up three coordinate axes in a 2-dimensional space. There are two perpendicular axes k1 and k2 representing two orthogonal k states, and a third axis A representing one A state. There are infinitely many ways to reach an arbitrary point P by travelling on paths which are always parallel to one or other of the three axes. Three such paths are shown. One of them OXP uses only k1 and k2 and demonstrates the completeness of the k-states, OZYP is a general path, and OQWP where Q is the foot of the perpendicular from P, is a special path which keeps to A as much as possible, and used k1 and k2 the least. For this special path, the resultant QP of the path sections parallel to k1 and k2 is perpendicular (orthogonal) to A. Exactly this notion can be used in (2) to define a unique expansion ; we require the second term in (2) to be orthogonal to every ϕ_A in the first,

$$\sum_k \langle A|k\rangle b_k = 0, \quad \text{all } A \tag{5}$$

If we substitute (2) into (1), multiply on the left by the complex conjugate of any member of the set {A,k} , and integrate over all space, we arrive at a system of linear equations for the coefficients a_A, b_k (see ref.5 for example). We write these equations in matrix form by collecting the coefficients a_A, b_k into a column matrix ;

$$\begin{vmatrix} H_A - 1\varepsilon & H_{AM} - \varepsilon S_{AM} \\ H_{MA} - \varepsilon S_{MA} & H_M - 1\varepsilon \end{vmatrix} \begin{vmatrix} a \\ b \end{vmatrix} = 0 \tag{6}$$

In (6) the matrix H of the one electron operator

$$\hat{H} = -(\hbar^2/2m)\nabla^2 + V \tag{7}$$

which appears in (1), is partitioned into an admolecule block H_A, a metal block H_M, and the coupling between them. Thus H_A has elements $\langle A|H|A'\rangle$, H_M has elements $\langle k|H|k'\rangle$, and H_{AM} has elements $\langle A|H|k\rangle$. S is the overlap matrix which differs from the unit matrix because the set {A,k} is overcomplete. S_{AM} has elements $\langle A|k\rangle$ so we can now write the subsidiary conditions (5) in the form

$$S_{AM}b = 0 \tag{8}$$

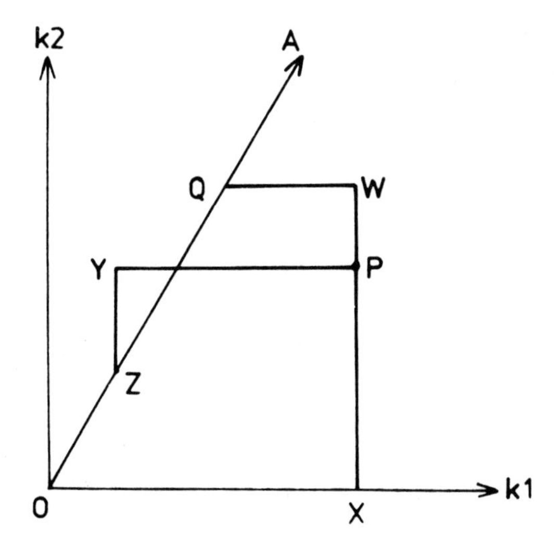

Fig. 2 : Illustration of overcompleteness

Because the set $\{\phi_k\}$ is complete, S has the property $S_{AM}S_{MA} = 1$.

From (6)

$$(H_A - 1\varepsilon)a + (H_{AM} - \varepsilon S_{AM})b = 0 \qquad (9)$$

$$(H_{MA} - \varepsilon S_{MA})a + (H_M - 1\varepsilon)b = 0 \qquad (10)$$

Using (8), equation (9) becomes

$$(H_A - 1\varepsilon)a + H_{AM}b = 0 \qquad (11)$$

and we use this to eliminate εS_{MA} from (10). Since

$$\varepsilon S_{MA}a = S_{MA}H_Aa + S_{MA}H_{AM}b$$

equation (10) becomes

$$(H_{MA} - S_{MA}H_A)a + (H_M - S_{MA}H_{AM} - 1\varepsilon)b = 0 \quad (12)$$

Equations (11) and (12) are replacements for (9) and (10), and the condition (8). To proceed we define

$$V_{MA} = H_{MA} - S_{MA}H_A \tag{13}$$

and then note that

$$\begin{aligned} H_{AM} &= H^{\dagger}_{MA} && \text{(H is Hermitian)} \\ &= S^{\dagger}_{MA}H_A + V^{\dagger}_{MA} && \text{(definition of } V_{MA}\text{)} \\ &= H_A S_{AM} + V^{\dagger}_{MA} && \text{(S is Hermitian)} \end{aligned}$$

Consequently,

$$H_{AM}b = H_A S_{AM}b + V^{\dagger}_{MA}\, b = V_{AM}\, b \text{ (using condition(8))},$$

and if we now define a pseudo-Hamiltonian for use with the k states

$$H^{ps}_M = H_M - S_{MA}H_{AM} \tag{14}$$

equations (11) and (12) can be written in matrix form :

$$\begin{vmatrix} H_A - 1\varepsilon & V^{\dagger}_{MA} \\ V_{MA} & H^{ps}_M - 1\varepsilon \end{vmatrix} \begin{vmatrix} a \\ b \end{vmatrix} = 0 \tag{15}$$

Comparing (15) with (6) we see that all trace of the overlap matrix signalling the overcompleteness of the basis has disappeared but we have instead a non-Hermitian pseudo-Hamiltonian H^{ps}_M and a pseudo-interaction V_{MA} between the metal and the admolecule[6]. A similar situation is found in the theory of transition metal impurities in simple metals[7].

1.3. The secular equation and the Greenian matrix

According to (15) the one-electron energies ε_μ are the roots of the secular equation

$$\det|F - 1\varepsilon| = 0, \qquad F = \begin{vmatrix} H_A & V^{\dagger}_{MA} \\ V_{MA} & H^{ps}_M \end{vmatrix} \tag{16}$$

This is an equation of ∞ order because there are an infinite number of metal atoms in the system, and the eigenvalue spectrum of F is therefore essentially continuous. In this situation there is no question of solving (16) for the eigenvalues ε_μ, we seek instead the distribution function $\rho(\varepsilon)$ governing the density of eigenstates as a function of energy. This density of states, and in fact much of what we need to know about the system can be calculated from a knowledge of the Green operator $G(\varepsilon)$ for the eigenvalue problem (15).

The Green operator is defined in terms of the one-electron states and energies of the system, admolecule + metal, i.e., in terms of the eigenfunctions ψ_μ and eigenvalues ε_μ of (1) ;

$$G(\varepsilon) = \sum_\mu \frac{|\mu\rangle\langle\mu|}{\zeta - \varepsilon_\mu} \quad , \zeta = \varepsilon + i0 \tag{17}$$

but of course every ψ_μ is expanded in the form (2), and every ε_μ satisfies (16). The matrix of G is diagonal in the representation afforded by the basis set $\{\psi_\mu\}$ with diagonal elements

$$\langle\mu|G|\mu\rangle = (\zeta - \varepsilon_\mu)^{-1} \tag{18}$$

Thus the imaginary part of $\langle\mu|G|\mu\rangle$ consists of a δ-function of strength $-\pi$ at ε_μ ;

$$\mathrm{Im}\langle\mu|G|\mu\rangle = -\pi\delta(\varepsilon - \varepsilon_\mu) \tag{19}$$

Since each eigenvalue ε_μ contributes a δ-function to the density of states we have

$$\rho(\varepsilon) = \sum_\mu \delta(\varepsilon - \varepsilon_\mu) = -\frac{1}{\pi}\,\mathrm{Im}\sum_\mu \langle\mu|G|\mu\rangle = -\frac{1}{\pi}\,\mathrm{Im\,Tr}\,G \tag{20}$$

To express $\rho(\varepsilon)$ in terms of the matrix elements of H, we improve the notation a little by collecting the eigenfunctions ψ_μ into a row vector ψ, and the basis states $\{\phi_A, \phi_k\}$ into a row vector ϕ. Then the eigenfunction expansions (2) are

$$\psi = \phi A \quad , \quad A^\dagger S A = 1 \tag{21}$$

with inverse

$$\phi = \psi B \quad , \quad AB = 1 \quad , \quad BA = 1 \tag{22}$$

From (22) we see that

$$B^\dagger B = S \tag{23}$$

but of course the inverse of S does not exist because the basis is overcomplete. From (18), the definition of G, the Green matrix in the ψ representation is

$$(1\zeta - \overline{H})G = 1 \tag{24}$$

where $\overline{H}$ is the matrix of the one-electron operator (7) in the ψ representation. Using the relations (21)-(23) for the matrices A and B, we can transform (24) into the form

$$(\zeta S - H)\tilde{G} = 1 \tag{25}$$

where

$$\tilde{G} = A\overline{G}A^{\dagger} \tag{26}$$

and H is as before, the matrix of the one-electron operator H in the ϕ-representation ($\overline{H} = A^{\dagger}HA$).

The Green matrix $\tilde{G}$ defined in (25) is the natural one of use for the eigenvalue problem (6). From (20), (23) and (26), remembering that the order of matrix multiplication is unimportant in the trace,

$$\rho(\varepsilon) = -\frac{1}{\pi}\,\mathrm{Im}\,\mathrm{Tr}\,(\tilde{G}S) \tag{27}$$

which is the desired expression for the density of states. We note that the matrix $\tilde{G}$ itself can give a function which would be known to quantum chemists as the net density of states, $\tilde{\rho}(\varepsilon)$,

$$\tilde{\rho}(\varepsilon) = -\frac{1}{\pi}\,\mathrm{Im}\,\mathrm{Tr}\,\tilde{G} \tag{28}$$

The function $\rho(\varepsilon)$ is the gross density of states ; the net/gross distinction can appear whenever the overlap matrix differs from the unit matrix[8] (but see below).

Equations (21)-(28) are general, and apply equally to overcomplete basis sets, and to incomplete but non-orthogonal basis sets, we only have to remember that, in the former case, S is singular. However we are concerned here with overcompleteness (see Section 1.9 for incomplete but overlapping basis sets), so we now examine how the subsidiary condition (8) can be used to simplify (25).

First we note that in our new notation, the subsidiary condition (8) becomes

$$\sum_k S_{Ak} A_{k\mu} = 0 \quad (\text{all } \mu) \tag{8}$$

and from (26)

$$\sum_k S_{Ak}\tilde{G}_{ks} = \sum_{k\mu} S_{Ak}A_{k\mu}(\zeta - \varepsilon_\mu)^{-1}A^\dagger_{\mu s} = 0$$

because of (8)

This gives the two matrix equations

$$S_{AM}\tilde{G}_{MA} = 0, \quad S_{AM}\tilde{G}_M = 0 \tag{29}$$

which are subsidiary conditions on the Green matrix in (25). Consequently, if we write (25) in block form

$$\begin{vmatrix} 1\zeta - HA & \zeta S_{AM} - H_{AM} \\ \zeta S_{MA} - H_{MA} & 1\zeta - H_M \end{vmatrix} \begin{vmatrix} \tilde{G}_A & \tilde{G}_{AM} \\ \tilde{G}_{MA} & \tilde{G}_M \end{vmatrix} = \begin{vmatrix} 1 & 0 \\ 0 & 1 \end{vmatrix} \tag{30}$$

and use the conditions (29) exactly as we used the condition (8) in (6), we arrive without difficulty at a formally simpler equation for $\tilde{G}$

$$\begin{vmatrix} 1\zeta - H_A & -V^\dagger_{MA} \\ -V_{MA} & 1\zeta - H^{ps}_M \end{vmatrix} \begin{vmatrix} \tilde{G}_A & \tilde{G}_{AM} \\ \tilde{G}_{MA} & \tilde{G}_M \end{vmatrix} = \begin{vmatrix} 1 & 0 \\ -S_{MA} & 1 \end{vmatrix} \tag{31}$$

involving the pseudo-Hamiltonian (14) for use with the k states and the pseudointeraction (13), and incorporating the subsidiary conditions (29). An alternative form of (31) is obtained by multiplying on the right with the matrix

$$K = \begin{vmatrix} 1 & 0 \\ S_{MA} & 1 - S_{MA}S_{AM} \end{vmatrix}$$

and using the subsidiary conditions (29) in the form

$$\tilde{G}_{AM}\,S_{MA} = 0, \quad \tilde{G}_M S_{MA} = 0$$

to establish that $\tilde{G}K = \tilde{G}$. Then

$$\begin{vmatrix} 1\zeta - H_A & -V_{AM} \\ -V_{MA} & 1\zeta - H^{ps}_M \end{vmatrix} \begin{vmatrix} \tilde{G}_A & \tilde{G}_{AM} \\ \tilde{G}_{MA} & \tilde{G}_M \end{vmatrix} = \begin{vmatrix} 1 & 0 \\ 0 & 1 - S_{MA}S_{AM} \end{vmatrix} \tag{32}$$

which was first derived by Newton[9] and by Gunnarsson and Hjelmberg[10].

We now examine the density of states (27). Using conditions (29)

$$\mathrm{Tr}\ \tilde{G}S = \mathrm{Tr}\ (\tilde{G}_A + \tilde{G}_M + \tilde{G}_{AM}S_{MA} + \tilde{G}_{MA}S_{AM})$$

$$= \mathrm{Tr}\ (\tilde{G}_A + \tilde{G}_M) = \mathrm{Tr}\ \tilde{G}$$

Consequently the subsidiary conditions effect a real simplification by eliminating the net/gross distinction ;

$$\rho(\varepsilon) = \tilde{\rho}(\varepsilon) = -\frac{1}{\pi}\ \mathrm{Im\ Tr}\ \tilde{G}S = -\frac{1}{\pi}\ \mathrm{Im\ Tr}\ \tilde{G} \qquad (33)$$

Generally, it is easiest to calculate $\tilde{G}S$ from (30), but for certain approximate models, it seems better to calculate $\tilde{G}$ from (32).

1.4. The electronic structure

We derive a formula for the density of states in the system admolecule + metal from (30) and (33). The first column of the matrix equations (30) is

$$(1\zeta - H_A)\tilde{G}_A + (\zeta S_{AM} - H_{AM})\tilde{G}_{MA} = 1$$

$$(\zeta S_{MA} - H_{MA})\tilde{G}_A + (1\zeta - H_M)\tilde{G}_{MA} = 0$$

and eliminating $\tilde{G}_{MA}$ we find

$$\tilde{G}_A = (1\zeta - H_A - q_A)^{-1} \qquad (34)$$

with

$$q_A = V_{AM}(1\zeta - H_M)^{-1}V_{MA} \qquad (35)$$

$$V_{AM} = \zeta S_{AM} - H_{AM} \qquad (36)$$

The second column of the matrix equation (30) is

$$(1\zeta - H_A)\tilde{G}_{AM} + (\zeta S_{AM} - H_{AM})\tilde{G}_M = 0 \qquad (37)$$

$$(\zeta S_{MA} - H_{MA})\widetilde{G}_{AM} + (1\zeta - H_M)\widetilde{G}_M = 1 \tag{37}$$

and eliminating $\widetilde{G}_M$, and using the notation (36)

$$\widetilde{G}_{AM} = \widetilde{G}_A V_{AM}(1\zeta - H_M)^{-1} \tag{38}$$

Using this in (37)

$$\widetilde{G}_M = (1\zeta - H_M)^{-1} [1 + V_{MA}\widetilde{G}_A V_{AM}(1\zeta - H_M)^{-1}] \tag{39}$$

From (33), (34) and (39) we now have a formula for the density of states ;

$$\begin{aligned}\rho(\varepsilon) &= -\frac{1}{\pi} \operatorname{Im} \operatorname{Tr} (\widetilde{G}_A + \widetilde{G}_M) \\ &= -\frac{1}{\pi} \operatorname{Im} \operatorname{Tr} [\widetilde{G}_A + (1\zeta - H_M)^{-1}(1 + \widetilde{G}_A q_A)]\end{aligned} \tag{40}$$

A more convenient form of this equation is obtained by noting that from (35) and (36)

$$\begin{aligned}\delta q_A/\delta\varepsilon = -V_{AM}(1\zeta - H_M)^{-2}V_{MA} - S_{AM}(1\zeta - H_M)^{-1}V_{MA} \\ - V_{AM}(1\zeta - H_M)^{-1}S_{MA}\end{aligned}$$

so that, using the subsidiary conditions (29) and equation (38), we establish

$$\operatorname{Tr} \widetilde{G}(\delta q_A/\delta\varepsilon) = -\operatorname{Tr} (1\zeta - H_M)^{-1}\widetilde{G}_A q_A$$

which enables us to cast (40) into the form

$$\rho(\varepsilon) = -\frac{1}{\pi} \operatorname{Im} \operatorname{Tr} \{\widetilde{G}_A(1 - \delta q_A/\delta\varepsilon) + (1\zeta - H_M)^{-1}\} \tag{41}$$

This result is also obtained directly by calculating GS from (30) and using $\rho(\varepsilon) = -\pi^{-1} \operatorname{Im} \operatorname{Tr} \widetilde{G}S$[11].

We are interested, not in $\rho(\varepsilon)$ itself, but in $\Delta\rho(\varepsilon)$, the contribution of the admolecule to the density of states ;

$$\Delta\rho = \rho - \rho_M^f \tag{42}$$

where ρ_M^f is the density of states in the free metal. $\Delta\rho$ is not obtained simply by dropping the last term in (41) because H_M differs from the free metal Hamiltonian matrix because of the disturbance due to the admolecule ;

$$H_M = H_M^f + X_M \tag{43}$$

Thus, in the coupled system there is an effective Green matrix for the metal, G_M^{eff}, defined by

$$G_M^{eff} = (1\zeta - H_M)^{-1} \tag{44}$$

which can be expressed in terms of the free metal Green matrix G_M^f by Dyson's equation

$$G_M^{eff} = G_M^f + G_M^f X_M G_M^{eff} \tag{45}$$

giving

$$G_M^{eff} = (1 - G_M^f X_M)^{-1} G_M^f \tag{46}$$

Consequently, if we use $\rho_M^f = -\pi^{-1}$ Im Tr G_M^f, we find from (41)-(46) that

$$\Delta\rho = -\frac{1}{\pi} \text{Im Tr } \{G_A (1 - \delta q_A/\delta\varepsilon) + G_M^f X_M G_M^{eff}\} \tag{47}$$

$\Delta\rho$ is required to calculate what I call the <u>one-electron</u> contribution D_{one} to the binding energy D of the admolecule to the metal (although whether D_{one} is really a one-electron contribution depends on the details of the model used),

$$D_{one} = \sum_\sigma \int_{-\infty}^{\varepsilon_F} d\varepsilon\ \varepsilon [\rho_A^f(\varepsilon) - \Delta\rho_\sigma(\varepsilon)] \tag{48}$$

where σ is the spin index, either $\uparrow$ or $\downarrow$, and ε_F is the Fermi level, here assumed the same in the coupled system as in the free metal, although this may not be so in all <u>models</u> of chemisorption (see section 2.5).

All realistic models of chemisorption involve some sort of self-consistent calculation to determine how electrons are shared between the admolecule and the metal in the chemisorption bond, and to carry out such calculations, the so-called charge-and-bond-order matrix (the density matrix in quantum chemistry) P is needed. This matrix is given by

$$P = -\frac{1}{\pi} \text{Im} \int_{-\infty}^{\varepsilon_F} d\varepsilon \, \tilde{G}(\varepsilon) \tag{49}$$

but in practical calculations one cannot achieve self-consistency over the whole system. In the simplest calculations one might aim at self-consistency only over the admolecule so that only $\tilde{G}_A$ is needed (see section 1.6). Better calculations extend the self-consistency into the metal, and achieve self-consistency over a cluster comprising the admolecule, and its first coordination shell of metal atoms (see section 2.5). Such calculations do not require a new formalism, we simply allow the subscript A to stand for the cluster, not just the admolecule. The overcompleteness of the basis set $\{\phi_A, \phi_k\}$ is now very apparent (see section 2.6).

Finally, we emphasize that the operator (17) is known one $\tilde{G}$ has been calculated,

$$G(r,r' ; \varepsilon) = \sum_{rs} \tilde{G}_{rs}(\varepsilon) \, \phi_r(r) \phi_s^*(r') \tag{50}$$

and

$$\rho(r,r' ; \varepsilon) = -\frac{1}{\pi} \text{Im} \, G(r,r';\varepsilon) \tag{51}$$

is a spectral density function required for Lehmann's representation. I do not however use these real space quantities in my lectures.

1.5. Surface and volume states

To implement the above formal scheme we need the states $\{\phi_k\}$, and so at this point we face the task of calculating the electronic structure, energy spectrum and wavefunctions, of a semi-infinite metal as described by Dr. Lang and Professor Allan. At present the information we need in our chemisorption theory is only obtainable for some simplified models, the Sommerfeld free electron (FE) model, the nearly-free electron (NFE) model, and the jellium model. So far the FE and NFE models have been used[9,12]. The FE model is straightforward, though of course wavefunctions for energies above, as well as below, the vacuum level are needed. With the NFE model, the possibility of surface states arises. These states are described by wavefunctions which are localized near the surface plane, and display periodic properties only in directions parallel to the surface (figure 3). In view of this localization, it might seem that these states would play a special rôle in chemisorption, and in other surface processes. However, this can only be discovered by detailed calculations. Consider a solid with one valency electron per atom. With periodic boundary conditions over N atoms parallel to the surface, the squared modulus of a surface state wavefunction near the surface plane is $\sim N^{-2}$, and since there can be only about

N^2 such bound states their physical effects are ~ 1. This is the same as that of the volume states ; N^3 volume states below the vacuum level each with squared modulus near the surface plane $\sim N^{-3}$.

The theory of surface states on NFE metals has been comprehensively discussed by Heine[13,14], Forstmann[15] and Hoffstein[16], assuming that the periodic crystal potential is terminated at the surface by a vertical step at the potential maximum (Figure 3). In a semi-infinite crystal there is only a 2-dimensional translation group $T_{\|}$ say, and Bloch's theorem is

$$\phi(k_{\|}, r + \tau_{\|}) = \exp(i\, k_{\|} \cdot \tau_{\|})\, \phi(k_{\|}, r) \quad (52)$$

where $\tau_{\|}$ is a translation vector of $T_{\|}$, and $k_{\|}$ is a 2-dimensional wave vector labelling the irreductible representations of $T_{\|}$. We use coordinate systems adapted to the exposed surface ; $r = (x,y,z) = (\rho,z)$, and denote by $k = (k_{\|}, k_z)$ the usual 3-dimensional wave vector of a Bloch state, and by $K = (K_{\|}, K_z)$ a 3-dimensional reference vector in the reciprocal lattice. Surface states do not propagate in the z-direction, instead they decay with $|z|$, but $k_{\|}$ is still a good quantum number, and the surface state wavefunctions satisfy (52). Their energies can only lie in gaps in the $k_{\|}$ subbands. These subbands form a 1-dimensional band system ; for given $k_{\|}$, the energy bands are exhibited along k_z in the usual 3-dimensional Brillouin zone (Figure 4).

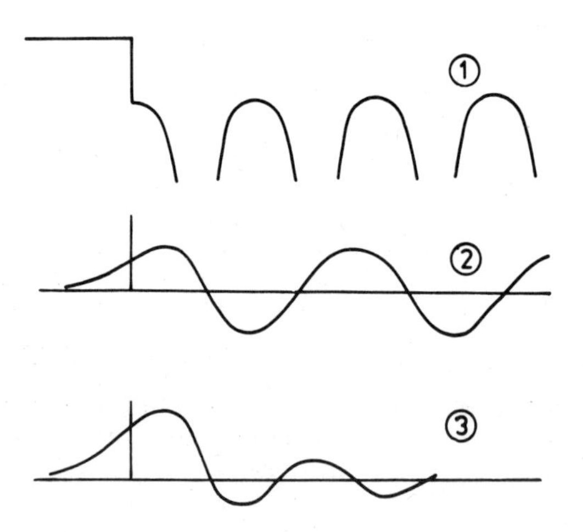

Fig. 3 : Surface and volume states : (1) the idealized potential, (2) volume state pseudowavefunction, (3) surface state pseudowavefunction.

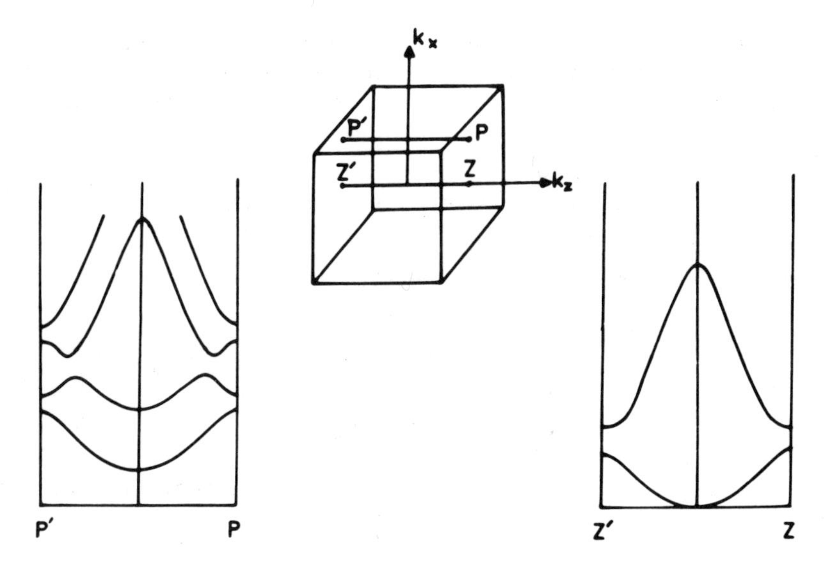

Fig. 4 : Energy band along Z'Z and P'P in a simple cubic crystal. These are the (0,0) and ($3\pi/4$,0) subbands respectively.

To find the energy spectrum of electrons in a semi-infinite crystal, we proceed as follows. For given $k_{\parallel}$ and energy ε, we write down the general solutions $\phi^{in}(k_{\parallel},r;\varepsilon)$ and $\phi^{out}(k_{\parallel},r;\varepsilon)$ of Schrödinger's equation for electrons in the metal ($z \geqslant 0$), and in free space outside ($z \leqslant 0$), and then find those values of ε for which the two solutions can be joined with continuous normal logarithmic derivative at the surface ($z = 0$) for all ρ. For given $k_{\parallel}$ and ε, there are an infinite number of solutions of Schrödinger's equation for electrons in a crystal, $\phi(k_z,k_{\parallel},\varepsilon;r)$ with different k_z. A small number may be propagating solutions with real k_z, the remainder are evanescent solutions with complex k_z; $k_z \to k_z+iq$. These evanescent solutions are discarded for the infinite crystal because they become arbitrarily large for $z \to \pm\infty$. But in the semi-infinite crystal we must not discard evanescent solutions which would only become infinite in the <u>missing half-crystal</u>. These solutions have $q > 0$, and without them we would not have a complete set of solutions with given $k_{\parallel}$. In a gap in the $k_{\parallel}$ subbands, there are <u>only</u> evanescent solutions. These have the form of the usual Bloch functions but k_z is complex. Thus they can be written

$$e^{-qz}\, v(k,\varepsilon;r) \qquad (53)$$

where v is a Block function with real $k = (k_{\parallel}, k_z)$. Labelling evanescent solutions by their complex k_z-values, propagating solutions by their real k_z-values, the general solution in the metal for arbitrary energy is

$$\phi^{in}(k_{\parallel},\varepsilon;r) = \sum_{k_z} A(k_z)\phi(k_z,k_{\parallel},\varepsilon;r) \qquad (54)$$

Outside the metal we have free electron solutions

$$\left.\begin{aligned} \chi(K_{\parallel},k_{\parallel},\varepsilon;r) &= \exp(\lambda z)\ \exp[i(k_{\parallel} + K_{\parallel}).\rho] \\ \lambda &= [2(V_o - \varepsilon) + |k_{\parallel} + K_{\parallel}|^2]^{1/2} \end{aligned}\right\} \qquad (55)$$

V_o is the height of the vacuum level above the mean crystal potential. For real λ these are evanescent free electron solutions. Consequently

$$\phi^{out}(k_{\parallel},\varepsilon;r) = \sum_{K} B(K_{\parallel})\chi(K_{\parallel},k_{\parallel},\varepsilon;r) \qquad (56)$$

The smooth joining of (54) and (56) on the plane z = 0 leads to a system of linear equations for the coefficients $A(k_z)$,

$$\sum_{k} [\lambda(K_{\parallel})\ F\ (K_{\parallel},k_z) - J(K_{\parallel},k_z)]\,A(k_z) = 0 \qquad (57)$$

where F and J are 2-dimensional Fourier transforms of

$\phi(k_z,k_{\parallel},\varepsilon;\rho,0)$ and $\delta\phi(k_z,k_{\parallel},\varepsilon;\rho,0)/\delta z$, for example

$$F = \int d\rho\ \exp[-i(k_{\parallel} + K_{\parallel}).\rho]\ \phi\ (k_z,k_{\parallel},\varepsilon;\rho,0) \qquad (58)$$

The following two statements are evidently true from the physics of the problem, but I do not know of a formal proof, although they are always apparent when one looks at particular examples of $k_{\parallel}$ subbands ;

(a) If ε lies in an allowed $k_{\parallel}$ subbands, there is, for each state, always one more k_z than $K_{\parallel}$ in (57) so that matched solutions always exist, and the system of equations in (57) determines the wavefunction coefficients $A(k_z)$.

(b) If ε lies in a gap in the $k_{\parallel}$ subbands, equal numbers of values of k_z and $K_{\parallel}$ are present so that (57) only has a non-trivial solution if

$$\det|\lambda F - J| = 0 \tag{59}$$

Since λ, F and J depend on ε explicitly or implicitly, (59) determines the energies of any surface states belonging to the $k_{\parallel}$ representation of $T_{\parallel}$, and lying in the band gap. The corresponding wavefunction coefficients are then found from (57).

For NFE metals, the crystal wavefunctions ϕ in (54) are expressed in terms of a few plane waves (with real or complex k_z)

$$\phi = \sum_K c(K) \exp[i(k + K).r] \tag{60}$$

Consequently

$$F = \sum_{K_z} c(K_z, K_{\parallel}), \quad J = i\sum_{K_z} (k_z + K_z) c(K_z, K_{\parallel}) \tag{61}$$

Since the coefficients $c(K_z, K_{\parallel})$ are known from band structure calculations on the infinite crystal, the states of the semi-infinite metal are now known in principle.

To see how this works in detail, consider the states with $k_{\parallel} = 0$ for Aℓ with a (100) surface. The subbands up to the top of the first gap (Figure 5) at least are adequately described by two plane waves with $K_{\parallel} = (0,0)$;

$$K = \frac{2\pi}{a}(0,0,0) \ , \ K = \frac{2\pi}{a}(\bar{2},0,0) \tag{62}$$

for k_z positive in (60). So from (61)

$$F(0,0) = c(0,0,0) + c(\bar{2},0,0) \tag{63}$$

$$J(0,0) = i[k_z c(0,0,0) + (k_z - \frac{4\pi}{a}) c(\bar{2},0,0)]$$

and from (55)

$$\lambda(0,0) = [2(V_o - \varepsilon)]^{1/2} \tag{64}$$

The Fourier coefficients $c(0,0,0)$ and $c(\bar{2},0,0)$ are known from the usual band structure eigenvalue problem

$$\begin{vmatrix} 1/2\, k_z^2 - \varepsilon & W_2 \\ W_2 & 1/2(k_z - \frac{4\pi}{a})^2 - \varepsilon \end{vmatrix} \begin{vmatrix} c(0,0,0) \\ c(\bar{2},0,0) \end{vmatrix} = 0 \tag{65}$$

where W_2 is the matrix element $\langle 0,0,0|W|\bar{2},0,0\rangle$ of the periodic potential between the two plane waves, $|W_2| = 0.055$ Ryd. for Aℓ[17]. Corresponding to this approximate band structure, there are only two terms on the right in (54), that at Δ and that at $\bar{\Delta}$, both propagating solutions. At $\bar{\Delta}$ we simply change k_z to $-k_z$ and $(\bar{2},0,0)$ to $(2,0,0)$. The system of equations (57) reduces to one equation

$$[\lambda(0,0)\,F(0,0,k_z) - J(0,0,k_z)]\,A(k_z) + [\lambda(0,0)F(0,0,-k_z) - J(0,0,-k_z)]\,A(-k_z) = 0 \qquad (66)$$

which determines $A(k_z)/A(-k_z)$ in the matched solution.

When we give ε a value in the band gap, (65) can only be satisfied for a complex k_z, in fact $k_z \to 2\pi/a + iq$. The first problem is to find the value of q (more generally k_z itself) for given ε. To do this, the eigenvalue problem (65) in ε is transformed to an eigenvalue problem in k_z. This doubles the size of the secular equation[16,18]. If we let c stand for the column vector in (65), and define

$$X = \begin{vmatrix} 0 & 0 \\ 0 & -4\pi k_z/a \end{vmatrix} \qquad Y = \begin{vmatrix} -\varepsilon & W_2 \\ W_2 & \dfrac{8\pi^2}{a^2} - \varepsilon \end{vmatrix} \qquad (67)$$

$$b = k_z c/\sqrt{2}$$

the eigenvalue problem in k_z is

$$\begin{vmatrix} 0 & 1 \\ -Y & -X \end{vmatrix} \begin{vmatrix} c \\ b \end{vmatrix} = \frac{k_z}{\sqrt{2}} \begin{vmatrix} c \\ b \end{vmatrix} \qquad (68)$$

which can be solved by standard techniques[19] to yield k_z (real or complex) and the vector c when ε is given. In the subband gap, the four eigenvalues of (68) are $2\pi/a \pm iq$, $-2\pi/a \pm iq$, but the second pair do not give new vectors c, and only the positive sign to q is physically acceptable for a metal in the half-space $z > 0$. Consequently, there is only one k_z-value in (54) and (57), $k_z = 2\pi/a + iq$, and (57) gives

$$\lambda(0,0)F(0,0,k_z) - J(0,0,k_z) = 0 \qquad (69)$$

which is the simplest version of (59). The question of course is whether (69) can be satisfied for an ε in the band gap. If not, there is no surface state with $k_{\parallel} = 0$ in this gap. Using (63), (64) and (68) in (69) it turns out that there is a surface state if $W_2 < 0$, but not otherwise. This is essentially a result proved long ago by

Shockley[20] for 1-dimensional systems; if $W_2 < 0$ the Bloch state at the bottom of the gap is bonding, at the top of the gap is anti-bonding. Since W_2 is negative† for the Aℓ pseudopotential, we see that there is indeed a surface state here. Of course, one does not need the general approach of equations (59) and (68) to investigate this state ; the general approach is needed however for complete investigations like those carried out for Aℓ by Boudreaux[18]; equation (68) generates the complex band structure.

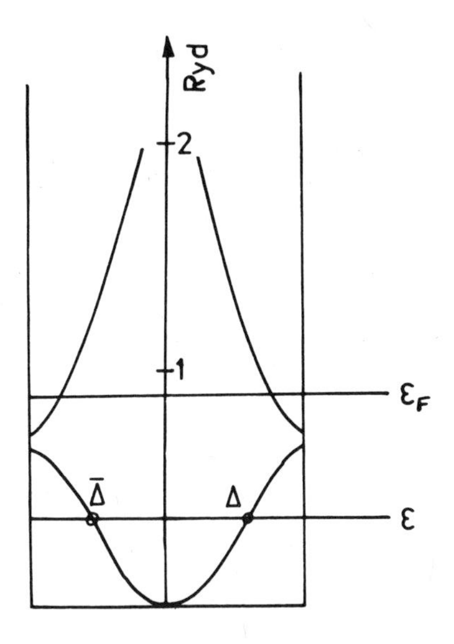

Fig. 5 : (0,0) subbands of aluminium

----

† *Our origin of coordinates is at the unit cell edge, not at its centre.*

1.6. An adatom on a free-electron metal

Consider hydrogen on aluminium. The energy levels are shown in Figure 6, and the hydrogen 1s level is far below the Fermi level ε_F so electron transfer to the hydrogen is expected. But if one electron is transferred, the 1s level is raised by something like 0.5 A.U. because of the Coulomb repulsion of the two electrons in it, and the extra electron is then returned to the metal causing the 1s level to fall again. This sort of situation is very familiar in chemistry, and what we do is to make ε_A, the position of the 1s level, depend on its occupancy $<n_A>$ and calculate $<n_A>$ self-consistently along with ε_A.

H_A in (30) is a single element $<A|-\frac{1}{2}\nabla^2 + V_A + V_M|A>$, and to allow for the 1s level shift with occupancy we write

$$<A| -\frac{1}{2}\nabla^2 + V_A + V_M|A> = E_A + J_A <n_A> = \varepsilon_A \tag{70}$$

The literal definition of J_A would be

$$J_A = \int dr_1\, dr_2 \frac{|\phi_A(r_1)|^2 |\phi_A(r_2)|^2}{r_{12}} \tag{71}$$

giving 0.625 A.U. for hydrogen, but to allow for orbital expansion in the H^- ion, one could take J_A as the difference between the ionization potential, and the electron affinity. This gives 0.456 A.U., and a further reduction through many body screening effects not included in a one-electron theory may be anticipated.

From (26) we see that, if we define an adatom density of states $\rho_A(\varepsilon)$ by

$$\rho_A = -\frac{1}{\pi} \operatorname{Im} \tilde{G}_A = \sum_\mu |A_{A\mu}|^2 \delta(\varepsilon - \varepsilon\mu) \tag{72}$$

then

$$<n_A> = \sum_{\mu<\varepsilon_F} |A_{A\mu}|^2 = \int_{-\infty}^{\varepsilon_F} d\varepsilon\, \rho_A(\varepsilon) \tag{73}$$

We note that generally ρ_A and $<n_A>$ in this way would be net quantities, but because of the condition (8) we have $A_{A\mu} = <A|\mu>$, and so they are also gross quantities. We now formulate the self-consistency condition.

The first column of the matrix equation (32) is

$$\left.\begin{aligned} (\zeta - \varepsilon_A)\tilde{G}_A - V_{AM}\,\tilde{G}_{MA} &= 1 \\ -V_{MA}\,\tilde{G}_A + (1\zeta - H_M^{ps})\,\tilde{G}_{MA} &= 0 \end{aligned}\right\} \tag{74}$$

and eliminating $\tilde{G}_{MA}$ we find

$$\tilde{G}_A = (\zeta - \varepsilon_A - q_A)^{-1} \tag{75}$$

with

$$q_A = V_{AM}\,(1\zeta - H_M^{ps})^{-1}\,V_{MA} \tag{76}$$

so the main problem is the inversion of the matrix $(1\zeta - H_M^{ps})$. From (14) we see that if we divide the self-consistent potential V(r) in (1) into that seen by an electron in the clean semi-infinite metal and the remainder, X, then

$$H_M^{ps} = \varepsilon_M - S_{MA}S_{AM}\varepsilon_M + X_M - S_{MA}X_{AM} \tag{77}$$

where ε_M is the diagonal matrix of the eigenvalues of the clean metal. Thus the adatom causes k-state scattering both directly through X, and indirectly through the overlap of the k states and the adatom state. No calculations retaining these scattering terms have been made at present. Instead the approximation

$$H_M^{ps} \simeq \varepsilon_M \tag{78}$$

has been used[6,11,12], although this has recently been criticized[10].

Using (78), the <u>chemisorption function</u> (76) is rather simple

$$q_A = \sum_k |V_{kA}|^2 / (\zeta - \varepsilon_k) = \alpha(\varepsilon) - i\,\Gamma(\varepsilon) \tag{79}$$

where

$$\Gamma(\varepsilon) = \pi \sum_k |V_{kA}|^2\,\delta(\varepsilon - \varepsilon_k) \tag{80}$$

and $\alpha(\varepsilon)$ is the Hilbert transform of $\Gamma(\varepsilon)$,

$$\alpha(\varepsilon) = \frac{1}{\pi}\int_{-\infty}^{+\infty} d\varepsilon'\,\frac{\Gamma(\varepsilon')}{\varepsilon - \varepsilon'} \tag{81}$$

If we divide the self-consistent potential V(r) in (1) into that seen by an electron in the 1s orbital in the free atom, and the remainder which we call V_M, then

$$V_{kA} = \langle k|V_M|A\rangle - \langle k|A\rangle\langle A|V_M|A\rangle \qquad (82)$$

The above equations have a certain generality, but for a simple free electron metal they are easy to handle because the chemisorption function does not have to be calculated self-consistently. Instead V_M is zero for $z < 0$, and is the inner potential, W_0 say, for $z > 0$, so that q_A is calculated once and for all at every adatom-metal distance s.Γ is calculated first,α then follows from (81). I mention only one point about such calculations. Using box normalization of the k states, one encounters terms in $|V_{Ak}|^2$ which oscillate infinitely rapidly as continuous functions k as the size of the box tends to infinity. In calculating Γ from (80) the summation over the actual eigenvalues is replaced by an integration over k, and now the rapidly oscillating terms must be replaced either by their mean values over the many eigenvalues which occur in one oscillation, or when there is only one eigenvalue per oscillation, by the slowly varying function defined only at the eigenvalues. The former situation is encountered with k states above the vacuum level, the latter with states below. Figure 7 shows the spectral density function $\Gamma(\varepsilon)$ calculated[12] for a hydrogen atom at distance 1 A.U. from free-electron Aℓ ;

$$\phi_A = \pi^{-1/2}\exp(-r) \ , \ W_0 = -0.5834 \text{ A.U.}$$

$$\varepsilon_F = - 0.1544 \text{ A.U.}$$

The unbound states, $\varepsilon_k > 0.5834$, are by no means unimportant. The spectral density of the overlap $|S_{Ak}|^2$ is also exhibited in Figure 7. This peaks in the unbound states, and demonstrates why the overcompleteness problem must be investigated.

The adatom density of states, and the occupancy of the hydrogen 1s orbital are calculated from (72), (73), (75) and (79). The theory allows the orbital occupancies to be different for ↑ and ↓ spins if this is the situation of lowest energy. We only need to exhibit the spin σ, either ↑ or ↓ in the orbital energy (70),

$$\varepsilon_{A\uparrow} = E_A + J_A \langle n_{A\downarrow}\rangle \ , \ \varepsilon_{A\downarrow} = E_A + J_A \langle n_{A\uparrow}\rangle$$

to see that the self-consistency conditions are

$$\langle n_{A\sigma}\rangle = -\frac{1}{\pi}\,\mathrm{Im}\int_{-\infty}^{\varepsilon_F} d\varepsilon/(\zeta - \varepsilon_{A\sigma} - q_A) \ , \ \sigma = \uparrow \text{ or } \downarrow . \qquad (83)$$

At s = 1 A.U., the ground state for hydrogen on Aℓ does have different occupancies for different spins (DODS), and[12] $\langle n_{A\sigma}\rangle = 0.9000$, $\langle n_{A-\sigma}\rangle = 0.1840$, which is similar to the situation in the free atom,

$<n_{A\sigma}> = 1$, $<n_{A-\sigma}> = 0$. As a matter of fact, the ground state is DODS for all $s > 0$ at least[9].

In the approximation in which we are working at present, all Coulomb interactions between the adatom and the metal are neglected, and the ground state energy of the system is given by

$$E = \sum_{\sigma\mu} \varepsilon_{\mu\sigma} <n_{\mu\sigma}> - J_A <n_{A\sigma}><n_{A-\sigma}> \tag{84}$$

The first term in (84) can be written as an integral over the density of states (33), and so we need the total contribution, $\Delta\rho$, which the adatom makes to the density of states in order to calculate D_{one} in (48). The calculation of $\Delta\rho$ parallels the work of section 1.4., although we have a different chemisorption function, and $1 - S_{MA}S_{AM}$ in (32) not 1 as in (30). The result is

$$\Delta\rho = -\frac{1}{\pi} \text{Im Tr} \{\tilde{G}_A - S_{AM}G_M^{ps}S_{MA} + G_M^f X_M^{ps} G_M^{ps}\} \tag{85}$$

with (compare (43) and (44))

$$H_M^{ps} = \varepsilon_M + X_M^{ps} \quad , \quad G_M^{ps} = (1\zeta - H_M^{ps})^{-1} \tag{86}$$

Equation (85) shows that the approximation (78) should be used with caution. It makes the last term in (85) zero and

$$\int_{-\infty}^{+\infty} d\varepsilon \; \Delta\rho(\varepsilon) = -\frac{1}{\pi} \text{Im} \int_{-\infty}^{+\infty} d\varepsilon \; \tilde{G}_A - 1 \tag{87}$$

and although the first term on the right in (87) is not unity because the approximation (78) is used in G_A, the normalization of $\Delta\rho$ is undoubtedly quite wrong. This may not be too serious if we do not discuss phenomena involving energies above ε_F because the spectral density of $|S_{Ak}|^2$, which gives the contribution -1 in (87), has its peak above the vacuum level (Figure 7). However, when approximation (78) is used, I prefer[11] to go back to (32), and drop also $S_{MA}S_{AM}$ on the right. This gives (cf.(47))

$$\begin{aligned} \Delta\rho &= -\frac{1}{\pi} \text{Im Tr} \{\tilde{G}_A(1 - \delta q_A/\delta\varepsilon)\} \\ \int_{-\infty}^{+\infty} d\varepsilon\Delta\rho\ (\varepsilon) &= 1 \end{aligned} \tag{88}$$

ρ_A is of course unaltered by this, but any approximation for H_M^{ps} in (76) destroys its normalization ;

$$\int_{-\infty}^{+\infty} d\,\varepsilon\,\rho_A(\varepsilon) \neq 1 \tag{89}$$

and one needs to be aware of this when using (78).

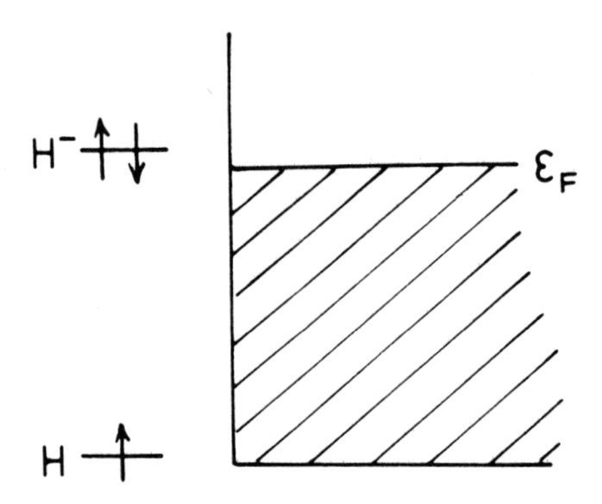

Fig. 6 : Energy levels for hydrogen and aluminium metal

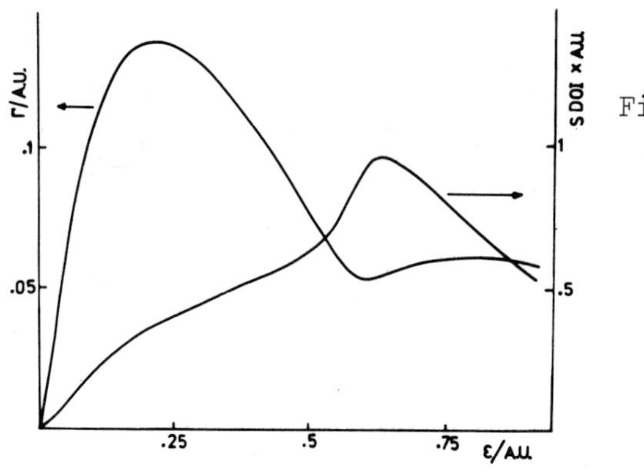

Fig. 7 :

The function $\Gamma(\varepsilon)$ for hydrogen at 1 A.U. from free-electron aluminium compared with the spectral density of the overlap integral (SDOI). Note the difference in scale.

1.7. An adatom on a nearly-free-electron metal

A self-consistent calculation like that described in the previous section has not yet been made for a NFE metal ; the calculation of the k-states (their wavefunctions and energies) at a large number of points in the surface Brillouin zone, of the matrix elements V_{Ak}, and finally of Γ from (80) is quite time consuming. What has been done so far[9] is to approximate the problem by a 1-dimensional one, "FE parallel, NFE perpendicular", by simply neglecting the periodic variation of the pseudopotential in the direction parallel variation to the surface (the ρ-direction). In this approximation the crystal wavefunctions which appear in (54) are written

$$\phi(k_z, k_\parallel, \varepsilon; r) \simeq f(k_z, \varepsilon; z)\exp[i(k_\parallel + K_\parallel).\rho] \qquad (90)$$

then f satisfies

$$[-\frac{1}{2}\frac{d^2}{dz^2} + V(z)]\, f = \varepsilon(k_z)f \qquad (91)$$

and

$$\varepsilon = \varepsilon(k_z) + \frac{1}{2}|k_\parallel + K_\parallel|^2 \qquad (92)$$

Equation (91) is a 1-dimensional crystal problem, and in our approximation the 3-dimensional band structure (92) has free-electron bands in 2-dimensions rising from each eigenvalue $\varepsilon(k_z)$ of (91). The (real) band structure $\varepsilon(k_z)$ can only have gaps at the centre, and at the boundary of the 1-dimensional Brillouin zone, and if there is a surface state in such a gap, there is, according to (92), a surface state on the semi-infinite crystal for every $k_\parallel$ and $K_\parallel$, i.e. a 2-dimensional free-electron band of surface states rising from each 1-dimensional surface state. Of course this is wrong for two reasons ; (a) the periodicity of V in the ρ-direction will break up the continuous spectrum of 2-dimensional free-electron bands into allowed regions, and forbidden gaps, (b) the correct band structure is much more complicated than (92), and there is no necessity for a gap in the $k_\parallel$ subbands to persist for all $k_\parallel$. Of these , (b) is the most serious, but even so we may expect to learn something about the rôle of surface states in chemisorption from this very simple model.

Newton[9] has calculated the chemisorption function for this model of hydrogen on (100)Aℓ using (79)-(82), and Fourier coefficients of the pseudopotential

$$W_0 = -0.5834 \text{ A.U.}, \qquad W_2 = \pm 0.0275 \text{ A.U.}$$

From this he has gone on to solve the self-consistency conditions (83), and to calculate the adatom density of states $\rho_{A\sigma}$ (the integrand in (83)). One of his results, that for an adatom-metal distance 1 A.U. is shown in Fig.8. This has the restriction $\langle n_{A\uparrow}\rangle = \langle n_{A\downarrow}\rangle$ imposed. If W_2 is negative (as it is for $A\ell$) there is a surface state in the first gap in $\varepsilon(k_z)$ at $k_z = 2\pi/a$, and correspondingly a complete band of surface states on the semi-infinite metal according to the present model. The striking feature of these results is the profound influence on ρ_A of changing the sign of such a small quantity like W_2. I do not think that this profound influence will be felt in the strength of the chemisorption bond ($\langle n_A\rangle$ for example does not alter profoundly being[9] 0.657 for W_2 positive and 0.575 for W_2 negative) but rather in the reactivity of the chemisorbed hydrogen towards other gases.

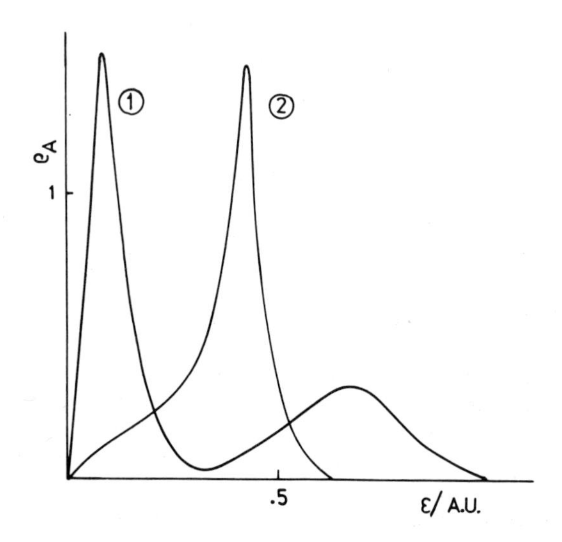

Fig. 8 : Adatom density of states ρ_A for hydrogen at 1 A.U. from nearly-free-electron aluminium, (1) with, and (2) without, surface states.

1.8. Anderson's Hamiltonian

An important question in chemisorption theroy concerns the limits of a one-electron (orbital) approximation. The one-electron theory presented in Sections 1.6 and 1.7 regards Coulomb interactions between two electrons in the localized state on the adatom as of vital importance in preventing large charge transfer between the adatom and the metal. Consequently it may not be good enough to treat these interactions in the simple self-consistent one-electron fashion. To examine this, and other problems, it is useful to have a theoretical model which will yield the results of Sections 1.6 and 1.7 in the Hartree-Fock approximation, but which is simple enough to be treated also in higher approximations. Anderson's[21] Hamiltonian is such a model, and it serves also to introduce the formalism of second quantization into chemisorption theory.

The Hamiltonian is a function of anti-commuting Fermion operators $c^{+}_{i\sigma}, c_{i\sigma}$, which create, and destroy, electrons with spin σ, either $\uparrow$ or $\downarrow$ in a chosen set of orthonormal orbitals $\{\phi_i\}$. For Anderson's model, we choose the set $\{\phi_A, \phi_k\}$, treat it as orthonormal, and write

$$H = \sum_{\sigma} E_A n_{A\sigma} + J_A n_{A\uparrow} n_{A\downarrow} + \sum_{k\sigma} \varepsilon_k n_{k\sigma} + \sum_{k\sigma} (V_{Ak} c^{+}_{A\sigma} c_{k\sigma} + V_{kA} c^{+}_{k\sigma} c_{A\sigma}) \qquad (93)$$

Here $n_{i\sigma}$ $(i = A,k)$ stands for $c^{+}_{i\sigma} c_{i\sigma}$, and is the operator for the number of spin σ electrons in the orbital ϕ_i.

The equation of motion of any operator $c(t)$ is

$$i\dot{c} = cH - Hc = [\, c, H \,] \qquad (94)$$

and for $c_{A\sigma}$ and $c_{k\sigma}$ we obtain

$$i\dot{c}_A = (E_A + J_A n_{A-\sigma}) c_{A\sigma} + \sum_k V_{Ak} c_{k\sigma}$$
$$i\dot{c}_k = \varepsilon_k c_{k\sigma} + V_{kA} c_{A\sigma} \qquad (95)$$

If we linearize these equations of motion by using

$$n_{A-\sigma} c_{A\sigma} \approx \langle n_{A-\sigma} \rangle c_{A\sigma} \qquad (96)$$

in the first, we obtain the Hartree-Fock approximation. Thus after using (96), and taking Fourier time transforms

$$c(\varepsilon) = \int_{-\infty}^{+\infty} dt\; c(t) \exp(i\varepsilon t)$$

we obtain

$$(\varepsilon - E_A - J_A \langle n_{A-\sigma} \rangle) c_{A\sigma}(\varepsilon) = \sum_k V_{Ak} c_{k\sigma}(\varepsilon)$$

$$(\varepsilon - \varepsilon_k)c_{k\sigma}(\varepsilon) = V_{kA}c_{A\sigma}(\varepsilon)$$

The matrix form of these linear equations is

$$\begin{vmatrix} \varepsilon_{A\sigma} - \varepsilon & V_{AM} \\ V_{MA} & \varepsilon_M - 1\varepsilon \end{vmatrix} \begin{vmatrix} c_{A\sigma} \\ c_{M\sigma} \end{vmatrix} = 0$$

which is like (15) with the approximation (78), except that the operators $c_{A\sigma}, c_{k\sigma}$ replace the coefficients a and b_k, and the spin σ is exhibited. Thus, the working equations of Sections 1.6 and 1.7, and equation (88) it may be noted, are re-derived by treating the model Hamiltonian (93) in the Hartree-Fock approximation.

To go beyond the Hartree-Fock approximation, we might look to improve on (96). However, because of its generality I prefer an analysis based on single particle (hole) propagators, and their Fourier time transforms (Green functions).

For any two orbitals ϕ_i and ϕ_j of the basis set, we define the propagator for spin σ particles by (see references 22 or 23 for example)

$$G^{\sigma}(ij;t) = \begin{cases} -\, i\langle c_{i\sigma}(t)c^{+}_{j\sigma}(0)\rangle & t > 0 \\ i\langle c^{+}_{j\sigma}(0)c_{i\sigma}(t)\rangle & t < 0 \end{cases} \tag{97}$$

where <.....> denotes the expectation value in the exact ground state. We form the equation of motion of $G^{\sigma}(ij;t)$ according to (94), take Fourier time transforms, and collect the resulting Green functions $G^{\sigma}(ij;\varepsilon)$ into a matrix $G^{\sigma}(\varepsilon)$. The result is

$$G^{\sigma} = (1\varepsilon - H - M^{\sigma})^{-1} \tag{98}$$

where H is the matrix of the Schrödinger Hamiltonian for a non-interacting electron in the system, and M^{σ}, the irreducible self-energy operator for the spin σ electrons, takes into account the electron interactions. Equation (98) is general, but for the Hamiltonian (93)

$$H = \begin{vmatrix} E_A & V_{AM} \\ V_{MA} & E_M \end{vmatrix} , \quad M^{\sigma} = \begin{vmatrix} M^{\sigma}_A & 0 \\ 0 & 0 \end{vmatrix} \tag{99}$$

so M^{σ} is particularly simple. From (98) and (99)

$$G(AA;\varepsilon) = (\varepsilon - E_A - M_A^\sigma - q_A)^{-1} \tag{100}$$

where q_A is given by (79).

In the Hartree-Fock approximation $M_A^\sigma = J_A\langle n_{A-\sigma}\rangle$, and an obvious improvement would be to use for M_A^σ the <u>exact</u> formula

$$M_A^\sigma = (\varepsilon - E_A)J_A\langle n_{A-\sigma}\rangle/[\varepsilon - E_A - J_A + J_A\langle n_{A-\sigma}\rangle] \tag{101}$$

for the <u>free</u> atom(11). No results using this formula in the FE and NFE models of Sections 1.6 and 1.7 are available.

1.9. <u>The atomic orbital basis set</u>

Noting that any system, adsorbate + metal is nothing more than a very large molecule, one should certainly develop the traditional "quantum chemistry" approach to chemisorption. This approach is characterized by the fact that, in the Hartree-Fock approximation, it degenerates to the familiar linear combination of atomic orbitals self-consistent field molecular orbital (LCAO-SCF-MO) theory.

Let $\{\psi_\mu\}$ be a complete orthonormal set of functions, which it is convenient, though not necessary, to think of as the ordinary Hartree-Fock molecular orbitals of the system adsorbate + metal. Let $c^+_{\mu\sigma}$, $c_{\mu\sigma}$ be anti-commuting Fermion operators which create and destroy electrons with spin σ in ψ_μ;

$$\begin{aligned} \{c_{\mu\sigma}, c^+_{\nu\sigma'}\} &= c_{\mu\sigma}c^+_{\nu\sigma'} + c^+_{\nu\sigma'}c_{\mu\sigma} = \delta_{\mu\nu}\delta_{\sigma\sigma'} \\ \{c_{\mu\sigma}, c_{\nu\sigma'}\} &= \{c^+_{\mu\sigma}, c^+_{\nu\sigma'}\} = 0 \end{aligned} \tag{102}$$

The Hamiltonian operator for the electrons in the system is (see reference 22 for example)

$$H = \sum_\sigma (c^+_\sigma \overline{H}_0 c_\sigma + \frac{1}{2} c^+_\sigma \overline{W}^\sigma c_\sigma) \tag{103}$$

where c^+_σ is a row matrix with elements $c^+_{\mu\sigma}$, c_σ is a column matrix with elements $c_{\mu\sigma}$, $\overline{H}_0$ is the matrix in the ψ_μ representation of $\hat{H}_0$, the Schrödinger operator for an electron moving in the field of the ion cores, and the matrix elements of $\overline{W}^\sigma$ are

$$\overline{W}^{\sigma}_{\mu\nu} = \sum_{\sigma'} \sum_{\kappa\lambda} \int dr_1 \, dr_2 \frac{\psi^*_\mu(r_1)\psi^*_\kappa(r_2)\psi_\nu(r_1)\psi_\lambda(r_2)}{|r_1 - r_2|} c^+_{\kappa\sigma'} c_{\lambda\sigma'} \tag{104}$$

$$= \sum_{\sigma'} \sum_{\lambda} <\mu\kappa|v|\nu\lambda> c^+_{\kappa\sigma'} c_{\lambda\sigma'}, \quad v = \frac{1}{|r_1 - r_2|}$$

We introduce a basis set $\{\phi_i\}$ of atomic orbitals on the adsorbate atoms, and on the atoms of the semi-infinite metal. Equations (21)-(23) hold, but S may not be singular because the basis ϕ may be incomplete. This would be the typical quantum chemistry situation. For the present however, we will allow S to be singular. New operators $a^+_{i\sigma}$, $a_{i\sigma}$ are defined in terms of $c^+_{\mu\sigma}$, $c_{\mu\sigma}$ by

$$a^+_\sigma = c^+_\sigma A^\dagger, \quad a_\sigma = Ac \tag{105}$$

so that (103) becomes

$$H = \sum_\sigma (a^+_\sigma H_o a_\sigma + \frac{1}{2} a^+_\sigma W^\sigma a_\sigma) \tag{106}$$

where H_0 is the matrix of the one-electron operator $\hat{H}_0$ in the ϕ representation, and W^σ has elements

$$W^\sigma_{rs} = \sum_{\sigma'} \sum_{tu} <rt|v|su> a^+_{t\sigma'} a_{u\sigma'} \tag{107}$$

The equations of motion of the operators $a_{r\sigma}$ are conveniently obtained from those of the $c_{\mu\sigma}$ by using (105) and (21)-(23). The equations of motion of the operators $c_{\mu\sigma}$ are obtained in the usual way by commuting them with H, and using the anti-commutation rules (102). We find

$$i\dot{c}_\sigma = (\overline{H}_o + \overline{W}^\sigma) c_\sigma \tag{108}$$

and noting that $c_\sigma = Ba_\sigma$, $B^\dagger B = S$

$$iS\dot{a}_\sigma = (H_o + W^\sigma) a_\sigma \tag{109}$$

This equation can be used to eliminate W^σ from (106), giving

$$H = \frac{1}{2} \sum_\sigma (ia^+_\sigma S a_\sigma + a^+_\sigma H_\sigma a_\sigma) \tag{110}$$

which is useful for expressing the ground state energy. The above equations have been given before(11), but I have been careful to derive them here when the inverse of S does not exist. To continue in this way, I note that, for any two operators A(t) and B(t) in Heisenberg's representation, the equation of motion of the propagator

$$\langle A;B\rangle = \begin{Bmatrix} -\, i\langle A(t)B(0)\rangle \ , & t > 0 \\ i\langle B(0)A(t)\rangle \ , & t < 0 \end{Bmatrix} \tag{111}$$

is

$$i\,\frac{\delta\langle A;B\rangle}{\delta t} = \langle\delta(t)\ \{A,B\}\rangle + \langle[A,H]\ ;\ B\rangle \tag{112}$$

The Fourier time transform of <A;B> is denoted $\langle\langle A;B\rangle\rangle_\varepsilon$, and

$$\varepsilon\langle\langle A;B\rangle\rangle_\varepsilon = \langle\{A,B\}\rangle + \langle\langle[A,H]\ ;\ B\rangle\rangle_\varepsilon \tag{113}$$

If we apply (113) to the propagators

$$G^\sigma(\mu,\nu;t) = \langle c_{\mu\sigma}(t)\ ;\ c^+_{\nu\sigma}(0)\rangle$$

for the Hamiltonian (103), we obtain the matrix equation

$$(1\varepsilon - H_o)G^\sigma = \langle\langle W^\sigma c_\sigma;\ c^+_\sigma\rangle\rangle_\varepsilon + 1 \tag{114}$$

Noting equations (21)-(23), (26), (104), (105) and (107), we convert (114) to an equation in the ϕ-representation

$$(\varepsilon S - H_o)\tilde{G}^\sigma = \langle\langle W^\sigma\ a_\sigma\ ;\ a^+_\sigma\rangle\rangle_\varepsilon + 1 \tag{115}$$

or

$$(\varepsilon S - H_o - M^\sigma)\tilde{G}^\sigma = 1 \tag{116}$$

where M^σ, the irreducible self-energy operator for spin σ electrons in the non-orthogonal ϕ-representation is defined by

$$M^\sigma G^\sigma = \langle\langle W^\sigma a_\sigma\ ;\ a^+_\sigma\rangle\rangle_\varepsilon \tag{117}$$

All electron interactions are contained in M^σ. The poles of G^σ are the single hole and single particle excitations (i.e., the ionization, and the affinity levels) and are obtained according to (116) by solving the secular equation

$$\det|\varepsilon S - H_o - M^\sigma| = 0 \tag{118}$$

This is evidently a very compact formulation of the problem of calculating, by configuration interaction techniques , the ground state of N electrons and the excited states of N-1 and N+1 electrons, in order to obtain the ionization and the affinity levels.

The ground state energy is determined by $\tilde{G}$,

$$E_o = \langle H\rangle = \frac{1}{4\pi i}\sum_\sigma \int_c d\varepsilon\ (\varepsilon \mathrm{Tr}\ S\ \tilde{G}^\sigma + \mathrm{Tr}\ H_o\ \tilde{G}^\sigma) \tag{119}$$

where the contour C consists of the real axis, and an infinite semicircle in the upper half-plane. Also

$$\langle a^+_{r\sigma}\ a_{s\sigma}\rangle = \frac{1}{2\pi i}\int_c d\varepsilon\ \tilde{G}^\sigma(s,r;\varepsilon) \tag{120}$$

serves to express the elements of the charge-and-bond-order matrix (briefly the density matrix of quantum chemistry) in terms of $\tilde{G}^\sigma$.

The well-known Hartree-Fock approximation is obtained with

$$M^\sigma_{rs} = W^\sigma_{rs} \simeq \sum_{\sigma'}\sum_{tu} (\langle rt|1/r_{12}|su\rangle - \delta_{\sigma\sigma'}\langle rt|1/r_{12}|us\rangle)\langle a^+_{t\sigma'}a_{u\sigma'}\rangle \tag{121}$$

1.10. Many electron wavefunctions

So far we have used the molecular orbital (MO) approach either explicitly, or, in the exact theory of Section 1.9 have deliberately made the Hartree-Fock MO theory rather easy to obtain. In Section 1.2 the system MO's are linear combinations of adsorbate and metal MO's, and having found these system MO's by solving the secular problem (6) or (15), many electron wavefunctions are obtained by assigning electrons to these MO's, i.e., by specifying the occupation numbers $\langle n_{\mu\sigma}\rangle$ of the orbitals ψ_μ for spin σ electrons. In this section I present the alternative approach of expressing the many electron wavefunction of the interacting system as a linear combination of the many electron wavefunctions of the non-iteracting system. The problem of finding the MO's of the interacting system is avoided, but the price is paid in requiring many electron wavefunctions describing the excited states of the non-iteracting systems.

Consider an adsorbate A and a metal M when the atom-metal distances are so large that there is no interaction between them. Let ψ_I, I = 0,1,2,... be the ground and excited states of the neutral system. Excited neutral states involve exciting electrons in the adsorbate, in the metal, or in both. Let $\psi_K(A^+,M^-)$ and $\psi_L(A^-,M^+)$ be charge transfer, i.e. polar states. Like the neutral states ψ_I, these polar states include all excited polar states. Thus the states $\psi_L(A^-,M^+)$ can have any number of electron-hole pairs in the metal in addition to the single hole formed by electron transfer to the adsorbate. For adsorbates where only single charge transfers are either possible or important (H, the alkali metals, CO etc.), the wavefunction for the interacting system can be written with sufficient accuracy as a linear combination of neutral and single charge transfer states,

$$\psi = \sum_I N_I\psi_I + \sum_K A_K\psi_K(A^-,M^+) + \sum_L C_L\psi_L(A^+,M^-) \quad (122)$$

but in general all the polar states will be needed,

$$\psi = \sum_I N_I\psi_I + \sum_P B_P\psi_P \quad (123)$$

The coefficients N_I and B_P are whose which minimize the expectation value of the Hamiltonian for the function (123). This requirement leads to the usual system of linear equations

$$\begin{vmatrix} H_N - S_N E & V - SE \\ V^\dagger - SE & H_B - S_B E \end{vmatrix} \begin{vmatrix} N \\ B \end{vmatrix} = 0 \quad (124)$$

Here N is a column matrix with elements N_I, and similarly B has elements B_P, and the Hamiltonian and overlap matrices in the representation afforded by the many electron basis $\{\psi_I,\psi_P\}$ are partitioned into neutral, polar, and neutral-polar blocks. This formulation is exact, but of course we can only make progress if we approximate. I shall mention two approximate versions here ; other approximate versions have recently been reviewed(6).

The fundamental approximation of the induced covalent bond theory of Schrieffer and Gomer(24) is

$$\psi = N_O\psi_O + \sum_{I \neq O} N_I\psi_I \quad (125)$$

i.e., the polar states are dropped from (123). This is exactly the approximation made[4] in treating the interaction between a radical (H) and a closed shell atom (Be) according to the theory of Heitler and London[3]. For two radicals, the first term in (125) is adequate but when one of the partners in the bond has no unpaired electrons in its ground state, the metal in our case, Be in BeH, the exchange of electrons with opposite spins, which characterizes the Heitler-London picture of the chemical bond, is impossible without violating the Exclusion Principle. The required electron exchange is only possible if the partner in the bond, lacking the unpaired electrons in its ground state, is excited to produce them. The covalent bond is induced, and to allow this, the excited states $I \neq 0$ are present on the right in (125).

In the approximation (125), the linear system (124) reduces to

$$(H_N - S_N E)N = 0 \tag{126}$$

and the lowest root $E_N(s)$ say of the secular equation gives the ground state energy as a function of the adsorbate-metal distance s. It is instructive to exhibit the effect of the excited states ($I \neq 0$) on the ground state ψ_0 in (125) since this is conveniently done by a method originally described by Löwdin[25]. After partitioning the matrixes in (126) into the $I = 0$ element, and the rest denoted by subscript R, we obtain the exact equation

$$E - E_0 = (V^{\dagger}_{OR} - S^{\dagger}_{OR}E)(S_R E - H_R)^{-1}(V_{RO} - S_{RO}E) \tag{127}$$

where $E_0(s)$ is the expectation value of the Hamiltonian for the state ψ_0,

$$E_0 = \int \psi_0^* H \psi_0 dr \tag{128}$$

and represents the exchange repulsion between the adsorbate and the metal when both are in their unperturbed ground states. This energy was calculated approximately by Toya[1] and by Wojciechowski[26] for hydrogen and a free electron metal, but a proper calculation for the model seems still to be lacking. Paulson and Schrieffer[27] calculated it for hydrogen and a tight binding s band solid. If we replace E on the right in (127) by E_0, and then neglect non-diagonal elements of $(S_R E_0 - H_R)^{-1}$, we derive a formula for E_N, the expectation value of H for the ground state function (125) which is correct through second-order terms :

$$E_N = E_0 + \sum_{I \neq 0} \frac{|V_{OI} - E_0 S_{OI}|^2}{E_0 - E_I} \tag{129}$$

where E_I is the expectation value of H for the excited (neutral) state ψ_I. Equation (129) leads directly to the weak interaction limit of Paulson and Schrieffer[27]. Their strong coupling limit will be discussed in Section 2.4.

The second approximate version of (123) I refer to is[1,2,28]

$$\psi = N_O\psi_O + \sum_K A_K\psi_K(A^-M^+) + \sum_L C_L\psi_L(A^+M^-) \qquad (130)$$

which drops the excited neutral states of the induced covalent bond theory in favour of the single charge-transfer states. Again we can study the effect of the charge-transfer states on the neutral state ψ_O by Löwdin's[25] perturbation theory. We partition all matrices according to the neutral (O) and polar (B) basis states, and so obtain the exact equation

$$E - E_O = (V^\dagger - SE)(S_BE - H_B)^{-1}(V - SE) \qquad (131)$$

To progress we neglect non-diagonal elements of $(S_BE - H_B)$ (no k-state scattering by the adsorbate) to get

$$E - E_O = \sum_P \frac{|V_{OP} - S_{OP}E|^2}{E - E_P} \qquad (132)$$

where E_P is the expectation value of H for the polar state ψ_P. The right-hand site of (132) is a many-electron chemisorption function (cf. equations (35) and (36)) and if we define

$$\Gamma(E,E') = \pi \sum_P |V_{OP} - S_{OP}E|^2\, \delta(E' - E_P) \qquad (133)$$

then

$$E - E_O = \frac{1}{\pi}\int_{-\infty}^{+\infty} dE' \,\frac{\Gamma(E,E')}{E-E'} \qquad (134)$$

Of course the function (133) is very difficult to calculate because all excited polar states are included in the basis, so either these excited states have been neglected[28], or they have been included[1,2] as fixed linear combinations in the expansion (130), i.e., in quantum chemistry language the basis has been contracted[6]. Because of these approximations, the essential <u>overcompleteness</u> of the basis (130) for hydrogen chemisorption by simple free-electron metals has not been noticable. To be clear that the basis <u>is overcomplete</u>, I need only remark that, for a hydrogen atom and N electrons in a metal, we have an (N+1)-electron system, and the cationic states $\psi_L(A^+M^-)$, being eigenstates of the isolated metal with

N+1 electrons in it, are themselves a complete set of (N+1)-electron states. Consequently we should introduce auxiliary conditions to go with the linear equations (124) along the lines of Section 1.2. This will introduce a pseudo-Hamiltonian for use with the cationic states, and pseudointeractions between these states and the neutral and anionic states. Since no calculations using these modifications have been made, I do not pause to give the details here.

2. CLUSTER CALCULATIONS

2.1. General

The idea that a variety of experimental observations in chemisorption, and catalysis can be understood in terms of the formation of a surface molecule involving the adsorbate, and a few substrate atoms to which it is strongly bonded, is widespread in the literature. Recently, experimental ion neutralization spectra (INS) of the chalcogens in Ni have been interpreted in such terms[29], and theoretical predictions of the angular dependence of energy resolved photoemission from surface molecules have been made[30] (see also Dr. Gadzuk's lecture). Of course the idea is an attractive one because, in its simplest form at least, it converts the problem of chemisorption into just another problem in molecular quantum mechanics, and so enables well-known,computational techniques to be employed. At a slightly more sophisticated level, one recognizes that one cannot simply cut off the surface molecule from the rest of the substrate ; at least a small coupling between the two remains, and this fixes the chemical potential of electrons in the surface molecule at that of the (extended) substrate[28]. A surface molecule is not therefore like an ordinary molecule, and this is one reason why heterogeneous processes occur which have no homogeneous countepart; another is that the positions of substrate atoms in a surface molecule are rather firmly fixed by forces originating outside the surface molecule. At the third level of sophistication, one accepts that the rest of the substrate might have a more profound effect on the surface molecule, and calculates this effect using some simple model, while treating the cluster of atoms comprising the surface molecule with a much better model[31]. Of course, if we make the cluster large enough, the second and third levels of sophistication are automatically encompassed. It seems however that, except perhaps for some simple systems (hydrogen on Li or graphite), one cannot do this, and still have realistic computing times.

2.2. The SCF-LCAO-MO calculation

The key papers here are those of Roothaan[32,33]but I obtain the equations I need quickly from the exact theory of Section 1.9. Using (121) in (109), (116) and (118) we have

$$i S \dot{a}_\sigma = F^\sigma a_\sigma \tag{135}$$

$$(\varepsilon S - F^\sigma)\, \tilde{G}^\sigma = 1 \tag{136}$$

$$\det|\varepsilon S - F^\sigma| = 0 \tag{137}$$

where F^σ is the Hartree-Fock matrix in the AO ϕ-representation

$$F^\sigma = H_0 + W^\sigma \tag{138}$$

W^σ involves the density matrix P^σ with elements

$$<a^+_{r\sigma} a_{s\sigma}> = -\frac{1}{\pi} \mathrm{Im} \int_{-\infty}^{\varepsilon_F} d\varepsilon\, \tilde{G}^\sigma(s,r;\zeta), \zeta = \varepsilon + i0 \tag{139}$$

so the familiar Hartree-Fock self-consistency problem is expressed by (136)-(139). Of course one does not normally encounter the Green function matrix $\tilde{G}^\sigma$ in molecular quantum mechanics. Instead P^σ is calculated from the matrix of the eigenvectors A^σ (equation 21)

$$(\varepsilon S - F^\sigma)A^\sigma = 0 \quad , \; (A^\sigma)^\dagger S A^\sigma = 1 \tag{140}$$

and

$$P^\sigma_{rs} = <a^+_{r\sigma} a_{s\sigma}> = \sum_\mu A^\sigma_{r\mu} A^{\sigma\dagger}_{\mu s} <n_{\mu\sigma}> \tag{141}$$

where $<n_{\mu\sigma}>$ is the occupancy (0 or 1) of the Hartree-Fock orbital ψ_μ in the ground state. Equation (141) follows immediately from (26) and (139). Finally the ground state energy is

$$E_0 = \sum_\sigma [\varepsilon_{\mu\sigma} <n_{\sigma\mu}> - \frac{1}{2} \sum_{rs} W^\sigma_{rs} <a^+_{r\sigma} a_{s\sigma}>] \tag{142}$$

where $\varepsilon_{\mu\sigma}$ are the orbital energies, i.e. the roots of (137).

In spite of its neglect of electron correlations, the above scheme (usually with no distinction between ↑ and ↓ values of σ, i.e. the restricted Hartree-Fock scheme) has been widely used to investigate the electronic structures and the bonding in medium sized molecules. For large molecular systems the bottleneck is the computation, and storage of the two-electron integrals $<rt|v|su>$ in W^σ_{rs}. As an example I quote from some data published by Clementi[34] For $C_{25}N_4O_7H_{14}$ with 50 atoms, 232 electrons, and 618 Gaussian orbitals in the basis set $\{\phi_r\}$ there are some 92.7 x 10^8 two-electron integrals. Their straight computation would require about 1800 hours on an IBM 360/195, so one can see the difficulty in trying to study chemisorption this way. However, it is possible to speed up the computations without loss of accuracy so that the above time is reduced to 6 1/2 hours[34]. Consequently, it seems to me that cluster model calculations for the chemisorption of say hydrogen and CO on (100)Ni using a cluster of four Ni atoms could be made. I do not know of any calculations of this sort. Of course cluster calculations have been made using semi-empirical versions of the SCF-LCAO-MO scheme like Extended Hückel (EH) or CNDO. Experience from chemistry is that EH is not quantitatively reliable, giving poor estimated of energies, geometries and electron distributions. The status of other semi-empirical methods in chemistry is controversial, but it seems that, with careful parameterization, MINDO can give quantitative information on thermodynamic properties, and on reaction paths for large molecular systems with realistic computing times(35). This is interesting because the increasingly used SCF-Xα-SW method, which I now describe, is also a carefully parameterized semi-empirical scheme.

2.3. The SCF-Xα-SW calculation

The key references here are the papers of Slater and Johnson(36) Slater(37) and Johnson(38). Instead of expressing the molecular orbitals in terms of a basis set of functions, the one-electron Schrödinger equation

$$[-\frac{1}{2}\nabla^2 + V(r)]\,\psi_\mu(r) = \varepsilon_\mu\psi_\mu(r) \tag{143}$$

is integrated numerically in different representations in different regions of the molecule, and the various solutions joined smoothly together where they meet, by a multiple-scattered-wave formalism. The smooth joining condition leads to a secular equation of lower order usually than that (equation (137)) encountered in the LCAO calculation. The potential function in (143) is taken as the sum of the Coulomb potential V_C due to the total electronic and nuclear charge, and the quantity $V_{X\alpha}$ which is Slater's Xα statistical approximation to the effective one-electron exchange and correlation energy (see also Dr. Lang's lectures).

$$V(r) = V_c(r) + V_{x\alpha}(r) \tag{144}$$

$$V_c(r) = -\sum_g Z_g/|r - Rg| + \int dr'\,\rho(r')/|r - r'| \tag{145}$$

$$V_{x\alpha} = -6\alpha[(3/8\pi)\,\rho(r)]^{1/3} \tag{146}$$

$$\rho(r) = \sum_\mu \langle n_\mu\rangle|\psi_\mu(r)|^2 \tag{147}$$

where $\rho(r)$ is the local electron density, and Z_g the charge on the nucleus g at R_g.

To see how the scheme works, consider the planar molecule BF_3. The molecular space is partitioned into three mutually exclusive regions (Figure 9). A sphere surrounds each atom giving region I, the atomic spheres. A single outer sphere surrounds the cluster of atomic spheres, and region III lies outside this sphere. This is the extramolecular region. The remainder of space in region II, the interatomic region. Inside a sphere in I, and for III, the potential V(r) is averaged over angles to give an approximate potential which is spherically symmetric in the centre of the particular sphere considered. In II, a volume average is taken to produce a constant potential. This is the "muffin tin" approximation. The disposable parameters are the radii of the atomic spheres R_B and R_F, the radius of the outer sphere R_{II}, and the values of α in $V_{X\alpha}$ for the boron and fluorine spheres, the interatomic region II, and the extramolecular region III. The usual choice of the radii is to take R_B and R_F proportional to the B and F atomic radii, and as large as possible without overlapping, and R_{II} as small as possible without overlapping the atomic spheres. The scaling parameter α is chosen for each atomic sphere by matching the $X\alpha$ total energy of the atom to the Hartree-Fock total energy. Such values are available[39] for the atoms hydrogen to niobium. Some average of these atomic α-values is used in regions II and III.

Having partitioned space into regions of spherically symmetric, and constant potential, the wavefunction is expanded in a partial wave representation in each region. In every atomic sphere, and in III, the expansion is made with respect to the sphere centre. In II, a multicentre expansion is appropriate. The smooth joining conditions lead to a secular equation for the partial wave expansion coefficients in region I and III whose order is therefore equal to the total number of partial waves used in these regions. For BF_3 with s and p waves in energy atomic sphere, and in III, the secular equation is 20 x 20. It is here the same size as that obtained in the LCAO scheme with a minimal basis set. The secular equation is smaller than might be expected at first sight because its matrix elements depend on ε as a result of the ε-dependence of the radial wavefunction in a spherical region. To put it differently : there are only s and p partial waves, not 1s, 2s and 2p partial waves.

To start the calculation, an initial potential formed by superposition of atomic potentials is used. From the initial set of orbitals obtained, a new potential is calculated from (144)-(147), spherically averaged in I and III, and volume averaged in II. A weighted average of the new, and the initial, potential is used for the first iteration. A new set of orbitals is obtained and the entire computation iterated to self-consistency.

Many molecules have been treated by the SCF-$X\alpha$-SW scheme, and ionization energies, and optical excitation energies calculated. The agreement with experimental is generally very good for these quanti-

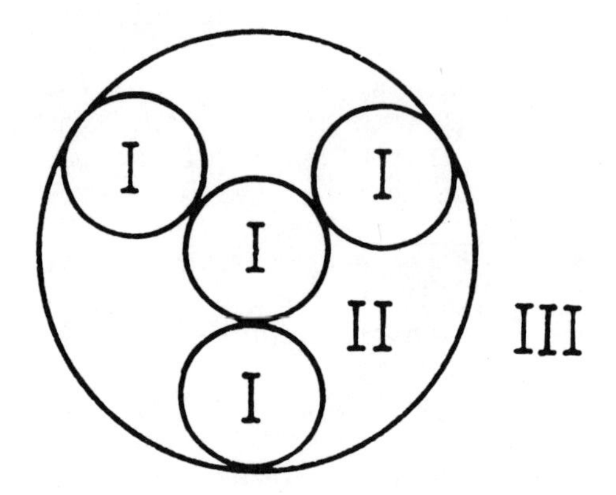

Fig. 9 : Division of space in the Xα scheme for BF_3

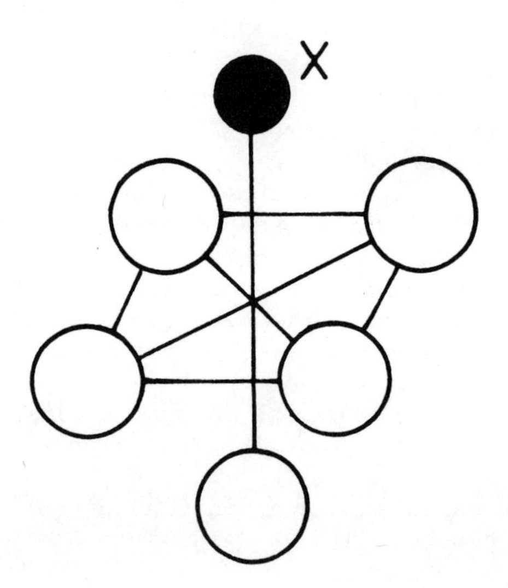

Fig. 10 : A cluster $X\text{-}Ni_5$ for chalcogens on (100)Ni

ties. On the other hand the total energies, binding energies, and equilibrium bond lengths are very bad. The calculations of Weinberger and Konowalow[40] on N_2, O_2 and F_2 provide good examples of this state of affairs. Of course efforts are being made to improve the total energy calculation, by using overlapping atomic spheres, and by using non-muffin tin potentials. It is not my purpose to review these developments here. Instead I mention the calculations of Niemczyk[41] on a cluster of five Ni atoms and a chalcogen in a configuration with C_{4v} symmetry representing chemisorption of the chalcogen X on (100)Ni (Figure 10), since this is one of many such calculations claiming to have some relevance to chemisorption by transition metals. The calculations were made in the manner outlined above with however, $\alpha = 2/3$ everywhere, and the 1s-2p states of Ni treated in the "frozen core" approximation. A minimal basis set of partial waves was used, and the total energy calculated as a function of the distance between X and the plane of the four Ni atoms. Only the $O\text{-}Ni_5$ system shows a minimum in the binding energy curve, but as no actual energies are given it is difficult to draw any conclusions from these calculations. The obvious criticism of this cluster calculation is that none of the five Ni atoms has even the correct nearest neighbour environment for a (100) surface. This is where chemisorption theory differs from the theory of small molecular systems, and although cluster calculations like those just described (and many others, reference 42 for example) will, and should, continue to be made, we must at the same time study the problem of joining the cluster to the rest of the semi-infinite substrate in some managable way, if only to see what the effect _is_ on the cluster. That this is a real problem even in catalysis where we are concerned with supported metal catalysts is evident when we note that the largest metal clusters handled by $X\alpha$ programs (not always self-consistently however) contain 13 atoms[43,44], whereas the small particles in the supported catalysis contain 100-10,000 atoms.

2.4. Formal theory of embedding

Consider the adsorbate-metal cluster C, and the rest of the metal D (Figure 11). This piece of metal is defective because of the metal atoms in C. The Hamiltonian is written $H = H_C + H_D + V = H_0 + V$, and its eigenstates ψ are expanded in terms of those of H_0 in the manner (123), i.e., in terms of the exact states of the non-interacting cluster, and defective metal, $\psi_I(C,D)$, $\psi_K(C^-,D^+)$, $\psi_L(C^+,D^-)$... These states are known in principle from a cluster calculation, and from the theory of metals, but in practice, since the metal is defective, its states will only be known for simplified models. However that may be, we have to solve a system of linear equations like (124). The ground state $\psi_0(C,D)$ of the non-interacting cluster and defective metal is our first approximation to the true ground state so we divide the basis into $\psi_0(C,D)$, and the remainder which we denote by R, then, using Löwdin's perturbation theory we have the exact equation (cf.(127))

$$E = E_O + (V^+_{OR} - S_{OR}E)(S_R E - H_R)^{-1}(V_{RO} - S_{RO}E) \qquad (148)$$

$$E_O = E_C + E_D + \langle\psi_O|V|\psi_O\rangle \qquad (149)$$

for the ground state energy, showing how the states R affect the state $\psi_O(C,D)$.

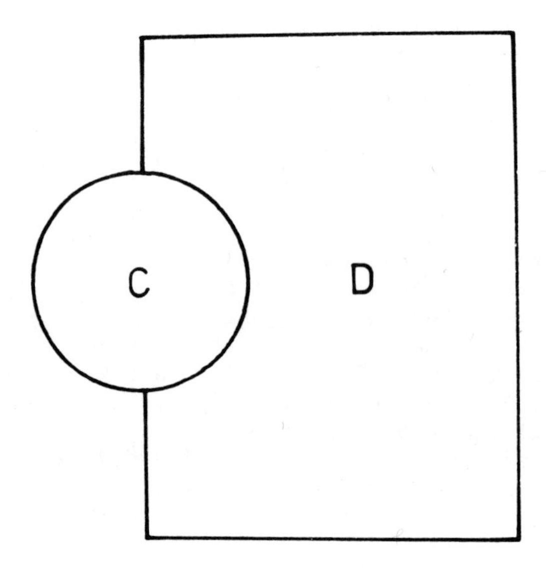

Fig. 11 : A cluster C and a defective metal D

To make progress we have to be prepared to approximate, but at this stage we can see that the formulation (148) is not very convenient for a large cluster C. Such a cluster may be expected to have a large number of low-lying excited states, both neutral, and ionic, so that a large number of terms make more or less equal contributions to the right hand side of (148). The same situation exists even without the adsorbate in the cluster ; a large number of terms are needed to join the cluster of metal atoms to the defective metal so as to recover the original semi-infinite metal. However, the present formulation is certainly useful for small clusters, and an important example of it is to be found in the work of Paulson and

Schrieffer[27] on hydrogen chemisorption by a simple cubic tight binding s band solid (the material often called simple cubium). Consider the local geometry of Figure 12 (on-site chemisorption). The cluster C is the diatomic molecule AB, and A is hydrogen. For a tight binding metal the coupling between C and D consists only of electron hopping between atom B and its five nearest neighbours in D, so that, if we use the Heitler-London description of the molecule AB, the only states coupled directly to ψ_0 by V are $\psi_K(C^-,D^+)$ and $\psi_L(C^+,D^-)$, i.e. only single charge transfer states. Moreover, since polar states with H^+ and H^- are not allowed in the Heitler-London description there are no double charge transfer states involving C^{2-} and C^{2+} anyway. Consequently good results are expected with the states R in (148) confined to single charge transfer states. Replacing E on the right in (148) by E_0, and neglecting non-diagonal elements of $(S_R E_0 - H_R)$ gives a formula correct through second order terms.

$$E = E_0 + \sum_R \frac{|V_{OR} - S_{OR}E_0|^2}{E_0 - E_R} \quad , \quad E_R = \langle\psi_R|H|\psi_R\rangle \tag{150}$$

This, with R limited as described above, leads directly to what Paulson and Schrieffer[27] call the strong limit of the induced covalent bond. The third term on the right in (149) is zero for the present model, and a simple formula exists for E_D ;

$$E_D = E_{metal} + 2\int_{-\infty}^{\varepsilon_F} d\varepsilon\,\varepsilon\,\Delta\rho(\varepsilon) \tag{151}$$

where $\Delta\rho$ is the change in the density of states when the orbital ϕ_B is removed from the semi-infinite metal basis set to produce the defective metal.

Fig. 12 :

On-site chemisorption on a cubic metal

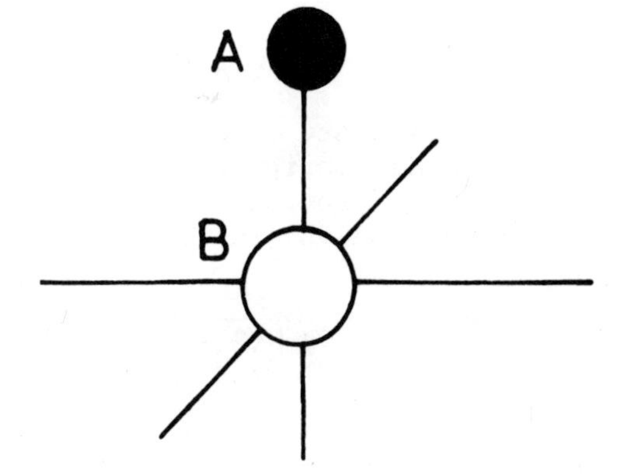

2.5. Embedding an SCF-LCAO-MO cluster

When we work with one electron states, and Fermion operators, the key equation for handling the coupling of the cluster C to the rest of the substrate D, and the responses of C and D to this coupling, is Dyson's[45,46] equation, first derived in connection with quantum electrodynamics. In this section we work with the Hartree-Fock versions of the equations of Section 1.9, i.e., equations like (135)-(142) apply now to the complete system C + D representing the adsorbate, and the whole metal.

The atomic orbital basis set $\{\phi_r\}$ is partitioned into those on the cluster C and those on the defective metal D. All matrices are partitioned accordingly. We drop the spin index σ for ease of writing, and divide the matrix $F - \varepsilon S$ into $F^o - \varepsilon S^o$ and X

$$F - \varepsilon S = F^o - \varepsilon S^o + X \tag{152}$$

X includes all interactions between C and D except those accounted for in the proviso that the density matrix which appears in the block F^o_C of F^o is that prevailing in the interacting system C,D, not that for the cluster C alone. If we define a Green function matrix $\tilde{G}^o$ for F^o,

$$(\zeta S^o - F^o)\,\tilde{G}^o = 1, \quad \zeta = \varepsilon + i0 \tag{153}$$

then from (136), (152) and (153) we derive Dyson's equation

$$\tilde{G} = \tilde{G}^o + \tilde{G}^o X \tilde{G} \tag{154}$$

Remembering that F^o and S^o are block diagonal, $\tilde{G}^o$ is block diagonal, so from (154)

$$\begin{aligned} \tilde{G}_C &= \tilde{G}^o_C + \tilde{G}^o_C (X_C\tilde{G}_C + X_{CD}\tilde{G}_{DC}) \\ \tilde{G}_{DC} &= \tilde{G}^f_D (X_D\tilde{G}_{DC} + X_{DC}\tilde{G}_C) \end{aligned} \tag{155}$$

$\tilde{G}^f_D$ is the free, but defective metal Green function matrix, and does not need a superscript zero because of the way X is defined. The cluster Green function matrix does, because it involves the density matrix prevailing in the interacting system. Solving (155) for $\tilde{G}_C$,

$$\tilde{G}_C = [\,(\tilde{G}^o_C)^{-1} - X_C - q_C]^{-1} = [\zeta S_C - F^o_C - X_C - q_C]^{-1} \tag{156}$$

$$q_C = X_{CD}\tilde{G}^{eff}_D X_{DC}\ , \quad \tilde{G}^{eff}_D = (1 - \tilde{G}^f_D X_D)^{-1}\,\tilde{G}^f_D \tag{157}$$

q_C is the cluster embedding matrix describing the response of C to its coupling to D. $\tilde{G}_D^{eff}$ is the defective metal Green function matrix corrected for the existence of the diagonal part X_M of the C-D coupling.

Equation (154) also contains the two equations

$$\tilde{G}_D = \tilde{G}_D^f + \tilde{G}_D^f (X_D \tilde{G}_D + X_{DC} \tilde{G}_{CD})$$
$$\tilde{G}_{CD} = \tilde{G}_C^o (X_C \tilde{G}_{CD} + X_{CD} \tilde{G}_D) \tag{158}$$

and from these we obtain

$$\tilde{G}_{CD} = \tilde{G}_C X_{CD} \tilde{G}_D^{eff} \tag{159}$$

$$\tilde{G}_D = \tilde{G}_D^{eff} [1 + X_{DC} \tilde{G}_C X_{CD} \tilde{G}_D^{eff}] \tag{160}$$

The Hartree-Fock self-consistency problem on the cluster C is defined by (153) together with the equation

$$P = -\frac{1}{\pi} \operatorname{Im} \int_{-\infty}^{\varepsilon_F} d\varepsilon \tilde{G}(\zeta) \tag{161}$$

for the density matrix. This equation with P and $\tilde{G}$, not P_C^o and $\tilde{G}_C^o$, is needed because F_C^o in (153) depends on the density matrix in the interacting system. This is the first thing to notice. The second is that the density matrix outside C is involved in the definition of F_C^o (see for example (121) and (138)). This is not peculiar to our problem ; it is always so. Because of this, all metal atoms, not just those in C are drawn into the self-consistency problem on C with the result that the cluster C is embedded self-consistently in the semi-infinite metal. Starting with an assumed initial P, F_C^o is calculated from (138), and then $\tilde{G}_C^o$ from (153). This enables the matrix $\tilde{G}$ to be calculated from (156), (157), (159) and (160), and so a new density matrix is obtained from (161). A weighted average of the new and the initial density matrix is used to define a new F_C^o, and the whole computation iterated to self-consistency.

If it could be performed, the above computation would give self-consistency over the whole system C + D, not simply over C. Of course it cannot be performed for a semi-infinite metal. But if we aim only at self-consistency over C, and use some other non-self-consistent model for D, and for the C-D coupling, the calculation is enormously simplified, firstly because we do not now go outside C in iterating to self-consistency, and secondly because the cluster embedding matrix q_C which determines everything, is not now subject to self-consistent adjustments, but is calculated once and for all.

The self-consistency conditions are in fact expressed by the symbolic equation

$$P_C = -\frac{1}{\pi} \text{Im} \int_{-\infty}^{\varepsilon_F} d\varepsilon \; (\zeta S_C - F_C^o - X_C - q_C)^{-1} \quad (162)$$

with F_C^o and X_C depending on P_C.

Having achieved self-consistency, the total energy is calculated from the usual Hartree-Fock expression. The electron interaction contribution is confined to C but the one-electron contribution is not. After some straightforward algebra, we find that $\Delta\rho$, the total adsorbate contribution to the density of states as defined in (42) is given by

$$\begin{aligned} \Delta\rho &= -\frac{1}{\pi} \text{ImTr} \; (\tilde{G}S - \tilde{G}_M^f S_M) \\ &= -\frac{1}{\pi} \text{ImTr} \; \{\tilde{G}_C(S_C - \frac{\partial q_C}{\partial \varepsilon}) + \tilde{G}_D^{eff} S_D - \tilde{G}_M^f S_M\} \quad (163) \end{aligned}$$

Here $\tilde{G}_M^f$ is the Green function matrix, and S_M the overlap matrix in the atomic orbital representation, for the perfect (i.e., not defective) semi-infinite metal. Equation (48) gives the one-electron contribution to D, the binding energy of the adsorbate to the metal.

The results of carrying through the above computational scheme for hydrogen on cubium in the local geometry of Figure 12, with the cluster consisting of only the diatomic molecule AB, will be exactly the same as those published by Grimley and Pisani(31) using a partitioning of the basis $\{\phi_r\}$ into adsorbate and metal orbitals. With this {A,M} partition, the self-consistency condition is formulated in the subspace A, but metal atoms are <u>always</u> drawn into the self-consistency problem.

Calculations have now been made in exactly the same way as before[31] for the on-site chemisorption of hydrogen by (100)Li and Na using a single Slater-type orbital on the metal atom B taken from free atom Hartree-Fock calculations[47]. Some of the results are given in Table 1. The contribution $\Delta n\varepsilon_F$ to D is present for the following reason : Since we do not achieve self-consistency over the whole system, but only over the cluster, the Fermi level ε_F changes with chemisorption to conserve electrons. This is unphysical. Conversely if we keep ε_F fixed, the number of electrons changes by Δn as the chemisorption bond is formed. Again this is unphysical. The contribution to D due to this effect is $\Delta n\varepsilon_F$, and this is included in our quoted D-values. One can argue that this spurious contribution should not be included in D, but however that may be, it should always be exhibited. It is a rather small quantity for hydrogen on the alkali metals ; not only is the cluster AB almost electrically neutral (though polar A^-B^+), but there is no significant charge accumulation, positive or negative, on the defective metal either.

TABLE 1 : Equilibrium distance d, binding energy D, gross occupancies $\langle n_{A\sigma}\rangle$ and $\langle n_{B\sigma}\rangle$, bond order $\langle a^+_{A\sigma}a_{B\sigma}\rangle$, and the contribution $\Delta n\varepsilon_F$ to D for hydrogen on (100)Li and Na in the on-site position.

	d/A.U.	D/Hartree	$\langle n_{A\sigma}\rangle$	$\langle n_{B\sigma}\rangle$	$\langle a^+_{A\sigma}a_{B\sigma}\rangle$	$-\Delta n\varepsilon_F$/Hartree
Li	1.45	0.220	0.947	0.042	0.022	0.006
Na	1.30	0.230	0.999	0.005	-0.012	0.003

2.6. Embedding a cluster into an arbitrary metal

LCAO-MO models are only generally used to describe the d bands of transition metals. Simple metals, and the conduction bands of transition metals, are described by non-localized states (k-states), which have not been constructed from a basis of localized states. Of course one can construct a localized basis from the k-states[48], and use these Wannier functions in the scheme described in the previous section. If we do not do this, the following problem arises : suppose we can solve the cluster problem and find the self-consistent MO's $\{\psi_c\}$ of the adsorbate-metal cluster (perhaps by the $X\alpha$ calculation), can we now embed this cluster calculation into a metal whose k-states are known ?

We formulate the chemisorption problem in the overcomplete basis $\{\psi_c,\psi_k\}$ of cluster states ψ_c, and states ψ_k of the semi-infinite metal, and so combine elements of Sections 1.2 and 2.5. We use F not H for the self-consistent one-electron operator, but this should not be taken to mean that the Hartree-Fock self-consistent scheme is to be used ; any self-consistent scheme is covered. We separate $\varepsilon S-F$ as in (152), and include in X all cluster-metal couplings (C-M couplings) except those accounted for by the proviso that in forming the block F^o_C, the self-consistent operator F appropriate to the actual coupled system, adsorbate-metal, is to be used. F^o defines a

Green function matrix as in (153), $(\zeta S^{o} - F^{o})\tilde{G}^{o} = 1$, but $S^{o} = 1$ now. $\tilde{G}^{o}$ is block diagonal because F^{o} is. The Green function matrix in the interacting system is defined as in (30), $(\zeta S - F)\tilde{G} = 1$, and Dyson's equation (154) holds. Thus we obtain (155)-(160) with the subscript D replaced by M. But because of overcompleteness we have (cf. (29))

$$S_{CM}\tilde{G}_{MC} = S_{CM}\tilde{G}_{M} = 0 \tag{164}$$

These auxiliary conditions affect the cluster embedding matrix. We now have

$$q_C = F_{CM}G_M^{eff}X_{MC} \tag{165}$$

(a similar change could have been made in (35)), but still (cf.(46) and (34))

$$G_M^{eff} = (1 - G_M^f X_M)^{-1} G_M^f = (1\zeta - \varepsilon_M - X_M)^{-1} \tag{166}$$

$$\tilde{G}_C = (1\zeta - F_C^o - X_C - q_C)^{-1} \tag{167}$$

Similarly, in place of (159) and (160) we have

$$\tilde{G}_{CM} = \tilde{G}_C F_{CM} G_M^{eff} \tag{168}$$

$$\tilde{G}_M = G_M^{eff}\,[1 + X_{MC}\tilde{G}_C F_{CM} G_M^{eff}\,] \tag{169}$$

There are no tildes over G_M^f and G_M^{eff} because the k-states are orthogonal. The self-consistent embedding of the cluster C is expressed by (167), the equation for the density matrix in the {C,M} representation

$$\langle a_{r\sigma}^{+} a_{s\sigma}\rangle = -\frac{1}{\pi}\,\mathrm{Im}\int_{-\infty}^{\varepsilon_F} d\varepsilon\; \tilde{G}^{\sigma}(s,r;\zeta)$$

equations (168) and (169) giving $\tilde{G}$ in terms of $\tilde{G}_C$ and other quantities, and the assumed dependence of the one-electron operator F on the density matrix. This formulation is exact.

To make progress we have to approximate, but as there is no generally valid approximation as simple as that used in the last section for a tight binding model, I shall not comment on practical schemes derived from the above equations until I, and others[10], have had the opportunity of testing the various approaches. Having achieved some approximate self-consistency, equation (47) with A replaced by C gives the adsorbate contribution to the density of states, and this determines the one-electron contribution to the binding energy of the adsorbate to the metal through (48).

REFERENCES

(1) T. Toya, J. Res. Inst. Catal. Hokkaido Univ., 6, 308 (1958)
(2) T. Toya, J. Res. Inst. Catal. Hokkaido Univ., 8, 209 (1960)
(3) W. Heitler and F. London, Z. Phys., 44, 455 (1927)
(4) C.E. Ireland, Phys. Rev. 43, 329 (1933)
(5) C.A. Coulson, Physical Chemistry, H. Eyring, D. Henderson and W. Jost, eds., Vol. 5, Academic Press, New York, (1970)
(6) T.B. Grimley, Prog. in Surf. and Membrane Sci., 9, (1975)
(7) W.A. Harrison, Solid State Theory, McGraw-Hill, New York, (1970) p. 480
(8) T.B. Grimley, J. Phys. C : Solid St. Phys. 3, 1934 (1970)
(9) C.P.J. Newton, Thesis, Liverpool, (1975)
(10) O. Gunnarson and H. Hjelmberg,Physica Scripta(Sweden)11,97 (1975)
(11) T.B. Grimley, Proc. Int. School of Physics "Enrico Fermi" Course LVIII, to be published
(12) T.B. Grimley, and C.P.J. Newton, Phys. Lett., to be published
(13) V. Heine, Proc. Phys. Soc., (London) 81, 300 (1963)
(14) V. Heine, Surface Sci., 2, 1 (1964)
(15) F. Forstmann, Z. Physik, 235, 69 (1970)
(16) V. Hoffstein, Surface Sci, 32, 149 (1972)
(17) W.H. Harrison, Pseudopotentials in the Theory of Metals, Benjamin, New York (1966)
(18) D.S. Boudreaux, Surface Sci., 28, 344 (1971)
(19) J.H. Wilkinson, The Algebraic Eigenvalue Problem, Clarendon Press, Oxford, (1965)
(20) W. Shockley, Phys. Rev., 56, 317 (1939)
(21) P.W. Anderson, Phys. Rev., 124, 41 (1961)
(22) S. Raimes, Many Electron Theory, North-Holland, Amsterdam (1972)
(23) A.L. Fetter and J.D. Walecka, Quantum Theory of Many-Particle Systems, McGraw-Hill, New York (1971)
(24) J.R. Schrieffer and R. Gomer, Surface Sci., 25, 315 (1971)
(25) P.O. Löwdin, J. Chem. Phys., 19, 1396 (1951)
(26) K.F. Wojciechowski, Act . Phys. Polon. 29, 119 (1966)
(27) R.H. Paulson and J.R. Schrieffer, Surface Sci., 48, 329 (1975)
(28) T.B. Grimley, Molecular Processes on Solid Surfaces, E. Drauglis, R.D. Gretz and R.I. Jaffee, eds., p. 181, McGraw-Hill, New York, (1969)
(29) H.D. Hagstrum and G.E. Becker, J. Chem. Phys. 54, 1015 (1971)
(30) J.W. Gadzuk, Phys. Rev., B10, 5030 (1974)
(31) T.B. Grimley, and C. Pisani, J. Phys. C., Solid State Phys. 7, 2831 (1974)
(32) C.C.J. Roothaan, Rev. Mod. Phys., 23, 69 (1951)
(33) C.C.J. Roothaan, Rev. Mod. Phys., 32, 179 (1960)

(34) E. Clementi, Proc. Nat. Acad. Sci. U.S.A., 69, 2942 (1972)
(35) M.J.S. Dewar, Wave Mechanics, W.C. Price, S.S. Chissick and T. Ravensdale, eds., p. 239, Butterworths, London (1973)
(36) J.C. Slater and K.H. Johnson, Phys. Rev., B5, 844 (1972)
(37) J.C. Slater, Advan. Quantum Chemistry, 6, 1, (1972)
(38) K.H. Johnson, Advan. Quantum Chemistry, 7, 143 (1973)
(39) K. Schwarz, Phys. Rev., B5, 2466 (1972)
(40) P. Weinberger and D.D. Konowalow, Internat. J. Quantum Chem., 75, 353 (1973)
(41) S.J. Niemczyk, J. Vac. Sci., Technol., 12, 246 (1975)
(42) I.P. Batra and O. Robaux, J. Vac. Sci. Technol., 12, 242 (1975)
(43) R.P. Messmer, Battelle Colloquium : The Physical Basis for Heterogeneous Catalysis, to be published
(44) R.O. Jones, Surface Sci., to be published
(45) E.J. Dyson, Phys. Rev., 75, 486 (1949)
(46) E.J. Dyson, Phys. Rev., 75, 1736 (1949)
(47) E. Clementi and D.L. Raimondi, J. Chem. Phys., 38, 2686 (1963)
(48) G. Wannier, Phys. Rev., 52, 191 (1937)

APPENDIX

If we multiply (31) on the right by the matrix

$$K' = \begin{vmatrix} 1 & 0 \\ S_{MA} & 1 \end{vmatrix}$$

for which $\tilde{G}K' = \tilde{G}$ also, we derive (32) again but with the unit matrix on the RHS. This seems to be the simplest form of (25) and the subsidiary conditions (29).

ELECTRONIC PROPERTIES OF CLEAN AND CHEMISORBED METAL SURFACES

Thor RHODIN[+]
Cornell University
Ithaca, New York, U.S.A. 14853
and
David ADAMS
Xerox Research Laboratory
Webster, New York, U.S.A. 14644

A. THEORETICAL APPROACHES TO CHEMISORPTION[(1-11)]

1. Introduction

The theory of chemisorption is based on an understanding of the atomic and electronic structure of clean metal surfaces and the extent to which the basic surface parameters are subsequently modified by adsorption or reaction processes resulting from both adsorbent-adsorbate and adsorbate-adsorbate interactions. Precise formulation of the electronic structure of metal surfaces in the clean state or with adsorbed atoms is complicated by the incomplete understanding of the nature of the metallic bond itself as well as by the lack of detailed experimental information on the electron structure specific to a metal surface and how this structure changes during a chemical transformation. At present major theoretical effort tends to be directed towards the modeling of rather simple molecular interactions on ideal atomically flat surfaces.

Reference is recommended to the monograph by Rhodin and Adams[(22)] for a detailed presentation of the concepts presented in this lecture. The status of chemisorption theory was also discussed recently[(12)] and independently reviewed in some detail by Gomer[(13)]. It seems de-

---

+ Support from National Science Foundation Grant DMR71-01769 A02, from the Advanced Research Agency through the Cornell Materials Science Center and from the NATO Advanced Study Institute is gratefully acknowledged (TNR).

sirable at this time to outline some of the characteristic features of the theory from the viewpoint of the experimentalist. The objective is to provide perspective on the adsorption process as well as a base for comparing experimental parameters and models with the corresponding variables critical to the theoretical models.

One of the main theoretical thrusts pertinent to adsorption phenomena at present is the calculation of the local density of states for one electron energy levels at the adsorbate. This in turn is used to predict atomic binding energies and charge transfer. A second objective is to formulate the microscopic models needed to interpret and analyze spectroscopic data on the electronic structure both of clean and chemisorbed metals[14]. It is therefore appropriate to note briefly here some of the main features of thc theoretical approaches and to follow this with a parallel discussion of three of the valence-level spectroscopies suited to measurement of the electronic properties of both clean and chemisorbed metal surfaces.

Two principal approaches are commonly used. One starts with an almost completely localized interaction involving a surface molecule or complex closely related to the quantum chemistry approach of chemical bond formation in molecular chemistry[14-17]. The other involves band structure concepts derived from methods of solid state physics[1-10,13,18-21]. In both cases the concept of orbitals or energy levels and their occupancy in terms of the interaction of the individual valence electrons, of the adatom with electron levels in the conduction band of the metal is used in chemisorption models[18]. One generally considers an energy level diagram (see Figure 1.) where the valence level of a monovalent atom is positioned relative to the vacuum and Fermi levels of simple free electron metal with a work function, ϕ and a Fermi energy, ε_f. A model involving a transition metal substrate would also contain a narrow d-band centered at some energy, ε_d with respect to the Fermi level.

The virtual state of the adatom corresponds to the sharp valence electron level of the isolated atom broadened by interaction with the substrate electrons when positioned at some stable position and with a broadened energy, ε_a in the proximity of the surface. The broadened and energy level position corresponding to this virtual state are basic parameters used to specify the interactions which describe the chemisorbed state. In the band theory approach the broadening of the narrow ionization level, ε_a of the isolated atom as it approaches the surface may be regarded as an effect of the uncertainly principle in which the electron associated with the adatom now spends less time on the adatom. The half-width, $\Gamma/2$ of the broadened level is approximately related to the rate, $\Gamma \approx 2\ h/\tau$ at which the electron hops back and forth between the surface and the adatom where τ = tunneling time of an electron (on the adatom) into the metal, $\Gamma(\varepsilon)$ is often referred to as the level-width function. In the surface molecule approach, to which reference was made briefly in pre-

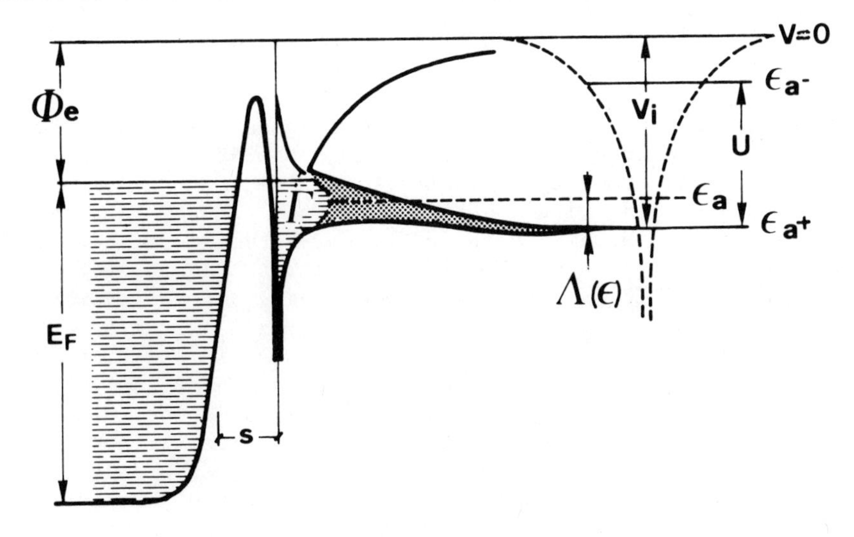

Figure 1. : Electron-level diagram relevant to a metal-surface impurity problem. The dashed curve is the ion-core potential for the particle at infinity. The solid curve is a schematic of the combined atomic and metal potential for the atom a distance s from the surface. V_i represents the ionization energy of the isolated atom, ΔE is the shift of the energy level, and Γ is the natural (lifetime) broadening due to the atom's interaction with the solid. The density-of-state is not indicated on the diagram. (After Gadzuk, refs. 18 and 19)

vious lectures, the broadened virtual state corresponds to the spectrum of molecular orbitals of the local complex.

The original broadened adsorbate level will also be shifted a net amount resulting from a downward shift related to the bonding interaction with the metal electrons as well as one upward shift due to the intra-atomic Coulomb repulsion, U and a second relative shift due to an attractive image potential interaction, V_{im} with the surface. The effective ionization potential, V_i is correspondingly reduced by these two upward-shifts. The degree of shift is sometimes referred to as the level-shift function, $\Lambda(\varepsilon)$. The characteristics of the broadened virtual state level defined in this manner describes the state of adsorbate resonance. In addition to the level width function $\Gamma(\varepsilon)$, and the level-shift function $\Lambda(\varepsilon)$, it is useful to describe the distribution of filling of the energy levels, i.e., the local density of states at the adsorbate, $\rho(\varepsilon)$.

The coupling of these chemisorption parameters with the spectra obtained experimentally occurs through the density of states $\rho(E)$. The density of occupied states is defined as the number of occupied orbitals lying within a specified small energy interval as a function of energy usually referred to the Fermi energy level. The density of states for an s-electron, $\rho_s(\varepsilon)$ with reference to a transition metal usually extends over a broad range of energy whereas that of d-electrons, $\rho_d(\varepsilon)$ tends to be localized around an energy, ε_d. These features are illustrated in Figure 2. In the simplest interpretation of electron emission from such surfaces, the observed spectra reflect essentially the coupling of a discrete valence level of the adatom at a specific chemisorption site, with the s-band-and d-band-density of states of the metal to give a broadened virtual level characterized by a local density of states. The latter may show a variety of forms involving highly delocalized density of electrons, or in some cases a split pair of localized densities corresponding to a low energy (occupied) and a high energy (unoccupied) level as illustrated in Figure 3 (to be discussed).

It should be noted that in the measurement of the local density of states by electron spectroscopic methods the spectra are not usually just a simple reflection of the characteristic local density of states of the surface molecule but contain additional features associated with the collective nature of electrons in the substrate as well as the electron emission process itself. The fact that the observed so-called optical density of states usually reflects a complicated transform of the original electron structure modified in some manner by the adatom as well as by the excitation and emission process complicate reduction of the observed spectra into the basic chemisorption parameters.

It is significant that the local density of states concept is directly referrable to models[14] (FEED, UPS, INS and ESCA) used to interpret measured electronic spectra of atoms chemisorbed on transition metal surfaces. By careful analysis and comparison of spectroscopic data it is possible to extract critical microscopic information on the chemisorption process. For a more detailed discussion of the relationship of these theoretical parameters to spectroscopic observations see Gadzuk[11].

Two approaches to the theoretical formulation of the chemisorption bond used in general to evaluate adsorbate resonance, in parallel with the methods of molecular structure calculations, are the molecular orbital, (MO) scheme[1-5,18,20,21] and the valence, (VB) scheme[15,16,23,24]. When repulsions between electrons are important, it is necessary to keep track of the individual spins in

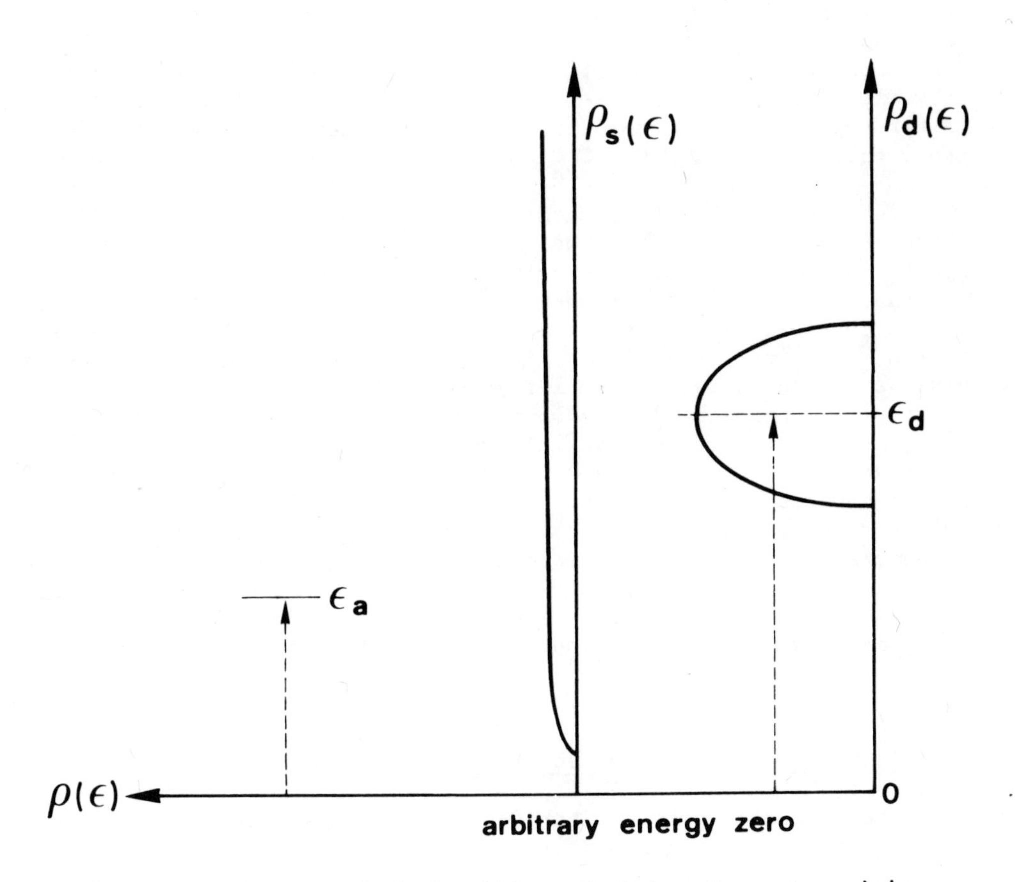

Figure 2. : Some typical densities of states for a transition metal showing from left to right, the isolated adatom valence level, the uncoupled distribution of s-electrons in the metal, and the unoccupied d-electrons in the metal. An adsorbate density-of-states can be calculated by combining these three components by providing appropriately for (1) position of ε_a relative to $\rho_d(\varepsilon)$, (2) strength of the metal adatom coupling, (3) coordination of adatom with surface atoms, (4) band structure of the substrate, (5) orbital symmetries.

the former scheme since the electrons belong to all the nuclei in the system. It has the advantage that all the atoms in the solid, including the adatom, are treated equally and the difficulty that description of the electron-electron correlation effects are often not readily reduced to manageable form. The VB-scheme on the other hand tends to over-account for electron correlation by considering only electron-pair or covalent bonds ant it is not clear how the latter approach can be used to account for energy band effects in the metal.

The present thrust of chemisorption theory attempts to define which of these factors are most critical for a given adsorption system and to formulate a description of the energy levels and their occupancies sufficiently rigorous to be precise, sufficiently clear in its physical significance, and sufficiently manageable in terms of its analytical formulation. This task is not always readily achieved. We will consider here briefly the salient features of some of the typical approaches to achieve a sense of analytical connection with results of valence level electron spectroscopy. The detailed mathematical formulations and approximations associated with the different theoretical approaches will be presented in other lectures directed specifically towards the theory. Emphasis is placed here on describing the main theoretical parameters and how they can be related to experiment. For more detailed comparative presentations of the various approaches, reference is recommended to recent reviews dealing with formulation and application of chemisorption theory[9-14].

2. Molecular Orbital Approches[1-5,14,20-26]

The molecular orbital approach that has been widely used is a one electron approximation which defines a basis set as a single free adsorbate wave function, ϕ_a plus the Bloch states of the metal in terms of wave functions, ϕ_m. This approach is derived mainly from the initial formulation originally employed by Anderson[27] to describe the local electronic interaction of an impurity atom in a dilute metal alloy.

The ratio of $\pi\Gamma/2$ to U determines whether it is essential in order to calculate the surface bonding, to correlate the spin of the hopping electron with the electrons in the metal to obtain a sufficiently accurate estimation of the energy of the system[1-5,10]. The electron correlation on the adsorbate becomes very important when the intra-atomic repulsion is larger than the broadened level, that is when $U > \pi\Gamma/2$. However the calculation of the adsorption interactions becomes correspondingly more difficult in this case. On the other hand, there are adsorption systems where the contrary is true, $U < \pi\Gamma/2$, that is the motion of the electrons back and forth to the adsorbate is relatively rapid. Under this circumstance the correlation contribution to the energy of the bonding electrons can be ignored and the energy of the system can be validly treated in the restricted Hartree-Fock approach.

This latter approach is commonly used and generally referred to as the self-consistent field (SCF) molecular orbital scheme. It is critical to define the shape and position of the adsorbate resonance to determine whether it is valid to apply the SCF molecular orbital approach. One considers that the one-electron wave function, Ψ_n extend over the whole electron system of metal plus adatom. The average charge density in the system is treated in terms of the electron density on the adatom as being composed of the sum of many small contributions from each of the one-electron states which may be occupied. Balanced up- and down-spins and absence of electron correlation are assumed in this approach.

The unrestricted HF approximation[10] on the other hand, considers both the correlation effects of the electron on the adsorbate as well as the fact that the electrons may not have balanced up- and down-spins. It is the effect of correlation and spin which is more complicated. It must however be considered in the more complete picture. In both approximations,it is essential that the charge distribution be evaluated in a manner which is consistent with the potentials of interaction used to describe the system. This requirement applies irrespective of what approximation is chosen to describe the contribution of the electron charge and spin. In his lectures, Grimley[2-4,21,25] discussed how to adapt this approach to the chemisorption problem using the restricted Hartree-Fock approximation. Some very important features resulting from this approach are emphasized in Figure 3. It can be shown[14] that one or more new states can be created by the adsorption interaction. The adsorbate density of states corresponding to increasing interaction strengths are illustrated by a, b and c. Strong chemisorption tends to form a surface molecule associated with the splitting off of two states, one above the band which is empty of electrons and one filled state below the band. These may be considered as antibonding and bonding levels respectively of a surface molecule formed between the adsorbate and the substrate. It results from the strong local nature of the interaction. It leads directly to the important conclusion that different adsorption states can exist not only on different crystal faces but even on the same face. These split-off states are associated with different bonding energies, characteristic orbital symmetries and atomic geometries.

The restricted Hartree-Fock one-electron approximation has been applied with reasonable success to metal atom adsorption on tungsten (28) involving relatively little transfer of charge as well as to alkali metal adsorption[29] involving rather large transfer of charge. Applications of the unrestricted Hartree-Fock approach to chemisorption systems have not been particularly successful.

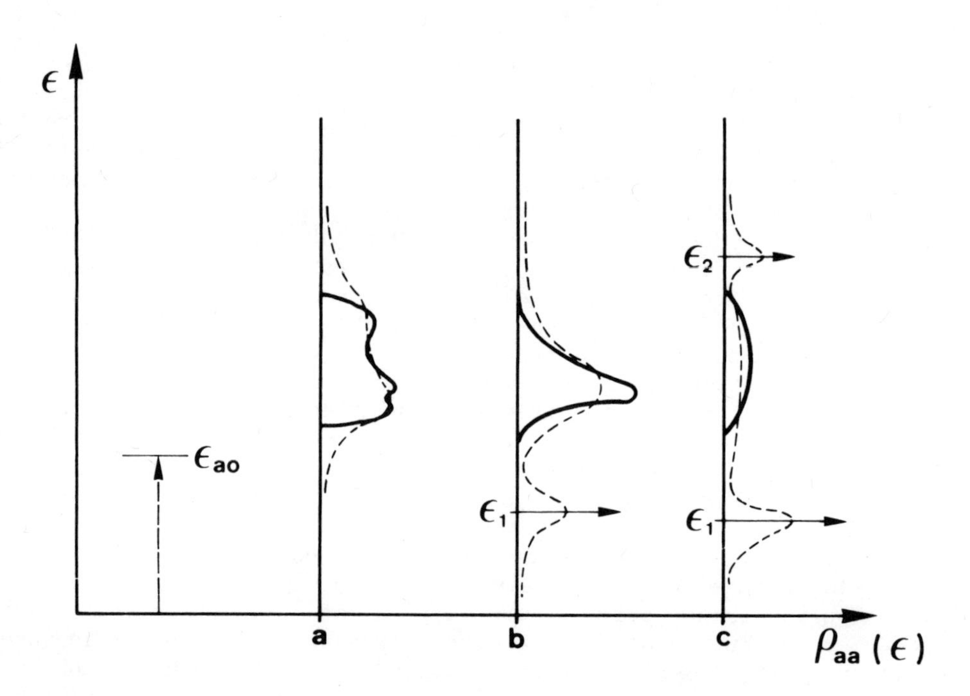

Figure 3. : Surface-molecule energy level diagram. Schematic examples of three possible adsorbate density-of-states for one adatom coupled respectively, with the d-band and the s-band density of states (dashed line) for chemisorption on an idealized transition metal surface. (a) A featureless energy level spectrum in which localized or characteristic bonding is absent. (b) Peaked distribution corresponding with formation of a virtual state (at the higher energy) and a localized surface molecule state (at the lower energy, ε_1). (c) Peaked distribution corresponding to localized distributions indicating formation of split-off states outside of the d-band at energies ε_2 and ε_1. Note : All three states a, b, and c are known to occur. State c is most readily analyzed in terms of spectroscopic measurements (After Gadzuk, ref. 14)

There are other important approaches to chemisorption theory within the molecular orbital scheme in addition to those of Newns (5,20) and of Grimley[1-4,21]. In analogy with the Korringa-Kohn-Rostoker approach to the band theory of solids, Johnson and co-workers[30,31] have developed a scattered wave scheme to solve approximately the local field problem on a self-consistent basis for a cluster of atoms. It is referred to as the self-consistent-field X_α-scattered-wave (SCF-X_α-SW) cluster method. It was originally developed by Slater and Johnson[32] and subsequently extended by Johnson and co-workers[30,31] to many special phenomena involving localized interaction among atoms in cluster configurations. For a review of such applications to adsorption the reader is referred to the recent work of Johnson and Messmer[17].

3. Valence Bond Approaches[15,16,23,24]

There are difficult problems associated with proper provision for the interelectron Coulomb repulsion and strong electron-electron correlation contributions on the adatom for the adatom-substrate system within the molecular orbital scheme. In cases of nonionic chemisorption such as hydrogen on transition metals, where the Coulomb term, U is large relative to the energy-broadening term $\Gamma(\varepsilon)$, it is desirable to construct a theory where correlation effects are treated *ab initio*. A new description of the adatom-substrate system analogous to the Heitler-London treatment of a diatomic molecule was proposed by Schrieffer and Gomer[23] with this objective in mind. It was developed in some detail by Paulson and Schrieffer[24] for hydrogen chemisorption.

The nonorthogonality of the wave functions of the adatom and those of the metal produce an antiferromagnetic exchange interaction between the adatom and the solid resulting in a significant spin density in the solid near the bonding site. It is postulated that the decrease in energy due to the local induced spin-pairing compared to the increase in energy required to produce an unpaired spin can be such as to result in the production of an induced covalent bond at the surface. Proper account of the correlation of electrons hopping between tha adatom and the metal is provided for by setting up an analog of valence-bond wave function. This considers only electron pair bonds and tends to exaggerate the correlation contributions and completely neglects the ionic contributions. Since no unpaired spins are available, at conventional adsorption temperature it is proposed that electron-hole pairs are created by promoting electrons above the Fermi energy to create free spins which can then pair with the spin on the adatom to form a valence bond. It is predicted that the binding energy increases as the band energy in the metal gets narrower[24]. This is qualitatively in agreement with the observation that the chemisorption of hydrogen is strongest on the transition metals characterized by narrower d-bands. Since this treatment is essentially a many body approach, the simpler concept of one-electron energy levels cannot be used. The fact that the quantitative formulation is difficult and that a local density of states is not readily derived from it so far have tended to restrict evaluation and application of this approach.

4. Density-Functional Approach[6,33-36]

The density-functional theory applied to the surface region treats electrons as a gas in which electrostatic self-consistency is strictly observed but where exchange and correlation contributions are treated approximately. The theory is without adjustable parameters and seems to account remarkably well for the charge density in the vicinity of the adsorbate. It thus explains dielectric response effects and work function changes upon adsorption rather effectively. The logical basis of this approach rests on a general variational scheme applied to the ground state of a nonmagnetic system of electrons of a given density distribution in a given external potential. It has been shown by Kohn and his co-workers[35,36] that there exist a universal function of the electron density in terms of which one may define an energy function from which minimization of the total energy of the adsorption system may be readily achieved.

The most extensive qualitative results for electron density distributions, surface energies and work functions are those of Lang and Kohn[6]. They are in good agreement with experiment. The same density-functional formalism has also recently been used to describe a model of hydrogen adsorbed on tungsten by Smith, Ying and Kohn[34]. Reasonable agreement with experiment is obtained.for the calculated ionic adsorption energy, the dipole moment, the resonance level position and the vibration frequency. The calculated binding energies for a proton imbedded in the electron gas of the metal on the other hand, are off. Detailed discussion of this apprcach to chemisorption is presented in the lectures by Lang.

The microscopic features of electron properties in these approaches can now be usefully considered in terms of modern methods of electron spectroscopy. It is clear from a general view-point that one very useful product of all these theoretical formulations of chemisorption is to focus attention on the theoretical variables which are most critical to the experimental parameters. Some of these experimental parameters will be specifically considered in terms of the following spectroscopic measurements of electrons at surfaces.

B. EXPERIMENTAL APPROACHES TO CHEMISORPTION[37-57]

1. Introduction

A significant developemnt of various methods of electron spectroscopy to the characterization of the electron structure of both clean and adsorbate-covered metal surfaces has occurred recently[37]. A variety of spectroscopies provide a wealth of detail on electron states and excitation processes. They involve either direct excitation of the valence electrons by photons, ions, electrons or high electric fields or indirect excitation in which the core level electrons are first excited.

Some common examples of the direct excitation methods include photoelectron emission excited in the vacuum ultraviolet (UPS)[38], ion neutralization spectroscopy (INS)[39], inelastic low energy electron-diffraction (ILEED)[40] and field electron energy distribution (FEED)[41]. These new surface spectroscopies (three typical ones are indicated in Figure 4) are developing rapidly in breadth and complexity. Reference is recommended to recent articles surveying the features of these various methods by Plummer[42], by Rhodin[43] and by Gomer[44].

The indirect excitation methods include techniques such as appearance potential spectroscopy (APS)[45-47], Auger electron spectroscopy (AES)[48-52] and X-ray photoelectron spectroscopy (XPS or ESCA)[53,54-56]. Core level electrons are excited by incident electrons in APS and AES, and by incident X-ray photons in XPS. Both the excitation and subsequent deexcitation of the core level (by electron or X-ray emission) may involve valence electron excitation (see Figure 1). Direct excitation of the valence electrons can also occur in XPS. The principles and applications of these various techniques for surface studies were outlined briefly in the first lecture and will be discussed in detail in the lectures by Gadzuk and by Brundle.

2. Significance of New Methods

We are interested in how these approaches can provide specific information on the microscopic nature of the adsorption process in terms of the electron structure at the surface. The different kinds of information obtained and the degree to which they supplement each other is of particular importance. All of these spectroscopies provide detailed information on the microscopic aspects of electrons at or near a metal surface. The present state of interpretation of the data however is often complicated. This results partly from the influence of band structure effects associated with the direct excitation methods and by the multiplicity of transitions associated

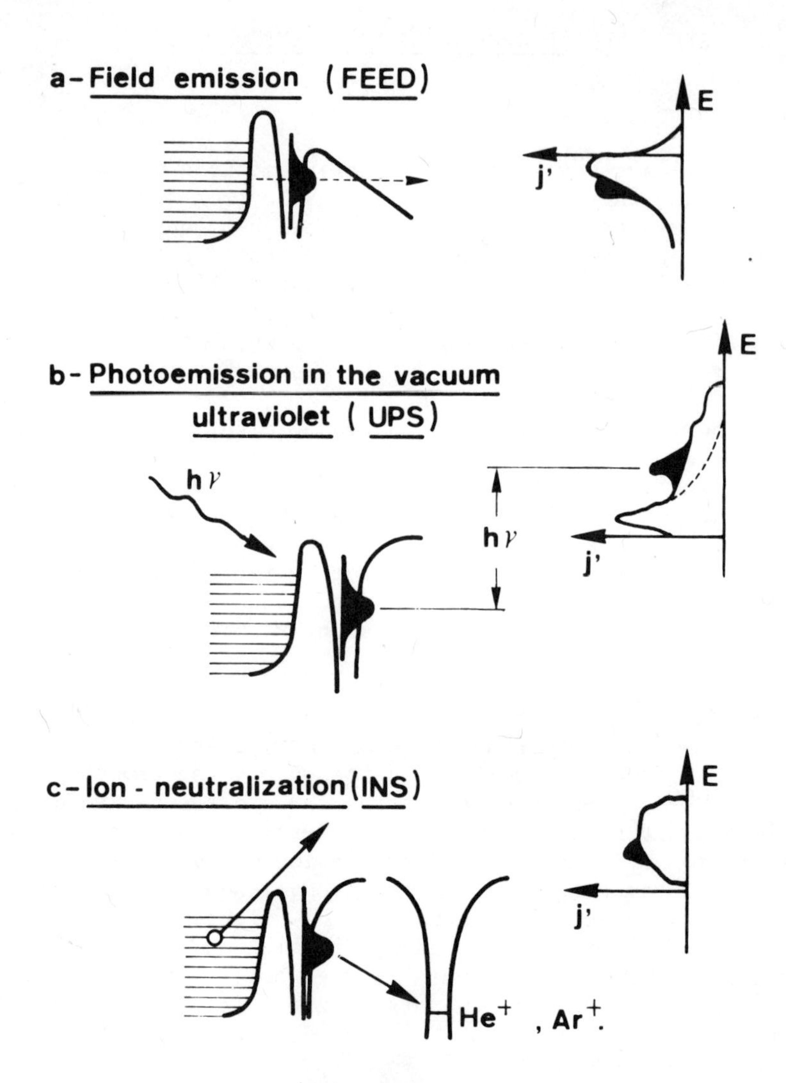

Figure 4. : Potential energy diagrams corresponding to one electron descriptions of electron emission promoted by high local field (FEED-a), by photon excitation (UPS-XPS-ESCA-b) or by ion neutralization (INS-c). The emission current vs energy plots on the right show the effect of chemisorption (dark area) against the current distribution characteristic of the clean surface.(Note: the spectra obtained directly from INS is the unfold. It must be deconvoluted to obtain a spectra corresponding to the other two methods).

with the indirect excitation methods as well as by uncertainties in defining precisely the role of the instrumental response function for the specific spectroscopic method employed.

Quantitative descriptions of specific features of the adsorption phenomena in terms of each spectral approach is relatively limited at present. The qualitative implications on the other hand, are interesting. Active development of the required theoretical background is in progress[12], and advances in the quantitative aspects can be expected. A detailed discussion of experimental and instrumental developments in the field of electron spectroscopy up to the end of 1971 is available[37]. Reference should also be made to specific reviews of the application of electron spectroscopy to adsorption on metals[42,43,48-56].

It is significant that electron spectroscopy measurements can provide information on the local density of states at the surface, characteristic of the chemisorption process. These data are essential to the evaluation of recently developed theoretical models. Depending on the specific type of electron spectroscopy employed, information is obtained on the atomic electron energy levels in terms of the position and lifetime broadening of the atomic band as a function of the physical parameters of adsorption. Of the many types of electron spectroscopies used for studies of metal surface reactions we deal here specifically with three examples which have proved particularly useful, tunneling-resonance field emission, ion-neutralization electron emission and photoelectron emission illustrated in Figure 4.

It should be reemphasized that the main link between theoretical models of chemisorption and electron emission spectra is in extraction of an adsorbate density of states to provide information on the critical parameters of chemisorption predicted by theory. The important question in every case is the extent to which an adsorbate density of states can be deduced from the experimental spectra. It seems that this can be most readily achieved for the FEED spectra, with considerable confidence for clean surfaces but with less quantitatively defined implications for chemisorbed surfaces. The photoemission and ion-neutralization spectra on the other hand for the present provide with some exceptions, primarily methods for "fingerprinting" the surface orbitals. This can be done by taking a difference curve between the clean surface spectra and that for the corresponding chemisorption system. At present, conversion of the measured optical density of states by either method to an adsorbate density of states involves some nontrivial uncertainties as indicated. However current theoretical efforts are likely to make useful contributions to this deconvolution process in the future[11]. Interesting interpretation of electron structure from INS spectra have also been achieved by Hagstrum[57] including experimental comparisons with UPS spectra. The spectroscopic methods are now in the process of being cross-compared with each other in some detail. Some of the more salient results of these measurements are now discussed with specific reference to each.

3. Field Emission Energy Distribution (FEED)[41,58,65]

The powerful experimental approach to surface studies based on field emission, field desorption and field ionization are derived largely from the original inventions and studies of E.W. Müller (58-60). It was subsequently predicted by Duke and Alferieff[61] that resonance transmission of tunneling electrons should occur in field emission when electrons in the metal have the same energy as a virtual level created by the adatom. The relationship between the relative characteristics of the resonance tunneling and the energy distribution of electrons for a clean surface compared to one modified by the presence of an adsorbed atom is illustrated in Figure 5.

The position and broadening of the atomic level can be obtained from the measurement of the energy distribution of the electrons field-emitted through the adsorbed atom in terms of resonance-enhanced tunneling demonstrated by Plummer, Gadzuk and Young[62,63]. Combining Gadzuk's tunneling-resonance theory[64] with the measured enhancement factors, Plummer and Young[63] obtained direct information on the lifetime broadening and energy shifts of the virtual surface impurity state of alkaline earth atoms on single crystal tungsten surfaces. The data for barium and calcium demonstrated clearly that the ground-state configuration of an adsorbed atom may be a mixture of the atomic ground-state and excited atomic states due to the atom-metal interaction, illustrated in Figure 6. Important features of adsorption such as the nature of the surface bond and the magnitude of the heat of adsorption are related directly to the number and character of the virtual levels below the Fermi surface of the metal.

The ability to probe molecular binding properties of single atoms establishes the unique applicability of tunneling field emission (FEED). The tunneling currents fall off rapidly however, below the Fermi level, E_F making it rather difficult to probe energy levels more than about 2.0 volts below this level. The potential applicability of field-emission energy-distribution (FEED) spectra as an effective probe of the electron structure of clean and adsorbed surfaces has been reviewed by Gadzuk ane Plummer[41]. Measurements from clean metal surfaces of tungsten by Plummer *et al.*[42] have also been shown to provide information about the density of metal states near the surface which can be compared to the relevant photoemission data.

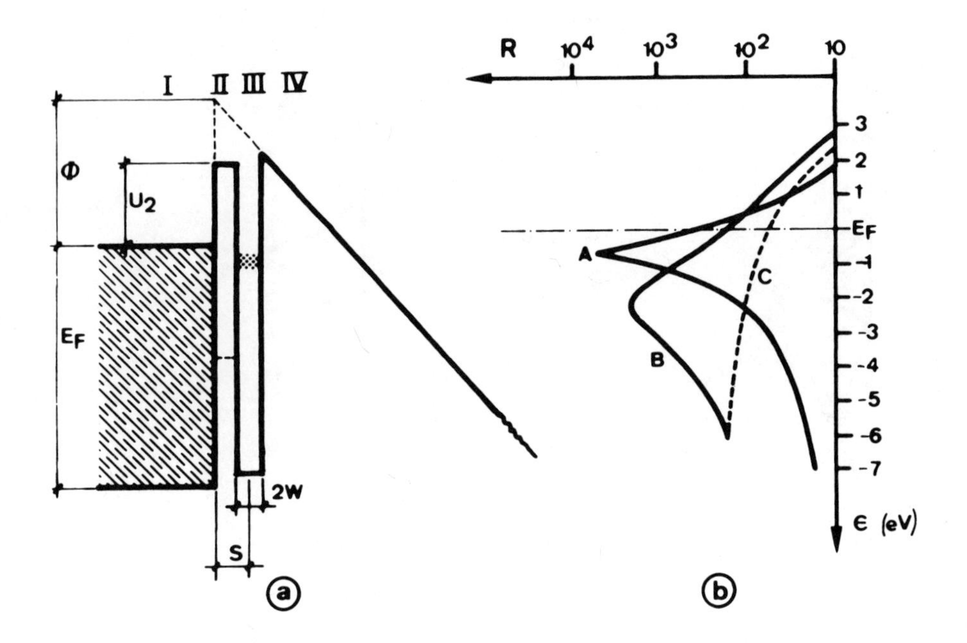

Figure 5. : (a) One-dimensional model for the metal (I) and the adsorbate (II) in the presence of an applied field. The width is 2W and depth of the well is -0.7 eV (shaded region).
(b) Enhancement factor for the potentials in (a); curve A (solid line potential), curve B (barrier height, is at -3.5 eV) and curve C (non-resonant contribution)
(After Plummer and Young, ref. 63)

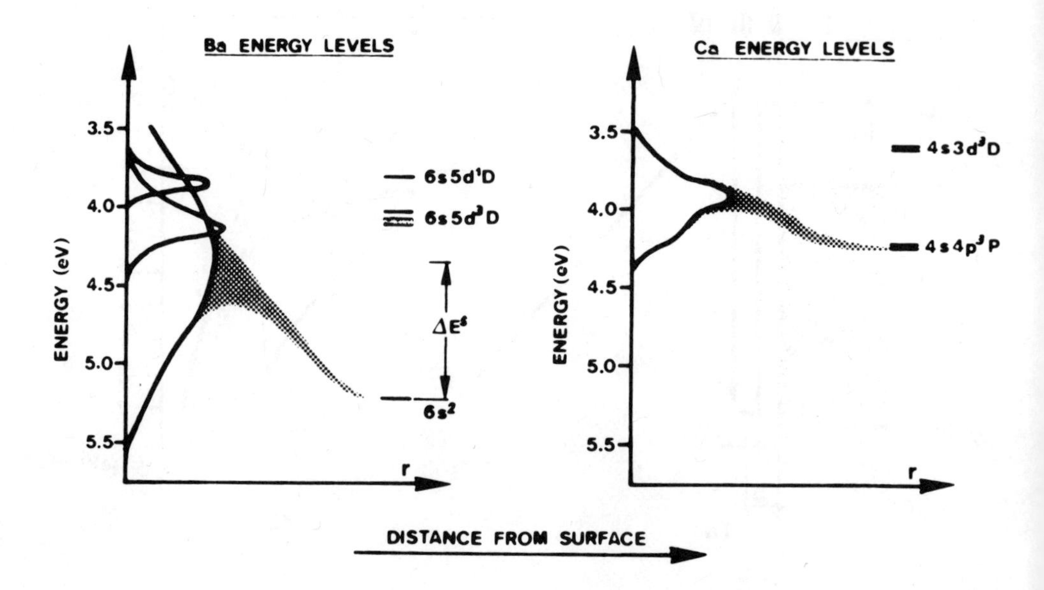

Figure 6. : Pictorial representation of the broadening and the shifting of the energy levels of barium and calcium as they interact with the surface. The shapes and position of the virtual levels at the surface are taken from the data on the low work-function planes of tungsten. (After Plummer and Young, ref. 63)

In addition, new FEED results[66,67] for many of the principal faces of the platinum-group metals (iridium, rhodium, palladium and platinum) indicate that such studies can be closely related to interpretation of band structure and density of state considerations for both the clean and adsorbate covered surfaces of these metals. The significance of these studies lies in the combination of the experimental features of field desorption, field emission and field ion microscopy. The structure contained in these energy spectra can be interpreted in terms of contributions among overlapping sub-bands of the electronic band structure of the metal as modified by the surface. The selective probe of the d-bands particularly at the surface is illustrated by the one-electron potential energy tunneling diagram of Figure 7 for a clean and chemisorbed 5-d transition metal such as platinum or iridium. A typical FEED spectra for Ir(111) in Figure 8 indicates clearly the electron structure associated with the d-band emission. (The enhancement curve is measure of the d-band emission relative to that of the sp-band contribution to the tunneling current.) It has been shown to be directly related to the one-dimensional directional density of states at the surface of the metal by Politzer and Cutler[91] and by Penn and Plummer[68]. The correlation with the corresponding calculated d-band structure (with some contradiction of the bulk band caused by the surface) is also indicated along the energy axis. Important details of the d-band surface structure can thus be interpreted from the high field tunneling spectra[41,66,67].

Chemisorption has a marked effect on the electron emission spectrum as indicated in Figure 9 for the associative chemisorption of nitrogen on Pt(001), Pt(110) and Pt(310) surfaces at 78 K. Although the initial chemisorption modifies the electron structure differently for different crystal faces, it is significant that the local density of state distributions are strikingly similar for the Pt(100) and Pt(310) surfaces after an exposure at 3.0 L. nitrogen. Chemisorption is clearly strongly modifying the electron structure of the surface differently for the various crystal faces. Quantitative interpretation of such effects in terms of the various hybridizations of the hydrogen molecular orbitals with the d-subband structure of the metal is possible in principle but has not as yet been achieved. Qualitative interpretations of these studies in combination with photoemission spectral data in terms of surface electron structure are described by Plummer _et al_.[41] for tungsten and by Rhodin _et al_.[66,67] for the platinum-group metals. Interpretation of the FEED emission spectra benefits mostly from a well developed theoretical understanding of the corresponding emission phenomena among these three electron spectroscopies (FEED, UPS and INS). All three methods as applied to the electronic interpretation of chemisorption have generally proved to be useful and productive in defining electron structure and excitations at metal surfaces.

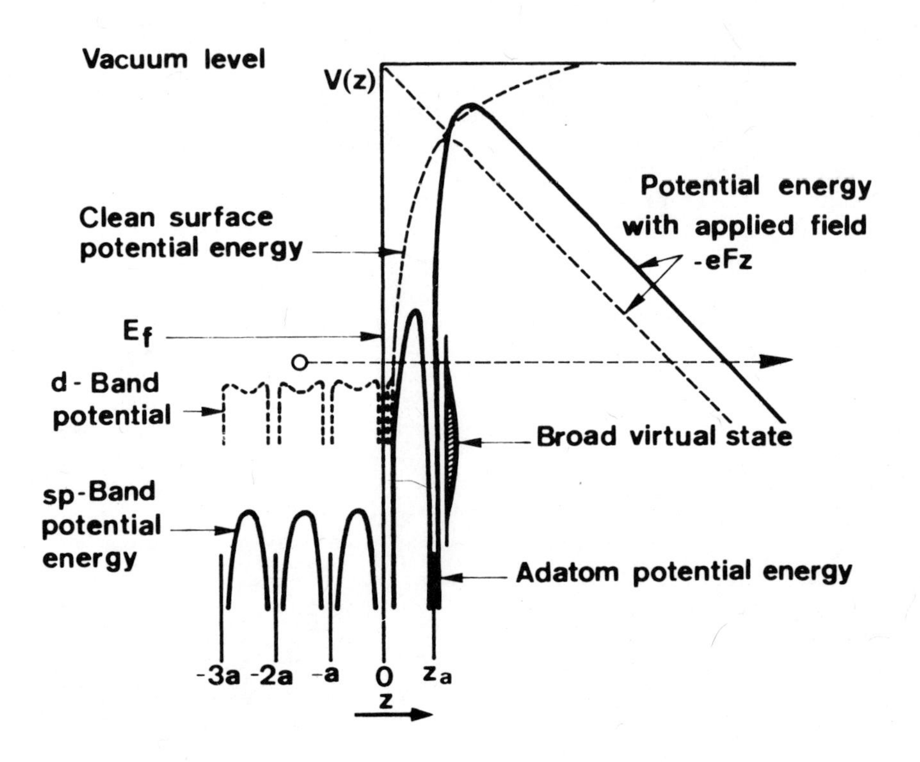

Figure 7. : One-dimensional potential energy diagram for a strongly bound adatom/substrate system. A broad virtual state of an adatom is illustrated relative to the sp- and d-bands for the surface of a metal under an externally applied field. Notice the difference in the barrier thickness for the two bands before and after a single adsorption event. (After Dionne and Rhodin, ref. 66)

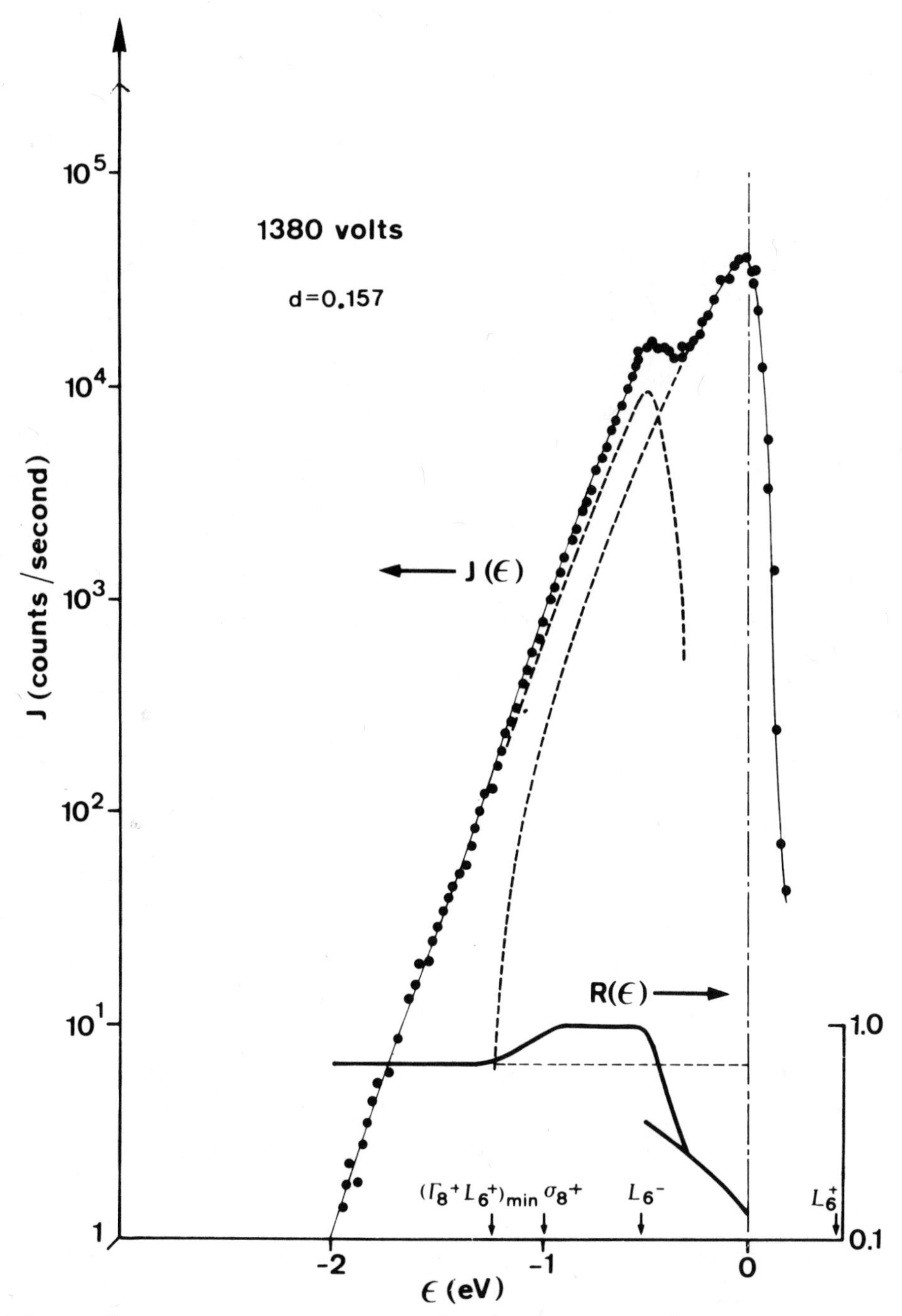

Figure 8. : FEED spectrum of Ir(111). The tunneling voltage and the tunneling parameter are indicated. The enhancement factor-plot and calculated bulk band structure notation are indicated below (After Dionne and Rhodin, ref. 66)

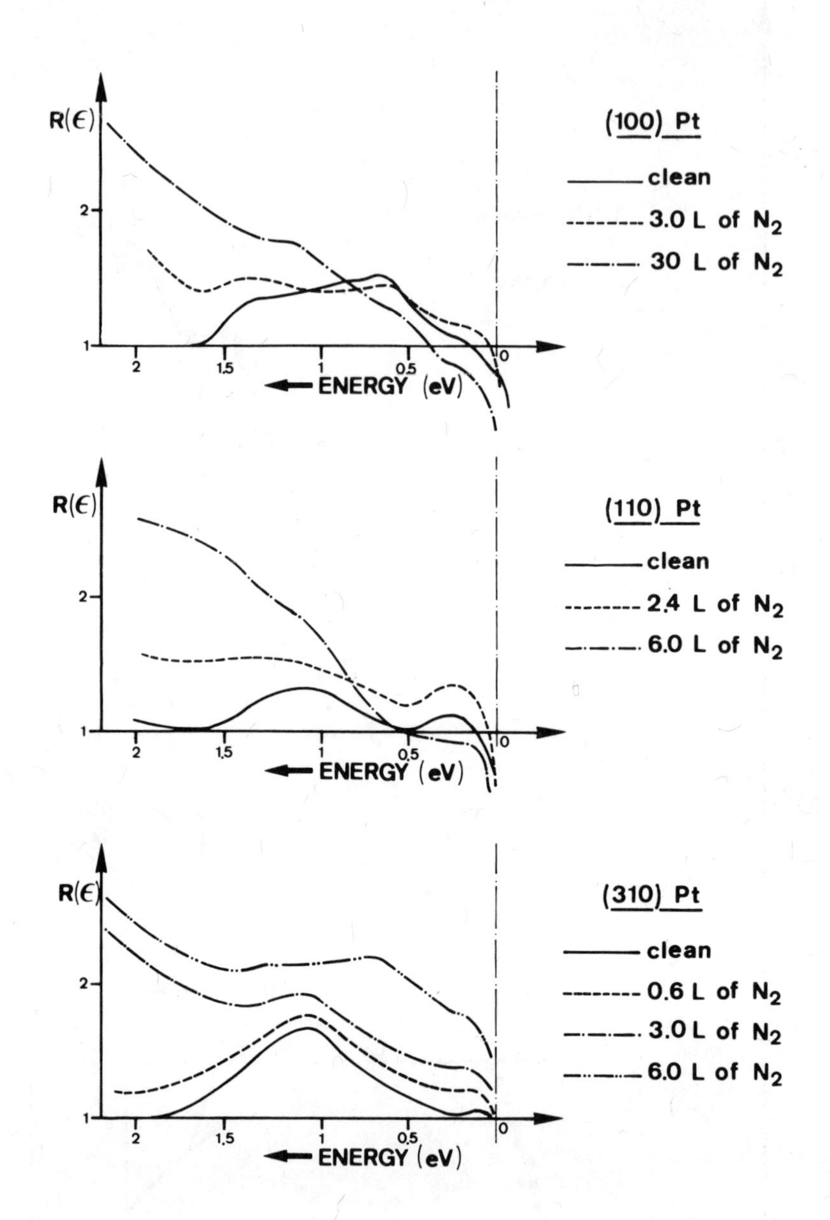

Figure 9. : Enhancement factor plots corresponding to density of state information from field emission energy distribution measurements of associative chemisorption of nitrogen on single crystal planes of platinum at 78 K.

4. Photoemission Spectroscopy[37,38,69-87]

For the study of adsorption on metals by photoemission, reference is recommended to typical studies by Eastman on simple gases and hydrocarbons on nickel[70,71], by Hagstrum on the chalcogens on nickel[57], by Plummer et al. on hydrogen and hydrocarbons on tungsten[42], as well as to studies by Linnett and co-workers[72,73], by Feuerbacher and Fitton[74-76] and by Rhodin et al.[77-79]. Preliminary systematic studies coupling FEED and UPS of reactive gases and hydrocarbons on platinum-group metals now in progress should provide a new and quantitative approach to the microscopic description of the surface physics of these catalytically important adsorption systems[12,78,79].

These studies make possible new microscopic approaches to the interpretation of chemical bonding at surfaces relevant to chemisorption and catalysis. Photoemission studies on elemental semiconductors by Rowe and Ibach[81,82], N.V. Smith[83] and associates on angular photoemission from layered semiconductor compounds, and by Spicer[69] and associates on a variety of both metals and semiconductors, have clearly indicated important connections with the occurrence of surface states and the unique features of chemical bonding on semiconductor surfaces. These are of significance with reference not only to modification of the band structure near the surface but to the symmetry of both the adsorption sites and bonding orbitals. The latter is associated with the potential applicability of angular resolved photoemission reported by Guastaffson[84] and by Smith et al.[83] and discussed by Gadzuk[85], and by Liebsch[86] and by Liebsch and Plummer[87].

The applicability of UPS to the study of gas adsorption and reaction on metals was demonstrated in the initial work of Eastman and Cashion[70]. The effect of chemisorbed carbon monoxide and oxygen on photoemission spectra from polycrystalline evaporated nickel films was observed. Energy distributions determined by the UPS-method as a function of oxygen exposure and heat treatment shown in Figure 10 are taken from the work. The energy level associated with the oxygen chemisorption increase in amplitude,broadens and shifts towards the Fermi level with increasing oxygen exposure. Upon further exposure, emission associated with the d-band of nickel decreases and the emission associated with the oxygen-bonding increases (see (3),(5) and (5) Figure 10). The level at 2 eV below E_F in curve 6b is attributed to the d-electrons of the oxidized nickel. Structure seen for the surface oxide is qualitatively similar to that observed on a nickel oxide crystal indicating occurrence of the predicted transition in chemical bonding from chemisorption to oxide formation. Heating the oxidized surface substantively reduces the oxygen levels and strengthens the nickel d-band emission consistent with the dissolution of oxygen in the metal as previously

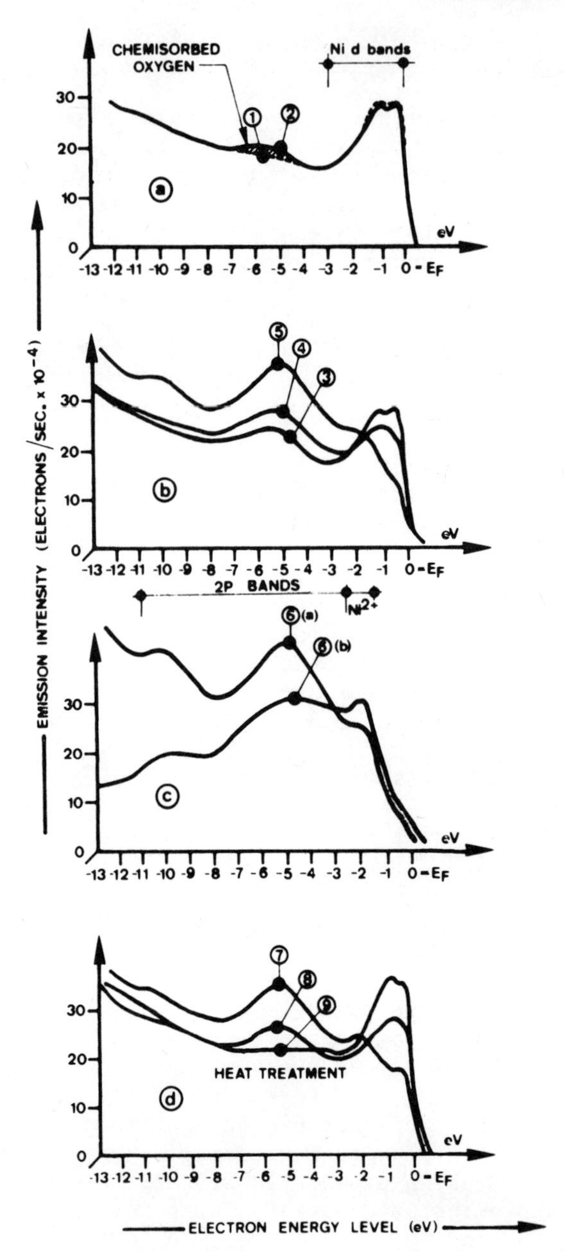

Figure 10. : Photoelectron emission spectra for adsorbed oxygen on nickel for various oxygen exposures (a), (b), (c) (one Langmuir (L) = 10^{-6} torr oxygen for 1 sec) and heat treatments (d) photon energy = 21.2 eV except for (6b) where hν = 41.8 eV (amplitude x 50) (After Eastman and Cashion, ref. 70)

reported from LEED observations[88]. More detailed UPS studies of the interaction of oxygen with clean surfaces of αFe(001) by Brucker and Rhodin[80] indicate clearly the different stages of the reaction as it progresses from simple superficial chemisorption, to two-dimensional oxide nucleation and then to bulk oxidation.

Interpretation of the photoemission spectra in terms of the surface electron structure of the clean, chemisorbed or chemically converted metal is always useful as a qualitative "finger-print". The basic objective, to identify the spectral peak structure in terms of the combined contributions of the molecular orbitals of the adsorbate and the spaced d-electron structure of the transition metal, is rendered difficult by the energy shifts, matrix effects and secondary excitations which complicate the spectra. In some special cases however such as the chemisorption of CO on polycrystalline Pd [89] and Ni [89] and on single crystal Ir(001)[77] it has been possible to index the peaks with some assurances in terms of the corresponding gas-phase spectra[90]. This is illustrated in Figure 11 from the UPS work of Brodén and Rhodin[77]. It is concluded that the peak A is mainly derived from the 4σ orbital of CO whereas the peaks B and C are derived from the 1σ and 5σ orbitals of CO respectively with an unknown component of d-electrons from the iridium orbitals. This interpretation is, of course, strong support for the surface molecule picture of chemisorption. Investigation of the photon excitation energy-dependence[89] of the spectra serves to reinforce these conclusions.

The final objective of interpretation of UPS-spectra is to obtain information on the nature of chemical bonding itself at the surface of the metal. It is of particular interest to monitor, for example, the rearrangement of chemical bonding of a hydrocarbon adsorbate on a catalytically-active surface. Photoemission studies of the dehydrogenation of chemisorbed ethylene and acetylene on Ni(111), first reported by Demuth and Eastman[91], indicated that the ethylene UPS-spectra assumed the appearance of the acetylene UPS-spectra upon heating. A similar study using photoemission by Brodén and Rhodin[79] on Ir(111) at 300 K show that the difference-spectra are quite different for the same two chemisorbed gases. Apparently, the dehydrogenated structure of ethylene on iridium represents some intermediate state between that of the two gases (see Figure 12). This is a good example of the use of UPS-spectra to give a "fingerprint" of the surface chemical bonding. The effect of temperature on the UPS-spectra of acetylene chemisorbed on Ir(001) is illustrated in Figure 13. It is quite unusual to observe a peak such as M shift to L, at a lower energy i.e. towards the Fermi level upon heating. The orbitals associated with this peak are thought to be the 3σ-ones joining the carbon atoms. One explanation is that the adsorbed acetylene molecule stretches with a decrease in bond energy when the temperature is increased. This is a tentative interpretation but the effect clearly indicates how UPS-spectra can be used to follow the electron changes associated with variation in chemical bon-

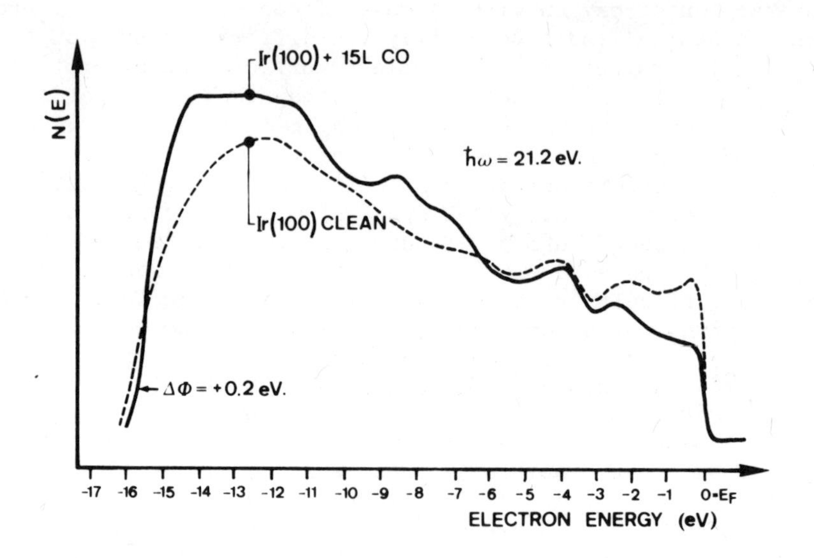

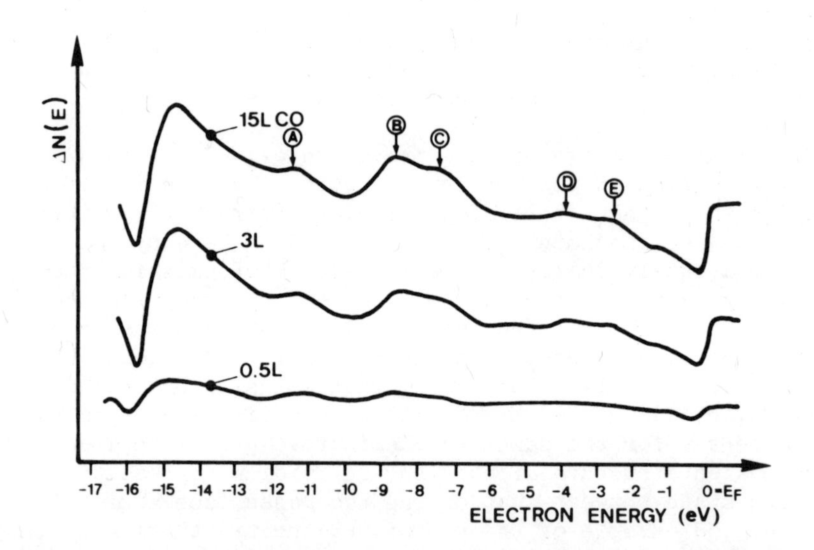

Figure 11. : UPS (photoemission) spectra for chemisorption of CO on Ir(001) as a function of exposure at 300 K. The actual spectra for the clean and chemisorbed surfaces (above) and the difference curves for different exposures (below). (From Brodén and Rhodin, ref. 77)

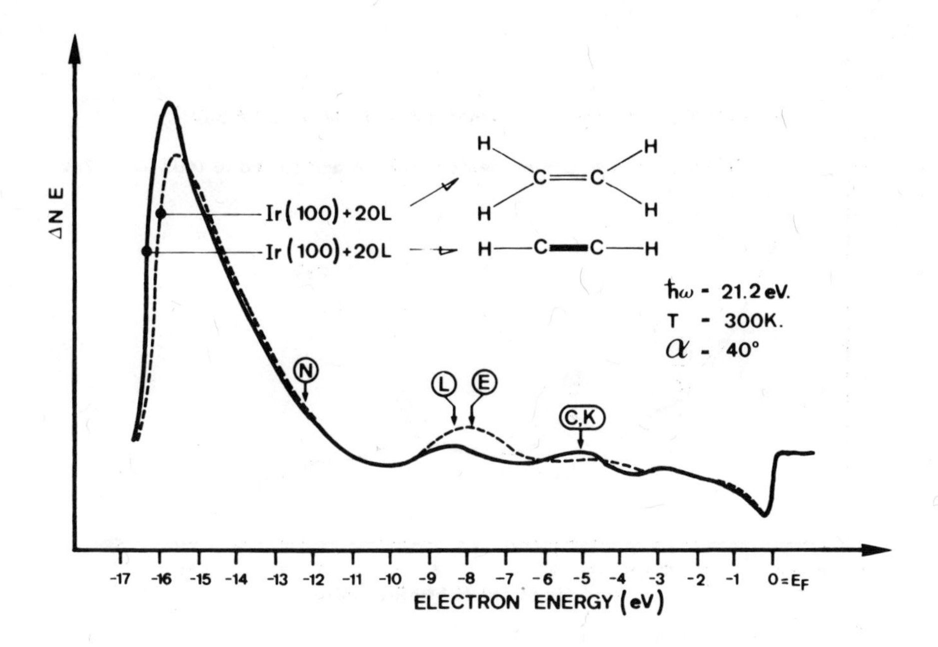

Figure 12. : UPS difference spectra of ethylene and acetylene chemisorbed on Ir(001) at 300 K. Note that the dehydrogenated ethylene spectra shows peaks at L and E not typical of the acetylene spectra and that the peaks C and K of the acetylene spectra are absent for the dehydrogenated ethylene. (After Brodén and Rhodin, ref. 79).

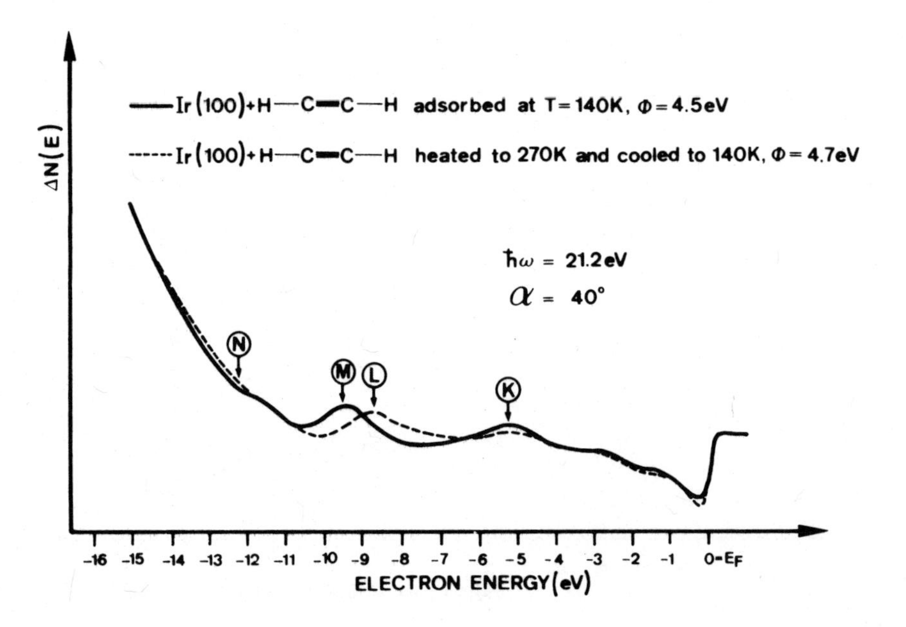

Figure 13. : UPS difference spectra of acetylene chemisorbed on Ir(001) showing effect of temperature on the peak positions. Note that the peak at M moves to a lower energy at L upon heating from 140 to 270 K. This may be attributed to a stretching of the double bond between the carbon atoms. (After Brodén and Rhodin, ref. 79).

ding of chemisorbed hydrocarbons.

In summary, these UPS studies demonstrate how the photoemission technique can be used to follow the changes in surface electron structure of the metal as it proceeds through chemisorption and chemical reaction stages. The present results although rather qualitative promise to provide more quantitative information on molecular orbitals and chemical bonding as the photoemission theory improves and as more systematic data become available.

5. Ion Neutralization Spectroscopy[39,57,92,94-96]

In addition to stimulating electron emission from surface atoms by high field vacuum-tunneling (FEED) and by ultraviolet photoexcitation (UPS), similar estimations of orbital energy spectra of electrons in chemisorption bonds have also been achieved by Hagstrum and Becker[93-96] using the previously described method of ion-neutralization spectroscopy (INS). Interpretation of the measurements from the ion-neutralization technique, may be complicated by the presence at the surface of the intense local field centered on the gas ion. In the INS technique, emission is achieved by generating an electron hole in a relatively low lying energy level in the conduction band of the metal through neutralization of the ion. A corresponding emission of an electron into the vacuum occurs through an Auger excitation process. The emission is very surface sensitive due to the dependence of the ion neutralization process on the detailed nature of the surface potential. It seems to be a more sensitive measurement of the adsorbate electron level structure than (UPS) because it selectively probes the outer surface. On the other hand, as previously indicated, interpretation in terms of electron structure of the adsorbed molecule is complicated by the nature of the excitation and emission processes. Hence, in comparison with (FEED) and (UPS), ion-neutralization spectroscopy (INS) is more difficult in its experimental implementation and mathematical interpretation. On the other hand, the uniqueness of (INS) makes it possible to probe the magnitude of the wave functions associated with bonding at the surface using local ion excitation without involving the use of high local electric fields or of ultraviolet photon excitation. Critical comparisons of UPS and INS appear to indicate that the two methods are evaluating closely related electron properties. An illustration of this is shown by the system c(2x2) oxygen on Ni(001). Spectra are given in Figure 14 from recent work of Hagstrum and Becker[93,95,96]. It is concluded that whereas electron structure in the subsurface region is accessible to UPS, ion-neutralization spectroscopy appears to show more sensitivity to bonding identified principally with the surface molecule. Comparison of the two methods in terms of the interpretation of adsorption behavior for chalcogens on nickel has been critically discussed[97].

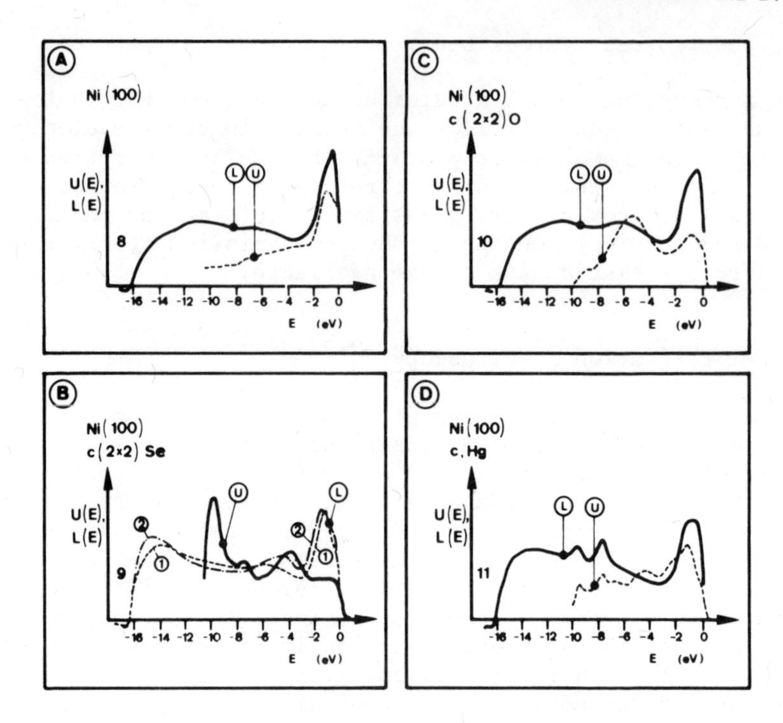

Figure 14. : Photoemission spectra (L) ($h\nu$ = 21.2 eV) compared to the ion neutralization unfold function (U) for the same clean surface and chemisorption systems indicating various significant similarities and differences for the two methods. (a) Atomically clean Ni(100).(Note extended range of UPS and somewhat higher sensitivity at very low energies corresponding to nickel d-band).(b) Ni(100) c(2x2)Se.(Note dependence of structure in UPS spectra on angle of incidence of photon beam (1) normal incidence and (2) 45° angle of incidence. In this case the electron photoemission is averaged over 2π sterradians. The increased structure in the INS spectra is associated with the greater sensitivity of the INS method to the chemisorbed selenium localized at the surface).(c) Ni(100) c(2x2)O.(Note essential similarity of spectra for both methods but enhanced sensitivity of INS for the O-p orbital at 5.5 eV).(d) Ni(100) containing carbon and mercury in the surface region but not necessarily localized at the surface.(Note that the different intensities for the UPS and INS spectra is partly associated with the location of the atoms with respect to the surface). (After Hagstrum and Becker, refs. 57 and 96)

As a general conclusion it is now established that these three particular methods (FEED, UPS and INS), in measuring specific features of the electronic state in similar but rather different ways at and near the surface region, provide unique and complementary information on the electronic details of adsorption on metals in terms of surface bond formation and the symmetry and orientation of adsorbed molecules. (For a more detailed comparison of the various spectroscopic methods see Hagstrum and McRae[92]).

There is no doubt that the continued refinement of experimental and theoretical methods for defining the quantum mechanical parameters describing the clean and chemisorbed metal surface will lead to more critical interpretations of the physics and chemistry of metals. This applies particularly to the application of electron spectroscopic methods and the concurrent development of more quantitative theories of surface electron structure and reactivity. Major advances in the field of electron properties and chemical behavior from the quantum mechanical viewpoint can be expected as these new approaches are exploited.

REFERENCES

(1) T.B. Grimley, Advances in Catalysis 12 (1960) 1
(2) T.B. Grimley, Proc. Phys. Soc. (London) 90 (1967) 751
(3) T.B. Grimley, Proc. Phys. Soc. (London) 92 (1967) 776
(4) T.B. Grimley, in : Molecular Processes on Solid Surfaces ; edited by E. Drauglis, R.D. Gretz, R.I. Jaffee, p. 299 (McGraw-Hill, 1969)
(5) D.M. Newns, Phys. Rev. 178 (1969) 1123
(6) N.D. Lang, Solid State Phys. 28 (1973) 225
(7) J. Horiuti and T. Toya, Solid State Surface Science 1 (1969)1
(8) K.F. Wojciechowski, Progress in Surface Science 1 (1971) 65
(9) T.B. Grimley, J. Vac. Sci. and Tech. 8 (1971) 31
(10) J.R. Schrieffer, J. Vac. Sci. and Tech. 9 (1972) 561
(11) J.W. Gadzuk, in : Surface Physics of Crystalline Materials, edited by J.M. Blakely (Academic Press, 1975)
(12) E. Drauglis and R.I. Jaffee, editors, The Physical Basis for Heterogeneous Catalysis, Proceedings of the Battelle Colloquium (Gstaad, Switzerland, 1974)
(13) R. Gomer, Chemisorption, Solid State Phys., to be published
(14) J.W. Gadzuk, Japan. J. Appl. Phys. Suppl. 2 (1974)
(15) A. van der Avoird, J. Chem. Phys. 47 (1967) 3649; Chem. Phys. Letters 1 (1967) 411
(16) L. Jansen, in ref. 4, p. 49
(17) K.H. Johnson and R.P. Messmer, J. Vac. Sci. and Tech. 11 (1974) 236; Surface Science 42 (1974) 341
(18) J.W. Gadzuk, Surface Science 43 (1974) 44
(19) J.W. Gadzuk, Surface Science 6 (1967) 133; Phys. Rev. B1 (1970) 2110

(20) D.M. Newns, J. Chem. Phys. 50 (1969) 4572; Phys. Rev. B1 (1970) 3304; Phys. Letters 38 (1972) 341
(21) T.B. Grimley and S.M. Walker, Surface Science 14 (1969) 395
(22) T.N. Rhodin and D.L. Adams, "Adsorption of Gases on Solids", Treatise on Solid State Chemistry, vol. 6, ed., N.B. Hannay, Plenum Press, New York (1975)
(23) J.R. Schrieffer and R. Gomer, Surface Science 25 (1971) 315
(24) R. Paulsen and J.R. Schrieffer, Surface Science (1975) in press
(25) T.B. Grimley and B.J. Thorpe, J. Phys. F 14 (1971)
(26) B.J. Thorpe, Surface Science 33 (1972) 306
(27) P.W. Anderson, Phys. Rev. 124 (1961) 41; in, Many Body Physics, edited by Dewitt and Balien (Gordon and Breach, 1969)
(28) D.N. Newns, Phys. Rev. 25 (1970) 1575
(29) J.W. Gadzuk, J.K. Hartman and T.N. Rhodin, Phys. Rev. B4 (1971) 241
(30) K.H. Johnson, Advances in Quantum Chemistry 7 (1973) 143
(31) K.H. Johnson, J.G. Norman and J.W.D. Connolly, in : Computational Methods for Large Molecules and Localized States in Solids, edited by Herman, McLean and Nesbit (Plenum, 1973)
(32) J.C. Slater and K.H. Jonhson, Physics Today 27 (1974) 34
(33) N.D. Lang and W. Kohn, Phys. Rev. B3 (1971) 1215
(34) J.R. Smith, S.C. Ying and W. Kohn, Phys. Rev. Letters 30 (1973) 610
(35) P. Hohenberg and W. Kohn, Phys. Rev. 136B (1964) 864
(36) W. Kohn and L.J. Sham, Phys. Rev. 140A (1965) 1133
(37) D.A. Shirley, editor, Electron Spectroscopy (North-Holland, 1972)
(38) D.E. Eastman, in : Techniques of Metals Research, Vol. 6, chapter 6, edited by R.P. Bunshah (Wiley, 1972) p. 411
(39) H.D. Eastman, in : Techniques of Metals Research, Vol. 6, chapter 4, edited by R.P. Bunshah (Wiley, 1972) p. 309
(40) C.B. Duke, Advances in Chem. Phys. to be published
(41) J.W. Gadzuk and E.W. Plummer, Rev. Mod. Phys. 45 (1973) 487
(42) E.W. Plummer, in : Topics in Applied Physics, edited by R. Gomer (Springer, 1975)
(43) T.N. Rhodin, in : Reactivity of Solids, Proceedings of the International Symposium, editors, J.W. Anderson, M.W. Roberts and F.S. Stone (Bristol, 1972) p. 651
(44) R. Gomer, Advances in Chemical Physics 27 (1974) 211
(45) R.L. Park and J.E. Houston, J. Vac. Sci. and Tech. 11 (1974)1
(46) A.M. Bradshaw, in : Surface and Defect Properties of Solids, vol. 3, chapter 5 (The Chemical Society, 1974)
(47) R.L. Park and J.E. Houston, Advances in X-Ray Analysis 15 (1972) 462
(48) C.C. Chang, Surface Science 25 (1971) 53
(49) N.J. Taylor, in : Techniques of Metals Research, vol. 7, edited by R.F. Bunshah (Interscience, 1972)
(50) J.C. Tracy, in : Electron Emission Spectroscopy, edited by W. DeKeyser, L. Fiermans, G. Vanderkelen and J. Vennik (Reidel, 1973)

(51) P.W. Palmberg, in : Electron Spectroscopy, edited by D. Shirley (North-Holland,1972)
(52) E.N. Sickafus, J. Vac. Sci. and Tech. 11 (1974) 299, and references therein
(53) K. Siegbahn, C. Nordling, A. Fahlman, R. Nordberg, K. Hamrin, J. Hedman, G. Johansson, T. Bergmark, S. Karlson, I. Lindgren, B. Lindberg, ESCA : Atomic,Molecular, and Solid State Structure Studies by Means of Electron Spectroscopy (Almqvist and Wiksells, 1967)
(54) C.R. Brundle, in : Surface and Defect Properties of Solids, vol. 1, chapter 6 (The Chemical Society, 1972)
(55) D.T. Clark, in ref. 50
(56) W.N. Delgass, T.R. Hughes and C.S. Fadley, Catalysis Rev. 4 (1971) 179
(57) H.D. Hagstrum, Science 178 (1972) 275
(58) E.W. Müller, Z. Physik 106 (1937) 541
(59) E.W. Müller, Z. Physik 131 (1951) 136
(60) E.W. Müller and T.T. Tsong, Field Ion Microscopy (Elsevier, 1969)
(61) C.B. Duke and M.E. Alferieff, J. Chem. Phys. 46 (1967) 923
(62) E.W. Plummer, J.W. Gadzuk and R.D. Young, Solid State Comm. 7 (1969) 487
(63) E.W. Plummer and R.D. Young, Phys. Rev. B1 (1970) 2088
(64) J.W. Gadzuk, Phys. Rev. 182 (1969) 945
(65) L.W. Swanson and A.E. Bell, Advances in Electronics and Electron Physics 32 (1973) 194
(66) N.J. Dionne and T.N. Rhodin, Phys. Rev. Letters 32 (1974) 1311; Bull. Am. Phys. Soc. 20 (1075) 359; N.J. Dionne, Ph.D. Thesis, 1974
(67) R.L. Billingston and T.N. Rhodin, Bull. Am. Phys. Soc. 20 (1975) 359
(68) D. Penn and E.W. Plummer, Phys. Rev. B9 (1974) 1216
(69) W.E. Spicer, Comments in Solid State Physics 5 (1973) 105
(70) D.E. Eastman and J.K. Cashion, Phys. Rev. Letters 27 (1971) 1520
(71) J.E. Demuth and D.E. Eastman, Phys. Rev. Letters 32 (1974)1123
(72) W.F. Egelhoff, J.W. Linnett and D.L. Perry, Disc. Farad. Soc. 58 (1974)
(73) W.F. Egelhoff and D.L. Perry, Phys. Rev. Letters 34 (1975) 93
(74) B. Feuerbacher and B. Fitton, Phys. Rev. Letters 29 (1972) 786
(75) B. Feuerbacher and B. Fitton, Phys. Rev. B8 (1973) 4890
(76) B. Feuerbacher and M.R. Adrianes, Surface Science 45 (1974)553
(77) G. Brodén and T.N. Rhodin, Solid State Comm. (to be published) 1975, vol. 17
(78) G. Brodén and T.N. Rhodin, Disc. Farad. Soc. vol. 60 (1975) in press
(79) G. Brodén and T.N. Rhodin, Surface Science (to be publised)
(80) C.A. Brucker and T.N. Rhodin, Bull. Am. Phys. Soc. 20 (1975) 304

(81) J.E.Rowe and H. Ibach, Phys. Rev. Letters 31 (1973) 102
(82) J.E. Rowe and H. Ibach, Phys. Rev. Letters 32 (1974) 421
(83) M.M. Taum, N.V. Smith and F.S. DiSalvo, Phys. Rev. Letters, 32 (1974) 1241
(84) T. Gustafsson, An experimental Investigation of the Electronic Structure of Silver and Berylium by Measurement of Photoelectron Spectra Including Variations with Emission Angle, Ph.D. Thesis, Chalmers University of Technology, Goteborg, 1973
(85) J.W. Gadzuk, Phys. Rev. B10 (1974) 5030
(86) A. Liebsch, Phys. Rev. Letters 32 (1974) 1203
(87) A. Liebsch and E.W. Plummer, Disc. Farad. Soc. 58 (1974)
(88) J.W. May, Ind. Eng. Chem. 57 (1965) 19; Advances in Catalysis 21 (1970) 244
(89) T. Gustaffson, E.W. Plummer, E. Eastman and J. Freeoff, Bull. Am. Phys. Soc. 20 (1975) 304 (abstract)
(90) D.W. Turner, Molecular Photoelectron Spectroscopy (Wiley, 1970)
(91) B. Politzer and P.H. Cutler, Phys. Rev. Lett. 28 (1972) 1330
(92) J.E. Demuth and D. Eastman, Phys. Rev. Letters 32 (1974) 1123
(93) H.D. Hagstrum and E. McRae, Treatise on Solid State Chemistry, edited by N.B. Hannay, vol. 6, Plenum Press, New York (1975)
(94) H.D. Hagstrum, Phys. Rev. 150 (1966) 495
(95) H.D. Hagstrum and G.E. Becker, Phys. Rev. 159 (1967) 572; Phys. Rev. B4 (1971) 4187
(96) H.D. Hagstrum and G.E. Becker, J. Chem. Phys. 54 (1971) 1015
(97) G.E. Becker and H.D. Hagstrum, Surface Science 30 (1972) 505
(98) H.D. Hagstrum, J. Vac. Sci. and Tech. 10 (1973) 264

APPROACH TO TWO PROTOTYPE METAL ADSORPTION SYSTEMS

Thor RHODIN[+]
Cornell University
Ithaca, New York, U.S.A. 14853
and
David ADAMS
Xerox Research Laboratory
Webster, New York, U.S.A. 14644

INTRODUCTION

The subject of metal surface properties has long been of great interest to scientists and a corresponding wealth of experimental and analytical information has been accumulated. In addition, it is clear from the use of more modern theoretical and experimental approaches discussed at this meeting that a great deal of additional information on atomistic and microscopic properties is now potentially available. It is instructive therefore to examine critically the degree to which a typical metal-gas interface system is understood in terms of basic physical and chemical principles. The conclusion is that there is no single such system about which we can claim a basic general understanding. Tungsten surfaces are by far the most extensively studied partly because of their refractory nature and partly because of historical precedent but serious gaps persist in the area of defining critical atomistic and microscopic properties. More recently, considerable useful information on the electron and atomic features of single crystal nickel surfaces has been gained. Although the basic properties of either metal-gas system are far from being adequately understood, it is instructive to discuss these two examples in some detail from the viewpoint of the

+ Support from National Science Foundation Grant No. DMR71-01769 A02, from the Advanced Research Projects Agency through the Cornell Materials Science Center and from the NATO Advanced Study Institute is gratefully acknowledged (TNR).

relevant concepts, models, and methods presented so far at this meeting. This presentation will indicate our present state of understanding in terms both of its achievements and its inadequacies. (Note : This material is drawn to a large extent from a more comprehensive treatment of the subject "Adsorption of Gases on Solids" by Rhodin and Adams[1]).

A. ADSORPTION OF DIATOMIC GASES (H_2, O_2, N_2, CO) on TUNGSTEN[1-17]

The relative ease of preparation of clean tungsten surfaces established[2-5] in recent years by Auger spectroscopy, enables contact to be made with a large body of previous work. Sticking probabilities for H_2,N_2,O_2 and CO are generally quite high (> 0.1), permitting adsorption experiments to be carried out under well controlled conditions. The major recent experimental studies have been the characterization of adsorption binding states, their kinetics of formation and desorption and binding energies and the influence of adsorbent crystal structure upon these properties. In the last few years low-energy electron-diffraction, field-emission energy distribution, UV photoemission and ESCA experiments have revealed interesting correlations between measurements of kinetics and surface structural and electronic properties.

1. Kinetics of Desorption and Binding States

The early thermal desorption experiments on polycrystalline tungsten[6-14] led to the conclusion that hydrogen, nitrogen and oxygen are dissociatively adsorbed at room temperature, but that carbon monoxide adsorbs without dissociation. In the cases of hydrogen and nitrogen, this followed from the observation of second-order desorption kinetics.

Recent thermal desorption experiments on single-crystal planes of tungsten[15-33] have revealed ambiguities in the interpretation of results from polycrystalline samples. Multiple binding states are found to occur on individual crystal planes within both the weakly bound (γ or α) and the strongly bound (β) categories respectively in the usual thermal desorption nomenclature. In addition, correlation between LEED and thermal desorption results has led to suggestions[19,34-37] that multiple binding states may result in part from lateral interactions between the adsorbed species rather than from distinct binding states. The consequent uncertainty concerning the mechanisms of desorption precludes the use of kinetic analysis of thermal desorption spectra in unambiguously deducing the state of association of the adsorbed species[36]. An additional complication in this respect is the possible occurrence of thermally induced conversion among binding states, generally believed to occur for adsorbed carbon monoxide[27,38].

A fairly clear distinction between the weakly bound (γ or α) and strongly bound (β) states is suggested by the results of isotopic mixing[15,25,39,40], infra-red absorption[41,42], electron-impact desorption[37,43,44], UV-photoemission[44-45], and ESCA[46,47] experiments. The absence of isotopic mixing in the α and γ states of hydrogen, nitrogen and carbon monoxide provides strong evidence that these states are essentially molecular in nature. This conclusion appears to be confirmed by more recent experiments using the above-mentioned techniques. The occurrence of isotopic mixing in the β-states of hydrogen and nitrogen has been attributed[15,25,40] to dissociatively adsorbed species.

In their pioneering study, Propost and Piper[48], using electron energy-loss spectroscopy observed characteristic energy losses upon adsorption of H_2,N_2 and CO upon W(100). These losses were attributed to vibrational excitation of the bonds between the adsorbed species and tungsten surface atoms. An energy level at 258 meV, quite close to the first vibrational level of free CO at 270 meV, could be associated with a weakly bound state which desorbed at room temperature. Propst and Piper[48] concluded that H_2 and N_2 were dissociatively adsorbed and that the CO bond was very much weakened,if not in fact broken by stimultaneous bond formation by C and O atoms to surface W atoms.

Adsorption of H_2 on W(100) produces a thermal desorption spectrum[16,25] (Figure 1) containing two peaks usually indentified as β_1 and β_2, which desorb with approximatively first and second order desorption kinetics, respectively. From observations of the vibrational fine structure in field-emission energy distribution, Plummer and Bell[49] have concluded that both β_1-and β_2-hydrogen on W(100) are derived from atomically bound species. Vibrational excitations characteristic of a molecularly bound species were observed for hydrogen adsorbed at cryogenic temperatures on W(111).

In summary, the available evidence at this time indicates that the weakly bound γ-and α-states of H_2,N_2 and CO are moleular in nature, whereas the strongly bound β-states of H_2,N_2 and O_2 result from dissociative adsorption. In the case of the strongly bound β-CO states, isotopic mixing studies[39] and vibrational spectra[48,49] appear to provide the strongest evidence in favor of dissociative adsorption. Comparative studies of CO and carbon and oxygen adsorption by UPS[45,87] and x-ray-photoemission[47] and by field-emission energy distribution[51] measurements have been interpreted in terms of considerable weakening of the CO bond but dissociation does not appear to be established. The exact origin of the multiplicity of binding states within the weakly bound and strongly bound categories is obscure. Distinction among different modes of bonding, effects of adsorbate-adsorbate interactions and thermally induced bonding conversions remains an area of active interest.

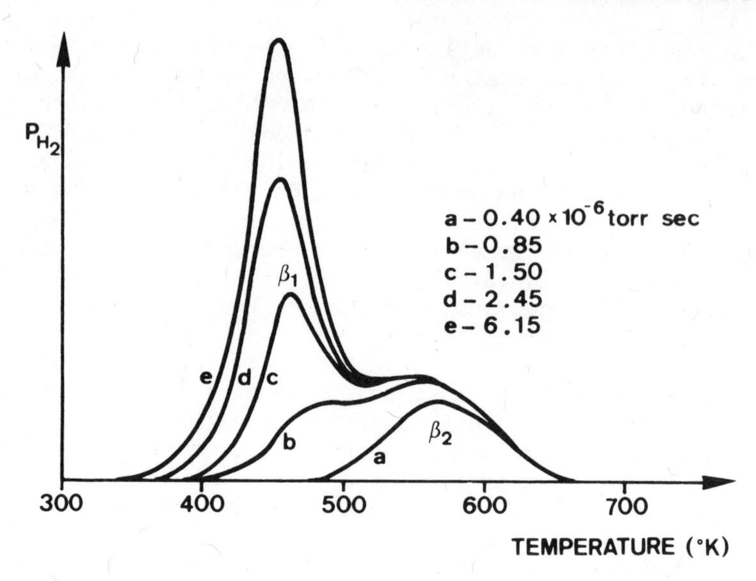

Figure 1 : Thermal desorption spectra obtained after different initial hydrogen exposures on W(100). The β_2-peak alone is observed in spectra taken after small exposures. With increasing exposure the β_1-peak at lower temperature develops (From Adams and Germer, ref. 17)

2. Kinetics of Adsorption

The adsorption kinetics for hydrogen, nitrogen, carbon monoxide and oxygen on tungsten are usually discussed in terms of the sticking probability for adsorption. (This can be defined as the ratio of the rates of adsorption and impingement). Such a concept is meaningful for irreversible adsorption in which the lifetime of the chemisorbed state is very long and the lifetime of any weakly adsorbed precursor state is very short compared to the duration of the experiment. For many combinations of adsorbate and crystal plane, the sticking probability is found to be independent of coverage, at least in the low coverage region. The implication is that the adsorption mechanism includes as a first step, formation of a weakly bound precursor state at either a filled or an empty site, followed by either migration of the precursor to an empty site or reflection of the adatom to the gas phase. For H_2-adsorption on W(100), W(111) and W(211), the sticking probability is independent of coverage at low coverage and the zero-coverage sticking probability is independent of adsorbent temperature. Tamm and Schmidt[86] have concluded from these observations that accomodation into the precursor state is determined by the energetics of bond formation rather than through phonon-assisted thermal accomodation.

Absolute values for sticking probabilities are subject to considerable experimental error. Accordingly, detailed comparisons of sticking probabilities for the different gases as a function of tungsten crystal plane must be made with caution. It is evident from the available data that large anisotropies exist for H_2- and N_2-adsorption, whereas CO- and O_2-adsorption appears to be less dependent upon crystal plane with sticking probabilities close to unity. A general explanation of the effect of crystal plane has been proposed for the case of N_2-adsorption in terms of a correlation between sticking probability and surface structure by Adams and Germer[21]. Adsorption on tungsten planes in the region of the stereographic triangle bounded by the [001], [$\bar{1}$11] and [01$\bar{1}$] zones occurs with relatively high sticking probability (> 0.1), whereas on the remaining planes sticking probabilities are low (< 10^{-2}), as shown in Figure 2. This correlation exists both for results on macroscopic single crystal planes and for FIM observations[52,53]. The reactive planes contain surface sites of a 4-fold symmetry, characteristic of those on the (100) plane[21]. The less reactive planes contain only the more close-packed sites as found on the (110) plane and shown in Figure 3. Since nitrogen adsorption from NH_3[54] for example does not involve similar anisotropies, the implication appears to be that dissociation of the N_2 molecule occurs more readily at the more open sites characteristic of the (100) plane.

3. Surface Structure

It is generally assumed that H_2-, N_2- and CO adsorption on tungsten occurs without geometrical rearrangement (reconstruction) of the metal surface. In the case of O_2-adsorption there seems to be little doubt that reconstruction and formation of surface oxides occur under specific conditions of coverage and heat treatment.

Quantitative surface structural analyses have not been performed on these systems however and the available LEED and FIM evidence is subject to conflicting interpretations. For example, Holscher and Sachtler[55] propose on the basis of FIM studies that adsorption of nitrogen and carbon monoxide, but not H_2, causes reconstruction of tungsten surfaces, whereas Ehrlich[6,58] has interpreted similar evidence in terms of adsorption without reconstruction.

Adsorption of nitrogen or carbon monoxide at room temperature produces an immobile layer without long range order. After heat treatment at appropriate coverages however, a variety of ordered structures are observed. In general it appears that the annealing temperatures required to produce long range order corresponds to the temperature for onset of surface mobility in agreement with earlier FEM experiments. It is frequently observed that heat treatment of

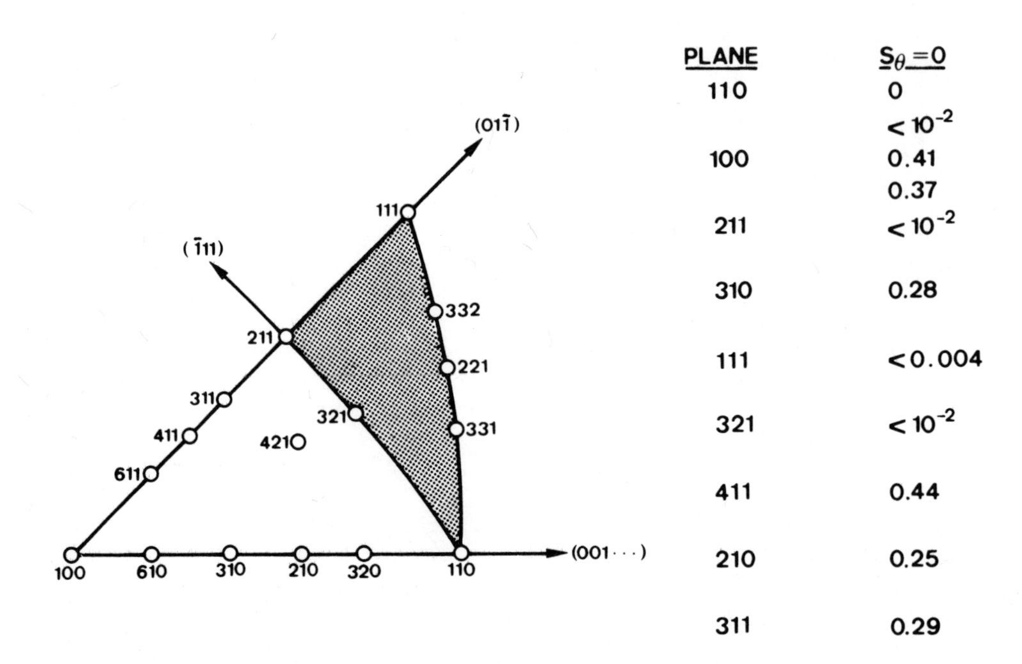

STEREOGRAPHIC UNIT TRIANGLE

Figure 2 : Stereographic plot of the normals to various planes in a cubic structure. The shades area indicates locations of tungsten planes which are predicted to be unreactive to nitrogen at 300 K. The sticking coefficient extrapolated to zero coverage for various crystallographic directions is indicated on the right (From Adams and Germer, ref. 21).

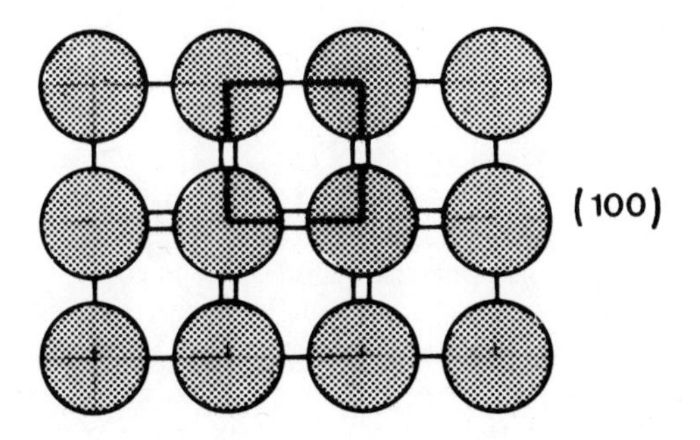

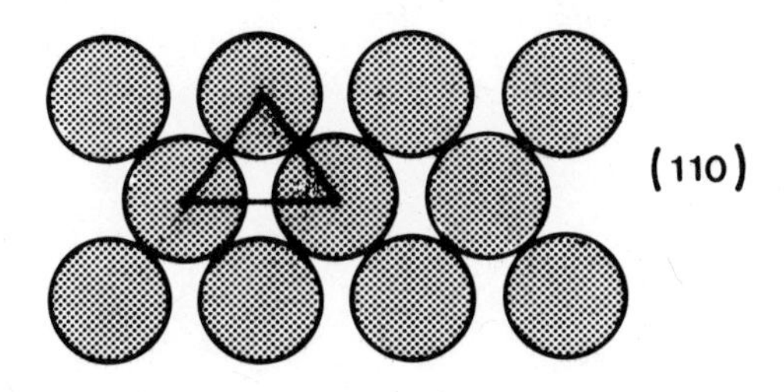

Figure 3. : Schematic models of the (100) and (110) planes of the bcc lattice. The high symmetry sites characteristic of each plane are indicated.

the adsorbed layer at constant coverage is accompanied by a decrease in work function. This has been variously attributed to,

a) reconstruction of the metal surface
b) in the case of carbon monoxide conversion from a "linear" to a "lying-down" bonding configuration[57]
c) conversion from a disordered to an ordered structure[19,21].

Complex LEED patterns are formed by CO-adsorption on W(110)[23] and (211)[24]. A reasonable explanation is that coincidence lattice formation occurs between an unreconstructed metal surface and a 2-D CO layer of larger unit mesh than the metal plane. This can result from repulsive interactions between CO molecules (or separate C and O atoms) along the close-packed <111> directions in the W(110) and (211) surfaces. This latter explanation is consistent with CO adsorption behavior on the W(100)[20], (310)[60] and (210)[19] planes which contain the more open <100> directions. At CO coverages less than one half monolayer, simple surface structures are observed with periodicity twice that of the metal plane along the <100> directions.

At higher coverages the double-spaced structures are succeeded by structures having the same unit mesh as the metal surface, except in the case of W(100) where more complex structures have been observed[60]. The observations have been interpreted[19] in terms of sequential formation of double-spaced and single-spaced structures controlled by the influence of repulsive adsorbate-adsorbate interactions in which CO molecules occupy the sites of 4-fold symmetry along the <100> rows on an unreconstructed surface.

LEED observations for H_2-and N_2-adsorption on tungsten planes tend to support the hypothesis that adsorption of these gases as well as that of CO occurs without reconstruction of the metal surface. For example, all three gases produce a c(2x2) structure at about half monolayer coverage on W(100)[17,18,20,61-63] and both nitrogen and carbon monoxide produce simple p(2x1) structures at about half monolayer coverage on W(310) and (210)[19,60]. This similarity in behavior for all three adsorbates and the simple periodicities observed tends to argue against the formation of unique surface compounds, at least for the planes of the <100> zone.

4. Chemical Bond Formation

A detailed picture of the chemical bonds formed by hydrogen, nitrogen, carbon monoxide and oxygen with tungsten surfaces requires as a minimum first stage, a quantitative understanding of the molecular constitution and geometrical arrangement of the adsorbed species in their various binding states. This is not currently available. However, some useful qualitative conclusions can be drawn.

The binding energies of the adsorbed species even in the most weakly bound states are indicative of chemical rather than physical bond formation. The changes in work function accompanying adsorption are indicative of covalent bond formation with net charge transfer between adsorbate and adsorbent of less than 0.1 electron charge per adsorbed atom. The vibrational spectra for the strongly bound β-states suggest that the local adsorbed complexes involve bridge-bonding with more than one tungsten atom in the surface[48]. In the case of carbon monoxide, bond formation by both the carbon and oxygen atoms to the surface is indicated. Infrared and vibrational spectra for the weakly bound α-CO states are consistent with single bond formation between a single tungsten atom and the carbon atom of a CO molecule, with the CO bond axis normal to the surface[44]. An additional α-CO state involving bonding between the carbon atom and two tungsten atoms with consequent weakening of the CO bond has been suggested by Yates and King[44] on the basis of their infrared and electron impact desorption studies[42,44] and of vibrational spectra by Propst and Piper[48].

The adsorption of hydrogen, nitrogen, carbon monoxide and oxygen on tungsten has also been the subject of a number of recent FEED[49,51,64,65] and photoemission[44,45,47,66-69] investigations.

Interpretation of adsorption-induced spectral changes often involves attempts to identify structure in the difference(UPS-spectra)or enhancement (FEED-spectra) curves in terms of the electron energy levels of the free atomic or molecular state of the adsorbate. This approach has been used successfully by Plummer and Young[70] in their FEED study of alkaline earth metal adsorption on tungsten and by Demuth and Eastman[71] in their UPS study of physical and chemisorption of hydrocarbons on nickel. This approach is useful but limited however for the adsorption systems discussed in this section because strong chemical bond formation is expected to produce large shifts and broadening of adsorbate levels, together with pronounced changes in the local density of states at adsorbent surface atoms.

A further complication which may lead to a more detailed understanding of surface electronic structure[72,73] is the occurrence of strong directional effects in both photoelectron emission as well as vacuum-tunneling spectroscopy. Thus although attempts have been made to interpret difference spectra in terms of broadened and shifted atomic levels, interpretation of spectra-changes is limited.

In the case of H_2-adsorption on W(100),FEED measurements by Plummer and Bell[49] have shown that with increasing hydrogen coverage, emission from the surface resonance[65,66,69] at 0.4 eV below the Fermi level is attenuated. New levels associated with adsorbed hydrogen are found at 0.9 eV and 1.1 eV below the Fermi level. These levels form in the regime of coverage where thermal desorption experiments show the occurrence of the β_2-peak alone and where LEED results indicate the development of a c(2x2) surface structure[17,62,74]. The half order LEED beams, characteristic of this struture first reach a maximum intensity and then split into four new beams which move apart continuously and decay in intensity with increasing coverage. In the regime of higher coverage, desorption spectra show the additional development of the β_1-peak. Correlated with these changes in the LEED pattern and desorption spectra, the characteristic levels of the β_2-state decay in intensity to be replaced by a broad level centered at about 2.5 eV below the Fermi level. The original FEED observations[49] have been confirmed by more recent UPS measurements[66-69,75], shown in Figure 4. Plummer and Bell[49] have concluded that the β_1- and β_2-desorption peaks do not correspond to distinct binding states which populate sequentially with increasing coverage. Feuerbacher and Fitton[67] agree with this conclusion and further suggest that two distinct binding states are not present at saturation coverage of hydrogen. A coverage-dependent change in adsorbate binding occurs in this system, is probably related to adsorbate-adsorbate interactions[35,36], however, the possibility of co-existing and mutually perturbing bonding types at high coverage cannot be ruled out.

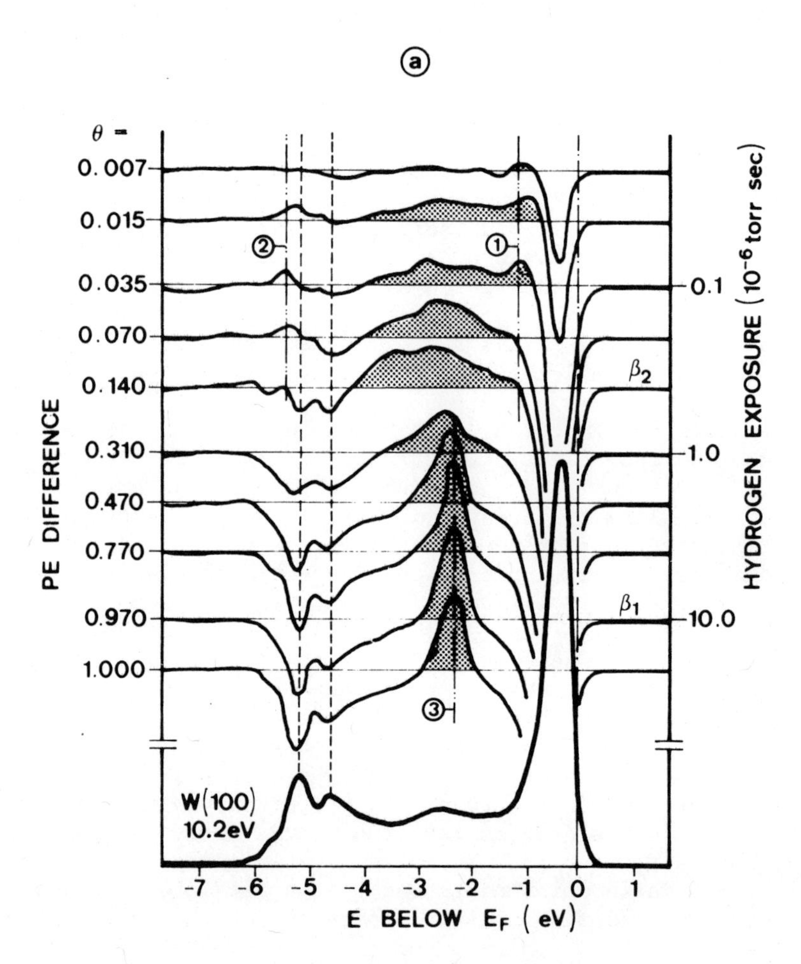

Figure 4. : Photoemission difference spectra for hydrogen absorption on W(100). The bottom curve shows the energy distribution spectrum of the photoelectrons emitted normal to the clean (100) face. θ gives the fractional coverage corresponding to the difference spectra. Photon energy 10.2 eV. (From Feuerbacher and Fitton, Ref. 66-68)

In the case of H_2-adsorption on W(100), Feuerbacher and Fitton[67] have reported sequential changes in UV-photoemission spectra with increasing hydrogen coverage as shown in Figure 5. These results can be correlated with the sequential occurrence of two thermal desorption peaks as observed by Tamm and Schmidt[15]. LEED observations have not revealed the occurrence of new diffraction features upon H_2-adsorption. It had been concluded[15,76] that coverage dependent effects found for H_2-adsorption on W(110) result from adsorbate-adsorbate interactions involving a single type of bonding.

Changes in work function with coverage have also been used in attempts to elucidate the origins of multiple peak desorption spectra. Barford and Rye[33] have suggested that multiple peak hydrogen desorption spectra result from multiple distinct bonding modes on W(211) and (111) but from lateral interactions on W(100) and (110). These conclusions are based on observations[33] of $\Delta\phi$ versus hydrogen coverage shown in Figure 6. The occurrence of distinct dipole moments of opposite sign is clearly a possible interpretation of the plot for W(211). The existence of two different states of adsorbed hydrogen can be correlated with double peak desorption spectra found for this plane.

Oxygen adsorption on W(110) has been studied using UV-photoemission by Baker and Eastman[87]. They reported that spectra taken after exposing the surface to oxygen at 1000°K, under conditions where a complex LEED pattern[77] forms are very similar to spectra obtained after saturation exposure at room temperature. On the assumption that the complex LEED pattern indicates that reconstruction of the surface has occurred it seems probable that saturation exposure at room temperature also produces reconstruction.

The interpretations of UV-photoemission results for CO-adsorption on W(100) illustrates the ambiguities involved in relating features in the difference spectra directly to known electron energy levels of the gas phase species. Baker and Eastman[87] have noted the similarities in difference spectra for adsorbed carbon monoxide and oxygen respectively, which they have suggested are consistent with dissociative adsorption of carbon monoxide. Waclawski and Plummer in contrast, have noted similarities between spectra for carbon monoxide adsorbed on W(100) and W(110) with spectra for $W(CO)_6$, presumably suggestive of a nondissociative adsorption of CO. It appears likely in view of all the measurements that the high temperature β-CO peak is characterized by CO-dissociation.

B. ADSORPTION OF CHALCOGENS (O_2,S,Se,Te) ON NICKEL[78-123]

A discussion of the adsorption of the Group 6 chalcogen elements upon nickel is particularly appropriate for this review since

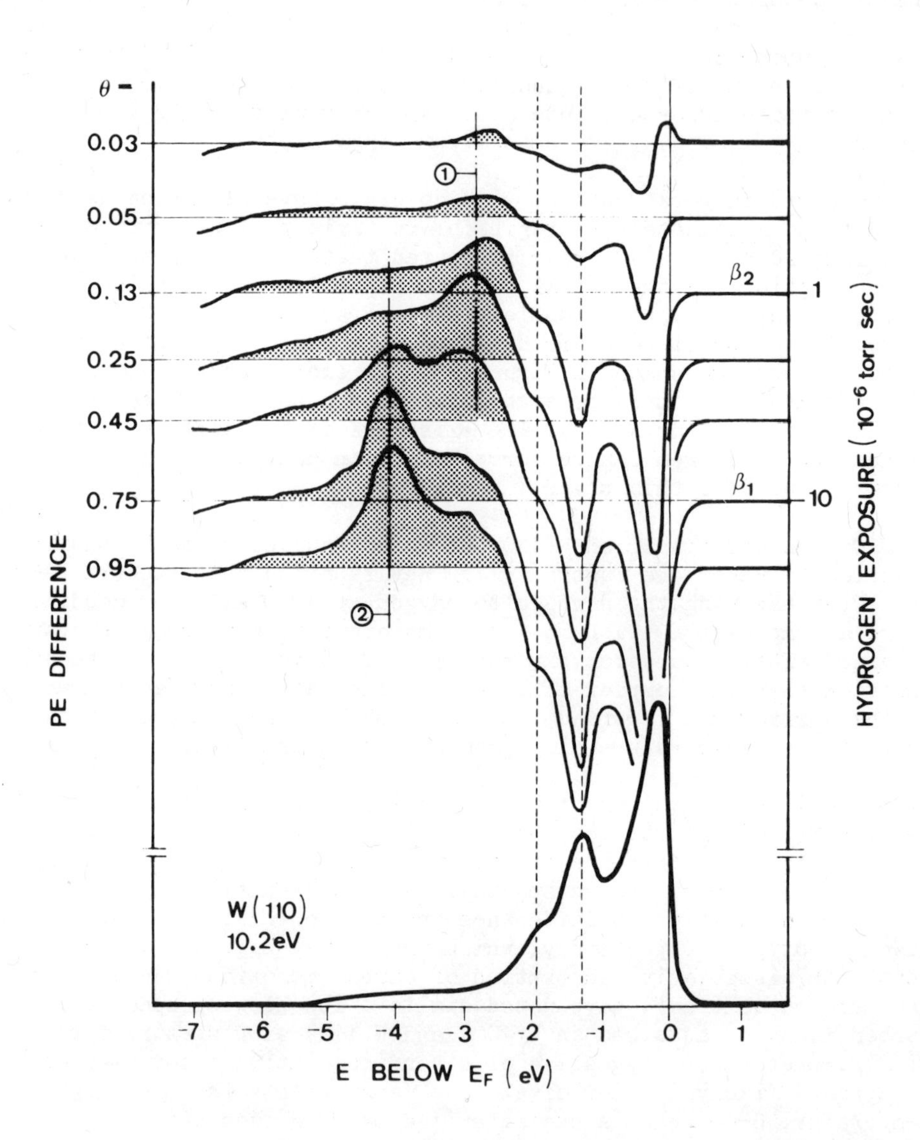

Figure 5. : Photoemission difference spectra for hydrogen adsorption on W(110). (From Feuerbacher and Fitton, refs. 66-68)

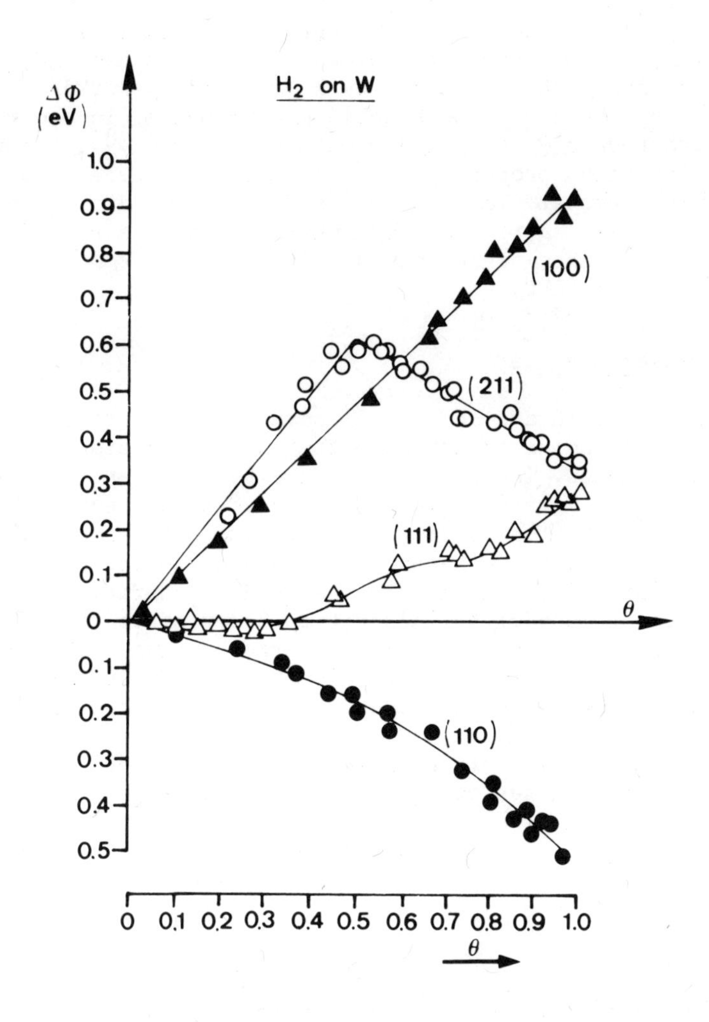

Figure 6. : Variation in change of work function, $\Delta\phi$ with coverage, θ for hydrogen adsorption on the close-packed planes of tungsten (From Barford and Rye, ref. 33)

reliable experimental information is available concerning both the structural and electronic properties of clean nickel surfaces containing chalcogen adsorbates. Comprehensive studies of oxygen adsorption on nickel have been made recently by Hagstrum and Becker using ion-neutralization spectroscopy, photoemission spectroscopy and LEED; by Demuth and Rhodin[78] using LEED, Auger spectroscopy and work function measurements; by Holloway and Hudson[79-80] using Auger spectroscopy and LEED, and by Horgan and co-workers[81,82] using Auger spectroscopy and kinetic measurements. In the work of Hagstrum and Becker[83-85], and Demuth and Rhodin[78], the variations in adsorption properties of the chalcogens were also studied on particular crystal planes. Studies of sulphur adsorption on nickel have also been carried out by Oudar and co-workers[88-93] using LEED, Auger and radio-tracer measurements, and by Edmonds and co-workers[94] using LEED.

1. Clean Nickel Surfaces

Recently Demuth and Rhodin[95] have derived surface interlayer spacings corresponding to not more than 2-3% expansion for all three close-packed planes. These analyses were based on an identification of the Bragg peak positions in intensity-voltage spectra, assuming a dominant contribution of single-scattering processes to the diffracted beam intensities. Recent multiple-scattering calculations (96-100) of intensity-voltage spectra tend to confirm that the nickel surfaces subject of the adsorption experiments to be described do not differ significantly from the corresponding bulk structure. Comparison between theory and experiment for the surface structure analysis of clean low index-planes of nickel is shown in Figure 7.

The preparation of clean nickel surfaces is more difficult than is the case for tungsten owing to the relative ease of solution and surface segregation of common impurities such as oxygen, carbon and sulphur. Procedures have been developed[82,95,101] however, which when used in conjunction with Auger spectroscopy to monitor the surface chemical composition, allow clean surfaces to be prepared quite routinely.

2. Kinetics at Adsorption of Oxygen Upon Nickel

Early measurements[102] of kinetics and change in work function led to the conclusion that the interaction between oxygen and nickel occurred in three distinct stages; rapid chemisorption, surface oxide formation, and slow growth of bulk oxide. This qualitative description has been substantiated by the results of recent investigations[78-82] which also provide additional quantitative information regarding the first two stages of the oxidation process. The different stages of interaction have been discussed by Horgan and King[81], by Holloway and Hudson[79,80] and by Demuth and Rhodin[78]

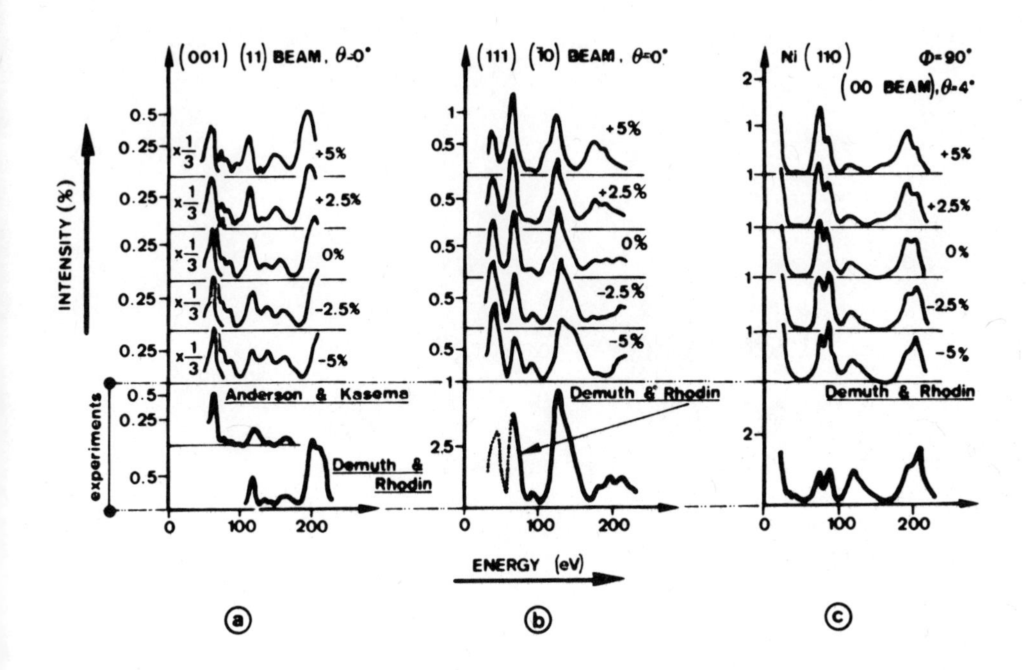

Figure 7. : Comparison of experimental measurements on clean surfaces by Andersson and Kasemo[124], Park and Farnsworth[125] and Demuth and Rhodin[95] to calculations by Demuth, Jepsen and Marcus[104,117] for the low index planes of nickel and the beams and diffraction observations as indicated. Relative intensities are on the ordinates. Percentage relaxation in the first layer is indicated for five possibilities in each case. Within experimental uncertainties of the comparison no significant relaxation is assumed to be indicated.

based on correlated observations of the change in sticking probability, surface structure and work function with oxygen coverage. The variation in sticking probability with coverage reported for oxygen adsorption on polycrystalline nickel films by Horgan and King[81] is shown in Figure 8. A similar dependence upon coverage has been found for the (100) and (111) planes by Holloway and Hudson[79,80]. The stages of interaction are discussed below.

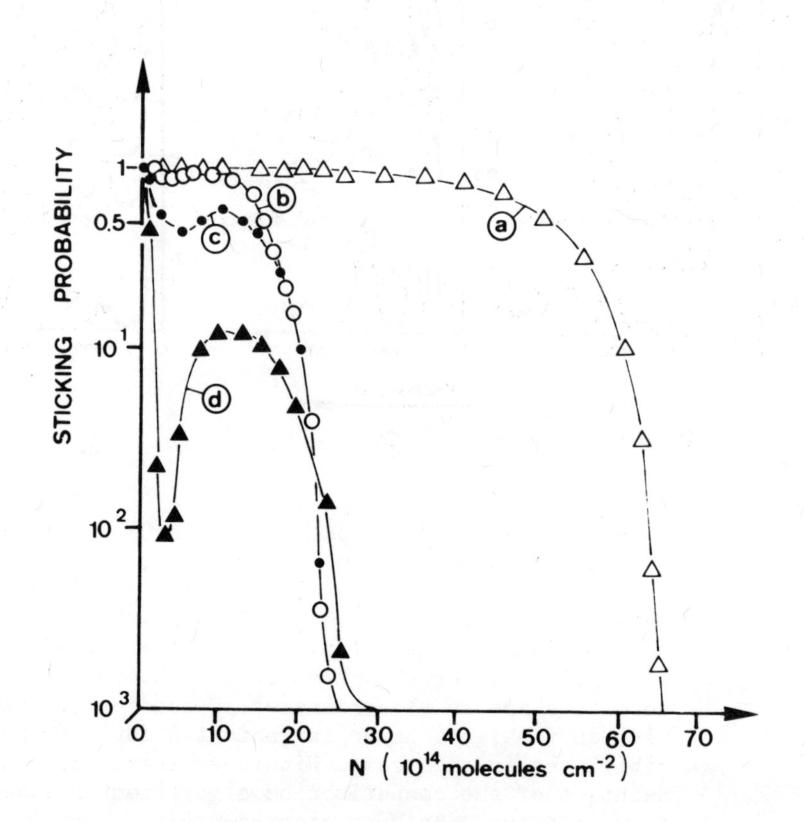

Figure 8. : Variation of sticking probability S with oxygen coverage N at
a) 77°K
b) 195°K
c) 290°K
d) 373°K
for oxygen adsorption on an evaporated nickel film (From Horgan and King, ref. 82)

CHEMISORPTION

In the first stage, chemisorption of oxygen is associated with the formation at room temperature of ordered surface structures having a simple relation to the underlying nickel plane. The work function increases with coverage. The sticking probability (Figure 8) decreases from unity by one or two orders of magnitude (depending upon the adsorption temperature) at room temperature or above, as the coverage is increased from zero to about half monolayer. The sticking probability decreases linearly with coverage in this region, consistent either with Langmuir adsorption of molecular oxygen[81,82] or more probably with dissociative adsorption with kinetics controlled by the growth of monolayer islands of the chemisorbed layer. Questions regarding the molecular constitution and detailed atomic arrangement in the surface region are not yet resolved.

SURFACE OXIDE NUCLEI FORMATION

The relatively simple structures observed in the chemisorption stage are superseded with increasing oxygen coverage by island growth of two-dimensional structures, associated with limited reconstruction of the surface. The work function decreases to a value below that of the clean surface due to incorporation of oxygen into the metal lattice. In this stage the sticking probability increases to a broad maximum then decreases with further coverage. In this stage the kinetics are[79,80,106] controlled by nucleation and growth of oxide islands, one or two layers thick. The initial increase in sticking probability with coverage in this stage, after the minimum in Figure 8, suggests that the availability of adsorption sites at island edges is rate-controlling[79,80]; thus the process is "auto-catalytic". At higher coverage, the decrease in sticking probability with coverage suggests that the increasing area occupied by oxide islands renders desorption as a precursor state before its entrapment at an island edge increasingly likely.

SLOW BULK OXIDE FORMATION

After the surface is covered by a very thin "oxygenated" layer, further thickening of this layer occurs with a bulk nickel oxide structure. The rate of this process is relatively slow compared to the rates of the first two stages of reaction. Continued growth of the oxide layer, requires transport of oxygen or nickel through the layer itself[107]. The sticking probability for oxygen adsorption on the oxide layer is very small at room temperature, but oxygen can be reversibly adsorbed at cryogenic temperatures giving an increase in work function[81].

Results found for individual crystal planes when considered in more specific terms permit a more detailed description of the surface structures associated with the first two stages of reaction.

3. Oxygen Surface Structures

On all three close-packed planes, the first stage of adsorption is characterized by the formation of surface structures with a simple periodic relation to the nickel plane in question. The sequence of structures formed with increasing oxygen coverage is shown in Figures 9, 10 and 11.

On the (100) plane (Figure 9) a p(2x2) structure is formed at low coverage with best development at about quarter monolayer. With further coverage a c(2x2) structure is formed with best development at about half monolayer. The two phases appear to coexist at coverages between quarter and half monolayer. The transition between p(2x2) and c(2x2) appears to occur by random filling of vacant sites in the p(2x2) structure. Transitions among these structures with exposure have been discussed in detail by Demuth and Rhodin[95]. We note that a coordinated movement of previously adsorbed oxygen is not necessarily required for this transformation.

On the (111) plane a p(2x2) structure is formed[108] at low coverage with best development at about quarter monolayer (see Fig. 13.10). At higher coverage a $(\sqrt{3} \times \sqrt{3})$ R30° structure has been reported[85,108] but it appears that heating the sample after adsorption at room temperature is required to produce this structure. In this case the structural transformation from p(2x2) to $(\sqrt{3} \times \sqrt{3})$R30° probably requires a coordinated reorganization of the previously adsorbed oxygen (see Figure 10). The apparent absence of structural transformations for this plane is probably associated with a weaker adsorbate interaction typical of the close-packed structure of the substrate.

On the (110) plane (see Figure 11) a complicated sequence of structures studied by MacRae[108] and in detail by Germer _et al_.[109] develops with increasing coverage. For adsorption at room temperature, well-ordered (3x1),(2x1) and (3x1) structures are formed with best development at about one-third, one-half and two-thirds monolayer coverage respectively. Two aspects of the structural evolution are of particular interest. Firstly, Germer _et al_.[109] observed that a reversible phase change between the (3x1) and (2x1) structures could be produced at a coverage of about one-third monolayer by heating to 300°C. Elongation of diffraction features in the (2x1) LEED pattern observed at 200°-300°C indicated partial disorder in the (2x1) structure. This is hardly surprising since segregation into (2x1) islands and bare surface regions presumably is involved[109]. It is noteworthy, however, that the disorder is predominantly along the close-packed $[1\bar{1}0]$ direction. Long-range order in the single-spaced periodicity along the [001] direction is preserved. The implication of this result is that a strong attractive interaction between the surface species along the [001] direction causes the struc-

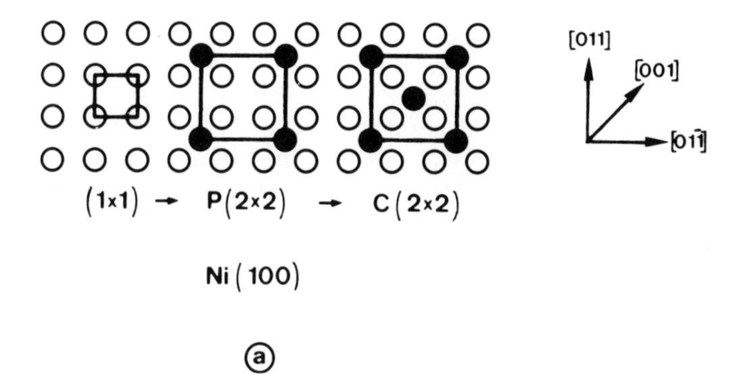

Figure 9. : Schematic of surface structures formed with increasing oxygen coverage on Ni(100)

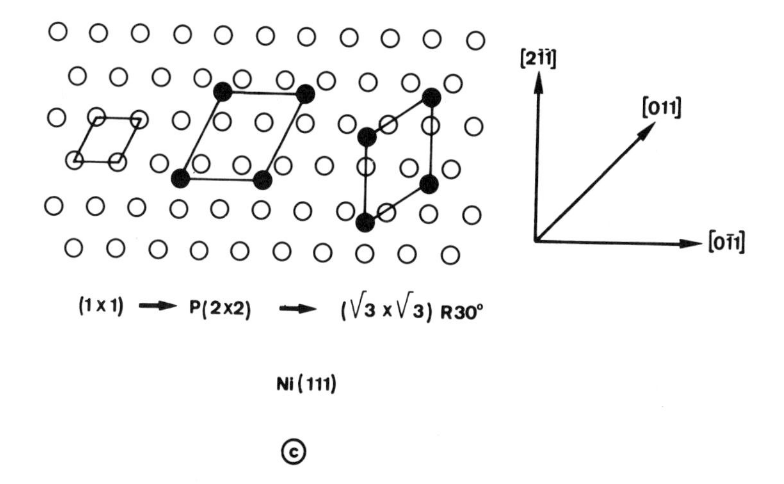

Figure 10.: Schematic of surface structures formed with increasing oxygen coverage on Ni(111).

tural modifications along adjacent $[1\bar{1}0]$ rows to be correlated.

The second set of interesting observations concern the transition between the (2x1) and (3x1) structures in the coverage range of about one-half to two-third mononlayer. For adsorption at room temperature this region is marked by elongation of the diffracted beams characteristic of the (2x1) structure. Again, the interpretation is that the (2x1) structure becomes disordered along the $[1\bar{1}0]$ rows with correlated behavior in adjacent rows. At about two-thirds monolayer coverage, the hitherto elongated beams are consolidated into well-resolved beams characteristic of the (3x1) structure. When the oxygen coverage is increased from one-half monolayer by adsorption at 300°C, however, the half-order beams of the (2x1) LEED pattern split into two sharp beams which move apart continuously with increasing coverage until the (3x1) LEED pattern is reached. Germer _et al._[109] showed that these results could be understood if it were assumed that the preferred surface structure as formed at 300°C contained an intimate mix of the (2x1) and (3x1) structures mutually arranged to produce the most uniform distribution of oxygen along the $[1\bar{1}0]$ rows at each intermediate coverage. After room temperature adsorption, evidently the distribution of oxygen is more random, yielding disordered structures.

Possible implications of these sequential changes in structure with oxygen coverage observed on the three close-packed planes can be summarized.

1. At oxygen coverage less than one-half monolayer the surface structures formed on each plane have periodicities greater than that of the corresponding nickel plane along the $\langle 110 \rangle$ rows of nickel atoms (Figures 9 and 10). Double periodicities are present in both structures formed on the (100) and (111) planes. Triple and double periodicities respectively are present in the (3x1) and (2x1) structures on the (110) planes. These results indicate the occurrence of repulsive interactions along the $\langle 110 \rangle$ surface directions.

2. Assuming that the optimum evolution of the structures with coverage is that which minimized repulsive interactions we note that this is probably achieved in the facile transformation from p(2x2) to c(2x2) on the (100) plane without reorganization of the previously adsorbed oxygen whereas the more demanding transformation from p(2x2) to $(\sqrt{3} \times \sqrt{3})$ R30° on the (111) plane presumably does require such a reorganization. The transformation with increasing coverage from the (2x1) to the (3x1) structure on the (110) plane presumably also requires a structural reorganization. In this case however the relatively minor adjustments required along the $[1\bar{1}0]$ rows may be facilitated by a low activation energy for diffusion along this close-packed direction.

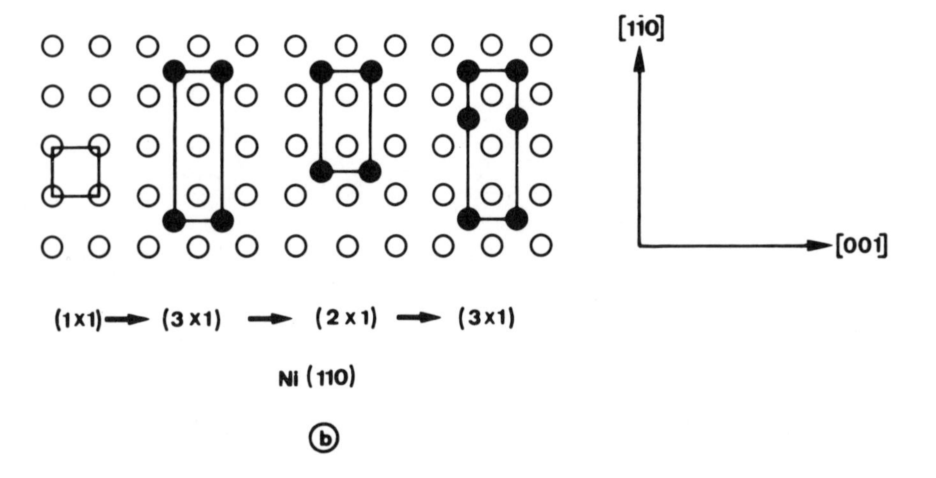

Figure 11. : Schematic of surface structures formed with increasing oxygen coverage on Ni(110)

3. The surface structures formed on the (110) plane all exhibit single-spaced periodicities along the [001] direction, even in the case of partial disorder, suggesting that strong attractive interactions occur in this direction. The formation of a p(2x3) rather than a c(2x2) structure at lowest coverages on the (100) plane, however, suggests that attractive interactions along the <001> surface directions are weak or non-existent on this plane.
4. All the structures formed in the first stage of adsorption are simply related to the adsorbent lattice. This observation suggests that the structure of the local adsorbed complex is determined primarily by short range adsorbate-adsorbent interactions rather than by interactions within the adsorbed layer.

4. Coordination and Bonding of Chemisorbed Oxygen

In our discussion of surface structures so far we have not dealt directly with the question of the actual atomic crystallography in the surface region. There is little doubt that in the second stage of adsorption,penetration of oxygen into the nickel lattice occurs i.e., reconstruction of the surface, to give a very thin oxygenated layer. On all three planes there is strong evidence that the oxygenated layer has a structure close to that of bulk NiO, although some distortion of the NiO lattice to reduce its mismatch with the nickel lattice probably occurs.

For the first stage of adsorption, the molecular constitution of the adsorbed oxygen and the precise atomic arrangement is now established with certainty. Most investigators of this system have assumed that oxygen is dissociatively adsorbed but we note that chemisorption of molecular oxygen has been concluded from kinetic studies on sintered polycrystalline films(81). Since these films might reasonably be expected to contain a large (111) component it is interesting to note that the possibility that nondissociative adsorption might be specific to the (111) plane has been proposed recently by Demuth and Rhodin(85). These authors have drawn attention to certain differences in adsorption behavior between the (111) and the (100) and (110) planes, namely that the change in work function is considerably larger on the (111) plane; the oxygen $KL_I L_{II}$ Auger line-shape more closely resembles that observed after carbon monoxide adsorption on the (111) plane than the line-shapes observed for adsorption on (100) or (110); both carbon monoxide (presumed to be molecularly adsorbed) and oxygen p(2x2) structures on the (111) plane undergo an order-disorder phase transition at about 160°C. These differences are shown in Figure 12. As noted by Demuth and Rhodin(78), however, these differences in adsorption behavior might also be consistently attributed to a different type of bonding of atomically adsorbed oxygen on the planes in question.

Although it seems quite certain that significant changes are produced in the electronic structure of surface nickel atoms we believe that the weight of the available evidence indicates that large alterations in nickel atom positions do not occur during the first stage of oxygen chemisorption. The results of recent studies by Eastman and Cashion(110) using ultraviolet photoemission spectroscopy, by Hagstrum(111) using ion-neutralization spectroscopy, by Horgan and Dallins(82) using Auger spectroscopy, and by Andersson and Nyberg(112) using appearance potential spectroscopy, in which spectra from surfaces containing low coverages of oxygen were compared with spectra from surfaces containing nickel oxide, all suggest that significant changes occur in the spectra characteristic of the transition from the chemisorbed structure to the oxygen reconstructed at the surface.

The detailed atomic arrangement in the various surface structures is accessible from measurements of the corresponding LEED intensities. Model calculations(103-105,113-115) via multiple-scattering theories have been made for the c(2x2) structure formed on the (100) plane, and more recently also for the p(2x2) structure(116). Structures suggested by different workers are now in agreement. Andersson _et al._(113) and Demuth _et al._(117) concur that the c(2x2)

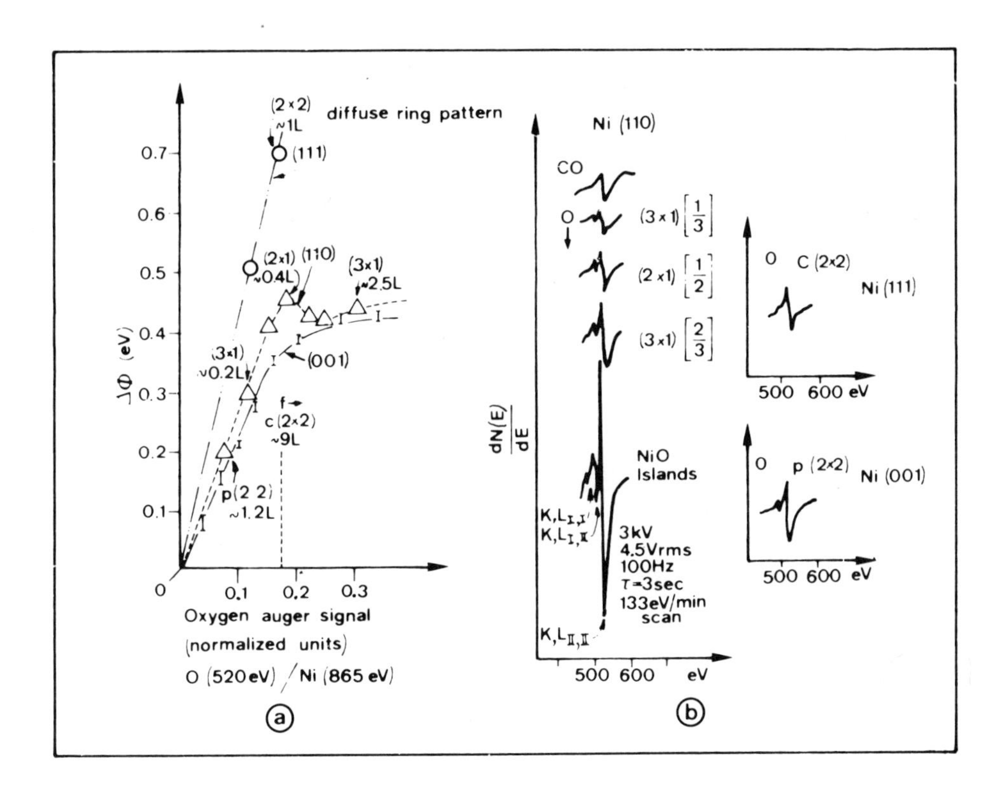

Figure 12. : (a) Change in work function versus detected oxygen Auger signal (Note : 0.18 Auger intensity units is 0.5 of a monolayer here). Observed LEED structures are indicated corresponding to the surface coverage as well as the exposure in Langmuirs. (I.L. = 10^{-6} torr-sec).

(b) Peak shape of the characteristic Auger signal for oxygen chemisorption on all three low index planes and for CO structure on Ni(110).

(From Demuth and Rhodin, ref. 78)

structure consists of oxygen atoms on the 4-coordinate sites on the (100) plane, with a spacing of 0.90 ± 0.1 Å between the plane of the oxygen atoms and the underlying nickel plane. Van Hove and Tong[116] have recently suggested a model for the p(2x2) structure on Ni(100) which in its local bonding configuration is identical to that proposed by Demuth *et al*.[117] for the c(2x2) structure. We note that good agreement has been obtained between experiment and theory, in terms of peak positions and peak shapes[116,117]. Additional grounds for confidence in the nonreconstructed model for oxygen adsorption proposed by these authors are to be found in the consistent surface crystallography results reported by Demuth *et al*.[103,104,117] and by van Hove and Tong[116] respectively from the LEED measurements of Demuth and Rhodin[78] for the c(2x2) and p(2x2) structures formed by sulphur, selenium and tellurium on Ni(100) in terms of similar nonreconstructed models.

It is interesting to note that early measurements of the change in work function produced by oxygen adsorption on nickel were interpreted in terms of essentially covalent nickel-oxygen bonding. From the nonreconstructed models proposed[116,117] for the p(2x2) and c(2x2) structures on Ni(100) we note that the effective radius of the oxygen atom may be determined as 0.73 Å assuming[117] no change in the metallic radius of surface nickel atoms. This is in close agreement with values quoted[118] for the single-bond covalent radius of oxygen and in agreement with the covalent radii for Na c(2x2) Al(001) reported by Hutchins, Rhodin and Demuth[119].

More direct information concerning the electronic character of nickel-oxygen bonding is available from the ion-neutralization studies of Hagstrum and Becker[83-85] and from the photoemission studies of Eastman and Cashion[110]. Although detailed interpretation of these measurements is not available, the interesting qualitative implications of these results are considered in the following section.

5. Adsorption of Oxygen, Sulphur, Selenium and Tellurium on Nickel: Geometry of Two-Dimensional Structures

Results of a variety of experimental approaches indicate that oxygen adsorption on nickel proceeds through at least three distinct stages; chemisorption, two-dimensional surface oxide formation and bulk oxide growth. Studies of sulphur adsorption suggest that a similar tendency may occur, although more severe conditions of exposure and temperature are required in this case. Quantitative information is available from the work of Hagstrum and Becker[83,84,120] and Demuth and Rhodin[78], for the low coverage region, concerning the structural and electronic properties of the adsorbed chalcogens. Sulphur adsorption on nickel has also been studied by Oudar *et al*.[88-93] and by Edmonds *et al*.[94] using LEED pattern analysis. In the former case, sulphur coverages were accurately determined by a radio-tracer technique.

The geometric surface structures observed for chalcogen adsorption on nickel are listed in Table 1, together with approximate coverages inferred in most cases from the LEED patterns. Considering the marked changes in chemical properties in the progression from oxygen to tellurium, the uniformity in structural behavior of the different elements at low coverage is rather remarkable. Only on the (110) plane is there a different structural sequence between oxygen on the one hand and sulfur and selenium on the other.

Apart from the differences in behavior found on the (110) plane, trends in the relative ease of formation of well-ordered structures have been observed. Thus formation of the p(2x2) structure on the (100) plane occurs readily for oxygen but a decrease in long-range order is apparently found in the progression from oxygen to tellurium. Conversely, the ($\sqrt{3}$ x $\sqrt{3}$) R30° structure observed on the (111) plane is best-developed for selenium and becomes increasingly disordered for sulphur and oxygen[85]. The relative ease of this transformation for selenium is perhaps indicative of a weaker nickel-chalcogen bond and a smaller activation energy for diffusion for the heavier chalcogens.

The most striking difference in the structures formed by the chalcogens is the absence of any evidence for formation of well-defined surface compounds of nickel with sulphur, selenium and tellerium as compared to oxygen for which the terminal structure appears to be NiO on all three close-packed planes.

It may be significant that unlike oxygen, the group of sulphur, selenium and tellurium are known to form nonstoichiometric bulk compounds with nickel based on the nickel-arsenide structure[121]. Accordingly, nonstoichiometric two-dimensional surface compounds may be responsible for the complex structures formed with sulphur, selenium and tellurium adsorption on nickel at higher temperatures.

6. Surface Structure and Chemical Bond Formation

Extensive LEED measurements by Demuth and Rhodin[78] have been analyzed by Demuth *et al.*[103-105,117] and by van Hove and Tong[116] in terms of comparison of calculated and experimental intensity-voltage spectra for a number of diffracted beams. The geometric parameters of the models resulting from these analyses for the c(2x2) and p(2x2) structures found on Ni(100) are given in Table 2. In all cases the proposed model structures have chalcogen atoms in the 4-coordinate sites at a distance, d from the underlying unreconstructed Ni(100) plane. With regard to the original differences between the proposed structures, a reevaluation by Andersson and Pendry[122] of the comparison between experimental and calculated intensities using a wider energy range of new data has led to agreement with

TABLE 1 : SURFACE STRUCTURES FORMED BY CHALCOGEN ADSORPTION ON NICKEL

Plane	Coverage (Monolayers)	Oxygen	Sulphur	Selenium	Tellurium
(100)	1/4	p(2x2)	p(2x2)	p(2x2)	p(2x2)
	1/2	c(2x2)	c(2x2)	c(2x2)	c(2x2)
	1	NiO(100)			
(111)	1/4	p(2x2)			
	1/3	$(\sqrt{3}x\sqrt{3})R30°$	$(\sqrt{3}x\sqrt{3})R30°$	$(\sqrt{3}x\sqrt{3})R30°$	
	1/2		Complex structures on heating		
	1	NiO(111)			
(110)	1/3	p(3x1)			
	1/2	p(2x1)	c(2x2)	c(2x2)	
	2/3	p(3x1)	p(3x2)		
	10/9	p(9x4)			
	1	NiO(100)			

TABLE 2 : MODEL STRUCTURES FOR CHALCOGENS ON Ni(100)

Covalent Radius		Demuth et al.		Van Hove and Tong	
		d	r	d	r
O	0.73	0.9	0.73	0.9	0.73
S	1.02	1.3	0.94	1.3	0.94
Se	1.16	1.45	1.04	1.55	1.10
Te	1.35	1.90	1.36	1.80	1.27

Interlayer spacing d and oxygen atom radius r for c(2x2) structures. All models have oxygen atom in site of 4-coordination on Ni (100) plane. Van Hove and Tong's analysis pertains to the p(2x2) structures.

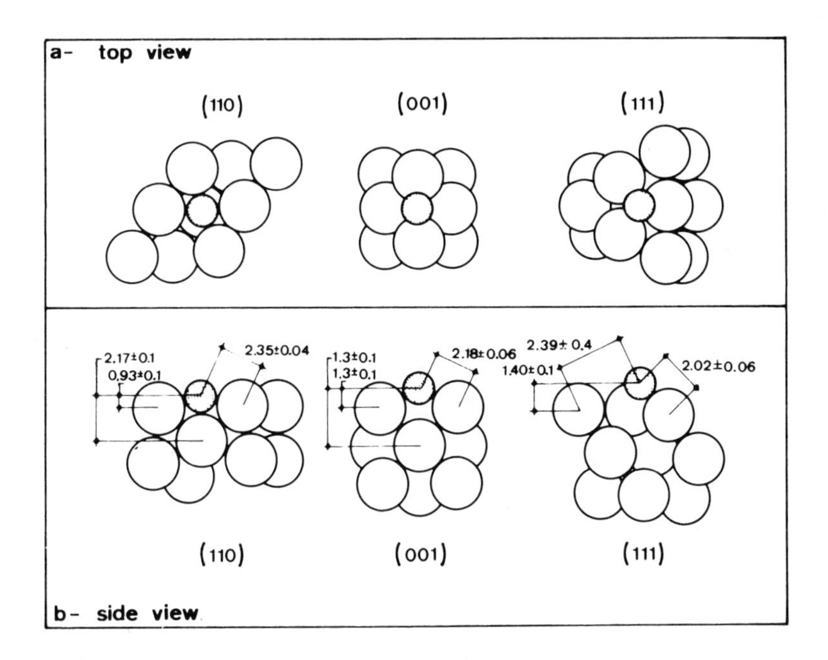

Figure 13.: Atomic location of adsorbed sulphur on the (110), (100) and (111) planes of nickel as determined from analysis of LEED intensity measurements. (From Demuth, Jepsen and Marcus, ref.104). Cross-hatched circles represent sulphur atoms, open circles represent nickel atoms.

the interlayer spacings d proposed by Demuth *et al.*[103-105,117]. The layer spacing of 1.7 Å originally proposed by Duke *et al.*[114] for the c(2x2) sulphur structure was apparently due in part to a misunderstanding in evaluation of Theeton's experimental data[114, 123]. A remaining conflict in interpretation, concerns the reconstructed structure proposed by Duke *et al.*[115] for the c(2x2) oxygen structure. We are not aware of any reevaluation of this structure at the present time.

The model structures proposed by Demuth *et al.* for the c(2x2) structures formed by sulphur on the (100) and (110) planes and the p(2x2) structures formed on the (111) plane are shown schematically in Figure 13. Again the model structures contain sulphur atoms in the sites of high coordination on an unreconstructed adsorbent.

In Table 2 are also listed the covalent radii for the chalcogens. For the model structures in which the nickel lattice is presumed to be unperturbed, we have calculated effective radii, r, of the adsorbed atoms from the values of the interlayer spacing d. It is evident that the model structures proposed by Demuth *et al.*[103-105, 117] and by van Hove and Tong[116] contain chalcogen atoms covalently bonded to the nickel surface atoms. From the quoted values (116) for the interlayer spacings in Table 2 and the changes in work function reported[78] for the lowest coverage, p(2x2) structures on Ni(100), we have derived dipole moments and charge transfer associated with the different nickel-chalcogen bonds, assuming the image plane to lie through the centers of the topmost Ni atoms and ignoring possible depolarization effects. These are listed in Table 3.

TABLE 3 : CHARGE TRANSFER IN NICKEL-CHALCOGEN BOND

Chalcogen	Interlayer Spacing (Å)	θ (monolayer)	$\Delta\phi$ (eV)	μ (Debye)	Δe
O	0.9	1/4	0.22	0.14	0.03
S	1.3	1/4	0.24	0.16	0.03
Se	1.45	1/4	0.08	0.05	0.01
Te	1.90	1/4	-0.29	-0.19	-0.02

The electronic aspects of bonding of oxygen, sulphur and selenium to single-crystal surfaces of nickel have been studied in detail by Hagstrum and Becker(83-85) using ion-neutralization spectroscopy. A detailed interpretation of the observed spectra and of recent ultraviolet photoemission spectra(110) must await the development of quantitative models of the adsorbed state and the emission processes in these spectroscopies.

C. SUMMARY - MICROSCOPIC AND ATOMIC PROPERTIES OF CLEAN AND CHEMISORBED SURFACES(78,83-85,103-105,117)

The most striking feature in the experimental approach to the study of metal surfaces is the capability to prepare, characterize and measure the electronic and atomistic properties of solid surfaces to a remarkable degree of sensitivity and precision. Significant progress has been made recently in the measurement of fundamental parameters associated with mainly static and some dynamic aspects of adsorption of simple molecules on single crystal metal surfaces.

Encouraging advances have been made for example in the determination of the surface crystallography of adsorbed layers using low-electron-diffraction. For simple structures, recent dynamical calculations of LEED intensity spectra using complete multiple-scattering formalisms have shown significant progress in terms of comparison with experimental spectra. Convergent iteration-perturbation methods for treatment of multiple-scattering are also being successfully applied and analysis of more complex surface structures using this approach should be possible.

Successful application of the transform-deconvolution method to a number of clean metal surfaces has also been achieved. Evaluation of the application of the method to more overlayer systems is in progress. Encouraging results have also been reported for clean Ni(100) using data averaging procedures. The application to more complex systems is being investigated.

The solid state electron spectroscopies based upon Auger electron emission, inelastic scattering of low-energy electrons, ultraviolet and x-ray photoemission, ion-neutralization, and vacuum tunneling spectroscopy are potentially able to provide detailed information on both the clean surface microscopic properties and the quantum mechanical parameters of adsorption and reaction on metals. Typical parameters relate to electron energy levels and occupancies, electronic charge distribution and electron transition probabilities. These studies are now being rapidly extended to many transition metals. Interpretation is based mainly on semiquantitative analysis. More quantitative theoretical formulations are in course of development.

It should be noted that the powerful methods of field emission, field ionization and field desorption invented by Müller are continuing to provide unique microscopic and atomistic adsorption information on metal surfaces. Not only is it possible to obtain quantitative information on atomic and electronic structure by various combinations of microscopic and spectroscopic features of these approaches but to do so on a scale of atomistic detail not accessible by other methods.

Techniques for measurement of thermodynamic and kinetic parameters also continue to provide a great deal of critical information. Thermal desorption and electron impact desorption are being widely used to investigate the nature of adsorbate binding states and to measure adsorbate binding energies and cross-sections for desorption processes. Interpretations of the occurrence of multiple binding states have been concerned with the alternative possibilities of distinct binding modes and induced surface heterogeneity due to adsorbate-adsorbate interactions. Combination of the desorption techniques with measurements of surface structure and change in work function using single-crystal surfaces has facilitated more detailed understanding.

Interpretation of measurements of adsorbate binding states and energies, and of kinetic parameters relating to rates of adsorption, desorption, surface diffusion and reactions between adsorbed molecules has provided a phenomenological basis for qualitative interpretation of the results of electron spectroscopic studies.

Studies involving correlation of static to dynamic measurements have been rather limited to date and directed to simple molecules adsorbed on single-crystal surfaces of a relatively limited number of metals. It is to be expected that expansion of the ways in which these methods may be combined and their extension to a broader class of adsorption systems characteristic perhaps of more complex structures will have a major impact on both the science and the engineering of solid surfaces.

REFERENCES

(1) T.N. Rhodin and D.L. Adams, "Adsorption of Gases on Solids", Vol. 6, Treatise on Solid State Chemistry, ed. N.B. Hannay, Plenum Press, New York (1975)
(2) N.J. Taylor, J. Vac. Sci. and Tech. 6 (1969) 241
(3) J.M. Chen and C.A. Papageorgopoulos, Surface Science 20 (1970) 1975
(4) R.G. Musket and J. Ferrante, J. Vac. Sci. and Tech. 7 (1970) 14

(5) R.W. Joyner, J. Rickman and M.W. Roberts, Surface Science 39 (1973) 445
(6) G. Ehrlich, Advances in Catalysis 14 (1963) 256
(7) T. Oguri, J. Phys. Soc. Japan 18 (1963) 1280
(8) G.E. Moore and F.C. Unterwald, J. Chem. Phys. 40 (1964) 2626
(9) F. Ricca, R. Medana and G. Saini, Trans. Farad. Soc. 61 (1965) 1492
(10) L.J. Rigby, Can. J. Phys. 43 (1965) 1020
(11) T.E. Madey and J.T. Yates, J. Chem. Phys. 44 (1966) 1675
(12) A.E. Bell and R. Gomer, J. Chem. Phys. 44 (1966) 1065
(13) V.J. Mimeault and R. Hansen, J. Chem. Phys. 45 (1966) 2240
(14) T.L. Rhodin, W.K. Warburton and T.N. Rhodin, J. Chem. Phys. 46 (1967) 665
(15) P.W. Tamm and L.D. Schmidt, J. Chem. Phys. 54 (1971) 4775
(16) T.E. Madey and J.T. Yates, in : Structure et Propriétés des Surfaces des Solides (Editions de C.N.R.S., Paris, 1970) No. 187, p. 155
(17) D.L. Adams and L.H. Germer, Surface Science 23 (1970) 419
(18) K. Yonehara and L.D. Schmidt, Surface Science 25 (1971) 238
(19) D.L. Adams and L.H. Germer, Surface Science 32 (1972) 205
(20) J. Andersson and P.J. Estrup, J. Chem. Phys. 46 (1967) 563
(21) D.L. Adams and L.H. Germer, Surface Science 27 (1971) 21
(22) T.A. Delchar and G. Ehrlich, J. Chem. Phys. 42 (1965) 2628
(23) J.W. May and L.H. Germer, J. Chem. Phys. 44 (1966) 2895
(24) C.C. Chang, J. Electrochem. Soc. 115 (1968) 355; Ph.D. Thesis Cornell University 1967
(25) P.W. Tamm and L.D. Schmidt, J. Chem. Phys. 51 (1969) 5352
(26) L.R. Clavenna and L.D. Schmidt, Surface Science 22 (1970) 365
(27) C. Kohrt and R. Gomer, Surface Science 24 (1971) 77
(28) J.T. Yates and T.E. Madey, J. Vac. Sci. and Tech. 8 (1971) 63
(29) J.T. Yates and T.E. Madey, J. Chem. Phys. 54 (1971) 4969
(30) R.R. Rye and B.D. Barford, Surface Science 27 (1971) 667
(31) L.R. Clavenna and L.D. Schmidt, Surface Science 33 (1972) 11
(32) R.R. Rye, B.D. Barford and P.G. Cartier, J. Chem. Phys. 59 (1973) 1693
(33) B.D. Barford and R.R. Rye, J. Chem. Phys. 60 (1974) 1046
(34) T.E. Madey and D. Menzel, Japan. J. Appl. Phys. Suppl. 2 (1974)
(35) T. Toya, J. Vac. Sci. and Tech. 9 (1972) 890, and private communication
(36) D.L. Adams, Surface Science 42 (1974) 12
(37) C.G. Goymour and D.A. King, J. Chem. Soc. Farad. I 69 (1973) 749
(38) L.W. Swanson and R. Gomer, J. Chem. Phys. 39 (1963) 2513
(39) T.E. Madey, J.T. Yates and R.C. Stern, J. Chem. Phys. 42 (1965) 1372
(40) J.T. Yates and T.E. Madey, J. Chem. Phys. 43 (1965) 1055
(41) J.T. Yates, R.G. Greenler, I. Ratajczykowa and D.A. King, Surface Science 36 (1973) 739
(42) J.T. Yates and D.A. King, Surface Science 30 (1972) 601
(43) D. Menzel, Surface Science 47 (1975) 370; for reviews of elec-

tron induced desorption see: T.E. Madey and J.T. Yates, J. Vac. Sci. and Tech. 8 (1971) 525
(44) J.T. Yates and D.A. King, Surface Science 32 (1972) 479
(45) W.F. Egelhoff, J.W. Linnett and D.L. Perry, Disc. Farad. Soc. 58 (1974)
(46) T.E. Maday, J.T. Yates and N.E. Erickson, Surface Science 43 (1974) 526
(47) J.T. Yates, T.E. Madey and N.E. Erickson, Surface Science 43 (1974) 257
(48) F.M. Propst and T.C. Piper, J. Vac. Sci. and Tech. 4 (1967) 53
(49) E.W. Plummer and A.E. Bell, J. Vac. Sci. and Tech. 9 (1972)583
(50) G. Ehrlich and F.G. Hudda, J. Chem. Phys. 35 (1961) 1421
(51) P.L. Young and R. Gomer, Surface Science 44 (1974) 277
(52) G. Ehrlich and F.G. Hudda, J. Chem. Phys. 36 (1962) 3233
(53) G. Ehrlich, in : Metal Surfaces (American Society for Metals, 1963) p. 221
(54) J.W. May, R.J. Szostak and L.H. Germer, Surface Science 15 (1969) 37
(55) Molecular Processes on Solid Surfaces, ed. Drauglis, Gretz and Jaffee, McGraw-Hill, New York, 1969, p. 317
(56) G. Ehrlich, Disc. Farad. Soc. 41 (1966) 54
(57) R.A. Armstrong, Can. J. Phys. 46 (1968) 949
(58) T.E. Madey and J.T. Yates, Suppl. Nuovo Cimento 5 (1967) 483
(59) T. Engel and R. Gomer, J. Chem. Phys. 50 (1969) 2428
(60) D.L. Adams, unpublished results
(61) D.L. Adams and L.H. Germer, Surface Science 26 (1971) 109
(62) P.J. Estrup and J. Andersson, J. Chem. Phys. 45 (1966) 2254
(63) P.J. Estrup and J. Andersson, J. Chem. Phys. 46 (1967) 567
(64) P.L. Young and R. Gomer, J. Chem. Phys. 61 (1974) 4955
(65) E.W. Plummer and J.W. Gadzuk, Phys. Rev. Letters 25 (1970) 1493
(66) B. Feuerbacher and B. Fitton, Phys. Rev. Letters 29 (1972) 786
(67) B. Feuerbacher and B. Fitton, Phys. Rev. Letters B8 (1973) 4890
(68) B. Feuerbacher and M.R. Adrianes, Surface Science 45 (1974) 553
(69) E.W. Plummer and B.J. Waclawski, Phys. Rev. Letters 29 (1972) 783
(70) E.W. Plummer and R.D. Young, Phys. Rev. B1 (1970) 2088
(71) J.E. Demuth and D.E. Eastman, Phys. Rev. Letters 32 (1974) 1123
(72) J.W. Gadzuk, Phys. Rev. B10 (1974) 5030
(73) A. Liebsch, Phys. Rev. Letters 32 (1974) 1203
(74) J. Pritchard and M.L. Sims, Trans. Farad. Soc. 66 (1972) 427
(75) B.J. Waclawski and E.W. Plummer, J. Vac. Sci. and Tech. 10 (1973) 292
(76) D.A. King, Surface Science 47 (1975) 384
(77) L.H. Germer and J.W. May, Surface Science 4 (1966) 452
(78) J.E. Demuth and T.N. Rhodin, Surface Science 45 (1974) 249
(79) P.H. Holloway and J.B. Hudson, Surface Science 43 (1974) 123
(80) P.H. Holloway and J.B. Hudson, Surface Science 43 (1974) 141
(81) A.M. Horgan and D.A. King, Surface Science 23 (1970) 259
(82) A.M. Horgan and I. Dalins, Surface Science 36 (1973) 526

(83) H.D. Hagstrum and G.E. Becker, Phys. Rev. Letters 22 (1969) 1054
(84) H.D. Hagstrum and G.E. Becker, J. Chem. Phys. 54 (1971) 1015
(85) G.E. Becker and H.D. Hagstrum, Surface Science 30 (1972) 505
(86) Tamm and Schmidt, J. Chem. Phys. 52 (1970) 1150
(87) J.M. Baker and D.E. Eastman, J. Vac. Sci. Tech. 10 (1973) 223
(88) J. Oudar, Compt. Rend. (Paris) 249 (1965) 91
(89) J.L. Domange, J. Oudar and J. Benard, Growth Mechanism and Structure of Adsorbed Layers, in ref. 55
(90) J.L.Domange and J. Oudar, Surface Science 11 (1968) 124
(91) M. Perdereau and J. Oudar, Surface Science 20 (1970) 80
(92) M. Perdereau, Surface Science 24 (1971) 239
(93) J. Oudar, J. Vac. Sci. and Tech. 9 (1972) 657
(94) T. Edmonds, J.J. McCarroll and R.C. Pitkethly, J. Vac. Sci. and T-ch. 8 (1971) 68
(95) J.E. Demuth and T.N. Rhodin, Surface Science 42 (1974) 261
(96) J.E. Demuth, P.M. Marcus and D.W. Jepsen, Phys. Rev. B11 (1975) 1460
(97) R.H. Tait, S.Y. Tong and T.N. Rhodin, Phys. Rev. Letters 28 (1972) 553
(98) S. Andersson and J.B. Pendry, J. Phys. C 5 (1972) 41
(99) S.Y. Tong and L.L. Kesmodel, Phys. Rev. B8 (1973) 3753
(100) G.E. Laramore, Phys. Rev. B8 (1973) 515
(101) R.L. Park and H.E. Farnsworth, J. Appl. Phys. 35 (1964) 2220
(102) See for example, T.A. Delcher and F.C. Tompkins, Proc. Roy. Soc. (London) A 300 (1967) 141; C.M. Quinn and M.W. Roberts, Trans. Farad. Soc. 61 (1964) 1775; 60 (1964) 899
(103) J.E. Demuth, D.W. Jepsen and P.M. Marcus, Solid State Comm. 13 (1973) 1311
(104) J.E. Demuth, D.W. Jepsen and P.M. Marcus, Phys. Rev. Letters 32 (1974) 1182
(105) J.E. Demuth, D.W. Jepsen and P.M. Marcus, Surface Science 45 (1974) 733
(106) T.N. Rhodin, W.H. Orr and D. Walton, Memoires Scientifiques Rev. Metall. 62 (1965) 67
(107) F.P. Fehlner and N.F. Mott, Oxidation of Metals 2 (1970) 52
(108) A.U. MacRae, Science 139 (1963) 379
(109) L.H. Germer, J.W. May and R.J. Szostak, Surface Science 7 (1967) 430
(110) D.E. Eastman and J.K. Cashion, Phys. Rev. Letters 27 (1971) 1520
(111) H.D. Hagstrum, J. Vac. Sci. and Tech. 10 (1973) 264
(112) S. Andersson and C. Nyberg, Solid State Comm. 15 (1974) 1145
(113) S. Andersson, J.B. Pendry, B. Kasemo and M. van Hove, Phys. Rev. Letters 31 (1973) 595
(114) C.B. Duke, N.O. Lipari, G.E. Laramore and J.P. Theeten, Solid State Comm. 13 (1973) 579
(115) C.B. Duke, N.O. Lipari and G.E. Laramore, J. Vac. Sci. and Tech. 11 (1974) 180

(116) M.A. van Hove and S.Y. Tong, J. Vac. Sci. and Tech. 12 (1975) 230
(117) J.E. Demuth, D.W. Jepsen and P.M. Marcus, Phys. Rev. Letters 31 (1973) 540
(118) L. Pauling, The Nature of the Chemical Bond, Third Edition, (Cornell University Press, 1960)
(119) B. Hutchins, T.N. Rhodin and J.E. Demuth, Surface Science (to be published) 1975
(120) H.D. Hagstrum and G.E. Becker, Phys. Rev. 159 (1967) 572; Phys. Rev. B4 (1971) 4187
(121) A.F. Wells, Structural Inorganic Chemistry, Third Edition, (Oxford University Press, 1962)
(122) S. Andersson and J.B. Pendry, Solid State Comm. 16 (1975) 563
(123) C.B. Duke, N.O. Lipari and G.E. Laramore, J. Vac. Sci. and Tech. 12 (1975) 222
(124) S. Andersson and B. Kasemo, Solid State Comm. 8 (1970) 1327; Surface Science 25 (1971) 273
(125) R.L. Park and H.F. Farnsworth, J. App. Phys. 35 (1964) 2220.

ATOMIC STRUCTURE AND THERMODYNAMICS

OF PURE AND TWO-COMPONENT METAL SURFACES

G.E. RHEAD

Laboratoire de Métallurgie et Physico-Chimie des Surfaces
E.N.S.C.P., Université Pierre et Marie Curie
11 rue Pierre et Marie Curie, 75005 Paris, France

This chapter is concerned with surface structure as described by 'hard-sphere' atomic models. It also examines the possibilities for predicting surface behaviour through the use of thermodynamics. In these approaches one tries to find answers to questions like : How do real surfaces differ from ideal bulk plane arrangements ? What defects are likely to be present on a surface ? What determines the topographical and chemical stability ? Is it possible to prepare a surface with a particular structure and composition ? These questions are fundamental for the design of catalysts. The atomistic and thermodynamic descriptions are necessary bridges in going from an understanding of electrons at surfaces to technological applications.

1. The surface structure of clean metals.

1.1. Ideal bulk plane arrangements.

The structure of the bulk crystal is the obvious starting point. We consider first the ideal structure of a surface obtained by cutting through the bulk lattice along a plane of any arbitrary crystallographic orientation and assuming that the atoms remain in their bulk positions. General geometrical descriptions have been given by Nicholas and his colleagues 1) and an atlas of models has been published 2). From these descriptions calculations have been made of the variation of different quantities with crystallographic orientation - for example the densities of broken bonds 3) and surface energies evaluated on the basis of pairwise interactions 4).

The importance of crystal anisotropy is immediately evident from this work. As an illustration fig.1 shows how the surface density of broken bonds varies with orientation in face-

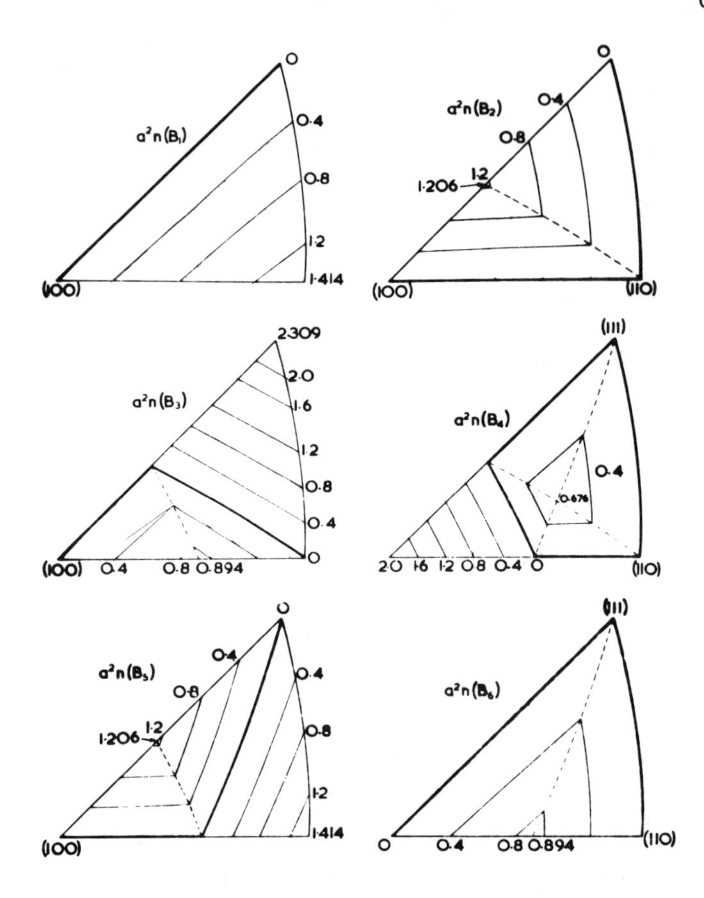

Fig.1 : Contour plots, for a face-centred cubic crystal, of the density of atoms which have 1, 2, ... 6 broken nearest neighbour bonds. The density is referred to area a^2 of surface, where a is the lattice parameter. Zero contours are plotted as thick lines. Dotted lines show where corners occur in the contours 3).

---

centred cubic crystals. In terms of first and second nearest neighbour bonds it is found that there are as many as nineteen crystallographically distinct types of surface atom for face-centred cubic crystals and fifteen for body-centred cubic crystals. Broken-bond models tend to overemphasize the effects of crystal anisotropy but nevertheless differences in surface orientation may be more important for certain properties than differences between one metal and another.

Simple ball models show that most structures can be analysed in terms of only three features: terraces, ledges (steps) and kinks in the steps. This, the so-called TLK model, is particularly useful for describing *vicinal* surfaces i.e. those with orientations close to low Miller-index orientations. The angle from the low-index pole determines the terrace width. Obviously this width will be a whole number of atomic units only for certain angles - in general it will be an average between two widths differing by one unit. The zone on which the orientation lies determine the ideal structure of the step - in particular the density of kinks.

The high Miller indices needed to specify a vicinal surface are cumbersome as a nomenclature and convey little information. Alternatively a surface may be designated by angular coordinates in the unit stereographic triangle or by defining the orientation of the terrace, its width and the orientation of the steps. Thus a surface of an f.c.c. crystal cut at about 9° from the (100) pole and along the $[01\bar{1}]$ zone can be designated as (911) ; $\alpha = 9°$, $\beta = 45°$; or [5(100)x(111)] , this last nomenclature indicating that there are (100) terraces five atoms wide and (111) oriented steps 5).

1.2. Real surfaces. Relaxation and reconstruction.

Because surface atoms are in a different, asymmetrical, environment they may take up equilibrium arrangements quite different from those in the bulk planes. One effect is that the interplanar spacing normal to the surface is changed (surface relaxation). Theoretically this may come about directly because of the asymmetry or through an electronic redistribution at the surface that changes the effective interatomic binding. In certain cases there may occur a total rearrangement within the surface plane and the formation of an entirely different structure (reconstruction).

Some theoretical work on relaxation has been done by using a pairwise interaction model and potential energy functions such as the Morse function. Starting with the perfect lattice positions, the atoms are relaxed by an iterative process until the calculated net force on each atom is zero [+]. Results for copper 6) show that there should be an *expansion* of the lattice by 5 to 15 per cent for the top layer. The calculated expansion decreases rapidly from layer to layer and is less than 0.1 per cent for the fourth plane. The fundamental reason for this expansion is that in the bulk crystal the nearest neighbour distance is smaller than the separation at the minimum energy in the pair potential - in the bulk crystal the force between nearest neighbours is repulsive.

Relaxation will evidently affect low-energy electron diffraction intensities since changes in lattice spacing must introduce changes in phase differences between waves scattered from

---

[+] A different approach employing the Green function technique of lattice statics has also been proposed 7).

different layers. However, only recently have the theoretical methods for interpreting the LEED data become sufficiently reliable. Results for aluminium 8), obtained using intensity averaging methods and pseudokinematic analyses, have shown that the first layer of the (110) surface is *contracted* by 3 to 5 per cent but that there is no significant relaxation for the (100) surface. Consideration of dynamic effects such as interlayer multiple scattering may lead to quantitative changes in the results but the *sign* of the relaxation is not in doubt and, very significantly, this points to the failure of calculations based on pairwise interactions and potentials derived from bulk properties.

LEED analyses involve comparison of experimental and theoretical plots of intensity versus energy for the various diffracted beams. The overall comparison is often assessed subjectively and it is difficult to set confidence limits.

An analysis of LEED data for the (100) face of molybdenum (the first results for a b.c.c. metal) shows an interplanar contraction of 11.5 per cent with respect to the bulk. This represents a reduction of 3.7 per cent in the nearest neighbour separation. The contraction does not persist when impurities are adsorbed 9).

An understanding of surface contraction requires more insight into the problem of electronic redistribution at the surface. Simple calculations for transition metals, using a tight-binding approximation, have been described 10). By explicity introducing electronic effects that alter the force constants at the surface it can be shown that contractions rather than expansions should indeed occur. In a different approach Finnis and Heine 11) have suggested an approximate answer for sp bonded metals. They consider that the electron charge density tends to be smoothed out parallel to the surface (as suggested by Smoluchowski 12)) and that the top surface layer can be represented by flattened asymmetrical Wigner-Seitz cells which are neutral objects that produce only small fields outside themselves. The ion core takes up the position where the electric field from the electron charge distribution is zero. Calculations on this basis for aluminium are in qualitative agreement with the experimental results. Alldredge and Kleinman 13) have shown, however, that for other metals crystalline effects can be as important as the electrostatic force from the uniform background density of electrons. Alldredge and Ma 14) have further shown that in d-band metals the mobile sp electron distribution has a dominant effect on relaxation which leads to a net contraction whereas the d electrons alone would give an outer expansion. The whole problem of relaxation is in the forefront of current theoretical work. There are uncertainties as to the relative magnitudes of different contributions to surface forces but, these aside, there is general agreement on the inadequacy of pair potentials derived from bulk data.

Surface relaxation causes rather subtle effects in the LEED intensities. Reconstruction, on the other hand, changes the surface mesh and so leads to extra diffraction spots. Several clean low-index metal surfaces are known to be reconstructed. Silicon and germanium also have reconstructed surfaces. Among metals, gold, platinum and iridium have been examined most closely and the gold (100) surface in particular. Fig.2 shows schematically the LEED pattern obtained from a clean, well annealed Au (100) surface and an explanation 15) of this pattern in terms of double diffraction is shown on the right. This interpretation is based on a model 18) in which the topmost layer is accurately hexagonal and has a laterally contracted interatomic spacing 4.4 ± 0.1 per cent smaller than that in the underlying bulk gold in the square lattice arrangement. The most remarkable aspect of this structure is that there is apparently *no* simple coincidence lattice with the underlying substrate. It appears that the lateral binding is sufficiently dominant that there is little tendency for the top overlayer to be bound to sites in the sublayer. A similar structure explains the pattern from the Pt (100) surface. Some attempts to give structural interpretations 16) of the LEED intensity data from these 'anomalous' surfaces seem to be wrong because they treat the patterns as if from a (5x1) structure thus mixing intensities from different spots and ignoring double diffraction.

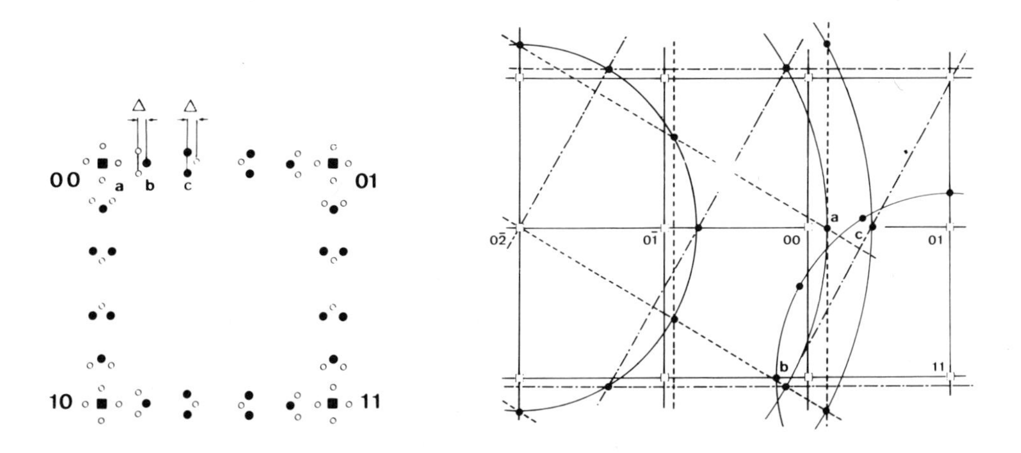

Fig.2 : Schematic representation of the LEED pattern from Au(100). A reciprocal lattice plot (right) in which two orientations of an hexagonal mesh (on arcs of circles), superimposed on a square mesh, simulates double diffraction from different substrate reflexions and explains the origin of the different features a, b, c. The splitting Δ is a sensitive measure of the contraction of the hexagonal overlayer 15).

Well resolved patterns from the Au(111) surface show satellite spots that indicate the same contracted hexagonal layer that occurs on the (100) surface. The patterns from Au(110) and Pt(110) have not yet been interpreted satisfactorily 15).

The reconstructed structures appear to be stable right up to the bulk melting point. A small amount of impurity, typically ∿ 0.1 monolayer, will, however, cause a transformation to the 'normal' (1x1) arrangement.

An important confirmation of the close-packed structure on the Au(100) surface has come recently from positive ion channeling spectroscopy 17). This technique is sensitive to the surface density of atoms and it has been shown that the reconstructed layer is denser than the (1x1) arrangement by an amount consistent with an hexagonal structure.

It is significant that for most metals surface reconstruction does not occur. For an explanation one may look to the special properties of the platinum series elements. Rhodin *et al.* 18) have invoked the concept of an enhanced surface valency accompanied by a contraction of the atomic radius. In gold, for example, the 'softness' of the 5d shell would make it easy to promote the $5d^{10}6s^1$ ground state to the higher valency configuration $5d^9 6s^2$. The full explanation is under active discussion. The anomalous structures of the (111) faces of silicon, (7x7), and germanium, (2x8), have been treated from quite different theoretical viewpoints via soft phonons 19), excitonic instabilities and surface vacancies 20).

Few studies have been made on relaxation and reconstruction on vicinal and high index surfaces, although the atomically rough nature of these planes might cause such effects to be even more important than on low-index faces. A calculation based on pairwise interactions and a bulk interatomic potential 21) shows that for the (112) plane of a b.c.c. crystal there should be a *tangential* relaxation much greater (13 per cent) than the normal relaxation (4 per cent). This effect would occur on planes that have less than two normal planes of symmetry. The effect of electron redistribution is not known and experimental data are not yet available. For reconstruction, a preliminary investigation has shown that gold surfaces cut at up to about 12° from the (100) orientation exhibit anomalous and complex LEED patterns suggesting that the clean atomic terraces retain the hexagonal arrangement 22).

It may be possible to study tangential relaxation by means of a new procedure in LEED 23) which is capable of determining lattice parameters in the surface plane to within 0.2 per cent. The method makes use of critical reflection along the crystal surface. Applied to the (100) face of α-iron it shows that the spacing between low index rows agrees accurately with the bulk value.

1.3. Surface steps

A most important difference between real and ideal surfaces is the presence of extra atomic steps. For a nominally low-index plane if the surface departs by only 1/2° from the ideal orientation then about 1 per cent of the surface atoms will be at step sites. Dislocations emerging at the surface may also introduce about the same step density. Such a surface concentration of defects is about the same as the concentration of impurity atoms that can be readily detected by Auger electron spectroscopy. The problem of steps and other structural imperfections is therefore replacing that of unwanted impurities as a main cause of uncertainty in many surface experiments.

Two kinds of experimental investigation are possible : studies of single isolated steps - such as can be made in the field ion microscope or by a decoration technique and studies of fairly large areas of vicinal surfaces for which the average step density can be varied systematically by varying the crystallographic orientation.

LEED studies of vicinal surfaces were started in several laboratories round about 1968, 24) 25) 26). The patterns obtained all show the same features : spots that were single in the pattern for the low-index plane are split into doublets due to the periodic array of steps (fig.3). The splitting is inversely proportional to the average distance between steps. The spot positions and their somewhat complex dependence on electron wavelength can be easily interpreted by simple kinematic theory. Essentially one applies the convolution theorem: convolution in real space corresponds to multiplication in reciprocal space. The patterns agree with the simple TLK model with step heights of one single atomic unit. Such patterns have now been obtained from several metals : copper, platinum, rhenium, tungsten and nickel. In the absence of impurities the vicinal surfaces are found to be topographically stable when heated in ultrahigh vacuum.

It is known that the apparent perfection of LEED patterns in general can "conceal an astonishing degree of structural anarchy" 27). This is particularly evident in the case of vicinal surfaces for, depending on the angle of cut of the crystal, the spot splitting may be very sharp and yet correspond to a terrace width that is a *non-integral* number of atomic units - the diffraction pattern gives an average width. Several authors have treated the problem of the effect of non-periodic defects on LEED patterns 27). The general results are well known in the theory of optical diffraction 28). Information on departures from periodicity appears only in the fine detail of the intensity profile of the diffracted beams. This is true for low-index as well as high-index orientations. In practice the resolution in LEED is not good enough to give significant information on non-periodic defects. This need not be a serious drawback since many properties can depend largely on the average

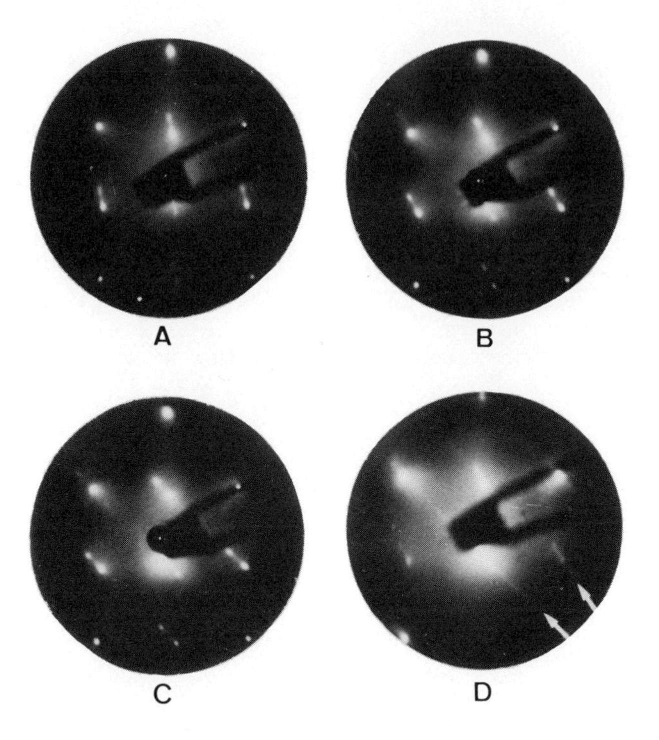

Fig.3 : LEED patterns (126 eV) from clean copper vicinal surfaces 12° from (100) and cut along zones at : (A) 15° from the [01$\bar{1}$] zone (slightly kinked steps); (B) 30° from [01$\bar{1}$] ; (C) along the [001] zone (fully kinked steps). D shows the pattern from orientation B after exposure to oxygen at ambient temperature: the steps tend to be rearranged into fully kinked orientations giving streaks (lower arrow 25).

step densities and their average crystallographic directions.

The observed stability of stepped surfaces agrees with results from surface work (energy) measurements (as discussed later). The anisotropy of surface work, γ , for clean surfaces is not sufficient to make faceting energetically favoured - the presence of adsorbed impurities is required. Heating a clean crystal should not produce extra steps - an important conclusion from the classic paper of Burton, Cabrera and Frank 29) on crystal growth, However, the structure of the step itself may well be strongly temperature dependent and heating may lead to the creation of new kink sites and even to a 'melting' of the step.

There is in fact evidence from LEED that kink sites can be randomly distributed along steps and their positions uncorrelated between neighbouring steps. Ellis 30) has shown from optical simulation that random kinks will produce patterns with doublets perpendicular to the average step direction and with no extra spots in other directions. The LEED patterns shown in fig.3 are of that type.

Little is known about the relaxation of atoms at step sites. In principle LEED intensity data may give information on this point. The problem is difficult because both the electronic redistribution associated with the step (Smoluchowski smoothing, see below) and changes in atomic positions will affect electron scattering. From the results on aluminium quoted above it seems probable that around the step there occurs an inward contraction rather than an expansion and there should also be lateral displacements. The terraces on vicinal surfaces should be slightly curved. Relaxation at steps can produce diffuse scattering features in RHEED 31). First experiments on W(110) surfaces suggest that clean steps cannot be relaxed by more than 3 to 10 per cent. More information on steps may ultimately come from ion scattering experiments 32). At present our knowledge of step structure is rudimentary.

Some of the effects of steps on surface properties may be used to detect these and other defects. A brief survey is appropriate.

Work function. The presence of steps introduces additional, positive-outward, dipoles and so reduces the work function. Observations supporting this conclusion have been reported by Körner 33) and Haas and Thomas 34) for tungsten. Typically a decrease of the order 0.1 eV is observed for a disorientation of about 10° from the (100) pole. Qualitatively this effect can be understood partly in terms of the smoothing of the surface electron distribution that leaves the atoms at step sites protruding and partially denuded of negative charge. A enhanced effect due to step roughening - greater denuding at kink sites - has been reported by Besocke and Wagner 35). The latter authors have deduced a value of about 0.3 D for the electric dipole moment of a tungsten atom at a step site. Some depolarising may occur if the steps are very close.
Surface states. UPS measurements on cleaved silicon (111) surfaces have revealed the existence of a surface state peak associated with steps at an energy 0.4 eV higher than the main surface state peak 36). Investigations on metals are underway in several laboratories.
Adsorption and reactivity. That the surface orientation affects adsorption has long been known. Studies with the field emission microscope 37) show very direct evidence. Some of the first LEED studies were made on copper 38). Various effects can occur: the rate of adsorption may be markedly increased by steps; for certain step densities adsorption can lead to a spontaneous regrouping of the steps - i.e. faceting; the adsorbed layer can have the same structure on the atomic terraces as on the corresponding low-index face - or, for narrow terraces, the structure may be modified ; if, because

of symmetry, several distinguishable orientations (domains) of a particular adsorbed structure occur on the low-index face, then steps can cause the preferential nucleation and growth of only certain domains (this can be an aid to LEED interpretations) ; binding to the step may produce modifications in the crystalline parameter of the adsorbed layer parallel to the step direction.

Somorjai and collaborators 5) have made extensive studies of stepped platinum surfaces. They have found notably that whereas molecular oxygen and hydrogen have very low sticking probabilities on (100) and (111) faces, dissociative chemisorption occurs readily in the presence of steps + x. Similarly, steps can decompose carbon monoxide. Chemisorption of hydrocarbons on stepped platinum surfaces can lead to hydrogen-containing 'carbonaceous' layers, important for catalytic reactions, whereas on low-index faces graphitic layers are formed. A study of the hydrogenolysis of cyclopropane on a vicinal platinum surface at atmospheric pressure 40) shows behaviour very similar to that of a highly dispersed supported catalyst and so suggests that stepped single crystals may serve as excellent models for supported catalysts.

Diffusion. It has been suggested 41) that diffusion along steps may be very rapid and so effectively short-circuit other surface diffusion processes. If this is the case the presence of steps due to emerging dislocations might make the usual TLK model invalid.

Electrochemistry on carefully oriented single crystal vicinal surfaces shows effects attributable to steps which might ultimately be used to check the residual surface defect concentrations on low-index faces 42).

1.4. Point defects and clusters

Little is known about the real, thermally created, defect structures at solid surfaces. Our ignorance stems partly from theoretical difficulties and the uncertainties regarding surface forces, partly from the lack of appropriate experimental methods.

+Since 'accidental' steps will be present on large areas of nominally low-index surfaces the very low reactivities are surprising. Possibly these steps are contaminated by residual impurities.

xA later investigation 39) shows that adsorption on the low-index planes may be inhibited by 'clean-off' reactions with carbon monoxide in the vacuum system. Nevertheless, steps produce a marked difference in behaviour.

The concentration of surface vacancies and self-adsorbed atoms (adatoms) will evidently increase with temperature and since the surface is in contact with an infinite source of vacancies one may expect the defect concentration to be higher than in the bulk. High defect concentrations tend to make it energetically easier to generate more defects and a complete atomic-scale roughening -surface melting- may occur. This cooperative phenomenon has been treated theoretically by several authors 29) 43) 44). The following simplified treatment due to Mullins 45) brings out a clear physical picture : the degree of roughening depends on the competition between low total bond energy and high configurational entropy.

A (100) surface of a simple cubic lattice is condidered and it is assumed that adatoms (surface level + 1) and vacancies (surface level - 1) are the only defects (fig. 4). The fraction of sites occupied by atoms in different levels are respectively X_1, X_{-1} and X_o (original surface level) ; $X_1 + X_o + X_{-1} = 1$. In a simple bond model, with a bond energy between nearest neighbours only, $\emptyset/2$ is the extra energy associated with a free bond. The energy associated with a particular defect concentration can therefore be calculated by counting the free bonds parallel to the original surface. (The number of free bonds normal to the surface is unaltered by roughening). Thus an additional energy will be associated with every pair of sites at which the atoms have different levels. Only the three combinations shown in fig. 4 contribute extra energy. It is clear from fig. 4 that the calculation will be unchanged if adatoms and vacancies are interchanged and therefore one may deal with only one variable : $X = X_1 = X_{-1} = (1 - X_o)/2$. From the concentrations of the different combinations one obtains the result that the total energy of any configuration, characterized by

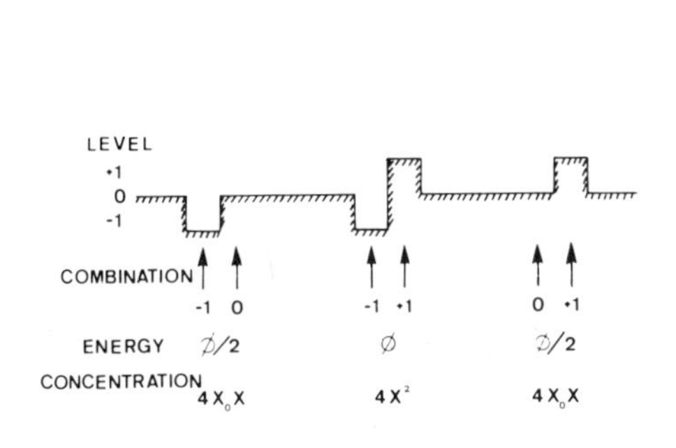

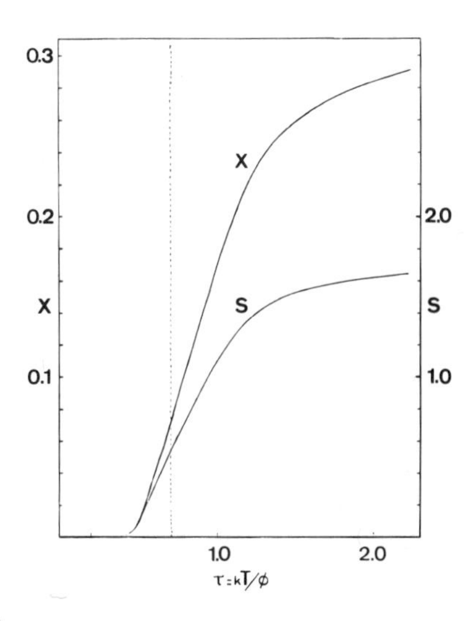

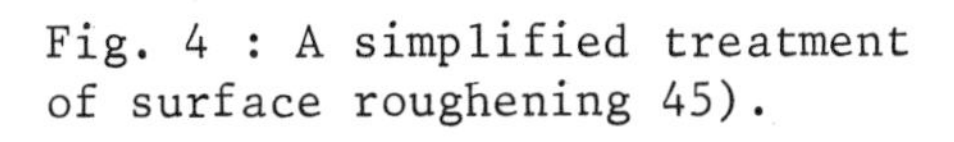
Fig. 4 : A simplified treatment of surface roughening 45).

a particular value of X, is $U = 4\,\emptyset\,X(1 - X)$. If the surface entropy arises only from the multiplicity of configurations (contributions from vibrational terms are ignored) then simple statistics gives for the entropy per site, n being the number of sites,

$$s = (k/n)\ \ln\left[(n!)/(nX!)^2(\left[1-2X\right]n)!\right]$$

$$s = k\left[2X\ln X + (1-2X)\ln(1-2X)\right] \quad (1)$$

The equilibrium condition $dU/dx - Tds/dx = 0$ then yields

$$X/(1-2X) = \exp\left[-2\emptyset/kT(1-2X)\right] \quad (2)$$

and at low concentrations

$$X \approx \exp\left[-2\emptyset/kT\right] \quad (3)$$

Equation (2), in the form $x = f(kT/\emptyset)$ is plotted in fig. 4, together with the roughness factor S, defined as the number of free bonds parallel to the surface per site. For this model $S = U/(\emptyset/2) = 8x(1 - X)$. The temperature scale may be interpreted by noting that for a rare gas solid -the type of solid best represented by this simple model- the melting point is given approximatively by $kT_m/\emptyset \sim 0.7$. It is seen that on this simple model complete roughening would begin to set in at temperatures very much higher than the melting point. However, the defect concentration 2X may be as high as 0.15 at the melting point.

Mullins' curves are close to those obtained by Burton, Cabrera and Frank 29) from a less approximate treatment. The results show no sudden changes of roughness that could be ascribed to surface melting below the bulk melting point, but the defect concentrations are quite high.

A more complete and rigorous theory of surface roughening presents formidable difficulties. There are many unknown quantities. Among the factors that must be considered are the effect of steps, vibrations, translational freedom (surface diffusion), cluster formation, the effect of including many atomic levels to describe the surface topography, the effect of a more realistic model for binding - including more neighbours and the changed interatomic potential at the surface⁺. Most of the theoretical progress to date has been made in establishing methods of calculation and in treating idealised models. There has emerged no clear picture of the probable

---

⁺ Estimations 6) of the relaxation around a surface vacancy, using pairwise bulk potentials, give an inward 'collapse' of up to 15 percent -but effects of charge redistribution on such calculations are unknown. A surface defect may itself change the charge distribution and the effective interatomic potential will differ from both the bulk and the 'surface terrace' potential. Understanding of the *interactions* between point defects is not very advanced.

defect structure of a metal surface at, say, above half the bulk melting point. A reasonable guess is that atomic steps so greatly facilitate the production of defects that their concentration exceeds 0.1 in this high temperature range. Dynamic factors, considered in the next section, suggest that just below the melting point a surface may be completely disordered.

On the experimental side field ion microscopy with its capability of imaging single atoms seems a potentially valuable technique for looking at surface defects. However, while adatoms and surface vacancies are detectable by FIM, the observations are generally limited to low, cryogenic, temperatures. The small dimensions of the emitter tip means that annealing of non-equilibrium defects occurs very rapidly by surface diffusion and it is rarely possible to observe quenched-in high temperature surface structures. A recent review by Seidman 46) emphasizes the applicability of FIM techniques to bulk rather than surface defects. Studies of deposited adatoms -their migrations and interactions- may ultimately lead to a better knowledge of surface defect structure, but probably only for quite low temperatures.

LEED, as we have noted, is relatively insensitive to surface defects. This is true insofar as the positions of the diffraction spots is concerned, but not as regards the intensities. Disordering reduces the diffracted intensities and increases the background. The effect is clearly shown in experiments in which amorphous silicon is formed on a silicon crystal substrate 47), -a fraction of a monolayer produces a noticeable reduction in intensity. Unfortunately it is difficult to use LEED to look at thermally created defects because the temperature-dependent atomic vibrations also affect the electron scattering processes. It is difficult also to separate the effects of different atomic layers. Observations have been reported on LEED patterns from lead, bismuth and tin crystals for temperatures right up to the bulk melting points. However, the persistence of the diffraction pattern does not mean that the top layer is not molten -crystalline second and third layers could produce the pattern.

Recent developments show that the sensitivity of work function measurements to surface imperfections may ultimately provide a very useful means of looking at point defects. It has been found 48) that a single tungsten atom deposited on a (110) tungsten substrate exhibits a dipole moment of about 1 D. For a coverage of about half a monolayer of adatoms there occur work function reductions of up to 0.6 eV (fig. 5). The effect is similar to that produced by atomic steps and can be understood in terms of the smoothing of the surface electron distribution. There is marked effect of temperature : it appears that even below 100°C the deposited atoms diffuse and form islands and so tend to restore the surface potential to its original value.

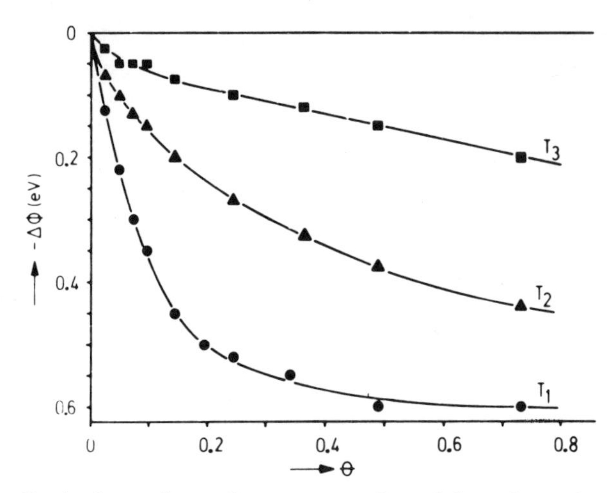

Fig. 5 : Work function changes produced by the adsorption of tungsten on tungsten (110) for different coverages Θ and for various substrate temperatures. T_1 = 20°C < T_2 < T_3 < 100°C. Ref. 48).

It is not known whether 'island clusters' are stable components of the equilibrium structure of metal surfaces at high temperature. Bonzel 49) has suggested that trimers stabilized by rotational motion are largely responsible for the high surface self-diffusivities of a metal near the melting point. The hypothesis is difficult to test. Observations of dimers and trimers have been made by FIM and there is currently great interest in their properties 50).

1.5. Dynamic aspects of surface structure

The static picture presented by simple surface models may be misleading, especially at high temperatures. A realistic description must include atomic vibrations - harmonic and anharmonic, diffusive jumps, possibly large-scale mobility and information not only on defect concentrations but also on their lifetimes. Such a complete dynamic picture is not yet available.

Theoretical methods have been developed 51) for treating surface vibrations, for examining the thermal dependence of surface expansion, mean square displacements of surface atoms and changes in the phonon spectrum. From various calculations the general conclusions are that normal mode frequencies are lowered for surface atoms and as a consequence there can be an increase in the mean square displacement of about 50 per cent. It is difficult to take into account possible electronic rearrangements at the surface.

Likewise, at least for temperatures above 0.1 Tm, there are doubts about the rates of various elementary diffusive motions on surfaces. While it is diffusing an atom may be regarded as a 'dynamic defect'. As has been commented elsewhere 52) the very high value of the self-diffusion coefficient of clean metals near their melting points ($D \sim 3x10^{-4} cm^2 \ s^{-1}$) suggests that *all* the surface atoms may be regarded as being in diffusive motion *all* the time - the surface is thus entirely defects. (At 0.9 T_m the diffusion data suggest that there may still be about 10 per cent 'dynamic defects'). There have been no theoretical attempts to treat this problem in more detail. At present an understanding of both the dynamic and defect properties of surfaces remains very limited.

2. Surface thermodynamics

2.1. Surface equilibrium : specific surface vork

Linford has written a very complete account of the surface thermodynamics of solids 53). He makes a special plea for rigour in nomenclature and symbols which, because of the confusion in the literature, is worth repeating. Following Lindford we call the most important quantity in surface thermodynamics the "specific surface work" and give it the symbol γ. It is defined as the reversible work required to form unit area of surface, at constant temperature, pressure and number of moles of each component, ideally by cleavage. It is a property of the whole system, bulk plus surface phases. The reader is urged to consult Lindford's excellent chapter⁺ for a full discussion of the distinctions between γ and the specific surface Helmholtz energy f_π (change in energy of the surface region or 'phase' as opposed to the change in energy of the whole system) and the surface stress tensor g_{ij} (the reversible work required to produce unit area of new surface by stretching). Different authors have called γ surface energy or surface tension. Much of the confusion arises because for a one-component liquid $\gamma = f_\pi = g_{ij}$ and is called surface tension. The importance of γ lies in the fact that it is a partial derivative of the Gibbs energy of the whole system and so determines equilibrium shapes and the configurations of intersecting surfaces.

Before discussing equilibrium it is important to consider the time required to reach an equilibrium situation. For changes of topography Herring has established scaling laws that show how changes produced by different mass transport mechanisms depend on distance 55). For small enough transport distances (normally up to several tens of microns for metals above 0.5 T_m) transport is predominantly

---

⁺Other important discussions of surface thermodynamics will be found in the ASM-AIME symposium 'Metal Surfaces' (1963) and papers by Winterbottom 54).

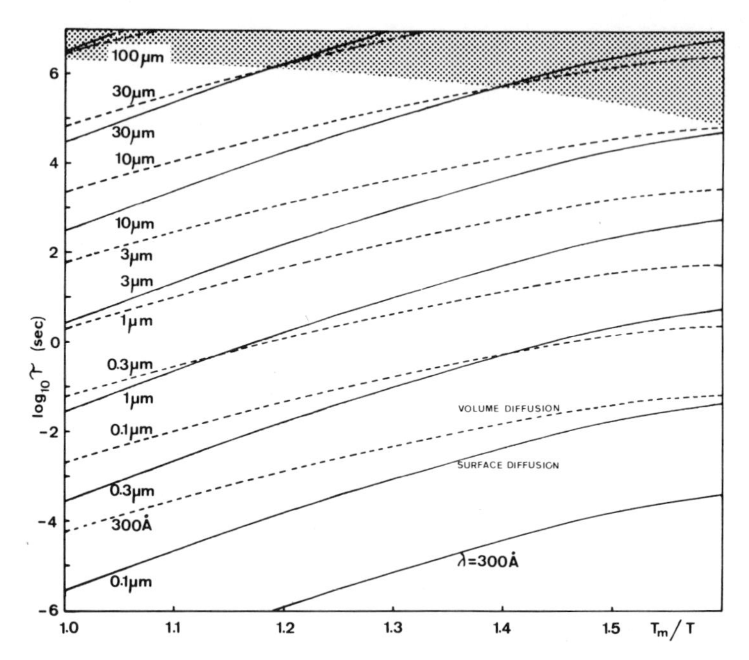

Fig.6 : Relaxation times for profile changes at different scales produced by surface or volume diffusion. Typical data for metals. Except in the shaded region surface diffusion dominates.

---

by surface self-diffusion. It is possible to make a simple plot of the characteristic relaxation time τ required for a change in topography over a distance λ. Such a plot is shown in fig.6. The relaxation times have been calculated for the exponential decay of a sinusoidal profile, wave length λ, using surface and bulk diffusion data normalized to the reduced temperature T/T_m.

The most significant relations in surface thermodynamics are:

$$\Delta \Sigma \gamma A_\pi = 0 \qquad \qquad (5)$$

and

$$d\gamma = - S_\pi dT - \sum_{i=1}^{n} \Gamma_i d\mu_i \qquad \qquad (6)$$

The first equation expresses the condition for topographical equilibrium - namely that the total surface work summed over all surfaces in the system (areas A_π) must be a minimum with respect to a virtual geometrical displacement. Eq.6 is the Gibbs adsorption equation for the variation of γ with temperature and with the adsorption of a chemical species i at a chemical potential μ_i and a surface excess (coverage in moles per unit area) Γ_i. S_π is the specific surface entropy.

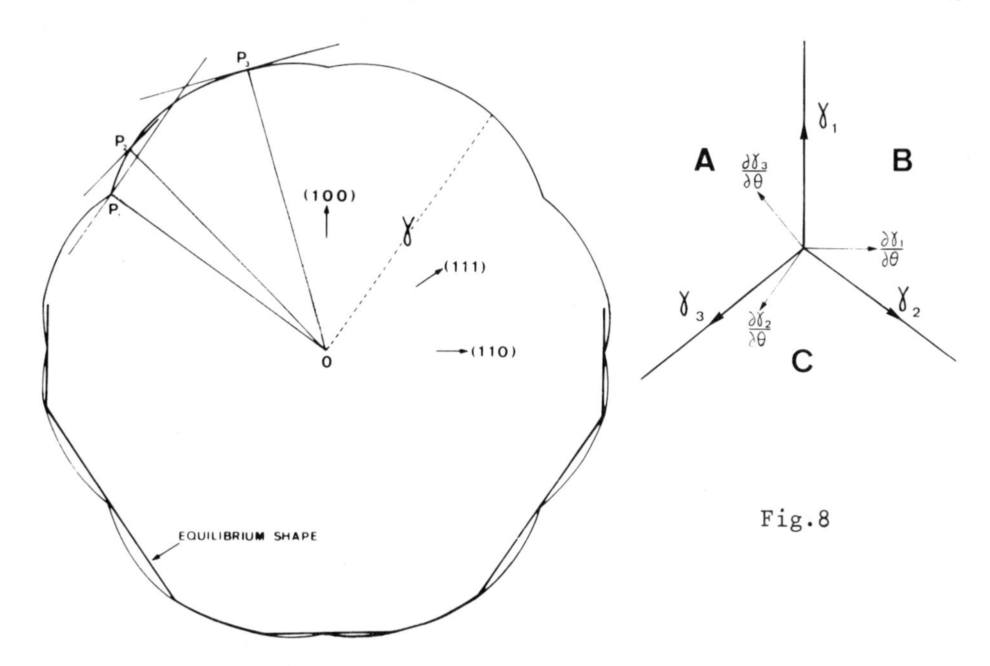

Fig.8

Fig.7 : The Wulff construction (two dimensional representation). Planes are drawn perpendicular to all radius vectors of the γ-plot. The equilibrium shape is geometrically similar to the body formed by all points reachable from the origin without crossing these planes.

For solid surfaces and interfaces γ is a function of the crystallographic orientation and the variation is plotted in a polar diagram known as the γ-plot. The γ-plot can be analysed in term of the terrace-ledge-kink model. The atomic steps and kinks add positive contributions to the surface work so that cusps of minimum γ occur at low-index orientations.

Eq.5 is the basis of both the Wulff construction (fig.7), relating the equilibrium shape of a crystal to the γ-plot, and the Herring equations 56) which describe equilibrium between intersecting surfaces or interfaces. These equations can be derived from the vector diagram of fig.8 which expresses the fact that each surface tends to reduce its area (vectors γ_1 , etc) and also tends to rotate toward crystallographic orientations having lower values of γ. The latter effect introduces 'torque terms' $\partial\gamma_1/\partial\theta$, etc.

Since the early 1960s many experiments have shown that such surface 'capillarity' phenomena as grain and twin boundary grooving, smoothing of surface asperities and undulations, faceting, blunting of field emitter tips, etc, are indeed governed by the above basic thermodynamic expressions. The Gibbs adsorption equation has been shown to be valid for solid surfaces 57).

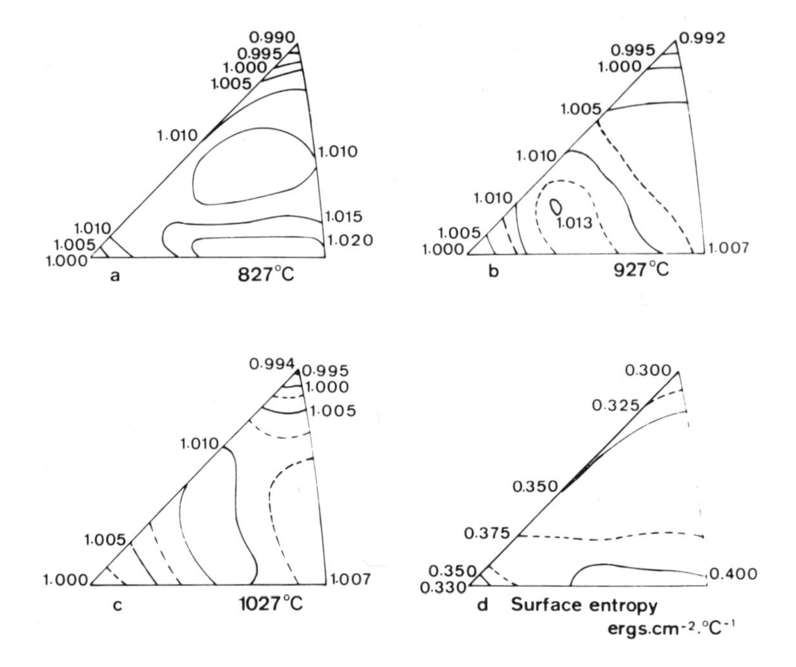

Fig.9 : γ-plots and surface entropy for clean copper surfaces 58).

A particularly detailed experimental study of γ-plots has been made by McLean 58) for clean copper. The results (fig.9) show that the variation with orientation is quite small (∿ 1 per cent). This means that all orientations are present in the equilibrium shape as determined by the Wulff construction. The same conclusion is believed to hold for most metals - at least for the temperature range where topographical equilibrium can be expected to be attained. The stability of vicinal and high-index surfaces, as observed by LEED, is not, therefore, surprising. An important aspect of McLean's results is that they cannot be explained by simple pairwise bonding models. Some experimentally determined orders of magnitude for copper are worth mentioning: typical surface entropies: 0.35 $mJ/m^2/°C$; contribution of atomic steps to γ : 2.5×10^{-11}J/m ; entropy associated with steps : 10^{-13}J/m/°C.

2.2 Topographical equilibrium

While the available evidence suggests that all orientations of clean metals should be topographically stable, it is known that adsorption can have a marked effect on the γ-plot and hence on surface stability. The change in anisotropy comes about through the general decrease of γ for all orientations and also because of the anisotropy of adsorption - the integral of the second term in eq.6 depends on the orientation.

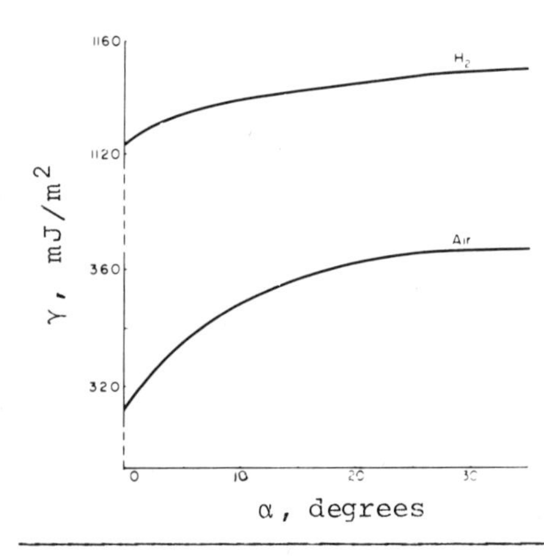

Fig.10 : Changes in γ (900°C) for silver surfaces near (111) that lead to faceting on adsorption of oxygen.
α, angle from the low-index orientation, 59).

The most striking effect of adsorption is faceting - the spontaneous break-up of a flat surface into a hill-and-valley structure comprising linear strips of certain crystallographic planes. The total area is increased but the total surface work is decreased by the development of planes with low γ.

Moore 60) gives a review of faceting on different metals but is overcautious in accepting the γ-plot explanation. The effect can be observed only if the atomic mobility is high enough and it has been examined in most detail for silver heated in oxygen at various pressures 59) 62). Specimens heated at 900°C develop either (100) or (111) facets which form a saw-tooth profile with ridges made by contacts with high-index surfaces. The angle across the ridge varies with the oxygen partial pressure; it can be predicted from the Herring equations - the cosine is approximately equal to the ratio of γ between the low and high-index faces.

Measurements 59) of the γ-plot for these faceted silver surfaces, determined from the equilibrium groove angles at twin boundaries, leads to the results shown in fig.10. Essentially the facets occur because of a very large reduction in the average value of γ and a small ($\sim$4 per cent) anisotropy of oxygen coverage. Together, these factors lead to very pronounced cusps at the low-index poles. The data of fig.10 permit a verification of Herring's criterion for the stability of a surface with respect to faceting 61). Part of the two-dimensional section of the γ-plot is drawn in polar coordinates (centre O) in fig.11. Considering any point A, Herring's criterion is : "A continuous curved surface with normals in the neighbourhood of OA will be stable with respect to the formation of a hill-and-valley structure if, and only if, the γ-plot passes nowhere inside the sphere passing through the origin and the tangent at A". Fig.11 illustrates this criterion for four different orientations. It is seen that the criterion involves the whole of the γ-plot and not just the local curvature. Simple analytical methods for predicting faceting from the shape of the γ-plot are not, there-

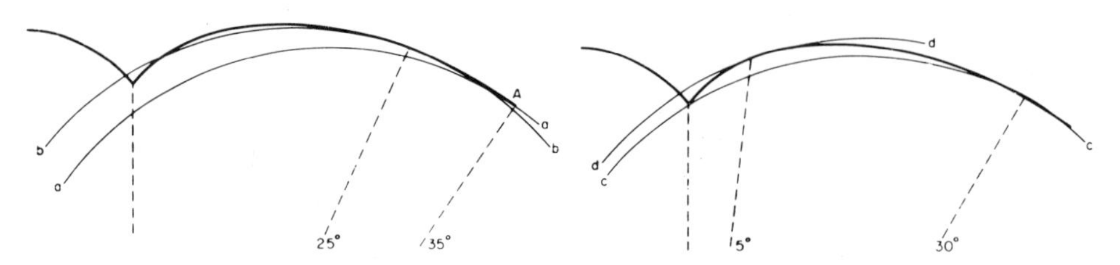

Fig.11 : The Herring criterion for faceting applied to γ-plots for silver in air at 900°C 59).

fore, easily applicable. For example at 25° from the pole the local curvature suggests stability, nevertheless the cusp penetrates the sphere and this orientation facets. Mullins has suggested that this type of faceting (as distinct from the faceting for the orientation at 5° in fig.11) is associated with a nucleation barrier - but there is no strong experimental evidence for this. It has been shown that facets grow by a surface diffusion mechanism 62). Generally facet densities are between 1 and 0.1 μm^{-1}. Probably growth starts at places where suitably oriented steps are more than one atom high. Facet growth must be accompanied by rapid equilibrium between the adsorbate and the newly created surface. This fact may explain some hot-stage microscopy observations of the early stages of growth that show fluctuations in size and even the occasional disappearance of small facets.

The γ-plot can vary markedly with temperature, becoming more sharply cusped at lower temperatures - especially if adsorbates are present. As a result *quenching* may produce new facet planes that are frequently seen as fine striations+. The presence or absence of these striations has sometimes been used as a criterion for marking a particular stage of adsorption and thus used to determine correspond ing heats of adsorption - a procedure that has little justification.

There have been many observations of faceting in LEED experiments on metal surfaces with various adsorbed layers. Facets produce new specular reflexions and associated diffraction spots. Since vacuum conditions usually prevail it is not certain whether the phenomenon is always a surface energy effect - it may be caused by preferential evaporation accentuated by the adsorbate.

The effect of adsorption on topographical stability can depend on rather subtle changes in the γ-plot. Facets appear when the cusps become sharply inward-pointing but another effect is also possible. Fig.12 illustrates schematically how the γ-plot changes continuously over the whole range of adsorption. A vicinal surface

+'Quenched striations', visible in the optical microscope, and not due to a high temperature anneal, can be produced in times of the order of a second or less. But because of the rapid reduction in relaxation time with temperature large-scale (> 1μm) surface topography is conserved during quenching.

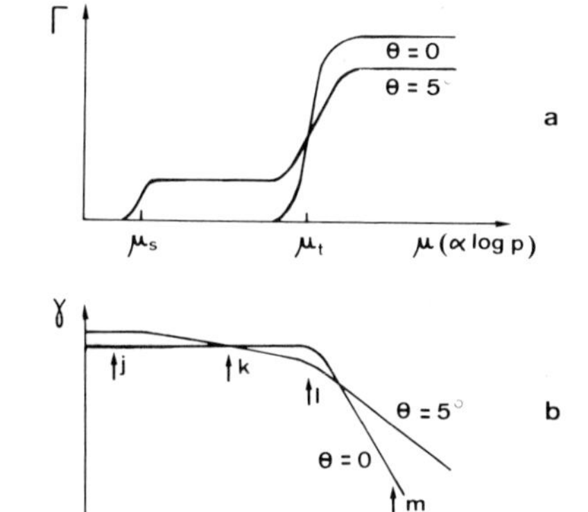

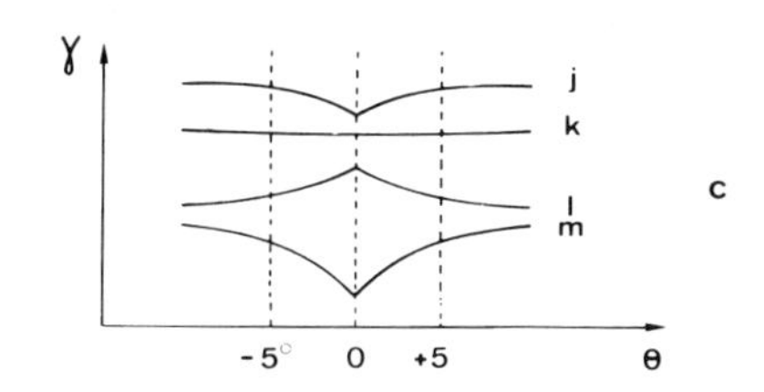

Fig.12 : γ-plot changes over the whole range of adsorption (schematic). Γ, coverage.

($\theta=5°$) is compared with a low index face. The characteristic chemical potential for adsorption on steps μ_s will generally be lower than that for terraces μ_t. As a result there exists a certain range of adsorption conditions for which the cusp can become *outward-pointing*. The Wulff construction then predicts that the low-index surface and a small range of vicinal orientations will be unstable. Such an effect has been observed 63) for silver surfaces with adsorbed sulphur in the temperature and pressure range 600-700°C and $0.08 < pH_2S/pH_2 < 0.20$. Orientations near (111) and (100) are unstable and break up into shallow conical pits with sides inclined at about 10°. The observations agree with measured adsorption isotherms. An outward-pointing cusp implies that the step energies (work) have become negative - a step will then tend to increase its length spontaneously. This process, accompanied by a regrouping at a different step density, leads to the pits. Negative contributions to γ are a direct consequence of the Gibbs adsorption equation ; there may exist conditions for which the total surface energy goes negative. The whole surface would then be unstable and the crystal could spontaneously disintegrate as an aerosol. Such an effect has not been observed for solid surfaces but spontaneous emulsification is known for liquid-liquid interfaces 63).

In connection with surface entropy Gruber and Mullins 64) have shown that γ for a vicinal surface can be written as

$$\gamma = \gamma_o \cos\theta + \frac{\emptyset\varepsilon}{h}\sin\theta \qquad \ldots\ldots\ldots\ldots\ldots \quad (7)$$

where γ_o refers to the low-index terrace, θ is the angle from the pole, ε is a step energy per unit length, h the step height and $\emptyset < 1$ is a number such that $\emptyset\varepsilon$ is a free energy that includes entropy effects. It can be shown that if $\emptyset$ is independent of θ all vicinal orientations are unstable - contrary to the observations for clean surfaces. Thus surface stability may be very sensitive to how the step entropy is affected by the step spacing - a higher configurational entropy may be associated with a greater freedom to 'meander'.

Winterbottom 65) has extended the theory of the Wulff construction to the equilibrium shape of a small crystal in contact with a substrate. At the orientation where there is contact the appropriate surface work is $\gamma_{sp} - \gamma_{sv}$ where the first term corresponds to the interface between the substrate and the particle and the second to the substrate-vapour interface. There is thus a singularity in γ for this direction: elsewhere γ_{pv} (particle-vapour interface) describes the γ-plot. The Wulff construction is still valid and can predict the degree of adhesion. 'Wetting' corresponds to an inward singularity (cf. faceting). But there can occur an outward singularity if the work of adhesion $\gamma_{pv} - (\gamma_{sp}-\gamma_{sv})$ is negative. There is then no surface of adhesion (cf. pitting).

2.3 Chemical equilibrium

For equilibrium with respect to changes of chemical composition the chemical potential for each element must be constant throughout the system. Usually the chemical potential is fixed by one of the bulk phases which acts as a chemical reservoir. The various processes by which a surface can reach equilibrium with the bulk phases are, of course, adsorption, desorption, segregation and dissolution. (If there is a chemical potential gradient *along* the surface changes in composition can also occur by surface diffusion).

A discussion of the thermodynamics of adsorption from the gas phase is outside our scope. We consider briefly the equilibrium of dissolved adsorbates that segregate from the bulk to the surface.

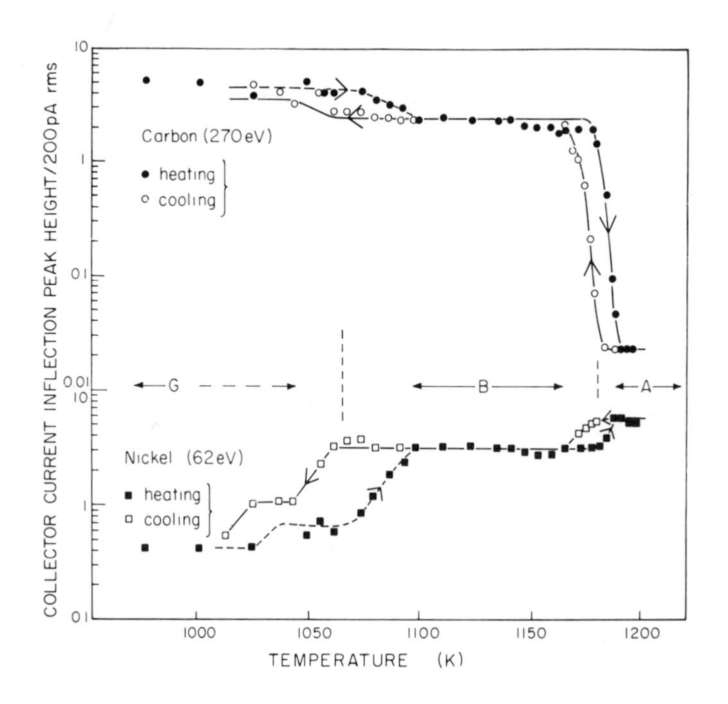

Fig.13

It is possible to achieve adsorbate-surface equilibrium via segregation even in the ultrahigh vacuum conditions required by many surface techniques. A recent comprehensive review of equilibrium adsorption and segregation is given by Blakely and Shelton 66).

Auger electron spectroscopy (AES) can be used to monitor surface concentrations under near-equilibrium conditions. Some results are shown in fig.13 for the segregation of carbon to a Ni(111) surface 66). The specimen had been previously doped at 0.3 per cent by treatment in CO. During slow heating the Auger peak heights change and one can distinguish three regions that can be correlated with LEED observations. In the range A there exists a very dilute carbon phase at the surface, probably with the bulk concentration. The range B corresponds to segregation of an epitaxial monolayer with the graphite structure. At lower temperatures bulk precipitation occurs (G). From the temperature range of the phase B it is possible to estimate the binding energy of carbon in the monolayer. The sharpness of the A-B transition near 1180 K indicates sharply varying adsorption isotherms of the Fowler rather than the Langmuir type : interactions within the adsorbate are important, as may be expected for a graphite structure. In other work, on polycrystalline surfaces 66) segregation data seem to be compatible with Langmuir isotherms but the effect of crystalline anisotropy may mask the real behaviour: a reliable comparison with theory generally requires the use of well defined single crystal substrates.

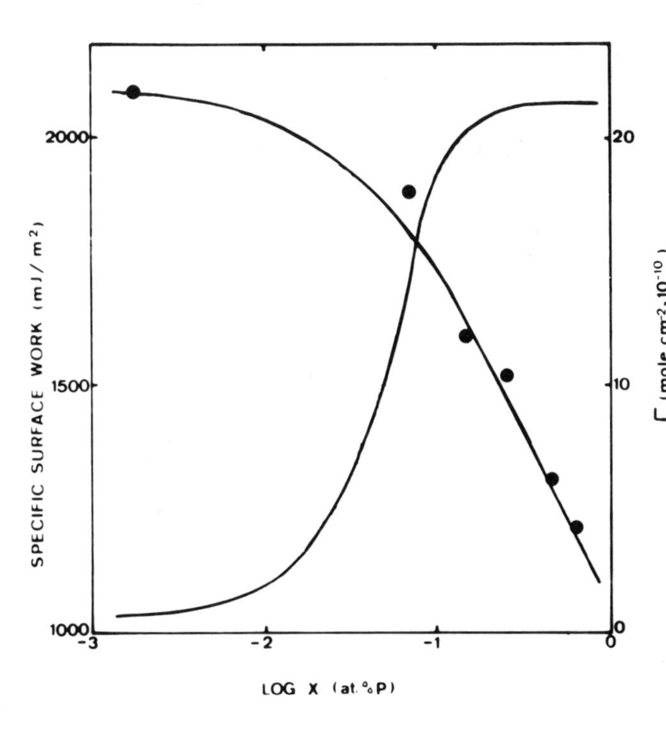

Fig.14 : Measured values of γ for iron at 1450°C, with different concentrations of phosphorus 57).

Measurements of the specific surface work γ, by the zero-creep method, have also been used to study segregation in various metal alloys 57). Fig.14 shows some results for phosphorus segregation in iron at 1450°C. The adsorption isotherm, derived by applying eq.6 to the curve for γ, shows a much sharper change of coverage than is predicted by the Langmuir isotherm.

While it is well understood that bulk and surface compositions can differ quite markedly at equilibrium and that any solute that lowers the surface energy is expected to segregate, a rigorous theoretical treatment of segregation is difficult. The general approach (comparable to treatments of surface roughening) is to calculate the Gibbs energy using, for example, nearest neighbour binding models and interaction energies derived from bulk properties. The limitations of such methods have already been mentioned. We will return to this point later.

The kinetics of segregation, or conversely the dissolution of a surface layer into the bulk, are important in many situations. For dissolution it is possible to relate the time variation of the surface concentration to the bulk diffusion coefficient D and the shape of the adsorption isotherm 67). Fig.15 shows such relations for different Fowler isotherms (w being the adsorbate interaction energy). The plots are normalized to a dimensionless reduced time, τ, equal to $Dt\,[X_b(\theta=1/2)\,/\,\Gamma_s(\max)]^2$. $X_b(\theta=1/2)$ is the equilibrium bulk concentration corresponding to half-monolayer coverage and Γ_s is the surface concentration. Thus from fig.15, using typical values $X_b \sim 10^{18}$ atoms cm^{-3} and $\Gamma_s(\max) \sim 10^{15}$ atoms cm^{-2}, one finds that to dissolve half of a complete monolayer in 1 hour requires a diffusivity of 10^{-10} cm^{2}s^{-1}.

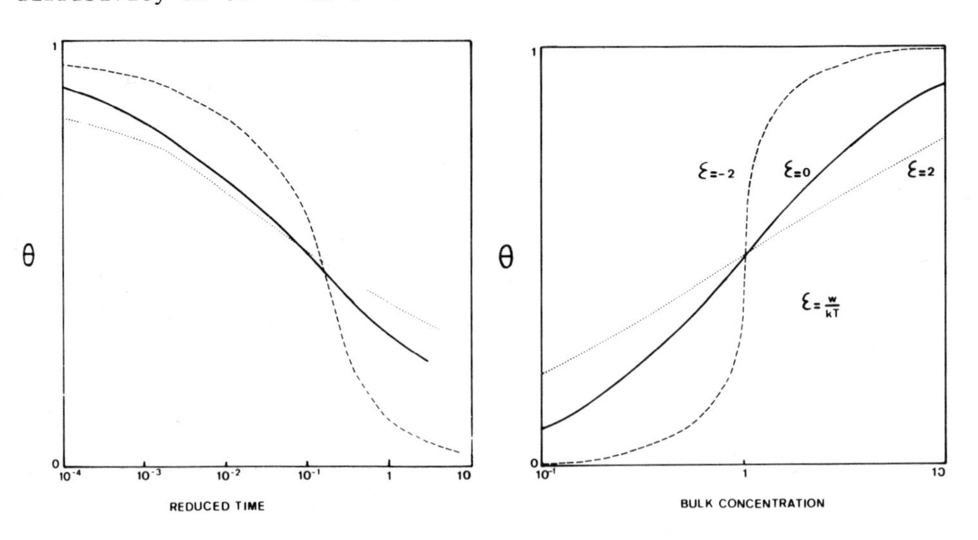

Fig.15 : The kinetics of monolayer dissolution (left) corresponding to different adsorption isotherms (right).

It is also possible for segregation to occur due to non-equilibrium effects involving defects or other impurity atoms. For example, the surface can act as a sink for vacancies. Any excess vacancies produced in the bulk by heat treatment or during solidification may migrate to the surface dragging solute atoms along with them. If this happens the segregated layer can extend over depths of tens of microns 68). Recrystallisation of surface layers and the annealing out of bombardment damage may produce such effects. Excess vacancies can in fact give rise to either surface enrichment or depletion, depending on the solute-to-solvent diffusivity ratio and on the magnitude of the solute-vacancy binding energy 69).

3. Binary metal systems

Combining metals in certain proportions so as to produce a material with desired properties is a traditional path of the metallurgist. With this approach and recently developed surface techniques a new *surface* metallurgy is emerging.

3.1 Metal overlayers

At pressures corresponding to undersaturation a metallic vapour can be adsorbed in monolayer quantities at thermodynamic equilibrium. An early study of the equilibrium adsorption of a metal was made by Bailey and Watkins 70) for lead on copper. These authors measured, for temperatures near 850°C, solid:solid and solid:liquid interfacial energies (surface works γ) with reference to liquid lead and ascribed the lowering of γ for the copper:vapour surface ($\sim$ 1.8 Jm^{-2} in hydrogen to $\sim$ 0.78 Jm^{-2} in lead vapour) to the formation of an adsorbed lead monolayer. The values of γ were deduced from simple microtopographical observations. This interesting method has not been much pursued. More recently the techniques of mass spectrometry and vapour beams have been used to obtain both thermodynamic and kinetic data on adsorbed metal phases 71).

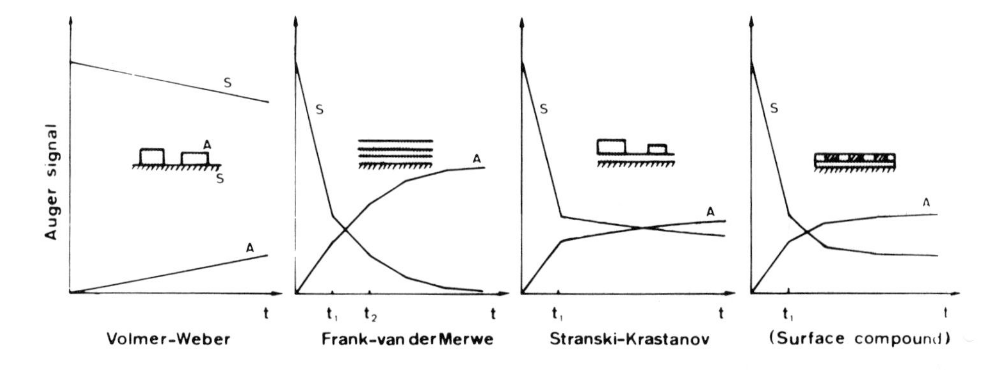

Fig.16 : Variations of Auger signals for the substrate and the adsorbate during film growth by different mechanisms.

Alternatively, monolayers can be obtained by interrupting the growth of a thin film as it forms from a supersaturated vapour. An adsorbed monolayer represents a first stage in both the Frank-van der Merwe and the Stranski-Krastanov mechanisms (Fig.16). The different growth modes are distinguishable by Auger electron spectroscopy and the completion of the first monolayer can be characterized by a sharp change in slope (knee) in the Auger signal versus deposition time (AST) plot.

An analysis 72) of the AST plots for layer-by-layer growth (valid for all growth modes up to the first monolayer) gives, for the adsorbate and substrate signals at the completion of the nth layer, the relations :

$$I_a^{(n)} = I_a^1 \left[1 + \alpha_a^a + (\alpha_a^a)^2 + \ldots (\alpha_a^a)^{n-1} \right] \qquad \ldots\ldots (8)$$

$$I_s^{(n)} = (\alpha_s^a)^n \, I_s^o \qquad \ldots\ldots\ldots\ldots\ldots\ldots (9)$$

and for $0 < \theta < 1$

$$\frac{I_a}{I_s} = \frac{I_a^1}{I_s^o} \frac{\theta}{1-\alpha_s^a\theta} \qquad \ldots\ldots\ldots\ldots\ldots\ldots (10)$$

where α_s^a represents the fraction of the Auger signal from the substrate transmitted by one monolayer of the adsorbate, etc... The transmission factors α are related to the inelastic mean free paths of the Auger electrons and to the geometry of the spectrometer 73). In the energy range 50 to 500 eV α is typically about 0.4.

Studies of metal adsorbates by LEED fall into two groups. First there are those concerned with low coverages in which binding is most probably at sites of high coordination : LEED intensity data for these layers have been used to test the usefulness of LEED theory as a tool for determining the type of site. Examples of this type of work are discussed by S. Andersson in another part of this course. Secondly, there are investigations of the epitaxy of dense overlayers.

Fig.17 shows a dense arrangement deduced from LEED patterns from copper with adsorbed lead. The coincidence mesh observed is too large to be explained by placing atoms only at corner sites but an hexagonal overlayer with a crystalline parameter within a few per cent of the bulk parameter fits neatly. The gross, averaged, features of the intensity distribution over the pattern can be explained by the modulation imposed by the overlayer coupled with double diffraction effects 74). A special feature of the overlayer on Cu (100) is that it melts at about 100°C below the bulk melting point, as shown by the intensity-temperature plots.

Similar close-packed hexagonal or pseudo-hexagonal arrangements have been deduced for different low-index faces for the following adsorbate: substrate systems - Cu:W, Pb:Pt, Bi:Cu, Pb:Au, Bi:Au, Na:Al, K:Al, K:Ni, Mn:Al. In the case of bismuth the

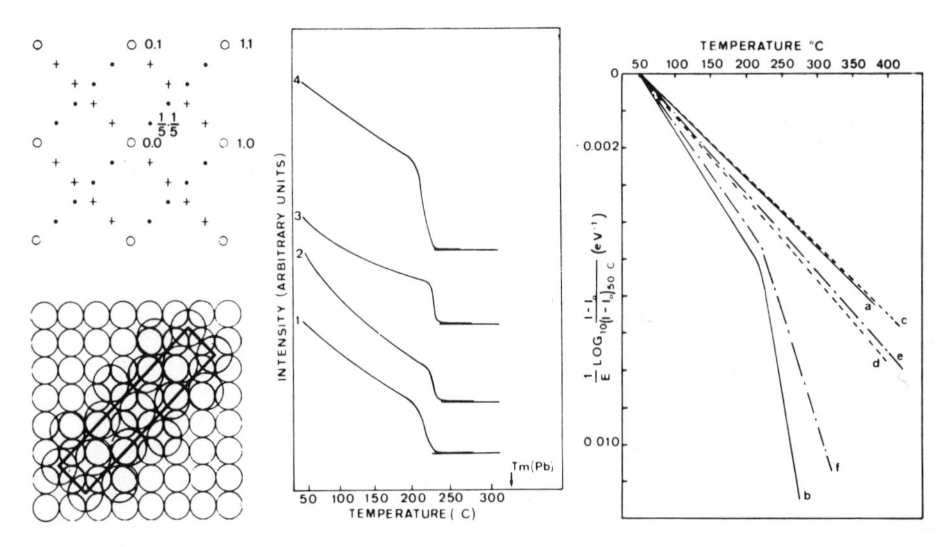

Fig.17 : Adsorbed lead on copper (100). Left : diffraction pattern (circles, substrate reflexions). Interpretation of the coincidence mesh. Centre : Intensity-temperature plots for various superlattice reflexions. Right : Intensities for the pattern from the clean (a, c, e) and lead-covered (b, d, f) substrate. a, b - 00 spot; c, d - 10 spot ; e, f - 11 spot. The sharp changes correspond to melting of the monolayer 74).

close-packed hexagonal structure is not one that exists in the bulk crystal. According to the models proposed, within the sometimes very large coincidence meshes containing tens of atoms, very few adsorbate atoms are at simple high-coordination sites : it appears that lateral interactions within the overlayer can dominate over the effects of substrate geometry. In extreme cases there may be no simple coincidence lattice and the mesh vectors of the overlayer can be irrational multiples of the substrate mesh vectors. Such structures are said to be *incoherent* : an example is the hexagonal layer that forms the anomalous structure of Au(100).

The importance of lateral interactions emerges clearly in studies of the highly ionic layers produced by adsorption of alkali metals. Potassium, for example, when adsorbed on a Ni(100) substrate 75) appears to take up the most uniform distribution compatible with the coverage : the corresponding LEED patterns show a continuously variable ring structure which 'condenses' into a pattern of sharp spots only when the average adsorbate atom spacing happens to coincide with the spacing between 4-fold sites of the substrate. At the highest coverages the ring pattern corresponds to an incoherent dense hexagonal overlayer with an interatomic spacing 8 per cent smaller than the bulk diameter. Similar ring patterns are observed for sodium on nickel 76) but it has been shown by laser simulation that the rings for low coverages can be produced by an overlayer adsorbed in sites but having a well defined *average* spacing determined by repulsive interactions. At high coverages incoherent hexagonal structures form to maximize the packing on (111) and (110) surfaces but a coherent $(\sqrt{2}x\sqrt{2})$ structure forms on Ni(100). For sodium on aluminium 77) on the other hand, where apparently the adsorbate-adsorbate interactions are weak, only coherent structures form.

Studies of the adsorption of cesium on tungsten go back to the work of Langmuir and are part of the historical origins of solid surface science. There has been some disagreement about the various adsorbed structures and their relation to the work function which, as the amount of adsorbed cesium increases, falls by some 3 eV and then slowly rises from a minimum value. LEED patterns for the (100) face progress from $(\sqrt{2}x\sqrt{2})$ to p(2x2) to an hexagonal pattern and the minimum in work function is associated with the transition between the last two patterns. Careful calibration of the adsorbate by AES 78) leads to the conclusion that the last pattern corresponds to a compact hexagonal monolayer while the minimum in work function occurs as this structure is nucleated at antiphase domain boundaries of the in-site p(2x2) structure. Random adsorption in sites can produce the $(\sqrt{2}x\sqrt{2})$ pattern.

Biberian and Huber 79) have recently reexamined compact metallic overlayer structures from the viewpoint of their symmetry properties. Assuming high rather than low symmetry they have established a scheme that leads to models with certain simple configurations, such as single and double atomic chains, and in which all the adsorbed atoms are in sites of high coordination. Modifications of the substrate atomic positions can be included if necessary. The general case of a square lattice with an $(n\sqrt{2}x\sqrt{2})45°$ overlayer mesh is instructive. The first arrangement, $(\sqrt{2}x\sqrt{2})45°$, placed in 4-fold sites, consists of equally spaced rows separated by rows of vacant sites. Increasing the density introduces antiphase domain boundaries (formation of double rows, fig.18 a). These boundaries give rise to a spreading of the 1/2 1/2 spots in the diffraction pattern (fig.18 b) The different $(n\sqrt{2}x\sqrt{2})45°$ structures develop naturally as the density is further increased - they correspond to different densities of double and single rows. Sharp diffraction patterns can be produced even if the structure is not strictly periodic. This description is closely related to Park's interpretation of diffraction from a disordered $(\sqrt{2}x\sqrt{2})45°$ structure with antiphase domains 80). The application of symmetry rules to a $(5\sqrt{2}x\sqrt{2})45°$ overlayer is illustrated in fig.18(c-j).

Laves' symmetry principle may have only limited application to metal surfaces and high-symmetry models may be criticized on the grounds that the distortions in atomic sizes can make them less plausible than low-symmetry models. Furthermore,high symmetry does not explain incoherent overlayers. The high-symmetry models may, nevertheless, have greater validity in the wider field of chemisorption (see Huber and Oudar 81)) and they provide useful procedures for obtaining 'trial structures'. They also demonstrate

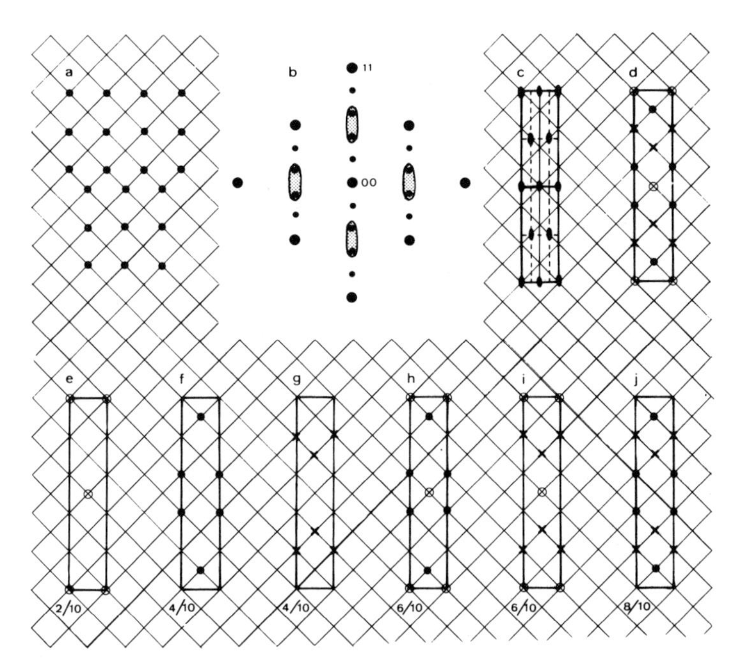

Fig.18 (a) A $(\sqrt{2}x\sqrt{2})45°$ overlayer with an antiphase boundary on a square lattice.

(b) Diffraction pattern for a disordered $(\sqrt{2}x\sqrt{2})45°$ structure (elongated 1/2 1/2 spots) and for a $(5\sqrt{2}x\sqrt{2})45°$ structure (one domain).

(c) The c2mm space group (corresponding to the highest symmetry for the $(5\sqrt{2}x\sqrt{2})45°$ mesh) superimposed on 4-fold sites.

(d) The three sets of sites related by symmetry

(e)-(j) Structures compatible with symmetry. The ratio gives the fraction of sites occupied. Deformation of the structure (h) would lead to the hexagonal overlayer shown in fig.17.

possible continuities between different structures and show how a change in a LEED pattern need not necessarily be due to a distinct change of structure or phase.

In a few cases metal adsorbates are known to react to form intermetallic compounds with the substrate. Gold, because of its electronegativity, is particularly interesting in this respect. When Au(100) substrates with deposited sodium are heated there appear various LEED patterns 82) that can be interpreted in terms of thin layers of Au_2Na. Similarly, lead deposited on gold can form $AuPb_2$. This last reaction occurs at ambient temperatures and has been followed in some detail by LEED and AES 83). The AST plots are of the type shown at the extreme right of fig.16. The signal from lead indicates a layer-by-layer type growth but the signal from the substrate element goes to a non-zero value due to migration of gold through the growing lead film. LEED observations show that at low coverages (about 1/10th of a full monolayer) the lead converts the 'anomalous' gold structures into p(1x1) arrangements. Subsequently there form simple structures (presumably adsorbed in sites) and then dense arrangements that are compatible with close-packed hexagonal structures. Further deposition beyond this point leads to a sudden change to a new hexagonal arrangement that corresponds to the (110) plane of the intermetallic compound $AuPb_2$. Thus the first lead layer alone does not alloy - alloying only occurs when there is sufficient lead to make up at least one unit cell of the bulk structure. The same alloy structure, appropriately oriented along low-index directions, is observed for all three low-index faces, (100), (111) and (110).

Some LEED patterns from Ni(111) with adsorbed molybdenum have also been ascribed to alloying 84) but the large meshes reported are possibly due to coincidence with a simple (unalloyed) overlayer. The system is worth reexamining by AES.

Field ion microscopy has been used - with limited success-to look for surface alloying. Gallium is found to alloy with molybdenum to form Mo_3Ga above 600°C but no similar reactivity is observed on tungsten emitters 85). There is, however, evidence for surface carbides and silicides on tungsten 86). Investigations have been made of adsorbed gold, silver and copper on tungsten emitters 87) and it has been shown that these metals can produce remarkable changes of structure with accompanying changes of work function. Modifications of the γ-plot, rather than surface alloying, is thought to be responsible.

As already noted, LEED can be used to examine the thermal stability of adsorbed layers. Melting leads to the sudden disappearance of 'extra' spots and is easily detected. Similarly, one can observe solid-solid phase changes in monolayers. For example, there occur two dense structures for bismuth on copper (111) - one slightly more compact than the other (fig.19). A reversible transformation takes place at about 250°C and it can be followed by recording the intensities of beams diffracted from each of the structures. First, the intensities of 'extra' spots show exponential Debye-Waller type behaviour. Then, at a temperature that depends on coverage, spots of one structure disappear and are replaced by others. At a slightly higher temperature the more compact structure undergoes a melting type transition - but this does not go to completion. The complexity of this type of thermal behaviour shows a danger of over-simplified models of adsorption: the possibility that the structure of a monolayer can depend quite markedly on both coverage and temperature is often ignored.

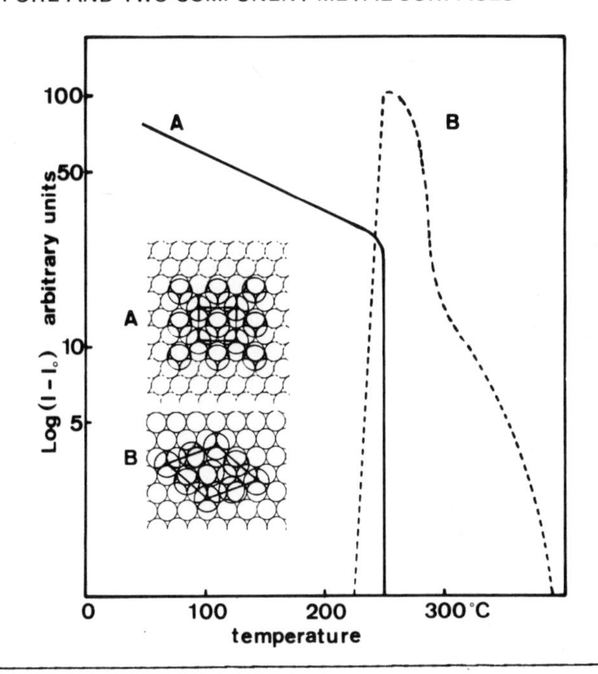

Fig.19 : Temperature dependence of LEED superlattice reflexions from two structures of bismuth on (111) copper, 88).

The whole field of surface phases is ripe for exploration with the new surface techniques. It would be of great interest to compare bulk and surface phase diagrams - to examine the effects of dimensionality and the influence of the substrate. Coadsorption of two elements and measurements of the properties of *binary* monolayers offers a great variety of experiments and the possibility of fabricating surface phases with special properties. LEED and AES are ideal for this type of work and it has been shown that binary metallic monolayers with variable composition can be produced in a controlled way using AES 72). Preliminary results for binary lead-bismuth monolayers on copper show that for a 'two-dimensional' system all the usual features of phase diagrams - solvus, solidus and liquidus curves, intermediate compounds, can be examined experimentally.

Ultimately it may also prove possible to use adsorbed metals to 'decorate' various surface defects and thus measure their concentrations.

3.2 Surfaces of binary alloys

In general the surface composition of an alloy will be different from the bulk. Provided the kinetics are fast enough for equilibrium to be reached the surface will be enriched by the component that lowers the free energy. Basically, the free energy is lowered when the element with weakest bonds is placed at the surface, since there it has fewer neighbours. Different thermodynamic expressions have been used to predict surface compositions of homogeneous binary solutions. The different structural factors and the necessary enthalpies and entropies are difficult to evaluate. Full accounts will be found in refs. 66) 89) 90) ; the following is a summary of the main points.

Two different types of structural model have been considered : one in which only the topmost layer is enriched (monolayer model) and the alternative multilayer model in which a number of successive layers are allowed to change. The alloy may be an ideal, regular or non-regular solution. For the monolayer, ideal solution, model one obtains by equating the chemical potentials of the bulk and surface (considered as distinct phases) and using the condition for ideality $\mu=\mu^{\circ} + RT\ln X$, the relation

$$\frac{X_B^s}{X_A^s} = \frac{X_B^b}{X_A^b} \exp\left[\frac{(\gamma_A - \gamma_B)a}{RT}\right] \qquad \text{.............. (11)}$$

where Xs are concentrations (superscripts, surface and bulk) for an alloy AB. γ is the specific surface work of the pure element and a is the specific atomic area. The component with the lowest γ is enriched at the surface.

For a regular solution the above expression is modified, the right hand side being multiplied by

$$\exp\left\{\frac{\Omega(\ell+m)}{RT}\left[(X_A^b)^2 - (X_B^b)^2\right] + \frac{\Omega\ell}{RT}\left[(X_B^s)^2 - (X_A^s)^2\right]\right\}$$

where Ω is the regular solution parameter given by

$$H_{AB} = \Omega + 1/2\ (H_{AA} + H_{BB})$$

where H_{AA} is the enthalpy associated with the A-A bond, etc ... ℓ and m characterize the crystal structure in that ℓ is the fraction of nearest neighbours in the same plane and m the fraction below the layer containing an atom.

Overbury *et al.* 90) have discussed the above expressions and have reviewed the available information on surface work that might be used to predict surface enrichment. Where experimental data is lacking it is possible to use semi-empirical expressions based on surface tensions of liquids or on heats of vaporization or sublimation. The above expressions show that quite small differences in γ can produce enrichments of several tenths of a monolayer. As the temperature is increased the surface enrichment decreases. Departure from ideality leads to greater enrichment for low bulk concentrations if Ω is positive (A-A and B-B bonds prefered over A-B bonds) and greater enrichment for high concentrations if Ω is negative (A-B bonds prefered).

Williams and Nason 89) have developed in some detail the multilayer model and have used a pair-bonding, quasi-chemical, approach to calculate specific surface works for both ideal and regular solutions. Four layers were found to be necessary and sufficient. The main conclusions are as follows :

- For an ideal solution, Ω=0, only the first layer has a different composition. The monolayer model is valid.
- In the first layer the enrichment (by the element with the lowest enthalpy per bond) is determined by $\Delta H_{sub}/RT$ (ΔH_{sub} = difference in the heats of vaporization of the two pure metals) and by the fraction of the nearest neighbour atoms missing for atoms in the first layer.
- The more atomically smooth the surface the smaller is the change in composition.
- Chemisorption of animpurity leads to preferential enrichment of the component that bonds most strongly to it.
- If, due to surface relaxation, bonds are actually stronger at the surface than in the bulk the enrichment is decreased. Changes also occur in the second layer.
- The regular solution parameter Ω can strongly influence the enrichment in the first layer, enhancing it if Ω is positive (for *all* bulk concentrations, in contrast with the results quoted above).
- In regular solutions with negative values of Ω one finds that because of preference for A-B bonds an enrichment of A in the first layer produces an enrichment of B (depletion of A) in the second layer (see fig.20).
- The results are modified in the case of small particles because segregation can alter the bulk composition and because a range of different crystal orientations are present[+].

Quantitatively these results have not yet been fully checked by experiment. That the quasichemical approach to surfaces may be rather approximate is well understood : it is known that the broken bond models do not always give good estimates of γ and other surface properties of pure metals. The effects of defects, thermal roughening, surface entropy terms and ordered surface phases are all difficult to assess.

These recent advances in the theory of surface enrichment in alloys have been stimulated by the availability of AES as a means for experimental confirmation. Two difficulties arise : that of calibration, which depends on the relative sensitivities of the technique to different elements, and the problem of knowing the depth analysed. As more data on Auger yields and inelastic mean free paths are becoming available these problems are being solved. The spectrum from a freshly ion-bombarded surface is sometimes taken as characteristic of the bulk composition but this assumes that no element is preferentially sputtered. In the case of Cu-Ni alloys, for example, argon ion bombardment preferentially removes copper and enriches the surface in nickel. A brief anneal at 650°C produces a copper-rich surface 92). Different studies of these alloys 92) 93), illustrate the importance of using Auger electrons in the 50 to 200 eV range

---

[+] Burton *et al.* 91) have discussed the application of regular solution theory to segregation in microclusters.

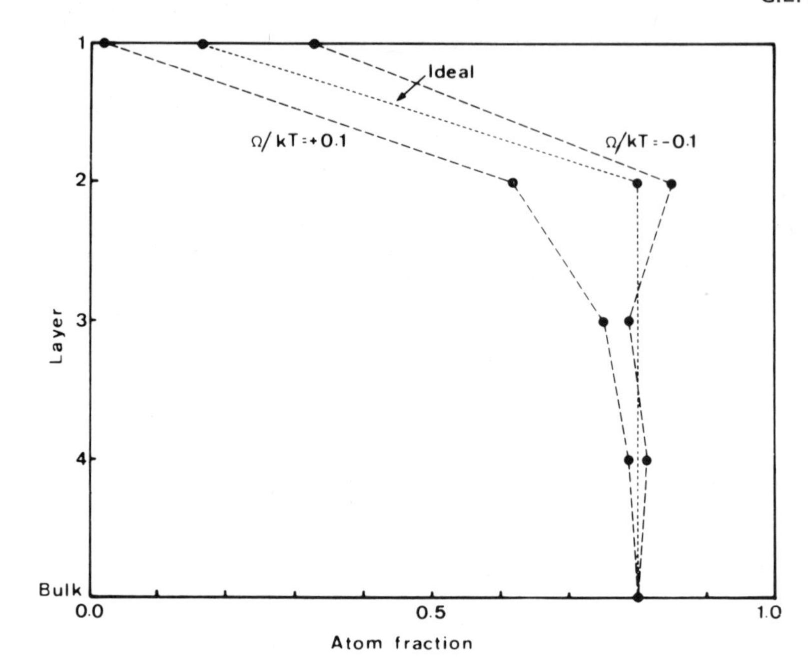

Fig.20 : Theoretical composition profiles for segregation in a binary alloy. f.c.c. (111) surface. Difference in heats of vaporization = 10 kT 89).

for true surface analysis. Auger spectra in the 700-900 eV range suggest no surface enrichment 93) but the low energy peaks near 100 eV indicate segregation of copper - as expected from its lower value of γ. The apparent discrepancy is resolved when escape depths are considered : up to eight atomic layers may contribute to the higher energy emission but only about two at 100 eV.

Concentration profiles at alloy surfaces are difficult to determine experimentally. One possibility is to exploit the effect described above by comparing the Auger signal ratios for different energies, thereby sampling different depths. Another approach would be to vary the energy of the primary excitation beam. Destructive methods such as 'ion-milling' or secondary ion mass spectrometer may not be fine enough to show up the subtle variations plotted in fig.20. Some results for platinum-tin alloys 94) do, however, show evidence for the second-layer depletion effect that has been predicted theoretically. This work was done on the intermetallic compounds Pt_3Sn and PtSn. Calibrations were effected by examining fracture surfaces created *in situ*. After annealing at 500°C surface enrichment of tin, by up to 20 per cent, is observed. This enrichment is enhanced by chemisorbing oxygen and reduced by hydrogen. Data obtained by X-ray photoelectron spectroscopy (high energy emission ∿ 1200 eV) when compared with AES (∿ 300 eV) suggest that there is a depletion of tin under the first layer.

Although the equilibrium surface composition of an alloy should depend quite markedly on temperature, there is, as yet, little experimental information on this point (apart from the work quoted earlier for carbon in nickel). Because of the possibility of fast changes during quenching, measurements should obviously be made with the specimen hot and held at the required temperature for sufficiently long to ensure equilibrium via diffusion. Significantly, the only confirmation of the exponential variation predicted by eq.11 has come from work on liquid alloys of lead and indium 95).

Very little is known yet about ordering at alloy surfaces. The copper-gold system, because of its well established order-disorder transformations in the bulk, has received some attention 96) 97) 98). In spite of gold having the lower γ(by about 300 mJ/m^2) AES indicates little or no surface segregation although sputtering, and also adsorbed sulphur, can cause the surface to be enriched with copper. The temperature dependence of long-range order at a (100) surface of Cu_3Au has been examined by LEED 96). The bulk alloy exhibits a slowly decreasing order parameter up to 390°C at which temperature there is a discontinuous first order transformation. A surface long-range order parameter, S, has been determined as a function of temperature from an analysis of intensities of superlattice reflexions. Basically, use is made of the expression $I \alpha S^2 e^{-2M}$, where M is the Debye-Waller factor, valid in the single scattering approximation. The experimental results show that the surface order parameter, unlike that for the bulk, is a *continuous* function of temperature. At about 60°C below the bulk transformation the surface begins to be appreciably more disordered than the bulk. This behaviour is comparable to the melting of monolayers mentioned above.

References

1) A.J.W. Moore and J.F. Nicholas, J. Phys. Chem. Solids, 20 222-229 (1961).
   J.F. Nicholas, J. Phys. Chem. Solids, 20 230-237 (1961).
   J.F. Nicholas, J. Phys. Chem. Solids, 23 1007-9 (1962).
2) J.F. Nicholas, *An Atlas of Models of Crystal Surfaces*. New York : Gordon & Breach (1965).
3) J.K. Mackenzie, A.J.W. Moore and J.F. Nicholas, J. Chem. Solids, 23 185-196 (1962).
4) J.F. Nicholas, Aust. J. Phys. 21 21-34 (1968).
5) G.A. Somorjai, R.W. Joyner and B. Lang, Proc. R. Soc. Lond. A 331 335-346 (1972).
6) P. Wynblatt and N.A. Gjostein, Surface Sci. 12 109-27 (1968).
7) J.W. Flocken, Phys. Rev. B 9 5133-43 (1974).
8) D. Aberdam, R. Baudoing, C. Gaubert and Y. Gauthier, Surface Sci. 47 181-2 (1975).

9) A. Ignatiev, F. Jona, H. D. Shih, D.W. Jepsen and P.M. Marcus, Phys. Rev. B 11 4787-94 (1975).
10) G. Allan and M. Lannoo, Surface Sci. 40 375-87 (1973).
11) M.W. Finnis and V. Heine, J. Phys. F 4 L37-41 (1974).
12) R. Smoluchowski, Phys. Rev. 60 661-74 (1941).
13) G.P. Alldredge and L. Kleinman, J. Phys. F 4 L207-11 (1974).
14) G.P. Alldredge and S.K.S. Ma, Phys. Lett. 51 A 155-6 (1975).
15) G.E. Rhead, J. Phys. F 3 L53-6 (1973).
16) T.A. Clarke, R. Mason and M. Tescari, Surface Sci. 40 1-14 (1973).
17) D.M. Zehner, B.R. Appleton, T.S. Noggle, J.W. Miller, J.H. Barrett, L.H. Jenkins and O.E. Schow III, J. Vac. Sci. Technol. 12 454-57 (1975).
18) T.N. Rhodin, P.W. Palmberg and E.W. Plummer, paper 22 in *The Structure and Chemistry of Solid Surfaces*, ed. G.A. Somorjai, Wiley (1969).
19) S.E. Trullinger, S.L. Cunningham, Phys. Rev. B 8 2622-38 (1973).
20) J.C. Phillips, Surface Sci. 40 459-69 (1973).
21) F. Bonneton and M. Drechsler, Surface Sci. 22 426-32 (1970).
22) J.P. Biberian, *Thèse*, Université de Paris VI (1975).
23) S. Nakanishi and T. Horiguchi, J. Phys. E 8 511-14 (1975).
24) W.P. Ellis and R.L. Schwoebel, Surface Sci. 11 82-89 (1968).
25) G.E. Rhead and J. Perdereau, in *Colloque Intern. sur la Structure et les Propriétés des Surfaces des Solides*, C.N.R.S. (1969).
26) M. Henzler, Surface Sci. 19 159-171 (1970).
27) J.E. Houston and R.L. Park, Surface Sci. 21 209-23 (1970).
28) G.W. Stroke, Progress in Optics 2 3-44 and 68-72 (1963).
29) W.K. Burton, N. Cabrera and F.C. Frank, Phil. Trans. Roy. Soc. Lond. A243 299-358 (1951).
30) W.P. Ellis, Surface Sci. 45 569-84 (1974).
31) K.J. Matysik, Surface Sci. 46 457-72 (1974).
32) E.P. Th. M. Suurmeijer and A.L. Boers, Surface Sci. 43 309-52 (1973).
33) W. Korner, Le Vide, 163 75-6 (1973).
34) G.A. Haas and R.E. Thomas, J. Appl. Phys. 40 3919-24 (1969).
35) K. Besocke and H. Wagner, Surface Science (in press) (1975).
36) J.E. Rowe, S.B. Christman and H. Ibach, Phys. Rev. Lett. 34 874-7 (1975).
37) G. Ehrlich, in *Metal Surfaces*, ASM-AIME Seminar (1963).
38) J. Perdereau and G.E. Rhead, Surface Sci. 24 551-71 (1971).
39) J.A. Joebstl, J. Vac. Sci. Technol. 12 347-50 (1975).
40) D.R. Kahn, E.E. Petersen and G.A. Somorjai, J. of Catalysis, 34 291-306 (1974).
41) C. Roulet, Surface Sci. 36 295-316 (1973).
42) A. Hamelin and J.P. Bellier, Electroanal. Chem. and Interfacial Electrochem. 41 179-192 (1973).
43) H.J. Leamy and K.A. Jackson, J. Appl. Phys. 42 2121-7 (1971).
44) R.A. Hunt and B. Gale, J. Phys. C 6 3571-84 (1973) ; J. Phys. C 7 507-15 (1974).
45) W.W. Mullins, Acta Metall. 7 746-7 (1959).

46) D.N. Seidman, J. Phys. F 3 393-421 (1973).
47) F. Jona, Surface Sci. 8 478-84 (1967).
48) K. Besocke and H. Wagner, Phys. Rev. B 8 4597-600 (1973).
49) H.P. Bonzel, Surface Sci. 21 45-60 (1970).
50) W.R. Graham and G. Ehrlich, J. Phys. F 4 L212-14 (1974).
51) D.J. Cheng, R.F. Wallis and L. Dobrzynski, Surface Sci. 49 9-20 (1975) and references therein.
52) G.E. Rhead, Surface Sci. 47 207-21 (1975).
53) R.G. Linford, Solid State Surface Science, 2 1-152 (1973).
54) W.L. Winterbottom, in *Surfaces and Interfaces* I. Syracuse Univ. Press (1967) ; and in *Structure and Properties of Metal Surfaces*, ed. S. Shimodaira et al., Maruzen (1973).
55) C. Herring, J. Appl. Phys. 21 301 (1950).
56) C. Herring in *Structure and Properties of Solid Surfaces* (ed. R. Gomer and C.S. Smith) Univ. Chicago Press (1952).
57) E.D. Hondros and D. McLean, in *Surface Phenomena of Metals*, Soc. Chem. Indust. Monograph (1968).
58) M. McLean, Acta Metall. 19 387-93 (1971).
59) G.E. Rhead and M. McLean, Acta Metall. 12 401-7 (1964).
60) A.J.W. Moore in *Metal Surfaces* ASM-AIME Symposium (1963).
61) C. Herring, Phys. Rev. 82 87-93 (1951).
62) G.E. Rhead, Acta Metall. 11 1035-42 (1963).
63) J. Perdereau and G.E. Rhead, Acta Metall. 16 1267-74 (1968).
64) E.E. Gruber and W.W. Mullins, J. Phys. Chem. Solids 28 875-87 (1967).
65) W.L. Winterbottom, Acta Metall. 15 303-10 (1967).
66) J.M. Blakely and J.C. Shelton, in *Surface Physics of Crystalline Materials* (ed. J.M. Blakely) Academic Press (1975).
67) M. Laguës and J.L. Domange, Surface Sci. 47 77-85 (1975).
68) K.T. Aust, R.E. Hanneman, P. Niessen and J.H. Westbrook, Acta Metall. 16 291-302 (1968).
69) R.E. Hanneman and T.R. Anthony, G.E.C. Report No 69-C-105 (1969).
70) G.L.J. Bailey and H.C. Watkins, Proc. Phys. Soc. B63 350-58 (1950).
71) J.B. Hudson and Chien Ming Lo, Surface Sci. 36 141-54 (1973).
72) C. Argile and G.E. Rhead, Surface Sci. in press (1975).
73) M. Seah, Surface Sci. 32 703-28 (1972).
74) J. Henrion and G.E. Rhead, Surface Sci. 29 20-36 (1972).
75) S. Andersson and U. Jostell, Solid State Comm. 13 829-32 (1973).
76) R.L. Gerlach and T.N. Rhodin, Surface Sci. 17 32-68 (1969).
77) J.O. Porteus, Surface Sci. 41 515-32 (1974).
78) J.L. Desplat, Japan J. Appl. Phys. Suppl.2, Pt.2, 177-80 (1974)
79) J.P. Biberian and M. Huber, Surface Sci. (in preparation, 1975).
80) R.L. Park, in *The Structure and Chemistry of Solid Surfaces*, ed. G.A. Somorjai (paper 28) Wiley (1969).
81) M. Huber and J. Oudar, Surface Sci. 47 605-21 (1975).
82) E. Bauer, in *Structure et propriétés des Surfaces des Solides*, C.N.R.S., Paris (1970).
83) J. Perdereau, J.P. Biberian and G.E. Rhead, J. Phys. F 4 798-806 (1974).

84) L.G. Feinstein and E. Blanc, Surface Sci. 18 350-56 (1969).
85) O. Nishikawa and T. Utsumi, J. Appl. Phys. 44 945-54 and 955-64 (1973).
86) P.C. Bettler, D.H. Bennum and C.M. Case, Surface Sci. 44 360-76 (1974).
87) A. Cetronio and J.P. Jones, Surface Sci. 40 227-48 (1973) and 44 109-28 (1974).
88) F. Delamare and G.E. Rhead, Surface Sci. 35 185-93 (1973).
89) F.L. Williams and D. Nason, Surface Sci. 45 377-408 (1974).
90) S.H. Overbury, P.A. Bertrand and G.A. Somorjai, Lawrence Berkeley Lab. Report 2746, to be published (1975).
91) J.J. Burton, E. Hyman and D.G. Fedak, J. of Catalysis 37 106-13 (1975).
92) C.R. Helms and K.Y. Yu, J. Vac. Sci. Technol. 12 276-78 (1975).
93) G. Ertl and J. Küppers, Surface Sci. 24 104-24 (1971).
94) R. Bouwman and P. Biloen, Surface Sci. 41 348-58 (1974).
95) S. Berglund and G.A. Somorjai, J. Chem. Phys. 59 5537 (1973).
96) V.S. Sundaram, R.S. Alben and W.D. Robertson, Surface Sci. 46 653-71 (1974).
97) H.C. Potter and J.M. Blakely, J. Vac. Sci. Technol. 12 635-42 (1975).
98) R.A. Van Santen, L.H. Toneman and R. Bouwman, Surface Sci. 47 64-76 (1975).

THEORY OF L.E.E.D.

B.W. HOLLAND

University of Warwick

Coventry, England

1. INTRODUCTION

In Low Energy Electron Diffraction (LEED) experiments one studies the elastic scattering of a beam of electrons of well defined energy by a solid surface. Low energy in this context means less than about 500 eV. Since electrons of such energy penetrate only a short distance (of order 10 Å) into the surface before inelastic scattering by valence electrons, the experimental data should yield information about the surface layers of the solid. In particular, the great hope was that LEED could be developed into a technique for the determination of surface structures. The main obstacle to the realisation of this hope was that the theory needed for the interpretation of the experimental data proved to be difficult to generate. The ion cores of the solid typically have cross-sections for elastic scattering of low energy electrons of the order of square angströms, which means that simple approximations used for neutron and X-ray scattering are inapplicable. One had to face up to the full multiple scattering problem in LEED, and after all the reasonable simplifications had been made, it was still necessary to develop very elaborate computer programmes in order to make useful calculations. However, we have now reached the point where it is established beyond doubt that several non-trivial surface structure problems have been solved.

The purpose of these lectures is not to give a detailed exposition of LEED theory ; this can be found for example in the monograph by Pendry (1) or the review by Duke (2), together with the literature cited. Rather we lay emphasis on the general structure of the theory, on the basic physics of the problem, on experimental corroboration of the theory, and on some recent developments in finding perturbation schemes that are fast and reliable. We first describe some general notions of scattering theory that are useful, before particularising to the LEED problem.

2. SCATTERING PROBLEMS AND THE T-MATRIX

We assume that our problem can be reduced to the solution of a one-particle Schrödinger equation :

$$\left[-\frac{\hbar^2}{2m}\nabla^2 + V(\underline{r})\right]\psi(\underline{r}) = E\psi(\underline{r}),$$

where $V(\underline{r})$ is the potential doing the scattering, and appropriate boundary conditions are imposed on $\psi(\underline{r})$, so that it describes an incident plane wave ($\phi_{\underline{k}}$) of wave vector $\underline{k}$, and a scattered wave (ϕ_s) propagating away from the scatterer. It is convenient to rewrite the Schrödinger equation in integral form thus :

$$\psi(\underline{r}) = \phi_{\underline{k}}(\underline{r}) + \int G(\underline{r}-\underline{r}')\, V(\underline{r}')\psi(\underline{r}')\, d^3\underline{r}', \quad (1)$$

where the free Green's function

$$G(\underline{r}-\underline{r}') = -\frac{2m}{\hbar^2}\cdot\frac{1}{4\pi}\frac{e^{i\kappa|\underline{r}-\underline{r}'|}}{|\underline{r}-\underline{r}'|}$$

with

$$\frac{\hbar^2}{2m}\kappa^2 = E.$$

We now write the scattered wave as

$$\begin{aligned}\psi_s(\underline{r}) &= \int G(\underline{r}-\underline{r}')V(\underline{r}')\psi(\underline{r}')d^3\underline{r}' \\ &= \int G(\underline{r}-\underline{r}')T(\underline{r}',\underline{x})\phi_{\underline{k}}(\underline{x})d^3\underline{r}'d^3\underline{x},\end{aligned} \quad (2)$$

which defines the T-matrix in the coordinate representation, having elements $T(\underline{r}',\underline{x})$. The T matrix is a kind of "black box" ; put in the incident wave and it generates the scattered wave. Clearly, finding the T matrix is equivalent to solving our original problem. We can easily find an equation determining the T matrix. From (1) and (2) we have

$$\psi(\underline{r}) = \phi_{\underline{k}}(\underline{r}) + \iint G(\underline{r}-\underline{r}^1)T(\underline{r}',\underline{x})\phi_{\underline{k}}(\underline{x})d^3\underline{r}'d^3\underline{x} \quad (3)$$

From (2) it follows that

$$V(\underline{r})\psi(\underline{r}) = \int T(\underline{r},\underline{x})\phi_{\underline{k}}(\underline{x})d^3\underline{x} \quad (4)$$

and from (3) that

$$V(\underline{r})\psi(\underline{r}) = V(\underline{r})\phi_{\underline{k}}(\underline{r}) + V(\underline{r})\iint G(\underline{r}-\underline{r}')T(\underline{r}',\underline{x})\phi_{\underline{k}}(\underline{x})d^3\underline{r}'d^3\underline{x} \tag{5}$$

so that comparing (4) and (5),

$$T(\underline{r},\underline{x}) = V(\underline{r})\delta(\underline{r}-\underline{x}) + V(\underline{r})\int G(\underline{r}-\underline{r}')T(\underline{r}',\underline{x})d^3\underline{r}' \tag{6}$$

The T matrix is the matrix representation of an operator that we call the transition operator T, and we can write (6) as an operator equation

$$T = V + V\,G\,T \tag{7}$$

where

$$G = \lim_{\varepsilon\to 0}\left(E - \frac{\hbar^2}{2m}\nabla^2 + i\varepsilon\right)^{-1}$$

It turns out that the relation between T matrix elements and experimental data is much more direct than we have suggested so far as we now show. We measure the rate at which scattered particles enter a detector, so that we need to calculate the scattered flux travelling in the direction of the detector. Since the distance from the scatterer to the detector is always much greater than the linear dimensions of the scatterer, we detect essentially the radial flux

$$J_s(\underline{r}) = \frac{\hbar}{2mi}\left(\psi_s^{*}\frac{\partial}{\partial r}\psi_s - \psi\frac{\partial}{s\partial r}\psi_s^{*}\right) \tag{8}$$

where $\underline{r}$ specifies a point on the window of the detector and the origin of coordinates is in the scatterer. Now

$$\psi_s(\underline{r}) = \int G(\underline{r}-\underline{r}')T(\underline{r}',\underline{x})\phi_{\underline{k}}(\underline{x})d^3\underline{r}'d^3\underline{x}$$

and it follows from (6) that $T(\underline{r}',\underline{x})$ vanishes when $\underline{r}'$or $\underline{x}$ lie outside the scatterer, hence at the detector $r>>r'$. Therefore

$$G(\underline{r}-\underline{r}') \triangleq -\frac{2m}{\hbar^2}\cdot\frac{1}{4\pi}\,\frac{e^{i\kappa r}}{r}\quad e^{-i\underline{k}'\cdot\underline{r}'}$$

where $\underline{k}' = \kappa\underline{r}/r$, so that

$$\psi_s(\underline{r}) \triangleq \frac{e^{i\kappa r}}{r}\quad f(\theta,\phi) \tag{9}$$

where the scattering amplitude

$$f(\theta,\phi) = -\frac{2m}{\hbar^2} \cdot \frac{1}{4\pi} \langle \underline{k}'|T|\underline{k}\rangle$$

The angles θ and ϕ specify the direction of the detector and

$$\langle \underline{k}'|T|\underline{k}\rangle = \iint e^{-i\underline{k}'\cdot\underline{r}'} T(\underline{r}',\underline{x})\, e^{i\underline{k}\cdot\underline{x}} d^3\underline{r}' d^3\underline{x}$$

is a simple matrix element of the T matrix in the momentum representation. Evaluating the radial scattered flux using (8) and (9) gives

$$J_s(\underline{r}) = \frac{\hbar\kappa}{m} \left| f(\theta,\phi)/r \right|^2$$

and if the detector subtends an element of solid angle $\delta\Omega$, the particle collection rate is

$$\frac{\hbar\kappa}{m} |f(\theta,\phi)|^2 \delta\Omega = \frac{\hbar\kappa}{m} \cdot \left(\frac{2m}{\hbar^2}\right)^2 \left(\frac{1}{4\pi}\right)^2 \left|\langle \underline{k}'|T|\underline{k}\rangle\right|^2 \delta\Omega$$

Thus the result of our scattering experiment is determined by a single T matrix element specified by the incident and scattered wave vectors. Our theoretical problem then is reduced to the calculation of such T matrix elements. Before going on to study how this is done for LEED however, let us first see what can be discovered from general considerations based on the symmetry of the surface.

3. USE OF SYMMETRY ARGUMENTS (4,5)

The crystal surface will have certain symmetries that allow one to make useful predictions without detailed calculation. If we adopt an idealised semi-infinite model for the surface, the symmetry will be described by one of the seventeen plane groups. Mathematically we express symmetry operations by coordinate transformations (see for example, the book by Heine (3)) thus :

$$S\underline{r} = \underline{r}'$$

where S is the symmetry operator. If for example we have a three-fold axis, S could be the operator that rotates the vector $\underline{r}$ by 120° to become the vector $\underline{r}'$. Now by definition, if S is a symmetry operator for the system, it leaves the Hamiltonian invariant, i.e.

$$SHS^{-1} = H$$

and it follows from (7) that T is also invariant under S ;

$$STS^{-1} = T$$

Therefore

$$\langle \underline{k}'|T|\underline{k}\rangle = \langle \underline{k}'|S^{-1}STS^{-1}S|\underline{k}\rangle = \langle \underline{k}'|S^{-1}TS|\underline{k}\rangle \quad (10)$$

In the coordinate representation

$$S\, e^{i\underline{k}\cdot\underline{r}} = e^{i\underline{k}\cdot(S\underline{r})} \quad (11)$$

If S induces translation by a Bravais net vector, $\underline{P}$, then according to (11)

$$S\, e^{i\underline{k}\cdot\underline{r}} = e^{i\underline{k}\cdot\underline{P}} e^{i\underline{k}\cdot\underline{r}}$$

and according to (10) therefore

$$\langle \underline{k}'|T|\underline{k}\rangle = e^{i(\underline{k}-\underline{k}').P} \langle \underline{k}'|T|\underline{k}\rangle$$

Hence $\langle \underline{k}|T|\underline{k}\rangle$ can be non zero, only if

$$e^{i(\underline{k}-\underline{k}')\cdot\underline{P}} = 1$$

or

$$(\underline{k}-\underline{k}')\cdot\underline{P} = 2\pi n$$

where n is an integer. By definition this implies that $\underline{k}_{11}-\underline{k}'_{11} = g$; a vector of the reciprocal net, where $\underline{k}_{11}$ is the component of the incident wave vector parallel to the surface. Thus the scattered beams will form a discrete set giving rise to a diffraction pattern of separate spots, from the geometry of which the Bravais net of the surface can immediately be deduced.

On the other hand, for a point operation (rotation on reflection)

$$\underline{k} \cdot (S\underline{r}) = (S^{-1}\underline{k})\cdot\underline{r}$$

i.e. a wave of wave vector $\underline{k}$ is transformed into a wave of wave vector $S^{-1}\underline{k}$. Hence from (10) we have

$$\langle \underline{k}'|T|\underline{k}\rangle = \langle S^{-1}\underline{k}'|T|S^{-1}\underline{k}\rangle \quad (12)$$

This leads to many intuitively obvious conclusions, such as for example that if we have a surface with a 3-fold axis, the diffraction pattern will be unchanged if we rotate the crystal by 120° about this axis, and that at normal incidence ($S^{-1}\underline{k} = \underline{k}$) the diffraction pattern itself will have at least the point symmetry of the surface.

A more interesting application occurs when the system has a glide plane, i.e. a symmetry operation consisting of a reflection with a translation by half a Bravais net vector in the reflection plane. Then (10) implies that

$$\langle \underline{k}'|T|\underline{k}\rangle = e^{i(\underline{k}-\underline{k}')\cdot\underline{a}/2}\langle R\underline{k}'|T|R\underline{k}\rangle$$

where R is the reflection operator and $\underline{a}$ is a primitive vector of the Bravais net. When the incident wave vector and the wave vector of the scattered beam both lie in the reflection plane, so that $R\underline{k} = \underline{k}$ and $R\underline{k}' = \underline{k}'$, then for non zero intensity of the scattered beam

$$e^{i(\underline{k}-\underline{k}')\cdot\underline{a}/2} = 1$$

or, since $\underline{a}$ is a Bravais net vector and $\underline{k}_{11} - \underline{k}'_{11} = \underline{g}$ a reciprocal net vector,

$$e^{in\pi} = 1$$

where n is an integer. Therefore n must be even for non zero intensity. It follows that if a glide plane is present, at normal incidence alternate spots along a line parallel to the glide plane and running through the centre of the diffraction pattern, will be missing.

Some of these points are illustrated in Figure 1. In 1(a) the substrate is denoted by filled circles and two possible relative positions of the same adsorbate structure are shown as crosses and open circles. 1(b) and 1(c) show respectively the reciprocal nets of the substrate alone and of the adsorbate alone. 1(d) shows a superposition of the two, but the reciprocal net of the combined substrate adsorbate structure is shown in 1(e). For the cross structure of 1(a) there is a glide plane and alternate spots on a vertical line would be missing in the diffraction pattern, while the diffraction pattern for the open circle structure, having no glide plane, would show no such feature.

Exploitation of time-reversal symmetry can also yield interesting results. Neglecting spin, for solutions of the time independent Schrödinger equation the time-reversal operation reduces to complex conjugation (3). For a plane wave therefore, it changes the wave vector $\underline{k}$ to $-\underline{k}$. Hence

$$\langle \underline{k}'|T|\underline{k}\rangle = \langle -\underline{k}\,|T|-\underline{k}'\rangle$$

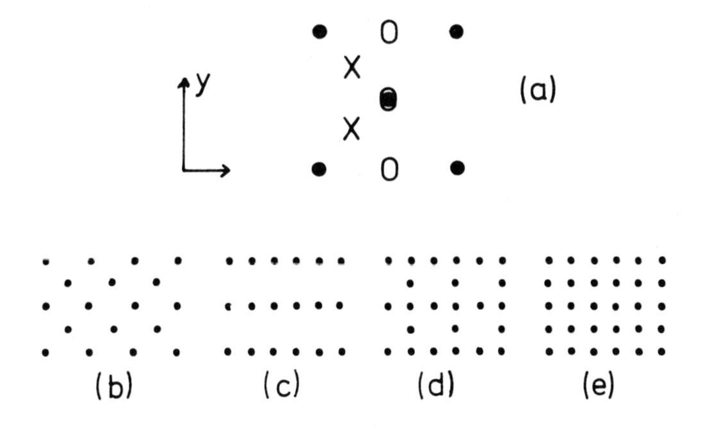

Figure 1

When combined with point symmetries, this can give some non-trivial predictions. For example consider the (111) face of a f.c.c. crystal. It has a 3-fold axis normal to the surface. We therefore expect a 3-fold symmetry of the $\underline{k}_{11} - \underline{k}'_{11} = 0$ or 00 beam as the crystal is rotated about the axis, but in fact this beam shows 6-fold symmetry. The reason is that for this beam alone time reversal is equivalent to a rotation of 180° about the normal, which combined with the 3-fold crystal symmetry gives a 6-fold symmetry. The diffraction pattern as a whole of course shows only a three-fold symmetry.

We see that general argument based on symmetry can give very useful non-trivial results. But they cannot give us actual numbers to compare with experiment. For that we need to make detailed calculations and we shall now proceed to see how this is done.

4. THE MODEL

We must first adopt a model of the physical problem that is sufficiently simple to be tractable and at the same time is capable of giving a good description of the experimental data. We view the crystal as a set of ion cores bathed in a valence electron gas, which for simplicity we take to be uniform. We assume that the Hamiltonian for the incident electron can be written as

$$H = H_o + V(\underline{r})$$

where

$$H_o = -\frac{\hbar^2}{2m}\nabla^2 + V_o$$

and

$$V(\underline{r}) = \sum_{\underline{R}} V_{\underline{R}}(\underline{r}-\underline{R})$$

where $V_{\underline{R}}$ is the potential due to the ion core at position $\underline{R}$. V_o describes the interaction between the incident electron and the valence electron gas within the quasi-particle model. Thus V_o is the self-energy of the incident electron due to the presence of the valence electron gas and is in general complex. The real part is negative and arises from polarisation of the electron gas by the incident electron. The imaginary part determines the life-time of the quasi-particle which is finite due to inelastic scattering by the valence electrons. For LEED, the energy is always large compared to Im V_o so that the quasi-particle model should be good.

To see the significance of the complex nature of V_o we note that the quasi-particle free states are determined by

$$\left(-\frac{\hbar^2}{2m}\nabla^2 + V_o\right)\phi(\underline{r}) = E\phi(\underline{r})$$

and there exist plane wave solutions

$$\phi_{\underline{k}}(\underline{r}) \sim e^{i\underline{k}\cdot\underline{r}}$$

such that

$$\frac{\hbar^2 k^2}{2m} = E - V_o \tag{13}$$

so that $\underline{k}$ is complex and the waves are damped. Estimates of V_o for electrons of LEED energies using say R.P.A. (6) for typical valence electron densities are of the order of a few eV, giving a characteristic damping length of a few Å. It is this damping that makes LEED a specific probe for the top few layers of the crystal only, and also profoundly affects the multiple scattering contributions to the scattering amplitude. The significance for LEED of inelastic scattering by the valence electrons was first fully recognised by Duke and Tucker (7), in a paper that represents a landmark in the development of LEED theory.

In fact R.P.A. is not reliable at valence electron densities and Re V_o and Im V_o are usually treated as adjustable parameters chosen to give the best fit to the experimental data. Typically $-\text{Re}V_o$ lies in the range 10-15 eV.

At the surface we assume that V_o vanishes abruptly on passing through a surface plane about half a layer spacing out from the top layer of atoms. We ignore scattering from the resulting potential step on grounds that an electron approaching the surface would in fact see a rather smoothly varying potential which would scatter very weakly.

We have already seen that the "free states" of the electron are modified by the presence of the uniform valence electron gas within the crystal. Any plane wave in the vacuum must join onto one in the solid having the same component of the wave vector parallel to the surface, since V_o is uniform over planes parallel to the surface so that the parallel component of momentum must be conserved. This condition together with energy conservation fixes uniquely the complex perpendicular component $k_\perp$, of the wave vector inside the crystal, given the wave vector outside, since (13) implies that

$$E - \frac{h^2}{2m}(k_{\|}^2 + k_\perp^2) - V_o = 0$$

For the ion core potentials the muffin tin model is adopted. The potential within each ion core differs from a constant value (the muffin tin zero) only within a certain radius (muffin tin radius) which is usually chosen to be as large as possible consistent with the condition that the potentials for different ion cores should not overlap. The potentials themselves might be constructed by using S.C.F. calculations for the neutral atoms and superposing the charge densities, placing the atoms on a lattice of the assumed structure and spherically averaging the charge densities about each lattice point. Sufficient uniform electron density is then added to make each muffin tin sphere electrically neutral. The radial Schrödinger equations for the appropriate energy and partial waves are then solved by the Hartree-Fock procedure or using the Slater method for the exchange interaction, and the phase shifts that determine the scattering properties found from the logarithmic derivative of the wave functions at the muffin tin radius.

We have in this section been concerned to give only a general description of the physical model on which LEED calculations are based. The details of the models used by different authors differ ; for further information and critical comparisons see the reviews by Pendry (1) and Duke (2).

Having chosen a specific model we know the free states for the scattering problem, together with the scattering properties of the ion cores and are in principle in a position to calculate numbers to compare with experiment. But before we can carry out this process in practice, we must first put the multiple scattering problem into a form that makes it tractable.

5. MULTIPLE SCATTERING THEROY (8,9)

We write the general scattering equation (3) symbolically as

$$\psi = \phi + GT\phi$$

ϕ representing the incident wave. Suppose our scattering system is a composite system of individual scatterers, then we can write

$$\psi = \phi + \sum_\beta F_\beta$$

where F_β is the scattered wave emanating from scatterer β. The wave incident on β, ϕ_β, is the incident wave plus scattered waves from all other scatterers :

$$\phi_\beta = \phi + \sum_{\alpha \neq \beta} F_\beta \tag{14}$$

By definition

$$F_\beta = G\, t_\beta \phi_\beta \tag{15}$$

where t_β is the transition operator for scatterer β alone. From (14) and (15) we find

$$F_\beta = Gt_\beta \phi + Gt_\beta \sum_{\alpha \neq \beta} F_\alpha \tag{16}$$

Define an effective transition operator T_β, such that

$$F_\beta = GT_\beta \phi \tag{17}$$

Then using (17) in (16) gives

$$GT_\beta \phi = Gt_\beta \phi + Gt_\beta \sum_{\alpha \neq \beta} GT_\alpha \phi$$

or

$$T_\beta = t_\beta + t_\beta G \sum_{\alpha \neq \beta} T_\alpha \tag{18}$$

a coupled set of equations for the effective transition operators for the different scatterers. These effective transition operators for the individual scatterers determine the transition operator for the composite scatterer. For the scattered

$$GT\phi = \sum_\beta F_\beta = G \sum_\beta T_\beta \phi$$

so that

$$T = \sum_\beta T_\beta \tag{19}$$

Our object is to calculate $<\underline{k}'|T|\underline{k}>$, for the LEED case. We shall not present the detailed derivation of the required equations (10,1,2) which involves a formidable amount of mathematical manipulation, but instead we shall sketch the structure of the equations so that some idea can be gained of what is involved in making detailed LEED calculations. In the simplest case where all ion cores in a layer are translationally equivalent, one treats the layers as the individual scatterers making up the composite scatterer, the crystal. Provided $\underline{k}'$ specifies an allowed scattered beam one finds that

$$<\underline{k}'|T|\underline{k}> = \sum_{\substack{\ell,m \\ \ell',m'}} Y_{\ell m}(\underline{k}')Y_{\ell'm'}(\underline{k}) \sum_{\nu} e^{i(\underline{k}-\underline{k}')\cdot\underline{d}_{\nu}} T_{\nu}^{\ell m\ell'm'} \tag{20}$$

where $Y_{\ell m}$ is a real sperical harmonic with arguments determined by the direction of $\underline{k}$, $\underline{d}_{\nu}$ is the position vector of the origin of layer ν. Comparing (19) and (20) one sees that $T_{\nu}^{\ell m\ell'm'}$ is essentially an element of the effective T matrix for layer ν in the angular momentum representation.

The effective T matrices themselves satisfy a set of coupled equations

$$\underset{\sim}{T}_{\nu} = \underset{\sim}{\tau}_{\nu} + \underset{\sim}{\tau}_{\nu} \sum_{\nu'(\neq\nu)} \underline{G}^{\nu\nu'}(\underline{k})\underset{\sim}{T}_{\nu'} \tag{21}$$

This set of equations is the direct analogue of the set of operator equations (18), $\underset{\sim}{\tau}_{\nu}$ representing the T matrix for layer ν alone.

$$\underline{G}^{\nu\nu'}(\underline{k}) = e^{-i\underline{k}\cdot(\underline{d}_{\nu}-\underline{d}_{\nu'})} \sum_{\underline{P}} e^{-i\underline{k}\cdot\underline{P}}\, \underline{G}(\underline{P}+\underline{d}_{\nu}-\underline{d}_{\nu'}) \tag{22}$$

where

$$G_{\ell m\ell'm'}(\underline{R}) = -4\pi\kappa i^{\ell-\ell'} \sum_{\ell'',m''} a_{\ell m\ell'm'\ell''m''} h^{+}_{\ell''}(\kappa R) Y_{\ell''m''}(\underline{R}) \tag{23}$$

$h^{+}(\kappa R)$ is a Spherical Hankel function and

$$a_{\ell m\ell'm'\ell''m''} = \iint Y_{\ell m}(\theta,\phi)Y_{\ell'm'}(\theta,\phi)Y_{\ell''m''}(\theta,\phi)\sin\theta\, d\theta d\phi$$

$$\underset{\sim}{\tau}_{\nu} = \left(1-\underset{\sim}{t}_{\nu}\underline{G}^{sp}(\underline{k})\right)^{-1} \underset{\sim}{t}_{\nu} \tag{24}$$

where

$$\underline{G}^{sp}(\underline{k}) = \sum_{\underline{P} \neq 0} e^{-i\underline{k}\cdot\underline{P}} \underline{G}(\underline{P}) \qquad (25)$$

and $\underline{t}_\nu$ is the T matrix for the ion core of type found in layer ν in the angular momentum representation, and is diagonal ; with elements determined by the phase shifts δ_ℓ, according to

$$t_\nu^{\ell} = -\frac{1}{2i\kappa}\left(e^{2i\delta_\ell} - 1\right) \qquad (26)$$

In order to specify the scattering amplitude for a chosen beam, given the energy and incident direction relative to the surface we need

(i) V_o
(ii) the ion core potential
(iii) the surface structure.

From the scattering potential we find the phase shift δ_ℓ, and using the fact that

$$\frac{\hbar^2}{2m}\kappa^2 = E - V_o$$

we find κ and hence from (26) the matrices $\underline{t}$ (if all ion-cores are identical there is of course only one such matrix). The internal incident wave vector $\underline{k}$, is determined uniquely as we have already seen, when V_o and the external wave vector are given, and knowing the Bravais net we can calculate $\underline{G}^{sp}(\underline{k})$ from (25) and (23). From $\underline{t}_\nu$ and $\underline{G}^{sp}(\underline{k})$ we find $\underline{\tau}_\nu$ from (24). Given the structure we can also now calculate the various structure factors $\underline{G}^{\nu\nu'}(\underline{k})$ from (22) and (23). We are then in a position to solve equations (21) for the $\underline{T}_\nu$ and then finally to find the scattering amplitude from (20).

We see that the solution of the multiple scattering problem involves some very hard work and the first question to raise is, is it all really necessary ? The fact is that it was recognised (11,12) immediately after the earliest LEED experiments of Davisson and Germer (13) that single scattering theory is quite inadequate for the interpretation of the data, and the large volume of work in recent years has substantiated that opinion. Consider for example the results shown in Figure 2. In the left hand panel we have the logarithm of the intensity of the 00 beam, for a series of angles of incidence in the same azimuth differing by 1°, for LiF(100), measured by McRae and Caldwell (14). The right hand panel shows the results of calculations (15) in which multiple scattering is neglected. There is practically no correspondence between theory and experiment. The middle panel shows what happens when the multiple scattering is "switched on". First the difference between the right hand and middle panels is very marked, and since the parameters used the calculation are reasonable, we must conclude that multiple scattering is in general very important. Secondly, the agreement between theory and experiment is dramatically improved by the inclusion of multiple scattering. We see that the concept of multiple scattering is quite fundamental to any attempt to understand LEED data, in particular the multiple scattering contribution cannot be regarded as a small correction to the single scattering approximation.

It is true that several attempts have been made to show that with suitable processing of the experimental data, for example by averaging over energy [16] or averaging at constant momentum transfer[17], or by the use of Patterson function [18], multiple scattering effects can safely be ignored. But none of these approaches has been given any reasonable theoretical basis, and until this is done they must be highly suspect. For success in a few specific cases is no guarantee of general reliability. Until one understands why a method of structure determination should work, one cannot know the circumstances in which it might fail, and hence one cannot reasonably have confidence in it.

Examination of Figure 2 shows that agreement between theory and experiment is far from perfect even when multiple scattering is included, but this is not surprising since in the calculations the ion cores were assumed to be pure S-wave scatterers. When more realistic models of the ion core scattering are adopted, detailed agreement can be obtained. We now have numerous examples in the literature which substantiate this statement. We show in Figure 3 an example in which Pendry's calculation [19] of the intensity of the 00 beam for Cu(001) is compared with the experimental data of Andersson [20].

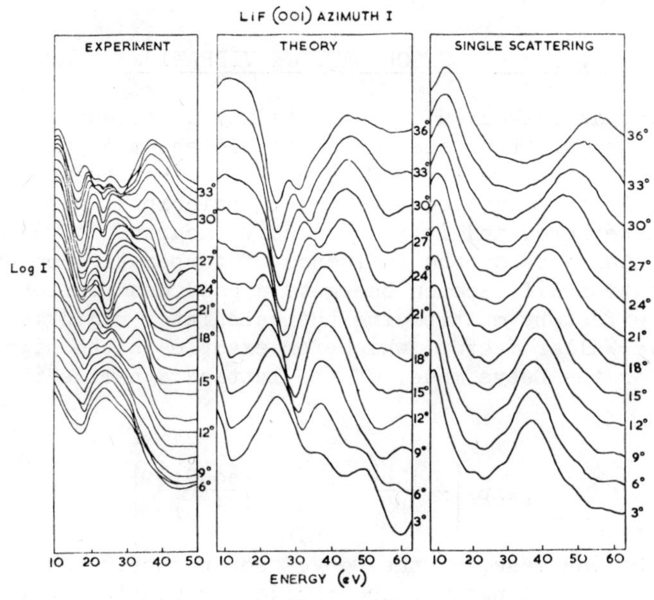

Figure 2

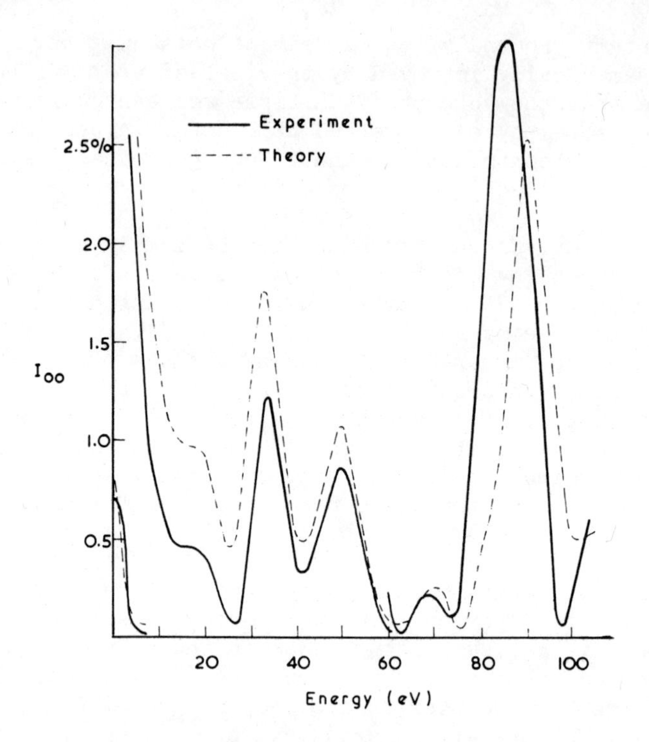

Figure 3

6. THE EFFECT OF LATTICE VIBRATIONS

The model described in §4 is certainly deficient in one important respect, namely that the ion cores are assumed rigidly fixed to the lattice sites, while in fact we know that they vibrate about these sites. A rigid lattice model is in principle inadequate even at the Absolute Zero, and it is obviously incapable of describing the experimental data on the influence of temperature on LEED. For X-ray and neutron diffraction one gets a good description of the temperature effect by multiplying the calculated rigid ion intensity by a Debye-Waller factor, which for example in the high temperature limit for the Debye model of the lattice dynamics is of the form

$$\exp\left(-\frac{12h^2}{Mk_B}\cdot\frac{T}{\theta_D^2}\left(\frac{\cos\theta}{\lambda}\right)^2\right)$$

where M is the ion mass, θ_D the Debye temperature, θ the scattering angle and λ the wavelength of the incident radiation. For LEED such a procedure is quite inadequate because of the presence of strong multiple scattering. If one plots the log of the intensity against temperature, from the slope one can determine an "effective Debye temperature". In the neutron and X-ray cases one finds a result independent of or varying very slowly with the angle of incidence and incident energy. But for LEED the effective Debye temperature often fluctuates violently with energy and angle of incidence. A striking example of the anomalous behaviour of LEED data with change in temperature is shown in the left-hand panel of Figure 4. This shows the results of Reid[21] on the 00 beam from Cu(001). Evidently varying the temperature actually causes a reversal of the relative intensities of the two peaks which are only separated in energy by a few eV.

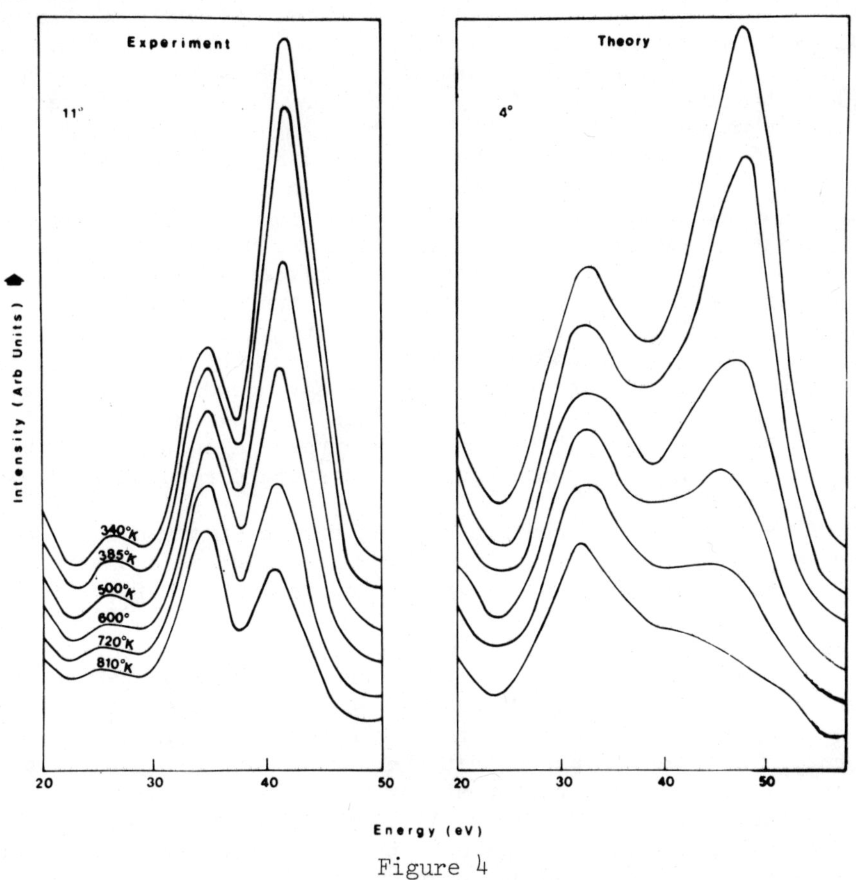

Figure 4

The theory of multiple scattering in a vibrating lattice is too complicated to present here. In the literature[22-25] it is shown that provided correlations between displacements of ions on different sites are neglected, and an isotropic model for the lattice vibrations used, the equations of the full theory given in section 5 are unchanged except that each of the ion core T matrix elements t_{ν}^{ℓ}, is modified by a temperature dependent factor, determined by the mean square displacement of the ion and the energy of the incident electrons. A crude version of this model[25] is capable of giving at least a qualitative account of the behaviour of Reid's doublet as can be seen from the right hand panel of Figure 4.

7. DEVELOPMENT OF FAST, RELIABLE PERTURBATION METHODS

The direct solution of the equations of section 5 makes very heavy demands on computing facilities, both of core storage and of time. For example, the solution of the set of equations (21) requires the inversion (or some equivalent procedure) of a matrix of order $N(\ell_{max}+1)^2$ where N is the number of layers (for the simplest case) and $\ell_{max}+1$ is the number of partial waves needed to describe adequately the ion core scattering properties. Thus the storage requirement is at least of order $N^2(\ell_{max}+1)^4$. For simple cases $\ell_{max} \sim 4$ and $N \sim 10$ so that even here we have a storage requirement of about 60K. For more complex cases the effective number of layers is increased, since within a single layer there will be several, say n inequivalent scatterers, giving an effective number of layers of nN. Further, since the time for matrix multiplication, or inversion scales as the order of the matrix to the third power, i.e. as $N^3(\ell_{max}+1)^6$, a small increase in complexity will be sufficient to make the direct solution impracticable, since the times taken for the simple cases so far treated are already uncomfortably long. It is clear therefore that there is a pressing need for the development of perturbation theories that allow accurate calculations to be made that are both fast and economic in storage requirements.

A variety of perturbation methods have been advanced ; see for example the recent study by Aberdam[26] et al where extensive references are given. With one exception however, they suffer from a basic weakness ; they are justified on the basis of showing that for a few chosen examples calculations to a certain order agree well with exact calculations. But such a procedure can never guarantee the reliability of a perturbation scheme, because for other cases it may be necessary to go to higher order in order to obtain a good enough approximation, or it may be that the series diverges. For example Pendry[27] showed that perturbation theory in the order of the ion core T matrices diverges for Ni(111), though it had previously been used successfully in other cases. It is therefore essential to be able easily to extend the calculations to higher order so as to be sure that discrepancies between theory and experiment arise from an incorrect choice of trial structure, or some other deficiency of the model, and not merely from the inaccuracy of the approximate calculations. In practice this implies that the calculational scheme must be iterative, so that the amount of work involved for each order is not much greater than that for the previous order.

The first iterative scheme to be put forward was the Renormalised Forward Scattering (RFS) theory of Pendry[28]. In this method the layer scattering amplitude is calculated exactly, but the calculation of multiple scattering between layers is based on a perturbation theory in the order of back scattering. The argument was that back scattering tends to be weak relative to forward scattering so it is natural to treat back scattering by perturbation theory. The most time consuming part of the calculation was in fact the evaluation of the layer scattering amplitude, and a great improvement in speed over the direct method of solution was obtained. The success of the method was demonstrated by calculations on Cu(100) and good agreement with exact calculations was already obtained in third order as can be seen in Figure 5. The intensity of the 00 beam is plotted against energy, the full curve represents the exact calculation. The dashed line corresponds to the first order result, and the crosses to third order. It can be seen that the convergence is very satisfactory. The calculational procedure is the same for every order, so it is very easy to establish convergence.

It was discovered by Zimmer and Holland[29] that if the Cu ion cores are replaced by "giant symmetric scatterers" having equal forward and backward scattering power and total cross sections much larger than that of Cu, that the RFS scheme still converges rapidly. This can be seen in Figure 6 in which the logarithm of the intensity

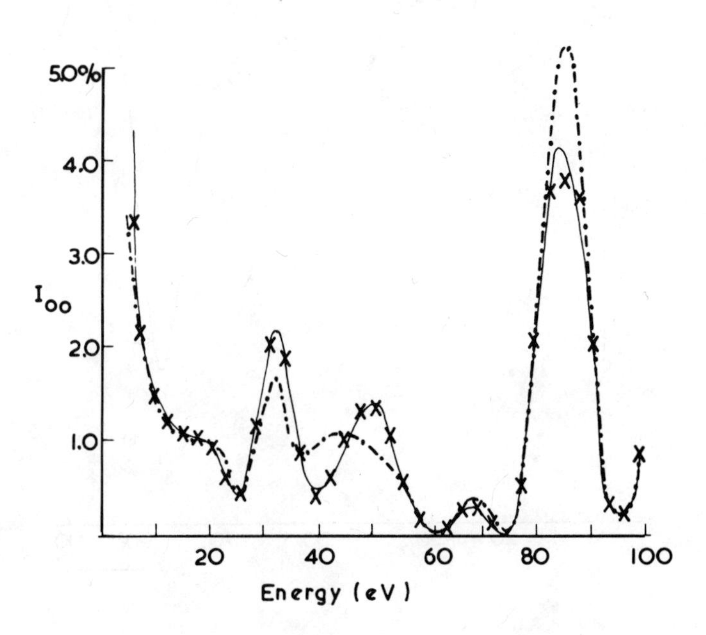

Figure 5

of the 00 beam for "quasi-copper" (100) is shown. Again the exact calculation is the full line. The first order calculation is represented by crosses and corresponds very closely with the exact result over almost the whole range. The higher order approximations are represented by the dashed lines and it is only when divergence occurs below 20 eV that they can be readily distinguished from the exact result. At divergence the cross-section for the giant symmetric scatterer is about 17($\mathring{A}$)^2, a huge value by LEED standards.

Evidently the weak back relative to forward scattering is not a necessary condition for the convergence of RFS. This suggests that a perturbation scheme of similar structure might work even for calculating the layer scattering amplitude, when there are several inequivalent scatterers in each layer. In such cases, which arise for any but the simplest surface structure problems, the calculation of the layer scattering amplitude might be a much more formidable task than the whole calculation for say a clean metal surface. Thus a perturbation scheme for the scattering amplitude of a complicated layer, that is as satisfactory as RFS is for the treatment of scattering between layers would be very valuable. Such a method has been devised by Zimmer and Holland[30], and we now describe it briefly.

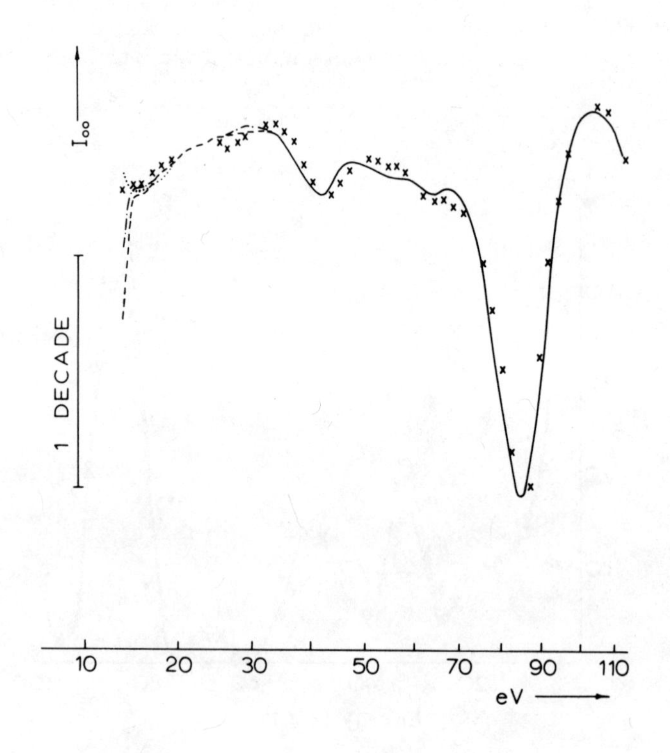

Figure 6

Our problem is basically to solve the equations (21) :

$$\underset{\sim}{T}_{\nu} = \underset{\sim}{\tau}_{\nu} + \underset{\sim}{\tau}_{\nu} \sum_{\nu'(\neq\nu)} \underline{G}^{\nu\nu'}(\underline{k})T_{\nu'}$$

Because of the damping of the electron waves we need consider only a finite number N of subplanes ν, i.e. of planes of inequivalent ion cores making up a layer (for the simplest case there is only one subplane per layer). It is in fact more convenient to solve a modified set of equations that follow from (21), namely

$$\underset{\sim}{B}_{\nu} = \underset{\sim}{\tau}_{\nu}\, \underline{Y} + \underset{\sim}{\tau}_{\nu} \sum_{\nu'(\neq\nu)} \underline{G}^{\nu\nu'} \underset{\sim}{B}_{\nu'}$$

where $\underline{Y}$ is a vector with elements $Y_{\ell m}(\underline{k})$ and

$$\underset{\sim}{B}_{\nu} = \underset{\sim}{T}_{\nu}\underline{Y}$$

for then the calculation involves repeated multiplication of matrices into vectors which scales in time as n^2 where n is the order of the matrix, instead of matrix-matrix multiplication which scales as n^3.

We assign a definite order to the subplanes by numbering adjacent subplanes in the top layer from one upwards, then moving to the second layer, and so on. Put

$$\underset{\sim}{B}_{\nu} = \underset{\sim}{\tau}_{\nu}\underline{Y} + \underset{\sim}{B}_{\nu}^{+} + \underset{\sim}{B}_{\nu}^{-}$$

where

$$\underset{\sim}{B}_{\nu}^{-} = \underset{\sim}{\tau}_{\nu} \sum_{\nu'<\nu} \underline{G}^{\nu\nu'} \underset{\sim}{B}_{\nu'}$$

and

$$\underset{\sim}{B}_{\nu}^{+} = \underset{\sim}{\tau}_{\nu} \sum_{\nu'>\nu} \underline{G}^{\nu\nu'} \underset{\sim}{B}_{\nu'}$$

Clearly $\underline{B}_{\nu}^{-} = 0 = \underset{\sim}{B}_{N}^{+}$ (27)

Let $\underset{\sim}{B}_{\nu}^{-(n)}$ represent the contribution to $\underset{\sim}{B}_{\nu}^{-}$ from scattering paths in which there are n reversals with respect to the chosen ordering of subplanes. Then

$$\underset{\sim}{B}_{\nu}^{-(n)} = \underset{\sim}{\tau}_{\nu} \sum_{\nu'<\nu} \underline{G}^{\nu\nu'} \left(\underset{\sim}{B}_{\nu'}^{-(n)} + \underset{\sim}{B}_{\nu'}^{+(n-1)}\right) \tag{28}$$

and similarly

$$\underset{\sim}{B}_{\nu}^{+(n)} = \underset{\sim}{\tau}_{\nu} \sum_{\nu'>\nu} \underline{G}^{\nu\nu'} \left(B_{\nu'}^{+(n)} + \underset{\sim}{B}_{\nu'}^{-(n-1)}\right) \tag{29}$$

Therefore to Mth order in path reversals,

$$\underset{\sim}{B}_{\nu}^{(M)} = \tau_{\nu}\underset{\sim}{Y} + \sum_{n=o}^{M} \left(\underset{\sim}{B}_{\nu}^{-(n)} + \underset{\sim}{B}_{\nu}^{+(n)}\right) \qquad (30)$$

Using the formal initial conditions

$$\underset{\sim}{B}_{\nu}^{+(-1)} = \tau_{\nu}\underset{\sim}{Y} = \underset{\sim}{B}_{\nu}^{-(-1)}$$

subject to (27), we can solve equations (28) and (29) to find the $\underset{\sim}{B}_{\nu}^{-(0)}$ and the $\underset{\sim}{B}_{\nu}^{+(0)}$. These solutions then provide the input for calculating the $\underset{\sim}{B}_{\nu}^{-(1)}$ and the $\underset{\sim}{B}_{\nu}^{+(1)}$ and so on up to any desired order of reverse scattering M. Then, to Mth order, the $\underset{\sim}{B}_{\nu}$ are found from (30).

To provide a severe test of this Perturbation Theory in the order of Reverse Scattering (RS scheme) it was used to calculate the scattering amplitude for a layer of Ni(111), the case for which perturbation theory in the order of the ion core T matrices is known to diverge. The results for the exact calculation and RS perturbation theory to fifth order were found to be identical when plotted out on the line printer. From further studies it appears that third order calculations will be sufficient for most purposes. For the Ni(111) case the RS calculations in third order were 24 times faster than the exact calculations. The time for RS scale as the fourth power of the number of partial waves instead of the sixth for exact calculations, and as N instead of N^2. Further the core storage requirement of the exact method scales as the square of that for the RS schemes. It seems therefore that the RS scheme extends the main advantages of RFS to structures with several subplanes per layer and hence opens the door to realistic LEED calculations for complex structures.

CONCLUSION

In summary we may conclude that LEED theory in its present state of development is adequate for the interpretation of data from clean metal surfaces and for some simple adsorbed layer structures on such surfaces. The recent development of fast reliable approximation schemes should allow more interesting and complex surface structure problems to be attacked in the near future. One may anticipate that further refinements of the model underlying LEED theory may then be required.

One area where some revision of the model may be needed is in the use of spherical ion core potentials and spherically symmetric models of lattice vibrations. Even in the bulk, the potentials in such materials as Si are likely to deviate considerably from spherical symmetry, and until more work is done on materials other than metals it will not be clear how important such effects are. Clearly the lattice vibrations at the surface are far from isotropic, and again work is needed to assess how serious the oversimplifications of the presently used isotropic model are. Could one develop the theory far enough to use LEED data for the elucidation of surface lattice dynamics ? The abandonment of the isotropic potential and isotropic lattice dynamics model would not require very major revision of the formalism ; it would mean simply that the matrices t_ν would no longer be diagonal.

A potentially more serious problem, and one that has been little explored concerns the model for the self-energy. Even if we assume that this can be represented by a local potential $V(\underline{r})$, are we not going too far in using the abrupt transition model in which $V(\underline{r}) = 0$ in the vacuum and $V(\underline{r}) = V_o$, a constant, in the crystal ? This model has worked quite well for clean surfaces, but as van Hove(31) has recently asked, is there any good reason for taking Re V_o to be the same in an adsorbed layer as it is in the substrate ? If one does not do so the extra freedom in the theoretical model gives rise to increased ambiguity and makes it more difficult to draw firm conclusions on the basis of comparisons between theory and experimental data.

REFERENCES

(1) J.B. Pendry (1974) Low Energy Electron Diffraction (Academic : London)
(2) C.B. Duke (1974) Advances in Chem. Phys. 27, 1
(3) V.Heine (1960) Group Theory in Quantum Mechanics (Pergamon : London)
(4) B.W. Holland and D.P. Woodruff (1973) Surface Sci. 36, 488
(5) D.P. Woodruff and B.W. Holland (1970) Phys. Letters 31A, 207
(6) J.J. Quinn (1962) Phys.Rev. 126, 1453
(7) C.B. Duke and C.W. Tucker (1969) Surface Sci. 15, 231
(8) M. Lax (1951) Rev. Mod. Phys. 23, 287
(9) E.G. McRae (1966) J. Chem. Phys. 45, 3258
(10) J.L. Beeby (1968) J. Phys. C. 1, 82
(11) H. Bethe (1928) Ann. d. Physik 87, 55
(12) P.M. Morse (1930) Phys. Rev. 35, 1310
(13) C.J. Davisson and L.H. Germer (1927) Phys. Rev. 30, 705
(14) E.G. McRae and C.W. Caldwell (1967) Surface Sci. 7, 41
(15) B.W. Holland, R.W. Hannum and A.M. Gibbons (1971) Surface Sci. 25, 561
(16) C.W. Tucker and C.B. Duke (1970) Surface Sci. 23, 411
(17) M.G. Lagally, T.C. Ngoc and M.B. Webb (1971) Phys. Rev. Lett. 26, 1557
(18) T.A. Clarke, R. Mason and M. Tescari (1972) Surface Sci. 30, 553

(19) J.B. Pendry (1971) J. Phys. C. 4, 2514
(20) S. Andersson (1969) Surface Sci. 13, 325
(21) R.J. Reid (1972) Surface Sci. 32, 139
(22) C.B. Duke and G.E. Laramore (1970) Phys. Rev. B2, 4765
(23) G.E. Laramore and C.B. Duke (1970) Phys. Rev. B2, 4783
(24) C.B. Duke, D.L. Smith and B.W. Holland (1972) Phys. Rev. B5, 3358
(25) B.W. Holland (1971) Surface Sci. 28, 258
(26) D. Aberdam, R. Baudoing and C. Gaubert (1975) Surface Sci. 48, 509
(27) J.B. Pendry (1971) J. Phys. C. 4, 3095
(28) J.B. Pendry (1971) Phys. Rev. Lett. 27, 856
(29) R.S. Zimmer and B.W. Holland (1975) Surface Sci. 47, 717
(30) R.S. Zimmer and B.W. Holland (1975) J. Phys. C. to be published
(31) M.A. Van Hove (1975) Surface Sci. 48, 406

APPLICATIONS OF LEED TO THE DETERMINATION OF SURFACE STRUCTURES ON METALS

S. ANDERSSON

Department of Physics, Chalmers University of Technology, Fack, S-402 20 Göteborg, Sweden

1. INTRODUCTION

Clean low-index single crystal surfaces of pure metals and ordered arrangements of adsorbed atoms and molecules on such surfaces constitute a most important class of idealized systems from which we expect to derive basic understanding about interactions in the surface region of metals. Even in the case of such simple systems at stationary state our present knowledge of conventional physical observables in solid state and molecular physics e.g. atomic arrangement, atomic vibrations and electron structure is relatively limited. This is so for experimental as well as for theoretical reasons. The surface region being an inhomogenity of matter is complicated to investigate experimentally comprising accurate control of a stationary state, sensitive measuring techniques and unfolding of surface and bulk information. Similarly the inhomogenity causing break-down of three-dimensional translational periodicity forces, in many cases, theoretical considerations of a formidable molecular problem encompassing a great number of atoms to be considered in order to account for the underlying bulk crystal.

The inhomogenity of the surface is one of the crucial points of surface crystallography. Actually the unit cell to be found extends straight through the semi-infinite crystal. In general it is just the outermost few atomic planes that is of prime importance for most systems of interest. These layers can be practically investigated by electron diffraction which is the experimental method that in principle can provide the same structural information about crystalline surfaces as the one X-ray diffraction has provided for bulk crystals. Several different techniques utilizing different diffraction parameters (electron energy, angle of incidence to the surface) are currently used; low-, medium-, and reflection high- energy electron diffraction (LEED, MEED and RHEED). Presently LEED is the most advanced tool of these. The experimental probe is a low-energy electron beam (< 1 keV) that diffracts from the surface under study. Elucidation of the surface unit will require some form of diffraction intensity analysis. The dynamical theory of LEED that describes the interactions of the probing electrons with the crystalline surface is the theoretical tool. It is now developed to such a level that a structure analysis of clean surfaces and simple chemisorption systems is attainable[1]. For the moment a structure analysis within this context has to be performed in a trial and error manner, that is, intensities for various trial structures are calculated and compared with experimental data. So far this method has been primarily applied to the kind of systems we are interested in here, namely, clean low-index surfaces of pure metals and ordered arrangements of adsorbates on such surfaces. Certainly there is no shortage of simple systems that can be tackled but the work has predominantly been concentrated on low-index surfaces of Al, Ni, Cu and Ag with simple overlayers containing one adsorbate atom like O, S, Se, Te, Na and I.

This indirect trial and error method for surface structure analysis can evidently be used with success for simple systems but some more manageable method must certainly be developed before the more complicated structures can be solved. Work on simplified intensity analysis schemes has for obvious reasons been guided by the simplicity and accuracy of the X-ray diffraction method for bulk crystals. Development along such lines has resulted in kinematic type data reduction techniques like energy averaging of experimental diffraction intensities[2], constant momentum-transfer averaging of experimental intensities[3] and Fourier transform methods[4].

2. LOW-ENERGY ELECTRON DIFFRACTION, LEED

2.1. Two-dimensionally periodic structures

Some concepts and notations related to surface structure work will be briefly presented here. A more thorough account of surface crystallography is given in a recent review by Strozier et al.[1]

Fig. 1 shows schematically some possible atomic arrangements in a surface. The substrate or bulk crystal is periodic

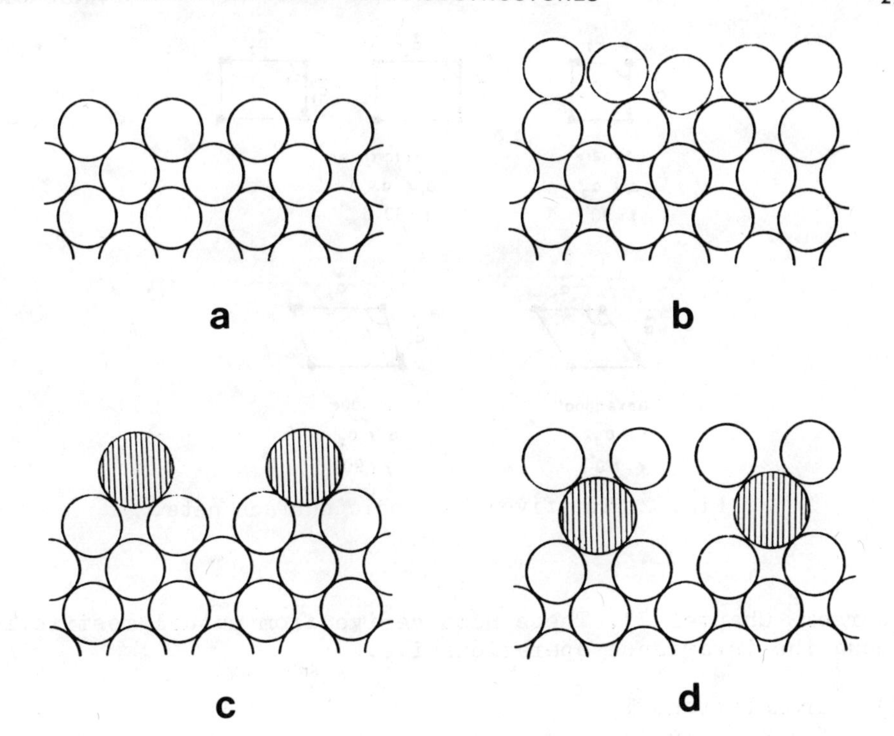

Fig. 1 Some possible model surface structures a, clean ideal surface, all atoms retain bulk positions, b, rearrangement in clean surface that yields a coincidence superposition of the first layer, c, adsorbed overlayer, d, surface compound formation.

in three dimensions and the surface (or selvedge, see Wood[5]) is the transition region between vacuum and the bulk crystal. The atomic arrangement in the surface is denoted surface structure and thus includes all the atomic layers which do not have the three dimensional periodicity of the bulk. In the crystalline case, which we will consider here, the surface structure is periodic in two dimensions i.e. diperiodic. Relations between the bulk structure and the surface structure will be discussed below.

Nets. In a diperiodic crystal surface the equivalent points form a net and the area units are unit meshes. The five nets or plane lattices are shown in Fig. 2. They correspond to the fourteen Bravais lattices for a triperiodic crystal (see

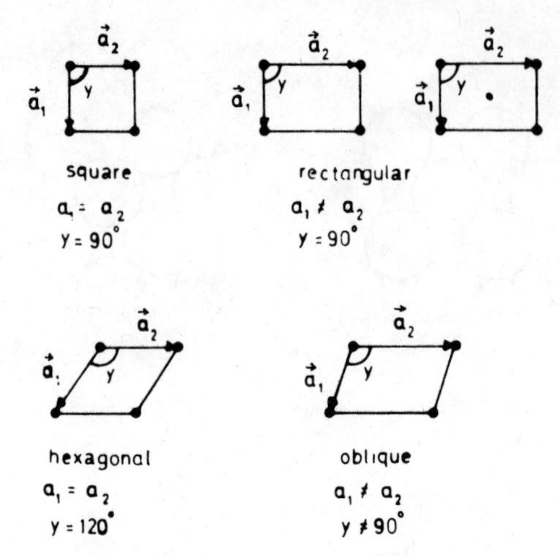

Fig. 2 The five diperiodic surface nets.

Buerger[6] Chapter 7). These nets derive from mutual restrictions among the invariance operations i.e.

i) translations T
$T = n_1 \underline{a}_1 + n_2 \underline{a}_2$
where $\underline{a}_1$ and $\underline{a}_2$ are the translation vectors along the side of the unit mesh.

ii) rotation by an angle
$R = \frac{2\pi}{n}$ $\quad n = 1, 2, 3, 4$ and 6

iii) mirror reflections.

The five nets describe just the symmetry of a plane point lattice. Each lattice point may consist of an arrangement of atoms which itself is invariant under certain symmetry operations. These are the 10 crystallographic point groups in a plane. They are shown in Fig. 3 and are five axial groups consistent with the permissible rotations (ii) and five combining the rotations (ii) and the mirror reflections (iii). The five nets are of course consistent with th 10 plane point groups. Combining the five nets and the 10 plans point groups yield the 17 strictly two-dimensional plane groups. It was pointed out by Wood[5] that another plane group namely the 80 diperiodic groups in three dimensions would be as useful when denoting diperiodic surface structures. We may anticipate the possible occurence of the surface structures described by the five nets. Fig. 4 and Fig. 5 show the symmetries of low-index (h k l) planes of the face- and body-centered cubic lattices and we notice square, rectangular and hexagonal nets.

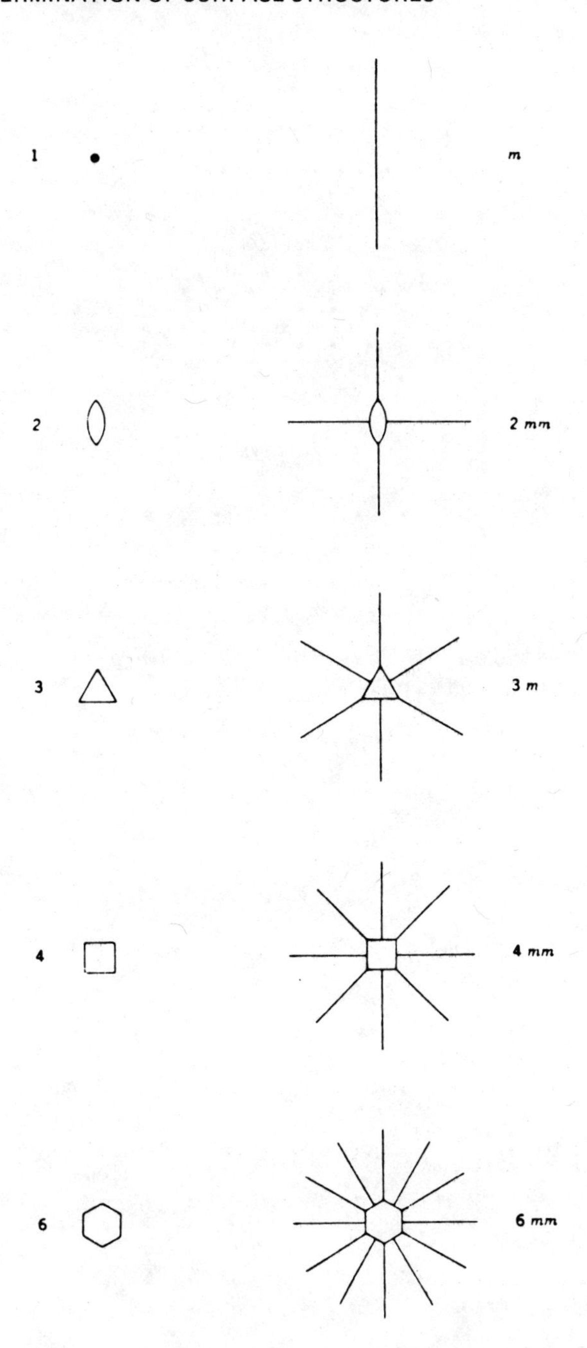

Fig. 3 The ten plane point groups.

fcc (111)
fcc (100)
fcc (110)

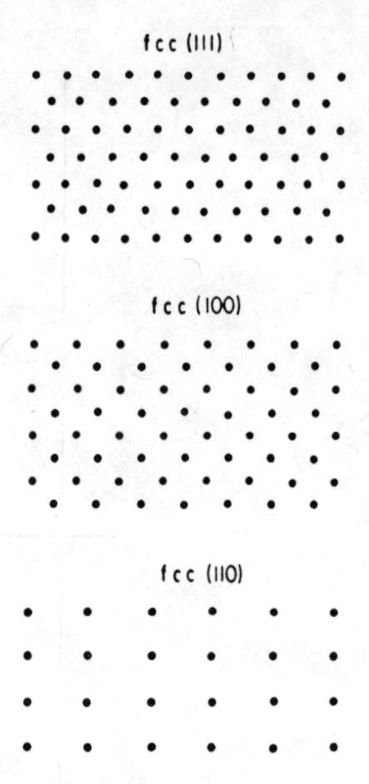

Fig. 4 Symmetries of low-index (h k l) planes of the f c c Bravais lattice.

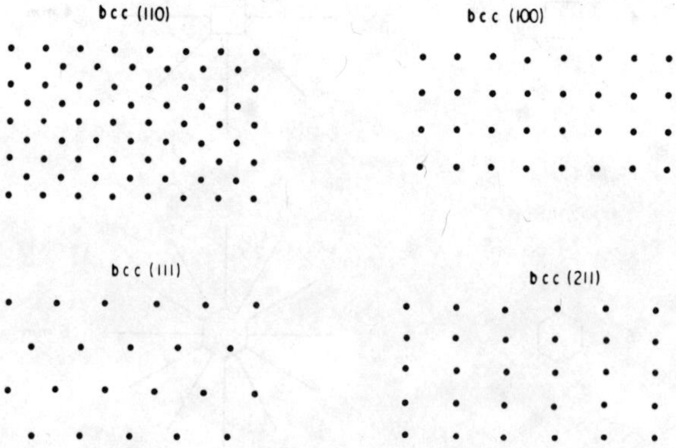

bcc (110)
bcc (100)
bcc (111)
bcc (211)

Fig. 5 Symmetry of low-index (h k l) planes of the b c c Bravais lattice.

Superposition of nets. When denoting the structure of rearranged surfaces and ordered adsorbate layers it is usually required that one expresses the translations $\underline{a}_i$ of one net of area A (e.g. the adsorbate layer net) to the translations $\underline{b}_j$ of another net of area B (e.g. the ideal substrate surface net). The transformation between the two nets may be described by the matrix[7]

$$\underline{\underline{M}} = \begin{bmatrix} m_{11} & m_{12} \\ m_{21} & m_{22} \end{bmatrix} \tag{2.1}$$

i.e. the translation vectors are related

$$\begin{aligned} \underline{a}_1 &= m_{11}\underline{b}_1 + m_{12}\underline{b}_2 \\ \underline{a}_2 &= m_{21}\underline{b}_1 + m_{22}\underline{b}_2 \end{aligned} \tag{2.2}$$

or short

$$\underline{\underline{a}} = \underline{\underline{M}} \cdot \underline{\underline{b}} \tag{2.3}$$

It is easy to show that the areas are related as

$$A = B \det \underline{\underline{M}} \tag{2.4}$$

The values of det $\underline{\underline{M}}$ define the character of the superposition of the two nets $\underline{\underline{a}}$ and $\underline{\underline{b}}$.

det $\underline{\underline{M}}$ integer:	$\underline{\underline{a}}$ and $\underline{\underline{b}}$ are simply related and the super- is simple
det $\underline{\underline{M}}$ rational fraction	$\underline{\underline{a}}$ and $\underline{\underline{b}}$ are rationally related and the superposition is coincidence-site
det $\underline{\underline{M}}$ irrational fraction	$\underline{\underline{a}}$ and $\underline{\underline{b}}$ are irrationallly related and the superposition is inchoherent.

The superpositioned nets (but for the incoherent) can be denoted by one net $\underline{\underline{c}}$ such that

$$\underline{\underline{c}} = \underline{\underline{M}} \cdot \underline{\underline{a}} = \underline{\underline{P}} \cdot \underline{\underline{b}} \tag{2.5}$$

It is usually suitable to choose the net $\underline{\underline{c}}$ with the smallest unit mesh. Thus det $\underline{\underline{M}}$ and det $\underline{\underline{P}}$ must be integers with no common factor.

Surface structure notation. There are several different ways of denoting surface structures in the literature. One obvious way of notation is in terms of the above described matrices[7].

i) If $\underline{\underline{b}}$ is the net of a reference surface structure the superposition is denoted by the matrix $\underline{\underline{P}}$ connecting the nets. Fig. 6 gives a few illustrative examples of the use of this notation for a square symmetric substrate. Fig. 6a shows a case where the translation periodicity is unaltered and thus

$$\underline{\underline{c}} = \underline{\underline{P}} \cdot \underline{\underline{b}}$$

$$\underline{\underline{P}} = \begin{pmatrix} 1 & 0 \\ 0 & 1 \end{pmatrix}$$

In Fig. 6b the superposed net has in one direction a translation vector that is twice that of the reference net while the other vector is unaltered. There are obviously two possible orientations. So

$$\underline{\underline{P}}_1 = \begin{pmatrix} 2 & 0 \\ 0 & 1 \end{pmatrix} \qquad \text{or}$$

$$\underline{\underline{P}}_2 = \begin{pmatrix} 1 & 0 \\ 0 & 2 \end{pmatrix}$$

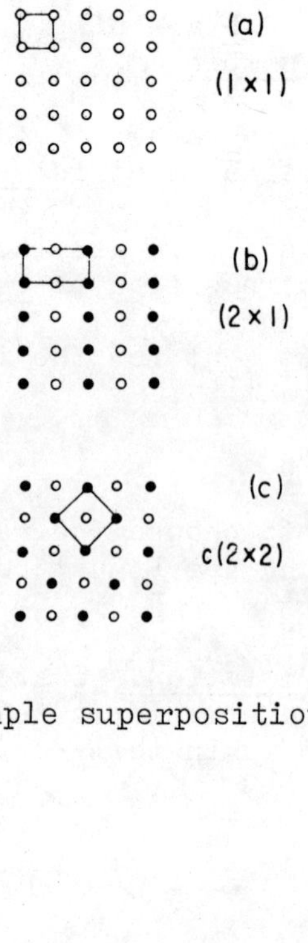

(a) (1×1)

(b) (2×1)

(c) c(2×2)

Fig. 6 Examples of simple superpositions on a square symmetric substrate.

Fig. 6c shows a situation where the superposed net is rotated by 45° and the modulus of the translation vector is $\sqrt{2}$ times that of the reference net. The notation is

$$\underline{\underline{P}} = \begin{bmatrix} 1 & -1 \\ 1 & 1 \end{bmatrix}$$

and the areas

$$C = A = B \det \underline{\underline{P}} = 2\,B$$

ii) The Wood[5] notation has been more frequently used. The superposition $\underline{\underline{c}}$ and reference net $\underline{\underline{b}}$ are related by the quotient of the lengths of the translation vectors and a rotation R

$$({}^{a_1}/{}_{b_1} \times {}^{a_2}/{}_{b_2})\ R$$

This notation can only be used if the unit meshes $\underline{\underline{a}}$ and $\underline{\underline{b}}$ have the same included angles.

The examples of Fig. 6 will be denoted

a) (1 x 1) or p(1 x 1) where p denotes primitive
b) (2 x 1) or p(2 x 1)
c) ($\sqrt{2}$ x $\sqrt{2}$) 45°

iii) In experimental literature a notation based on the choice of non-primitive unit meshes is commonly used. Returning to Fig. 6 the notation will be

a) p(1 x 1)
b) p(2 x 1)
c) c(2 x 2) where c denotes centered

<u>The reciprocal net</u>. An important item when describing physical phenomena related to periodic objects is the construction of a related reciprocal space. This is born out by the fact that periodic functions are expressed as Fourier series in such a space. In this text we will in particular be concerned with reciprocal space when describing diffraction from diperiodic objects.

<u>The reciprocal net unit vectors</u> $\underline{a}_1^*$, $\underline{a}_1^*$ are related to the <u>direct net unit vectors</u> $\underline{a}_1$, $\underline{a}_2$ by

$$\underline{a}_i^* \cdot \underline{a}_j = 2\pi\delta ij \tag{2.6}$$

or in matrix form

$$\tilde{\underline{\underline{a}}}^* \ \tilde{\underline{\underline{a}}} = 2\pi \ \underline{\underline{I}}$$

where $\underline{\underline{I}}$ is the 2 x 2 unit matrix.

An arbitrary reciprocal net vector is given by

$$\underline{g}_{hk} = h\,\underline{a}_1^* + k\,\underline{a}_2^* \tag{2.7}$$

previously we had

$$\underline{\underline{a}} = \underline{\underline{M}} \cdot \underline{\underline{b}}$$

for the superposition of two nets. We will now prove that for the reciprocal nets we have

$$\underline{\underline{a}} = \underline{\underline{\tilde{M}}}^{-1}\,\underline{\underline{b}}^* \tag{2.8}$$

postmultiply both sides by $\underline{\underline{a}}$ and use $\underline{\underline{\tilde{a}}} = \underline{\underline{\tilde{b}}}\,\underline{\underline{\tilde{M}}}$

$$\underline{\underline{a}}^*\,\underline{\underline{\tilde{a}}} = \underline{\underline{\tilde{M}}}^{-1}\,\underline{\underline{b}}^*\,\underline{\underline{\tilde{a}}}$$

left side $\qquad \underline{\underline{a}}^*\,\underline{\underline{\tilde{a}}} = 2\pi\,\underline{\underline{I}}$

right side $\qquad \underline{\underline{\tilde{M}}}^{-1}\,\underline{\underline{b}}^*\,\underline{\underline{\tilde{a}}} = \underline{\underline{\tilde{M}}}^{-1}\,\underline{\underline{b}}^*\,\underline{\underline{\tilde{b}}} \cdot \underline{\underline{\tilde{M}}} =$

$$\underline{\underline{M}}^{-1}\,2\pi\,\underline{\underline{I}}\,\underline{\underline{\tilde{M}}} = 2\pi\,\underline{\underline{I}}$$

The reciprocal net has some elementary properties

i) each vector of the reciprocal net is normal to a set of net rows of the direct net.

ii) $\underline{g}$ is inversely proportional to the spacing of the net rows normal to $\underline{g}$ provided the components h, k of $\underline{g}$ has no common factor, i.e. the spacing between the rows is just

$$d_{(hk)} = \frac{2\pi}{|\underline{g}|} \tag{2.9}$$

iii) the area of the reciprocal unit mesh is inversely proportional to the area of the direct mesh.

$$A = \frac{4\pi^2}{A} \tag{2.10}$$

It is obvious that a particular set of net rows can be denoted by the appropriate low-index reciprocal vector. This notation has a close resemblance to the Miller indices in crystallography.

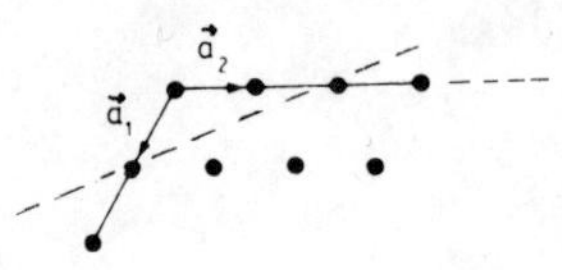

The Miller indices are determined by finding the intercepts with the two unit mesh axes, take the reciprocals of these numbers and reduce to the smallest integers with the same ratio. The notation is (h k).

Interrow spacing for the five nets. As we have seen the interrow spacings are important when we want to describe the properties of a net, we can easily derive expression for this spacing for the five nets.

oblique: $$\frac{1}{d^2_{(hk)}} = \frac{h^2}{a_1^2 \sin^2\gamma} + \frac{k^2}{a_2^2 \sin^2\gamma} - \frac{2hk\cos\gamma}{a_1 a_2 \sin^2\gamma}$$

rectangular p and c: $$\frac{1}{d^2_{(hk)}} = \left(\frac{h}{a_1}\right)^2 + \left(\frac{k}{a_2}\right)^2$$

hexagonal: $$\frac{1}{d^2_{(hk)}} = \frac{4}{3} \cdot \frac{h^2 + hk + k^2}{a^2}$$

square: $$\frac{1}{d^2_{(hk)}} = \frac{h^2 + k^2}{a^2}$$

2.2 Electron diffraction from diperiodic structures

The obvious means to investigate the relative positions of atoms in a periodic arrangement is diffraction. Depending on the properties of the periodic object different methods are employed. Concerning surface structure studies, electron diffraction has been found to be a sensitive tool. Different techniques are used but we will exclusively deal with the method of low-energy electron diffraction, LEED. A schematic LEED experiment using a fluorescent screen display of the diffraction pattern is shown in Fig. 7 together with a photograph of the diffraction pattern from a clean Cu(100) surface (fcc, square symmetry). An electron beam of small energy spread impinges on the specimen surface at normal or nearly normal incidence. The electron energy, E, is typically < 1 keV, usually 50-250 eV. The diffracted electrons are 'back-reflected' from the crystal surface and the main contribution comes from the outermost atomic planes. This surface sensitivity of LEED is due to two courses of events in the mechanism (i) large elastic scattering cross-sections of the low-energy electron on the ion cores and (ii) strong attenuation of the electron in the solid because of inelastic collisions (in metals primarily

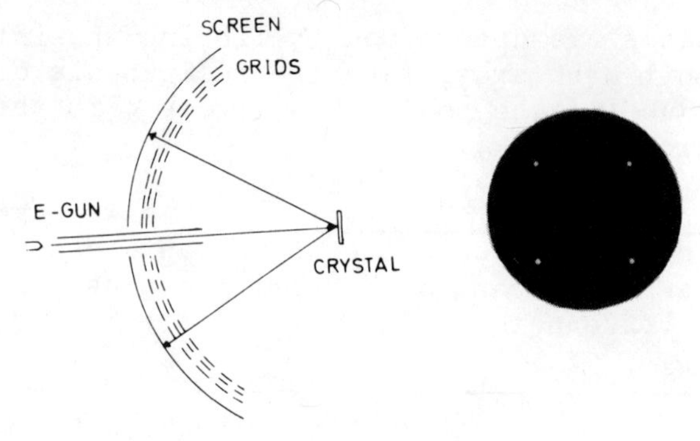

Fig. 7 Schematic LEED display equipment and a diffraction pattern from a Cu(100) surface (50 eV).

due to plasman excitations). Mechanism (ii) yields mean free paths of about 5-10 Å for electrons in the range 50-250 eV. Most of the inelastically scattered electrons are filtered off by using one or several of the grids as a velocity filter. Fig. 8 shows an energy distribution, N(E), of the back-scattered electrons from the Cu (111) specimen obtained by modulation technique by using the two grids close to the screen as retarders and the screen as detector. The diffraction pattern in Fig. 7 is due to the electrons that yield the 'quasi-elastic'

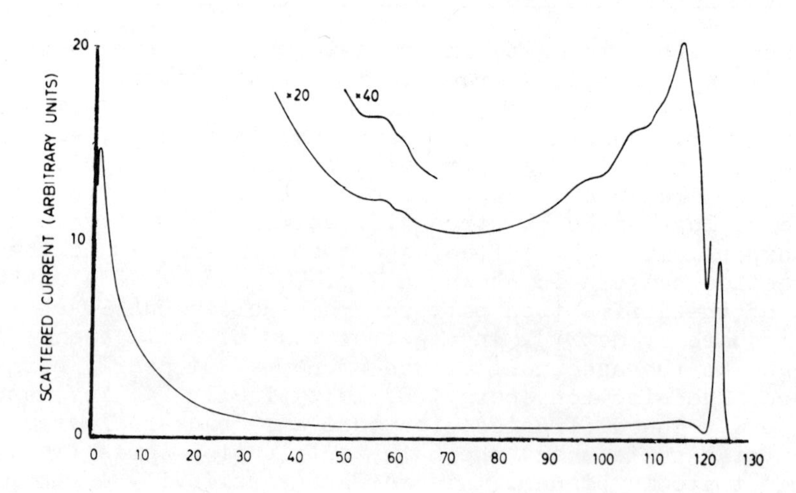

Fig. 8 Energy distribution of electron back-scattered from a clean Cu(111) surface for an incident beam of 125 eV electrons.

high-energy peak. The resolution, $\Delta E/E$, is typically 1 % for $E > 50$ eV. The energy range close to (< 30 eV) the 'quasi-elastic' peak, exhibits a set of discrete loss lines due to inelastic collisions and around 60 eV is a double line due to Auger electrons related to the Cu3p core levels. Auger lines can be further emphasized by differentiating the back-scattered current twice recording $dN(E)/dE$ rather than $N(E)$ and using higher energy in the incident electron (1-2 keV). Auger electron spectroscopy, AES, is frequently used in conjuction with LEED to monitor surface composition.

Fig. 9 shows the exterior of a display type LEED system (Varian 120) equipped with a goniometer for positioning of the specimen, a quadropole masspectrometer for gas analysis, an ion gun for noble gas ion bombardment cleaning of the specimen and a shielded watercooled evaporation cell for various metal vapour sources, (resistive or electron bombardment heating). The diffraction pattern from the Cu(111) surface is observable through the front viewing port. More thorough presentations of the method and its technical details can be found in the literature[8].

Fig. 9 The exterior of a Varian LEED system of display type equipped with a masspectrometer and an evaporator unit. The LEED pattern from a clean Cu(111) surface is visible through the front window.

Diffraction conditions. The LEED experiment is characterized by the diffraction parameters which together specify the energy, E, of the primary electron beam and the direction of this beam relative to the crystal axes. The directions are the angle of incidence, θ , relative to the surface normal and the aximuthal angle, ψ, relative to a unit vector $\underline{a}$ parallel to the crystal surface.

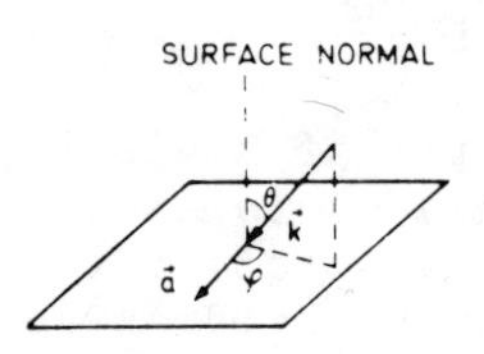

The energy is related to $\underline{K}$ (the propagation vector) by

$$E = \frac{\hbar^2}{2m} k^2 \tag{2.11}$$

$$k = \frac{2\pi}{\lambda} \tag{2.12}$$

and thus

$$\lambda = 12.26 / E^{1/2} \qquad (\lambda \text{ in Å, E in eV}) \tag{2.13}$$

$\underline{k}$ is conveniently expressed in terms of its projection $\underline{k}_{\parallel}$ and $\underline{k}_{\perp}$ onto the surface and the surface normal, respectively.

$$\underline{k} = (\underline{k}_{\parallel}, \underline{k}_{\perp})$$

by using θ , ψ and $\underline{a}$ we get

$$\underline{k}_{\perp} = k \cos$$

$$\underline{k} \cdot \underline{a} = k_{\parallel} \cos \psi$$

The observed quantities are the back-diffracted beams, their directions and intensities. The wave vector $\underline{k}'$ of a diffracted beam may as above be given in terms of angles θ', ψ'. The diffraction process is elastic i.e. energy and the surface component of momentums are conserved. These conditions can be vizualized by the Ewald sphere construction in the reciprocal space. The construction is shown in Fig. 10. A set of reciprocal net normals are drawn through the points g_{hk} of the reciprocal net in the direction normal to the net (which is also normal to the direct net since $\underline{a}_i$ and $\underline{a}_i^*$ are coplanar). Let the

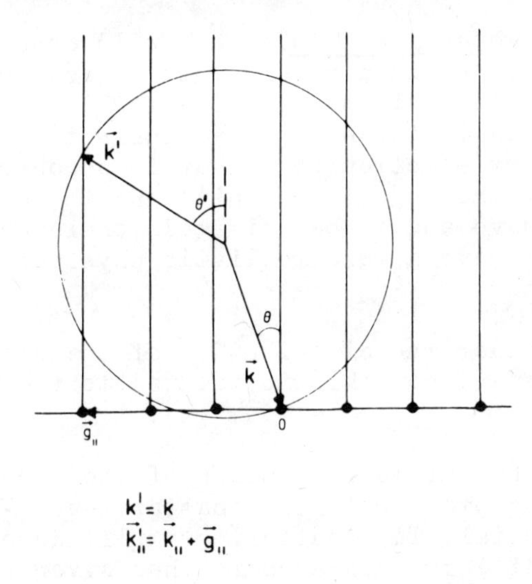

Fig. 10 Ewald sphere construction for a two-dimensional reciprocal net.

propagation vector $\underline{k}$ terminate at the origin of the reciprocal net and let $\underline{k}$ be the radius vector of the Ewald sphere. The intersection of the net normals with the sphere gives the directions of forward and backward diffracted beams such that

$$k' = k \tag{2.14}$$

i.e. $E = E$ energy is conserved

$$\underline{k}'_{\parallel} = \underline{k}_{\parallel} + \underline{g}_{hk} \tag{2.15}$$

i.e. the surface component of momentum is conserved.

The propagation vectors are thus given by

$$\underline{k}^{\pm} = (\underline{k}_{\parallel} + \underline{g}_{hk}, \pm \underline{k}_{\perp}) \tag{2.16}$$

$$\underline{k}_{\perp} = (k^2 - |\underline{k}_{\parallel} + \underline{g}_{hk}|^2)^{\frac{1}{2}} \; (k_{\perp} \text{ real}) \tag{2.17}$$

where the back-diffracted beams, $\underline{k}^-$, are observed in the LEED experiment. These beams will generate a diffraction pattern which reveals the reciprocal unit mesh. From this we can deduce the symmetry and size of the direct unit mesh. The kinds and relative positions of atoms in the unit mehs have to be found from the diffraction intensities.

The interference function. The Ewald sphere construction in the reciprocal net is a purely mathematical construction and cannot be realized in nature. The requirements would be a perfectly periodic in two dimensions infinitely extended object and a primary electron beam that is a coherent pure plane wave. In reality the object is finite and the electron beam is not a plane wave and it has limited coherence. It is certainly useful to have a more realistic physical picture of the diffraction process.

Let us consider the distribution of the diffracted low-energy electrons and how this is related to the surface structure.

Let $\Psi(\underline{r}')$ be the wave function of the electron involved in the scattering process in the scatterer and $V(\underline{r}')$ be the scattering potential. The scattering amplitude of the scattering volume Ω at large distance is then given by

$$A = -\frac{1}{4\pi} \int_{\Omega} e^{-i\,\underline{k}'\cdot\underline{r}'}\, V(\underline{r}')\, \Psi(\underline{r}')\, d\underline{r}' \tag{2.18}$$

The laterial periodicity is given by the unit mesh vectors $\underline{a}_1$, $\underline{a}_2$ and thus for any point $\underline{r}' = \underline{r}'_o + n_1\underline{a}_1 + n_2\underline{a}_2$

$$V(\underline{r}') = V(r'_o) \tag{2.19}$$

The wave function at equivalent points of the net differs by only a phase factor. For an incident plane wave $\Psi = e^{i\,\underline{k}\cdot\underline{r}}$ we have

$$\Psi(\underline{r}') = e^{i\,\underline{k}(n_1\underline{a}_1+n_2\underline{a}_2)} \cdot \Psi(r'_o) \tag{2.20}$$

We can break up the integration (2.18) over the whole scattering volume into a summation over the net and integration over the unit cell.

$$A = \sum_{n_1 n_2} e^{-i(\underline{k}'-\underline{k})(n_1\underline{a}_1+n_2\underline{a}_2)} \times \left(-\frac{1}{4\pi}\int_{\Omega_o} e^{-i\underline{k}\,\underline{r}'_o}\, V(\underline{r}'_o)\,\Psi(\underline{r}'_o)\, d\underline{r}'_o\right)$$

$$= S(\underline{k}', \underline{k}) \cdot F(\underline{k}', \underline{k}) \tag{2.21}$$

The intensity distribution of the scattered wave is proportional to $|A|^2$

$$I(\underline{k}', \underline{k}) = |S|^2\, |F|^2 \tag{2.22}$$

$|S|^2$ is called the interference function, denoted J, and determines the directions of the diffracted waves. The dynamical structure amplitude F determines their relative intensities and carries information about the relative positions of the atoms in the unit cell. The integration in $F(\underline{k}', \underline{k})$ extends over all points within the "unit cell" which is column of layers normal to the surface with an area $|\underline{a}_1 \times \underline{a}_2|$ and a height $z \cdot a_3$ determined by the penetration depth of the wave field.

Performing the summation over n_1, n_2 for a perfect net of dimensions $N_1\underline{a}_1 \times N_2\underline{a}_2$ yields

$$J = \frac{\sin^2(\frac{1}{2} N_1 \delta_1)}{\sin^2(\frac{1}{2} \delta_1)} \cdot \frac{\sin^2(\frac{1}{2} N_2 \delta_2)}{\sin^2(\frac{1}{2} \delta_2)} \tag{2.23}$$

$$\begin{aligned} \delta_1 &= (\underline{k} - \underline{k}')\cdot\underline{a}_1 = (\underline{k}_{\parallel} - \underline{k}'_{\parallel})\cdot\underline{a}_1 \\ \delta_2 &= (\underline{k} - \underline{k}')\cdot\underline{a}_2 = (k_{\parallel} - \underline{k}'_{\parallel})\cdot\underline{a}_2 \end{aligned} \tag{2.24}$$

J yields the symmetry, spacings and beam profiles in the diffraction pattern. J has maxima of height $N_1^2 \times N_2^2$ whenever

$$\delta_i = 2\pi h_i \qquad (h_i \text{ integer}) \tag{2.25}$$

This statement is equivalent to the Laue conditions for diffraction from a two-dimensional grating.

The first zero in the interference function may be taken as a measure of the beam width. Consider $\sin(\frac{1}{2} N\delta^0) = 0$

$$\sin(\tfrac{1}{2} N\delta^0) = 0 \tag{2.26}$$

$$\tfrac{1}{2} N\delta^0 = \tfrac{1}{2} N\, 2\pi h + \pi$$

first zero from the maximum at $\delta = 2\pi h$
thus

$$\Delta\delta = \delta^0 - \delta = (\underline{k}'_{\parallel} - \underline{k}'^{0}_{\parallel})\underline{a} = \frac{2\pi}{N}$$

For simplicity consider $\underline{k}'_{\parallel}$ to be parallel to $\underline{a}$ and $\underline{k}$ to be normal to the surface

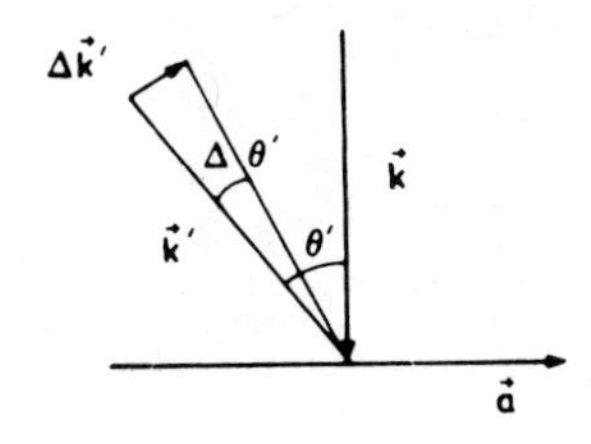

$$\Delta k' = k' \Delta \theta$$

$$\Delta k_{\parallel}' = \Delta k' \cos \theta = k' \Delta \theta \cos \theta$$

furthermore

$$\underline{k}_{\parallel}' \cdot \underline{a} = k' \sin \theta' a = 2\pi h$$

this gives

$$\Delta \delta = k' \Delta \theta' \cos \theta' a = \frac{2\pi h}{\sin \theta'} \cos \theta' \Delta \theta' = \frac{2\pi}{N}$$

thus

$$\Delta \theta' = \frac{\tan \theta'}{h N} \tag{2.27}$$

is a measure of the angular broadening of the diffraction beam due to limited periodicity in the surface net.

Coherence. According to (2.27) the width of the diffraction beams provide information about the dimensions $N_1 \times N_2$ of the coherent scattering domains. The instrumental resolution, however, sets a lower limit to the area over which the primary wave field is coherent, the coherence zone. The character of the primary wave is determined by the electron source. For an extended, thermally heated source as used in LEED optics the coherence zone will be limited by time incoherence due to the thermal energy spread ΔE and by space incoherence which derives from divergence of the electron beam as measured by the half-angle β_s subtended at the crystal surface by the electron source[9]. Let the electron propogation vector $\underline{k}$ be normal to the crystal surface.

$$E = \frac{\hbar^2}{2m} k^2 \quad \text{gives the spread}$$

$$k = \frac{2m}{\hbar^2} \frac{\Delta E}{2k} = k \frac{\Delta E}{E} \quad (\text{along } \underline{k})$$

Due to the divergence there is a spread in the surface component of $\underline{k}$, $2\underline{k}\beta_s$ combining the uncertainties in quadrature gives

$$\Delta k_{\parallel} \approx 2 [(k\beta_s)^2 + (\Delta k \beta_s)^2]^{\frac{1}{2}} =$$

$$2\beta_s \frac{2\pi}{\lambda} [1 + (\frac{\Delta E}{2E})^2]^{\frac{1}{2}} \tag{2.28}$$

Let the coherence criterion be given by

$$\Delta X \cdot \Delta k_{\parallel} = 2\pi$$

where ΔX is the coherence zone diameter

$$\Delta X \approx \frac{\lambda}{2\ \beta_s\ [1 + (\Delta E/2E)^2]^{\frac{1}{2}}} \qquad (2.29)$$

No area larger than $\approx (\Delta X)^2$ can contribute coherently to the diffraction pattern.

We may estimate ΔX for an ordinary LEED optics. Assume $E \approx \frac{3}{2}$ k T, i.e. the average velocity of the thermally emitted electrons and putting $\beta_s \approx 0.005$ (for a source radius 0.5 mm at 100 mm distance from the crystal surface). $\Delta E \approx 0.2$ eV for a low-temperature source at 800°C. It is obvious from (2.29) that the time-incoherence due to ΔE will only be of importance at the lowest energies. At E = 150 eV, $\lambda \approx 1$ Å gives

$\Delta X \approx 100$ Å

A similar figure is found experimentally[10].

Surface disorder. There are of course situations applying to clean single crystal surfaces when limited order in the surface net gives rise to beam broadening. The surface order can be detoriated during noble gas ion bombardment cleaning. In this way disorder may accumulate until the diffraction pattern is completely obliterated e.g. Si, Ge kept at room temperature. The disorder can usually be leaked out by careful annealing (suitable temperatures for Si and Ge are 700°C and 300°C respectively) and the surface becomes perfect over regions wide enough compared to $(\Delta X)^2$.

Antoher kind of disorder may occur when clean single crystal surfaces are produced by cleavage. Usually such surfaces are perfectly periodic and flat over large areas but cleavage steps may appear in a very large number breaking long range order in the surface direction perpendicular to the step direction. High step densities can also be introduced by deliberatly cutting the crystal face at a slightly wrong angle.

In the case of adsorbed layers on single crystal surfaces, a number of different mechanismes may give rise to beam broadening e.g. effects of domain size, different kinds of one-dimensional and two-dimensional disorder (out-of-phase domains) such effects have been discussed at several places in the literature[11,12].

Fig. 11 illustrates disordering of the Ni(100) surface by noble gas ionbombardment (200 eV organ ions, 1 μA/cm^2, 2 hours). The observed beam broadening reveals that the size of the ordered domains has been reduced below the size of the coherence

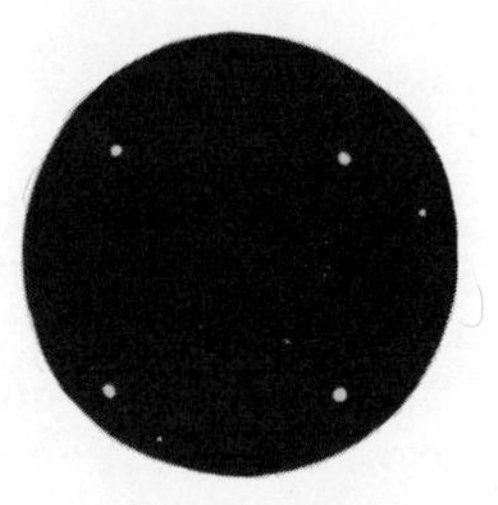

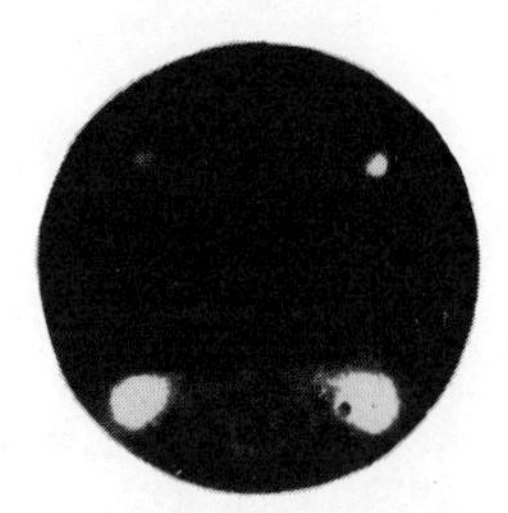

Fig. 11 Diffraction patterns from an ordered and a disordered Ni(100) surface respectively (LEED).

zone $(\Delta X)^2$. We may estimate the domains size from

$$\Delta\theta = \frac{\tan\theta}{h \cdot N}$$

$$\Delta\theta = 0.04 \text{ radians}, \; h = 1, \; \theta = 34^\circ$$

which gives N = 18 i.e.

$$Na = 18 \cdot 2.50 \text{ Å} = 45 \text{ Å}$$

compare $\Delta X \approx 150$ Å at $\lambda = 1.5$ Å

2.3 LEED intensity analysis

As in any diffraction study of a crystalline object a

determination of the kinds and relative positions of the atoms in an ordered surface structure has to be based on an analysis of the LEED intensities, $I(\underline{k}',\underline{k})$. The serious problem of LEED is that the dynamical structure amplitude $F(\underline{k}',\underline{k})$ is most involved quantity because of the strong multiple scattering of the low-energy electrons on the scatters. Thus the simple geometrical phase relations among the atoms in the unit cell are buried and there is no obvious Fourier transform of $I(\underline{k}',\underline{k})$ as in the X-ray case. A diffraction intensity analysis within the framework of the dynamical theory of LEED then in general implies a comparision of experimentally determined intensities with those calculated for a postulated trial structure. The practical procedure is described in sections 3 and 4.

Any dynamical calculation of $I(\underline{k}',\underline{k})$ amounts to calculating $\Psi(r')$ in Eq. 2.18 for the appropriate arrangement of scatters. It has been found preferable to regard the semiinfinite crystal as a pile of layers parallel to surface[13-17] assuming the layers to be infinite perfect nets. Domain size effects parallel to the surface are thus disregarded. Because of the attenuation of the electron wave field inside the crystal only a finite number of layers are considered. This division of the problem has natural advantages. The two-dimensional periodicity of the layers implies a Bloch wave representation of the waves utilizing the symmetry properties of the layers and it allows for a variation of position and or composition of the layers e.g. contraction or expansion of the surface layer of a clean low-index metal surface or superposition of a chemisorbed layer on such a surface.

Detailed account of the dynamical LEED theory utilizing the layer scheme mentioned above are given in the literature [1,13-17,18] where also the computational procedure and explicit results can be found[16,17]. A brief presentation will be given here. The theory utilizes the muffin-tin model of the crystal potential which is assumed to be composed of perfect two-dimensionally periodic layers of mono-atomic thickness parallel to the surface. The electron waves are represented in the constant potential, V_0, between layers in terms of a discrete set of plane waves that are diffracted by each layer. The incident electron beam is represented by a plane wave which connects to the plane waves between layers via the two-dimensional reciprocal lattice vectors of each layer. The plane-wave diffraction properties of a single layer are calculated from those of a single atom and are then used to generate the diffraction properties of the pile of layers considered. Fig. 12 illustrates a scattering situation with an overlayer on a substrate. The diffraction properties of the whole system is calculated from the reflectivition coefficients $R_{\underline{gg}'}$ of the substrate and the reflection and transmission coefficients of the overlayer

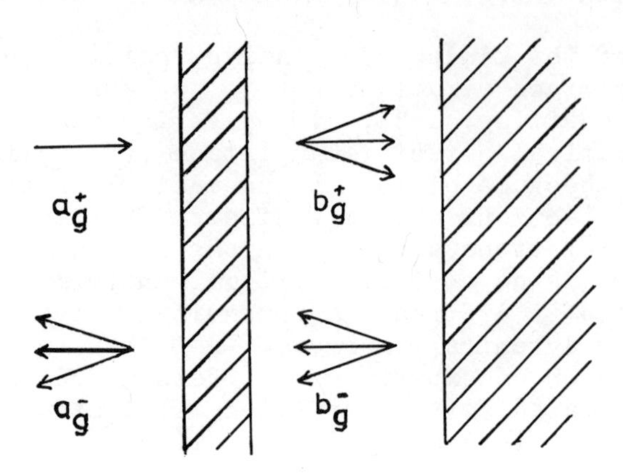

Fig. 12 An overlayer on a substrate. The arrows denote waves outside the system and between overlayer and substrate.

R^{-+}, R^{+-}, T^{++} and T^{--}, respectively. If $b_{\underline{g}}^{+}$ are the amplitudes of waves incident on the substrate and $b_{\underline{g}}^{-}$ are the amplitudes of waves reflected by the substrate we have

$$b_{\underline{g}}^{-} = \sum_{\underline{g}'} R_{\underline{gg}'} b_{\underline{g}'}^{+} \tag{2.30}$$

Including the overlayer the various amplitudes are related as

$$b_{\underline{g}}^{+} = \sum_{\underline{g}'} T_{\underline{gg}'}^{++} a_{\underline{g}'}^{+} + R_{\underline{gg}'}^{+-} b_{\underline{g}'}^{-} \tag{2.31}$$

$$a_{\underline{g}}^{-} = \sum_{\underline{g}'} T_{\underline{gg}'}^{--} b_{\underline{g}'}^{-} + R_{\underline{gg}'}^{-+} a_{\underline{g}'}^{+} \tag{2.32}$$

where $a_{\underline{g}}^{+}$ are amplitudes of waves incident on the system and $a_{\underline{g}}^{-}$ are the amplitudes of diffracted waves we wish to find. From equations (2.30-2.32) we find

$$a_{\underline{g}}^{-} = \sum_{\underline{g}'} \left| T^{--}(1-RR^{+-})^{-1} RT^{++} + R^{-+} \right|_{\underline{gg}'} \times a_{\underline{g}'}^{+} \tag{2.33}$$

The reflection coefficients $R_{\underline{gg}'}$ of the substrate are generated in a similar way as $a_{\underline{g}}^{-}$ from the reflection and transmission coefficients of the atomic layers constituting the substrate crystal. These coefficient can for the case of a two-dimensional Bravais lattice (one atom per unit cell) be obtained from the following expression[16] which yields the single-layer diffraction amplitudes between an incident wave $\exp(i\,\underline{k}_{\underline{g}}^{+}\cdot\underline{r})$ and the wave $\exp(i\,\underline{k}_{\underline{g}'}^{+}\cdot\underline{r})$

$$M^{\pm\pm}_{\underline{g}'\underline{g}} = \frac{8\pi^2\cdot i}{A|\underline{k}_{\underline{g}}^{\pm}|\underline{k}^{+}_{\underline{g}'\perp}} \sum_{\substack{l'm'\\ lm}} |i^{l}(-1)^{m}Y_{lm}(\Omega(\underline{k}_{\underline{g}}^{\pm}))|$$

$$x\ (1-X)^{-1}_{lm,l'm'}\ |i^{-l'}Y_{l'm'}(\Omega(\underline{k}_{\underline{g}'}^{\pm})|e^{i\delta_{l'}}\sin\delta_{l'}. \qquad (2.34)$$

The origin of coordinates is an atomic centre. A is the area of the layer unit mesh, $\Omega(\underline{k})$ is the direction of the wave-vector $\underline{k}$ and δ_l is the set of phase shifts that describes the electron-atom scattering. The matrix X describes multiple scattering within the layer. The wave vectors $\underline{k}_{\underline{g}}^{\pm}$ of the plane waves are given (see equations 2.16, 17) in terms of parallel ($\parallel$) and normal ($\perp$) components with respect to the surface

$$\underline{k}_{\underline{g}}^{\pm} = (\underline{k}_{\underline{g}\parallel}^{\pm}\ ,\ k_{\underline{g}\perp}^{\pm})$$

$$\underline{k}_{\underline{g}\parallel}^{\pm} = \underline{k}_{\parallel} + \underline{g} \qquad (2.35)$$

$$k_{\underline{g}\perp}^{\pm} = \pm\,[2(E-V_o) - |\underline{k}_{\underline{g}\parallel}^{\pm}|^2]^{1/2}$$

The forward and backward transmission matrices are $T^{++} = M^{++} + I$ and $T^{--} = M^{--} + I$ where I is the unit matrix. The forward to backward and backward to forward reflection matrices are $R^{-+} = M^{-+}$ and $R^{+-} = M^{+-}$.

An intensity analysis within the framework of this dynamical diffraction theory starts with the postulation of a model, which is specified by a number of parameters. The parameters that specify the positions of atoms are conveniently called <u>structural parameter</u> while the <u>non-structural</u> parameters are those that describe the potentials chosen to calculate the scattering of electrons, the absorption and refraction of electrons and the vibrations of surface and bulk atoms. As an example of the application of the method Fig. 13 shows a comparision of calculated and experimental LEED intensities for the three lowest index diffraction beams from the clean Cu(100) surface. The calculation is due to Pendry[16] and assumes rigid bulk atomic position (no thermal vibrations). The ion-core scattering was obtained from a Hartree-Fock potential for Cu, the real part of the uniform potential, V_{or}, was deduced to be

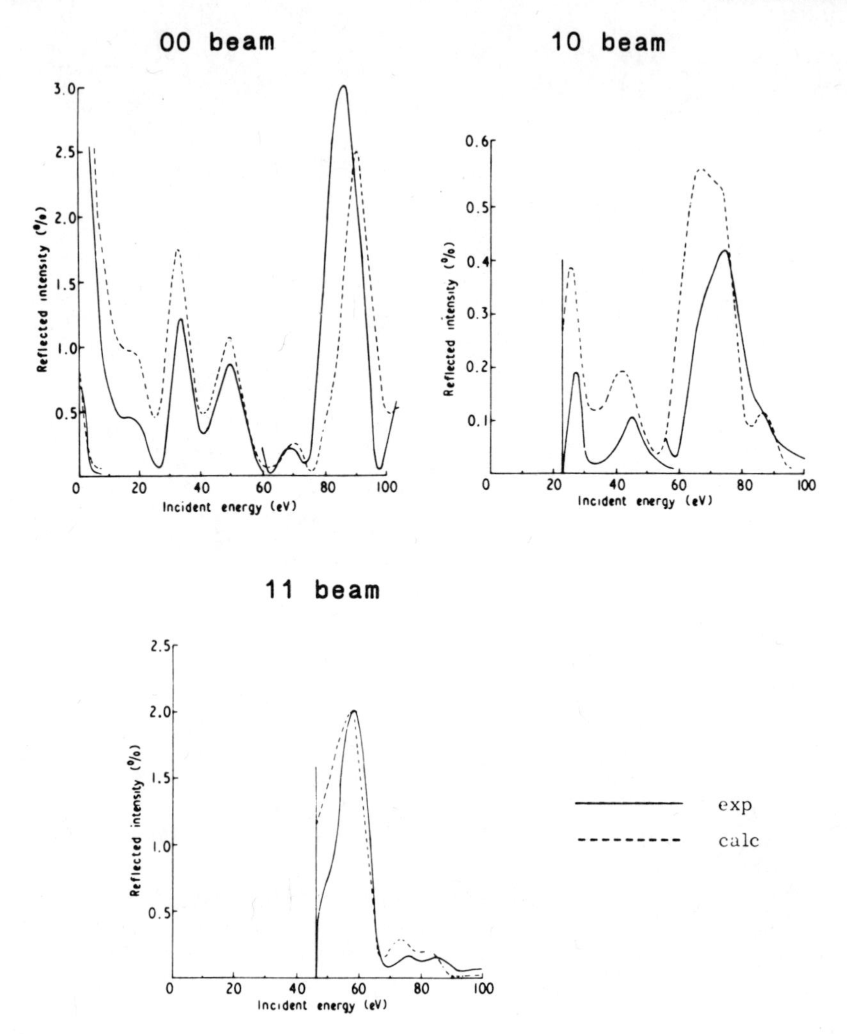

Fig. 13 Comparison of experimental and theoretical LEED beam intensities for Cu(100) (Pendry[16]).

-15 eV from a comparison with a bandstructure calculation for Cu and the imaginary part, V_{oi}, describing the damping of the electron flux was adjusted empirically to the experimental intensity data, V_{oi} = 4 eV for E > 10 eV, V_{oi} = 1 eV for E < 10 eV. Reflection from the surface potential barrier was neglected. It is obvious from Fig. 13 that the calculation reproduces position, shape and relative magnitude of experimental intensity peaks well. It was found that the theoretical results were very sensitive to the Cu ion-core potential used[16].

3. STRUCTURE OF CLEAN LOW-INDEX METAL SURFACES

3.1 Surface preparation

In general LEED studies of clean low-index metal surfaces has a dual purpose in surface structure work

(i) to determine atomic positions within the clean surface and to establish a suitable dynamical diffraction treatment

(ii) to produce a well-defined substrate for a study of adsorbate structures.

An important step in a surface structure investigation is the preparation and cleaning of the single crystal specimen. In the LEED experiment suitable specimens should have an area considerably larger than the area of the electron beam (about 1 mm^2). The bulk material should be high purity since contaminants often tend to segregate at the surface during heating of the specimen. Some commonly used preparation and cleaning techniques are listed below.

a) The specimen may be prepared by <u>cleavage in situ</u> at UHV conditions (10^{-10}Torr). The surface obtained in this way are usually free of contamination and can be of very high structural perfection. The technique can only be applied to few metals like Be and Zn[19].

b) <u>Epitaxial growth</u>, from the vapour phase onto a suitable single crystalline substrate, at UHV conditions, can yield excellent specimens. The substrate surface may be prepared by cleavage (e.g. alkali halides, mica). Contamination from the substrate may be a problem. Thus for example Ag(100) can be grown on KCl(100)[20], Al(111) on Si(111)[21,22], Cu(111) on W(110)[23], Ag(111) on Cu(100)[24], Na(110) on Ni(100)[25].

c) In most cases the specimens are prepared from high purity bulk single crystal outside the UHV system by means of <u>cutting and polishing</u>. The surface orientation is obtained by X-ray back reflection technique to within 0.1^o-1°. Cutting may be performed by spark erosion, chemical erosion or by a diamond saw depending on the material. Mechanical and electrolytical (or chemical) polishing is applied to obtain an undistroted flat surface. These specimens have to be cleaned in the LEED UHV system. Commonly used methods are

heating
ion bombardment-annealing
chemical reaction.

The appropriate technique again depends on the material. W surfaces, for instance, may be cleaned by heating since segregated carbon can be flashed off. Carbon segregated on Ni surfaces has to be removed by ion bombardment or chemical reaction.

The characterization of the surface condition is an important problem. A combination of the LEED observations and a surface sensitive spectroscopic technique as grazing angle Auger electron spectroscopy (AES) yield in most cases a decent diagnostics of surface order and purity. LEED patterns that show sharp diffraction spots, high contrast and low background reveal in general good crystallinity. AES spectra free from impurity lines correspond in general to impurity concentrations lower than 0.1% of a monolayer.

3.2 Diffraction patterns

LEED patterns have been recorded from a great number of clean low-index metal surfaces. In most cases the primary aim has been to produce suitable surfaces for adsorption studies[26]. In some cases a sequence of low-index surfaces have been investigated like (111), (100) and (110) of fcc metals (e.g. Cu, Ni, Pt) and (110), (100), (111) and (112) of bcc metals (e.g. W). A number of metals have also been investigated with the particular emphasis on obtaining diffraction intensities useful for structure analysis and some examples of this will be given in the next paragraph.

In general the lateral periodicity of most clean low-index metal surfaces has been found to be identical to that of a corresponding plane in the bulk crystal. This is in most cases established from the observation that the diffraction pattern does not contain any other diffraction spots than those expected from ideal periodicity. LEED patterns from the clean (111) and (100) surfaces of Cu (fcc) are shown in Fig. 14 and we recognize hexagonal and square symmetries as expected. The clean (110), (100) and (211) surfaces of W yield the LEED patterns in Fig. 15 which also correspond to ideal lateral periodicity. The unit mesh dimensions can of course be determined from the diffraction pattern provided the spot positions on the photographs have been calibrated in terms of diffraction angle θ', At normal incidence we have

$$\underline{k}'_{\parallel} = \underline{g}_{hk} \qquad \text{or}$$

$$\frac{2\pi}{\lambda} \cdot \sin\theta' = \frac{2\pi}{d_{(hk)}}$$

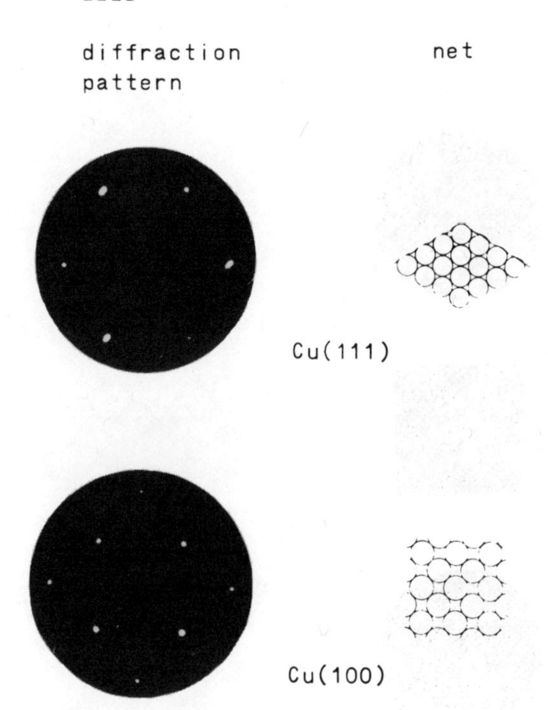

Fig. 14 Diffraction patterns and direct net models for Cu(111) and Cu(100) surfaces (LEED).

$$d_{(hk)} \sin \theta' = \lambda$$

This is called the plane grating formula. From 2.13 we get

$$E = V - V_c = \frac{150.4}{d^2_{(hk)}} \cdot \frac{1}{\sin^2 \theta'}$$

where V is the accelerating potential and Vc is the contact potential between the cathode and the specimen

$$V_c = \phi_s - \phi_c$$

where ϕ_s and ϕ_c are the workfunctions of the specimen and cathode respectively. The experimental relation V versus $1/\sin^2\theta'$ for Ni(100) is shown in Fig. 16. The observed lateral spacing is (2.46±0.03)Å to be compared with the X-ray value

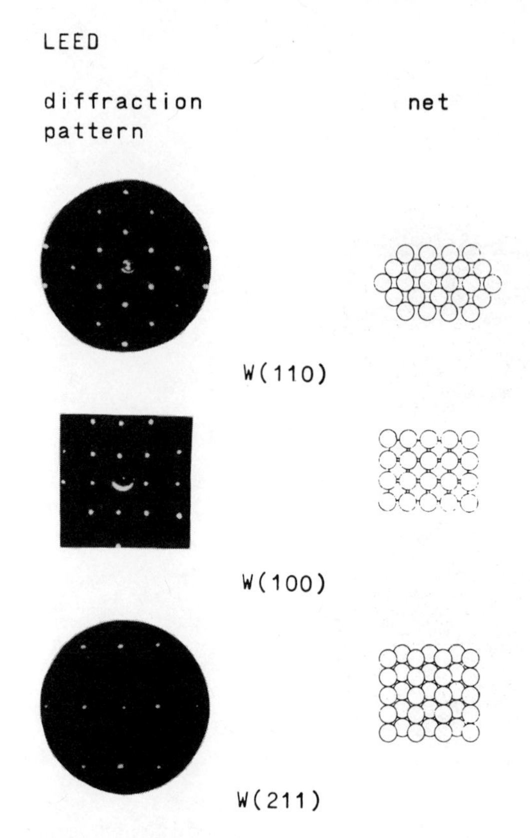

Fig. 15 Diffraction patterns and direct net models for W(110), W(100) and W(211) surfaces (LEED).

2.49 Å. The accuracy of this kind of measurements is typically ≈1% provided stray electric and magnetic fields are well cancelled. Further improvements can be made by careful calibrations and is of interest in particular applications[27].

Clean surfaces of the fcc metals Pt[28], Au[29,30] and Ir[31] have been thoroughly investigated since the LEED patterns indicate surface rearrangement. The (100) surfaces of these metals yield diffraction patterns indicating a (5x1) unit mesh. Such a pattern from Au(100) is shown in Fig. 17 together with a model structure that assumes a hexagonal arrangement of Au atoms superpositioned coincidently on the underlying square symmetric (100) substrate. There has been suggestions that these rearrangements are associated with contamination (e.g. O[32]) but the evidence seems unclear and the (5x1) structures have indeed been observed by a number of different investigators.

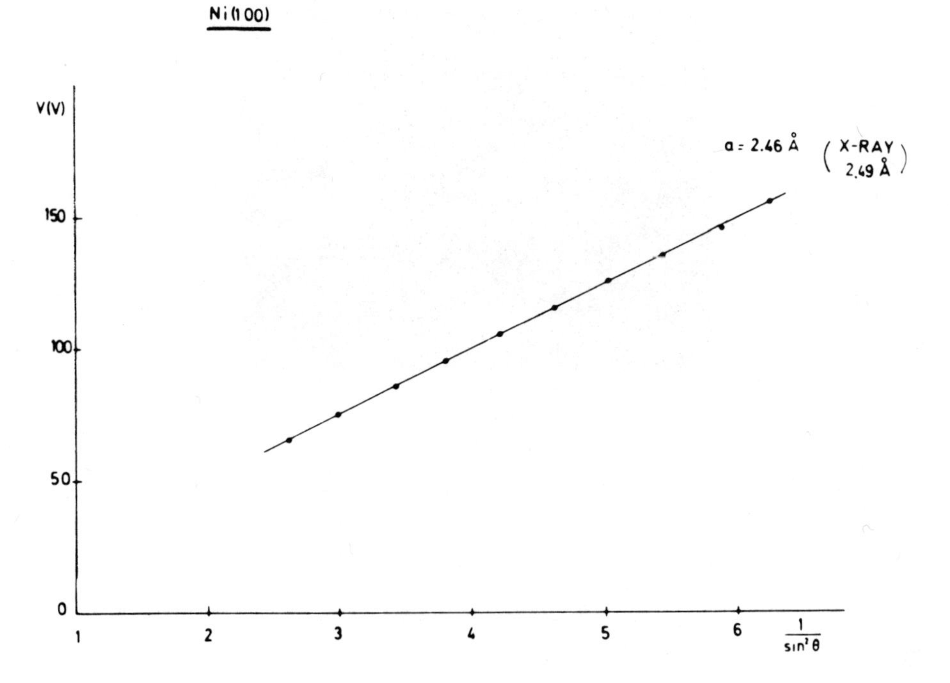

Fig. 16 Surface net spacing determination for Ni(100). Electron potential, V, versus $1/\sin^2\theta'$, θ' diffraction angle.

3.3 Intensity analysis

In a fluorescent screen display type LEED system (see Fig. 7) the diffraction beam intensities can be recorded either by a movable Faraday cage or by a spotphotometer that records the intensity of the diffraction spots on the screen. Both methods discriminate against the inelastically scattered electrons with an energy resolution of $\Delta E/E \approx 1\%$, E is the energy of the incident electrons. While the Faraday cage directly reads the electron current in the hk beam, i_{hk}, the spotphotometer reading has to be calibrated. This is usually achieved by illuminating the screen with an electron beam of known intensity. The homogenity of the screen has to be checked in this procedure. The recorded beam intensities are normalized to the current, i_o, in the incident electron beam and displayed like

$$I_{hk} = i_{hk}/i_o$$

as a function of electron energy, E, for one or several angles of incidence, θ, along one or several azimuths, ψ.

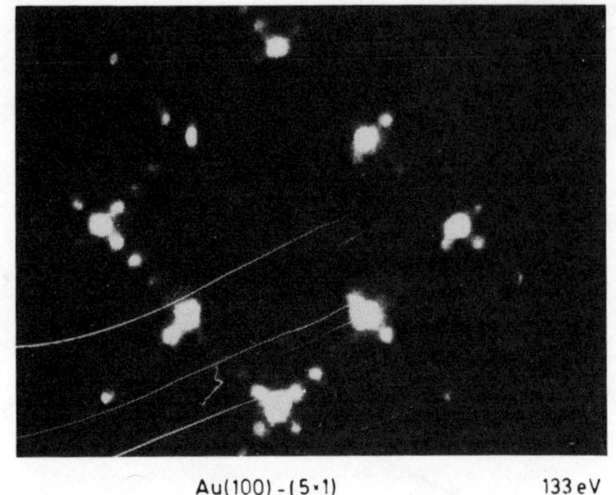

(a)

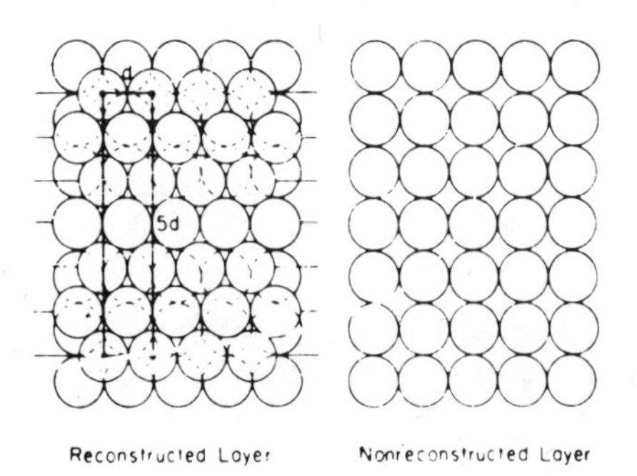

(b)

Fig. 17 a. Diffraction pattern from Au(100)-(5x1).
b. Model structure of coincidental hexagonal Au layer superpositioned on the underlying (100) substrate. (Palmberg and Rhodin[30]).

Diffraction intensity analysis as outlined in section 2.3. has been carried out for a number of low-index metal surfaces. A list of examples is given in Table I (a more complete list is in Ref. 1). They mainly refer to a comparison of experimental LEED intensities with those calculated within the framework of the dynamical LEED theory. A surface structure determination within this context is for the time being exclusively performed in an indirect trial and error manner adjusting the atomic positions (the structural parameters) (see section 2.3). The diffraction intensities are calculated for a trial structure that is compatible with the space group symmetry revealed by the LEED pattern. For the surfaces listed in Table I the obvious initial trial structure is the bulk structure, since these

TABLE I

LEED intensity analysis of low-index metal surfaces.

Element	Structure	Bulk lattice constant	Surface		
Al	fcc	4.05 Å	(111) expanded ~2.5% Ref. 33, 34	(100) see e.g. Ref. 17, 33, 35	(110) contracted 5 ~ 15% Ref. 33, 34, 36
Ni	fcc	3.52 Å	(111) Ref. 37, 38, 39	(100) Ref. 37, 39, 40	(110) contracted ~5% Ref. 39
Cu	fcc	3.61 Å	(111) Ref. 41	(100) Ref. 16, 17	
Ag	fcc	4.09 Å	(111) Ref. 42, 43	(100) Ref. 17	
Na	bcc	4.23 Å	(110) Ref. 44		
Mo	bcc	3.15 Å		(100) contracted ~11% Ref. 45	
Be	hcp	a 2.27 Å b 3.59 Å	(1000) Ref. 46		
Ti	hcp	a 2.95 Å b 4.68 Å	(1000) Ref. 45		

surfaces exhibit ideal lateral periodicity. If the calculated and experimental intensities agree for some range of diffraction parameter ($0 < E < 200$ eV, one or a few angles of incidence θ) then the trial structure is probably correct. Improvements may then be achieved by modifying the non-structural parameters (see section 2.3). If the experiment and theory disagree strongly then the choice of structural parameters is most probably wrong and a new set has to be tried. Unfortunately there is no satisfactory definition of a "reliability factor", R, for surface structure analysis. In general the models are evaluated by doing a semiquantitative comparison of experimental and calculated intensities emphazising on positions, shapes and relative magnitudes of peaks in the intensity curves. Effects of surface roughness e.g. cannot be accounted for so a quantitative agreement of absolute intensities is not expected. It has been proposed[47] that a possible "reliability factor" could be

$$R = \Sigma \ (cI_{calc} - I_{obs})^2 / \Sigma (I_{obs})^2$$

where I_{calc} and I_{obs} are the calculated and observed intensities respectively. The sums are taken for each beam and over a sufficiently dense net of energy points (e.g. every eV) over the experimental energy range. The scaling constant c is calculated for each beam so that R is minimized.

A few examples of intensity analysis work for clean surfaces will be presented here. In some instances several workers have considered the same problem and achieved good agreement. In section 2.3 and Fig. 13 we considered calculated LEED intensities for the clean Cu(100) surface obtained by Pendry. His results can be compared with those due to Jepsen, Marcus and Jona [17] shown in Fig. 18 together with the same experimental results as in Fig. 13. The latter calculation also assumes bulk atomic positions but are corrected for lattice vibrations at 298 K assuming a Debye temperature θ_D = 339 K. The Cu ion-core potential is the Chodorov potential used for bulk band structure calculations and the damping was described by V_{oi} = 3.4 eV. The two calculations agree reasonable well. Similar results have been achieved for several other (100) surfaces of fcc metals as e.g. Al, Ni, Ag.

For the dense (111) and (100) surfaces of fcc metals one has in general found that the postulation of bulk atomic positions is good to within about ~2.5% variation of the interlayer spacing between the first two layers. For the less dense Ni(110) surface Demuth, Marcus and Jepsen[39] have observed a contraction of about 5% in this spacing. A comparison of calculated and experimental LEED intensities for this system is shown in Fig. 19. The calculation assumes an effective Debye temperature of 335 K, and inner potential V_o=11 eV and an energy (E) dependent imaginary potential of the form V_{oi}=0.85 $E^{1/3}$ (E in eV).

Several bcc materials have also been investigated and data for Na(110) and Mo(100) will be discussed here. Fig. 20 shows

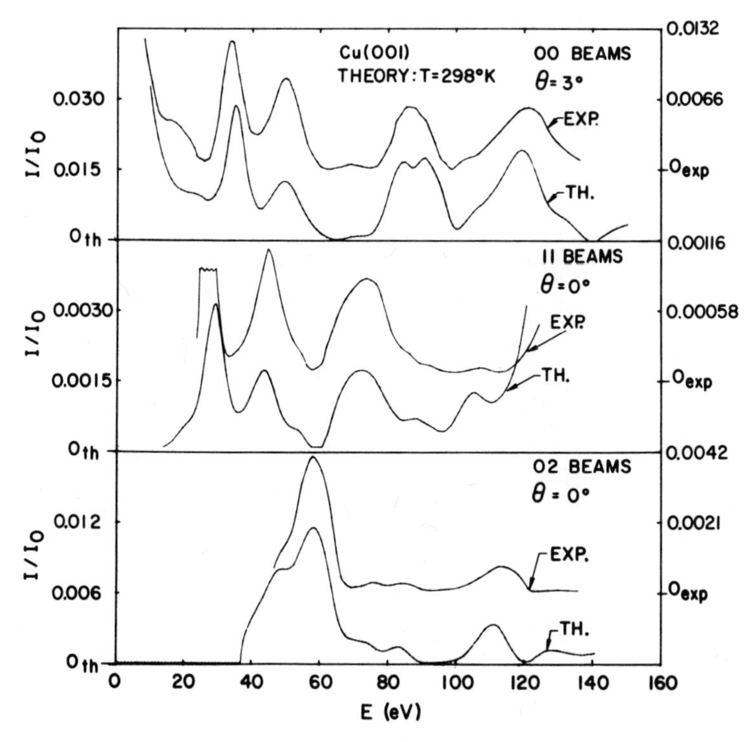

Fig. 18 Comparison of experimental and theoretical LEED beam intensities from Cu(100). (Jepsen, Marcus and Jona[17]). Compare with Fig. 13.

results for the celan Na(110) surface[44]. The Na single crystal is grown epitaxially at 173 K on a Ni(100) substrate[25] and is rotationally disordered around the surface normal. All experimental diffraction intensities were recorded for normal incidence of the primary electron beam. The calculation assumes bulk atomic positions and is corrected for lattice motions at 173 K assuming an effective Debye temperature of 114 K. Both the calculated sets assume the imaginary potential, V_{oi}, derived for a free electron system of Na density[48] ($r_s \approx 4$). One set is calculated with a Slater exchange potential plus a uniform shift of 2.5 eV and the other set with a Hartree-Fock potential for the ioncore plus a local energy dependent Hedin-Lundqvist exchange and correlation potential for the rest of the electrons. Evidently the last set gives a very good match to the experimental data.

The Mo(100) surface has been investigated by Jona et al.[49] and is reported by Jona[45]. A contraction of about 11% of the

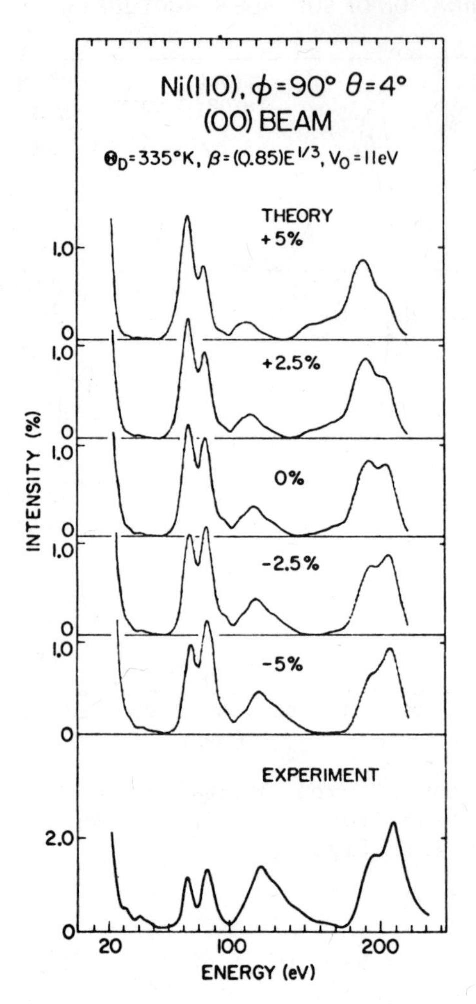

Fig. 19 Comparison of experimental and theoretical LEED 00 beam intensities from Ni(110) (Demuth, Marcus and Jepsen[39]) assuming expansion (+) or contraction(-) between the two first Ni(110) layer is varied in the calculation.

interlayer spacing between the first and second atomic layers was found to yield better agreement between theory and experiment than bulk atomic positions. The results are shown in Fig. 21 and are corrected for a surface Debye temperature (top layer) of 150 K to be compared with a bulk value of 360 K.

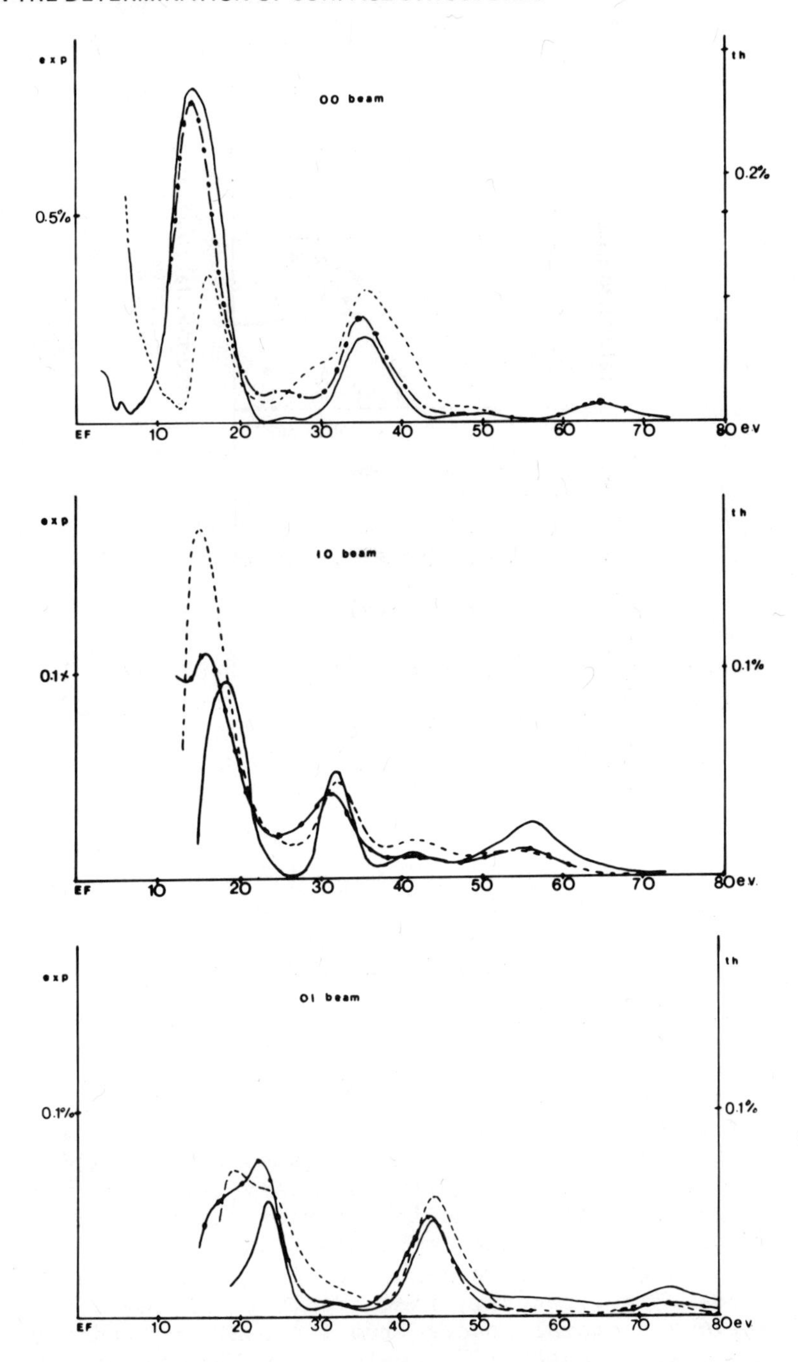

Fig. 20 Comparison of experimental (-) and theoretical (---, -o-o-) LEED beam intensities for Na(110) (Andersson, Echenique and Pendry[44]) normal incidence, θ = 0.

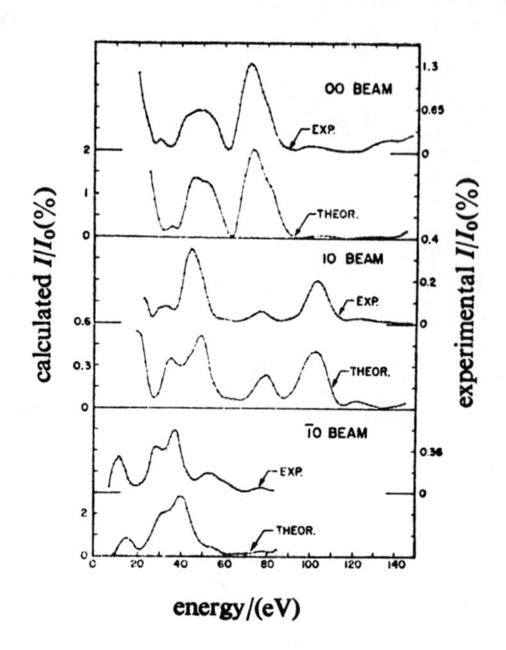

Fig. 21 Comparison of experimental and theoretical LEED beam intensities for Mo(100). Calculated intensities for a contraction of 11% between the first two Mo(100) layers. (Jona[45])

4. STRUCTURE OF ADSORBED LAYERS ON LOW-INDEX METAL SURFACES

4.1 Adsorption

Already 40 years ago Germer and Farnsworth observed by LEED that gas adsorption on Ni(111)[50] and Cu(100), Ag(100)[51] surfaces produced new lateral periodicities in the surface. Since then, and particularly after 1960, a large number of different adsorption systems utilizing low-index metal surfaces as substrates has been investigated by LEED and many has been found to yield two-dimensionally periodic structures (see Ref. 26). Reaction kinetics, structural transformations as a function of substrate temperature and adsorbate exposure, work function changes have been recorded in many cases. Until about 5 years ago, however, it was only possible to deduce qualitative structural models from the LEED spot pattern. Nowadays a number of adsorption system have been investigated using the same kind of trial and error LEED intensity analys as described in section 3.3 for clean metal surfaces.

Adsorbate layers may be formed in different ways. The

conventional method is adsorption from the gas phase either by leaking the appropriate gas into the UHV system via an adjustable leak-value (H_2, N_2, O_2, CO, CO_2, NH_3, hydrocarbons etc.) or by deposition from a suitable kind of vapour source. Thus alkali and alkaline earth atoms may be deposited from break-seal glass ampoulles containing high purity metal and broken in the UHV system. Transition metal atoms like Ti, V, Cr, Fe, Ni, Cu, may be evaporated from helical coil heavy gauge W or Mo sources, Ag and Au from alumina crucibles or W coil evaporators etc. S and Se can be adsorbed via the decomposition of H_2S and H_2Se, Sr and Ba via decomposition of $SrCO_3$ and $BaCO_3$. Adsorbate layers may also be formed by segregation of a bulk impurity at the surface (C on W, S on Ni). Characteric for all these methods is that it is in general difficult to control the amount of material on the surface. In some instances it may be possible to calibrate Auger electron spectroscopy(AES), soft X-ray emission spectroscopy (SXE) or X-ray photoelectron spectroscopy (XPS) to allow a quantitative determination of the amount of material adsorbed. Radioactive tracer methods may be used in some cases (e.g. S^{32}).

4.2 Diffraction patterns

As already mentioned above there is a vast literature on LEED studies of various adsorption systems and some examples will be given here that illustrate the different kinds of observations made. They will also serve as examples of the various superposition structures discussed in section 2.1.

Fig. 22 shows the LEED patterns observed after O_2 adsorption on Ni(100), the sequence is clean Ni(100), Ni(100) p(2x2)O and Ni(100) c(2x2)O, and the pattern clearly illustrates how the size of the reciprocal net changes as a consequence of oxygen adsorption. The p(2x2)O structure nucleates in small ordered p(2x2) domains already at low exposures and the structure is complited for exposures around 3L ($1L = 1 \cdot 10^{-6}$ Torr-sec) after which the intensity of some of the diffraction beams start to decline indicating the formation of the c(2x2)O structure. Fig. 23a shows the LEED pattern after 1 L exposure and Fig. 23b shows the intensity of the 1/2 0 beam as a function of exposure as measured by a spotphoto meter with an aperture of 1.5° seen from the crystal. The initial rise is linear i.e. $I_{1/2\ 0} \propto$ exposure which is due to the fact that the 1/2 0 spot is large relative to the measuring aperture and $I \propto N$, the number of adsorbed atoms, rather than N^2 as expected for the growth of ordered domains (see section 2.2).

Adsorption of S, Se and Te on Ni(100) yields the same sequence of LEED patterns as O_2 adsorption did. Table II lists a

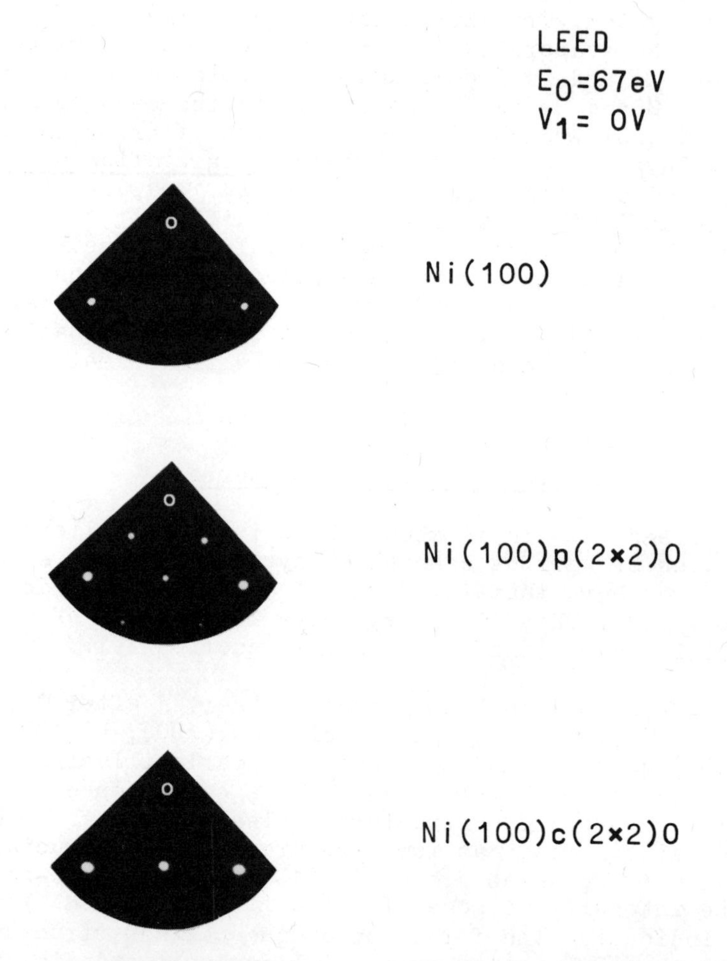

Fig. 22 LEED patterns from the ordered p(2x2)O and c(2x2)O structures on Ni(100) formed during oxygen adsorption.

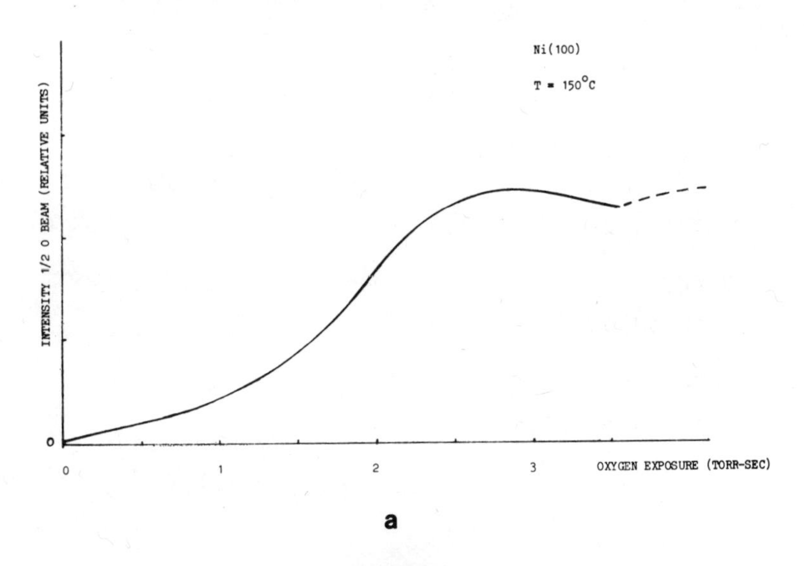

a

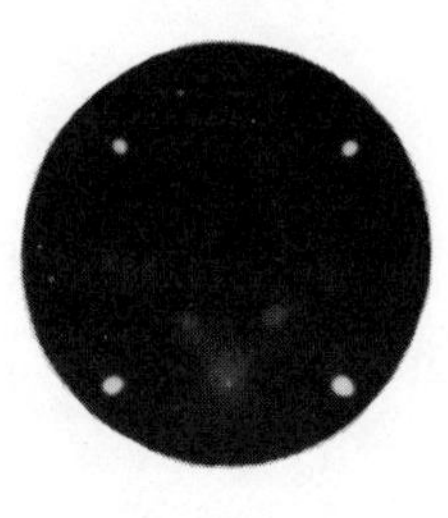

b

Fig. 23 a. Intensity of first ordered nonspecular LEED beam (1/20) from the Ni(100p(2x2)O structure during oxygen exposure.
b. LEED pattern from poorly ordered Ni(100)p(2x2)O at an oxygen exposure of 1.5 L (1 L = $1 \cdot 10^{-6}$ Torr-sec). 150°C. (Ordered p(2x2)O see Fig. 22).

number of structures (notation derived from LEED patterns)[26] observed after chalcogen (O, S, Se, Te) adsorption on low-index surfaces of Ni and Cu. It is clear that different adsorbate structures may form on these two metals indicating differencies in the nature of the adatom-metal bond. Fig. 24 shows diffraction patterns from Te adsorbed on Cu(100). Initially on ordered p(2x2)Te structure is formed (Fig. 24a), which with increased Te density transforms to a coincidence lattice (probably a dense Te layer) that yields the pattern of Fig. 24b.

TABLE II

Chalcogen overlayer structures on Ni and Cu surfaces.

Ni(111)		Cu(111)	
	(2x2) O		(O_2 lattice?)
	$(\sqrt{3}x\sqrt{3})$ 30° O		$(\sqrt{3}x\sqrt{3})$ 30° O
	(2x2) S		(2x2) S
	$(\sqrt{3}x\sqrt{3})$ 30° S		$(\sqrt{3}x\sqrt{3})$ 30° S
	(2x2) Se		
	$(\sqrt{3}x\sqrt{3})$ 30° Se		
			$(2\sqrt{3}x2\sqrt{3})$ 30° Te
			$(\sqrt{3}x\sqrt{3})$ 30° Te
Ni(100)	p(2x2) O	**Cu(100)**	oblique O
	c(2x2) O		p(2x4) 45° O
	p(2x2) S		p(2x2) S
	c(2x2) S		p(2x1) S
	p(2x2) Se		
	c(2x2) Se		
	p(2x2) Te		p(2x2) Te
	c(2x2) Te		coincidence Te

Alkali atoms adsorb on Ni surfaces[53,54] in a manner quite different from the chalcogens. Fig. 25 show LEED pattern obtained during Na and K adsorption on Ni(100). At low surface density of alkali atoms the LEED patterns exhibit a diffraction ring which is interpreted to be due to a uniform distribution of alkali atoms with a mean separation R of small spread. At higher densities a c(2x2)Na is formed and a hexagonal K structure (two kinds of domains yield 12 fold symmetry) coincidently matched to Ni(100) along the [10] and [01] Ni surface directions.

As a final example a set of LEED patterns from the deposition of Cu on a W(110) surface[23] is shown in Fig. 26. During the early stage of growth surface alloying seems to occur and after about ten Cu layer a wellordered Cu(111) single crystal has developed.

LEED

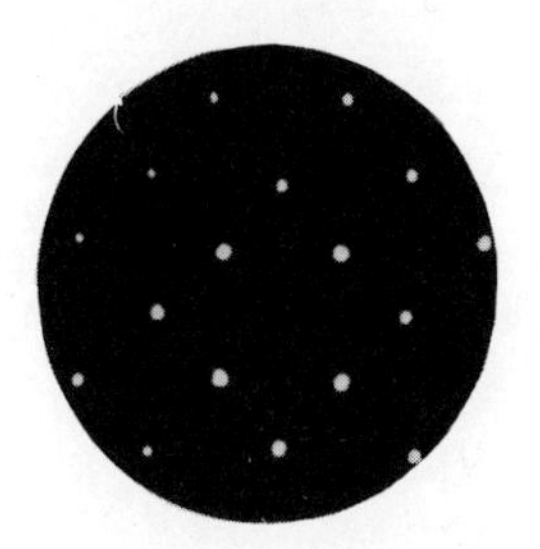

Cu(100)p(2×2)Te

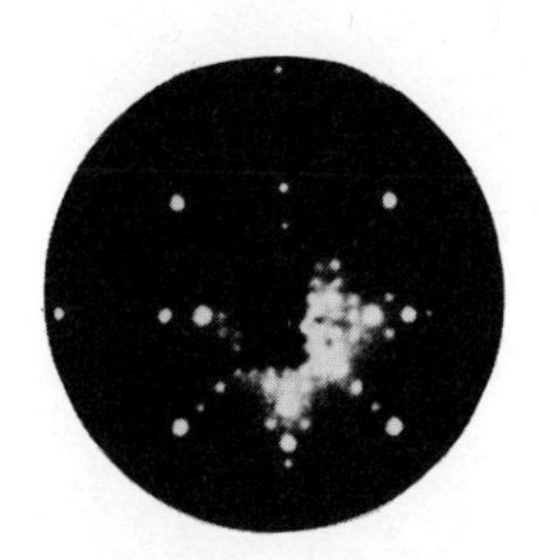

Cu(100) coincidence Te

Fig. 24 LEED patterns from the ordered p(2x2)Te and coincidence Te structures on Cu(100) formed by tellurium vapour deposition.

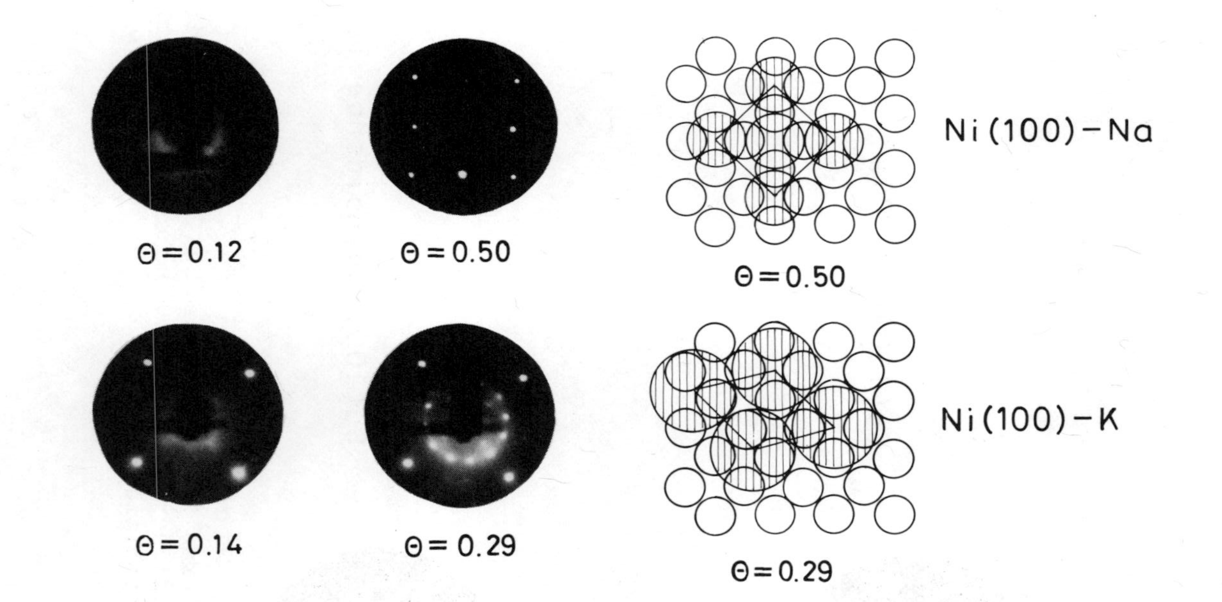

Fig. 25 LEED patterns from Na and K adsorption on Ni(100) and model structures for the c(2x2)Na and hexagonal K structures.

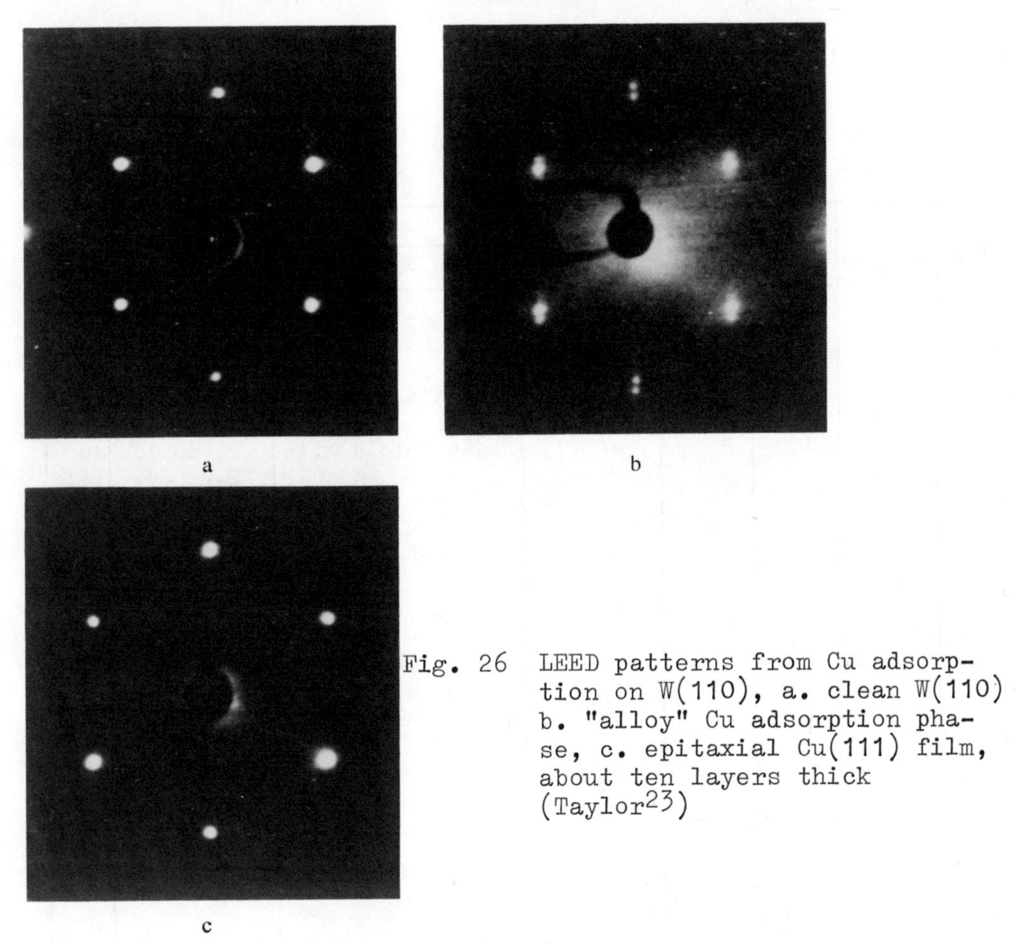

Fig. 26 LEED patterns from Cu adsorption on W(110), a. clean W(110) b. "alloy" Cu adsorption phase, c. epitaxial Cu(111) film, about ten layers thick (Taylor[23])

4.3 Intensity analysis

For the study of overlayer structures the LEED intensities are recorded experimentally and calculated dynamically along the same lines as was described for the work on clean low-index metal surfaces. Control of crystallinity and stoichiometry is for obvious reasons less quantitative than in studies of pure metals, though in order to obtain reliable intensity data preparation of the overlayer structure is as critical as the substrate surface preparation. In gas exposure work one establishes for example a range of exposures and substrate temperatures for which the recorded intensity data is essentially identical. AES is used to establish that the only species at the surface are the substrate and adsorbate atoms.

Table III lists a number of overlayer structures studied within the intensity analysis scheme described. Only the simplest overlayer structures have been investigated so far since

TABLE III

LEED intensity analysis of adsorbed layers on low-index metal surfaces.

substrate	surface	adsorbate	notation	Surface structure: site symmetry	Surface structure: dÅ	Ref.
Al	(100)	Na	c(2x2)	Na in 4-fold site	2.01±0.1	55
Ni	(111)	S	p(2x2)	S in 3-fold site ABC stacking	1.40±0.1	63
Ni	(100)	Na	c(2x2)	Na in 4-fold site	(2.9±0.1)[x] 2.2±0.1	40 56, 57
		O	p(2x2)	O in 4-fold site	0.9±0.1	58, 59
			c(2x2)	" " "	(1.5±0.1)[xx] 0.9±0.1	60 61, 59
		S	p(2x2)	S in 4-fold site	1.30±0.1	58, 59
			c(2x2)	" " "	1.30±0.1	61, 62, 59
		Se	p(2x2) c(2x2)	Se in 4-fold site " " "	1.55±0.1 1.45±0.1	58 61
		Te	p(2x2) c(2x2)	Te in 4-fold site " " "	1.80±0.1 1.90±0.1	58 61
Ni	(110)	S	c(2x2)	S in 4-fold site	0.93±0.1	63
Ag	(111)	I	$(\sqrt{3}x\sqrt{3})30^{o}$	I in 3-fold site ABC stacking	2.25	64
	(100)	Se	c(2x2)	Se in 4-fold site	1.9±0.1	
Mo	(100)	Si	(1x1)	Si in 4-fold site	1.21±0.2	65

[x] revised to 2.2±0.1 Å, see Ref. 56

[xx] revised to 0.9±0.1 Å, see Ref. 59

such structures involve relatively few extra diffraction beams and calculations are managable. The Ni(100) surface happens to be particularly favourable in this respect since a number of adsorbates form the simple c(2x2) structure (Na, O, S, Se, Te, CO) which involves just one extra (fractional order, 1/2 1/2) beam with respect to the substrate reciprocal unit mesh (see Fig. 22 and 25). There has been dispute about some of the structures reported in Table III and it may be that some of the structures will have to be revised with increasing experience about the various non-structural parameters and other features of the method.

Ni(100)c(2x2)Na. As an instructive example of the method and some of the complications one may run into we will firstly consider the Ni(100)c(2x2)Na structure. This was the first overlayer structure to be studied within the dynamical intensity analysis scheme[40]. It has the simple property that other physical information indicates the Na layer to constitute a separate layer on top of Ni(100) which considerably reduces the number of structures to be tried. The initial analysis proposed the Na atoms to occupy the fourfold symmetric Ni sites (see Fig. 25) and the Na layer to be separated (2.9±0.1) Å from the first Ni-layer. A reanalysis using extended experimental results of improved quality is shown in Fig. 27. The fourfold coordination is found to hold but there is some uncertainty with respect to the Na-Ni layer spacing. This spacing is most sensitively determined from the specular 00 beam which emphasises interferences between layers parallel to the surface (in particular at normal or close to normal incidence). The most plausible spacing seems to be d=2.2 Å though 1.5 Å and 2.9 Å still yield reasonably good agreement with experiment. The reason is that when the adsorbate is a weak scatterer of electrons approximate multiple coincidences may occur. Interferences between reflections from the substrate and the overlayer will importantly depend on the phase difference

$$(\underline{k} - \underline{k}') \cdot \underline{d} = \phi$$

where $\underline{k}$ and $\underline{k}'$ are the vectors of waves incident on and reflected from the substrate. If $\underline{d}$ is increased by $\Delta \underline{d}$ such that

$$(\underline{k} - \underline{k}') \cdot \Delta \underline{d} = 2\pi n$$

almost the same intensity will arise. To distinguish between $\underline{d}$ and $\underline{d} + \Delta\underline{d}$ it is necessary to change $(\underline{k}-\underline{k}')$ by considering different beams or a different incident energy.

E(eV)	d_{00} Å	d_{10} Å
40	0.8	1.0
100	0.6	0.6

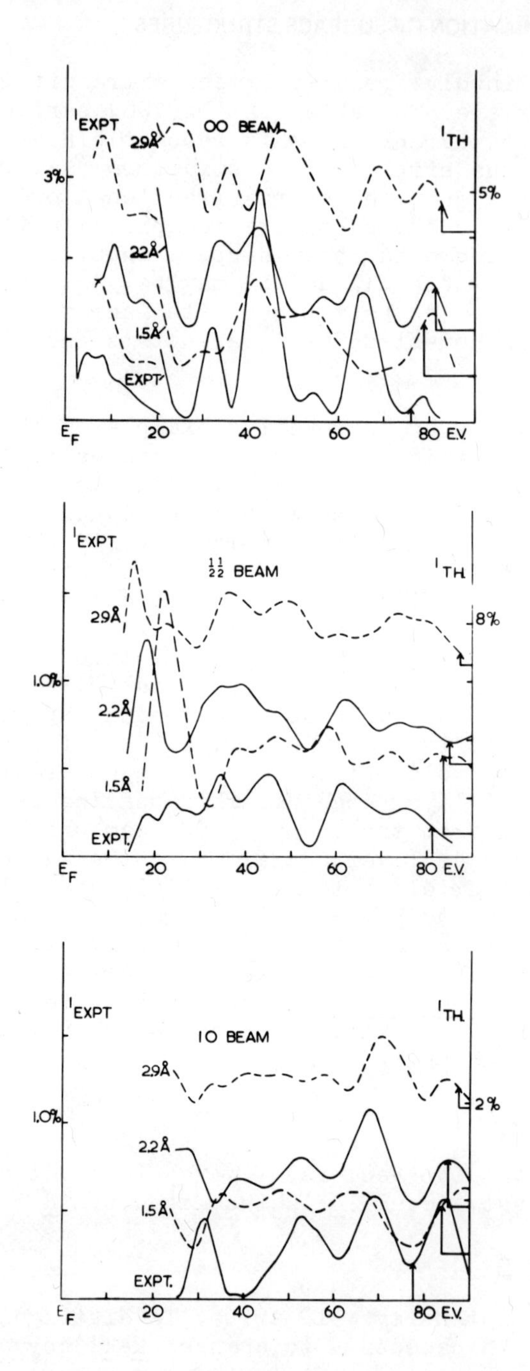

Fig. 27 Comparison of experimental and theoretical LEED beam intensities for Ni(100)c(2x2)Na. The calculated intensities correspond to Na in the 4-fold hollow site in the top nickel layer for three different Na layer displacements.

Some values of $\Delta \underline{d}$ for two different beams and two electron energies are given here. They illustrate that the problem may be rather difficultsince the resolution claimed is about 0.1 Å. If one can rely more on the shape and relative intensities of the data a more confident analysis may be achieved. This is illustrated by a calculation due to Demuth, Jepsen and Marcus[57] for the same system and is shown in Fig. 28. The comparison of the specular, 00, beam in Fig. 28a is in favour of the spacing 2.2 Å, Na in 4-fold hollows V_{or} = 11 eV for the whole system. In Fig. 28b it is seen that using a suitable mean potential V_o = 5.7 eV in the Na layer (Fermi level match) and an adsorbate Debye temperature of 200 K improves agreement between the calculated and the experimental intensities considerably. The comparison of the 00, 1/2 1/2, and 10 beams in Fig. 28c now shows an agreement as good as that obtained for clean surfaces.

Ni(100) - p(2x2) S

The O, S, Se and Te adsorbate structures on Ni(100) offers a chemically more interesting situation than Na does. Particularly what concerns the p(2x2) and c(2x2) oxygen structures there has been a lot of discussion about whether these structures consists of a pure oxygen overlayer or mixed nickel-oxygen layers (reconstruction). LEED intesity structure analysis by two different groups[60,61] found that a plausible structure was a separate O layer, O occuping the 4-fold symmetric Ni sites but different O-Ni layer spacing were prefered by the two groups. Demuth, Jepsen and Marcus[61] found that a spacing 0.9 Å gave better agreement than 1.5 Å and Andersson, Kasemo, Pendry and Van Hove found the reverse. The experimental data agreed well in the overlapping energy range. A similar multiple coincidence mechanism as discussed above for Na is probably responsible for this controversary[56]. Better knowledge about the "non-structural" parameters and chemical arguments may solve it. For a separation of 0.9 Å one finds a Ni-O band length of 1.97 Å which is within the range of plausible values 1.84-2.08 Å for different character Ni-O band lengths while 1.5 Å yield the Ni-O band length of 2.30 Å. A later reanalysis by Andersson and Pendry[59] using Demuth's oxygen potential supports 0.9 Å strongly. This value is also found in the analysis of the p(2x2)O structure[58,59]. In order to illustrate the kind of results that is obtained for the chalcogen overlayers from different workers a comparison of experiment and theory for the p(2x2)S structure is shown in Fig. 29. Fig. 29a, b shows results from the p(2x2)S obtained by Andersson and Pendry[59] and Fig. 29c shows results obtained by Van Hove and Tong using Demuth's experimental data. Both groups agree on the structure, S constitutes a separate layer, the S atoms sits in the 4-fold symmetrical Ni-sites and the S-Ni layer separation is 1.3 Å which is the same as Demuth et al.[62] found for the Ni(100)c(2x2) S structure. The 00 beam in Fig. 29a indicates a value in the range 1.2 - 1.3 Å.

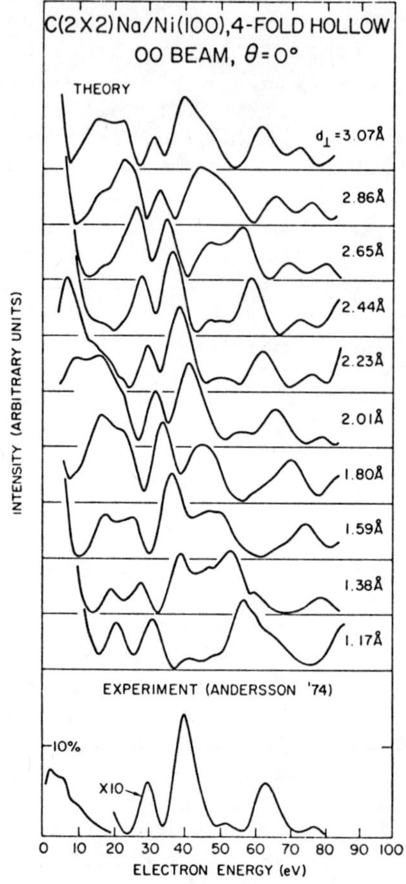

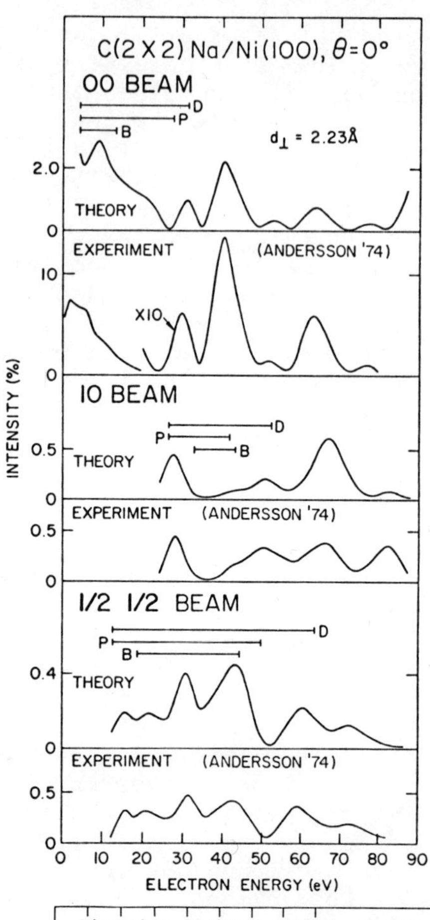

Fig. 28 Comparison of experimental and theoretical LEED beam intensities for Ni(100)c(2x2)Na (Demuth, Jepsen and Marcus[57]). a. 00 beam intensity for series of Na layer displacements, d , with Na in 4-fold hollows. b. Dependence of 1/2 1/2 beam intensity on the adsorbate potential (d =2.23 Å 4-fold hollow). c. Optimal agreement between experimental and calculated 00, 1/2 1/2 and 10 beam intensities for d = =2.23 Å, 4-fold hollow. A metallic Na potential, an innerpotential V_{or}= =5.7 eV, an absorptive potential V_{oi}=2.5 eV and a Debye temperature of 200 K were used for the Na-layer.

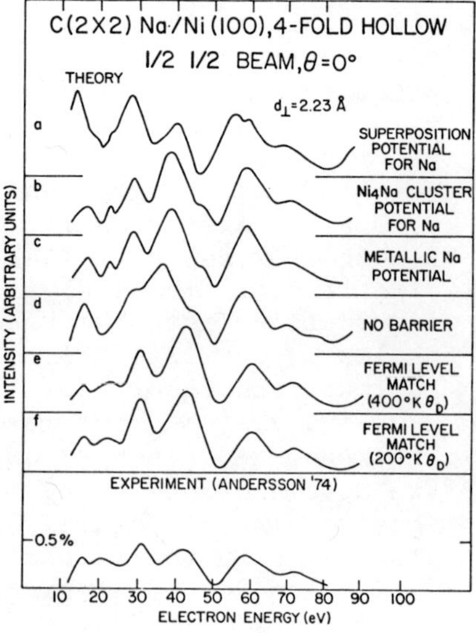

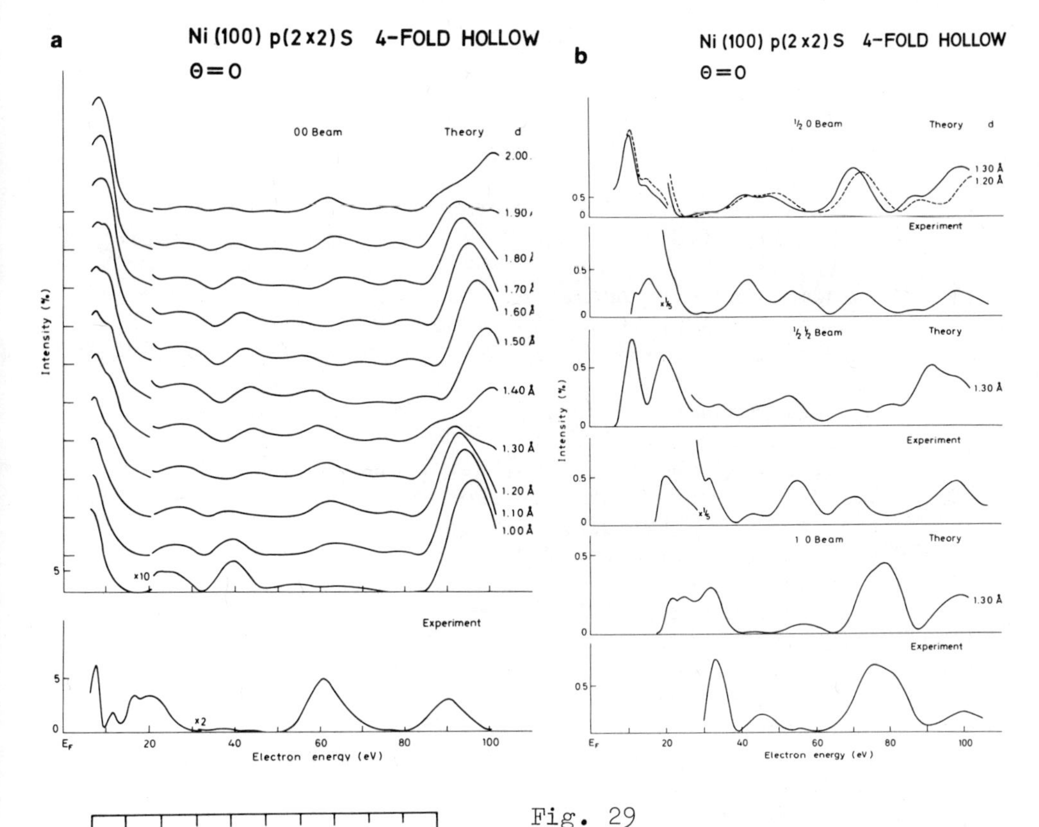

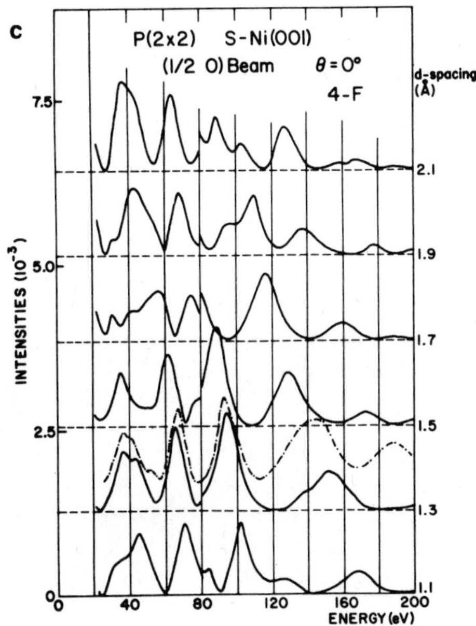

Fig. 29
Comparison of experimental and theoretical LEED beam intensities for Ni(100)p(2x2)S (a, and b, Andersson and Pendry[59], c, Van Hove and Tong using Demuth's experimental data[58]).
a. 00 beam intensity for different S layer displacements (d) with S in 4-fold hollow. Normal incidence, θ = 0.
b. 1/2 0,1/2 1/2 and 10 beam intensities for S in 4-fold hollow with d= 1.3 Å (1.2 and 1.3 for 1/2 0). Normal incidence, θ =0.
c. 1/2 0 beam intensities for different S-layer displacements. (d) with S in 4-fold hollow.

REFERENCES

(1) For a recent review of surface crystallography see J.A. Strozier Jr., D.W. Jepsen and F. Jona in Surface Physics of Crystalline Materials, J.M. Blakely ed. (Academic Press, New York, 1975)
(2) C.W. Tucker and C.B. Duke, Surface Sci., 23, 41 (1970)
(3) M.G. Lagally, T.C. Ngoc and M.B. Webb, Phys. Rev. Letters, 26, 1557 (1971)
(4) U. Landman and D.L. Adams, J. Vac. Sci. Techn., 11, 195 (1974)
(5) E.A. Wood, J. Appl. Phys. 35, 1306 (1964)
(6) M.J. Buerger, Elementary Crystallography (Wiley, New York, 1963)
(7) R.L. Park and H.H. Madden, Surface Sci., 8, 426 (1967)
(8) See e.g. E. Bauer in Techniques for Metals Research Vol. II, part 2, ch. 15 and 16 (Wiley, New York, 1969)
(9) T.D. Heidenreich, Fundamentals of Transmission Electron Microscopy, p. 104 (Interscience Publishers Inc., New York, 1964)
(10) R.L. Park, J. Appl. Phys., 37, 295 (1966)
(11) J.J. Lander, Surface Sci., 1, 125 (1964)
(12) P.J. Estrup and E.G. McRae, Surface Sci., 25, 1 (1971)
(13) K. Kambe, Z. Naturf., 22a, 322, 422 (1967)
(14) J.L. Beeby, J. Phys. C, 6, 601 (1968)
(15) E.G. McRae, Surface Sci., 11, 479 (1968)
(16) J.B. Pendry, J. Phys. C 4, 2501, 2514 (1971)
(17) D.W. Jepsen, P.M. Marcus and F. Jona, Phys. Rev. B5, 3933 (1972)
(18) J.B. Pendry in Low-energy electron diffraction (Academic Press, London and New York, 1974)
(19) J.M. Baker and J.M. Blakely, Surface Sci. 32, 45 (1972)
(20) P.W. Palmberg, T.N. Rhodin and C.J. Todd, Appl. Phys. Letters 10, 122 (1967)
(21) J.J. Lauder and J. Morrison, Surface Sci. 2, 553 (1964)
(22) J.O. Porteus and W.N. Faith, Phys. Rev. B8, 491 (1973)
(23) N.J. Taylor, Surface Sci., 4, 161 (1966)
(24) E. Bauer, Surface Sci., 7, 351 (1967)
(25) S. Andersson and U. Jostell, to be published
(26) For a review see G.A. Somorjai, Surface Sci., 34, 156 (1973)
(27) C. Leygraf and S. Ekelund, Surface Sci., 40, 609 (1973)
(28) H.B. Lyon Jr. and G.A. Somorjai, J. Chem. Phys., 46, 2539 (1967)
(29) D.G. Fedak and N.A. Gjostein, Surface Sci., 8, 77 (1967)
(30) P.W. Palmberg and T.N. Rhodin, Phys. Rev., 161, 586 (1967)
(31) J.T. Grant, Surface Sci., 18, 288 (1969)
(32) J.T. Grant and T.W. Haas, Surface Sci., 18, 457 (1969)
(33) G.E. Laramore and C.B. Duke, Phys. Rev., B5, 267 (1972)
(34) D.W. Jepsen, P.M. Marcus and F. Jona, Phys. Rev. B6, 3684 (1972)
(35) S.Y. Tong and T.N. Rhodin, Phys. Rev. Letters 26, 711 (1971)
(36) F. Jona, J.A. Strozier Jr., and C. Wong, Surface Sci., 30, 255 (1972)
(37) G.E. Laramore, Phys. Rev., B8, 515 (1973)
(38) M.G. Lagally, T.C. Ngoc, and M.B. Webb, J.Vac.Sci.Technol., 9,

645 (1972)
(39) J.E. Demuth, P.M. Marcus, and D.W. Jepsen, Phys. Rev. B11, 1460 (1975)
(40) S. Andersson and J.B. Pendry, J. Phys. C, 6, 601 (1973)
(41) G.E. Laramore, Phys. Rev., B9, 1204 (1974)
(42) D.W. Jepsen, P.M. Marcus, and F. Jona, Surface Sci., 41, 223 (1974)
(43) T.C. Ngoc, M.G. Lagally, and M.B. Webb, Surface Sci., 35, 117 (1973)
(44) S. Andersson, P. Echenique, and J.B. Pendry, to be published (1975)
(45) F. Jona, Proceedings of the Faraday Discussion 60 (1975)
(46) J.A. Strozier and R.O. Jones, Phys. Rev. B3, 3228 (1971)
(47) A. Ignatiev, F. Jona, D.W. Jepsen, and P.M. Marcus, LEED 7 seminar notes, p. 74 (San Diego, California, March 19-21 1973)
(48) B.I. Lundqvist, Physica Status Solidii, 32, 273 (1969)
(49) A. Ignatiev, F. Jona, H.D. Shih, D.W. Jepsen, and P.M. Marcus, to be published (1975)
(50) L.H. Germer, Z. Physik 54, 408 (1929)
(51) H.E. Farnsworth, Phys. Rev. 34, 679, (1930); 40, 634 (1932)
(52) J.L. Domange and J. Oudar, Surface Sci. 11; 124 (1968)
(53) R.L. Gerlach and T.N. Rhodin, Surface Sci. 19, 403 (1970)
(54) S. Andersson and U. Jostell, Surface Sci. 19, 403 (1974)
(55) B.A. Hutchins, T.N. Rhodin, and J.E. Demuth, to be published (1975)
(56) S. Andersson and J.B. Pendry, Solid State Commun., 16, 569 (1975)
(57) J.E. Demuth, D.W. Jepsen, and P.M. Marcus, J. Phys. C, 8, L25 (1975)
(58) M. Van Hove and S.Y. Tong, J. Vac. Sci. Technol. 12, 230 (1975)
(59) S. Andersson and J.B. Pendry, to be published
(60) S. Andersson, B. Kasemo, J.B. Pendry, and M. Van Hove, Phys. Rev. Letters 31, 595 (1973)
(61) J.E. Demuth, D.W. Jepsen, and P.M. Marcus, Phys. Rev. Letters, 31, 540 (1973)
(62) a. J.E. Demuth, D.W. Jepsen, and P.M. Marcus, Solid State Commun. 13, 1311 (1973)
b. C.B. Duke, N.O. Lipari and G.E. Laramore, J. Vac. Sci. Technol. 12, 222 (1973)
(63) J.E. Demuth, D.W. Jepsen and P.M. Marcus, Phys. Rev. Letters, 32, 1182 (1974)
(64) A. Ignatiev, F. Jona, D.W. Jepsen, and P.M. Marcus, Surface Sci., 40, 439 (1973)
(65) A. Ignatiev, F. Jona, D.W. Jepsen, and P.M. Marcus, to be published (1975)

ELECTRON SPECTROSCOPY OF SURFACES VIA FIELD AND PHOTOEMISSION

J.W. GADZUK

National Bureau of Standards

Washington, D.C. 20234, U.S.A.

1. INTRODUCTION

The purpose of these lectures is to convey a sense of what information on the electronic and geometrical properties of surfaces, with and without adsorbed monolayers of gases, is currently being obtained using field and photoemission spectroscopy. These notes are not a comprehensive review of the field as many such articles already exist or are in preparation now[1-6]. The emphasis here will be on some of the theoretical principles underlying the physics of the measurement process.

In particular, we will be interested in understanding :

(a) What makes a spectroscopy surface sensitive,
(b) What are some of the fundamental differences between atomic, molecular, and solid state bulk or surface photoelectron spectroscopies,
(c) How many body effects alter data interpretation based on a non-interacting electron picture. (Relaxation effects, satellites, softening of single electron selection rules, etc.)
(d) Why angle resolved measurements are useful.

First, we outline some of the useful features of Field Emission Energy Distributions(FEED) and discuss the way in which such measurements can be related to electronic properties of the surfaces. The major portion of these lectures however will be devoted to the various aspects of photoelectron spectroscopy in the ultra-violet to soft x-ray range mentioned above.

The basic ideas of the emission spectroscopies are illustrated in Fig.1. The solid is represented by a potential well of depth $\phi+E_F$ where ϕ is the workfunction and E_F, the kinetic energy of an electron at the Fermi level. Also shown is the potential associated with a chemisorbed atom that has either a virtual atomic or bonding surface molecule level centered at an energy ε below the Fermi level[3].

In the case of field emission, the sample is formed into a wire with a sharp (radius of curvative ∿ 1000 Å) point. Thus the application of a moderate (∿2000 volt) potential between the tip and an anode spaced ∿1-10 cm from the tip creates a field ∿.2 - .4 V/Å at the surface. As a result, electrons in occupied states of the solid can tunnel through the barrier and then be energy analyzed. An energy distribution (ED) for a clean, free electron metal is shown as the dashed curve on the lower righthand side of Fig.1. The basic features of this curve are the sharp Fermi level cutoff (for $T = 0°$) and an exponential attenuation with decreasing energy[4,7]. In addition, fine structure due to band or surface state effects appears on ED's from non-free electron metals[4,8,9,10] and this will be discussed in Section II. If an adsorbed atom has a virtual resonance near the Fermi level, then additional structure, the shaded area in the FEED, will appear due to resonance tunneling and this structure can be related to the local density of states associated with the chemically bonded adatom and metal[4,11].

Photoemission of electrons occurs when a solid is irradiated with photons of energy $\hbar\omega > \phi$ which are adsorbed by electrons, some of which are then emitted from the solid. Presumably the energy distribution of the ejected electrons is related to the electronic states of the solid, both before and after photon adsorption (hence hole creation), and thus information about the initial state can be extracted from such spectra. A characteristic ultraviolet photoemission spectrum from a clean solid is shown as the dashed curve in the upper righthand corner of Fig.1. Again a Fermi edge cutoff (rounded by the finite resolution of the energy analyzer) is seen, but now at $\hbar\omega$ above the emitter Fermi level. A modest peak slightly below the Fermi level could be due to d-band emission and the large increase in current at low energies is due to inelastically scattered electrons[5]. A similar type of distribution would occur for x-ray photoemission, but in addition sharp peaks in the ED at kinetic energies less than the valence level electrons would be observed, due to photoemission from core levels[6]. As in field emission, the presence of a virtual level in a chemisorption complex results in additional structure in the ED (shaded area). The surface sensitivity of a photoemission probe depends upon the value of λ, the inelastic electron attenuation length, all other things being equal. The attenuation length in effect is a measure of the depth within the solid (and thus number of substrate atoms per unit area) from which photoexcited electrons emerge without undergoing energy loss to plasmon or single electron modes. Shown in Fig.2 is a compilation of "experimentally determined" values of λ due to Powell[12]. (See the article by Powell for the sources of the data points.) In the UV and soft x-ray energy range, $\lambda \lesssim 15$ Å and thus only the first few layers of substrate are sampled. The degree to which the electronic properties of the first four or five layers mimic the bulk properties determines the usefulness of photoemission as a tool for studying volume effects[13]. Fortunately, most geometrical reconstruction and lattice spacing relaxation[14] is restricted to the first layer or two and the electron healing length due to the surface perturbation is scaled by k_F or κ (the Fermi or screening wavenumbers), so by 15 Å, bulk properties for elemental materials may be measured. Due to concentration gradients, this may not be true for alloys. On the otherhand, the total elastic current density from the substrate roughly equals $n_s \sigma_s \lambda v_s$ where n_s is the volume density of substrate atoms σ_s is some average photoionization cross section, and v_s is the number of valence electrons (including d electrons) per atom whereas the curve from an adsorbed monolayer goes as $n_{ads} \sigma_{ads} v_{ads}$ so the ratio $\eta \equiv \sigma_{ads} v_{ads} / (\sigma_s v_s \lambda / d)$, with d the substrate innerplanar separation, determines the strength of the adsorbate to substrate signal. In most published studies it appears that $\eta \sim 10\%$ is experimentally realized although there is no reason why this must always be. In fact Gustaftsson et al.[15] have performed experiments using synchrotron radiation in the range 25 eV $\leq \hbar\omega \leq$ 105 eV to study CO adsorption on Pd and Ni. They took advantage of the differential variation of σ_{ads} and σ_s with $\hbar\omega$ in order to identify the particular bonding levels and in this range of $\hbar\omega$, η varies from a few % to a tens % (by rough eyeball determination). Fortunately it appears that by judiciously choosing the right combinations of experimental variables such as $\hbar\omega$, angle of incidence[16], polarization, etc. one can obtain both bulk and surface information in UPS and XPS.

2. FEED

Historically, field emission was the first experimental confirmation of tunneling theory. Since its theoretical beginnings, in 1928, field emission experiments have been used to obtain values of workfunctions and workfunction changes of clean and adsorbed surfaces by measuring the total emission current as a function of applied field. In 1959, Young derived an expression for the FEED of a free electron metal and demonstrated that this was a reasonable approximation to the observed distribution from a W field emitter. Since then, FEED studies have been refined to the point where both band sturcture effects[4,8-10,14] and chemisorption energy levels[4,11] are being measured. The effects of the breakdown in free electron behavior of real materials on combined total current and energy distribution determinations of workfunctions have also been handled[17].

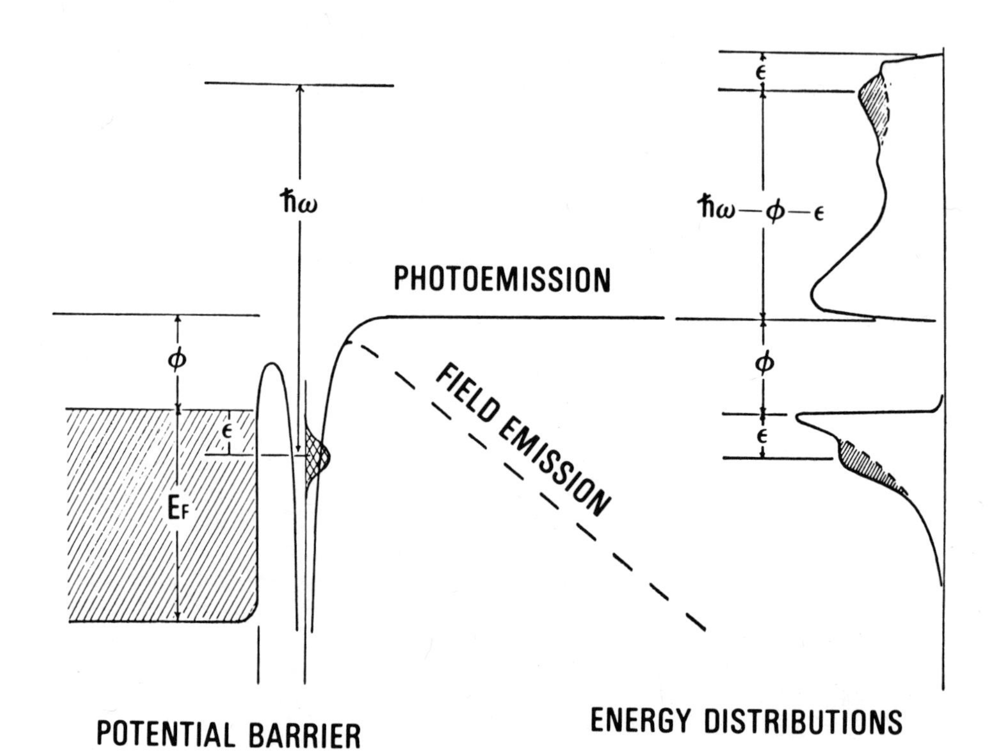

Fig.1 : Schematic potential energy diagram for a metal with an atom adsorbed on the surface. The dashed line depicts the potential when a large electrostatic field is applied, thus allowing for electrons to tunnel or be field emitted from the metal or adsorbed atom. The energy distribution on the bottom right is a FEED whereas that on the upper right is a photoemission spectrum. The shaded regions in the energy distribution indicate the additional current due to emission from the adsorbed atom.

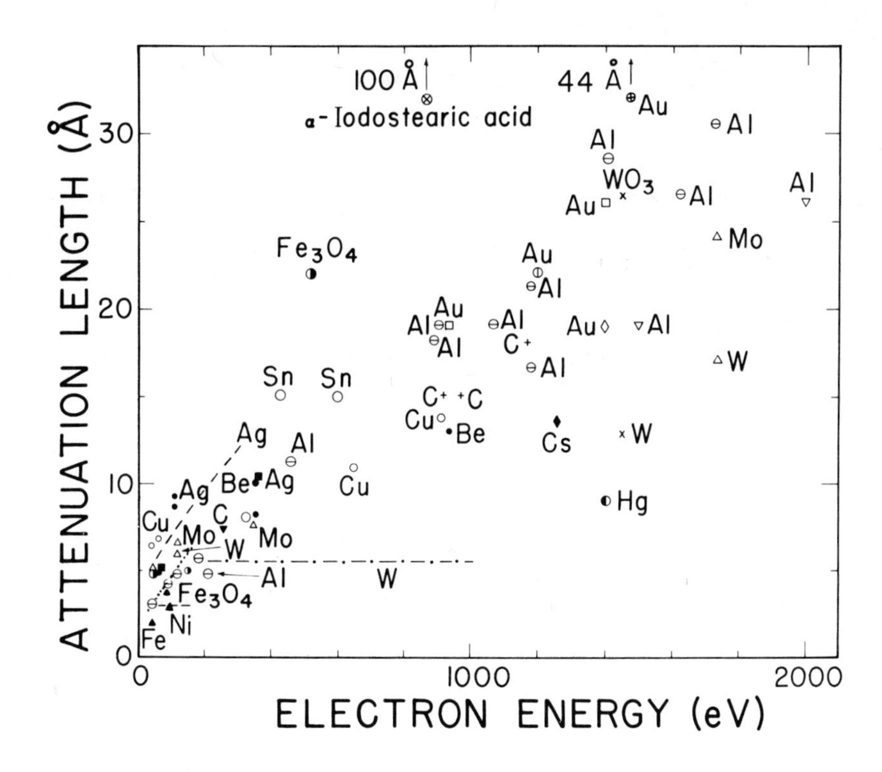

Fig. 2 : Measured or derived values of electron attenuation lengths for different materials as a function of electron energy. (Ref.12, Powell)

Let us first consider the free electron energy distribution, given by

$$\frac{dj}{d\varepsilon} \equiv j'(\varepsilon) = 2e \sum_{\mathbf{k}_{\parallel}, k_z > 0} \delta(\varepsilon - \varepsilon(\mathbf{k})) j_{\ell}(\mathbf{k}) D_{\ell r}(\mathbf{k}) f(\varepsilon(\mathbf{k})) \quad (1)$$

where $j_{\ell}(\mathbf{k})$ is the flux of electrons in state $\psi_{\mathbf{k}}$ incident upon a mathematical surface, say at the classical turning point nearest the surface, $D_{\ell r}(\mathbf{k})$ is the tunneling probability through the barrier, $\varepsilon(\mathbf{k})$ is the energy of state $\psi_{\mathbf{k}}$, f is the Fermi function, and the delta function picks out only those $\mathbf{k}$ states with energy ε. The quantum mechanical flux associated with the electron in state $\psi_{\mathbf{k}}$ incident on ℓ is given by

$$j_{\ell}(\mathbf{k}) = \frac{i\hbar}{2m} [\psi_{\mathbf{k}}(\mathbf{r}) \frac{d}{dz} \psi^*_{\mathbf{k}}(r) - \psi^*_{\mathbf{k}}(\mathbf{r}) \frac{d}{dz} \psi_{\mathbf{k}}(r)] \Big|_{z=\ell} \quad (2)$$

Consider first the specific case of a free electron metal in which $\varepsilon(\mathbf{k}) = \hbar^2 k^2/2m$ and $\psi_{\mathbf{k}}(\mathbf{r}) = e^{i\mathbf{k}\cdot\mathbf{r}}$. From Eq.(2), $j_{\ell}(\mathbf{k}) = \frac{\hbar k_z}{m}$. The WKB tunneling probability, in which the "usual" field emission approximations (such as $|\varepsilon| \lesssim 1$ eV) have been invoked, is a function of $k_z = k\cos\theta$ only and has been shown(4) to be

$$D_{\ell r}(\mathbf{k}) = D_{\ell r}(k_z) = \exp[-c + (W - E_F)/d] \quad (3)$$

where(4) $c \propto \phi^{3/2}/F$, $1/d \propto \phi^{1/2}/F$, F is the applied field, $W = \hbar^2 k^2 \cos^2\theta/2m$ is the so-called normal energy, and E_F is the Fermi energy. Replacing the summation in Eq.(1) by the integral

$$\sum_{\mathbf{k}_{\parallel}, k_z > 0} \rightarrow \frac{1}{8\pi^2} \left(\frac{2m}{\hbar^2}\right)^{3/2} \int \varepsilon(k)^{1/2} d\varepsilon(k) \int_0^1 d(\cos\theta),$$

it is a straightforward exercise to show that Eq.(1) reduces to

$$\frac{dj}{d\varepsilon} = \frac{J_o}{d} f(\varepsilon) e^{\varepsilon/d} \equiv j_o'(\varepsilon) \quad (4)$$

where $J_o \propto d^2 e^{-c}$. As previously mentioned, Eq.(4) describes the gross features of the data although many intriguing departures from free electron behavior have been reported, some of which may be due to band structure effects(4). Most recently, this prompted Penn and Plummer(9) to note explicitly the manner in which local density of states effects might alter the FEED. It should be pointed out that the calculations of Obermair(18) and Politzer and Cutler(8) had previously included such effects although they did not explicitly identify a surface local density of states in their formalisms. The theory of Penn[9] is rather detailed and for present purposes can be simulated as follows.

Within the solid, the electron eigenfunctions are Bloch functions

$$\psi_{\underset{\sim}{k}}(\underset{\sim}{r}) = e^{i\underset{\sim}{k}\cdot\underset{\sim}{r}} u_{\underset{\sim}{k}}(\underset{\sim}{r}) \tag{5}$$

Near the surface, the standing wave character of the wave functions is obtained by taking linear sums of $\psi_{\underset{\sim}{k}_{\|},k_z}$ and $\psi_{\underset{\sim}{k}_{\|},-k_z}$ (neglecting surface Umklapp[19]) with appropriate phase shifts to account for surface scattering and with $u_{\underset{\sim}{k}}(\underset{\sim}{r})$ modified in accord with the surface potential. The flux at the classical turning point that is due to the incident part of the standing wave is from Eqs.(2) and (5),

$$j_\ell(\underset{\sim}{k}) = [\frac{\hbar}{m} k_z|u_{\underset{\sim}{k}}(\underset{\sim}{r})|^2 + \frac{i\hbar}{2m}(u_{\underset{\sim}{k}}(\underset{\sim}{r})\frac{d}{dz}u^*_{\underset{\sim}{k}}(\underset{\sim}{r}) - c.c.)]|_{z=\ell} \tag{6}$$

where c.c. denotes complex conjugate and the expression is evaluated on the $z = \ell$ plane where it is assumed that a surface averaged value is appropriate. If u is pure real or imaginary, or if in the spirit of the WKB tunneling approximation (in which complex pre-exponential phase factors have been neglected) the second term in Eq.(6) is dropped, then Eqs.(1) and (6) yield

$$\frac{dj}{d\varepsilon} \simeq \frac{2e\hbar}{m} \sum_{\underset{\sim}{k}} \delta(\varepsilon-\varepsilon(\underset{\sim}{k}))f(\varepsilon)k_z|u_{\underset{\sim}{k}}(\underset{\sim}{r})|^2_{z=\ell} D_{\ell r}(k_z) \tag{7}$$

which, up to some multiplicative constants, is similar to the result of Penn and Plummer[9]. The presence of the modulus squared wave function provides the "band structure" sensitivity and the steep exponential drop off of $D_{\ell r}(k_z) = D_{\ell r}(\varepsilon,\underset{\sim}{k}_{\|})$ as $\underset{\sim}{k}_{\|}$ increases from zero, for constant ε, strongly weights only states $u_{\underset{\sim}{k}}$ with $k_z \gg o$, $\underset{\sim}{k}_{\|} \approx 0$; hence the terminology that a FEED is a measure of the one-dimensional states at a surface has been introduced. To emphasize the departure from free electron behavior, it is standard practice to present data in the form of an "enhancement factor" $R(\varepsilon) \equiv j'(\varepsilon)_{obs}/j'_o(\varepsilon)$ as this removes the dominating exponential from the data display. While it is true that different crystal faces of the same material or same faces of different materials yield R(ε) curves with quite different structure, as shown in Fig.3, this type data has so far received little quantitative utilization, mainly due to our present theoretical inability to meaningfully calculate electronic surface properties of the refractory transition metals that can be formed into field emission tips. Furthermore, due to some fundamental limitations in the field emission process[20], it appears unlikely that field emission spectroscopy can provide information on any states other than those within about 1-2 eV of the Fermi level. Thus FEED of clean surfaces may serve as a backup to other more wide ranging probes, but it seems doubtful that it will become as versatile a tool as photoelectron spectroscopy.

We now turn briefly to the use of FEED in chemisorption studies. Several years ago, Duke and Alferieff pointed out that if there is a resonance or virtual level associated with a chemisorbed atom or surface molecule complex which is in the vicinity of the emitter Fermi level, then the field emission current can be enhanced or diminished to a much greater degree than would be inferred from a work function change. In addition, new structure, related to the local density of states of the surface complex, would appear on the FEED[4,11,21]. Detailed theories, from several different viewpoints[11] have been presented but invariably they all lead to similar results, that the ratio of the change of the FEED divided by the free electron expression takes the form :

$$R(\varepsilon) \equiv \Delta j'(\varepsilon)/j_o'(\varepsilon) \propto \rho_a(\varepsilon)\, e^{2\kappa s} \qquad (8)$$

where $\rho_a(\varepsilon)$ is the local density of states on the adatom, s is the atom-metal separation, and $\kappa \approx [\frac{2m}{\hbar^2}(\phi+\varepsilon)]^{1/2}$ is roughly the WKB decay constant in the region between the atom and the metal. The physical content of Eq.(8) can be seen as follows. In analogy with Eq. (7), the FEED is considered to be proportional to a local density of states multiplied by a tunneling probability. Frequently in chemisorption systems $\rho_a(\varepsilon)$ is similar to a Lorentzian with a full width at half maximum ∿1eV which varies much more rapidly than the clean surface density of states. Consequently, the density of states ratio is well approximated simply by $\rho_a(\varepsilon)$. Furthermore, tunneling from the adatom is much more probable than direct tunneling from the substrate since that part of the tunneling barrier between the metal and adatom is no longer present. This enhanced tunneling probability is accounted for by the exponential in Eq.(8). The resulting experimental line shape thus reflects the adatom density of states skewed by the exponential which increases with decreasing energy ε. Although within the context of the Anderson-Grimley-Newns theory of chemisorption, the adatom density of states is given by

$$\rho_a(\varepsilon) = \frac{1}{\pi}\,\frac{\Delta_a(\varepsilon)}{(\varepsilon-\varepsilon_a-\Lambda(\varepsilon))^2+\Delta_a(\varepsilon)^2} \qquad (9a,b)$$

with

$$\Lambda(\varepsilon) + i\Delta_a(\varepsilon) = \sum_{\underset{\sim}{k}} \frac{|V_{a\underset{\sim}{k}}|^2}{\varepsilon-\varepsilon(\underset{\sim}{k})+i\delta}$$

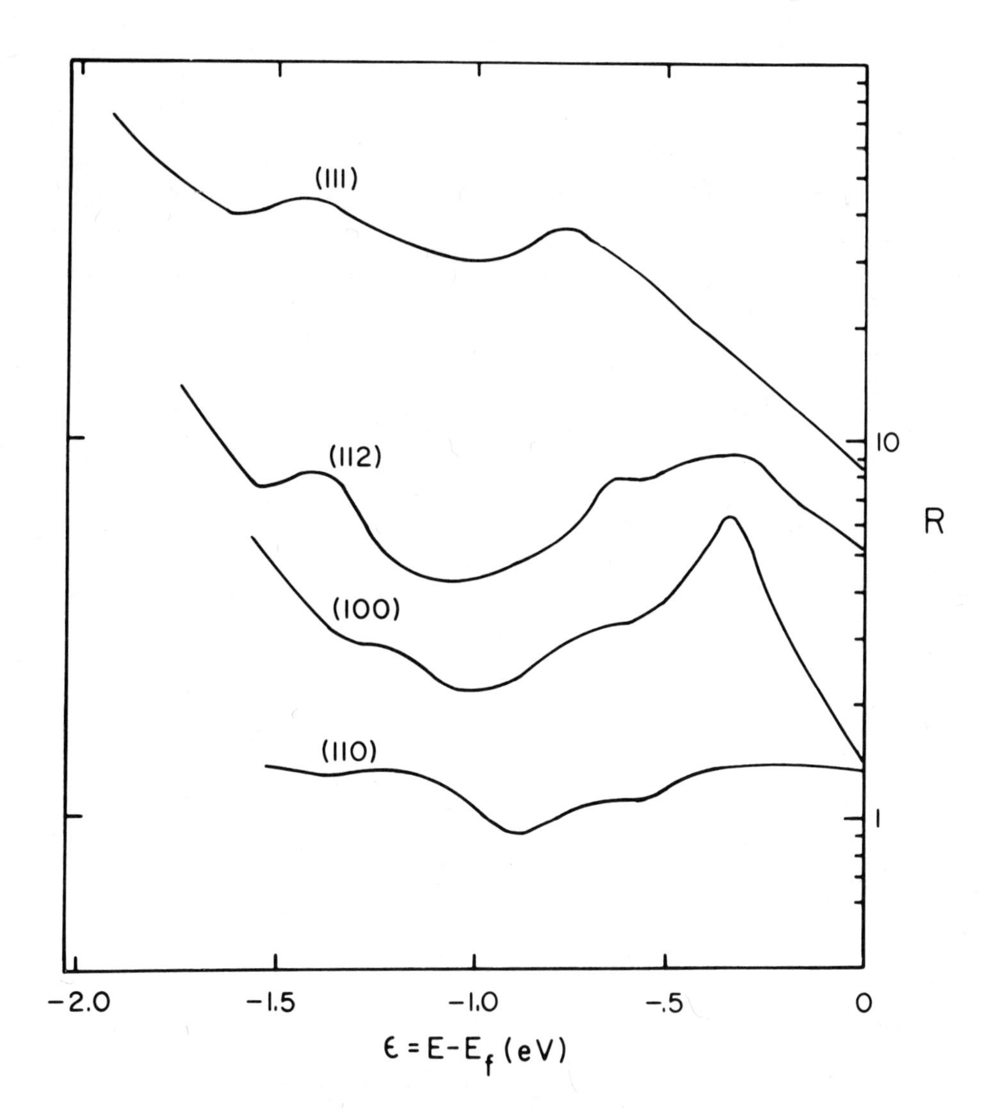

Fig.3 : The enhancement factor R(ε) for the four low index planes of tungsten. The free-electron energy distribution was calculated using ϕ(110)= 5.25 eV, ϕ(100) = 4.64 eV, ϕ(112) = 4.90 eV, and ϕ(111) = 4.45 eV. (Ref. 4, Gadzuk and Plummer)

where $V_{a\underset{\sim}{k}}$ is the $\underset{\sim}{k}$ dependent atom-metal hopping integral and $\Lambda(\varepsilon)$ and $\Delta_a(\varepsilon)$ are complicated functions of ε, Eq.(9) is often approximated by a two parameter Lorentzian (by ignoring the energy dependence of Λ and Δ_a). These parameters, the level shift and width are then obtained by fitting Eqs.(8) and (9) to experimentally determined $R(\varepsilon)$ curves. Typically shifts and widths of the order of 1 eV are required[22,23]. These values are in accord with current ideas in chemisorption theory[3]. As with the clean surface studies, such measurements are restricted to states near the Fermi level. The advantages of a field emission experiment are that adsorption occurs on a perfect single crystal facet, the field (and thus level position and effective adatom charge[24]) can be varied, and single adatom events can be monitored[4,25].

3. SINGLE ELECTRON PHOTOEMISSION

The most difficult task in writing lecture notes on photoelectron spectroscopy, both ultraviolet (UPS) and x-ray (XPS), is discovering new ways to improve upon the excellent notes of Hedin[26] and S. Lundqvist[27] to which the reader is referred for a very lucid description of bulk effects in photoemission. Recent theoretical papers which are influencing current thinking on UPS (including surface effects) should also be checked[28-39] in addition to the review articles cited[5]. In this section we will then first present a brief exposition of generally accepted wisdom and problems and then move on to some new thoughts in which the photoemission event is considered within a localized rather than customary $\underset{\sim}{k}$-space representation. In this way, photoejection of electrons from atoms, molecules, core and valence states of solids with surfaces, and chemisorption complexes can all be described within the same basic formulation. The unifying concept is the degree of coherence between electron emission from each of the constituent atoms in the ensemble being considered. This will be clarified in the second part of this section.

1. General Background

Before proceeding into detailed theory, consider the qualitative features of various photoemission examples shown in Fig.4. In all three cases, an electron with an initial state wave function ψ_i is excited by a photon with energy $\hbar\omega$ to a final state $\phi_>$. Independent of one's personal preferences (Green's functions[28,31], quadratic response[29], Kubo formulae[30,39] or Golden rule[33-35]) all single electron theories seem to give equivalent expressions for the energy and angle resolved photocurrent, namely[36] :

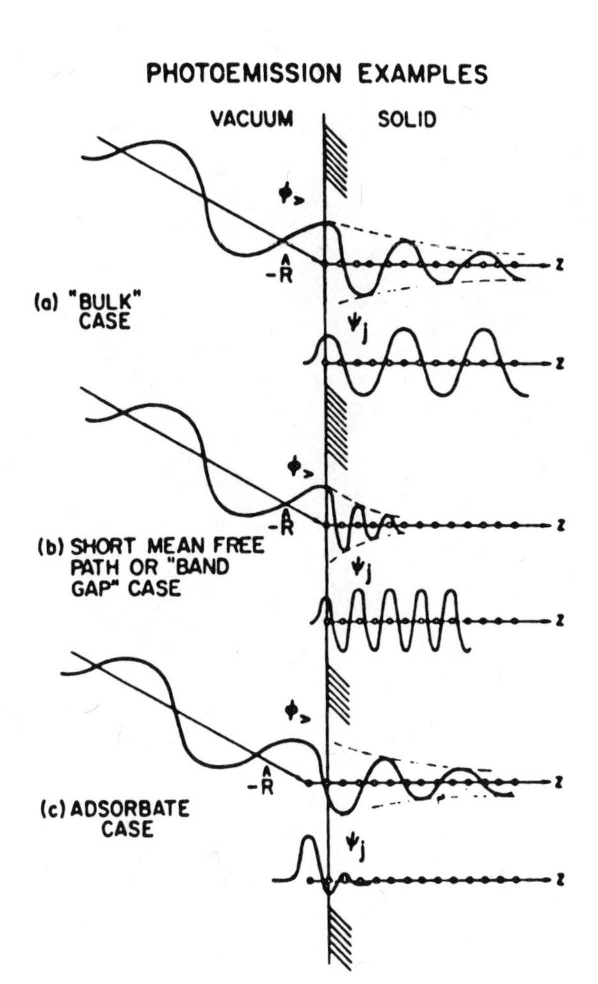

Fig.4 : (a) Schematic diagram of the initial state wavefunction ψ_j and final state wavefunction $\phi_>$ involved in direct interband transitions for the case of "bulk" photoemission from a semi-infinite solid.
(b) Schematic diagram for the short-mean-free path or the "band gap" cases in which the final state wave function $\phi_>$ within the solid consists only of rapidly decaying surface evanescent waves.
(c) Schematic diagram of ψ_j and $\phi_>$ for the case of an adsorbate with localized initial state wavefunctions ψ_j. (Ref.36, Feibelman and Eastman)

$$\frac{d^2j}{d\varepsilon d\Omega} \equiv j'' = 2e\,\frac{\hbar k}{m}\left(\frac{e}{2mc}\right)^2\left(\frac{m}{2\pi\hbar^2}\right)^2 \sum_j f(\varepsilon_j)\delta(\varepsilon-\hbar\omega-\varepsilon_j)$$

$$\times \left|\int d^3r\ \phi_>^*(\underset{\sim}{r};\hat{R},\varepsilon)\tau(\underset{\sim}{r})\psi_j(\underset{\sim}{r})\right|^2 \qquad (10a,b)$$

where $\tau(\underset{\sim}{r}) = \frac{1}{2}\,(\underset{\sim}{A}(\underset{\sim}{r})\cdot\underset{\sim}{p}_{op} + \underset{\sim}{p}_{op}\cdot\underset{\sim}{A}(\underset{\sim}{r}))$

with $\underset{\sim}{A}(\underset{\sim}{r})$ the total vector potential of the photon field at $\underset{\sim}{r}$, $\underset{\sim}{p}_{op}$ the electron momentum operator, and k the outgoing electron wave number far from the solid. The final state $\phi_>(\underset{\sim}{r};\hat{R};\varepsilon)$ is , in the confusing language of formal scattering theory(40), an outgoing wave satisfying incoming wave boundary conditions, the time reversed familiar scattering wave functions given by(37)

$$\phi_>(\underset{\sim}{r};\hat{R};\varepsilon) = e^{i\underset{\sim}{k}\cdot\underset{\sim}{r}} + \int d^3r'\ e^{i\underset{\sim}{k}\cdot\underset{\sim}{r}'}\ V(\underset{\sim}{r}')G^r(r',r;\varepsilon) \qquad (11)$$

with $\underset{\sim}{k} = -\ (2m\varepsilon/\hbar^2)^{1/2}\hat{R}$, $\hat{R}$ a unit vector pointing towards the detector, $V(\underset{\sim}{r}')$ the complete semi-infinite potential of the solid, and G^r the retarded electron Green's function. Note that if j" is treated in the Born approximation, then $\phi_> \simeq e^{i\underset{\sim}{k}\cdot\underset{\sim}{r}}$ and no subtleties arise. The physical significance and implications of the particular final state $\phi_>$ are the following. In Eq.(10), we are only interested in those transitions which produce an electron with energy ε moving in the direction $\hat{R}$ as the time interval after scattering (or more precisely, after photon absorption) approaches + ∞. But this is just the time reversed usual scattering wave function satisfying outgoing wave boundary conditions in which at t = − ∞, a plane wave with energy ε moves towards the scattering region and produces scattered waves as time moves forward to + ∞. For the present case in which the scattering is due to the lattice potential $V(\underset{\sim}{r}')$, the usual scattering state is a LEED wave function(41), hence the identification of the final state in the photoemission matrix element as a time reversed LEED wave function(36) in which at t = +∞ we have a plane wave moving away from the solid rather than at t = −∞ a plane wave moving towards the solid. One might wonder why such questions do not arise in the theory of Extended X-ray Absorption Fine Structure (EXAFS)(42) The reason seems to be that in an absorption measurement, one is looking at transitions to all possible states, not just those with energy ε moving in direction $\hat{R}$. Thus a simple spherical outgoing wave final state is adequate. As a final aside, it is interesting to note that the usual definitive reference to this problem is the paper by Breit and Behte(43). In my opinion, a more lucid description is given in an earlier paper in which Bethe discussed both the intuitive and counter-intuitive notions required to understand this point.(44)

Returning to Eq.(10) and Fig.5, we note that since in the uv and x-ray range, the photon attenuation length is long ($\gtrsim$ 100 Å), the envelope function of $\underset{\sim}{A}(\underset{\sim}{r})$ can be taken to be constant on the scale of either electron attenuation or simple molecular lengths and thus $\underset{\sim}{A}(\underset{\sim}{r})$ can be removed from the matrix element. Consequently the photoemission current is governed by momentum matrix elements. In the so-called bulk case shown in Fig.4a, the initial state ψ_j is a sum of extended Bloch functions forming a standing curve. The final state $\phi_>$ is also a sum of Bloch waves within the crystal matched on to appropriate plane waves outside in a way which satisfies the LEED requirements just discussed. Within the solid, $\phi_>$ is shown as an attenuated wave to account for the effects of inelastic electron scattering. This approximation is usually invoked in an ad hoc manner and some of its implications have recently been discussed by Feibelman(37). Problems arise because a description of the interacting many-body system is given in terms of a single electron theory in which the many-body interactions are simulated by a complex optical potential in the final state which is different than that of the initial state. Due to the non-hermitian nature of the final state optical potential, some steps in the derivation of Eq.(10) are not strictly valid. The degree to which this leads to serious consequences is not presently known. If attenuation and surface effects are neglected, then Fig.4a depicts the usual k-conserving bulk photoemission process(45,46).

Another situation which has drawn attention recently is what Feuerbacher has called "directional bandgap" emission(47). This is shown in Fig.4b. In this case, the energy and direction of the final state electron are such that there are no propagating Bloch states within the solid. Of course in the field free vacuum half space, all energies and directions are allowed. Thus the vacuum plane waves match onto evanescent waves in the solid(48) whose envelope decays rapidly, within a few layers. Thus in the momentum matrix element, overlap between ψ_j and $\phi_>$ is large only in the surface region which gives this type of photoemission extreme surface sensitivity.

The last example, Fig.4c, shows the case in which the initial state ψ_j is a localized extrinsic surface state due to an adsorbed atom(3). If the degree of localization of the surface state is small relative to the attenuation envelope of $\phi_>$, then the photoejected current from the adsorption complex is likely to form a small portion of the total elastic current, as mentioned in the introduction. Aspects of the theory of photoemission from adsorbed atoms such as line shapes(33,38) and angular distributions(34,35) have been treated.

B. Theory

The basic philosophy of the present approach has arisen from

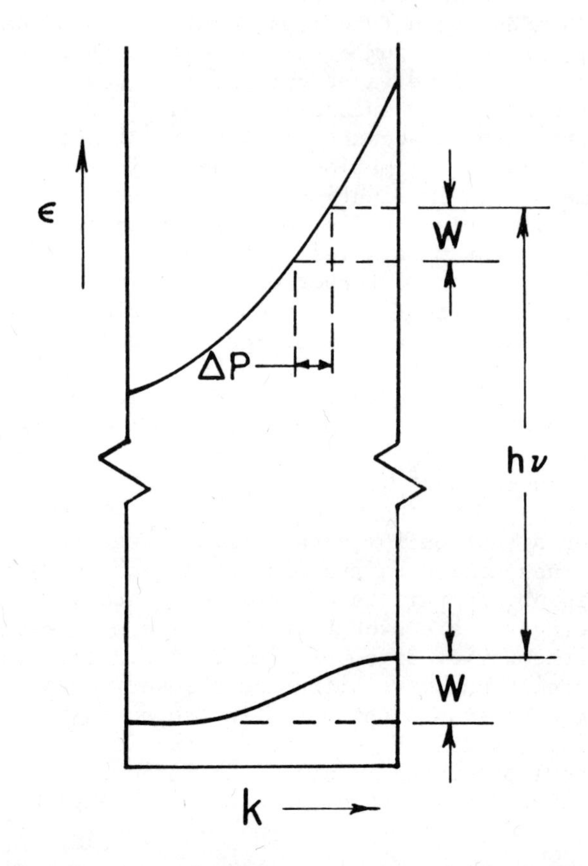

Fig. 5 : Energy band diagram for photoemission of electrons initially in a narrow band of width W and finally in a free electron band.

an attempt to answer the oft asked question, "in solid state photoemission is an electron excited from one atom or is 1/N of an electron excited from each of the N atoms in the solid?"[49] As it turns out, the question is more one of which quantum mechanical representation one feels comfortable thinking in. Certainly any predicted physical consequences must be representation independent. In the case of molecules and solids, a hole created on one atom can hop around to the other centers within the total structure (equivalent to the spreading of a wavepacket) and the speed at which this occurs is $\tau_h \sim \hbar/\Delta E$ where ΔE is of the order of either the bonding-antibonding state separation in molecules or the bandwidth in solids. If the energy discrimination in an experiment is much less than ΔE, (implying an observation time $>> \tau_h$) it is not possible, in principle, to independently determine from which atom the observed electron came from[50]. Consequently the observed currents are given as a coherent sum of currents from each atom in the system. The interference effects give rise to the characteristic features which differentiate atomic photoionization from solid state photoemission. The localized wave functions on one atom in a molecule or solid are not eigenfunctions of the system but instead have nonvanishing projections on each eigenstate. If one observes photoemission involving a specific final eigenstate, then this additional constraint (a <u>given</u> final state rather than the <u>spectrum</u> of final states allowed by the projection of the localized initial function onto the initial eigenstates) determines the relative phases of the individual atom currents needed to compute the coherent sums over all sources, the atoms in the system.

First let us consider some of the qualitative distinctions between gas phase and solid state photoejection.

(a) Hole localization. In the gas phase, the hole created in the photoejection process remains localized on one atom. Since a "record' of the ejected electron's origin is left behind, there are no interference effects between electrons ejected from various atoms[50]. Under certain conditions to be discussed later, a similar situation appears to prevail in the photoejection of core electrons from solids[51-55] or of valence electrons in flat bands[31,56]. In contrast, "good" band effects in a solid imply that a hole created on one atom rapidly delocalizes throughout the entire crystal. Interference effects and consequently, $\underset{\sim}{k}$ conservation selection rules then follow.

(b) Screening and final state interaction. An electron which is photoejected still interacts with the hole potential via "final state interactions" (FSI)[57,58]. The gas phase hole potential is unscreened and thus assumes a long range 1/r form. Consequently, the correct final state wave function which includes all FSI is a Coulomb function. In solids, FSI, often referred to as exciton effects, can be reduced since the free electrons at the Fermi surface respond to the sudden creation of the hole to produce a screened hole potential[51-55,59]. Due to the short range of the screened potential, the final state wave function is often taken to be an eigenfunction of a Hamiltonian which does not include the hole potential. This will not produce any serious problems provided that one does not equate hole screening with hole delocalization.

(c) Orientation. In the solid state, the atoms are arranged in a periodic array whereas in the gas phase, the arrangement is random. Furthermore, due to Ligand effects, the atomic orbitals in the solid array have a preferential orientation whereas in the gas phase, the orbitals are randomly oriented. This feature is currently being exploited in photoemission angular distribution studies[34,35,60].

(d) Lattice effects in the final state. In a solid there are weak final state interactions between the excited electron and the periodic lattice potential which give rise to diffraction effects in the outgoing electron wave (the secondary cones discussed by Mahan[28]). Moving on to some quantitative notions, the general expression for the photocurrent has been given in Eqs.(10a,b) and (11). As previously discussed, the final state should be taken to be a scattering state which satisfies ingoing wave boundary conditions. If final state interactions are neglected (as they will be in this treatment), then the final state is a plane wave $e^{i\mathbf{k}\cdot\mathbf{r}}$ and the consequences of ingoing versus outgoing waves are irrelevant. In the case of solid state photoemission, it will be assumed that the final state, although definitely not the initial state, is free electron like and thus $\varepsilon = \frac{\hbar^2 k^2}{2m}$ in an extended zone scheme or $\varepsilon = \frac{\hbar^2|\mathbf{k}_f+\mathbf{G}|^2}{2m}$ in the reduced zone scheme, where $\mathbf{k}_f$ is within the first Brillouin zone and $\mathbf{G}$ is the reciprocal lattice vector required to fold the extended band back into the Brillouin zone. To include surface or LEED effects, multiply the plane wave by a factor $T(k_z,\varepsilon)$, the transmission amplitude for an electron with energy ε and normal component of wave number k_z, to pass through the surface region from the vacuum into the solid. Aside from the pathological cases discussed by Feuerbacher and Christenson[47] in which ε falls in a band gap, T is expected to vary slowly with energy over the width of a narrow d-band as ε becomes large.

Recent attention has been directed onto the fact that at least for $\omega <$ the solid plasmon frequency, the normal component of $\mathbf{A}$ in Eq.(10b) varies rapidly within a few angstroms at the surface[32,37] and thus $[p_{op},A] \neq 0$. We will however ignore this effect and take the commutator = 0. Then the symmetrized perturbation is just $\tau = \mathbf{A}\cdot\mathbf{p}_{op} = \mathbf{p}_{op}\cdot\mathbf{A}$. Since the final state wave function is a plane wave, the optical matrix element in Eq.(10), can be written

$$\underset{\sim}{A} \cdot \int d^3r\, e^{-i\underset{\sim}{k}\cdot\underset{\sim}{r}}\, \underset{\sim}{\nabla}\psi_j(\underset{\sim}{r}) = -iA\cos\gamma\, \psi_j(\underset{\sim}{k}) \tag{12}$$

where γ is the angle between $\underset{\sim}{A}$ and $\underset{\sim}{k}$ and $\psi_j(\underset{\sim}{k})$ is the Fourier transform of the initial state wave function. Consequently, Eq.(10) reduces to

$$\frac{d^2j}{d\varepsilon d\Omega} = C\, A^2 \sum_j \cos^2\gamma k^3 |\psi_j(\underset{\sim}{k})|^2 \delta(\varepsilon - \hbar\omega - \varepsilon_j) \tag{13}$$

where C is just a group of constants. In this study, we will consider initial state wave functions which are written in the molecular orbital (MO) form (this includes the tightbinding Bloch functions of solids)

$$\psi_j(\underset{\sim}{r}) = \sum_i e^{i\delta_i(\varepsilon_j)}\, Q_i\, \phi_i(\underset{\sim}{r} - \underset{\sim}{R}_i) \tag{14}$$

where the sum i is over the position of all ion core centers in the ensemble which are located at $\underset{\sim}{R}_i$, $\delta_i(\varepsilon_j)$ specifies the energy dependent phase of the localized atomic-like orbital ϕ_i at position $\underset{\sim}{R}_i$, and Q_i is a positive real coefficient. Surface effects in the initial state can be included by proper choice of δ_i an Q_i. If surface effects are included, $Q_i = Q_i(\varepsilon_j)$. It is straightforward to show that the Fourier transform of Eq.(14) is

$$\psi_j(\underset{\sim}{k}) = \sum_i Q_i \phi_i(\underset{\sim}{k}) e^{i(\delta_i(\varepsilon_j) - \underset{\sim}{k}\cdot\underset{\sim}{R}_i)}$$

with $\phi_i(\underset{\sim}{k}) \equiv \int d^3r\, e^{-i\underset{\sim}{k}\cdot\underset{\sim}{r}}\, \phi_i(\underset{\sim}{r})$. Thus Eq.(13) can be written :

$$\frac{d^2j}{d\varepsilon d\Omega} = C\, A^2 \sum_j \cos^2\gamma k^3 \delta(\varepsilon - \hbar\omega - \varepsilon_j) \Big| \sum_j Q_i \phi_i(\underset{\sim}{k}) e^{i(\delta_i(\varepsilon_j) - \underset{\sim}{k}\cdot\underset{\sim}{R}_i)} \Big|^2 \tag{16}$$

It must be kept in mind that in the sum on initial states, $k \sim \varepsilon^{1/2}$ is an implicit function of ε_j through the energy conserving delta function. As Eq.(16) stands, for the case of solid photoemission, the $d\Omega$ is specified in terms of a coordinate system within the solid. In order to relate the angular distribution in terms of internal variables to variables outside the solid, the transformation outlined by Mahan(28) must be made in order to account for refraction effects at the surface.

Finally note that if hole lifetime effects due to radiative or Auger decay are important, the delta function in Eq.(7) should be replaced by a Lorentzian whose width is dictated by the hole lifetime.

C. Specific Applications

1. ATOMS

The first specific application of Eqs.(16) is to the case of photoionization of an atom centered at $\underset{\sim}{R}_s$. From Eq.(14), $Q_i = \delta_{i,s}$ and the value of the phase factor is irrelevant. The sum on initial states in Eq.(16) includes only the ground state valence electron with ionization potential V_i. Consequently, Eq.(16) is simply

$$\frac{d^2j}{d\varepsilon d\Omega} = C\ A^2\ \cos^2\gamma|\phi_i(\underset{\sim}{k})|^2\ k^3\delta(-V_i + \hbar\omega - \varepsilon) \qquad (17)$$

The single atom matrix element $\sim \cos^2\gamma|\phi_i(\underset{\sim}{k})|^2$ in Eq.(17) contains the angular momentum selection rules $\Delta\ell = \pm 1$, $\Delta m = 0, \pm 1$. In addition, standard gas phase photoionization theory(61) gives $\frac{dj}{d\Omega} = a + b\cos^2\gamma$ with $0 \leqslant |a/b| \leqslant 1$ when an average is performed over all atom orientations. Equation (17) is in accord with this result.

2. MOLECULES

Now consider a photoejection experiment from an oriented H_2 molecule(34,62). The molecular state wave functions are taken to be molecular orbitals,

$$\psi_{\mp} = \frac{1}{\sqrt{2}}\,(\phi_i(\underset{\sim}{r} - \underset{\sim}{R}_1) \pm \phi_i(\underset{\sim}{r} - \underset{\sim}{R}_2))$$

where ψ_- is the bonding orbital with energy $\varepsilon_{in} = -V$, ψ_+ the anti-bonding orbital with energy $+V$, and $V = \langle 1|H|2\rangle$ is the "hopping matrix element"(39). In a resonance theory of chemical bonding, $\hbar/V$ is the time scale for an electron (or hole) to remain localized on a given center. If the hole created on one center, say at $\underset{\sim}{R}_1$ is rapidly delocalized** to the other center at $\underset{\sim}{R}_2$ then the current arri-

---

* *Assuming that $k_{\parallel}$, the component of $\underset{\sim}{k}$ parallel to the surface is conserved in going from the metal to vacuum, the angles are related through $\sin\theta_{out} = (\varepsilon_f/(\varepsilon_f-\phi_e))^{1/2}\sin\theta_{in}$ and $\phi_{in} = \phi_{out}$ when ε_f is referenced with respect to a zero at the bottom of the conduction band and ϕ_e is the work function.*

** *The criteria of rapid is that $\tau_{hopping} = \hbar/2V \ll \hbar/\Delta\varepsilon$ where $\Delta\varepsilon$ is the energy discrimination or acceptance of the particular experiment.*

ving at the limited solid angle detector appears to have been generated from two coherent sources at $\underset{\sim}{R}_1$ and $\underset{\sim}{R}_2$. Consequently, the amplitudes add and interference effects are observed in the angular distribution. This is obtained from Eq.(16) by summing over i = 1 and 2. With $Q_1 = Q_2 = 1/\sqrt{2}$, $\underset{\sim}{R}_1 = -a/2\ \hat{i}_z$, $\underset{\sim}{R}_2 = +a/2\ \hat{i}_z$, $\delta_1 = \delta_2$ for the bonding orbital initial state, and $\delta_1 = \delta_2 + \pi$ for the antibonding orbital initial state, the angular distribution from each possible state are

$$\frac{dj}{d\Omega}\mp = C\ A^2 \cos^2\gamma|\phi_i(\underset{\sim}{k})|^2\ k^3 \begin{matrix}\cos^2\\ \sin^2\end{matrix}\left(\frac{k_z a}{2}\right) \qquad (18)$$

where j_-', the AD from the bonding orbital, is given by the $\cos^2$ term, and j_+', the AD from the antibonding orbital, by the $\sin^2$ term. In Eq.(18), $k = [\frac{2m}{\hbar^2}(\varepsilon_j + h\nu)]^{1/2}$ by virtue of the dε integration. In the limit in wich V → 0, the bonding and antibonding states become degenerate so both must be included in the sum on initial states. From Eq.(18), $j_{Tot}' = j_-' + j_+' = j_{atom}'(\cos^2 + \sin^2) = j_{atom}'$ and we recover the independent atom result. It can thus be seen that the possibility of spatial coherence in the emission redistributes the individual atom emission into preferred directions at particular energies but if a sum over all the possible energies is performed, then the coherence effects all cancel. An equivalent, although not so transparent, situation prevails in photoemission from band states in solids.

3. SOLIDS

The basic role of spatial coherence in solids does not differ, in principle, from that which was demonstrated in molecules. For illustrative purposes, consider photoemission within the band picture shown in Fig.5. The initial states are those of a narrow band $\varepsilon_j = \varepsilon_j(\underset{\sim}{k}')$ with width W and it is assumed that the final states are in a free electron band $\varepsilon = \hbar^2(\underset{\sim}{k}_f + \underset{\sim}{G})^2/2m$ with $\underset{\sim}{G}$ the reciprocal lattice vector which folds the extended zone dispersion relation back into the first zone. By energy conservation alone, the possible combinations of initial and final states are related by

$$\frac{\hbar^2 k^2}{2m} = \hbar^2|\underset{\sim}{k}_f + \underset{\sim}{G}|^2/2m = \hbar\omega + \varepsilon_j(\underset{\sim}{k})$$

Since $\frac{\hbar^2}{2m}((k + \Delta k)^2 - k^2) \equiv W$, the span of energy allowed wave vectors in the final state is

$$\Delta k \simeq \frac{mW}{\hbar^2 k} \simeq \frac{W}{4\varepsilon^{1/2}}\ \text{Å}^{-1} \qquad (19)$$

where W and ε are given in eV, Δk in Å^{-1}, and ε is measured from the bottom of the free electron bands. Equation (19) suggests that Δk can be made arbitrarily small as either the bandwidth W goes to zero or the excitation energy $\hbar\omega$ and thus ε become very large. The relevant criteria is that if the slope of the final state band becomes much larger than the slope of the initial state band, the spread in final state wave numbers decreases and it might be suspected that the importance of so-called momentum conservation selection rules and final state band effects diminishes. This in fact is the justification for assuming that only initial state band structure is important in x-ray photoemission[63].

The quantitative implications of these ideas are made clear by rewriting Eq.(16) as

$$\frac{d^2j}{d\varepsilon d\Omega} = \frac{CA^2}{N} \sum_{\underset{\sim}{k}'} \cos^2\gamma k^3 \delta(\varepsilon-\hbar\omega-\varepsilon(\underset{\sim}{k}'))|\phi(\underset{\sim}{k})|^2 \sum_{i,\ell} e^{-i\underset{\sim}{k}\cdot(\underset{\sim}{R}_i-\underset{\sim}{R}_\ell)} \times e^{ik'\cdot(\underset{\sim}{R}_i-\underset{\sim}{R}_\ell)} \quad (20)$$

where in Eq.(15) $\delta_i(\varepsilon_j) = \underset{\sim}{k}' \cdot \underset{\sim}{R}_i$ for tightbinding Bloch functions, $Q_i = Q_\ell = 1/\sqrt{N}$, and we have taken ϕ_i to be site independent. By taking Q_i = constant, surface effects are implicitly neglected. The sum on initial states can be replaced by an integral on $\underset{\sim}{k}'$:

$$\sum_{\underset{\sim}{k}'} = \Omega \int \frac{d^3k'}{(2\pi)^3}$$

where Ω is the volume of the crystal and $\underset{\sim}{k}'$ must lie within the first zone. To proceed, more specific statements of the models being considered are required.

i) Narrow bands or core states

In the narrow band (including core state "bands") or large $\hbar\omega$ limit, $\Delta k \to 0$ and thus both Δk and ε are independent of k'. Consequently, Eq.(20) can be written as

$$\frac{d^2j}{d\varepsilon d\Omega} = CA^2\delta(\varepsilon-\hbar\omega-\varepsilon')\Omega_a \cos^2\gamma|\phi(\underset{\sim}{k})|^2 k^3 \sum_{i,\ell} e^{-i\underset{\sim}{k}\cdot(\underset{\sim}{R}_i-\underset{\sim}{R}_\ell)} \times \int \frac{d^3k'}{(2\pi)^3} e^{i\underset{\sim}{k}'\cdot(\underset{\sim}{R}_i-\underset{\sim}{R}_\ell)} \quad (21)$$

where Ω_a is the volume per atom, ε' is some energy within the initial state band, and the limits on the k' integration must still be specified. The criterion of narrowness is once again energy resolution or acceptance (either that in the actual measurement process or that due to finite lifetime effects) compared to the bandwidth which is a measure of the hole localization time. If the energy acceptance is larger than the bandwidth, then one cannot determine, in principle, the initial state within the Brillouin zone. Consequently the integral on $\underset{\sim}{k}'$ in Eq.(21) extends over the complete zone (for a filled band) and is equal to a delta function on site indices. Thus

$$\frac{d^2j}{d\varepsilon d\Omega} = CA^2\delta(\varepsilon-\hbar\omega-\varepsilon')\Omega_a \cos^2\gamma|\phi(\underset{\sim}{k})|^2 k^3 \sum_{i,\ell} e^{-i\underset{\sim}{k}\cdot(\underset{\sim}{R}_i-\underset{\sim}{R}_\ell)}\delta_{i,\ell}$$

$$= NCA^2\delta(\varepsilon-\hbar\omega-\varepsilon')\Omega_a \cos^2\gamma k^3|\phi(\underset{\sim}{k})|^2 \qquad (22)$$

which is just the independent atom result. As with the V = 0 limit of the H_2 molecule, interferences due to coherent emission from various centers appear in the photocurrent only if one does an experiment in which the spectrum of possible states is energy resolved. In the case just treated, in which the angular distribution of the total yield from a filled band was obtained, the interferences from spatial coherence sum to zero and the equivalent of incoherent atom emission follows.

ii) k-conserving direct transitions

If the $\underset{\sim}{k}'$ integration runs over only part of the zone, the site index delta function does not appear and interference effects due to hole hopping and consequently, the indistinguishability of source atoms result. In this case the sum on initial states is restricted to one or a few states with $\varepsilon_j = \varepsilon(\underset{\sim}{k}')$. From Eq.(20), the angular distribution from one such state is

$$\frac{dj(\varepsilon_j(\underset{\sim}{k}'))}{d\Omega} = CA^2 \cos^2\gamma|\phi(\underset{\sim}{k})|^2 k^3 \sum_{i,\ell} e^{i(\underset{\sim}{k}'-\underset{\sim}{k})\cdot(\underset{\sim}{R}_i-\underset{\sim}{R}_\ell)} \qquad (23)$$

For crystalline ordering

$$\sum_{i,\ell} e^{i(\underset{\sim}{k}'-\underset{\sim}{k})\cdot(\underset{\sim}{R}_i-\underset{\sim}{R}_\ell)} = \sum_G \delta_{\underset{\sim}{k}'-\underset{\sim}{k},\underset{\sim}{G}}$$

where $\underset{\sim}{G}$ is a reciprocal lattice vector, so Eq.(23) becomes

$$\frac{dj(\varepsilon_j(\underset{\sim}{k}'))}{d\Omega} = CA^2 k^3 \sum_{\underset{\sim}{G}} \cos^2\gamma|\phi(\underset{\sim}{k})|^2 \delta_{\underset{\sim}{k}'-\underset{\sim}{k},\underset{\sim}{G}} \qquad (24)$$

The Kronecker delta expresses the so-called momentum ccnservation selection rule which is modulated by the atomic-like matrix element. If the initial state is written as a coherent sum of plane waves which satisfy the Bloch theorem

(i.e. $|j\rangle = \sum_{G} u(\underset{\sim}{k}',\underset{\sim}{G})e^{i(\underset{\sim}{k}'+\underset{\sim}{G})\cdot\underset{\sim}{r}}$)

then it is an easy matter to show that $\phi(\underset{\sim}{k}) = u(\underset{\sim}{k}'.\underset{\sim}{G})$ and Eq.(24) becomes precisely the condition for the Mahan cones[28]. From the study in this section, the fundamental origin of $\underset{\sim}{k}$-conservation in photoemission is apparent. Again one must be able to resolve the split levels or bands formed when the degenerate atomic levels are allowed to interact via hopping. If this resolution is possible, then hole delocalization results in observable photoemission characteristics of an array of phase coherent electron sources at each atom center. If the atoms are arranged in a lattice, the interference effects manifest themselves in $\underset{\sim}{k}$-conservation.

To make contact with the standard angle integrated theoretical energy distributions, Eq.(16) together with the results of Eq.(24) can be cast in the form

$$\frac{dj}{d\varepsilon} \propto A^2 \sum_{\underset{\sim}{k}',\underset{\sim}{G}} \int k d\Omega\ \delta(\varepsilon-\hbar\omega-\varepsilon_j(\underset{\sim}{k}'))k^2\cos^2\gamma|\phi(\underset{\sim}{k})|^2\ \delta^{(3)}_{\underset{\sim}{k}'-\underset{\sim}{k},\underset{\sim}{G}} \quad (25)$$

The usual approximation is to assume constant matrix elements (i.e., $Ak\cos\gamma\phi(\underset{\sim}{k})$ = const.) which then eliminates the $\underset{\sim}{G}$ sum. However note that $\underset{\sim}{k}$ conservation came out of a calculation of the matrix element when energy resolved spatial coherence was considered and was not something which was mysteriously entered into the theory. Under this assumption, Eq.(25) becomes

$$\frac{dj}{d\varepsilon} \propto \sum_{\underset{\sim}{k}'} \int k d\Omega\ \delta(\varepsilon-\hbar\omega-\varepsilon_j(\underset{\sim}{k}'))\delta^{(3)}_{\underset{\sim}{k},\underset{\sim}{k}'}$$

or alternatively

$$\frac{dj}{d\varepsilon} \propto \sum_{\underset{\sim}{k}'} \int k d\Omega d\varepsilon(\underset{\sim}{k})\delta(\varepsilon-\varepsilon(\underset{\sim}{k}))\delta(\varepsilon(\underset{\sim}{k})-\hbar\omega-\varepsilon_j(\underset{\sim}{k}'))\delta^{(3)}_{\underset{\sim}{k},\underset{\sim}{k}'} \quad (26)$$

For the free electron final state that has been assumed, $\int k d\Omega d\varepsilon(\underset{\sim}{k}) = \sum_{\underset{\sim}{k}}$ and when this sum is performed, using up the Kronecker delta, the final result :

$$\frac{dj}{d\varepsilon} \propto \sum_{\underset{\sim}{k}'} \delta(\varepsilon-\varepsilon(\underset{\sim}{k}'))\delta(\varepsilon(\underset{\sim}{k}') - \hbar\omega-\varepsilon_j(\underset{\sim}{k}')) \quad (27)$$

is the energy distribution of the "joint density of states". The expression given by Eq.(27) has formed the basis for interpretation of large quantities of photoemission data[5,45,46,64]. Since surface effects have been neglected a description in terms of direct transitions might be expected to be most valid for final state electron energies where the electron mean free path is long.

As the best example of a quantitative comparison between experiment and direct transition theory, energy distribution curves for photoemission from Cu calculated by Janak, Williams, and Moruzzi[64] are shown in Fig.6 together with experimental curves obtained by Eastman and Grobman[65]. In these calculations, constant matrix elements were not assumed, but instead momentum matrix elements involving states with $\varepsilon_j(\underset{\sim}{k}')$ and $\varepsilon(\underset{\sim}{k}')$ were included with the sum in Eq.(27). Whether such a theory satisfactorily accounts for the features of the experimental curves is left to the taste of the reader.

iii) Indirect transitions

If for some reason, translational invariance is broken, then the final state cannot be a momentum eigenfunction and thus k conservation in optical excitation is no longer meaningful[30,56]. Spicer has suggested that a hole in a narrow band provides a localized potential which breaks the symmetry. In the present framework we would say that the localized hole represents a record of which atom the electron came from. Thus there are no coherence effects, and as a result, the momentum conserving delta function of Eq.(24) does not arise. Noting that in Eq.(26), $\int k d\Omega d\varepsilon(k)$ is equivalent to a sum on final states and neglecting all k or k' dependences,

$$\frac{dj}{d\varepsilon} \sim \sum_{j,fin} \delta(\varepsilon-\varepsilon_{fin})\delta(\varepsilon_{fin}-\hbar\omega-\varepsilon_j) \sim \rho_{fin}(\varepsilon)\,\rho_{in}(\varepsilon-\hbar\omega)$$

which means that if k-conservation is not an applicable selection rule, then the energy distribution is just the product of the system density of states evaluated at the initial and final energies. This is a convenient way to interpret spectra, but unfortunately it does not appear to be experimentally substantiated.

iv) Surfaces

To date surface effects have usually been neglected in photoemission calculations which include optical matrix elements. Formulation of the photoemission problem within the localized picture lends itself naturally to the inclusion of such effects. As a simple example, consider a tight binding standing wave initial state

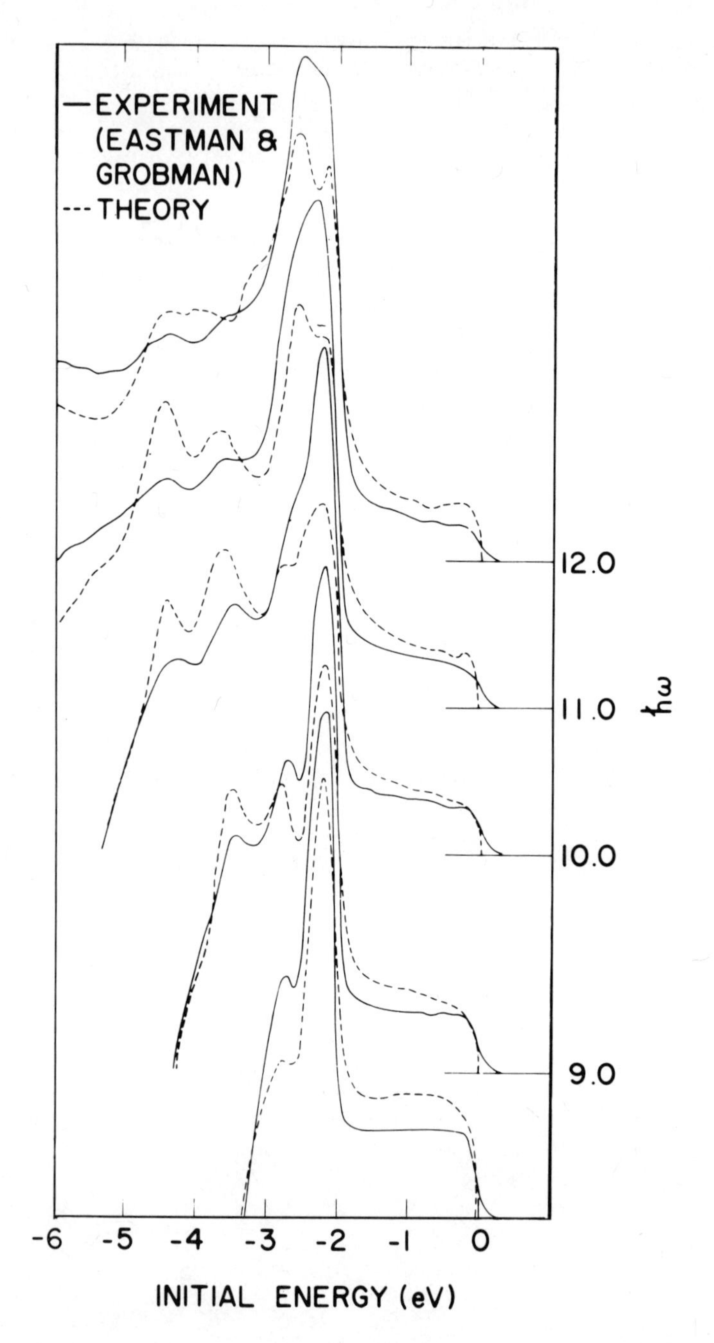

Fig. 6 : Comparison of experimental and theoretical photoemission energy distributions for $\hbar\omega$ between 8 and 12 eV. (Ref.64, Janak, Williams, and Moruzzi)

$$\psi_j(\underset{\sim}{r}) = \sum_{i(i_z>0)} e^{i\underset{\sim}{k}'_{\parallel}\cdot\underset{\sim}{R}_i} \sin(k'_z R_{iz} + \eta(\varepsilon_j, \underset{\sim}{k}'))\phi_i(\underset{\sim}{r}-\underset{\sim}{R}_i) \quad (28)$$

where the sum is restricted to the solid half space and $\eta(\varepsilon_j, \underset{\sim}{k}')$ is a phase shift due to the surface potential. Assuming that the atomic-like function is identical on all sites, the Fourier transform of Eq.(28) is

$$\psi_j(\underset{\sim}{k}) = \frac{\phi(\underset{\sim}{k})}{2i} \sum_i e^{i(\underset{\sim}{k}'_{\parallel}-\underset{\sim}{k}_{\parallel})\cdot\underset{\sim}{R}_i}\{e^{i\eta} e^{i(k'_z-k_z)R_{iz}} - e^{-i\eta} e^{-i(k'_z+k_z)R_{iz}}\}$$

which yields

$$|\psi_j(\underset{\sim}{k})|^2 = \sum_{G_{\parallel}} \frac{|\phi(\underset{\sim}{k})|^2}{4} \delta_{\underset{\sim}{k}'_{\parallel}-\underset{\sim}{k}_{\parallel},\underset{\sim}{G}_{\parallel}} \left|\sum_{i_z} e^{i\eta} e^{i(k'_z-k_z)R_{iz}} - e^{-i\eta} e^{-i(k'_z+k_z)R_{iz}}\right|^2 \quad (29)$$

Equation (29) is to be inserted in Eq.(13) to yield an expression for the photoemission current. In order to account for the surface localization due to inelastic damping, standard procedure is to replace k_z by $k_z - i\kappa$ (with $\kappa = 1/2\lambda$) as this exponentially favors emission from the layers nearest the surface. With this ad hoc substitution, the summation on i_z (neglecting lattice relaxation) can be performed analytically and the result obtained from Eqs.(13) and (29) for the elastic photoemission current (with $\eta = 0$) from state $\varepsilon_j(\underset{\sim}{k}')$ is

$$\frac{dj}{d\Omega} = CA^2k^3 \sum_{\underset{\sim}{G}} \cos^2\gamma|\phi(\underset{\sim}{k})|^2 \delta^{(2)}_{\underset{\sim}{k}'_{\parallel}-\underset{\sim}{k}_{\parallel},\underset{\sim}{G}_{\parallel}} e^{-2\kappa a} \sin^2 k_z a$$

$$/(1+e^{-4\kappa a}+4e^{-2\kappa a}[\cos^2 k'_z a+\cos^2 k_z a - \frac{1}{2} - 2\cosh\kappa a \cos k'_z a \cos k_z a]) \quad (30)$$

where a is the inner planar spacing. In analogy with Eq.(24), Eq. (30) displays selection rules from both atomic-like aspects ($\phi(\underset{\sim}{k})$) and also from the spatial coherence (the two dimensional delta function due to translational invariance in the transverse plane). However since the initial state was a standing wave and the inelastic damping localized the real space source region of the photocurrent near the surface, an inherent uncertainty or spread of k_z occurs and this enters the theory through the relaxed restrictions on k_z conservation, given by the denominator in Eq.(30). I am not aware of any calculations in which such surface effects have been explicitly included in realistic calculations, and thus cite this case as an example of what will probably be treated in the next generation of photoemission numerology.

v) Adsorption

A natural consequence of the surface sensitivity of UPS is its usefulness as a tool for studying both electronic and geometric properties of adsorbed (physisorbed and chemisorbed) atoms or molecules. As already suggested in Fig.1, new structure in a photoemission energy distribution due to an adsorbed atom or molecule can be related to the electronic states involved in either the chemisorption bond or the physisorbed species. Penn[33] demonstrated that the change in the ED upon chemisorption[3] (within the Anderson model) is given by

$$\Delta j'(\varepsilon) = a\ \mathrm{Im}G_a(\varepsilon) + b\mathrm{Re}G_a(\varepsilon) \qquad (31)$$

where a and b are functions of system parameters with b usually less then a, $G_a(\varepsilon) = (\varepsilon-\varepsilon_a-\Lambda+i\Delta_a)^{-1}$ is the adatom Green's function in the energy independent Λ and Δ_a approximation discussed in conjunction with Eq.(9), and the energy dependence of matrix elements has been neglected. The noteworthy feature of Eq.(31) is the presence of the term $\sim \mathrm{Re}G_a(\varepsilon) = (\varepsilon-\varepsilon_a-\Lambda)/((\varepsilon-\varepsilon_a-\Lambda)^2+\Delta_a^2)$ which adds an antisymmetric contribution to the chemisorbed atom line shape. This term thus displaces the energy position of the adsorption induced peak and this effect should be considered when analyzing experimental data. Whether such asymmetric line shapes have been observed is still open to debate as a very delicate procedure is involved in obtaining photoemission difference curves. The absolute normalization of both the before and after adsorption curves is important and unfortunately, the "observed" line shape and thus degree of asymmetry due to this interference effect are crucially dependent on the choice of normalizations.

Two examples are now given to illustrate the scope of UPS studies. In Fig.7 a photoemission spectrum of CO chemisorbed on the (100) face of Ni, obtained by Eastman and Cashion[67] is shown. The broad peak between -3 eV and the Fermi level is due to emission from the Ni d bands. Two rather weak peaks at -7.5 and -10.7 eV result from the chemisorbed CO. Considerable controversy has arisen over the interpretation and labeling of these peaks. It is likely that this problem has not yet seen its final resolution. Batra and Bagus[68] have performed X-α scattered wave calculations on the $CO(Ni)_5$ cluster shown in the inset, obtaining the cluster energy levels displayed also in Fig.7. Possible interpretations and implications of these numbers within the context of previous studies of CO on Ni is fully discussed in the excellent paper of Batra and Bagus[68].

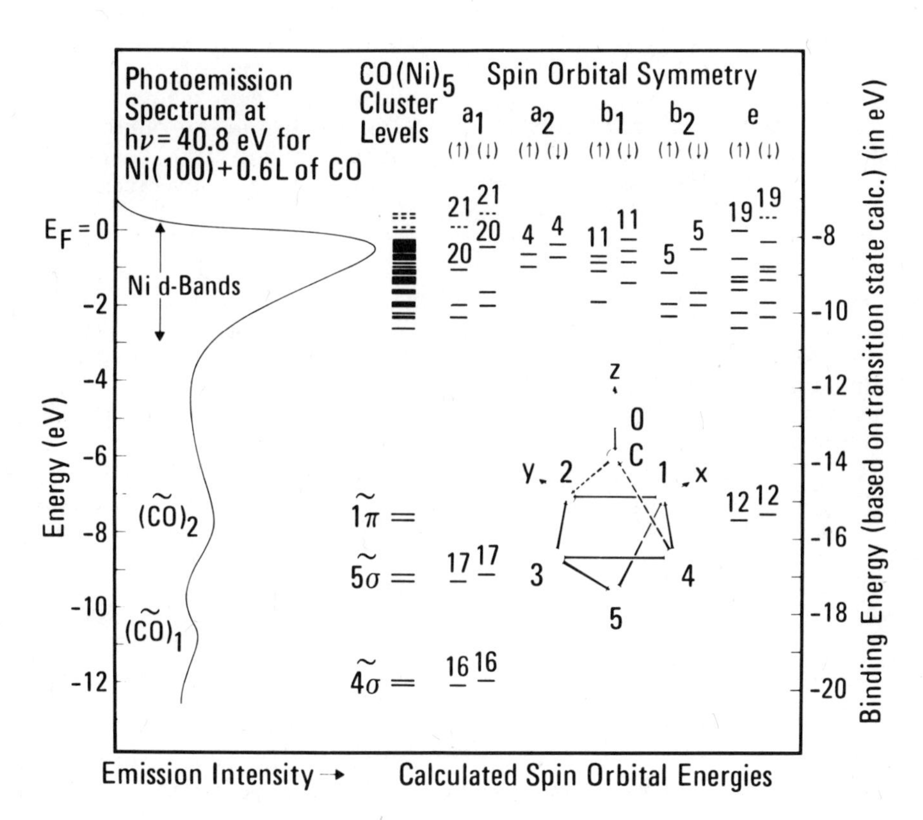

Fig. 7 : Comparison of theoretical spin-unrestricted energy levels for the $CO(Ni)_5$ cluster and experimental photoemission spectrum of CO chemisorbed on Ni(100) (reproduced from reference 67). Occupied levels are denoted by solid lines, unoccupied levels by dashed lines. In the ten rightmost columns spin orbital symmetry for all levels is given. For each symmetry species, the levels are labeled by serial numbers in order of increasing energy. Only a few of these serial numbers are actually shown. Insert shows the model cluster (Ref. 68, Batra and Bagus)

UPS is also being used to study physisorption. In Fig.8, UPS spectra and difference curves obtained by Waclawski and Herbst[69] for Xe physisorbed on the (100) face of W are shown. The two large Xe induced peaks at -6.76 and -8.06 eV are believed to be emission from the spin-orbit split $5p_{1/2}$ and $5p_{3/2}$ states of Xe since the splitting in the physisorbed case shown is identical with that in the gas phase. Furthermore the integrated intensity of the 3/2 to 1/2 peak is 1.4-1.6 as in the gas phase. It has been shown[69] that the unexpected broadening of the 3/2 peak is due to an unresolved crystal field splitting. Lastly the peak positions with respect to the vacuum potential are shifted by 1.6 eV due to extra-atomic relaxation energies, in good agreement with expectations[70].

Another area of recent interest has been in angle resolved photoemission spectra (ARPS) from chemisorbed atoms[34,35] as it is hoped that geometrical information on chemisorption bonds can be obtained. For example consider Fig.9 in which an adatom A bonds in an out of plane bridge site between two substrate atoms S_1 and S_2. The detector D can be moved to obtain the ARPS. At least two possible effects should result in geometrical information. The first, referred to as initial state interferences in Fig.9a, is due to the coherent emission from the source atoms in the surface molecule complex[34]. This is completely analogous to the coherence effects discussed in the sections on H_2 and $\underset{\sim}{k}$-conservation. Final state effects shown in Fig.9b are due to the interferences between the direct wave at the detector and the diffracted or backscattered waves[35]. We will ignore this second effect here but suggest that the interested reader consult Ref.71 for a detailed comparative discussion of the two.

In the case of initial state effects, if the coordinate origin is taken at the adatom center, then Eq.(16) can be written as

$$\frac{dj}{d\Omega} \sim \cos^2\gamma \left| \phi_a(\underset{\sim}{k}) + \phi_s(\underset{\sim}{k}) \sum_i Q_i\, e^{i\delta_i}\, e^{-i\underset{\sim}{k}\cdot\underset{\sim}{R}_i} \right|^2 \qquad (32)$$

where $\phi_a(\underset{\sim}{k})$ and $\phi_s(\underset{\sim}{k})$ are the Fourier transforms of the adatom and substrate orbitals respectively and the sum on i is restricted to the near neighbour substrate atoms involved in the chemical bond. Equation (32) is precisely the form first presented by Gadzuk[34]. It is satisfying that this example of photoemission can also be trivially handled within the single theoretical framework. As an example of the type of information expected in an ARPS, Eq.(32) has been evaluated for an adsorbate with an s valence electron bonding in the flat fourfold site with substrate d orbitals, as shown in Fig.10. Normal incidence unpolarized photons are assumed. One of the four equivalent quadrants of an azimuthal scan, with polar angle $\theta = 45°$, $r' \equiv \phi_s(k)Q_i/\phi_a(k) = 1$, and electron binding energy $\varepsilon_B = 10$ eV (with respect to vacuum) are shown in Fig.10, treating

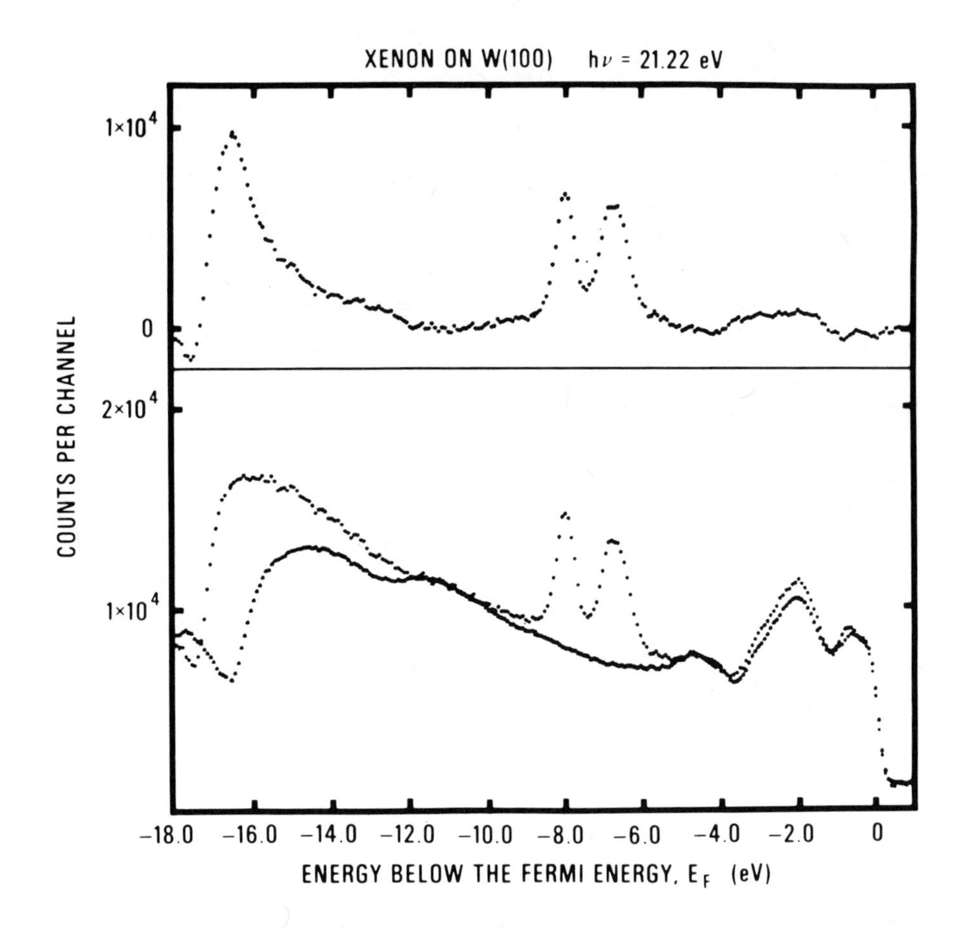

Fig. 8 : Photoelectron energy distributions. The lowermost curve is for W(100) at a temperature of ∿80°K, and directly above it is the distribution after exposure to 5 Langmuirs of xenon. The difference of the two distributions is given at the top of the figure (Ref. 69, Waclawski and Herbst)

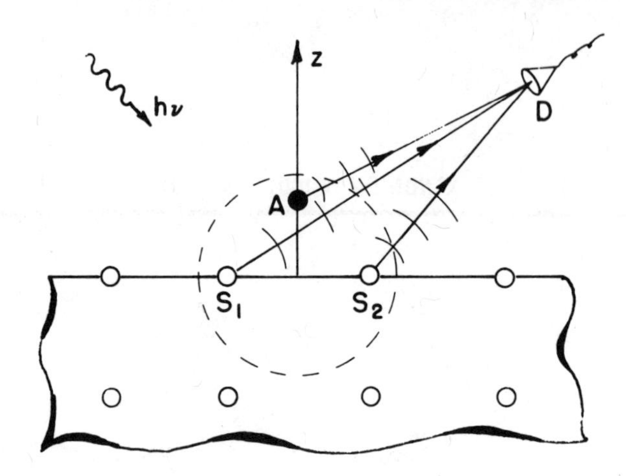

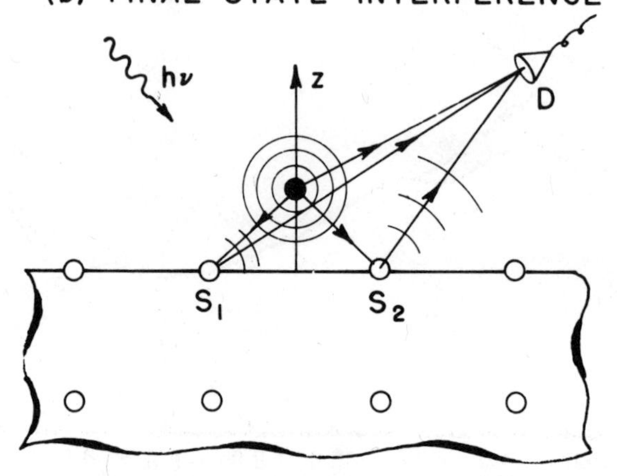

Fig. 9 : Schematic surface molecule photoemission diagram which depicts the origins of interference effects. The adatom is labeled A, the nearest neighbour substrate atoms S_1 and S_2, and the detector D.

(a) In the initial state, A, S_1, and S_2 act as coherent sources of electrons

(b) In the final state, electrons generated at A proceed directly to D, or propagate to S_1, S_2 where they are scattered back to the detector (Ref. 71, Gadzuk)

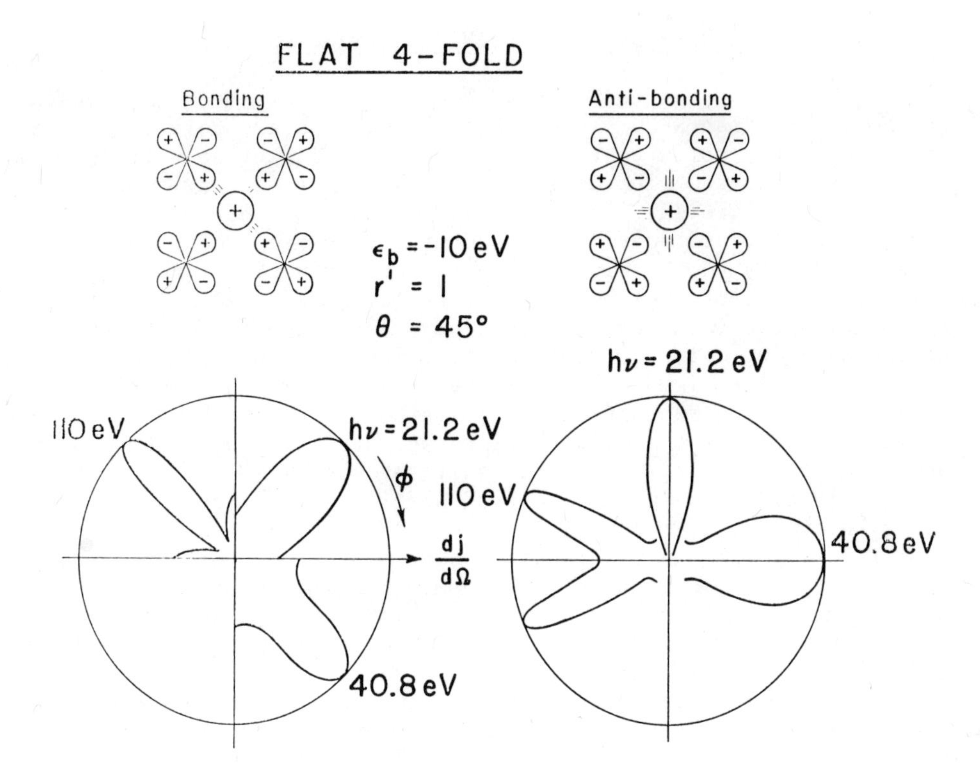

Fig. 10 : Azimuthal PAD for an s adatom in a flat fourfold bond site for various labeled photon frequencies, a polar angle of 45°, and a binding energy of 10 eV. The parameter r' (defined in the text) = 1. The orbital geometries and the bonding or anti-bonding charge densities are shown in the top row. The appropriate azimuthal PAD's are shown in the bottom row, where it should be noted that only one of the four equivalent quadrants is shown for each value of hν. (Ref. 71, Gadzuk)

the photon energy $h\nu = \hbar\omega$ parametrically. It is clear that geometrical information is contained in such distributions and studies of ARPS, as a function of $h\nu$, should be very fruitful in providing new and needed information on chemisorption.

4. MANY-BODY EFFECTS IN PHOTOEMISSION

Up to this point most of our considerations have been focused on UPS rather than XPS. Since valence electrons are relatively weakly bound to the solid they may be extracted with uv as well as x-ray photons and the theories do not differ in any basic way. However x-rays have sufficient energy to excite localized core electrons and in these events some new fundamental processes occur which require a theory beyond the single electron approximation. The original hope of many people has been that core level spectra should be easy to use since the spectra should be a simple series of narrow peaks at characteristic energies of the constituent atoms. The interpretation of such experimental data is relatively straightforward if the conditions of the excitation process are such that Koopmans' theorem is valid ; that is the difference between the single electron initial and final state energies is precisely the energy of the adsorbed photon, as is the case in valence band spectroscopy of spatially extended eigenstates. If Koopmans' theorem is not valid (as is usually the case for localized initial states) then the ejected electron emerges at an energy ε_r higher than the Koopman result. The term ε_r has been called the relaxation energy[7,26,27,70,72-76]. The sum $\varepsilon_r + \varepsilon_{orb}$, the orbital energy of the localized state is rigorously defined as the difference in total ground state energies of the initial N and final N-1 particle target systems, a time independent quantity. The "relaxation energy" is the adiabatic energy decrease of the N-1 electrons, due to the presence of the localized hole.

Another class of problems is concerned with the time dependence of the relaxation process and thus any possible observable consequences of the finite relaxation time[26,27,51-55,76-79]. Best[78] has given an excellent qualitative summary of the role of time scales in core level spectroscopies. Meldner and Perez[77] have presented a theory of gas phase atomic relaxation energies and relaxation times in the adiabatic and sudden limits, as follows. An electron in a Hartree-Fock single particle orbital with orbital energy ε_0 is excited by a photon with energy $h\nu$. The time scale by which the remaining electrons of the atom feel and thus respond to the deficiency of an electron charge (a positive hole) is determined by the kinetic energy of the excited electron. As the kinetic energy approaches zero (the photoionization threshold, if delayed onsets due to centrifugal barriers are neglected[80]), the excited electron moves very slowly from the ion core, and the resulting hole potential can thus be imagined to be turned on adiabatically. The electrons in the ion lower their energy by slowly relaxing around the extra positive charge, while always remaining in the ground state of the instantaneous self-consistent potential. In order that energy is conserved, the ejected electron must pick up this "intra-atomic relaxation energy" ε_a, and thus it emerges with a total energy ε_a greater than that inferred from a picture in which the orbitals in the ion are frozen in the same configuration they had in the atom. The photoionization energy spectrum would thus be a delta function (neglecting hole lifetime decay processes) at the energy $\varepsilon_t = h\nu + \varepsilon_a - \varepsilon_0$, as shown in Fig. 11a. In the other extreme of very large kinetic energy, the ejected electron leaves quickly, and the hole potential appears to be switched on instantly. In this sudden limit, the wave functions of the electrons in the ion core are continuous functions of time, but now the eigenstates of the ion are eigenfunctions of a new Hamiltonian, the atomic Hamiltonian with a hole potential. Consequently, the ion wave functions have non-vanishing projections onto the excited eigenstates of the ion. There is thus the possibility of leaving the ion in an excited state[73,74,81] and to conserve energy, the ejected electron must then emerge at discrete energies below the adiabatic energy as shown in Fig.11a. These peaks are called shake-up peaks. There is also the possibility of ejecting a second electron in this relaxation process, and the resulting continuous photoelectron spectrum is called a shake-off satellite. The adiabatic limit is obtained when the switching on time of the hole potential $\equiv \tau \gg \hbar/E_x$ with E_x a typical ion excitation energy. Likewise, the sudden limit is valid when $\tau \ll \hbar/E_x$.

In solid state physics, considerable attention has been focused on the long time response or relaxation of an electron gas when a localized potential such as a core hole is instantly switched on or off[51-55,59,73,76,82,83]. It has been shown that for certain simple hole potentials the sudden switching triggers a chaotic rearrangement of electrons in the conduction band, resulting in the excitation of many low-energy electron-hole pairs, the low energy excited states of the metal. In fact, an infrared catastrophe should occur in which an infinite number of zero energy pairs are created. The experimental ramifications of this effect are that x-ray emission or absorption spectra (obtained when core holes are switched off or on) should show singularities or exceptional rounding off at the threshold energies, where the spectra are predicted to have the functional form $D(\varepsilon) \sim 1/(\varepsilon_t - \varepsilon)^{1-\alpha}$ where α is related to the phase shifts of Fermi level electrons scattering from the screened hole potential.

Now consider the case of core level XPS of metals. We shall follow the same arguments used to discuss the Meldner and Perez[77] relaxation theory of atoms. If it were possible to switch on the hole potential adiabatically, then the energy level diagram shown in Fig. 11b would apply. In addition to the intra-atomic relaxation energy shift another shift, the extra-atomic relaxation energy $\equiv\Delta\varepsilon_r$, occurs due to the fact that the ion core is embedded in a polarizable electron gas. The conduction band electrons lower their energy by screening the positive ion core, and this energy is picked up by the excited electron. Now consider the sudden limit. In analogy with atomic shake-up peaks, it is expected that the solid might be left in some excited states, and thus the photoejected electrons could emerge with energies < ε_T. For a solid of inifinite extent, the electron-hole pair excited states form a continuum with $0 \leq \varepsilon_{e-h} < \infty$. If the atomic shake-up peaks are then allowed to merge together the photoelectron spectrum shown in Fig.11c would result. Here shake-up plasmon satellites are also shown[73,76]. Doniach and Šunjić[51] (DS) noticed that the actual line shape of a photoemitted electron should include the (assumed symmetric) broadening due to the finite lifetime of the core hole, not just the shake-up or Mahan structure shown in Fig.11c. Working in the sudden limit, they presented a theory in which asymmetric line shapes, skewed towards the high binding energy (low kinetic energy) side, were predicted. Such line shapes have since been observed[54]. Gadzuk and Šunjić have extended the theory in a way which extrapolates continuously from the adiabatic to sudden limits[55].

From this discussion our attention is directed towards the following many-body effects which are to be understood in XPS : (1) adiabatic relaxation energies, (2) sudden approximation shake-up effects such as line asymmetries and satellites, and (3) the connections between these effects.

A. Relaxation Energies

Consider first the intra-atomic relaxation energy labeled ε_a in Fig.11a. When one electron is removed from the core level ε_{HF}, the nucleus is less shielded and consequently the remaining atomic electrons readjust to this additional attraction, thus lowering their energy[72]. This purely single atom effect is usually assumed either to be identical or simply related in the gas and solid phase. Brute force calculations of total energy differences between N and N-1 electron systems (atoms or ions) in principle yield the exact value of ε_a. In practice this small number ($\sim$ 1 Rydberg) is the difference between two very large numbers and thus accurate calculations are difficult or more time consuming than practical. Another approach is the so-called transition state calculation[84] in which essentially the initial and final state potentials (determining single electron eigenstates) are replaced by some average potential of the N and N-1 electron system. There is then no need to calculate a relaxation energy as this is already accounted for in the particular choice of effective potential. The disadvantage is that a different effective potential is required for each possible transition. In any event, the principles involved in ε_a are well understood and the only problem is finding efficient and accurate computational tech-

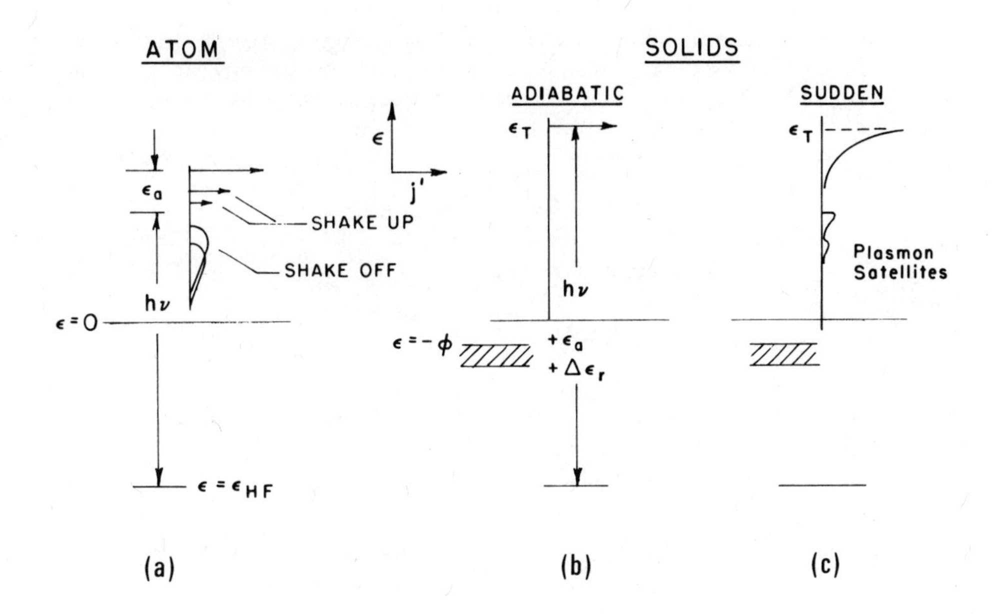

Fig. 11 : Energy level diagrams and photoemission energy distributions for core level XPS.

(a) Atomic photoionization showing the leading peak, shake-up peaks, and shake-off satellites. For adiabatic excitation, the intensity in all but the leading peak is zero.

(b) Solid state photoemission in a hypothetical adiabatic limit. Note that the leading peak has an additional relaxation energy shift $\Delta\varepsilon_r$ compared to the gas phase.

(c) Solid state photoemission in the sudden limit. The Mahan "shake-up" structure or the leading peak and intrinsic plasmon satellites are shown. (Ref. 55, Gadzuk and Šunjić)

niques.

In the case of core states in metals, the resulting ion is embedded in a polarizable medium, the conduction band electron gas and the interaction between the hole and the induced polarization or screening charge is called the extra-atomic relaxation energy $\equiv \Delta\varepsilon_r$.

Within a linear response theory the extra-atomic relaxation energy is related to the classical polarization energy as follows (26,27,70,76). The hole charge density (for a core state wave function $\psi(\underset{\sim}{r})$

$$\rho_h(\underset{\sim}{r}) = |\psi(\underset{\sim}{r})|^2 = \int \frac{d^3q}{(2\pi)^3}\, \rho_h(\underset{\sim}{q})\, e^{i\underset{\sim}{q}\cdot\underset{\sim}{r}} \tag{33}$$

induces a screening charge density

$$\rho_{in}(\underset{\sim}{r}) = \int \frac{d^3q}{(2\pi)^3}\, \rho_{in}(\underset{\sim}{q})\, e^{i\underset{\sim}{q}\cdot\underset{\sim}{r}} \tag{34}$$

with

$$\rho_{in}(\underset{\sim}{q}) = \left[\frac{1 - \varepsilon(q,0)}{\varepsilon(q,0)}\right] \rho_h(\underset{\sim}{q}) \tag{35}$$

and $\varepsilon(q,0)$ is the static wavenumber dependent dielectric function of the initially uniform electron gas. The classical polarization energy $\equiv\Delta\varepsilon_r$ due to the Coulomb interaction between the hole and the induced charge density is

$$\Delta\varepsilon_r = \frac{e^2}{2} \int d^3r\, d^3r' \, \frac{\rho_{in}(\underset{\sim}{r})\, \rho_h(\underset{\sim}{r}')}{|\underset{\sim}{r} - \underset{\sim}{r}'|}$$

which when combined with Eqs.(33)-(35), can be written as

$$\Delta\varepsilon_r = \frac{1}{2} \int \frac{d^3q}{(2\pi)^3}\, v(q) \left[\frac{1-\varepsilon(q,0)}{\varepsilon(q,0)}\right] |\rho_h(\underset{\sim}{q})|^2 \tag{36}$$

where $v(q) = 4\pi e^2/q^2$ is the Fourier transform of the Coulomb potential. For a spherically symmetric hole charge distribution, Eq.(36) becomes

$$\Delta\varepsilon_{rs} = \frac{e^2}{\pi} \int_0^\infty dq |\rho_h(q)|^2 \left[\frac{1-\varepsilon(q,0)}{\varepsilon(q,0)}\right] \tag{37}$$

As a preliminary exercise, consider a delta function hole charge density $\rho(\underset{\sim}{r}) = \delta^{(3)}(\underset{\sim}{r})$ (thus $\rho(\underset{\sim}{q}) = 1$ for all q) and a Fermi-Thomas dielectric function $\varepsilon(\underset{\sim}{q},0) = 1 + \kappa^2/q^2$ with the inverse screening length $\kappa = 2.95/r_s^{1/2}$ Å^{-1}, and r_s the electron gas density parameter. For this model it is straightforward to show that Eq.(37) leads to a polarization energy

$$\Delta\varepsilon_r = \frac{e^2\kappa}{2}$$

Next consider a hole in a 1s state in which the hydrogenic function is approximated by a single optimum Gaussian wave function(85)

$$\phi_{1s}(\underset{\sim}{r}) = \left(\frac{2\alpha}{\pi}\right)^{3/4} e^{-\alpha r^2}$$

with $\alpha = .96$ Å^{-2}. The hole charge density Fourier components associated with this state are

$$\rho_{1s}(q) = \int d^3r\, e^{i\underset{\sim}{q}\cdot\underset{\sim}{r}} |\phi_{1s}(\underset{\sim}{r})|^2 = e^{-q^2/8\alpha} \tag{38}$$

Consequently, in the Fermi-Thomas approximation, the relaxation energy obtained from Eqs.(37) and (38) is

$$\Delta\varepsilon_r(\text{atom}) = \frac{e^2\kappa^2}{\pi} \int_0^\infty \frac{dq\, e^{-q^2/4\alpha}}{q^2 + \kappa^2} = \frac{e^2\kappa}{2}\, e^{\kappa^2/4\alpha}(1-\text{erf}(\kappa/2\alpha^{1/2})) \tag{39}$$

For the Gaussian orbital, α is related to the mean radius as $\langle r\rangle = \sqrt{2/\pi\alpha}$. From Eq.(39), the ratio $2\Delta\varepsilon_r(\text{atom})/e^2\kappa$ as a function of $\kappa\langle r\rangle$ has been calculated treating r_s parametrically and the results are given in Table 1. As expected, the relaxation energy is reduced significantly when the spatial extent of the hole charge density is comparable to or greater than $\sim 1/\kappa$. Similar results are obtained for RPA dielectric functions and more complicated hole wave functions(86). In fact, as the spatial extent of the hole becomes unlimited (as with Bloch functions), $\Delta\varepsilon_r$ goes to zero and thus Koopmans' theorem is applicable.

Although the results just discussed apply to the homogeneous electron gas with no surfaces, it has been demonstrated in several papers(70,86-89) that the bulk expression for $\Delta\varepsilon_r$, Eq.(36), goes smoothly to an image potential shift,

$$\lim_{s \gg \langle r\rangle} \Delta\varepsilon_r(s) = e^2/4s$$

TABLE 1 : Ratio of the single Gaussian hydrogen 1s screening energy to that of a point hole as a function of hole size in units of inverse screening lengths.

$\kappa\langle r\rangle$	$2\varepsilon_r(\text{atom})/e^2\kappa$
.0	1.000
.2	.874
.4	.770
.6	.679
.8	.617
1.0	.555
1.2	.510
1.4	.453
1.6	.432
1.8	.392
2.0	.370
2.4	.326
2.8	.282

as s, the atom-metal separation becomes much larger than the radius of the hole which is on an atom outside the surface. For typical separations involving adsorbates ($s \sim 1$-2 Å), $\Delta\varepsilon_r \sim 1$-$3$ eV. Such shifts are generally observed experimentally, as for instance in the Xe on W spectrum shown in Fig.8.

B. Shake-up effects

Before an x-ray is absorbed, the many-body wave function of the solid $\equiv \Phi_i$ is an eigenfunction of some Hamiltonian H_o. Upon absorption, an additional localized hole potential V_h is rapidly switched on. This time dependent potential can excite electron-hole pairs or plasmons and this results in some photoejected electrons emerging at lower energies than would occur in adiabatic switching (55,59,73,76, 77,84,89). Within the sudden approximation, the wave function Φ_i is continuous in time after hole creation but is then a nonstationary state of the new Hamiltonian $H_o + V_h$ (which changed discontinuously as a function of time) and should be expanded in terms of ψ_n, the complete set (including excited states) of eigenstates of $H_o + V_h$.

The implications of this fact in an XPS experiment are made more clear by considering the Golden rule photocurrent in which the initial and final many electron states are written in the product form

$$\Phi_i = |c\rangle|N-1;\ c;\ 0\rangle \tag{40a,b}$$

and

$$\psi_n = |\underset{\sim}{k}\rangle|N-1;\ h;\ n\rangle$$

where $|c\rangle$ and $|\underset{\sim}{k}\rangle$ are single electron core and continuum states of the electron which is photoexcited, $|N-1;\ c;\ 0\rangle$ is the many-body ground state of the remaining N-1 electrons when the core state is occupied (i.e., when $H=H_o$) and $|N-1;\ h;\ n\rangle$ is the n'th excited state of the N-1 electrons when the core state is vacant (i.e., when $H = H_o + V_h$). The excitations can include pairs, plasmons, and localized atomic-like electron-hole pairs on adsorbed species. The expression for the photoemission spectrum is then

$$\frac{dj}{d\varepsilon} \sim \sum_{\varepsilon_k,\varepsilon_n} \delta(\varepsilon-\varepsilon_k)\delta(\varepsilon_k+\varepsilon_n-h\nu-\varepsilon_o-\varepsilon_r)|\langle\underset{\sim}{k}|\underset{\sim}{p}_{op}\cdot\underset{\sim}{A}|c\rangle|^2$$

$$\times\ |\langle N-1;\ h;\ n|N-1;\ c;\ 0\rangle|^2 \tag{41}$$

where the total energy of the final state, the sum of ε_k the photoelectron energy and ε_n the excitation or shake-up energy equals the initial core state energy ε_o plus the photon and relaxation energy. Within the typical ranges of ε_n, ($\lesssim$ 1 eV for pairs, $\sim$ 10 eV for plasmons), the single electron optical matrix element is usually slowly varying and is taken to be constant. Then the all important factor in determining the relative strengths of the shake-up satellites is the overlap between the initial and final N-1 electron states given in Eq.(40a) and (40b). Satellites or low energy tails generated in this way are called intrinsic[87-91] because they are produced in the actual photoabsorption process and not in the less interesting but ubiquitous electron transport process out of the solid[92]. Within the constant matrix element approximation, Eq.(41) is (dropping the N-1 label)

$$\frac{dj}{d\varepsilon} \sim \sum_n \delta(\varepsilon+\varepsilon_n-h\nu-\varepsilon_o-\varepsilon_r)|\langle h;\, n|c;\, 0\rangle|^2 \tag{42}$$

To allow for the finite lifetime of the hole $\equiv \tau_\ell$, the delta function can be replaced by a Lorentzian of width $\gamma = \hbar/\tau_\ell$. Then defining $|\langle h;\, n|c;\, 0\rangle|^2 \equiv D(\varepsilon_n)$, Eq.(42) is simply the convolution integral

$$\frac{dj}{d\varepsilon} \sim \frac{\gamma}{\pi}\int_0^\infty d\varepsilon_n \frac{D(\varepsilon_n)}{(\varepsilon_n+\varepsilon)^2+\gamma^2} \tag{43}$$

where for convenience, the zero of the energy scale is taken at the adiabatic threshold energy $\varepsilon_t \equiv h\nu + \varepsilon_o + \varepsilon_r$.

As suggested in the introduction to this section and also in Fig.11, the electron-hole pair spectrum results in an overlap function $D(\varepsilon_n) \sim 1/\varepsilon_n^{1-\alpha}$. In fact Müller-Hartman et al. have shown[59] that for a localized hole potential that is exponentially switched on at a rate η, the resulting pair spectrum is $D(\varepsilon_n) \sim e^{-\varepsilon_n/\eta}/\varepsilon_n^{1-\alpha}$. Gadzuk and Šunjić[55] have recently pointed out that the XPS line shape, from Eq.(43) and the just cited form for $D(\varepsilon_n)$, is

$$\frac{dj}{d\varepsilon} \sim \frac{e^{\varepsilon/\eta}}{(\varepsilon^2+\gamma^2)^{\frac{1-\alpha}{2}}} \quad \cos[(\alpha-1)\tan^{-1}(\varepsilon/\gamma) - \frac{\pi\alpha}{2} - \gamma/\eta] - \delta I(\varepsilon) \tag{44}$$

where δI is a small correction equal to zero for $\eta = 0$ (the adiabatic limit) and $\eta = \infty$(the sudden limit). This result in the sudden limit was first obtained by Doniach and Šunjić[51] in a much different way. Generally in experimentally realized situations, $\eta \to \infty$ is a good approximation. The experimental implications of Eq.(44) are twofold. In the adiabatic limit (no shake-up), with $\gamma = 0$, the line shape is a delta function at $\varepsilon = 0$. In the sudden limit, the observed line shape is an asymmetric Lorentzian skewed to lower kinetic energies ($\varepsilon < 0$) which is due to the impossibility of resolving the continuous pair shake-up structure from the $\varepsilon_n \rightarrow 0$ structure. Such asymmetric line shapes have been observed[54,93] although certain procedures in "background subtraction" make these observations subject to some arbitrariness. A second point is that the peak maximum of Eq.(44) is displaced downwards from $\varepsilon = 0$ to $\varepsilon_{max} = -\gamma \cot(\pi/(2-\alpha))$. Since $0 < \alpha \leqslant 1$, this shows that $\gamma < \varepsilon_{max} < 0$ and thus the "apparent" binding energy (defined in terms of the position of the peak maximum) depends on the parameters of the shake-up and relaxation process.

Recently Brenig[94] has discussed a similar process involving paramagnon satellites in photoemission of chemisorbed hydrogen on high susceptibility substrates. Breaking a covalent bond (by photoemission) leaves a localized unpaired spin which can then excite paramagnons, again a shake-up-like process. The resulting photoemission line shape is a Lorentzian convolved with an overlap function similar to the $D(\varepsilon_n)$ used in Eq.(44), although the physical significance of the parameter α is quite different. This may explain why H adsorbed on Pd (high susceptibility) shows a very broad peak whereas H on W (low susceptibility) shows a much narrower peak. Gumhalter and Newns[95] have also considered similar satellite problems with respect to photoemission from adsorbed atoms.

Plasmon satellites are also possible[26,27,73,76,87-92,96]. From Eq.(42), if $\varepsilon_n = n\hbar\omega_o$, with n an integer and ω_o the plasmon (or surface plasmon) frequency, then the photoelectron spectrum is a Poisson distribution of peaks at integral multiples of the relevant plasmon energies below the adiabatic threshold $\equiv \varepsilon = 0$. In the extensive list of references just cited, it has been demonstrated that Eq.(42) can be written

$$\frac{dj}{d\varepsilon} = \sum_n \delta(\varepsilon - n\hbar\omega_o)\, e^{-\beta}\, \frac{\beta^n}{n!}$$

with $\beta = \sum_{\underset{\sim}{k}} g_{\underset{\sim}{k}}^2/\hbar^2\omega_o^2(\underset{\sim}{k})$, $g_{\underset{\sim}{k}}$ the electron-plasmon coupling constant and $\omega_o(\underset{\sim}{k})$ the plasmon dispersion relation. Harris has recently shown[89] that for a localized hole created in an adsorbed atom a distance s from the surface,

$$\beta_s \approx (\text{Image energy})/(\text{Surface plasmon energy}) = e^2/4s\hbar\omega_s$$

for dispersionless surface plasmons. Bulk plasmon satellites due to intrinsic processes in adsorbed atoms are expected to be small. As a test of the surface plasmon shake-up theory, XPS experiments of chemisorbed atoms on substrates which have well defined surface plasmons would be useful.

Fuggle, Madey, Steinkilburg and Menzel[97] have obtained XPS from O and CO adsorbed on the basal (001) face of Ru and have observed satellite peaks below the adiabatic peaks of the 1s O level characteristic of both the CO and the Ru. Quantitative interpretation of these results must still be done. There is much work remaining before the full significance of shake-up effects in XPS on chemisorption systems is realized. However, as pointed out by Fuggle et al., (97) there is a good possibility that measurements of the relative strength of such satellites can be used as an additional piece of data for analytical purposes.

C. Relation between relaxation and shake-up

Although the theory of the extra-atomic relaxation energy given in Sec.A and of the shake-up effects given in Sec.B appear quite different, there is a rather profound connection between the two. B. Lundqvist[73,76] first recognized this although Manne and Aberg[94] are usually cited in the chemically oriented literature. Essentially the relaxation energy is due to the static response of the electron gas to the core hole potential whereas the shake-up satellites are due to the dynamic response. However the static and dynamic responses are related by Kramers-Kronig transformations,

$$R(\omega=0) = -\int \frac{d\omega'}{\omega'} \text{ Im } R(\omega')$$

where $R(\omega)$ is the electron gas density-density response function. Lundqvist proved that the following sum rule is satisfied (in the sudden limit)

$$\int \frac{d\omega}{2\pi} \omega D(\omega) = \varepsilon_{frozen}$$

where $D(\omega)$ is the hole spectral function (or shake-up function discussed in B) and ε_{frozen} is the energy that the ejected electron would emerge with if the N-1 electrons were frozen into the configuration they had when the core state was occupied. In other works, the many-body response redistributes the spectral weight of the hole, raising the maximum kinetic energy of the ejected electron by the "relaxation energy" but by also putting some spectral weight in low energy satellites in a way in which the average energy of the spectral function remains unchanged. Doniach has provided another form of the sum rule

$$\varepsilon_{rel} - \varepsilon_{frozen} = \Delta\varepsilon_r = -\int \frac{d\omega}{2\pi} \omega B(\omega) \qquad (45)$$

which defines the function $B(\omega)$ called the "satellite generator"[49]

As noted by Lundqvist and Wendin(27) and by Langreth(76), Eq.(45) can be reduced to

$$\Delta\varepsilon_r \sim \int \frac{d\omega}{\omega} \text{ Im } R(\omega)$$

from which it is seen that $\Delta\varepsilon_r$ the extra-atomic relaxation or polarization energy of Sec.A is simply related to the Kramers-Kronig transform of the shake-up satellites. Consult the excellent papers cited in this section for a more complete discussion of these points. Again because of background subtraction problems and instrumental transmission and response functions, it has not yet been possible to check and use the relations which exist between the static and dynamic responses as either an analytic tool or as an internal consistency check. But there is hope that in the future such added sophistications will be put to advantage in this fast developing field of electron spectroscopy.

REFERENCES

(1) For an introductory survey, see "Physics Today", April, 1975

(2) For surveys of the theory of the electronic structure of clean surfaces see : N.D. Lang, Solid State Physics 28, 225 (1973); J.A. Appelbaum, in "Surface Physics of Crystalline Materials", ed. by J.M. Blakely (Academic, N.Y., 1975)

(3) Surveys of chemisorption theory have been given by : T.B. Grimley, J. Vac. Sci. Technol. 8, 31 (1971); J.R. Schrieffer, J. Vac. Sci. Technol. 9, 561 (1972); R. Gomer, Critical Reviews in Sol. State Sci. 4, 247 (1974); J.W. Gadzuk, in "Surface Physics of Crystalline Materials", ed.by J.M. Blakely (Academic Press, N.Y., 1975); various aspects of the surface molecule concept in chemisorption have been discussed by : D.M. Newns, Phys. Rev. 178, 1123 (1969); T.B. Grimley and M. Torrini, J. Phys. C 6, 868 (1973): B.J. Thorpe, Surface Sci. 33, 306 (1972); D.R. Penn, Surface Sci. 39, 333 (1973); J.W. Gadzuk, Surface Sci. 43, 44 (1974); M.J. Kelly, Surface Sci. 43, 587 (1974); F. Cyrot-Lackman, M.C. Desjonqueres, and J.P. Gaspard, J. Phys. C 7, 925 (1974); T.E. Einstein, Surface Sci. 45, 713 (1974)

(4) Reviews of field emission spectroscopy : J.W. Gadzuk and E.W. Plummer, Rev. Mod. Phys. 45, 487 (1973); L.W. Swanson and A.E. Bell, Adv. Electronics and Electron Phys. 32, 194 (1973)

(5) Reviews or surveys concentrating mainly on UPS include : N.V. Smith, Crit. Rev. Solid State Sci. 2, 45 (1971) ; D.E. Eastman, in "Techniques of Metals Research VI, ed. by E. Passaglia (Interscience, New York, 1972) ; W.E. Spicer, Comments in Solid State Phys. 5, 105 (1973); D.T. Pierce, Acta Elec. 18, 69 (1975); B. Feuerbacher, Surface Sci. 47, 115 (1975); Surface Sci. 48, 99 (1975) ; E.W. Plummer in "Topics in Appl.Phys." Vol.4, "Interactions on Metal Surfaces", ed. by R. Goner (Springer-Verlag, N.Y. 1975) ; A.E. Bradshaw (to be published) ; M.L. Glasser and A.

Bachi (to be published).
(6) Conference proceedings, some published as special issues of journals, which contain up to date articles on both UPS and XPS include : "Electron Spectroscopy", ed. by D.A. Shirley (North Holland, Amsterdam, 1972); J. Elec. Spectroscopy 5, 1-1136 (1974); "Vacuum Ultraviolet Radiation Physics", ed. by E.E. Koch, R. Haensel, and C. Kunz (Pergamon, Braunschweig, 1974); Proc. of 2nd Intl. Conf. on Solid Surfaces, Japan J. Appl. Phys. Suppl. 2, Part 2 (1974)
(7) R.D. Young, Phys. Rev. 113, 110 (1959)
(8) B. Politzer and P.H. Cutler, Phys. Rev. Lett. 28, 1330 (1972)
(9) D.R. Penn and E.W. Plummer, Phys. Rev. B 9, 1216 (1974)
(10) N. Nicolaou and A. Modinos, Phys. Rev. B 11, 3687 (1975)
(11) C.B. Duke and M.E. Alferieff, J. Chem. Phys. 46, 923 (1967); J.W. Gadzuk, Phys. Rev. B 1, 2110 (1970); D. Penn, R. Gomer, and M.H. Cohen, Phys. Rev. B 5, 768 (1972)
(12) C.J. Powell, Surface Sci. 44, 29 (1974)
(13) P.O. Nilsson, in "Elementary Excitations in Solids, Molecules, and Atoms", ed. by J.T. Devreese, A.B. Kunz, and T.C. Collins (Plenum, New York, 1974)
(14) G. Allan, M. Lannov, Surf. Sci. 40, 375 (1973) ; D.J. Cheng, R.F. Wallis, and L. Dobrzynski, Surf. Sci. 40, 400 (1974)
(15) T. Gustafsson, E.W. Plummer, D.E. Eastman, and J.L..Freeouf Solid State Commun. 17, 391 (1975).
(16) C.S. Fadley, J. Electron Spectr. 5, 725 (1974)
(17) T.V. Vorburger, D. Penn, and E.W. Plummer, Surface Sci. 48, 417 (1975)
(18) G. Obermair, Z. Physik 217, 91 (1968)
(19) R.M. More, Phys. Rev. B 9, 392 (1974)
(20) J.W. Gadzuk and E.W. Plummer, Phys. Rev. Lett. 26, 92 (1971); J.W. Gadzuk and A.A. Lucas, Phys.Rev. B 7, 4770 (1973)
(21) J.W. Gadzuk, Japan J. Appl. Phys. Suppl. 2, Part 2, 851 (1974)
(22) J.W. Gadzuk, E.W. Plummer, H.E. Clark, and R.D. Young, in "Electronic Density of States", ed. by L.H. Bennett (NBS Spec. Publ. No. 323, Washington, 1971)
(23) P.L. Young and R. Gomer, J. Chem. Phys. 61, 4955 (1974); A. Bagchi and P.L. Young, Phys. Rev. B 9, 1194 (1974)
(24) A. J. Bennett and L.M. Falicov, Phys. Rev. 151, 512 (1966) ; J.W. Gadzuk, J.K. Hartman, and T.N. Rhodin, Phys. Rev. B 4, 241 (1971); A.J. Bennett, Surf. Sci. 50, 77 (1975)
(25) H.E. Clark and R.D. Young, Surf. Sci. 12, 385 (1968)
(26) L.Hedin, Arkiv Fysik 30, 231 (1965); L. Hedin and A. Johansson, J. Phys. B 2, 1336 (1969); L. Hedin and S. Lundqvist, Solid State Phys. 23, 1 (1969); L. Hedin, in "Electrons in Crystalline Solids", International Atomic Energy Agency, Vienna (1973) L. Hedin, in "X-ray Spectroscopy", ed. by L.V. Azaroff (McGraw Hill, N.Y., 1974)
(27) S. Lundqvist, in "Elementary Excitations in Solids, Molecules, and Atoms, Part A", ed. by J.T. Devreese, A.B. Kunz, and T.C. Collins (Plenum Press, London and N.Y., 1974) ;

S. Lundqvist and G. Wendin, J. Elec. Spect. 5, 513 (1974)
(28) G.D. Mahan, Phys. Rev. B 2, 4334 (1970)
(29) W.L. Schaich and N.W. Ashcroft, Phys. Rev. B 3, 2452 (1971)
(30) S. Doniach, Phys. Rev. B 2; 3898 (1970)
(31) C. Caroli, D. Lederer-Rozenblatt, B. Roulet, and D. Saint-James, Phys. Rev. B 8, 4552 (1973)
(32) J. Endriz, Phys. Rev. B 7, 3464 (1973)
(33) D.R. Penn, Phys. Rev. Letters 28, 1041 (1972)
(34) J.W. Gadzuk, Solid State Commun. 15, 1011 (1974); Phys. Rev. B 10, 5030 (1974)
(35) A. Liebsch, Phys. Rev. Lett. 32, 1203 (1974)
(36) P.J. Feibelman and D.E. Eastman, Phys. Rev. B 10, 4932 (1974)
(37) P.J. Feibelman, Surf. Sci. 46, 558 (1974)
(38) T.E. Einstein, Surf. Sci. 45, 713 (1974)
(39) S. Doniach and E.H. Sondheimer, "Green's Functions for Solid State Physicists", W.A. Benjamin (Reading, Mass., 1974)
(40) P. Roman, "Advanced Quantum Theory" (Addison-Wesley, Reading, Mass., 1965) p. 293
(41) M.B. Webb and M.G. Lagally, Solid State Phys. 28, 302 (1973) C.B. Duke, Adv. Chem. Phys. 27, 1 (1974); J.B. Pendry, "Low Energy Electron Diffraction", Academic (London, 1974)
(42) W. Schaich, Phys. Rev. B 8, 4028 (1973); C.A. Ashley and S. Doniach, Phys. Rev. B 11, 1279 (1975); E.A. Stern, D.E. Sayers, and F.W. Lytle, Phys. Rev. B 11, 4836 (1975)
(43) G. Breit and H.A. Bethe, Phys. Rev. 93, 888 (1954)
(44) H.A. Bethe, L. Maximon, and F. Low, Phys. Rev. 91, 417 (1953)
(45) C.N. Berglund and W.E. Spicer, Phys. Rev. 136, A1030 (1964); 136, A1044 (1964)
(46) N.V. Smith and L.F. Mattheiss, Phys. Rev. B 9, 1341 (1974)
(47) N.E. Christensen and B. Feuerbacher, Phys. Rev. B 10, 2349 (1974); 10, 2373 (1974)
(48) E.A. Stern, Phys. Rev. 162, 565 (1967)
(49) J.C. Slater, in "Computational Methods in Band Theory", ed. by P.M. Marcus, J.F. Janak, and A.R. Williams (Plenum Press, New York; 1971); S. Doniach, ibid
(50) R.P. Feynmann, R.B. Leighton, and M. Sands, "The Feynman Lectures on Physics, Quantum Mechanics, Vol. III" (Addison Wesley Reading, Mass.; 1965)
(51) S. Doniach and M. Sunjic, J. Phys. C 3, 285 (1970)
(52) G.D. Mahan, Phys. Rev. 163, 612 (1967); Solid State Phys. 29, 75 (1974)
(53) J.D. Dow, J.E. Robinson, J.H. Slowik, and B.F. Sonntag, Phys. Rev. B 10, 432 (1974)
(54) L.Ley, F.R. McFeely, S.P. Kowakczyk, J.G. Jenkin, and D.A. Shirley, Phys. Rev. B 11, 600 (1975)
(55) J.W. Gadzuk and M. Sunjic, Phys. Rev. B 12, 524(1975)
(56) W.E. Spicer, Phys. Rev. 154, 385 (1967)
(57) M.L. Goldberger and K.M. Watson, "Collision Theory" (Wiley, New York, 1964), p. 540

(58) J.J. Hopfield, Comm. on Solid State Phys. 1, 198 (1969)
(59) E. Müller-Hartman, T.V. Ramakrishnan, and G. Toulouse, Phys. Rev. B 3, 1102 (1971)
(60) M.M. Traum, N.V. Smith, and F.J. DiSalvo, Phys. Rev. Lett. 32, 1241 (1974); J.E. Rowe, M.M. Traum, and N.V. Smith, Phys. Rev. Lett. 33, 1333 (1974); W.F. Egelhoff and D.L. Perry, Phys. Rev. Lett. 34, 93 (1975); J. Anderson and G.J. Lapeyre, to be published
(61) J. Cooper and R.N. Zare, J. Chem. Phys. 48, 942 (1968); and "Lectures in Theoretical Physics, Atomic Collisions Processes", Vol.XI C(Gordon and Breach, N.Y., 1969); D.J. Kennedy and S.T. Manson, Phys. Rev. A 5, 227 (1972); T.A. Carlson, G.E. McGuire, A.E. Jonas, K.L. Cheng, C.P. Anderson, C.C. Lu, and B.P. Pullen, in "Electron Spectroscopy", ed. by D.A. Shirley (North Holland, Amsterdam, 1972); J.W. Rabalais and T.P. Debies, J. Elect. Spect. 5, 847 (1974)
(62) I.G. Kaplan and A.P. Markin, Dokl. Akad. Nauk SSR 184, 66 (1969) (Engl. Transl. : Soviet Phys. Dokl. 14, 36 (1969))
(63) C.S. Fadley and D.A. Shirley, in "Electronic Density of States", NBS Special Publication 323, Washington , 1971. S. Hüffner,G.K. Wertheim, and J.H. Wernick, Phys. Rev. B 8, 4511 (1973)
(64) J.F. Janak, A.R. Williams, and V.L. Moruzzi, Phys. Rev. B 11, 1522 (1975)
(65) D.E. Eastman and W. Grobman (unpublished)
(66) D.C. Langreth, Phys. Rev. B 11, 2155 (1975)
(67) D.E. Eastman and J.K. Cashion, Phys. Rev. Lett. 27, 1520 (1971)
(68) I.P. Batra and P.S. Bagus, Solid State Comm. 16, 1097 (1975)
(69) B.J. Waclawski and J.F. Herbst (submitted)
(70) J.W. Gadzuk, J. Vac. Sci. Technol. 12, 289 (1975)
(71) J.W. Gadzuk, Surf. Sci. (in press)
(72) P. Bagus, Phys. Rev. 139, A619 (1965)
(73) B. Lundqvist, Phys. Konders. Materie 9, 236 (1969)
(74) R. Manne and T. Åberg, Chem. Phys. Lett. 7, 282 (1970)
(75) D.A. Shirley, Chem. Phys. Lett. 16, 220 (1972); D.W. Davis and D.A. Shirley, J. Elect. Spect. 3, 137 (1974); P.H. Citrin and D.R. Hamann, Chem. Phys. Lett. 22, 301 (1973); Phys. Rev. B 10, 4948 (1974)
(77) H.W. Meldner and J.D. Perez, Phys. Rev. A 4, 1388 (1971)
(78) P.E. Best, in "X-ray Spectroscopy", ed. by L.V. Azaroff (McGraw-Hill, N.Y., 1974)
(79) J.T. Yue and S. Doniach, Phys. Rev. B 8, 4578 (1973)
(80) U. Fano and J.W. Cooper, Rev. Mod. Phys. 40, 441 (1968); J.W. Gadzuk, Phys. Rev. B (submitted)
(81) C.S. Fadley, Chem. Phys. Lett. 25, 225 (1974)
(82) P. Nozieres and C.T. de Dominicis, Phys. Rev. 178, 1097 (1969)
(83) J.D. Dow and D.R. Franceschetti, Phys. Rev. Lett. 34, 1320 (1975)
(84) J.C. Slater and K.H. Johnson, Phys. Rev. B 5, 844 (1972)
(85) R.F. Stewart, J. Chem. Phys. 52, 431 (1970)
(86) J.W. Gadzuk, Chem. Phys. Lett. (submitted)

(87) A.C. Hewson and D.M. Newns, Japan J. Appl. Phys. Suppl. 2; part 2, 121 (1974)
(88) G.E. Laramore and W.J. Camp, Phys. Rev. B9, 3270 (1974)
(89) J. Harris, Solid State Comm. 16, 671 (1975)
(90) D.C. Langreth, Phys. Rev. Lett. 26, 1229 (1971)
(91) A.M. Bradshaw, S.L. Cederbaum, W. Domcke, and U. Krause, J. Phys. C 7, 4503 (1974)
(92) W.J. Pardee, G.D. Mahan, D.E. Eastman, R.A. Pollak, L. Ley, F.R. McFeely, S.P. Kowalczyk, and D.A. Shirley, Phys. Rev. B 11, 3614 (1975)
(93) S. Hüffner, G.K. Wertheim, D.N.E. Buchanan, and K.W. West, Phys. Lett. 46A, 420 (1974); N.J. Shevchik, Phys. Rev. Lett. 33, 1336 (1974)
(94) W. Brenig, Z. Physik B 20, 55 (1975)
(95) B. Gumhalter and D.M. Newns, Phys. Lett. 53A, 137 (1975)
(96) A.A. Lucas and M. Šunjić, Prog. in Surf. Sci. 2, 75 (1972); M. Šunjić and D. Sokcevic, J. Elec. Spect. 5, 963 (1974)
(97) J.C. Fuggle, T.E. Madey, M. Steinkilberg, and D. Menzel, Chem. Phys. Lett. 33, 233 (1975)

ELECTRON SPECTROSCOPY FOR THE INVESTIGATION OF

METALLIC SURFACES AND ADSORBED SPECIES

C.R. BRUNDLE

IBM Research Laboratory

San Jose, California 95193

I. INTRODUCTION

These lecture notes cover the experimental use of Ultraviolet Photoemission (UPS) and X-ray Photoemission(XPS) for solid-state and surface adsorption studies and applications. As such they are complementary to the largely theoretical notes of J. W. Gadzuk (JWG)[1] which provide a unifying theoretical framework of photoemission. Examples of some of the phenomena discussed by J. W. Gadzuk are given (in particular the detection and use of adsorbate resonances, the phenomena of relaxation and "shake-up" structure, the sum rules connecting the two, and, briefly, some aspects of angular photoemission). Practical consideration is given to their ability to provide information on bonding and analysis at surfaces. Prior to the more esoteric considerations of electronic structure effects during adsorbate/substrate interactions, the more mundane aspects of (a) quantitative elemental analysis at surfaces, and (b) "fingerprinting" methods of molecular and chemical analysis at surfaces are covered.

In addition to UPS and XPS, some aspects of Auger Electron Spectrocopy, AES, Ion Neutralization Spectrocopy, INS, and Energy Loss Spectroscopy, ELS, are discussed. The coverage of these techniques is very cursory.

Detailed descriptions of the physical principles involved in the electron spectroscopic techniques have been given many times already and are not repeated here. Reviews by Fadley (XPS)[2]; Bradshaw (UPS)[3]; Tracy (Auger)[4]; and Brundle (general),[5-7] form a useful basis from which to work. In addition, several textbooks[8-11] on general aspects of UPS and XPS are available.

The aspects of the techniques which will be emphasized here are their relative abilities to:

(a) provide an elemental analysis
(b) provide a molecular analysis
(c) provide electronic structure and hence bonding information

For a) one ideally requires measurement of a phenomenon which is entirely an atomic property. In fact, core-level Binding Energies, (B.E), as measured by XPS or AES are sufficiently characteristic of an atom to provide element identification because the chemical shift effects on the core-levels occurring when the individual atom is bonded to others are small. Photoionization cross-sections for core-levels are also an atomic property which means that XPS and Auger Core Level transition intensities may be used for a quantitative element analysis, provided one measures the appropriate total experimental photoelectron intensity. The small chemical shifts observed for core-levels indicate that extra-atomic effects (both initial and final state) are of some importance, however (fortunately for electronic structure and bonding studies), and it is these same effects which cause a redistribution of the spectral intensity between the principal core-line of an element (in practice, the one of lowest B.E. - apparently not a necessary restriction in theory) and the so-called shake-up and shake-off satellite structure associated with that core-line. All this "lost" intensity in the satellite structure should be included in the intensity measurements for quantitative analysis, though in practice it is difficult to estimate.

Since ELS, INS, and UPS do not detect the deep core-levels, their use for elemental analysis is very limited (to those cases where levels in the accessable energy region are, in fact, core-like).

Molecular identification can be obtained from two sources - either directly from examining the valence level B.E's (characteristic of the ensemble, rather than the atom) or the chemical shifts in the core-levels occurring as a response to changes in the valence electron distribution (both initial and final states).

All the techniques are therefore capable in principle, of providing a molecular identification, and XPS and AES can attempt to do it from both valence and core measurements. For analysis purposes it is necessary either to know the M.O. pattern and chemical shifts of a given molecule to be able to identify it, or to have confidence in calculations of these values. This is no problem in the gas phase and homogeneous bulk solid situations. It is not so straight forward for surfaces where confirmation that a particular molecular species can exist may be a prime objective. The by now famous case of CO on Ni is an example of how the correct molecular identification (undisociated CO present) was originally arrived at from a wrong M.O. pattern identification in the UPS (which stood for 4 years).[12]

All the techniques are capable of providing electronic structure and bonding information, for the same reason as for b) above. Now that electron spectroscopy and electron spectroscipists are reaching maturity, it is realized that the electronic structure effects observed are as much a property of the final hole-state formed by ionization as of the initial state. This fact has been realized by people in other areas for some years. Whether this is to be considered a disaster or another exciting facet of electron spectroscopy depends on one's aims (or if one is interested in reactivity, whether the initial or final state best approximates the situation of a transition-state complex). Certainly the separation of initial and final state effects (relaxation phenomena) is proving problematical, though attempts are being made.[2]

II. SURFACE SENSITIVITY CONSIDERATIONS

The surface sensitivities of the techniques may be compared from the standpoints of:

(1) the depth of surface examined
(2) the time in which data (of sufficiently good statistics) may be obtained.

(1) is partly determined by the inelastic mean free path length, L_e, of the electron being detected. The electron spectroscopic techniques can be used for the study of surface composition and electronic structure because L_e is usually quite small. For UPS and XPS the impacting photons always penetrate to a considerable depth, but the L_e value of the ejected photoelectron is a function of its kinetic energy, the relationship and approximate absolute (for metals) values being shown in Figure 1. Some experimental numbers are given in JWG's notes, and elsewhere.[6,7,13] Thus it can be seen that UPS spectra might consist of electrons with L_e values as low as 5Å, whereas an Alk$_\alpha$ XPS spectrum of the same valence levels will consist of electrons with L_e some 25-30Å. Only those electrons which do not undergo inelastic scattering will contribute to the primary photoelectron spectrum, the rest appearing in the scattered electron background at lower kinetic energies. In practice, this results in the less surface-sensitive XPS valence level spectrum being "cleaner" than the UPS because multiple scattering results in the majority of scattered electrons having energies near zero, thus leading to a high electron

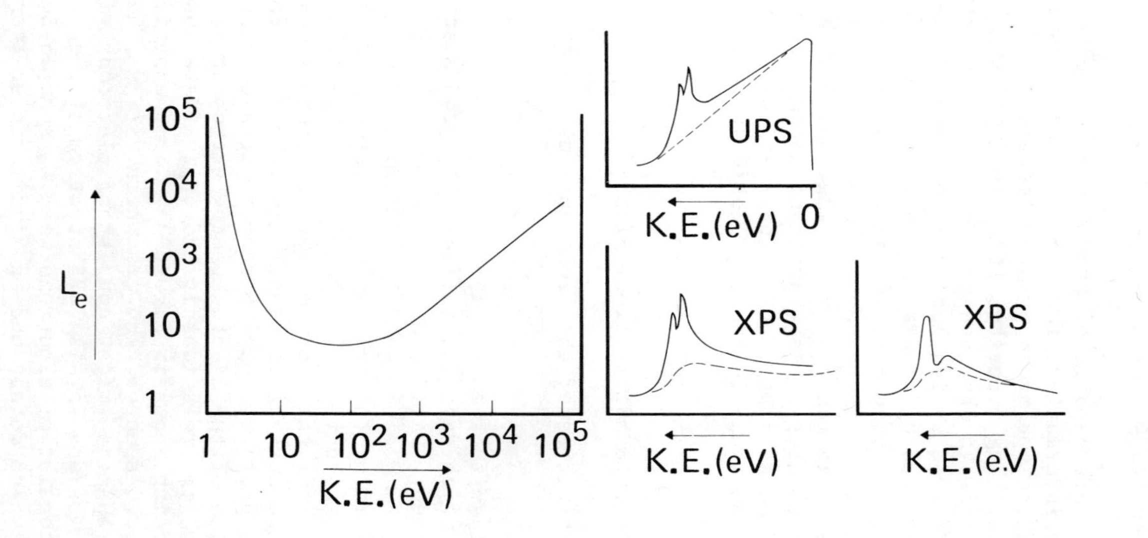

Figure 1. (left) Mean free path length, L_e, of an electron through a metal as a function of K.E.
(center) Schematic shape of solid–state valence band spectra in XPS and UPS, showing background scattered electron contribution.
(right) XPS core–level schematic spectrum.

background for the UPS spectrum but a lower one for XPS (Figure 1). XPS of core-levels will have a similar scattered electron background shape (Figure 1).

In AES the primary electron beam has a much shorter stopping length than for hν, but since energies of 3-5 KeV are usually used, it is still the shorter L_e of the ejected Auger electron which determines the depth of surface examined. Auger electron K.E.'s typically fall in the 100-600eV range so that the depth of surface examined is usually smaller than for XPS (Figure 1). This is not always the case, however, since sometimes the Auger K.E.'s of a particular element are higher than AlK_α photoejected core electrons.

In ELS,[14,15] one detects an inelastically scattered primary beam. In experiments specifically designed for ELS (which are not at all common) beam energies usually vary from a few 10's of eV to a few hundred, thus minimizing the L_e values. Most ELS data are obtained as additional information to AES by looking at the energy losses associated with the AES primary beam. Since this is generally not below 1 KeV, surface sensitivity will be reduced accordingly.

INS[16,17] is the most surface sensitive technique in terms of thickness of surface examined since the technique does not depend on the penetration of hν or e into the bulk to create core-holes, but on the tunnelling out of electrons from the surface to neutralize an incoming He^+ ion which never penetrates the surface. (The differences between the AES and INS processes are schematically shown in Figure 2).

The second parameter which influences the thickness of surface examined is the experimental geometry of the spectrometer. Making the assumptions of a uniform solid with completely flat surface; an exponential attenuation (inelastic scattering) of electrons by the solid; and an initial isotropic angular distribution of photoelectrons; the photoelectron intensity (in a particular photoelectron peak) escaping normal to the surface and originating within a depth, d, of the surface, is given by

$$I_d = I_\alpha (1 - e^{-d/L_e})$$

where I_α is the total photoelectron intensity emitted normal to the surface from an infinitely thick sample (since L_e is short, several hundred Å is usually a good approximation to "infinite"). The experimental geometry is illustrated in Figure 3(a). Putting in some realistic numbers for L_e,[18] say, for the $3d_{5/2}$ level of a clean Mo sample, allows us to estimate that about 18% intensity of the $3d_{5/2}$ observed XPS signal would originate from the top atomic layer of Mo atoms.

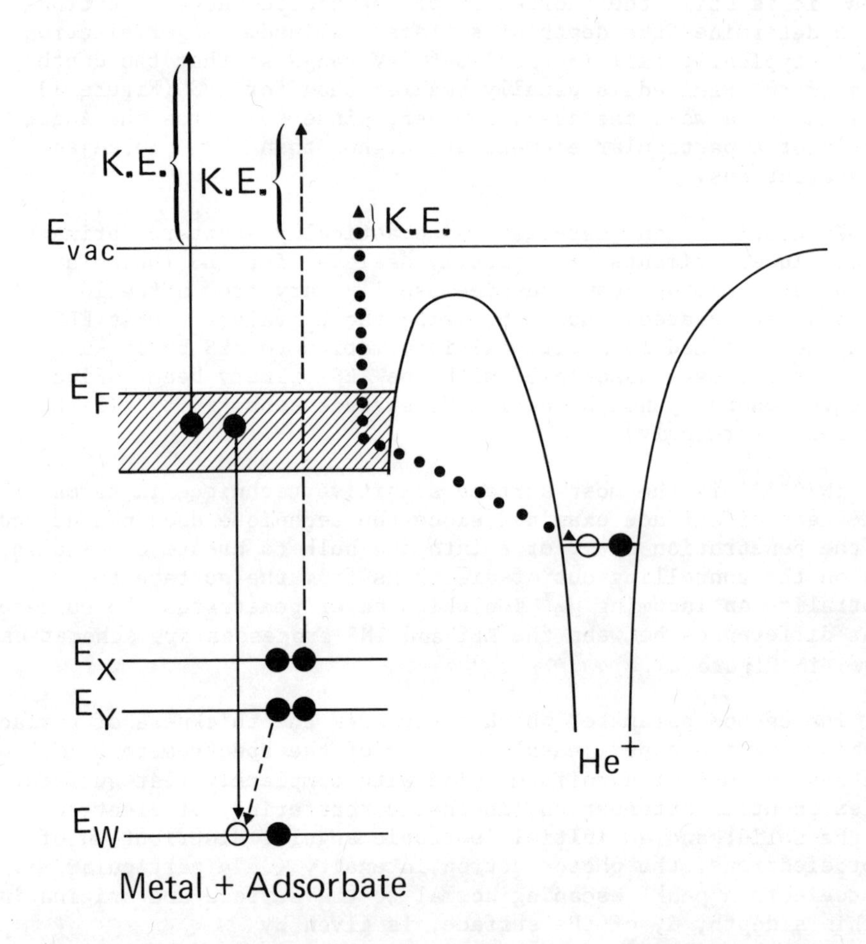

Figure 2. Core–valence–valence Auger transitions (→); core–core–core–Auger transitions (– –→); and the INS process (···), schematically represented.

If one alters the detection angle to $\theta < 90°$ (Figure 3b) the experimental path length in the solid that the photoelectrons must traverse before escaping to the vacuum is given by $d/\text{Sin}\theta$ and Equation (1) is modified to

$$I_d = I_\alpha(1 - e^{-d/L_e \text{Sin}\theta})$$

Thus reducing θ increases the percentage of the observed signal originating from the top atomic layer. $\theta = 45°$ is common experimentally. At this angle ca. 25% of the Mo $3d_{5/2}$ signal originates from the top atomic layer. The argument applies equally well for the case of an adsorbed layer on a substrate, so that the relative intensity of the adsorbate to substrate XPS (or UPS or AES) signals can be increased by an order of magnitude by going to grazing angle θ (5-10°). In addition to increasing surface sensitivity in this fashion, following the relative intensities of XPS (or AES) signals as a function of θ affords a qualitative depth distribution measurement of the elements present. Figure 4 shows a trivial example for the case of CO adsorption on Mo.[19] Part of the XPS spectrum taken at two angles, 45° and 15°, is shown. The O(1s)/Mo(3s) ratio increases by a factor of about 3 on going to the smaller angle telling us at once that the oxygen atoms are located at the surface and not distributed through the bulk material.

Analysis based on the above equations are the simplest approximations one can make. In practice, of course, surfaces are not completely flat, there may be anisotropy in the photoemission process (particularly for single crystal surfaces - see JWG's notes), and it is often an experimental constraint that changing θ also changes the angle of incidence of the photon beam (the case in the CO/Mo example given). C. S. Fadley, and co-workers,[20-22] have done a considerable amount of work in considering the influence of some of these complications. In particular, they have investigated the role of surface roughness and microscopic geometry rather throughly. Suffice it is to say here that all forms of surface roughness reduce the amount of surface enhancement obtainable by going to low θ.

(2) is governed by the quantity of material present; the intensity of the primary photon or electron beam; the ionization cross-section concerned; and the transmission and detection properties of the particular spectrometer used.

A few years ago 100 μ amps was a typical primary beam current used in AES. Today with the cylindrical mirror analyzer (CMA),[23] becoming standard in AES, beam currents of 5 μ amps are more typical. This is still two orders of magnitude larger than the photon fluxes obtainable in typical Alk$_\alpha$ XPS sources, but probably comparable to the vacuum UV flux from an HeI UPS source.

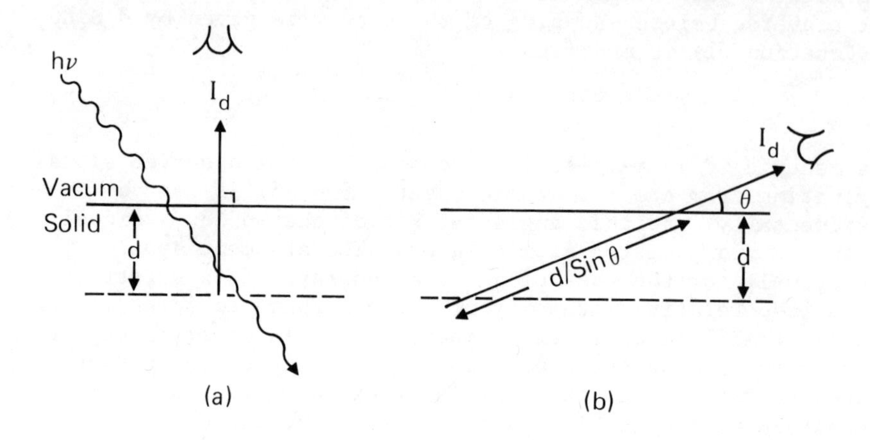

Figure 3. Possible experimental geometries for XPS or UPS.

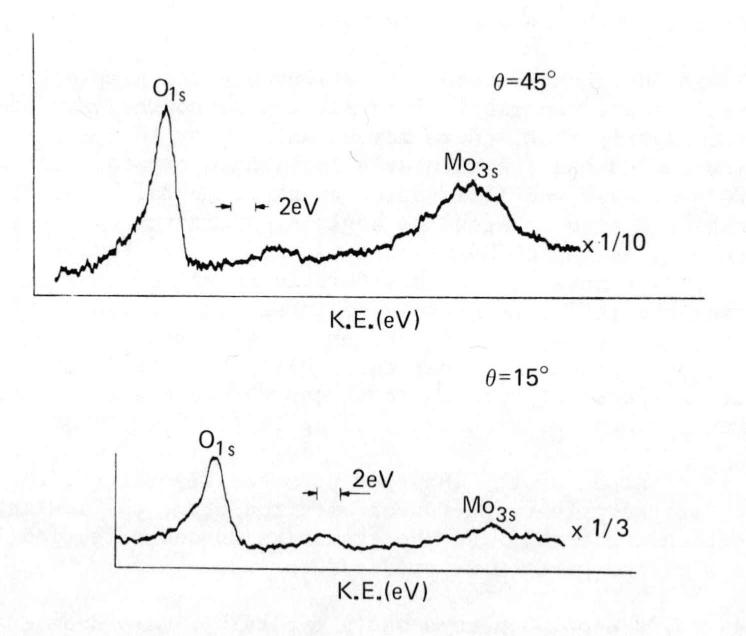

Figure 4. Mo (3s) and O(1s) XPS of CO adsorbed on an Mo film, taken at different electron ejection angles.[19]

The photoionization cross-section, $\sigma_{h\nu}$, for core-levels with Alk$_\alpha$ radiation varies from element to element and core-level to core-level. Taking the core-level with highest cross-section in each case, the variation in $\sigma_{h\nu}$ across most of the periodic table is no more than a factor of 20.[24-27] The ionization cross-section by electron impact, σ_e, for the same core-levels and at the primary electron beam energies typical of AES can be as much as two order of magnitude higher than $\sigma_{h\nu}$, but for some elements is actually lower than $\sigma_{h\nu}$.[28] (the variation of $\sigma_{h\nu}$ with hν is step-like, with a maximum near ionization threshold, whereas the variation of σ_e with e is a slow rise from threshold to a maximum at about 3-5 times threshold energy. Thus the ratio $\sigma_{h\nu}/\sigma_e$ depends intimately on the values of hν, e, and the B.E. of the core electron concerned).

Valence levels typically have very low $\sigma_{h\nu}$ at Alk$_\alpha$ photon energies[24] and high $\sigma_{h\nu}$ at UPS photon energies. For adsorption studies knowledge of the relative cross-sections for adsorbate and substrate levels, and the relative variation of cross-sections with change in hν can be important for both analysis and electronic structure studies, as will be seen later.

The analyzer now usually used in AES, the CMA[23], has a high transmission ($\sim$ 10%) by virtue of its large solid angle of acceptance and the fact that it is usually set for relatively low resolution (no pre-retardation). The analyzer used in XPS, (a hemispherical sector usually)[29] is set for the highest resolution possible since detection of small chemical shifts is a primary objective, and has a low transmission (0.1%). UPS data are recorded on both types of analyzer. The very high instrument absolute resolution of the XPS analyzer is not necessary for UPS because of the low electron energies involved, but the high UPS signal strengths make its use quite adequate in terms of count rate. When CMA's are used, far higher count-rates are obtained.

Owing to all the above factors typical operating conditions and detection limits might be: For XPS, a 30eV wide scan over the O(1s) region taken in 5 minutes with about 2% monolayer of oxygen being detectable; for AES, a 500eV scan in 1 minute (covering many element Auger positions) with the detection limit maybe 0.2% monolayer oxygen. It must be stressed that the main reasons for the higher sensitivity of AES are the high intensity electron beams used and the analyzer used. Unfortunately the highly destructive nature of electron beams (desorption, dissociation) means that in many cases the adsorbate/substrate interaction under study is altered before the required information can be obtained, so that the apparent speed advantage of AES is in such cases worthless. One factor, nevertheless, ensures that AES equipment outsells XPS commercially, and that is the ease with which electron beams can be focussed to a small spot size

(25μ) and rastered across a surface.[30] Though this is of no concern in the subject matter of this conference it is obviously of extreme technological importance in the semiconductor, thin film, and metallurgy based industries.

III. QUANTITATIVE ELEMENT ANALYSIS

XPS is simpler to consider and use than AES because only the core-ionization itself is involved. Assume for the moment that all the photoelectron intensity for each core-ionization goes into one well-defined peak in the experimental spectrum and let us consider a gaseous sample. For an absolute quantitative analysis of the elements present in our gaseous sample what are required are the ionization cross-sections for the core-levels concerned, the X-ray photon flux and the absolute detection efficiency of the spectrometer (which will probably vary with K.E. of detected electrons).

If all these are known an absolute analysis of all elements present could be made from the peak areas observed in the spectrum. A relative element concentration analysis would normally be all that is required, in which case a knowledge of the relative photoionization cross-sections and the variation of detection efficiency with K.E. is all that is required. In practice the normal experimental procedure would probably be to calibrate using gas mixtures with known partial pressures and establish a set of inter-element core line relative sensitivities.

Unfortunately, the basic premise of the whole procedure - that all the photoelectron intensity goes into one line - is incorrect. Though the photoionization process removes an electron from an orbital of well-defined energy, E_i, in the initial non-ionized on electron system, the act of removing it may result in a number of possible n-1 electron final states, which have slightly differing energies, resulting in a number of photoelectron lines with differing energies. The most obvious forms of line splitting are spin-orbit interaction, for example the Ar $2p_{3/2}$ and $2p_{1/2}$ states formed by removal of an Ar 2p electron, and spin-spin interaction (multiplet splitting) which is common in transition metal spectra where there are often several unpaired valence electrons to which the remaining unpaired core-electron may couple. These two phenomena are well-understood and the energy splittings and relative intensities of the component peaks can be calculated.[31,32] The effect of the "relaxation" phenomena[1,33] which is directly responsible for the so-called shake-up and shake-off satellites in XPS spectra is more difficult to handle. The theoretical basis of relaxation and shake-up satellite structure, and their inter-relationship, is lucidly described by JWG.[1] Owing to the redistribution of the system's valence electrons, caused by the influence of the positive core-hole creation during the photoionization process, there is finite probability that the valence electron configuration of the final state after the ionization process is complete will be different from the initial state. For a gaseous molecule this can be phenomenalogically viewed as the simultaneous excitation of a valence electron together with ionization of the core-electron (Figure 5a), though it should be emphasized that the correct theoretical description is as an n to n-1 electron process. In Figure 5a if the probabilities of causing valence excitations A, B, or C in conjunction with the core-ionization are 50%, 10% and 25% of the simple core-ionization process, then satellite peaks of there relative intensities will appear at kinetic energies ΔE_A, ΔE_B, and ΔE_C below the main line. In a shake-off process the valence electron is ejected (process D in Figure 5a) instead of merely excited, leading to a step followed by a continuum at kinetic energy, ΔE_D, below the main line (a step because the excess kinetic energy is partitioned between <u>two</u> ejected electrons). A large number of these shake-up and shake-off produced final states can carry finite spectral intensity. It is difficult to estimate these experimentally because (a) the populated shake-up states may spread up to about 100eV from the main line, and therefore get mixed up with states from other ionization processes and (b) because of the continuum nature of the shake-off structure. Unfortunately, for quantitative analysis, the total intensity should all be included in the measured photoelectron peak area because the distribution of intensity among the possible final states and the description of the final states themselves is chemical dependent, which means, for instance, that the fraction of the total photoelectron intensity going into the main O(1s) peak in CO will be different than for CO_2. Thus using peak intensity ratios based on the main lines only will introduce errors which may be small or large depending on the systems involved.

A homogeneous solid sample presents additional problems for quantitative analysis. Variation in the K.E. of detected photoelectrons results in variable L_e values and therefore signals originate from different depths for different elements. Second, though it is usual to try and include satellite intensities one has even less hope for a bulk solid of estimating the intensity correctly because of the contribution from the inelastically scattered electrons which lie in the same region. As an example of the strong chemical dependence of satellite structure, Figure 6 shows the XPS $2p_{3/2}$ region for Fe in iron metal and Fe_3O_4[34] and Cu in Cu_2O and CuO.[35] Many attempts have been made to establish experimentally an inter-element core-line relative sensitivity table for bulk chemical analysis.[24-27,36-38] In addition to the problems mentioned above it is probable that many of the "stoichiometric" samples used for calibration were (a)

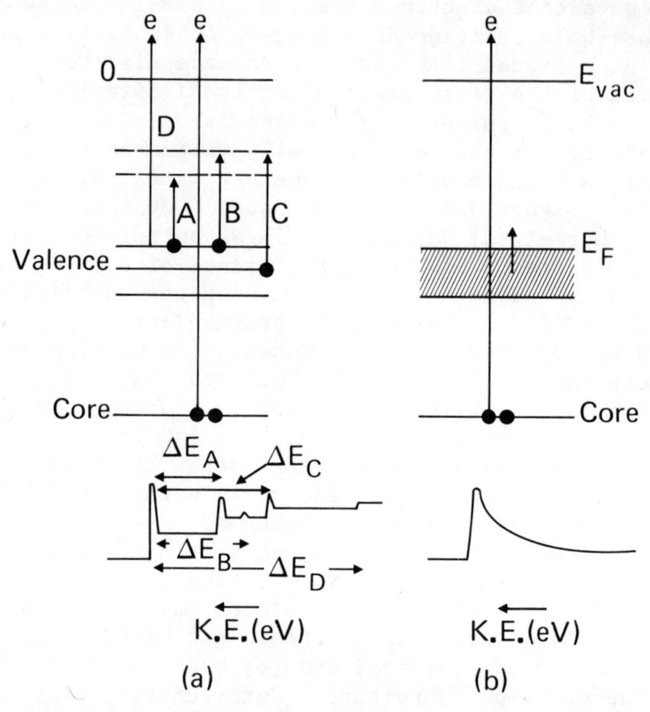

Figure 5. Shake–up and shake–off processes for (a) a free molecule, and (b) a metal.

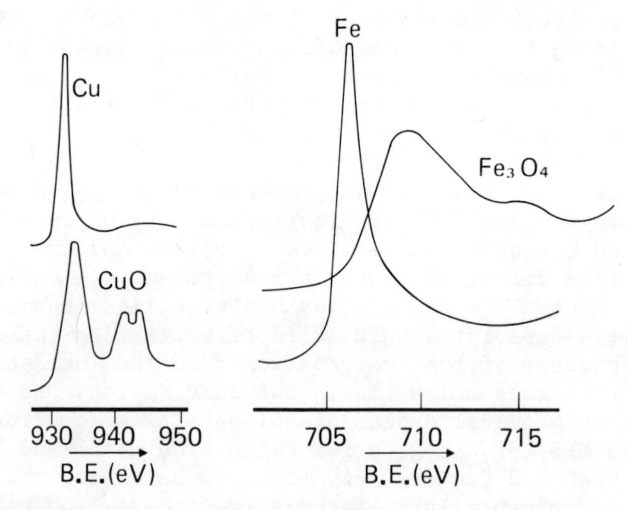

Figure 6. $2p_{3/2}$ XPS region for Cu, Cu_2O,[35] Fe, and Fe_3O_4,[34] illustrating the effect of shake–up structure.

not stoichiometric in the surface region, and (b) heavily surface contaminated. A selection of some of the results is given in Table I, together with a set of calculated values based on the atomic photoelectron cross-section calculation of Scofield[24] and an allowance for escape depth corrections by assuming $L_e \alpha \sqrt{K.E.}$ The calculated values are therefore appropriate to bulk solids. At present it is not clear how much of the discrepancy is due to genuine instrumental differences and how much to the problems mentioned above.

The specific case of a metal is of particular interest at this conference. JWG[1] has explained how, for an infinite conductor, electron-hole pair excitation processes, which are the equivalent of the discrete shake-up processes in insulators or free molecules, form a continuum from zero energy to infinity such that instead of a main XPS core-line followed by discrete satellite structure, one should observe an asymmetric core-line, (Figure 5b). Wertheim and co-worker[39] have observed such asymmetries and have suggested that they make it impossible to use XPS quantitatively for analysis of conductors because the asymmetric tail theoretically spreads to zero kinetic energy making it experimentally impossible to estimate the area under the peak. This actually seems rather unreasonable since the Sum Rules discussed by JWG[1] make it clear that the amount of intensity which can go into the asymmetric tail is restricted. We will come back to this point in Section V. For the case of an adsorbate monolayer quantitative analysis is actually in better shape than for the homogeneous bulk case because there is little attenuation of the adsorbate core-level photoelectrons by inelastic scattering and therefore no large scattered electron background from which to separate genuine satellite structure. The situation is, in fact, more like the gas phase. Table I, column 5 shows some inter-element core-level relative sensitivities established by condensing monolayer quantities of gases at inert surfaces (low temperature, Au substrate).[7] These should be more appropriate to adsorption studies than the bulk determined numbers.

Estimation of the coverage by adsorbate atoms can be done by estimating the percentage signal originating from the first atomic layer of the substrate (see Section I) and using adsorbate atom substrate atom core-level relative sensitivities. A more common procedure, typical of all types of surface coverage determinations, not just XPS, is to calibrate by relating the adsorbate signal intensity at some coverage point established by another technique (LEED, flash desorption, ellipsometry, radiotracers, or volumetric measurement).

In AES, Auger electron intensities depend both on the core-level ionization cross-sections (cf. XPS) and on the

TABLE I

Relative Intensities* of Photoelectron Peaks at $h\nu = \mathrm{Alk}_{\alpha}$

Element	Level	Experiment (Bulk Compounds)			Expt. (Adsorbate)	Calculation
		Ref 36	Ref 37	Ref 25	Ref 7	Ref 37
F	(1s)	1.0	1.0	1.0	1.0	1.0
C	(1s)	0.24	0.29	0.24	0.18	0.277
O	(1s)	0.61	0.53	0.35	0.48	0.522
Na	(1s)	2.09	1.44	1.89	--	1.32
Si	$(2p_{3/2})$	0.17	0.23	0.15	--	0.161
P	$(2p_{3/2})$	0.26	0.18	0.12	--	0.167
S	$(2p_{3/2})$	0.33	0.30	0.18	--	0.232
Cℓ	$(2p_{3/2})$	0.46	0.43	0.25	0.49	0.312
K	$(2p_{3/2})$	0.85	1.03	0.83	--	0.723
Ca	$(2p_{3/2})$	1.01	1.06	0.98	--	0.903
Pb	$(4f_{7/2})$	4.10	4.12	--	--	3.74

*Normalized to F (1s).

subsequent Auger transition rate to fill the core-hole. The alternative mode of de-excitation of the core-hole, X-ray fluorescence, is not a significant competitor to the Auger process for the majority of Auger transitions with which surface analysis is concerned (high energy transitions are required before fluorescence becomes significant) and therefore the total Auger cross-section for filling a given hole will be nearly the same as the ionization cross-section. This may be split between several possible Auger processes, however. In some cases individual transition rates have been calculated[40] and one could go after an absolute gas analysis as for XPS.

For solids the quantitative analysis problems are worse than for XPS. The largest proportion of detected current in solid-state AES is from the scattered primary beam (Figure 7), upon which are superimposed the small Auger signals. To aid in the detection of these signals and to suppress the rapidly-changing background it is normal to record in the dN(E)/dE mode (Figure 7). It is not possible to measure Auger peak areas in this mode so peak-to-peak heights in the derivative curve are usually used as a measure of Auger signal intensity. This is one stage worse than using just the main core-line intensity, without satellites, in XPS, since it is only strictly valid if an element Auger peak shape (in the N(E) curve) does not differ from compound to compound - a situation known to be untrue.

AES has been used in surface studies for much longer than XPS, and the words "quantitative surface analysis" are often associated with it. In fact, quantitative analysis in the manner attempted by XPS - the determination of the relative concentrations of different atoms at a surface - is not often practiced in AES. Most "quantitative analyses" consist only of coverage determinations by calibration against another technique (cf. XPS), or of following relative changes in concentrations during depth profiling experiments. The reasons for the lack of quantitative work on absorbates, I believe, can be traced to the difficulty of using differential Auger traces quantitatively, and, more important, to the very destructive nature of electron beams. It should be emphasized that for absorbate/substrate systems electron beam damage is <u>the rule</u> rather than the exception. Chang[41] has attempted to put Auger spectroscopy on a quantitative footing by experimentally determining the relative sensitivities of Auger transitions from pure elements and bulk compounds. He quotes the results in terms of inverse sensitivity factors, α, as shown in Table II. He notes that the values apply only to the analyzer and analysis parameters he uses and also that an elemental α can change in different compounds. This is probably mostly the result of peak shape changes and the dN/dE mode of recording. The relative sensitivities work reasonably well for estimating dopants in SiO_2 films, for bulk compounds,

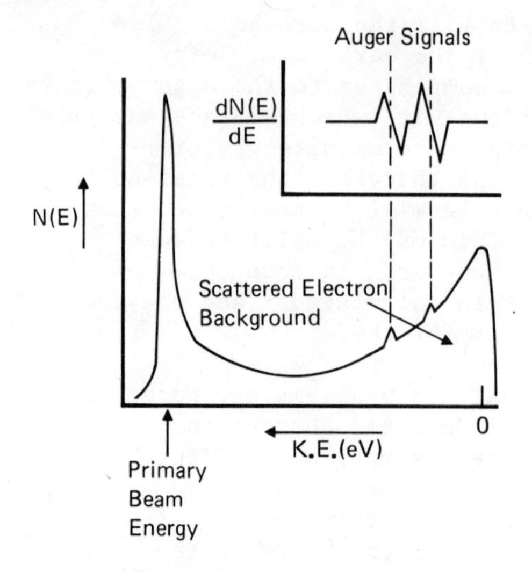

Figure 7. The electron emission Spectrum during AES.

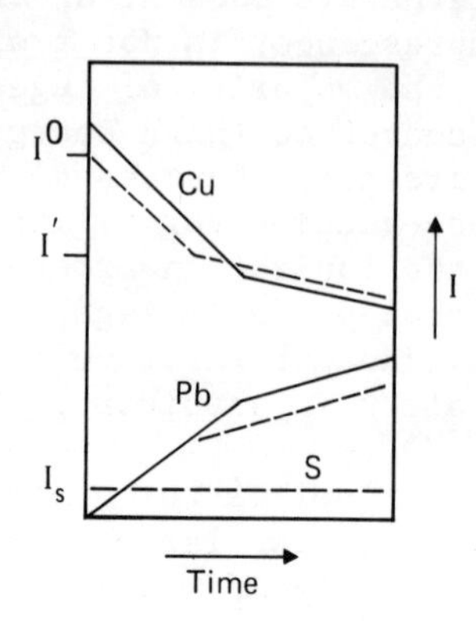

Figure 8. AES intensities during the evaporation of Pb onto clean Cu (solid curves), and S contaminated Cu (dashed curve).

TABLE II

Inverse Auger Sensitivity Factors Normalized to the Si 92eV Peak*

Element	Auger Energy (eV)	α	Element	Auger Energy (eV)	α
B	179	2.1	Fe	703	1.9
C	273	3.0	Cu	920	1.9
N	386	1.4	Ga	84	14
O	495	6.0	Ga	1070	3.7
Aℓ	66	1.7	Aℓ	95	16.6
Si	92	1.0	Pd	330	0.70
Si	1620	10	Tn	1680	10
P	120	0.86	W	169	3.5
S	150	0.70	Pt	168	8
Ti	418	1.1	Au	150	10

* From Reference 41. These values are only appropriate to the experimental conditions used in Reference 41.

and for carbon overlayer thickness estimates. Strong evidence for sample decomposition was observed in some cases, and no attempts were made to apply the analysis to adsorbate systems.

Morabito[42] has suggested the following approximations to obtain a "first approximation to quantitative Auger analysis": (1) L_e is proportional to $\sqrt{E}$ where E is the Auger electron kinetic energy; (2) ionization cross-sections, σ_e, are proportional to $1/E^2$ for K and L shell ionization; (3) the transmission of a CMA is proportional to E. Making the further assumptions that the backscatter correction ratios are near unity, the relative sputtering yields for different elements approach unity, and that one Auger transition is dominant for each core-ionization (so that σ_e may be used instead of the Auger yield x σ_e) he used approximations (1) - (3) to estimate the composition of MgO, Cu_2O, CdS, GaP, and K Cl. Considering the crudeness of the approximations the agreement obtained with stoichiometry (within 15%) is remarkable. The model also worked for ion-implanted nitrogen, oxygen, and carbon deposits. No attempts were made to apply it to adsorbates.

There has been one attempt at establishing calibration procedures in AES which deserves special attention because of its unique nature and its specific relevance to adsorption studies. This is the "Co-adsorption Calibration" described by Argile and Rhead.[43] In work on the adsorption of metals on metals, they established that both the increase in adsorbate signal and the decrease in substrate signal as a function of evaporation time exhibited a sharp break at the monolayer point (Figure 8). The actual form of the curves and the break depend on which of the several overlayer growth mechanisms are appropriate (see the lecture notes of G. Rhead in this volume). In this way a calibration for Pb on Cu was established. However, if the original Cu surface was partly contaminated with sulphur, the break-points for the Pb and Cu signals were observed to come earlier (Figure 8). Making the assumptions that (a) adsorption of Pb occurs only at free Cu sites, and (b) that a close-packed adsorption arrangement was adopted, a simple equation relating I' the Cu Auger intensity at the break point, I° the original Cu intensity of the sulphur contaminated surface, I_s, the S Auger intensity, and α_s and α_{Pb}, the fractional transmission coefficients of the Cu Auger electron through a close-packed layer of sulphur or of lead, was derived. The equation is:

$$I'/I^\circ - I'/I^\circ\,(1 - \alpha_s)\,I_s/I^\circ = I_s/I^\circ\,C\,(\alpha_s - \alpha_{Pb}) + \alpha_{Pb}$$

where C is an instrumental constant. Measurement of I°/I° as a function of I_s (i.e., for different initial coverages of sulphur) allows α_{Pb} and α_s to be determined. Use of this data allows one to establish the relative sensitivities of the Pb and S Auger signals, and, when repeated with oxygen contamination, Pb and O sensitivities (break-point calibrations cannot be used for S or O as for Pb because layer growth does not occur in the same fashion). It was found that the relative sensitivity of the S LMM Auger transition to the O KLL on copper is 30:1, and an O KLL spectrum of only 1% intensity of the Cu MMM substrate peak would correspond to monolayer coverage.

IV. MOLECULAR AND CHEMICAL ANALYSIS

To use any spectroscopic technique for a molecular or chemical analysis, as opposed to an elemental one, it is first of all necessary to be detecting transitions which are sufficiently affected by the chemical bonding involved - i.e., each molecule should have a unique spectrum. In different ways and to different extents all the electron spectroscopic techniques satisfy this requirement. The second requirement is to have a bank of available known "fingerprinting" spectra of pure compounds available so that molecule A can be distinguished from B by simple comparison. Thirdly, if the techniques are actually going to have <u>practical</u> value in analysis, they must have sufficient resolution so that A can be distinguished in mixtures of A, B, C and the amount present estimated. It is not <u>necessary</u> to be able to understand all the details of the spectra being recorded to be able to use them in a fingerprinting way, but it helps avoid errors if one is not entirely ignorant either!

Let us consider, for the gas phase first, the various chemical dependent features that can be used for a molecular analysis, and the chances of achieving that analysis in a mixture of molecules for the different techniques.

<u>UPS</u> The valence level M.O. pattern of a molecule is unique and may therefore be used to distinguish one molecule from another. The technique is a high-resolution one, often allowing the resolution of vibrational fine-structure. Unfortunately even for small molecules there are usually several MO's with I.P.'s less than 21eV so the possibility of successfully distinguishing A in mixtures of A, B, and C rapidly becomes untenable as the size of the molecules grow, owing to the overlapping of photoelectron bands.

<u>ELS</u> Energy loss spectroscopy can have equally good resolution but since there are several filled M.O. → empty M.O. transitions possible for each I.P. of a UPS spectrum, the potential for molecular analysis of mixtures is less than UPS.

<u>XPS</u> A range of chemical effects can be utilized in XPS. The valence levels can be examined directly, but the lower resolution

of the technique (X-ray line-width limited - ca. 1.1eV without monochromator, typically 0.5eV with monochromator) and the low ionization cross-section for valence levels largely rule this out as a practical method. For core-level studies chemical shifts, spin-orbit splitting, multiplet-structure, and satellite structure are all chemical dependent and usually they can be studied for several core-lines for each element present. In addition the initial elemental analysis, establishing the atomic ratios of elements present is an advantage entirely missing in UPS and ELS studies. If all these factors were to be used together a rather good molecular analysis of gases could be made. One can usually find cheaper ways of performing a gaseous analysis, but as we shall see it is exactly the above procedure that is used for adsorbates where there usually is no alternative method.

AES In principle AES offers the greatest possibility for molecular analysis. For an Auger transition which involves a core-level hole and valence level electrons (WVV transitions) the general energy region of the ejected Auger electrons is defined by that of the core-hole and therefore provides an elemental identification, but the composition of the structure within that region is determined by the valence level pattern of the molecule. Higher resolution may be achieved than in XPS because the K.E. of an Auger electron is independent of the ionizing electron or photon beam creating the core-hole and is not, therefore, line-width limited. As an example the C KLL Auger spectra of gaseous CO_2 and CO[44] are shown in Figure 9. The general energy spread of the two spectra is the same, but the detailed structure is quite different. The O KLL Auger region could also be used for confirmation of identification. In mixtures the overlapping of structure causes a similar problem as in UPS and ELS, but the element analysis and the availability of a "fingerprinting" region for each element is a great advantage over UPS and ELS.

Moving on to the solid state all the valence-region studies, by UPS, ELS, or XPS, become very valuable in delineating electronic structures, as will be seen in Section V, but because of the band nature of the spectrum their use for analytical purposes is restricted to very simple systems (two-component alloys for instance - see Section V). The core-level techniques retain their usefullness because core-levels are only slightly broadened by solid-state effects, and in fact, come into their own because of the lack of other techniques for analyzing the surface region of solids. As an example of the combined use of chemical shifts, spin-orbit splitting, multiplet splitting, and shake-up structure for a chemical analysis, Figure 10 shows the $Co2p_{3/2,1/2}$ spectra of a pure CoO surface and a pure Co_3O_4 surface.[45] One can see that (a) there are chemical shifts between

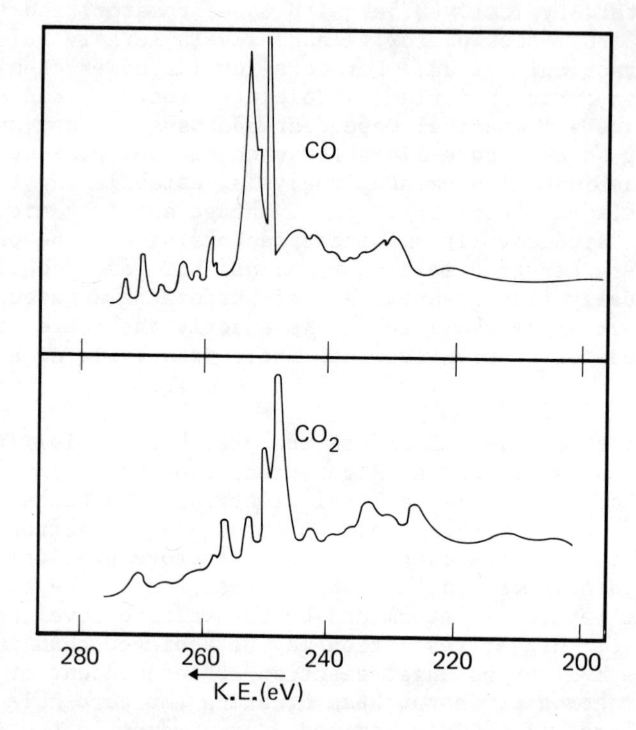

Figure 9. C KLL Auger spectra for gaseous CO and CO_2.[44]

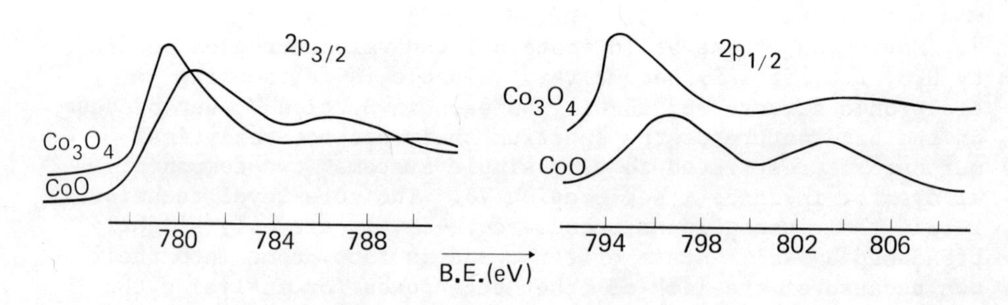

Figure 10. $2p_{3/2,\,1/2}$ XPS spectra for CoO and Co_3O_4.[45]

the two compounds, (b) the spin-orbit splitting is different in the two cases, (c) the line-widths are different in the two compounds (different multiplet splitting), and (d) the shake-up structure is quite different. Utilizing all these differences between the spectra of the pure compounds, it becomes possible to detect 10% concentrations of either in the other over the analysis depth region of the technique. Some interesting results are obtained on commercial samples. CoO comes covered in a few layer of Co_3O_4 and "Co_2O_3" bought commercially does not contain any Co_2O_3 on the surface or in the bulk, but is almost pure Co_3O_4.

In AES, core-level chemical shifts can also be used for diagnostic purposes. One should distinguish between Auger processes where all the levels involved in the Auger process are core-levels (often called WXY processed and those in which valence levels are involved (WXV or WVV processes). Auger electrons originating from the former will have chemical shifts which are a composite of shifts in the individual core-levels. Their interpretation is therefore more complex than in XPS, but they may be used for chemical identification purposes, nevertheless. To generalize, Auger lines are usually broader and more complex than XPS lines and therefore less diagnostically usefull. There are quite a few cases, however, where Auger chemical shifts are much larger than the equivalent XPS shifts (owing to the one hole → two hole nature of the transition), more than compensating for broader line-widths.[46-48] In other cases, line-widths are no broader than the XPS core-levels and shifts are also larger. Figure 11 illustrates one such case, Ga/Ga_2O_3.[49] It can be seen that the shift between metal and oxide is much larger in the Auger case than in the XPS $2p_{3/2}$ core-level. Another example is the Cu/Cu_2O system. The $Cu2p_{3/2}$ XPS core-level of Cu_2O is almost superimposable with that of Cu and so cannot be used for distinguishing between Cu^o and Cu^I. The Cu - Auger transitions are quite well separated in the two cases, however (about 2eV), and so may be used.[50]

For the WVV transitions, as we have seen for the gases, information is contained on the valence level patterns, as well as on the chemical shift of the core-level involved. For solids the valence level structure is band-like, of course. The superimposition of the two effects, core-level shifts and valence level changes, may give a very complex overall effect on the experimental spectrum, which can, nevertheless, be used in a "fingerprinting" manner.

When considering adsorbate/substrate interactions, a primary piece of information one would often like to have is whether the molecule retains its molecular entity or not during adsorption. If the adsorbate molecular electronic structure is not seriously perturbed compared to its free (gas phase) state - i.e., for

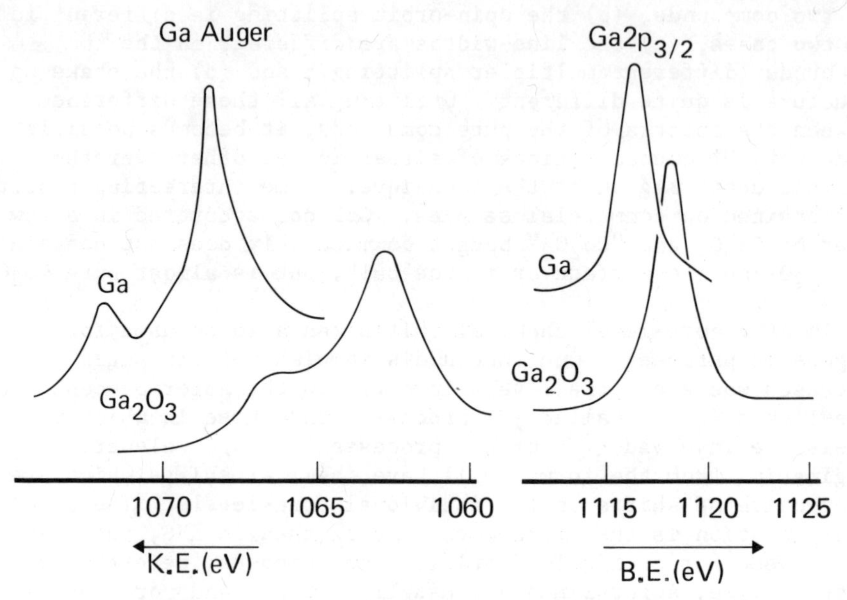

Figure 11. XPS and Auger spectra of Ga and Ga_2O_3.[49]

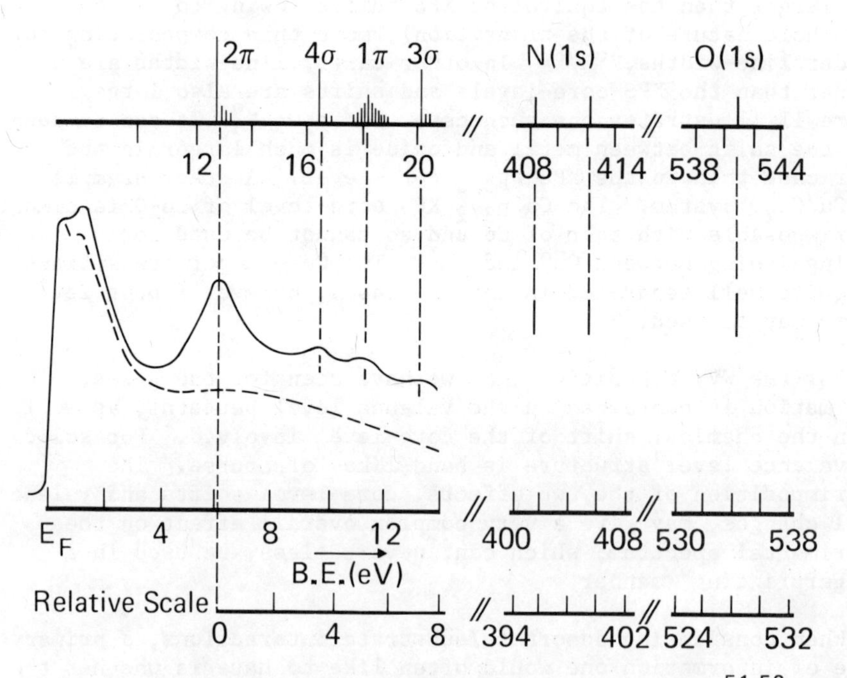

Figure 12. HeI spectrum and core–levels of N_2O on Ni at 80K,[51,52] Dashed trace represents Ni UPS. Gas phase N_2O spectrum shown above.

condensation, physical adsorption, or possibly weak chemisorption, a molecular identity can often be inferred directly from the UPS valence region spectrum in conjunction with an XPS identification of the core-levels. XPS itself cannot be used effectively in the valence region for adsorbed species for the same reason that it is less effective than UPS in the gas phase - low cross-section and resolution.

As an example of molecular identification of adsorbates by UPS and XPS, the spectrum of N_2O adsorbed at 80K on an Ni film is shown in Figure 12 and compared to the gas phase spectrum of N_2O.[51] For the moment we shall not consider the relationship between the adsorbant B.E. values with its inherent difficulties (see Section V), and so a second energy scale zeroed at the first I.P. for both gaseous and adsorbed situations is shown in the figure. It can be seen that (a) all the valence and core-levels of N_2O are identified in the adsorbate case; (b) the core-level relative intensities confirm the stoichiometry of 2N:O; and (c) the relative B.E.'s of all orbitals, valence and core, are the same in gaseous and adsorbed states. Thus it is convincingly demonstrated that the adsorbate is present at the surface in a largely undisturbed molecular state. A large number of adsorbate molecules have been identified in the condensed or physisorbed state in this manner.[51-54] Not all have been identified by both UPS <u>and</u> XPS. Some studies have been UPS only. This is unlikely to produce mis-identification for simple single-component condensates, but it is dangerous in the case of mixtures, or in chemisorption situations, where the valence and core-levels might be modified considerably, or dissociation may take place. In such cases, which are the subject matter of the following section, it is imperative to use a core-level method to provide an element analysis and to assist and confirm, by use of chemical shifts, assignments made in the valence region. In principle, AES could be used as an alternative to XPS. In practice for the condensation and physical adsorption studies this has not happened, probably because of the severe electron beam effects that would result. In the many chemisorption studies by UPS, AES has sometimes been used as an auxiliary tool, but so far only to the end of determining surface cleanliness and detecting the elements present during adsorption, not for additional electronic structure information.

The two other valence level techniques, ELS and INS, have never been used for "fingerprinting" adsorbate molecular identification purposes, as for as I know. In principle, there is no reason why they should not, but the practitioners of the techniques have reserved their use for detailed studies of a few dissociative chemisorption reactions.

V. SURFACE ELECTRONIC STRUCTURE STUDIES

In this section the information available from electron spectroscopy concerning (a) the electronic structure of adsorbate free metal (and other) surfaces and (b) the electronic structure and bonding of adsorbates to metal surfaces, will be considered.

For definitive information concerning adsorbate bonding it is, of course, usual to work with well-characterized single crystal surfaces to reduce the effects of surface heterogeneity and multiple adsorption sites. In a few cases LEED intensity analysis has supplied definitive geometry information for those surface adsorption atoms which contribute to the LEED pattern analyzed.[55,56] In the more common LEED without intensity analysis at least the repeat surface structure is determined, which usually limits the feasible possibilities for the adsorption sites actually adopted. In neither case, however, is information usually available on the percentage of the adsorbate present which contributes to the ordered LEED analyzed structures, or where the remainder is, thus reducing the definitive nature of the single crystal study.

The majority of UPS and XPS (but not AES, INS or ELS) adsorbate studies have been performed so far on polycrystalline surfaces. Whereas this certainly leaves room for ambiguity of interpretation in some cases, such studies are more than adequate to indicate the use of those phenomena available in UPS and XPS data for providing bonding information on the adsorbate - surface interaction.

(a) Surfaces Free From Adsorbates

It is not my brief, nor am I qualified, to consider the way in which the details of the ε,k diagram of the band-structure of a crystalline material may be studied by electron spectroscopy. I shall assume that the quantity of interest here is the one-electron Density of States (DOS) description of the surface material and the way the electron spectroscopies data may be related to this description.

Figure 13 (a) and (b) show the UPS spectra of polycrystalline gold using HeI (21.2eV) and HeII (40.8eV) photons.[57] Unlike the situation in the UPS spectra of gaseous molecules for which only relative intensity changes due to variations in cross-section would be expected, the two spectra are quite dissimilar. The dissimilarity reflects an important modifying effect of final-state structure for crystalline solids. If we consider the hypothetical E,k diagram of Figure 14(a) the photoionization process may be considered as raising an electron occupying an

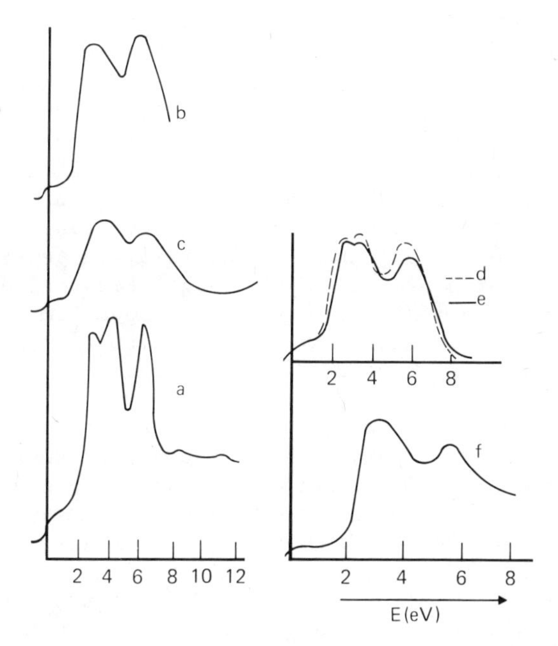

Figure 13. Photoemission of Au valence level region: (a), (b), and (c) HeI, HeII, and AlKα spectra of polycryst. Au;[57] (d) calculated DOS of Au;[59] (e) AlKα XPS (monochromatized) of single crystal Au;[59] and (f) HeI spectrum of molten Au.[61]

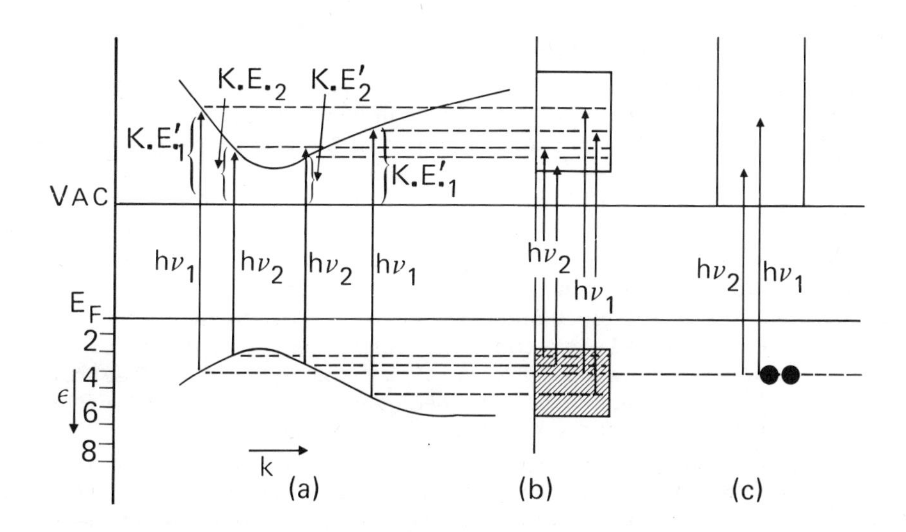

Figure 14. Hypothetical (E,k) diagram (a) and DOS (b) illustrating the effects of k–conservation and final state structure on UPS variation in hν; (c) situation for a free molecule.

allowed ε level in k space by hν. However, unlike the situation in the free molecule, the electron is still within the crystal at this point and must end up occupying another allowed ε level. If the quantum restriction of k-conservation is maintained a transition can only occur at those values of k for which there is a gap of exactly hν between initial and final states. Having reached the final state (the first step of the three-step model)[58] the electron is then transported through the crystal, and ejected across the surface. Thus, changing hν will, in general, change the transitions which are allowable (e.g., Figure 14a in which $h\nu_1$ is changed to $h\nu_2$) unlike a gas phase spectrum where a continuum is always available to connect with the initial state, no matter what the value of hν (Figure 14c).

From what has been said above the UPS spectrum of crystalline Au would be expected to vary with hν and not be representative of the initial DOS. This is apparently so up to about 35eV for the case of Au. At higher hν, variation with hν becomes less dramatic. For example the HeII spectrum of Figure 13b is quite similar to the XPS spectrum (1486.6eV) of Figure 13c,[57] and both are quite similar to the calculated DOS[59] (Figure 13d). This comes about because at high enough hν the final states are so numerous and folded back on each other in the ε,k, diagram that they approximate continuum behavior in that a level is always available for connection with an initial state for any value of hν and k. Even in this situation, there is still the possibility that the experimental spectrum is a modified version of the DOS owing to differences in the ionization cross-sections of electrons with different quantum character (n,ℓ). These cross-sections themselves are a function of hν. In the case of Au, the variation in cross-section between the sp region and the d region is apparently not dramatic or there would not be good agreement with the calculated DOS and the relative intensities of the sp and d regions would change significantly on going from HeII to AlK_α. More dramatic effects are found when n = 2,3 or 4 atomic orbitals make up the valence levels concerned. For example, the C2s/2p ratio increases by more than an order of magnitude on going from HeII to Mg Kα. Since the changes in relative cross-sections of the various atomic levels of the different elements is by now fairly well-known, it can be utilized to help determine electronic structure. This has been demonstrated most convincingly for the DOS of Ge and Si, as reported by Ley, et al.[60] and described here. The experimental XPS spectrum for crystalline Ge and Si are shown in Figure 15, together with the calculated DOS. One can see that all the major features in the calculated DOS are reproduced in the experimental structure, but there are intensity discrepancies. Whereas the calculated intensity of the band nearest E_F relative to the two other bands is almost the same for Ge and Si, the experimental ratio decreases considerably on going from Ge to Si. From data available on 3s, 3p, 4s, and 4p

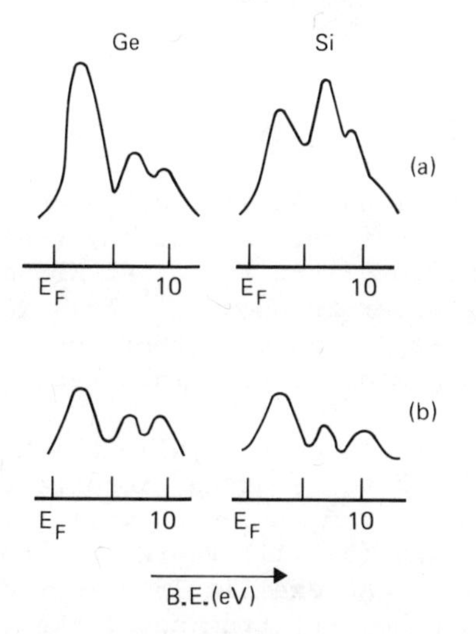

Figure 15. (a) XPS valence band structure of crystalline Ge and Si[60] (b) calculated DOS of crystalline Ge and Si[60]

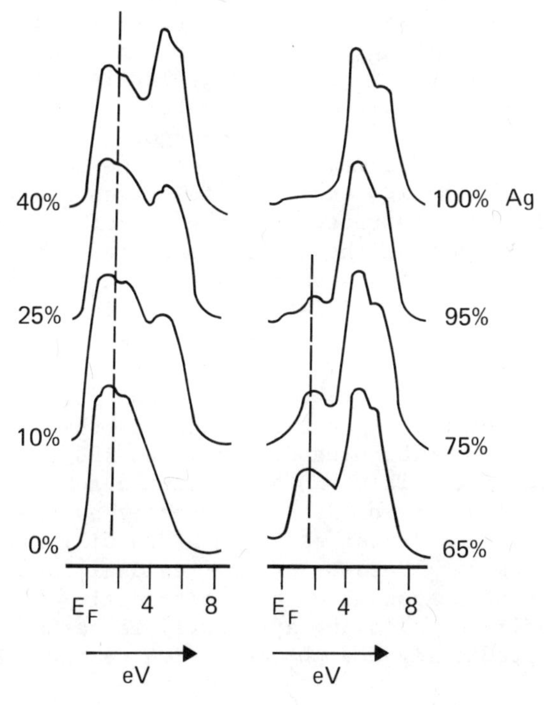

Figure 16. XPS spectra for Ag/Pd alloys [62].

core-levels in other elements it can be shown that the cross-section ratios are given by:

$$3s/3p : 4s/4p \simeq 2.3$$

The valence levels in Si and Ge are 3s, 3p and 4s, 4p, respectively. One, therefore, expects the the photoelectron ratio of s type bands to p type in Si, at XPS energies, should be about 2.3 times greater than in Ge. This is so if one assigns the band near E_F as p-type and the other two as s-type in both Ge and Si. This assignment is also in agreement with theory.

For amorphous materials or alloys in which there might be insufficient order for k to be of any importance, it is more likely that the UPS spectrum at low $h\nu$ will (a) not alter with small changes in $h\nu$, and (b) will represent the initial DOS, cross-section modified. An example is observed with molten Au[61] (Figure 13f) where the HeI spectrum now looks like the HeII and the DOS.

For alloys the advent of photoemission data has provided data that allowed a choice to be made between the several electronic-structure models of alloying which were available. As explained by Ponec in this volume the rigid-band (or common-band model), which provides a DOS invariant of component composition but filled to different E_F levels as a function of composition, was the basis of much postulation concerning catalytic mechanisms some 30-40 years ago. Photoemission data disproved this model for many alloys and one could be forgiven for thinking that the XPS and UPS data suggest an alloy DOS which is just a weighted construction of the DOS of the pure components. This would indicate that the d-electrons within the valence region behave in a predominantly atomic fashion and that the influence of neighboring atom potentials is not drastic, a premise of the tight-binding description. Figure 16 shows the XPS valence region for a range of Ag Pd alloys.[62] One can see that it is an oversimplification to describe the spectrum just as a weighted superimposition of pure Ag and Pd since, as the concentraton of one of the components becomes diluted, its XPS features narrow as well as decrease. This can be accounted for by the loss of some band-character in the dilute component as the separation of its individual atoms increases. The width will never become atomic-like, however, because of the residual interaction with the free electron component of the host metal. The coherent potential model[63] of alloying apparently is capable of successfully predicting the photoemission features described above.

UPS alloy data provides similar information on the valence levels to XPS, with the qualifications that it will be measuring much more of the surface electronic structure and at low hν there may be possible k-dependent modification. The Cu/Ni system has been studied in detail.[64] The results indicate that the Cu and Ni d electrons form largely independent levels and there is no transfer of electrons from Cu to Ni to fill the Ni 3d shell, as would be expected from rigid band theory. XPS data on the Cu/Ni system leads to similar conclusions.[62]

Some of the transition metal chalcogenides have excited a lot of experimental and theoretical interest recently because of their layer structure (chalcogen-metal-chalcogen; weak bonds between chalcogen atoms of successive layer) and because their electrical properties range from insulating (ZrS_2) to superconducting ($NbSe_2$). Figure 17 shows the UPS spectra of a number of such compounds[65] to illustrate the general form of their DOS which approximately conforms to the Wilson and Yoffe rigid-band model[66] for the series. The model predicts a filled σ-bonding level, largely of chalcogen p-character, a non-bonding metal d_{z^2}, $d_{x^2-y^2}$' and d_{xy} group and an empty σ* antibonding level well above E_F. It is the position and occupancy of the non-bonding metal d levels which determines the electrical properties. In Figure 17 we observe the progressive filling of a d band near E_F from Zr to Mo. ZrS_2 is an insulator because there are only sufficient electrons to populate the σ bonding band (p states) which is well below E_F. TiS_2 should be the same but there is a weak band at E_F which, it has been suggested, is due to electron donation from excess Ti into the non-bonding d level. Nb and Ta have an additional electron, and so their chalcogenides have a partly occupied d band, and thus become semiconductors or metals depending on its exact position. Mo compounds have two electrons available for the d band, but the band is lower lying and now overlaps the p band.

The XPS spectra of the transition metal chalcogenides[67] confirm most of the features of the UPS studies, though the resolution is lower. However, they reveal clearly the where-abouts of the chalcogen s levels, which is not obvious from the UPS data, and also indicate that part of the structure ascribed to chalcogen p states in the UPS data must be due to scattered electron density. The monochromatized AlK_α XPS results for $NbSe_2$ and MoS_2 are included in Figure 17 for comparison with UPS. Note that, probably as a result of cross-section and (for UPS) scattered background effects, the relative intensities of the different sets of fine structure differ in the two spectra.

The angular resolved UPS spectrum of $TaSe_2$ (type 1T) has been discussed in the chapter by JWG and will not be discussed again here. In an extension to their original work, Smith and Traum[68] showed that the angular resolved UPS of Ta Se_2 (type 2H) was rather different and that the difference is apparently a simple

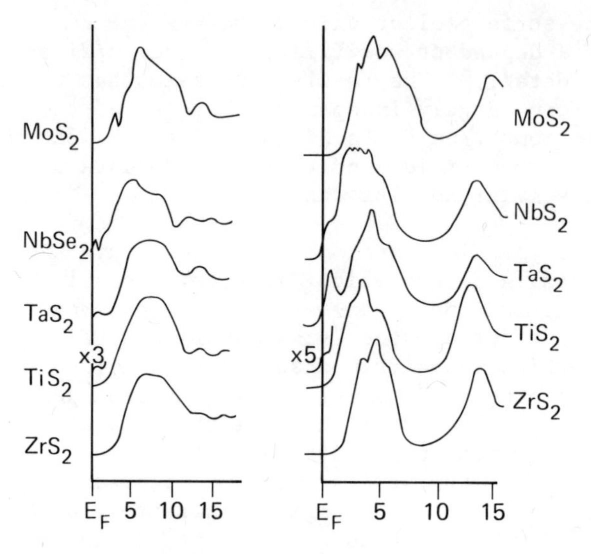

Figure 17. (left) HeII UPS of some layer chalcogenides[65] (right) XPS of some layer chalcogenides [67].

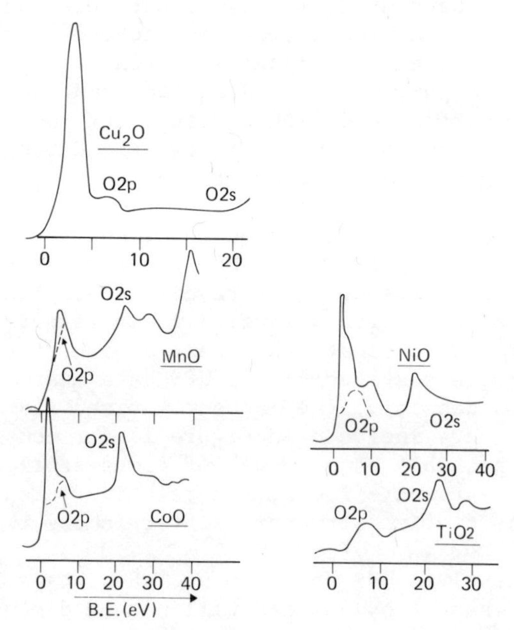

Figure 18. XPS of some mono and dioxides of transition metals.[69]

consequence of the symmetries of stacking in the two types. Each layer of $TaSe_2$ has 3-fold rotational symmetry. In the 1T type successive layers have the same orientation so that 3-fold symmetry is retained throughout the bulk. In the 2H type successive layers are rotated by 180° so that the overall symmetry is hexagonal. The trigonal symmetry of 1T - $TaSe_2$ shows up in the angular resolved photoemission, whereas that of 2H - $TaSe_2$ is close to 6 fold. However, because the photoemission from the second layer of $TaSe_2$ is weaker than from the top layer (attenuation) the radial plot of the azimuthal photoemission dependence is, in fact, the superimposition of two 3-fold patterns, one having a weaker intensity than the other. Smith and Traum conclude that a symmetry analysis of UPS data offers a powerful way of separating surface contributions from bulk contributions in those cases where the surface atoms are arranged with different rotational symmetry from the bulk (reconstructed clean metal surfaces, for instance).

The electronic structures of bulk metal oxides are of interest in the context of metal surface and adsorption studies because they may serve as models or at least provide assistance in the analysis for metal/O_2 interactions.

The XPS measurements on the valence levels of transition metal oxides all exhibit the same general pattern: if there are d-electrons in the compound one finds the d-electron density of the metal component near E_F, and O(2s) bands lying deeper (cf. the metal d, p and s levels of the layer chalcogenides). The separation between O(2p) and (2s) levels is always about 16eV,[69] close to the free atom value. This is what would be expected in a fully ionic model, but in fact cannot be considered as good evidence for the validity of such a model since various properties of the transition metal oxides indicate that hybridization between metal d and oxygen 2p levels (covalency) is important in some cases. The XPS spectra of some monoxides and dioxides are shown in Figure 18.[69] One can see that for MnO the levels derived from 3d and 2p overlap strongly, suggesting that hybridization may be significant. NiO presents an instructive example since although it is the oxide for which most experimental work (XPS and many other techniques) and theoretical work has been done, yet it is not clearly understood. In the XPS spectrum it is not at all obvious where the O(2p) levels lie, though O(2s) is clear enough. There are four possible explanations for the structure between 0 and 14eV. The first is that only the feature at 2eV represents the Ni3d levels, the peaks at 3.8 and 9eV being satellites due to multi-electron excitation and the O(2p) level being weak and buried in the 6eV region. Support for this view comes from the Ni 2p core level spectrum which shows similar satellite structure. However, in other Ni compounds where satellites are observed in the (2p) level, similar structure is not observed in the valence region, which suggests that the structure in the NiO valence region may be due to something else. The second possibility is that the entire structure from 0-14eV is Ni3d, with O(2p) being weak and buried at 6eV. There are theoretical models which fit either of these cases, i.e., narrow 3d level and separate O(2p) level; or 3d and O(2p) strongly hybridized to spread the 3d band considerably. A third possibility is that the first two features are genuine Ni3d structure and only the 9eV peak a satellite, perhaps a surface plasmon. Finally, a fourth and more unlikely possibility is that the 9eV feature represents O(2p) and that the O(2p)-(2s) separation is not 16eV in NiO. If the latter explanation is discounted, why is the O(2p) level so weak so as not to be observed? Here we return to arguments about relative cross-sections. The startling effect that gross differences in cross-section can have on a spectrum is illustrated by the case of ReO_3.[70] Covalency effects are very strong in ReO_3 so that the calculated DOS in the 0-10eV region has a combination of Re 5d and O(2p) character throughout (Figure 19a). It is clear from a comparison of Figures 19a and b, representing the XPS valence spectrum of ReO_3, that the strong features in the DOS at ca. 3eV, almost entirely O(2p) in character, make little contribution to the XPS spectrum. This suggests that the relative cross-section for O(2p) is very low (perhaps twenty times less than for Re(5d)) and that the observed XPS spectrum really represents only the Re(5d) character in the total DOS. It has been suggested that the experimental XPS spectra of other oxides (particularly NiO) should be reconsidered in the light of the ReO_3 results.

Perhaps because of instrumental difficulties (e.g., charging effects) or from difficulty in obtaining samples which are stoichiometric in the surface region, UPS studies of bulk metal oxides have not as yet been very common, particularly in the HeI, HeII photon energy range. More common are experiments which trace the effect of adsorption of oxygen and the later stages of oxidation of metal surfaces. Figure 20 shows the HeII and XPS spectra of Cu_2O grown from the metal substrate,[71] and the HeI[72] and XPS[69] spectra of NiO produced in a similar manner. It must be emphasised that there is no guarantee that these are stoichiometric samples, and in fact, for NiO there is strong evidence for two oxygen species, one of which is probably adsorbed atoms.[73] Apart from the improved resolution for the HeII case and a difference in the relative intensities of the Cu(3d)/O(2p) features reflecting the lower relative cross-section of the 2p level at XPS photon energies, the HeII and XPS spectrum of Cu_2O are quite similar. Just as for the XPS results discussed above, the interpretation of the UPS of NiO is still not definitely settled. Relative O(2p) cross-sections at 21.2eV photon energy are even higher than at 40.8eV and so the O(2p) level centered at about 5eV below E_F, shows up strongly. This adds support to the idea that in the XPS, O(2p) is present at about 5eV but is

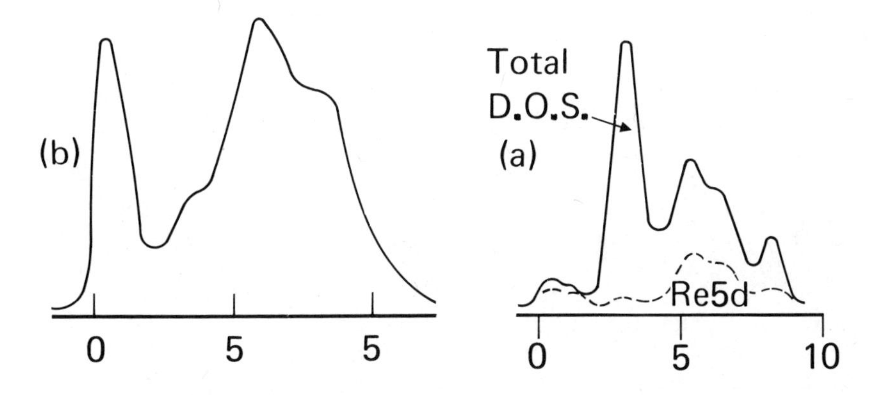

Figure 19. (a) Calculated DOS for ReO_3; (b) XPS of ReO_3[70].

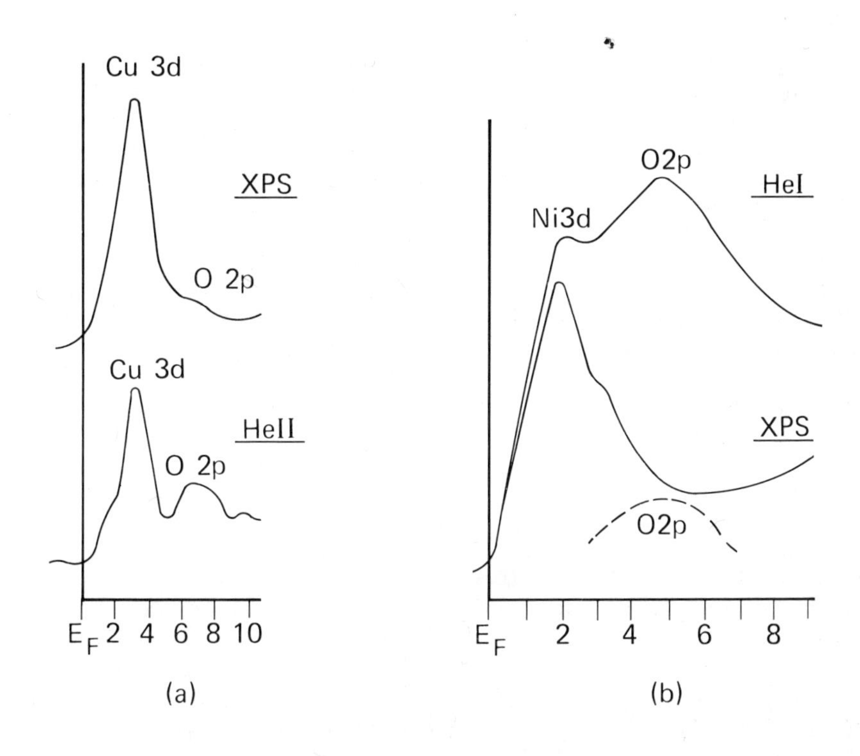

Figure 20. (a) HeII and XPS spectra of Cu_2O[71]
(b) HeI[72] and XPS[69] spectra of NiO.

not observed because of its low cross-section and the superimposition of satellite structure (see above).

Returning to the electronic structure and spectra of metals and alloys, it is, of course, to be expected that the DOS in the surface region may be different from that of the bulk structure. It may be different in two distinct ways. First, the actual atomic structure might be different in the surface region (expansion, contraction, or reconstruction of a metallic surface - variation in atomic composition for alloys). Second, even if the atomic structure and composition stay regular up to the surface layer, the bulk must terminate at that point, and that itself will change the DOS in the surface region. It has been postulated, and in some cases calculated, that the modification will take the form of a band narrowing owing to the lower co-ordination of the surface atoms. In addition, new surface states may be formed. For further discussion relating to the theoretical treatments of metal surface DOS the reader is referred elsewhere in this volume.[74,75]

Difference between bulk and surface DOS are not easily detectable by XPS or UPS because the escape depths are such that the experimental spectra are always a mixture of surface and bulk features. One can switch from HeI to HeII to reduce the escape depth and in some cases observe a narrowing of the experimental spectrum (e.g. in Ni), which may be attributable to a narrow DOS at the surface, but it may also be due to k-dependent effects. Azimuthal angular resolved photoemission, as described earlier for $TaSe_2$, offers the possibility of separating experimental surface and bulk contributions in the special case when there is different rotational symmetry between the top layer and the bulk. Strong surface states are more easily assigned since they are often destroyed by adsorption, which may therefore be used as a probe for their presence. For instance the W(100) surface exhibits a strong surface state which is destroyed by adsorption of 1/4 monolayer of H_2.[76]

The use of AES to obtain DOS information on metal surfaces, by looking at WVV transitions, has been suggested several times in the literature.[77-79] Since many of the transitions for metals results in low K. E. Auger electrons the technique does probe surface DOS more than XPS does, though not more than HeII UPS.

Attempts to extract a DOS from a WVV Auger peak are based on the premise that the experimental peak shape is proportional to N(E) * N(E) of the valence bond. There are several difficulties in applying such an analysis. If the spectrum is recorded in the dN(E)/dE mode it must be re-integrated. A rather subjective subtraction is then required to separate the small Auger peaks from the large and rapidly varying scattered electron background.

(It has been suggested that this can be quantitatively be removed by a dynamic background subtraction scheme).[79] Since the Auger process involves the transition of an electron from valence band to core-level hole, the overlap integral for this transition is likely to vary strongly across the width of the valence band, the variation being dependent on the electronic character across the band and the nature of the core-hole. Thus the experimental spectrum may be strongly modified by a varying transition probability across the peak width. Finally the fact that the initial state for the Auger process has a core-hole may itself modify the valence-level DOS from that characteristic of the neutral system.

An example of an attempt to relate a WVV Auger transition to a DOS is given in Figure 21.[78] The experimental LVV Auger peak of Al was corrected by the background subtraction shown, an additional correction for electron scattering out of the Auger peak itself was made, and the result compared to an N(E) * N(E) curve generated from a bulk DOS calculation. The width of the experimental peak is in reasonable agreement with the calculated curve, but the peak shape is badly skewed. This could be due to one or several of the aforementioned problems (the author suggested transition probability variation, but subsequent work indicated that attempts to include transition probabilities worsened the agreement),[80] or it could be due to a genuine difference between surface and bulk DOS. Recently Houston,[79] using the dynamic background subtraction scheme[81] on essentially the same data, produced a corrected experimental curve whose shape was in good agreement with the theoretical curve, but which was somewhat narrower. He attributed the difference to a genuine narrowing of the surface DOS. The differences between the two analyses of the same transition highlight the difficulty of obtaining DOS information by this method.

The WVV Auger spectra of Ag,[78] Cu,[82] and Zn[82] cannot be adequately represented by a bulk N(E) * N(E). They contain fine-structure remeniscent of atomic spectra. This may be due to localization effects caused by the core-hole initial-state in these non free-electron like materials.[79]

The INS process, being equivalent to a WVV Auger process, suffers from similar problems as AES in extracting DOS information. A very complex mathematical procedure for unfolding the DOS from an experimental INS curve has been developed by the authors,[83] though it could probably be simplified without losing much accuracy. INS does have two advantages which stem from the same source. Since the incoming He^+ ion does not penetrate the surface the neutralization reaction proceeds by a tunneling process from the surface, thus reducing the probing depth to a couple of layers at most. Any DOS information obtained therefore

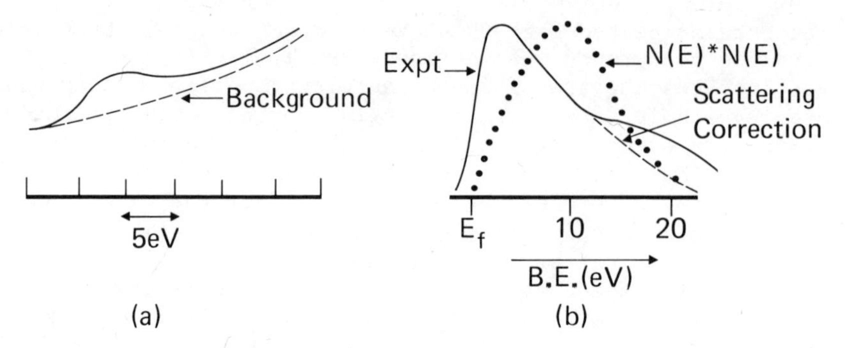

Figure 21. (a) Electron induced LVV Auger spectrum of Al (b) Comparison of N(E)*(E) curve generated from a theoretical DOS to the background subtracted LVV Auger spectrum[78].

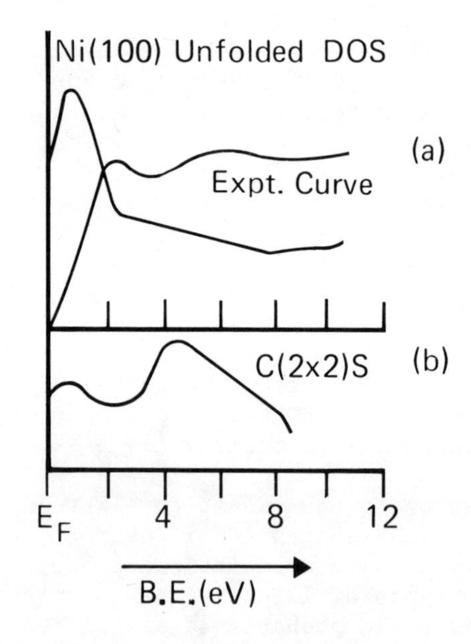

Figure 22. INS spectra of clean Ni(100) (upper traces) and that of the C(2x2)S surface formed on adsorption of sulphur (lower trace)[17].

relates specifically to the surface. A second consequence of the fact that no electrons originate below the surface is a lack of a large scattered electron background thus removing the subjective background subtraction necessary in AES. Figure 22(a) shows both the raw INS data for Ni(100) and the unfolded DOS curve.[17] The width of the d-band is slightly narrower than is the HeII UPS, presumably due to a genuine narrowing of the surface DOS.

Adsorbate - Surface Interactions

Adsorption of a gas phase molecule will induce changes in the surface DOS which for convenience can be split into two categories. The DOS of the substrate may undergo changes owing to the depletion of electron density for those electrons which become involved in bonding with the adsorbate. These changes are likely to the subtle (except in the special case of supression of a strong surface state) and difficult to quantify because a separation of bulk and surface contributions to the experimental spectrum is still required.

Effects are certainly observed in INS, for instance the d-bands of Ni(100) are almost completely depressed by the adsorption of half a monolayer of S[17] (Figure 22b), but the effect is likely to be a combination of genuine change in the Ni d-band electron density and a strong attenuation of the emission from these levels by the S overlayer. In UPS d-band electron density changes have also been observed[84] and the use of "difference curves" (subtraction of the clean surface spectrum from the covered surface spectrum) in principle, helps separate mere attenuation effects from electron density depletion. The regions of depletion shows up as small dips in the difference curve. They are small, except again in the case of strong surface states, because a lot of the total photoemission in UPS comes from deeper than the outer layer, and so far little interpretive effort has been directed towards them.

The adsorbate will also induce <u>additional</u> features in the DOS which are directly or indirectly related to the valence levels of the gas phase molecule. These features are more striking and more interpretable than the substrate electron density changes and so our attention will be directed here. As explained in Section IV, for very weak interactions, such as condensation or physical adsorption, the relationship between gaseous and adsorbate induced features is very close - in fact, the latter are merely broadened versions of the former. Stronger interactions will be expected to modify one or more of the adsorbate gas phase valence level orbitals. If it can be established as to which levels are affected, the orbitals involved in the chemisorption bonding can be identified and (possibly) some information about the strength of the interaction may be obtained. Once one has moved away from the simple condensate overlayer situation one must exercise discretion in making simple analogies to gas phase spectra, particulary in situations where ordered structures are found. The effects of the periodicity[1] may be manifested by strong angular dependent and $h\nu$ dependent (k-conservation) structure in the adsorbate induced levels in a similar fashion to that appearing for the clean substrate (see earlier). Experimentally this could result in peak splittings and shiftings as a function of angle or $h\nu$ so that a comparison with the gas phase at any one angle or $h\nu$ could be misleading. In practice effects which split or significantly shift peaks have not yet been observed for molecular adsorption, though changes in relative intensities of bands occur. In the case of $h\nu$ variation, this is due to the expected cross-section variation[85,86] and for angular variation can apparently be explained[87] on the basis of the angular distributions expected for the oriented free molecule itself.

In the case of dissociative adsorption, e.g., W/O_2, stronger angular effects have been observed,[88] and split peaks which may correspond to orbitals derived from the different symmetry O2p levels (p_xp_y or p_z) are clear. Again, however, only relative intensity changes of these levels as a function of angle are observed.

Since the valence level electron density at the substrate surface is altered by the chemical process, an effect should be observable on the substrate core-levels, as evidenced by XPS or AES. There has been some controversy over the magnitude and the interpretation of such effects which will not be regenerated here.[18, 89] The facts are that the difference in B.E. of the surface region substrate atoms from the bulk is slight (a few tenths of an eV). Whatever the explanation, which must include consideration of (a) the terms going to make up the observed "chemical shift" and (b) the number of substrate layers through which the effect is spread, the experimental situation is that at the usual spectrometer geometry ($\theta = 45°$) it is impossible to observe more than a slight asymmetry developing in the substrate core-level XPS corresponding to the surface component with shifted B.E.[18] Going to low θ (5 or 10°) and thus increasing the percentage of the signal coming from the outermost substrate layer makes detection easier,[89] but interpretation in terms of the chemisorption bonding is still lacking.

As is the case for the valence levels, the study of the B.E. of the adsorbate core-levels is more fruitful in revealing the nature of chemisorption reactions.

TABLE III

$B.E._g$ [(a)] and $B.E._{ads}$ [(b)] Values(eV) for N_2O (80K on Ni)

Level	2π	4σ	1π	3σ	N(1s)	O(1s)
$B.E._g$	12.9	16.4	18.2	20.1	408.5,412.5	541.2
$B.E._{ads}$	6.3	9.7	11.5	13.4	402.0,406.0	534.6
$B.E._g-B.E._{ads}$	6.6	6.7	6.7	6.7	6.5,6.5	6.6

(a) Referenced with respect to the vacuum level

(b) Referenced with respect to E_F of the substrate

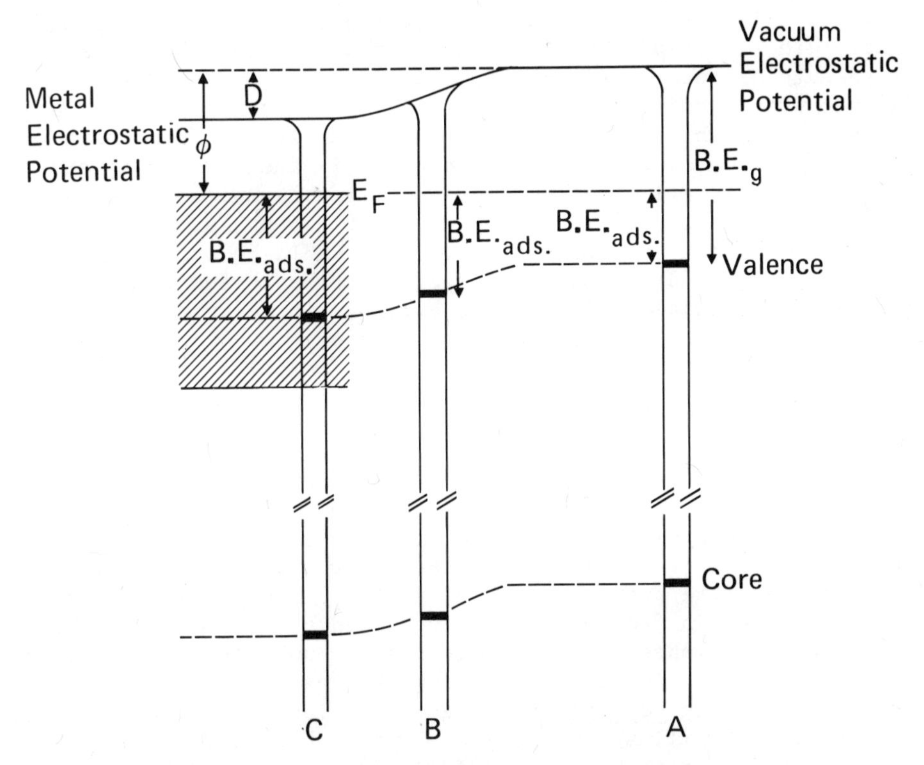

Figure 23. Adsorbate energy level behavior as a function of distance of the adsorbate from the metal surface in the limit of zero bonding interaction between molecule and surface. (see script)

Before going on to give specific examples of XPS and UPS studies of adsorbate/substrate systems to illustrate the information obtained by a comparison of gas phase and adsorbate B.E.'s it is necesary to consider first exactly how the comparison is experimentally made and what terms are involved.

The simplest situation to consider is one in which there is expected to be no bonding between adsorbate and substrate. Condensed layer are the best approximation to this, so we return to the XPS/UPS spectra of N_2O on Ni at 80K (Figure 12). It was noted in Section IV that the relative B.E.'s of all the N_2O levels remained unchanged in the condensed situation, making a "fingerprinting" identification of molecular N_2O easy. Here we will consider the absolute B.E. values. The gas phase values, $B.E._g$, which are experimentally determined with respect to the vacuum level, and the condensed state values, $B.E._{ads.}$, which are all experimentally determined with respect to the Fermi level of the substrate, are given in Table III. The difference between the values, $B.E._g - B.E._{ads.}$ varies between 6.5 and 6.7eV which is, within experimental error, constant, as noted above. What is responsible for this difference, remembering that no bonding effects are involved? In Figure 23 we consider what happens to the valence or core-level orbital energies, ε_j, of the N_2O molecule when adsorbed at different distances from the Ni surface. At position A, which is defined as sufficiently far from the surface to be outside the surface dipole, in the absence of any bonding effects, all orbital energies, ε_j, referenced to the vacuum level are unchanged compared to the gas phase. If UPS or XPS B.E.'s directly measured ε_j, $B.E._{ads.}$ would be simply related to $B.E._g$ by the addition of the experimental work function of the adsorbed situation, ϕ, because $B.E._{ads.}$ is measured experimentally with respect to E_F substrate.

$$B.E._g - B.E._{ads.} = \phi \tag{1}$$

The photoemission process does not measure ε_j, however, but the difference between the initial state of the total system before ionization and the final state of the total system after ionization. This is only equal to ε_j if, when an electron is removed from the jth orbital, all the other orbital energy levels remain frozen (the approximation of Koopmans' Theorem). In fact, they do not remain frozen and, as explained by JWG,[1] there is a relaxation of valence electrons towards the positive hole created by the photoionization process, which reduces the energy of the final state by the Relaxation Energy, E_j^R, less than ε_j. This is schematically illustrated to the left of Figure 24 for an atom or molecule. In gas phase work E^R is often, for calculation and explanation purposes, divided into two components, the atomic relaxation relating to the relaxation of the valence levels of the atom which posesses the core-hole, and the extra-atomic

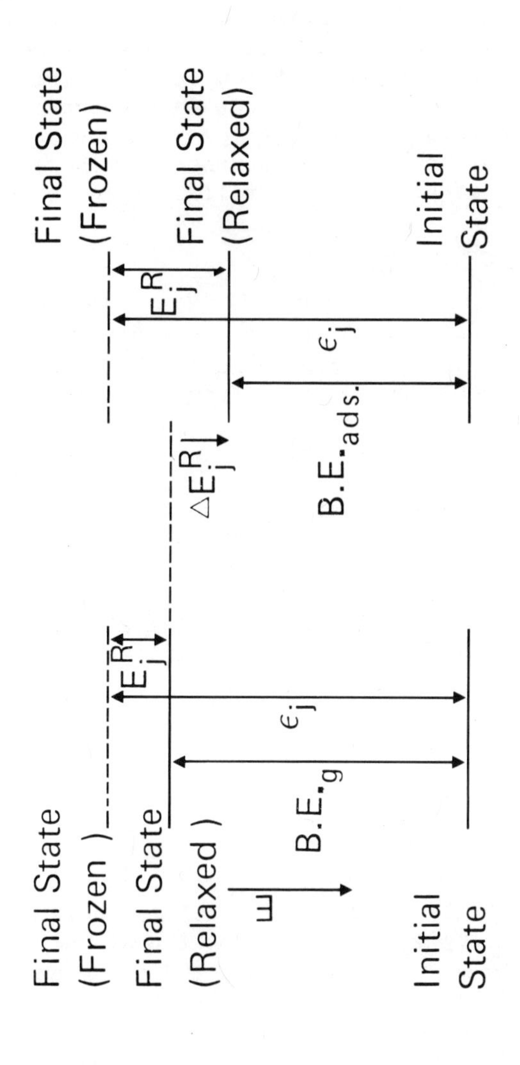

Figure 24. Relationship between orbital energy, ϵ_j, experimental binding energy, B.E. and relaxation energy E_j^R, for a free molecule (left), and for an adsorbed molecule (right) in the limit of zero bonding interactions between molecule and surface.

relaxation relating to the additional relaxation possible in molecules from the valence levels of the other atoms in the molecule.

If the molecule is now adsorbed or condensed at a surface the amount of extra-atomic relaxation occurring in the final state after photoemission is likely to be larger than for the free molecule because of the availability of valence electrons both from the substrate and from the other adsorbed molecules present. (A similar situation is, of course, true in a comparison between free atoms or molecules and bulk compounds). This is illustrated to the right of Figure 24. Thus on adsorption, the increase in extra-atomic relaxation, ΔE_j^R, lowers the experimental binding energy of the adsorbate by that amount. We therefore have a corrected version of Equation 1, still for the no bonding situation:

$$(B.E._{g.} - B.E._{ads.})_j = \phi + \Delta E_j^R \qquad (2)$$

In many condensed situations, ϕ, $B.E._{g.}$, and $B.E._{ads.}$ are known so that if one assumes the condensed molecule to be at position A in Figure 23, ΔE_j^R may be found from Equation 2. For N_2O/Ni ϕ is about 5.5eV (not accurately determined) so that ΔE_j^R for the various orbital ionizations varies between 6.5 and 6.7eV minus 5.5eV - i.e., about 1eV. For many molecules, the values of ΔE_j^R obtained on condensation are constant for all j and lie between 1 and 3eV (usually nearer 1eV). It has been suggested by JWG[1] that ΔE^R can be accounted for by an image charge potential model for a point charge above a metal surface. In the case of a condensate molecule, it seems likely that relaxation from surrounding condensate molecules (solvent effect) would be more important.

We now have a diagnostic test for an adsorbate with little or no bonding to the surface. All the adsorbate's gas phase orbitals should be recognizable, their relative B.E.'s should be the same as in the gas phase, and the value of $(B.E._{ads.} - B.E._{g} - \phi)$ should be small (< 3eV).

There is one further interest for a molecule in position A in that $B.E._{ads.}$ should vary with coverage by the amount the surface dipole (and hence ϕ) varies, so that the B.E. remains fixed with respect to the vacuum level[125]. The only case I know of where the UPS B.E.'s of condensates have been studied as a function of coverage and variation in ϕ, is for a MoS_2 substrate.[54] Here it was observed that the absorbate levels were tied to the vacuum level, indicating that the condensed molecules are effectively outside the surface dipole. On the other hand it appears that the $3p_{3/2,\,1/2}$ B.E.'s for Xe physisorbed on W(100) remain constant with respect to E_F over a wide range of coverage and change in ϕ.[90] In terms of Figure 23 this is consistent with the Xe atom being at position C, completely inside the surface dipole, D, since in position C a molecule with no bonding interaction will have its ε_j values increased by D compared to a molecule at position A. Thus:

$$(B.E._{g} - B.E._{ads.})_j = (\phi - D) + \Delta E_j^R \quad (3)$$

As the coverage changes, D changes, but ϕ - D remains constant so that $B.E._{ads.}$ remains constant. This does not necessarily imply that $B.E._{g.}$ - $B.E._{ads.}$ is smaller than at position A because though the term ϕ - D is involved instead of ϕ, ΔE^R will probably be larger than in position A because of the proximity of the available electron density of the metal surface.

For a non-interacting molecule in position B, which is merely defined as somewhere between A and C (Figure 23), then some part of D must be included in the equation:

$$(B.E._{g.} - B.E._{ads.})_j = (\phi - fD) + \Delta E_j^R \quad (4)$$

In this case $B.E._{ads.}$ will neither be tied to E_F nor the vacuum level, as a function of coverage. Equation (4) is a generalized form which also covers the situation for Equations (2) and (3) where f would be 0 and 1, respectively.

Let us now consider the case where a bonding interaction (chemisorption) taken place. Some of the initial state ε_j values of the adsorbate will now be modified compared to the free molecule values (those of the valence orbitals involved in the bonding, and the core levels which respond to the change in electron density in the valence shell) so that the generalized Equation (4) now becomes

$$(B.E._{g} - B.E._{ads.})_j = (\phi - fD) + \Delta E_j^R \pm \Delta E_j^B \quad (5)$$

where $\pm \Delta E_j^B$ represents the initial state chemical shift of ε_j due to adsorption, the quantity we are really trying to determine. Whereas for the non-interacting situation it was noted that ΔE_j^R is experimentally observed to be constant for all j, it is not to be generally expected that this will be so in the chemisorbed situation because those orbitals which are involved in the chemisorption bond with the metal surface might be expected to have larger ΔE^R values than the rest. In practice one cannot experimentally separate the terms ($\pm \Delta E_j^B + \Delta E_j^R$); values for f and D are unknown (it is unrealistic in the chemisorbed situation to expect f to be zero, i.e., the molecule to be at position A in Figure 23); and both ΔE_j^B and ΔE_j^R may change with coverage. Experimentally it has often been observed that $B.E._{ads.}$ for a chemisorbed species either remains constant with respect to E_F

as a function of coverage, or varies only slightly (few tenths of an eV).[91] This could mean that the molecule or atom is within the surface dipole at position C(f=1) and ΔE_j^B and ΔE_j^R do not vary with coverage, or that f ≠ 1 and changes in D, ΔE_j^R and ΔE_j^B approximately cancel. For the physisorption of Xe considered above as an example of a non-interacting situation, it should be remembered that there will, in fact, be a small $\pm\Delta E_j^B$ term and the constancy of the $5p_{3/2,1/2}$ B.E.s with respect to E_F may also be due to a cancellation of term. The Xe core-levels have been reported as undergoing small changes in $B.E._{ads.}$ as a function of coverage.[92]

In general terms, then, we can expect that the chemisorbed situation may be distinguished from the non-interaction situation by the following: $B.E._{g.} - B.E._{ads.}$ will not be constant for all j; for the valence levels not involved in bonding $B.E._{g.} - B.E._{ads.}$ will be expected to be larger than for the situation where the same molecule is condensed at the surface, because ΔE^R will be larger; and for the other energy levels $B.E._{g.} - B.E._{ads.}$ may be larger or smaller than the non-bonding levels depending on the sign of ΔE^B. There is unlikely to be any simple relationship of $B.E._{ads.}$ to the Fermi level of the substrate or the vacuum level as a function of coverage.

Some Specific Examples of Adsorbate/Substrate Studies

The examples given here come from three categories of surface reaction: (1) Chemisorption without dissociation; (2) Chemisorption accompanied by dissociation; and (3) Chemisorption accompanied by dissociation and extensive reaction (> 1 monolayer) with the surface. Whether a particular adsorbate/substrate interaction falls into categories (1), (2), or (3) often depends on temperature. It is therefore not sensible to try and split the discussion up into sections dealing with the three categories separately.

<u>C_6H_6, C_2H_4, and C_2H_2 on Ni, Cu, Fe.</u> Figure 25 shows the UPS spectra (HeI) for C_6H_6 adsorbed on nickel at 77K and 300K.[49] A schematic representation of the gas phase spectrum is also shown. At 77K the intensities of the absorbate induced features indicate that several layers are present. The gas phase and adsorbate B.E. scales are matched using Equation 2

$$(B.E._{g.} - B.E._{ads.})_j = \phi + \Delta E_j^R$$

where a value of ϕ of 5.5eV leads to a ΔE_j^R of about 1.3eV for all j. Thus we have identified a non-interacting condensed

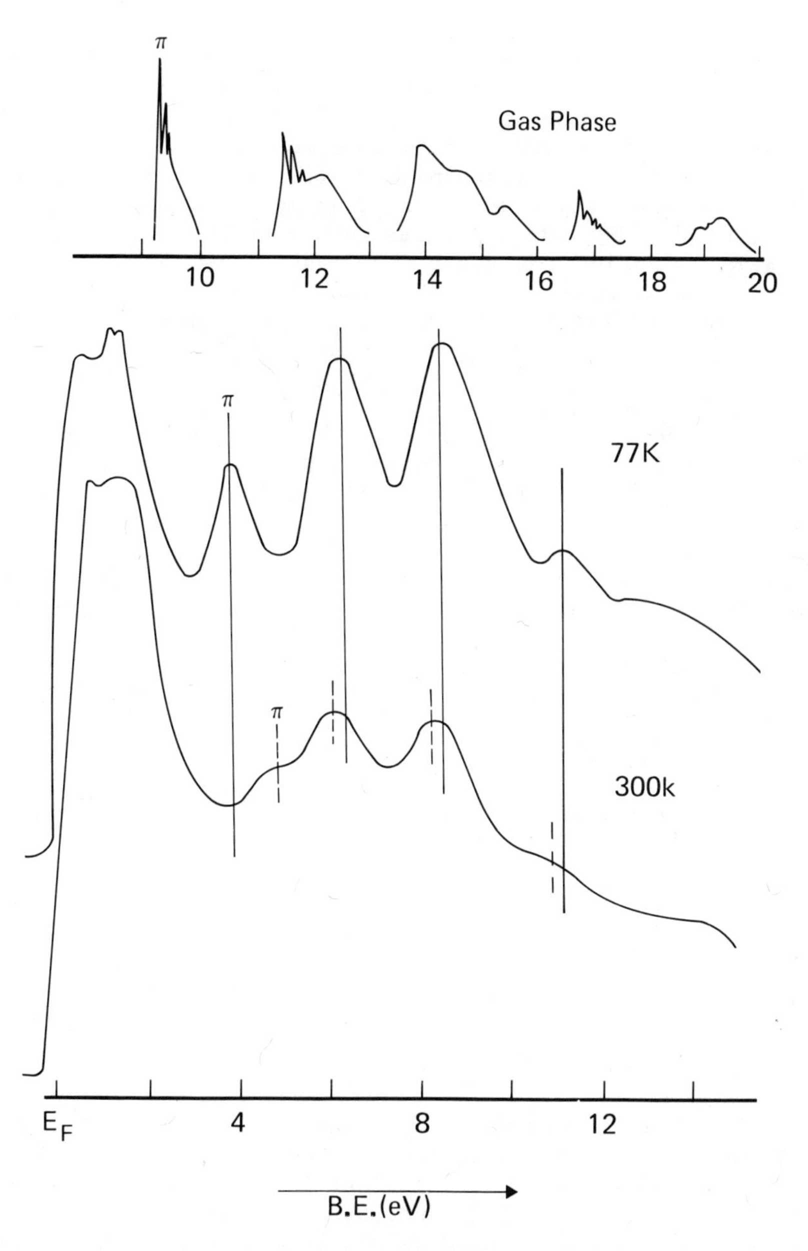

Figure 25. HeI UPS spectrum of C_6H_6 adsorbed on Ni 77K and warmed to 300K.[49] The gas phase spectrum is shown above.

species (surprise!). Warming to 300K desorbs the multilayer of C_6H_6. All the B.E.$_{ads.}$ of the remaining C_6H_6 features are shifted compared to the 77K situation, but only the top π level has shifted relative to the others, from which it is concluded that this is the orbital involved is the chemisorption bonding. Considering the comments previously made concerning comparison of condensed to chemisorbed situations we can go to Equation (5), and if we assume f = 1, explain the shifts in B.E.$_{ads.}$ for those valence levels not involved in the bonding by increasing ΔE^R from 1.33V to 1.7eV. For the top level we have $(\Delta E^R_\pi \mp \Delta E^R_\pi) = 0.5$eV. This is as far as one can get using Equation 5. Demuth and Eastman,[53] who studied C_6H_6, C_2H_4, and C_2H_2 adsorption on Ni(111) and were the first to try and extract bonding information from B.E.$_{ads.}$ shifts, make the further assumption that ΔE^R_π is the same as for the other levels (cf. above where I have stated that generally this is not to be expected), and therefore obtain a value of $\Delta E^B_\pi = -1.2$eV. They then use this value to calculate the chemisorption energy, ΔE^C, using the chemisorption model of Grimley[93] which is based on Mulliken's theory for a Donor Acceptor Complex.

Figure 26 presents a similar set of data for C_2H_4/Ni.[49] Adsorption at 80K produces condensed multilayers and a ΔE^R of about 1.5eV. Warming to 200K desorbs the multilayer and increases ΔE^R of the non-interacting orbitals to ca. 2.1eV uniformly. $\Delta E^R \pm \Delta E^B$ for the top π orbital is different at ca. 1.2eV, identifying this orbital as the one involved in the bonding. Assuming that ΔE^R_π is also 2.1eV gives a ΔE^B_π of -0.9eV from which a ΔE^C of 1.0 was calculated.[53] The same treatment for C_2H_2 yielded a ΔE^B_π of 1.5eV and a ΔE^C of 4.2eV.[53] These results are summarized in Table IV. The reaction

$$C_2H_4 \rightarrow C_2H_2 + H_2$$

is 1.8eV endothermic in the gas phase. From the figures in Table IV it would be expected to be exothermic in the chemisorbed state on Ni since

$$\Delta H^C_r = +1.8\text{eV} - \Delta E^C\ (C_2H_4) + \Delta E^C\ (C_2H_2) = -1.4\text{eV}$$

Trace C of Figure 26 shows what happens when the substrate temperature for C_2H_4/Ni is raised to 300K. All resemblence to the gas phase spectrum of C_2H_4 is lost and the spectrum becomes identical to that of the 300K interaction of acetylene with Ni. Thus, it appears that the experimentally determined ΔE^B_π values for C_2H_4 and C_2H_2 correctly predict the dehydrogenation reaction.[53]

Subsequent work has been carried out on Fe and Cu by Yu, et al.[94] Figure 27 shows the HeI UPS (difference spectra) of C_2H_4

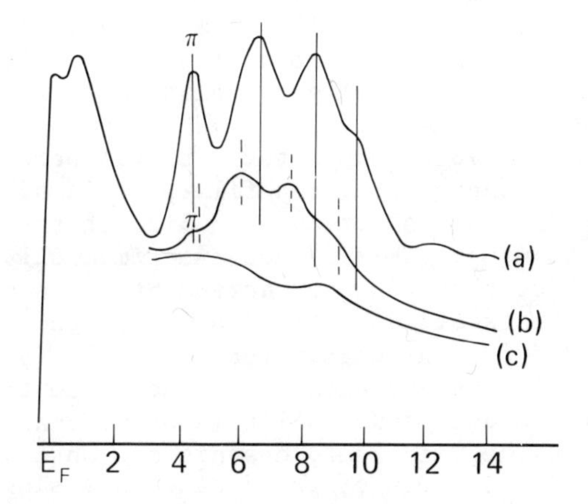

Figure 26. HeI UPS spectrum of C_2H_4 adsorbed on Ni at 77K (trace a), warmed to 200K (trace b), and warmed to 300K (trace c).[49]

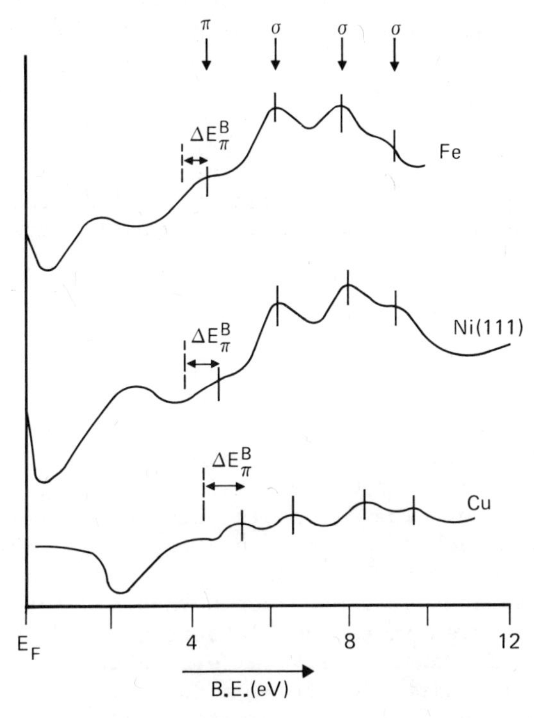

Figure 27. HeI UPS difference spectra of C_2H_4 adsorbed on Fe, Ni, and Cu.[94]

in the chemisorbed but undissociated state, together with Demuth and Eastman's results for Ni(111). Table IV gives the ΔE_{π}^{B} and ΔE^{C} values, and also the experimental thermochemical values for the heats of adsorption.[94] The correlation between ΔE^{C} and the thermochemical values is very poor. While the thermochemical data may be questionable, since the work was performed on spectroscopically uncharacterized films, it is significant that the calculated ΔE^{C} values for Cu also predict that C_2H_4 will dehydrogenate to C_2H_2. This is not experimentally observed. Yu et al.,[94] consider that the poor agreement between ΔE^{C} and the thermochemical heats of adsorption is due to the inappropriateness of the model used for calculating ΔE^{C} from ΔE_{π}^{B} (plus the fact that ΔE_{π}^{B} itself is suspect because of the assumptions necessary about ΔE_{π}^{R}). The model predicts little or no change in the substrate d-band when, in fact, changes are observed. It also requires the approximation of the d-band to a single d level for the calculation, and the calculated ΔE^{C} is therefore sensitive to the choice of position of the d level.

One is forced to the conclusion that at the present time whereas the identification of the orbital(s) involved in chemisorption is on a reasonably sound footing in UPS, a quantitative description of the bonding cannot be obtained from the measured binding energy shifts.

CO Interactions with Ni, Cu, W, Mo, and Ru. The Ni/CO interaction was the first adsorbate - substrate system to be investigated under adequate UHV and surface cleaniness conditions.[12] The HeI spectrum was recorded at 300K and an assignment made on the basis of Equation 6.

$$(B.E._{g.} - B.E._{ads.})_j = \phi \pm \Delta E_j^B \qquad (6)$$

The possibility of ΔE^{R} terms was not considered at that time. An HeI spectrum,[52] similar to the original, is shown in Figure 28 together with the assignment made then. The low $B.E._{ads.}$ feature was associated with the 5σ level and the higher feature with 1π. Tentative arguments were made concerning the nature of the bonding involved. The difficult point to explain with this assignment is the lack of any evidence for the 4σ level which should lie a few eV above 1π. For the HeI spectrum an argument could be advanced that it comes just beyond the cut-off point, but it is not observed in the HeII spectrum either which extends the accessible B.E. range by several eV. Alternatively, the cross-section for ionization for 4σ might have been reduced dramatically, but there is no theoretical basis to such a suggestion. Finally, possible strong angular effects might be invoked, but subsequent work has shown that the level is not detected with a hemispherical grid system collecting over all

TABLE IV

Data[(a)] for Chemisorbed Hydrocarbons on Ni, Fe, and Cu

Molecule	Parameter(eV)	Ni	Fe	Cu
	$-\Delta E_{\pi}^{B}$	1.2	-	-
C_1H_6	$-\Delta E^{C}$	1.7	-	-
C_6H_6	$-\Delta H$(expt.)	-	-	-
	$-\Delta E_{\pi}^{B}$	0.9	0.6	1.0
C_2H_4	$-\Delta E^{C}$	1.0	0.7	0.8
	$-\Delta H$(expt.)	2.5	3.0	0.8
	$-\Delta E_{\pi}^{B}$	1.5	41.9	1.9
C_2H_2	$-\Delta E^{C}$	4.2	5.6	3.7
	$-\Delta H$(expt.)	2.9	-	0.8

(a) This table is reproduced from Reference 94. The Ni data is taken from Reference 53.

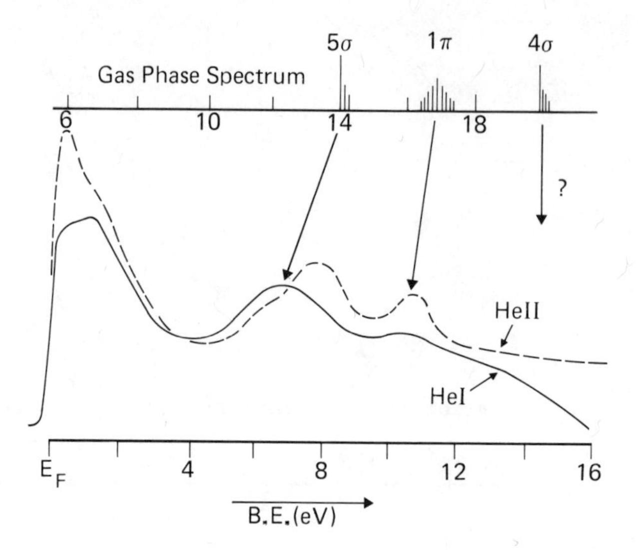

Figure 28. UPS of free CO and Adsorbed on Ni.[52]

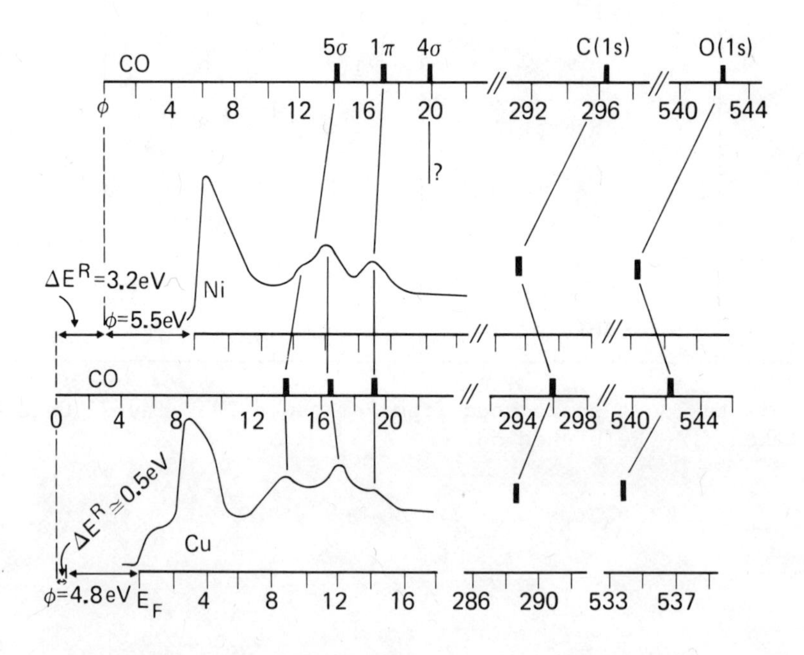

Figure 29. HeII and XPS spectra for Co on Ni and Cu, with alternative correlations to the gas phase spectrum.[52]

angles.[95] The original assignment was, in fact, incorrect, as was appreciated when (a) it was realized that large ΔE^R terms could be involved; (b) the HeI and HeII spectra were carefully compared; and (c) when the C(1s) and)(1s) core-levels were examined.[52] The HeII spectrum is shown in Figure 28 and again in Figure 29 where the core-level positions in the adsorbed state are added.[52] The HeII to HeI comparison reveals that the first band contains two, not one level, the relative ionization cross-sections of which change with $h\nu$. The adsorbate core-level values indicate that the assignment based on Equation 6, with a ϕ value of 5.5eV, cannot conceivably be correct because it implies a ΔE^B value of about 5.2eV for the C(1s) and O(1s) levels - far too large for any chemisorption interaction. With this information the natural assignment is to correlate both 5σ and 1π with the lower double band, and the missing 4σ with the higher band.[52,96] The changes in relative cross-section with $h\nu$ are in accord with such an assignment since they follow those of the free CO molecule. If the reasonable assumption that 4σ is not involved in the bonding is made a value of ΔE^R of 3.2eV is obtained for that level. Taking the approach of Demuth and Eastman and assuming all the other ΔE^R values to be 3.2eV also allows a match of gas phase and adsorbate scales (Figure 29) and leads to ΔE^B values of about - 1eV for 5σ and + 2eV for the core-levels.

The core-level shifts still seem rather large and it may be that ΔE^R is larger than 3.2eV for these levels. Gustaffson et al.[86] have extended the variation of $h\nu$ between 25 and 105eV both for CO in the gas phase and in the absorbed state on Ni. They also conclude that the relative intensity changes are only compatible with the lower B.E. band being a superimposition of 5σ and 1π and the higher B.E. represents 4σ. A detailed angular resolved study of Ni/CO has not yet been reported, but that of Ru(100)/CO has. (The Ru/CO UPS spectrum is quite similar to Ni/CO, and the same assignments are appropriate on the basis of HeI, HeII and XPS core-level comparisons). Fuggle and Menzel[87] have measured the changes in relative intensities of the lower BE band to higher B.E. band as a function of polar angle variation. The variations can be explained in terms of the angular variations expected for free CO molecules oriented perpendicular to the surface, carbon down, provided the $5\sigma + 1\pi$ low B.E. and 4σ upper B.E. assignment is followed.

The HeII spectrum of CO on polycrystalline Cu at 80K[57] is also shown in Figure 29. The heat of adsorption is much lower than for Ni and adsorption does not occur at 300K. One might intuitively expect, therefore, that this weaker bonding would be reflected in the UPS/XPS spectra by a smaller ΔE^R for all orbitals and a smaller ΔE^B term for the bonding 5σ orbital. This appears to be what happens as the valence levels and the core-levels are moved to higher B.E. for CO/Cu compared to CO/Ni and the separation between valence levels is now similar to that in the gas phase. This Cu/CO data should be regarded as preliminary, however, since it has been reported that different spectra can be obtained on single crystal surfaces.[97] Essentially the same data as that shown in Figure 29 has also been interpreted in terms of the first B.E. band being a composition of 1π and 5σ, the second being 4σ (cf. Ni) and the third being a spurious energy loss feature.[98]

The shake-up structure associated with the O(1s) of Ru/CO has been studied. The spectrum is reproduced in Figure 30.[89,91] A full interpretation has not yet been given but it was noted that the 16.6eV shake-up peak was at similar energy to a satellite observed for free CO and the 7eV shake-up peak was similar to one observed for gaseous metal carbonyls which has been assigned to a metal-to-Ligand transition. A detailed interpretation should contain information on the nature of the CO-Ru bond, or alternatively a theoretical description of the bonding must be able to reproduce the shake-up satellite structure and intensity. In principle, the sum-rule relating spectral weighting to relaxation energy, E^R (total E^R, not ΔE^R) ought to be useful also. E^R is the difference in energy between the Koopmans' Theorem (KT) core-binding energy (frozen orbital situation) and the experimentally observed lowest binding energy main peak which corresponds to the relaxed final ionized state. The sum rule operates like a lever-arm principle with the KT B.E. position as the pivot. The intensity of the main peak multiplied by E^R must be balanced by the intensity of the satellite structure on the high B.E. side of the pivot multiplied by the energy separation of that structure from the pivot. In addition, in the sudden approximation limit, the total intensity is unity. In principle, the rule implies that for a free molecule the pivot position - i.e., the KT B.E., which is equal to ε_j, can be found from an analysis of the satellite intensity. In practice this is almost impossible to apply because the sum rule involves intensity X energy separation terms, which means that very weak satellite structure a long way from the pivot is important, but it is just this structure which is difficult to evaluate experimentally. In the adsorbed situation a greater amount of relaxation occurs in the final state (greater by ΔE^R) which should therefore imply, from the sum rule, that the percentage intensity in the main line is decreased and that in the satellite-structure on the other side of the pivot increased. In a non-interacting situation an intensity analysis should just give us back the same KT B.E. and, therefore, ε_j as for the free molecule. In addition, the satellite structure should not alter much in character. Whether an increase in the proportion of satellite structure intensity can experimentally be observed is as yet unchecked, but will depend upon the relative sizes of E^R and ΔE^R. In a bonding

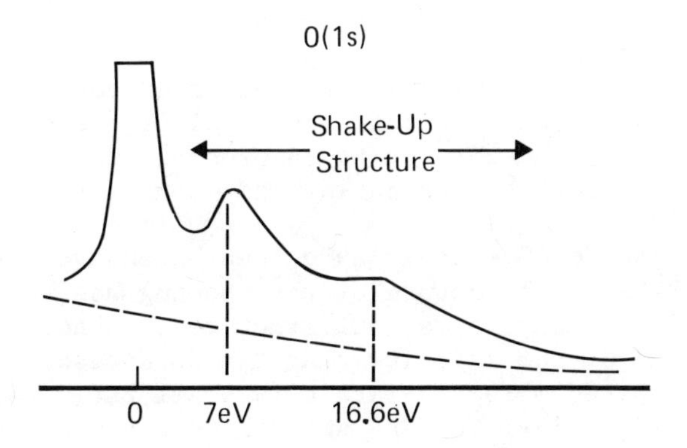

Figure 30. O(1s) XPS spectrum of CO adsorbed on Ru.[89]

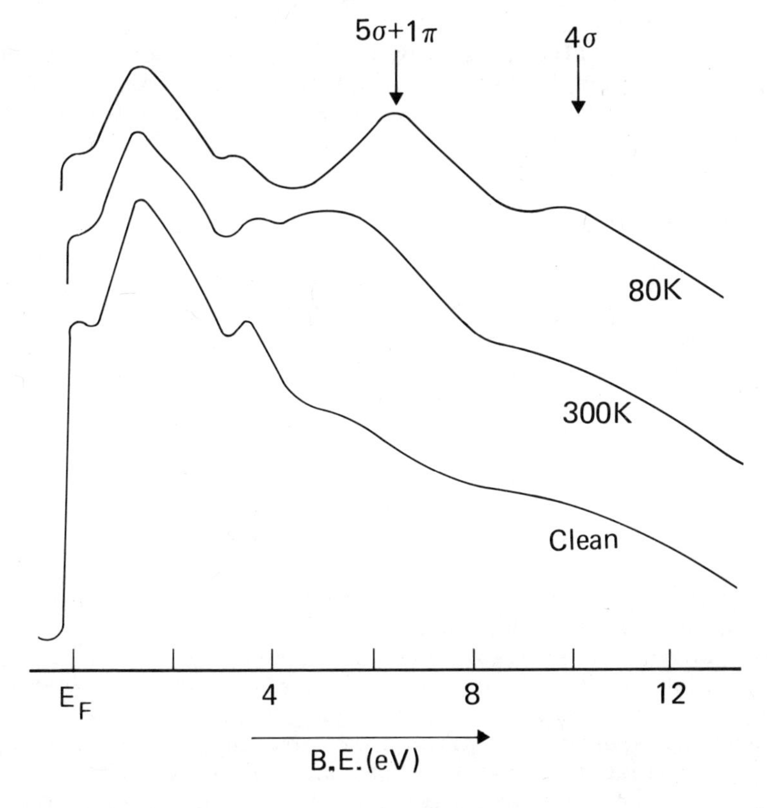

Figure 31. HeI spectrum of CO adsorbed on Mo at 80K and 300K.[102]

situation an intensity analysis will theoretically yield ε_j (adsorbate) which is different from the free molecule value by the amount $\pm\Delta E_j^B$ that we have been trying so hard to establish up to now. No attempts to derive $\pm\Delta E_j^B$ values of core-levels in this way have been made yet, and I am sceptical of the possibility of success because of the difficulty in properly separating the satellite structure from the background to make the analysis.

The W/CO and Mo/CO systems have been extensively studied by UPS and XPS.[99-102] The nature of the bonding has been controversial for many years, but prior to the photoelectron work it was accepted by the majority that the 300K adsorption (β-state) was an undisociated molecular state (a small but vocal minority thought otherwise).[103] The UPS and XPS work on the two systems clearly indicates that the β-state is dissociated. In the UPS there is no resemblence to the Ni/CO spectrum and it is impossible to assign the observed structure to CO molecular levels. The spectrum (Figure 31 for Mo), however, closely resembles a superimposition of those of Mo/O and Mo/C. The core-level C(1s) and O(1s) values are also the same as for Mo/O and Mo/C. In addition, no satellite structure that can be associated with molecular or carbonyl-like CO can be observed accompanying the O(1s) line, either for Mo/CO[49] or W/CO.[104] In the latter case the shake-up structure observed is the same as for the W/O_2 interaction.

Adsorption at 80K on W or Mo populates a state long known as the "virgin state"[105] because on warming it converts irreversibly to the β-state. The UPS and XPS reveal this to be a strongly chemisorbed but molecular state by the presence of the familar CO UPS molecular spectrum (Figure 31) and XPS O(1s) and C(1s) values which are at higher B.E. than those appropriate for dissociated O and C. An investigation for the expected shake-up structure accompanying O(1s) has not yet been made. On warming from 80K to 300K the virgin state dissociates and, as anticipated, yields UPS and XPS spectra identical to those produced by β-state adsorption at 300K.

<u>NO Interaction with Ni, W, and Pd.</u> The HeI, HeII, N(1s) and O(1s) spectra for the interaction of NO with a Ni film under various conditions are reproduced in Figures 32,33.[51,52] At 80K the gas phase UPS B.E.'s can be recognized in the adsorbate spectra. A comparison between the gas phase spectrum[106] and the 80K adsorbate spectrum is given in Figure 34 and the $(\text{B.E.}_{g.} - \text{B.E.}_{ads.})_j$ values in Table V. They are not at all constant and some of them are very large, in contrast to the 80K adsorption of N_2O given in Figure 12, and Table III. We can immediately conclude that the 80K adsorption is a strongly chemisorbed state. The gas phase B.E. scale of Figure 34 has been matched to that

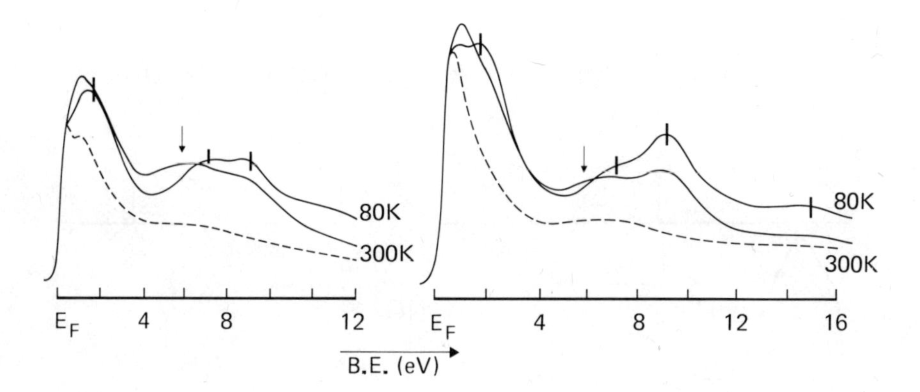

Figure 32. HeI and HeII spectra of NO adsorbed on Ni at 80K and 300K.[51,52]

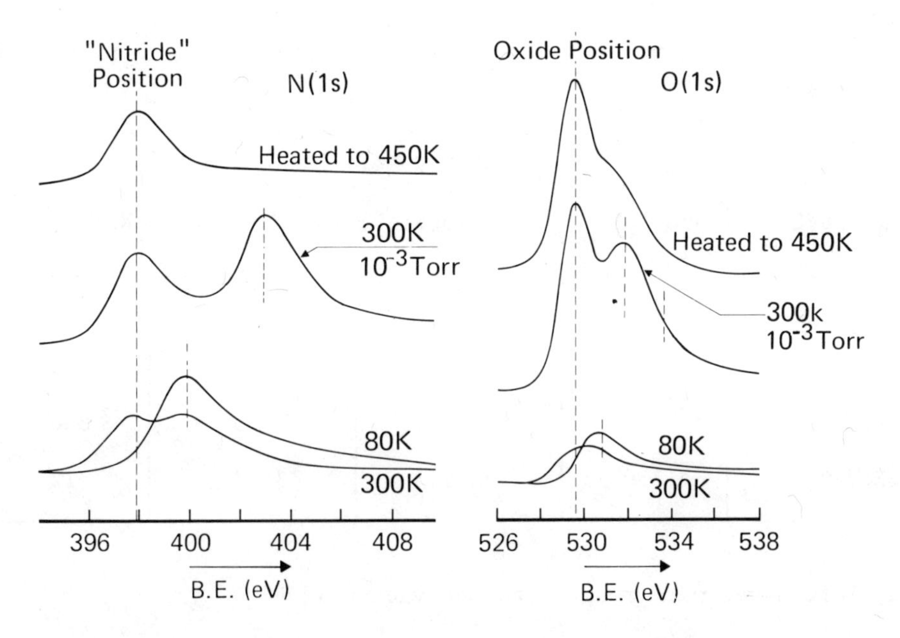

Figure 33. XPS N(1s) and O(1s) spectra for NO adsorption on Ni under a variety of conditions.[51,52]

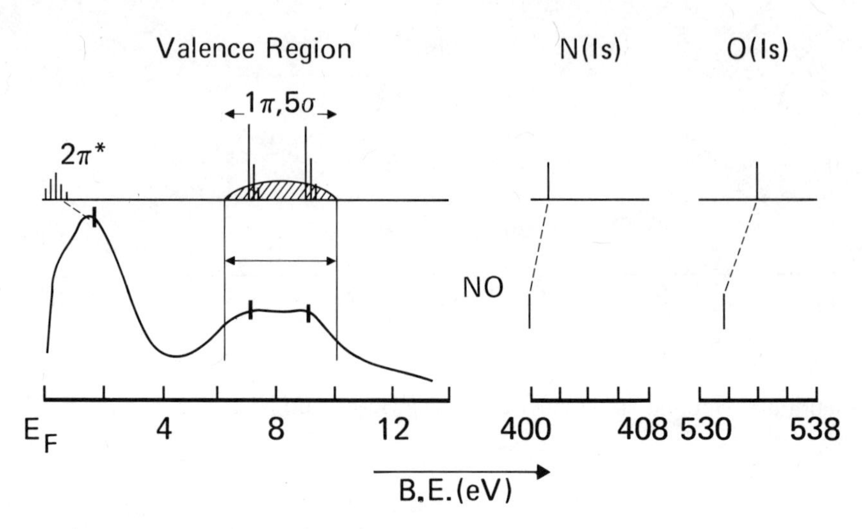

Figure 34. Comparison of NO molecular adsorbate energy levels to gas phase NO values.[51]

TABLE V

$B.E._g$ [(a)] and $B.E._{ads}$ [(b)] Values(eV) for NO (80K on Ni)

Level	2π*	1π, 5σ	N(1s)	O(1s)
$B.E._g$	9.5	15.5 → 20	410.3	543.3
$B.E._{ads}$	1.9	6 → 10.5	399.9	530.9
$B.E._g - B.E._{ads}$	7.6	ca 9.5	10.4	12.4

(a) Referenced with respect to the vacuum level.

(b) Referenced with respect to E_F of the substrate.

of the adsorbate spectrum by assuming that the 1π and 5σ levels are non-bonding, yielding a ΔE^R value of about 3.5eV for a ϕ of 5.5eV. If we make the usual, but probably inappropriate, assumption that ΔE_j^R for the other levels is also 3.5eV we derive ΔE^B values of -1.4eV, +1.5eV, and +3.5eV, respectively for the $2\pi^*$, N(1s) and O(1s) levels. The experimental assignment that the bonding is being done largely through the singly occupied antibonding $2\pi^*$ level is in agreement with an SCF X-α cluster calculation which reproduces the experimental adsorbate valence levels remarkably well.[107] On warming to 300K partial dissociation occurs as judged by the increase in the O2p and N2p derived region in the UPS (Figure 32) and the appearance of N(1s) and O(1s) features in characteristic atomic position in the XPS (Figure 33). At this point no atoms have been lost from the surface (core-level intensity measurement), but on subsequent exposure to NO at 300K and high pressure (10^{-3} Torr) further extensive adsorption occurs with considerable loss of nitrogen to yield a thick oxide layer (2-3 monolayer) containing a little "nitride," plus a weakly adsorbed NO state.[51] The NO state is identified as such by its N(1s) B.E. (Figure 33) which falls within the range for bulk nitrosyls, and the N(1s) and O(1s) intensity decreases caused by desorbing at an elevated temperature. In this final complex extensive reaction state UPS is of little assistance because of the complexity of the overlapping valence bond structures. It is worth noting that extensive interaction of NO_2 with Ni produces a very similar end product though molecular NO_2 can be detected from the UPS at 80K.[51]

The XPS results for NO/Ni are compared to those obtained for W/NO,[108] and Pd/NO[109] in Figure 35. It can be seen that W is more active towards dissociation than Ni since partial dissociation occurs even at 120K and is complete at 300K, whereas for Pd dissociation doesn't even occur at 300K.

<u>H_2S Interaction with Ni, Fe.</u> Figure 36 shows the HeI, HeII and S(2p) spectra for the interaction of H_2S with Ni at 80K and then warmed to 300K.[52] At 80K the three M.O.'s of gaseous H_2S are clearly observed with unchanged separations compared to the gas phase. A constant ΔE^R of about 0.8eV is found using Equation 2. We therefore have condensed multilayers present. The HeII spectrum shows the top M.O. only very weakly, if at all (Figure 36), owing to the large drop in ionization cross-section for S(3p) character (the top orbital is the sulphur lone pair orbital) at that photon energy. On warming to 300K the multilayer desorb and the remaining chemisorbed H_2S dissociates to leave a sulphide-like species at the surface as judged by the position of the S(2p) peak and the loss of the characteristic H_2S orbital structure is the HeI spectrum. The one element XPS cannot detect

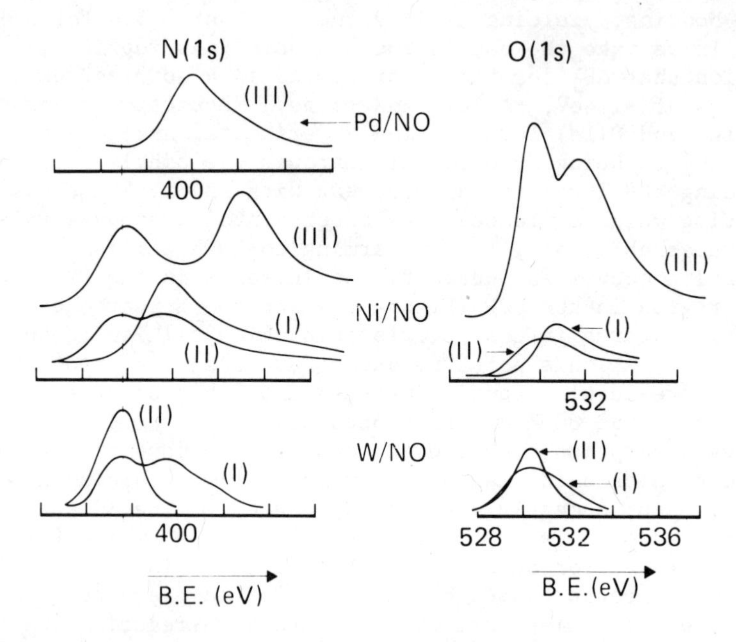

Figure 35. Comparison of N(1s) and O(1s) XPS spectra for NO adsorbed on W, Ni, and Pd.[51]

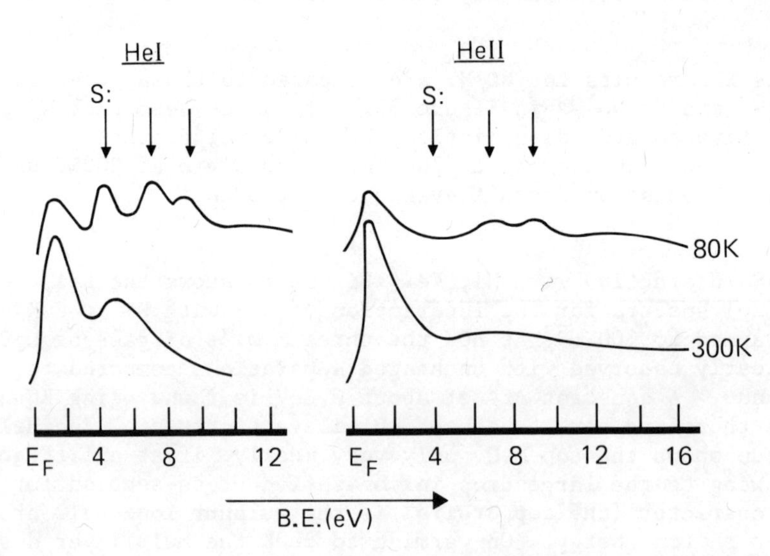

Figure 36. HeI and HeII spectra of H_2S adsorbed on Ni at 80K and 300K.[52]

is hydrogen, of course, so one is assuming that the hydrogen is lost from the system. In fact, there is good indication of this from the HeII spectrum, which on warming from 80K to 300K returns almost to its clean Ni appearance. One knows from the HeI and S(2p) that sulphur is present, however. The reason for the absence of any adsorbate feature in the HeII must therefore be because the adsorbate is entirely sulphur which has the very low 3p cross-section. If the species were H-S- or any other combination involving H, the H(1s) orbital contributions to the valence-level orbital structure would produce an observable cross-section (cf. the two MO;s for condensed H_2S which are observed at in the HeII spectrum).

Kishi and Roberts[110] have recently noted an interesting co-adsorption effect for CO and H_2S on Fe. They found that at 300K CO was initially adsorbed molecularly on Fe films judged by the UPS and XPS spectra, but that a slow dissociation process occurred yielding atomic C and O at the surface. At slightly elevated temperatures the process could be made to go to completion rapidly. They than found that the pre-adsorption of a fraction of a monolayer of H_2S at 300K, which, like Ni, yields dissociated sulphur atoms at the surface, completely stopped the dissociation of subsequently adsorbed CO. Sulphur doping to a small extent is quite common in commercial catalytic processes to maintain yields and specificity. Too much sulphur poisons the catalyts. The possible significance of the H_2S,CO/Fe work in this area should be self-evident. This piece of work is the only example given in this chapter of co-adsorption or sequential reaction studies by UPS and XPS. Many others are being studied, however, e.g., the reduction of "O" covered surfaces by H_2;[111] the replacement of O at surfaces by S;[49] and the replacement of CO by O_2,[49] or O_2 by CO.[92] This is an area likely to develop rapidly once sufficient is known about the characteristics of the single species reaction.

Oxygen Chemisorption and Oxidation of Transition Metals. Oxygen chemisorption and oxidation of transition metals have received a lot of attention in XPS[35,91(112-116)] and certain aspects of the results and opinions generated by them can best be described as provocative. This subject is therefore a suitable one with which to close the chapter. Since the author is involved in the controversy, it should not be taken for granted that the following discussion is un-biased.

Some fifteen or sixteen metals show, on the interaction with pure oxygen at low pressures, an O(1s) feature which is at or close to 529.5eV B.E. with respect to E_F. Other O(1s) peaks often also appear between 1eV to 4eV to higher binding energy. The relative intensities of these higher B.E. peaks are a function of temperature, pressure, and exposure. At 300K the 529.5eV peak is usually the dominant feature but at low temperatures (80K) its growth can be inhibited and in some cases suppressed altogether. Only a few of the metal/oxygen systems have been studied in detail but it is clear that at 300K for Ni, Fe, Zn and possibly Cu the 529.5eV peak and the higher B.E. O(1s) structure grows together (though not necessarily at the same rate) during the adsorption from the lowest detectable coverage (for Ru (100), however, no higher B.E. features are reported).[92] The 529.5eV peak continues to grow at a measurable rate at 300K, at low oxygen pressure (10^{-6} Torr or less), as a function of exposure until the sticking probability is reduced to a very low value ($< 10^{-4}$). For Ni this is known to correspond to an uptake equivalent of about 2 layers of NiO.[117] The uptake is greater for Fe (slightly)[49] and Zn[49] and lower for Cu.[49] Under these conditions the higher B.E. features never exceed an intensity appropriate to monolayer atomic coverage. At high oxygen pressure ($> 10^{-4}$ Torr, sometimes atmospheric) and often under conditions of poor sample purity (air, high pressure of oxygen from cylinders), the oxygen uptake can be increased and sometimes a higher B.E. feature at about 531.5eV becomes dominant. These facts are schematically illustrated in the spectra of Figure 37. Also shown is a typical bulk oxide spectrum. The better defined the oxide surface (e.g., clean single crystal NiO) the less the intensity of the high B.E. O(1s) component. It is clear that the O(1s) feature near 529.5eV is characteristic of an oxide and to a first approximation one might write the oxygen as O^{2-} and suggest that the potential felt by the oxygen atom is very similar for all the oxides with O(1s) peaks near 529.5eV. This argument has been criticized[118] as being far too simple since it ignores the (supposedly different) Madelung Potential terms for the different oxides. Whatever the charge assigned to the O atom in the individual oxides, it is not in doubt, however, that the 529.5eV peak observed for the bulk oxides is characteristic of oxide oxygen atoms. The controversial point then is whether the 529.5eV peak observed during oxygen "chemisorption," can therefore, also be characterized as oxide oxygen. My own opinion[119] is that this is the most sensible conclusion unless and until it can be disproved. The oxygen is electronically indistinguishable from bulk oxide oxygen (and therefore is probably also close to O^{2-}). The argument has been criticized on the same grounds as before, i.e., an expected difference in Madelung potential terms for surface oxygen,[118] plus the fact that chemisorbed O is considered in at least some theoretical treatments to have delocalized bonding states with the metal surface and can not in such a description be considered O^{2-}.[120] These two criticisms miss the point of the argument being advanced, namely that the 529.5eV peak is not chemisorbed oxygen in the traditional sense at all, but well on the way to being oxide. It may therefore be in or below the surface rather than

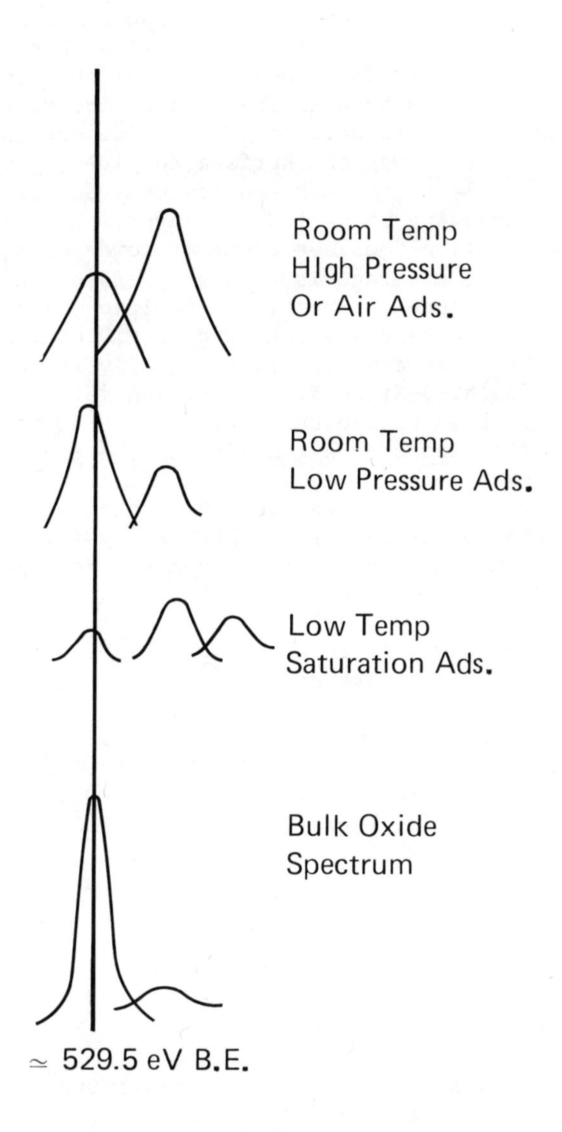

Figure 37. Schematic representation of O(1s) XPS characteristics during O_2 chemisorption and oxidation of metals.

above it and it may be present in small nucleated oxide islands. The higher B.E. O(1s) features would then be assigned as representing traditional chemisorbed overlayer oxygen.[121] Support is given to this idea by the fact that low temperature adsorption suppresses the oxide growth (529.5eV peak) with respect to the high B.E. features. It is also well-established in Ni and Fe single crystal studies by LEED that O atom incorporation and island nucleation of a double-layer of oxide occurs and that this subsequently spreads over the surface to give a complete double-layer.[122,123] The LEED studies only detect the nucleation at coverages exceeding about 1/3 monolayer, but this does not prove that nucleation does not occur at lower coverage, since a significant island coverage is required before the NiO or FeO LEED spots will become apparent. In addition, a certain domain size is required to be characteristic of NiO by LEED whereas it is unknown what size cluster electronically is required to resemble NiO (is Ni-O-Ni an NiO nucleation island)? In the case of Ni/O_2 it has been well-established by a LEED intensity analysis that some of the adsorbed oxygen sits in the four-fold coordination site about 1Å above the surface.[55] No information is provided on how much oxygen is not in this position or where it is under the conditions of the LEED analysis. It is known that more oxygen has been adsorbed than can be accounted for by the four-fold overlayer site adsorption.[124]

Summarizing, then, it is clear that much further work must be done before the controversy is resolved. In particular, the XPS and LEED work needs to be done simultaneously to establish which features correlate. Angular (polar) studies in the XPS would also help to establish the depth distribution of the various O(1s) features. Preliminary work on the Zn/O_2 system has been reported[116] and a model suggested in which ZnO islands nucleate on the surface (the 529.5eV O(1s)), with the higher B.E. O(1s) feature representing atoms randomly dissolved in lone patches of Zn surface between the islands.

REFERENCES

1. J. W. Gadzuk, this volume.

2. C. S. Fadley in "Electron Emission Spectroscopy" Ed. Dekeyser, Fiermans, Van der Kalen, and Vennik. D. Reidel, 1973.

3. A. Bradshaw, L. Cederbaum, and W. Domke in "Photoelectron Spectroscopy," Springer-Verlag (in press).

4. J. C. Tracy, as reference 2.

5. C. R. Brundle, in "Surface and Defect Properties of Solids," Vol. 1, Ed. J. M. Thomas and M. W. Roberts, Specialist Periodical report of the Chemical Society (London) 1972.

6. C. R. Brundle, J. Vac. Sci. Tech. 11, 212, 1975.

7. C. R. Brundle, Surface Science, 48, 99, 1975.

8. "Photoelectron Spectroscopy" by A. D. Baker and D. Betteridge, Pergamon Press, 1972.

9. "Photoelectron Spectroscopy" by J. H. D. Eland, Butterworths, 1974.

10. "Electron Spectroscopy for Chemical Analysis" by K. Siegbahn et al., AFML-TR-68.189.

11. "ESCA applied to Free Molecules," by K. Siegbahn et al., North-Holland (Amsterdam), 1969.

12. D. E. Eastman and J. K. Cashion, Phys. Rev. Lett. 27, 1520, (1971). 1520, 1971.

13. C. J. Powell, Surface Science 44, 29, 1974.

14. J. E. Rowe and H. Ibach, Phys. Rev. Letters 32, 421, 1974.

15. F. M. Propst and T. C. Piper, J. Vac. Sci. Tech. 4, 53, 1966.

16. H. D. Hagstrum and G. E. Becker, Proc. Roy. Soc. A331, 395, 1972.

17. H. D. Hagstrum and G. E. Becker, Phys. Rev. Letters 22, 1054 (1969); J. Chem. Phys. 54, 1015, 1971.

18. C. R. Brundle and A. F. Carley, Chem. Phys. Letters 33, 41, 1975.

19. C. R. Brundle, J. Elect. Spectr. 5, 291, 1975.

20. C. S. Fadley, R. J. Baird, W. Siekhaus, T. Novakov, and S.A.L. Bergström, J. Elect. Spectr. 4, 93, 1974.

21. C. S. Fadley, J. Elect. Spectr 5, 725, 1974.

22. C. S. Fadley, Disc. Faraday Soc. No. 60, to be published.

23. H. E. Bishop, J. P. Coad and J. C. Rivieŕe, J. Elect. Spectr. 1, 389, 1973.

24. J. H. Scofield, Lawrence Livermore Laboratory Report UCRL-51326, 1973.

25. V. I. Nefedov, N. P. Serguishin, I. M. Band and M. B. Trazhaskovskaya, J. Elect. Spectr. 2, 393, 1973.

26. Ibid 7, 175, 1975.

27. B. L. Henke, and R. L. Elgin, Adv. X-ray Anal. 13, 639, 1970.

28. T. E. Gallon, and J.A.D. Mathew, J. Phys. D. 4, 269, 1972.

29. C. R. Brundle, M. W. Roberts, D. Latham and K. Yates, J. Elect. Spectr. 3, 241, 1974.

30. Available from Physical Electronics Industries, Varian, and V. G. Scientific Ltd.

31. C. S. Fadley, D. A. Shirley, A. J. Freeman, P. S. Bagus and J. V. Mallow, Phys. Rev. Letters 23, 1397, 1969; A. J. Freeman, P. S. Bagus and J. V. Mallow, Int. J. Mag. 4, 35, 1973.

32. P. S. Bagus, A. J. Freeman and F. Sasaki, Phys. Rev. Letters 18, 850, 1973.

33. T. Robert, Chemical Physics 8, 123, 1975.

34. J. P. Coad and J. P. Cunningham, J. Elect. Spectr. J. Elect. Spectr. 3, 435, 1974.

35. S. Evans, Faraday Trans. II, 1044, 1975.

36. C. D. Wagner, Anal. Chem. 44, 1050, 1972.

37. W. J. Carter, G. K. Schweitzer and T. A. Carlson, J. Elect. Spectr. 5, 827, 1974.

38. C. K. Jorgensen and H. Berthou, Faraday Disc. No. 54, p. 269, 1972.

39. G. K. Wertheim and S. Hüfner, J. Inorg. Nucl. Chem., to be published.

40. E. McGuire, Phys. Rev. A 10, 32, 1974.

41. C. C. Chang, Surface Science, 48, 9, 1975.

42. J. M. Morabito, Surface Science, 49, 318, 1975.

43. C. Argile and G. E. Rhead, J. Phys. C. 7, L261, 1974.

44. W. E. Moddeman, T. A. Carlson, M. O. Krause, B. P. Pullen, W. E. Bull and G. K. Scheveitzer, J. Chem. Phys. 55, 231, 1971.

45. T. J. Chuang, to be published.

46. C. D. Wagner and P. Biloen, Surface Science, 35, 82, 1973.

47. S. P. Kowalczyk, R. A. Pollak, F. R. McFeely, L. Ley, and D. A. Shirley, Phys. Rev. B. 8, 2387, 1973.

48. C. D. Wagner, Analytical Chemistry, to be published.

49. C. R. Brundle, unpublished data.

50. P. E. Larson, J. Elect. Spectr. 4, 213, 1974.

51. C. R. Brundle, J. Vac. Sci. Tech., to be published.

52. C. R. Brundle and A. F. Carley, Faraday Discussions No. 60, to be published.

53. J. F. Demuth and D. E. Eastman, Phys. Rev. Letters 32, 1123, 1974.

54. K. Y. Yu, J. C. McMenamin, and W. E. Spicer, Surface Science 50, 149, 1975.

55. J. E. Demuth, D. W. Jepsen, and P. M. Marcus, Phys. Rev. Letters 31, 540, 1973.

56. P. M. Marcus, J. E. Demuth and D. W. Jepsen, Surface Science, to be published.

57. C. R. Brundle and M. W. Roberts, Proc. Roy. Soc. (London) A 331, 383, 1972.

58. C. N. Berglund and W. E. Spicer, Phys. Rev. A136, 1030; 1044, 1964.

59. D. A. Shirley, Phys. Rev. B5, 4709, 1972.

60. L. Ley, S. Kowalczyk, R. Pollak and D. A. Shirley, Phys. Rev. Letters, 29, 1088, 1972.

61. D. E. Eastman, Phys. Rev. Letters, 26, 1108, 1971.

62. S. Hüfner, G. K. Wertheim and J. H. Wernick, Phys. Rev. B8, 4511, 1973.

63. G. M. Stocks, R. W. Williams and J. S. Faulkner, Phys. Rev. B4, 4390, 1971.

64. D. H. Seib and W. E. Spicer, Phys. Rev. B2, 1676, 1970; 2, 1694, 1970.

65. P. M. Williams and F. R. Shepherd, J. Phys. C6, L36, 1973.

66. J. A. Wilson and A. D. Yoffe, Adv. Phys. 18, 193, 1969.

67. G. K. Wertheim, F. J. Di Salvo and D.N.E. Buchanan, Solid State Comm. 13, 1225, 1973.

68. N. V. Smith and M. M. Traum, Surface Science 45, 745, 1974.

69. S. Hüfner and G. K. Wertheim, Phys. Rev. B8, 4857, 1973.

70. G. K. Wertheim, L. F. Mattheiss, M. Campagna and T. P. Pearsall, Phys. Rev. Letters 32, 997, 1974.

71. S. Evans, Farad. Trans. II, to be published.

72. C. R. Brundle, unpublished data.

73. C. R. Brundle and A. F. Carley, Chem. Phys. Letters 31, 423, 1975.

74. N. Lang, this volume.

75. G. Allan, this volume.

76. B. J. Waclawski and E. W. Plummer, Phys. Rev. Letters 29, 783, 1972.

77. G. F. Amelio and E. J. Scheibner, Surface Science, 11, 242, 1968.

78. C. J. Powell, Phys. Rev. Letters 30, 1179, 1973.

79. J. E. Houston, J. Vac. Sci. Tech. 12, 255, 1975.

80. J. W. Gadzuk, Phys. Rev. B 9, 1978, 1974.

81. J. E. Houston, Rev. Sci. Sistr. 45, 897, 1974.

82. L. Yin, I. Adler, T. Tsang, M. H. Chen, and B. Craseman, Phys. Letters 46A, 113, 1973.

83. H. D. Hagstrum, Phys. Rev. 150, 495, 1966; H. D. Hagstrum and G. E. Becker, J. Chem. Phys. 54, 1015, 1971.

84. B. Feuerbacher and M. R. Adriens, Surface Science, 45, 553, 1974.

85. C. R. Brundle, J. Elect. Spectr., to be published.

86. T. Gustafsson, E. W. Plummer, D. E. Eastman, and J. L. Freeouf, Solid State Comm. 17, 391, 1975.

87. J. C. Fuggle and D. Menzel, to be published.

88. B. J. Waclawski, T. V. Vorburger, and R. J. Stein J. Vac. Sci. Tech. 12, 301, 1975.

89. J. C. Fuggle and D. Menzel, Chem. Phys. Letters 33, 37 1975.

90. J. Herbst and B. J. Waclawski, to be published.

91. J. C. Fuggle and D. Menzel, Surface Science, to be published.

92. T. B. Grimley in "Molecular Processes on Solid Surfaces," Ed. E. Drauglis, R. D. Gretz and R. I. Jaffee, McGraw-Hill, N.Y., 1969.

93. J. T. Yates and N. E. Erickson, Surface Science, 44, 489, 1974.

94. K. Y. Yu, W. E. Spicer, I. Lindan, P. Pianetta and S. F. Lin. J. Vac. Sci. Tech., to be published.

95. D. E. Eastman and J. E. Demuth, Phys. Rev., to be published.

96. See the discussion remarks of Faraday Discussions, no. 58, 1975; I.P. Batra and P.S. Bagus, Solid State Comm., 16, 1097, 1975

97. J. E. Demuth, personal communication.

98. G. Ertl, personal communication.

99. T. E. Madey, J. T. Yates, and N. Erickson, Chem. Phys. Letters 19, 487, 1973.

100. J. T. Yates, T. E. Madey, and N. E. Erickson, Surface Science, 43, 257, 1974.

101. S. J. Atkinson, C. R. Brundle and M. W. Roberts, Chem. Phys. Letters 2, 105, 1973.

102. S. J. Atkinson, C. R. Brundle and M. W. Roberts, Chem. Phys. Letters 24, 175, 1974.

]03. D. A. King, C. G. Goymour and J. T. Yates, Proc. Roy. Soc. A 331, 361, 1972.

104. J. C. Fuggle and D. Menzel, to be published.

105. R. R. Ford, Adv. Catalysis 21, 51 (1970) and references therein.

106. D. W. Turner, A. D. Baker, C. Baker and C. R. Brundle, "Molecular Photoelectron Spectroscopy," Wiley and Sons, London, 1970.

107. I. P. Batra, to be published.

108. T. E. Madey, J. T. Yates, N. E. Erickson, Surface Science, 43, 526, 1974.

109. K. Kishi and S. Ikeda, Bull. Chem. Soc. Japan 47, 2532, 1974.

110. K. Kishi and M. W. Roberts, to be published.

111. H. Conrad, G. Ertl, J. Küppers and E. E. Latta, Surface Science 50, 296, 1975.

112. C. R. Brundle and A. F. Carley, Chem. Phys. Letters 31, 423, 1975.

113. R. W. Joyner and M. W. Roberts, Chem. Phys. Letters 29, 447, 1974.

114. K. S. Kim and N. Winograd, Surface Science, 43, 625, 1974.

115. J. J. Braithwaite, R. W. Joyner, and M. W. Roberts, Faraday Discussion No. 60, to be published.

116. D. Briggs, Faraday Discussion No. 60, to be published.

117. A. M. Horgan and D. A. King, Surface Science, 23, 259, 1970.

118. See the remarks of S. Evans and J. M. Thomas in Faraday Discussion, No. 58, 1975.

119. See the remarks in Faraday Discussions, No. 58 (1975), and 60, to be published.

120. See the remarks of M. Quinn in the Faraday Discussions, No. 58 (1975) and No. 60, to be published.

121. Obviously under the conditions when the high B.E. O(1s) peak is too intense to represent monolayer proportions of oxygen (high pressure O_2 or air exposure) it cannot be all chemicosbed overlayer, and may then include contributions from non-stoichiometric oxide bulk oxygen (e.g., Ni_2O_3).

122. R. H. Holloway and J. B. Hudson, Surface Science, 43, 123, 1974; 43, 141, 1974.

123. G. W. Simmons and D. J. Dwyer, Surface Science 48, 373, 1975.

124. S. Anderson, personal communition; J. E. Demuth personal communication.

125. P.H. Citrin, to be published; J.W. Gadzuk, personal communication.

[illegible] to be published.

[illegible] to be published.

[illegible]

[illegible]

[illegible] to be published.

[illegible]

[illegible]

[illegible]

[illegible]

[illegible]

S.I.M.S. STUDIES AT METAL SURFACES

M. BARBER

University of Manchester Institute
of Science and Technology
Department of Chemistry
MANCHESTER M60 1QD

1. INTRODUCTION

Although considerable advances have been made during the past decade in our knowledge of many aspects of surface reactivity, a complete understanding of the processes is hampered by lack of detailed information upon the composition and electronic properties of the very uppermost layers of the solid, and the mode of interaction between those layers and surface species such as reactant molecules.

Answers to some fairly obvious and basic questions must be sought in any research in surface chemistry before a real understanding of mechanism etc. of the surface processes under investigation can be attained, e.g.

(i) What is the composition and electronic state of the uppermost surface layer(s) of the solid, before, during, and after any treatment to which the solid may be subjected ?

(ii) What kind of surface species can be formed by reactant molecules at the surface, which surface atoms are attacked, how are the species bonded to the surface and what is the surface coverage of such species ?

(iii) What is the reactivity of the surface species, and how is this related to the electronic state of the uppermost layers ?

(iv) How can surface species leave or be removed from the surface ?

In an effort to answer some or all of these questions existing physical techniques have been exploited, such as I.R., visible and U.V. spectroscopies, with success in some restricted fields, and new ones, such as ion reflection spectroscopy and the various electron spectroscopies (Auger, P.E.S., etc.) have been developed. Due to the pioneering work of Fogel[1] in Russia and Benninghoven[2] in Germany, the new techniques of secondary ion mass spectrometry (S.I.M.S.) has emerged, alongside those mentioned above, as a powerful addition to the armoury of methods available to the surface scientist. It would seem, both from previous workers' results and our own, to offer the exciting possibility for the chemist of obtaining information concerning the surface "compounds" formed during surface/molecule interactions.

2. THE METHOD OF SECONDARY ION MASS SPECTROMETRY

The method is based on the well-known phenomenon of the sputtering of particles from a solid surface by bombardment with energetic particles. When a surface is subjected to bombardment by say argon ions in the KeV range, provided that the ion is not reflected, then it will penetrate to a certain depth into the solid and will transfer its energy to the lattice atoms by collision processes, each of which will of course in turn set up its own collision chain and so on, so that an extended zone in the lattice is affected. Some of these collision chains will however terminate at the surface (see Fig. 1) and induce emission of a surface particle which may be an atom or a cluster of atoms in a charged or uncharged state. S.I.M.S., however, only concerns itself with the positive or negative ions which are produced.

The mechanism of these processes have been studied for many years now, both theoretically and experimentally; indeed, the earliest observations of sputtering from surfaces were carried out in the 1890s ! Two types of theoretical treatments have been reported, one in which the total process of surface bombardment and sputtering is treated classically; this has had considerable success in predicting the gross features of the kinetic energy distribution of the ions emitted and a second, investigating probably a much more intractable problem, of considering the ionization process itself as a quantum mechanical phenomenon[4,5]. An excellent source book on this subject is that quoted in Ref.3 which describes in some detail the current theories together with an extensive bibliography. Although these theoretical treatments have had considerable success in predicting or accounting for many of the effects observed in sputtering, there are a number of unknowns which must be borne in mind before using S.I.M.S. as a surface technique.

(i) On impact the primary bombarding particle will be incorporated into the lattice. At the same time the collision cascades set up by it will induce lattice disorders and defects and possibly chemical reactions at the surface.

3. INSTRUMENTATION

In essence the instrument consists of a vacuum enclosure, some means of producing a beam of high energy primary ions, the sample and a mass spectrometer. The arrangement is shown in Fig.2. The vacuum vessels follow standard U.H.V. practice, are made out of stainless steel, and pumped with oil diffusion pumps and liquid N_2 cold traps. The whole system can be baked out to 250°C. With this system base pressures of 10^{-11} can be routinely obtained. The primary ion beam (usually Ar^+) is produced by electron bombardment of the gas in a "sealed" source, followed by mass analysis of the ions emerging to give us a "pure" $^{40}Ar^+$ beam which is then accelerated to the required energy (this is variable from a few hundred eV to about 5 KeV). The mass analysis of the primary ion beam is carried out using a small R.F. quadrupole mass filter. The intensity of the ion beam is maintained at a constant preset value (which can be varied over the range 10^{-6} - 10^{-11} amps) by a sampling and feedback system. The whole ion gun assembly is differentially pumped so that the pressure in the analysis vessel does not rise about 10^{-9} even when the highest ion current is required.

The sample is mounted on a moveable shaft and can be heated to 1000°C or cooled to liquid N_2 temperature. The primary ion beam strikes it at an angle of about 20°. The ions sputtered from the surface are focussed into the main mass spectrometer and crudely energy filtered by means of a lens system. This mass spectrometer is a large, high mass R.F. quadrupole, having a mass range of 1 - 800 a.m.u.

An isolatable sample preparation chamber is also included in the instrument for cleaning and pretreating the specimens which are transported from this vessel to the analysis chamber on the moveable shaft by means of a stainless steel bellows system.

A typical example of a mass spectrum which may be obtained from an instrument such as the one described above is shown in Fig.3. This is the positive ion spectrum observed from surface oxidized Ti metal, which was sputtered in the instrument with 1.5 keV Ar^+ at an ion current of 10^{-9} amps cm^{-2} sec^{-1}. The features of the spectrum are fairly straightforward; at the low mass end an intense set of peaks occur around mass 48 due to Ti^+ and its isotopes. These are followed at higher mass by peaks of much lower intensity due to clusters of the general form TiO_x^+. Note that at mass 96 there is the possibility of both Ti_2^+ and TiO_3^+ being present, since they both have the same nominal mass number. Above mass 96 clusters due to $Ti_2O_x^+$ occur. The (x1) range on the intensity scale has full scale deflection of 10^5 counts sec^{-1}.

(ii) The technique relies on particles being removed from the surface. However the yields of sputtered particles do vary from element to element and from material to material so that preferential enrichment can occur for one element at the expense of another at the surface of a complex sample.

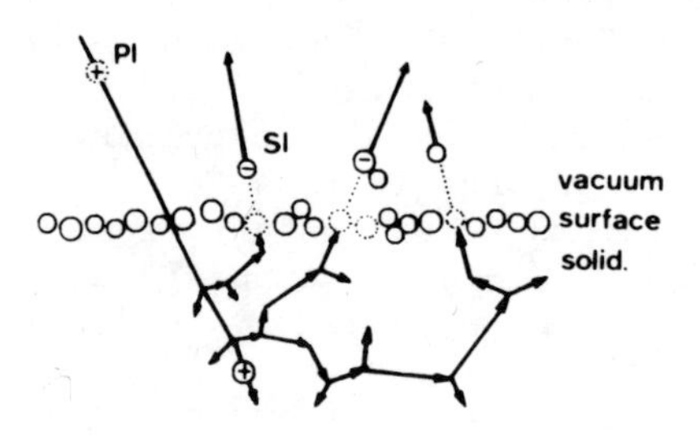

Fig.1 : A schematic representation of sputtering

Until recently, however, secondary ion mass spectrometry had only been applied to the analysis of the atomic composition of bulk materials and films. Use had also been made of the capability of focussing the primary ion beam to spot sizes of the order of 2μ or less, so that determining atomic concentration topology of samples became possible, and a number of commercial instruments were even developed. The primary ion beam currents used in this form of application completely precluded its use in surface analysis, since in order to maintain high bulk sensitivity etc. high incident primary ion fluxes were used (typically 10^{13} - 10^{16} ions cm^{-2} sec^{-1}) which inevitably produced a rapid surface destruction (tens of monolayers per second being sputtered). For the method to be usable in surface investigations the surface erosion rate had to be reduced such that the time for the removal of 1 monolayer is long in comparison to the time taken to carry out a mass spectral analysis. This is achieved by the simple but effective expedient of reducing the primary ion flux to the order of 10^{9} - 10^{10} ions cm^{-2} sec^{-1} and using modern counting and integrating techniques at the mass spectrometer detector and recording end. Under these primary ion fluxes it takes several hours to remove 1 monolayer, the time for a mass spectrometric scan from say m/e 2 - 300 being under these conditions of the order of 5 mins.

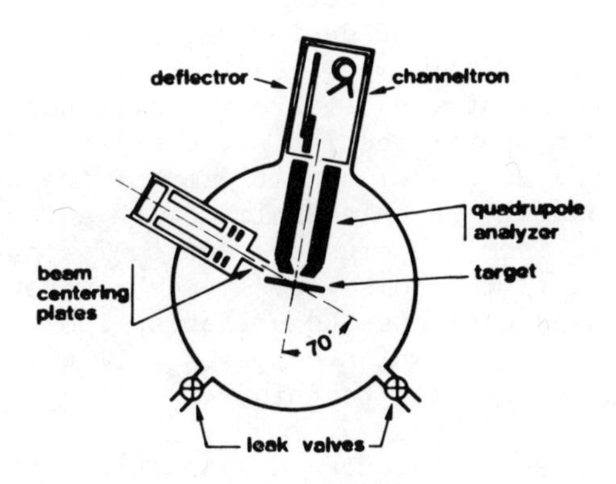

Figure 2 : Schematic layout of instrument

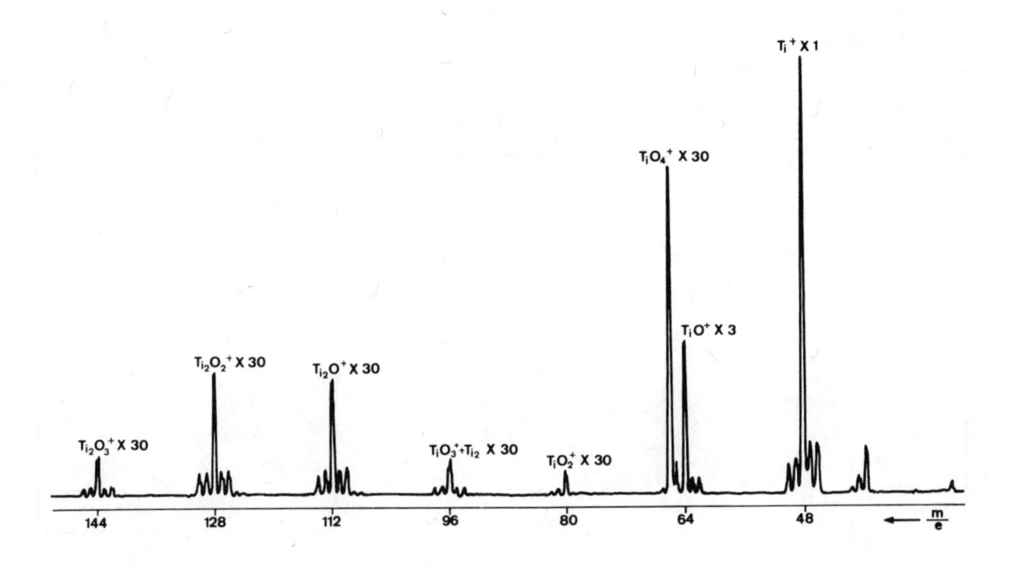

Figure 3

A number of inferences can be made from such spectra. For example, because of the inherent mass resolution of the large quadrupole, it has the capability of separating the isotopes of any element even down to hydrogen and deuterium, so that the technique offers the possibility of following reactions on surfaces by isotopic labelling of the reactants, a feature which is not available with the various electron spectroscopies. It is also obvious from the count rates that the monolayer sensitivity of the technique is high. In the example shown we are only dealing with a few layers of oxide and yet the signals from the clusters due to Ti plus O are very large in comparison to the noise level and background. It has been claimed that for many elements the detection limit is less than 10^{-6} of the monolayer, a figure which compares well with the claimed sensitivity of any other surface technique.

There are however disadvantages with this method, one of the main ones being the problems associated with attempting to carry out quantitative measurements of species present on the surface. As was mentioned above, the sputtering yield does vary with element, but it also varies with the state of the surface. Table 1 shows some figures quoted by Benninghoven(2) for the absolute yields of positive ions from a variety of clean metals and also from the same metals in the oxidised form, note the enormous differences in going from element to element. With this sort of variation of a fundamental sensitivity parameter it can be easily appreciated that quantification of any surface reacting system becomes very difficult. However, for a detailed listing of the advantages of this technique I would refer the reader to Ref.2.

Metal	$S(Me^+)$ Cleaned surface	$S(Me^+)$ Oxidized surface
Mg	0.0085	0.65
Al	0.02	2.0
V	0.0013	1.2
Cr	0.005	1.2
Fe	0.001	0.38
Ni	0.0003	0.02
Cu	0.00013	0.0045
Sr	0.0002	0.13

4. APPLICATIONS OF S.I.M.S. TO SOME "SIMPLE" SYSTEMS

Most of the early applications of the technique were devoted to the study of the primary oxidation processes of metal surfaces[6,7]. A typical example is the investigation of the surface oxidation of clean polycrystalline Cr. The types of secondary ions produced at various stages in oxide formation, how they vary with oxygen doseage and the model used to interpret the results are worth describing in some little detail[6].

The secondary ion mass spectrum of oxidised Cr shows a number of ions of the general form $Cr_xO_y^{\pm}$ many of which show very different dependencies on oxygen dosage, but all show saturation behaviour at about 600 L. In particular the species CrO_2^-, CrO_3^- and CrO^+ are worth careful scrutiny, since they are considered to be representative of the different oxide structures. Fig.4 shows the change in relative intensity of these species with oxygen exposure. Benninghoven then made several simplifying assumptions about the secondary particle emission from metals and oxides in order to interpret the results on a coherent model of oxide formation. Firstly the atoms of an emitted secondary cluster must have come from adjacent sites on the surface. Fig.5 shows schematically how this can happen. Secondly, an emitted cluster will tend to preserve the sign of the charge given by the sum of the charges of its constituent atoms in the lattice before emmission. Bearing these in mind, the following rationalization of the behaviour of the various species can be made with references to Fig.6. The initial oxidation step gives rise to CrO_2^- after an initial oxygen dose of about 50 - 100 L, which rapidly builds up to a monolayer.
As the thickness of the oxide phase increase CrO_3^- species are emitted and as the oxidation ends a higher oxide is formed at the very topmost layer, resulting in the emission of CrO^+. Further work on the sputtering behaviour of the various species from a completed oxide layer tended to confirm this model. Thus by considering in a simple manner how clusters can arise, considerable detail can be obtained as to the various stages in the oxidation process.

If we now consider the application of the technique to the investigation of molecular chemisorbed species on surfaces, then the question arises as to how much fragmentation of the species is produced by the bombardment process itself, and whether or not chemical reactions are induced by it. Two simple molecular chemisorption systems will now be described and the results compared with those obtained from other techniques. Let us consider first of all the adsorption of CO on Ni. Fig. 7 shows the mass spectrum of a clean Ni surface. The lack of any peaks corresponding to carbon or oxygen-containing species should be noted. The peaks at mass numbers 23 and 39 are due to sodium and potassium respectively, and are produced from the primary ion gun which unfortunately was not at this time mass filtered. The temperature of the clean sample was then reduced to 77 K, and CO admitted in measured doses. After the addition of some 14 L of CO, the surface seemed to saturate and the spectrum shown in Fig.8 was obtained. It will be noted that in addition to the peaks at 58 and 60 due to Ni^+ isotopes, a large pair of peaks

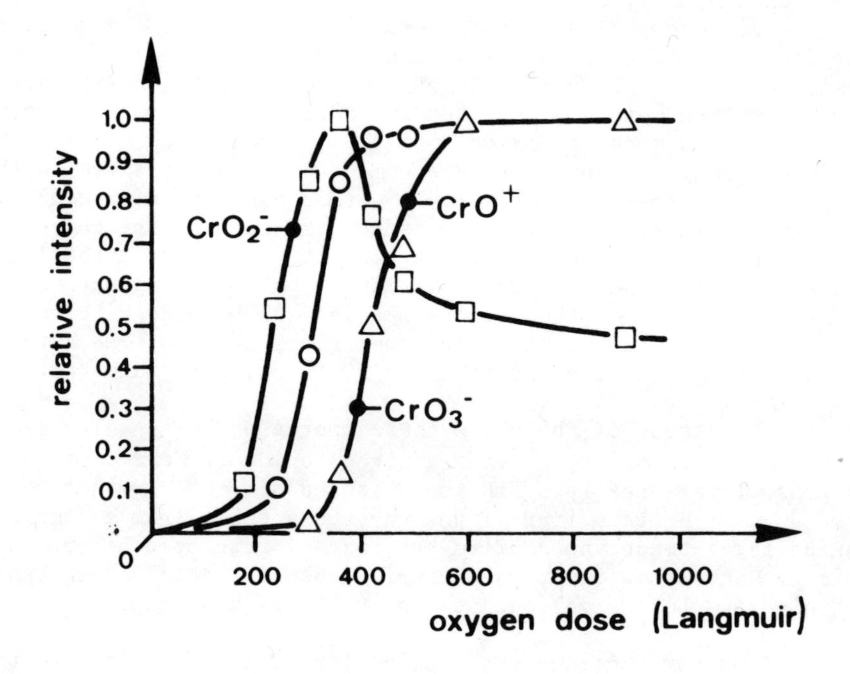

Fig. 4 : Changes in relative intensity of various secondary ions with O_2 dosage for the oxidation of Cr

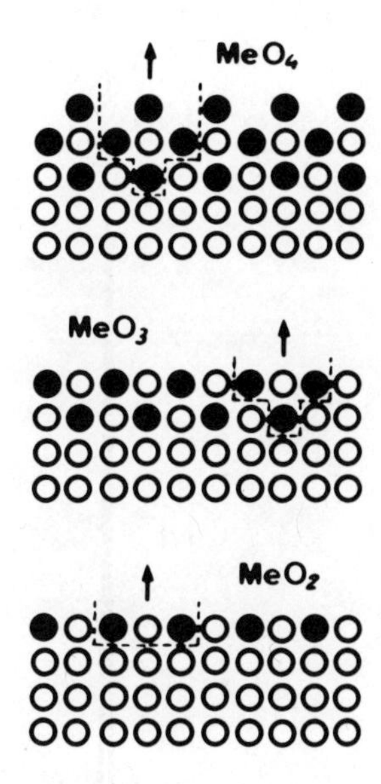

Fig. 5 : Cluster formation in oxide layers

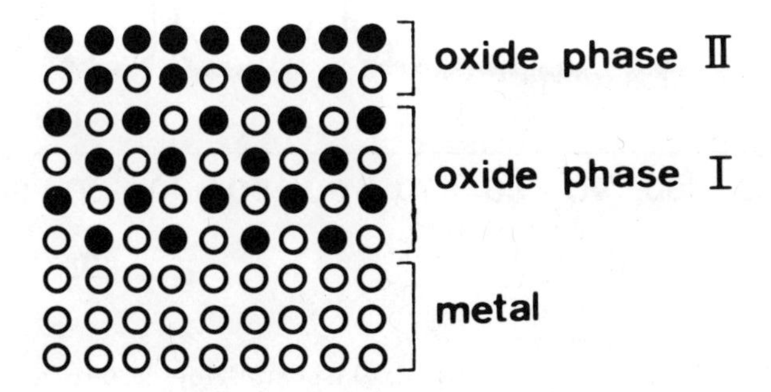

Fig. 6 : Model for oxide formation on Cr

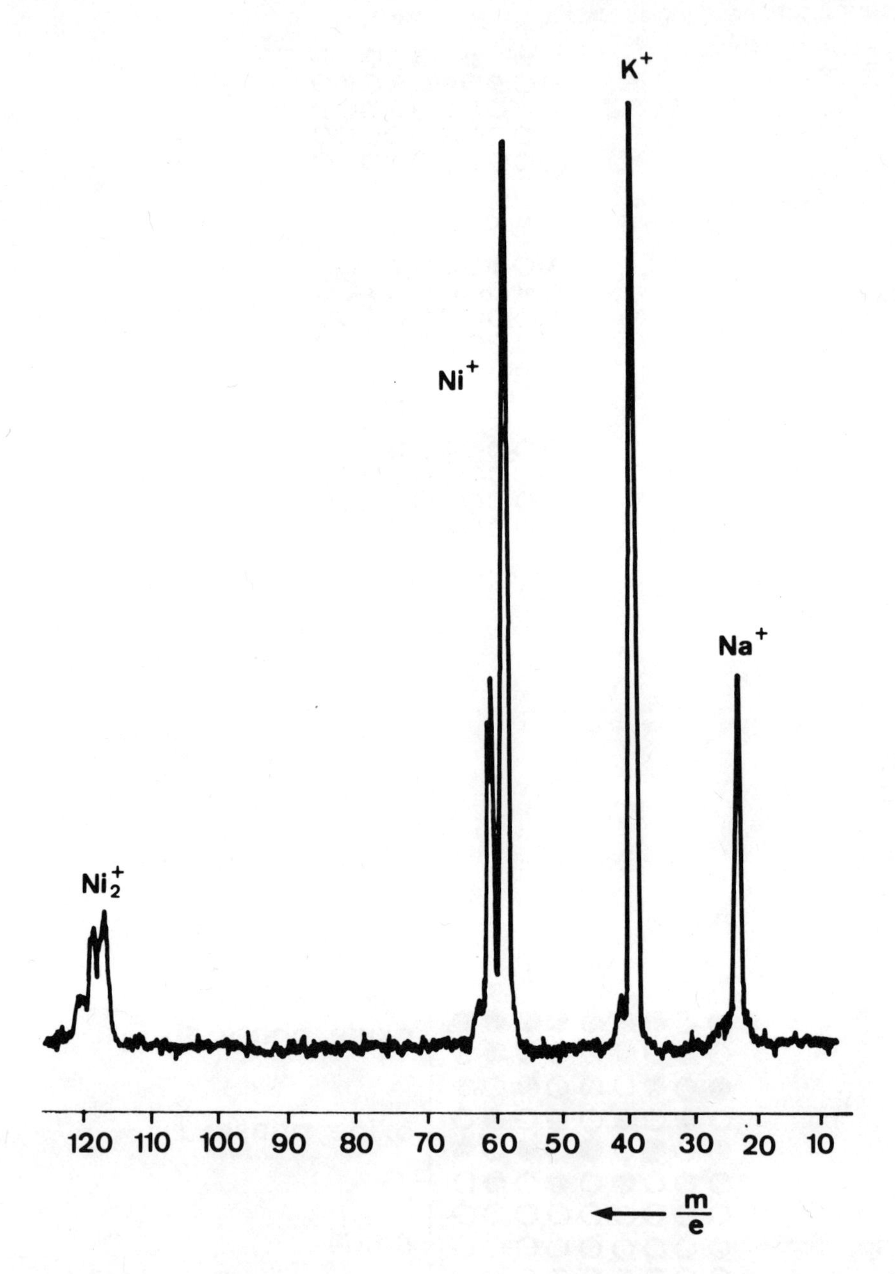

Fig. 7 : SIMS spectrum of nickel surface after argon ion etching

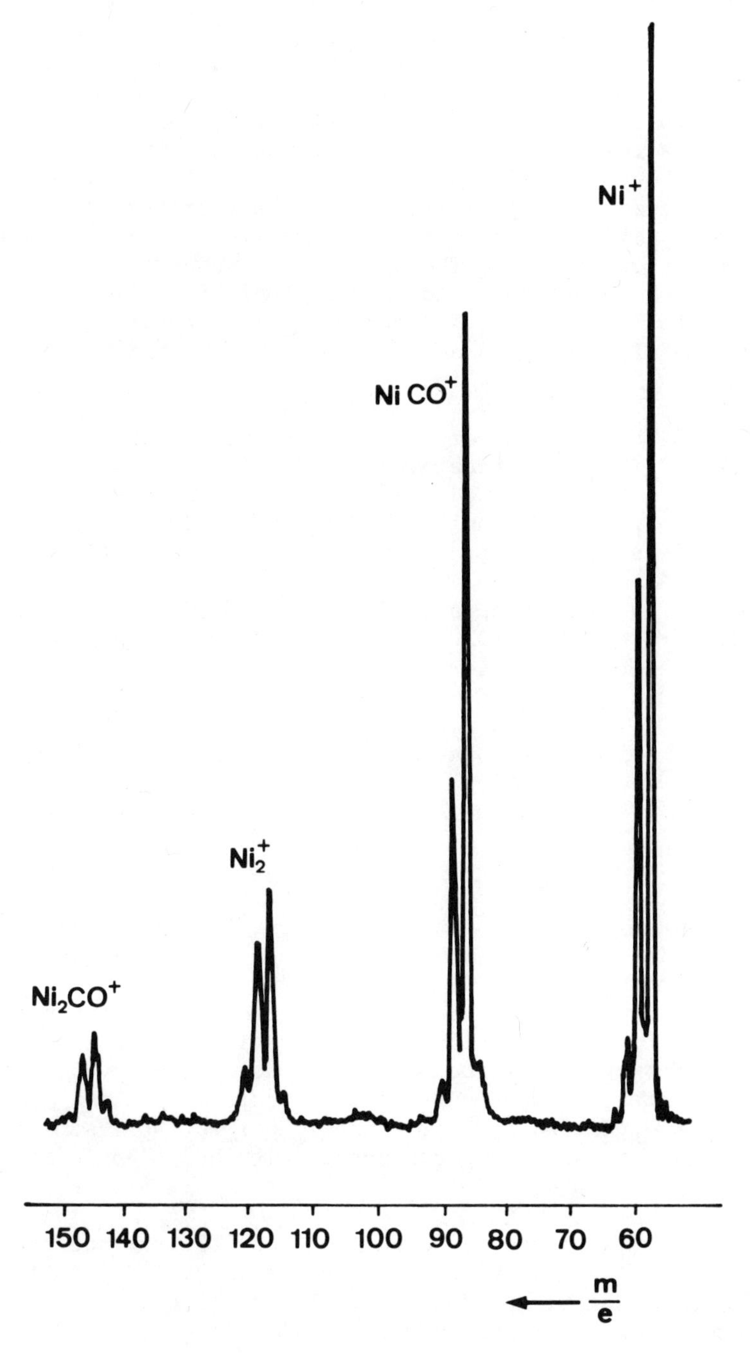

Fig. 8 : SIMS spectrum after saturation with carbon monoxide at 77K

occur at mass numbers 86 and 88 due to $NiCO^+$. It will also be noted that fairly intense peaks appear at 116, 118 and 120 due to Ni_2^+ and small peaks at 144, 146 and 148 due to Ni_2CO^+. On allowing the sample to warm to room temperature the intensities of the $NiCO^+$ and Ni_2CO^+ signals decreased until, at room temperature, their intensities were only about 10 % of those at 77K. Similar spectra were obtained starting from a clean nickel surface at room temperature (295K) except that saturation did not occur until about 40 L of carbon monoxide had been admitted. The spectrum is shown in Fig. 9. It will be noted, however, that under these conditions, small peaks corresponding to Ni_2C^+ and Ni_2O^+ appeared, the intensities of these were approximately equal and about 10 % of the intensity of the Ni_2CO^+ peaks.

Upon heating the sample to 390 K, after saturation of the surface at 295 K, the intensities of the $NiCO^+$ peaks quickly diminished and could not be detected at about 370 K. There was no significant alteration in the intensities of the peaks due to Ni_2^+, Ni_2C^+, Ni_2O^+ or Ni_2CO^+ and no new species appeared.

Upon the admission of a further 40 L of carbon monoxide at this temperature the Ni_2^+, Ni_2C^+, Ni_2O^+ and Ni_2CO^+ signals doubled their intensities but a further 20L made no significant difference. Prolonged heating at 390 K and several large doses of carbon monoxide (each dose being about 0.5 torr for 5 mins) caused a build-up of the Ni_2C^+ relative to that of the Ni_2CO^+ until no Ni_2CO remained on the surface. The final spectrum is shown in Fig. 10. This is in agreement with the results obtained from P.E.S. by Joyner and Roberts(8).

A number of points arise from the above observations. Firstly, S.I.M.S. can obviously detect chemisorbed species, but secondly the question can be asked as to whether the $NiCO^+$ species is simply a fragment formed from Ni_2CO^+ or whether there are two distinct surface species. If the former were the case, then it would be expected that this ratio would remain virtually constant over the range of temperatures considered. This was not so, however; the $NiCO^+$ peak had completely disapeared from the spectrum at about 370 K while the Ni_2CO^+ remained. Thus, it seems correct to assume that these two species are distinct surface species and the one does not derive from the other.

It is now necessary to suggest possible surface structures from which these secondary ions are derived. If we follow the methods and assumptions proposed by Benninghove, then Fig. 11 gives possible surface processes which would give rise to $NiCO^+$ and Ni_2CO^+. Since there is no evidence to say that adsorbed carbon monoxide is bonded to a metal surface via the oxygen atom, it seems likely that $NiCO^+$ is formed from process (a).

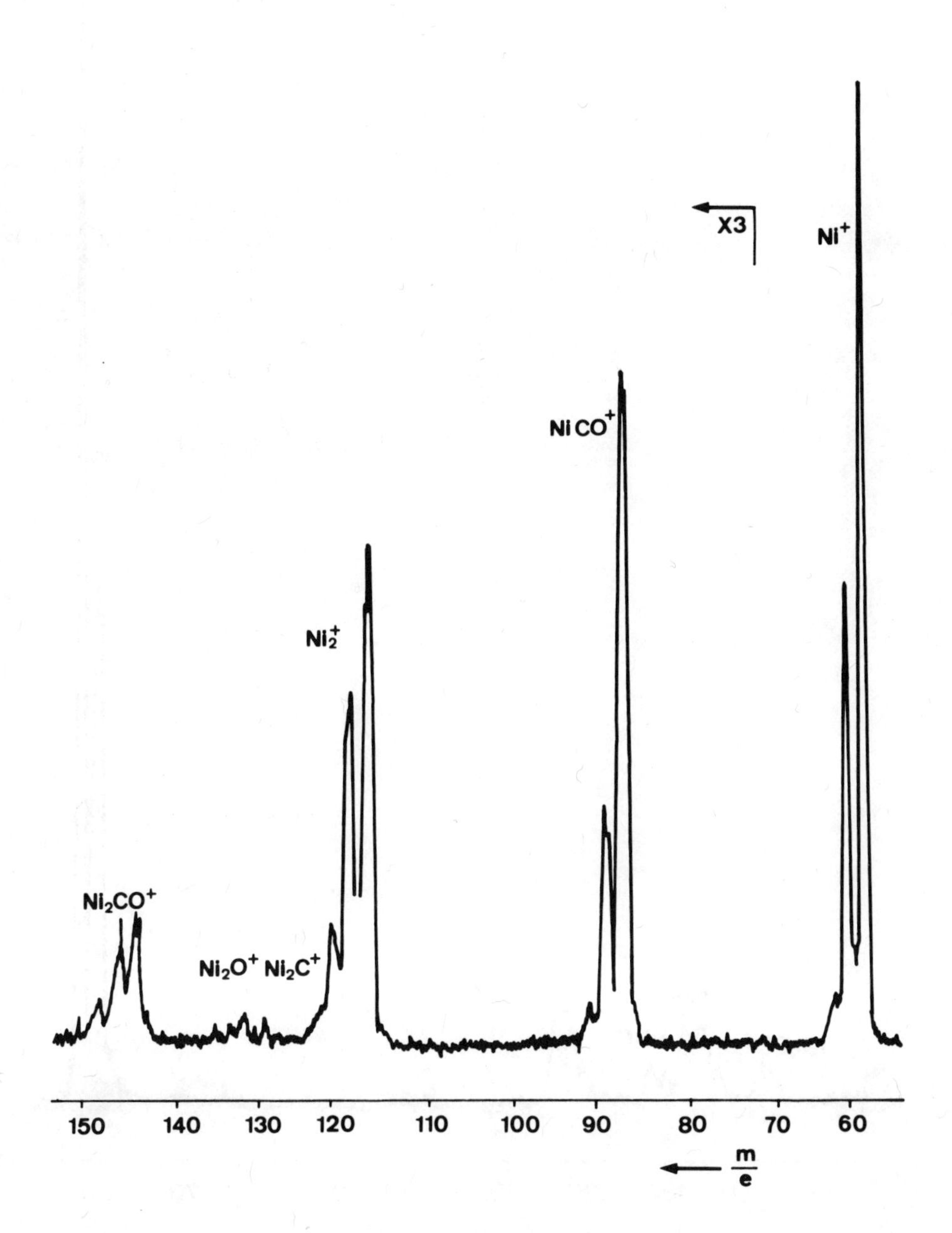

Fig. 9 : SIMS spectrum after saturation with carbon monoxide at room temperature

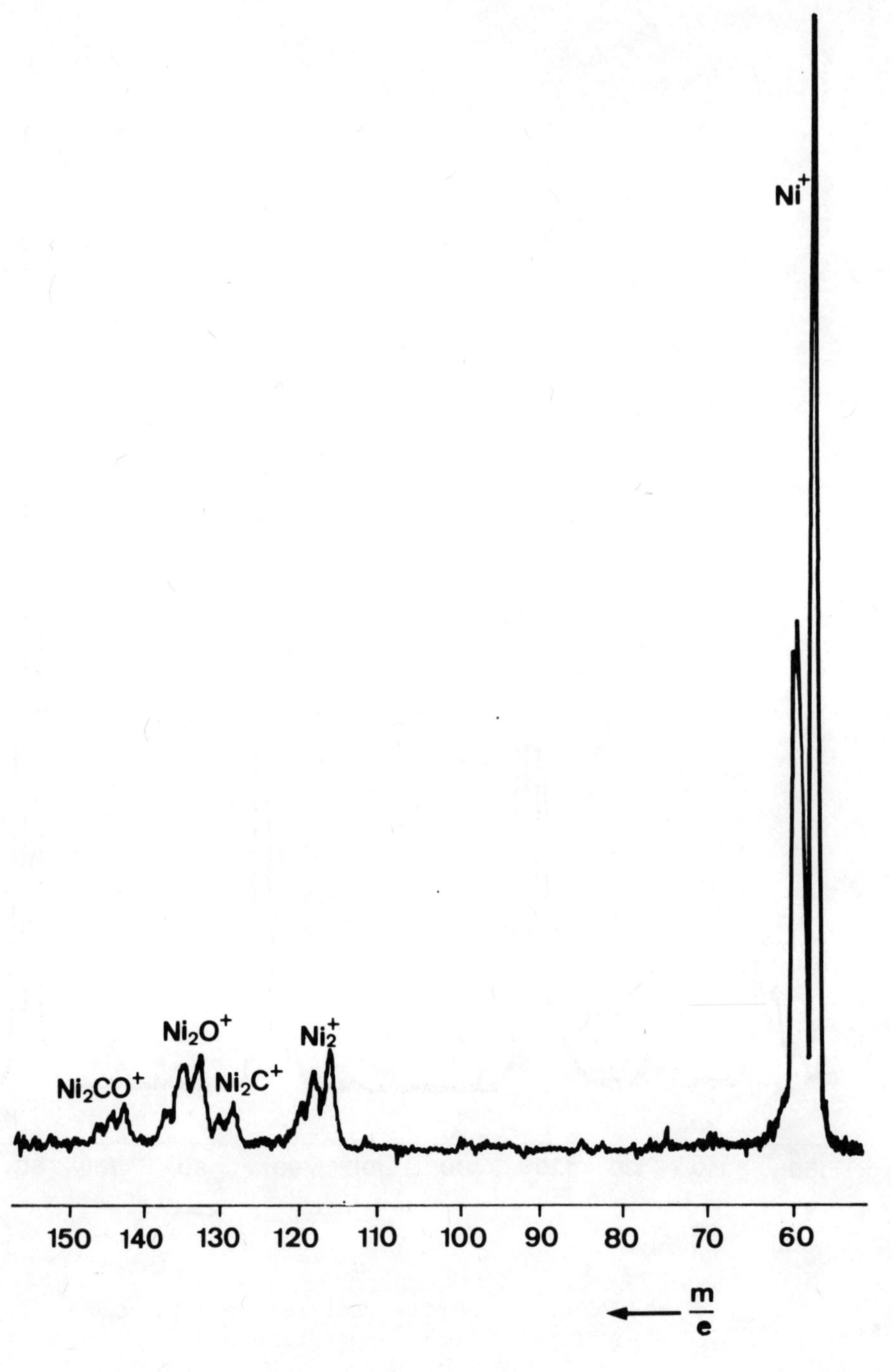

Fig. 10 : SIMS spectrum after large doses of carbon monoxide at 390 K

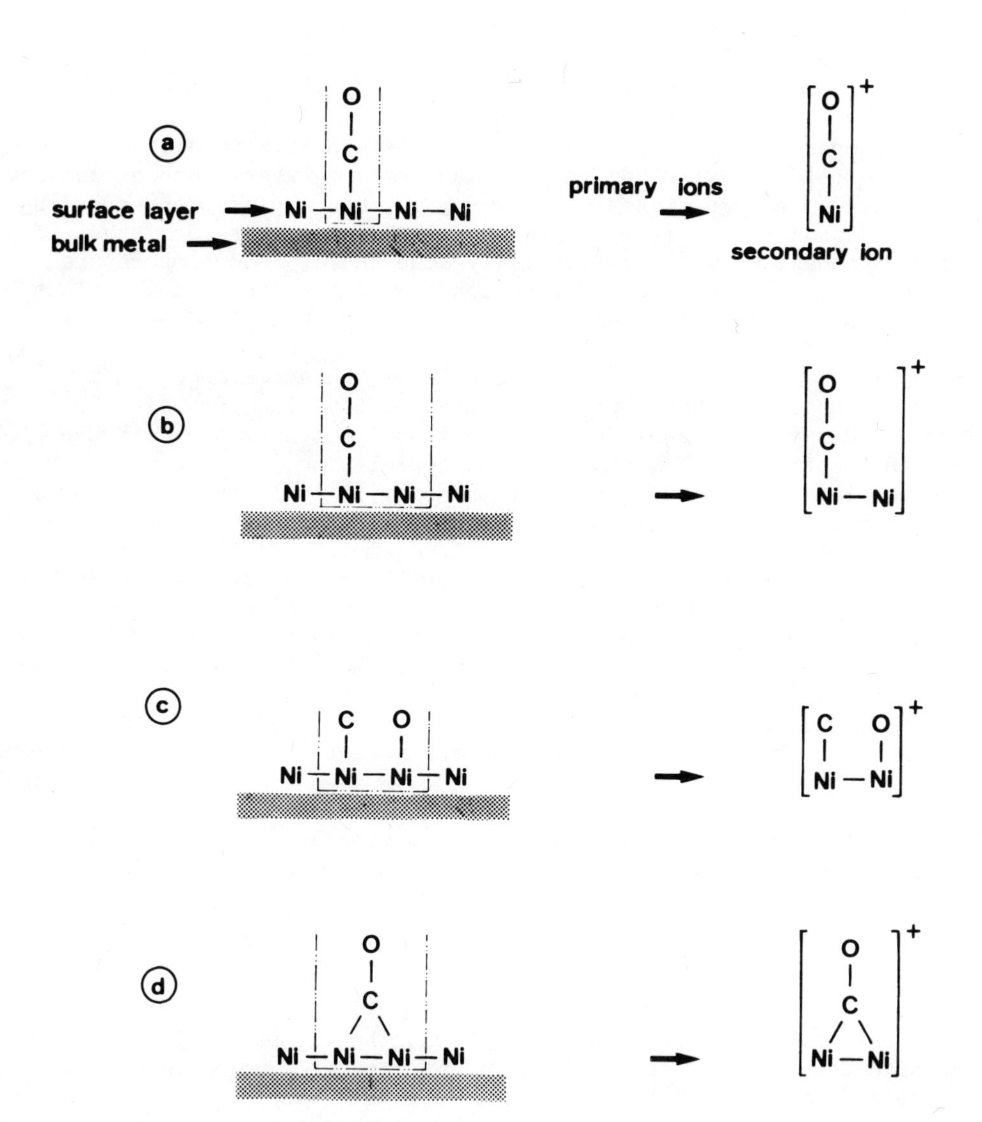

Fig. 11 : Possible surface processes which would give rise to the appearance of $NiCO^+$ and Ni_2CO^+

The surface structure illustrated by (b) appears to be identical with that in (a). It is, perhaps, possible that there are two types of site on the nickel surface and on both of these sites carbon monoxide is bonded in a linear manner. From one of these sites sputtering gives rise to only $NiCO^+$ and from the other only Ni_2CO^+, but this seems unlikely.

Process (c) suggests that the carbon monoxide is dissociatively adsorbed. It was found that at temperatures where carbon monoxide is known to adsorb dissociatively(8), we obtain peaks corresponding to Ni_2C^+ and Ni_2O^+. It is reasonable to assume, therefore, that if carbon monoxide was dissociatively adsorbed at room temperature then we would obtain peaks corresponding to the same species. Thus it is unlikely that process (c) gives rise to Ni_2CO^+.

It is logical, therefore, to ascribe the appearance of Ni_2CO^+ to a structure of the type illustrated in (d). The idea of a "bridge" bonded carbon monoxide molecule is supported by the evidence of IR spectroscopy(9). However, the correlation is not complete since the initial adsorption of carbon monoxide is IR inactive whereas we observe "linear" and "bridged" forms even at low exposures.

Our observations that, on heating the adsorbed carbon monoxide at 390 K in the presence of gaseous carbon monoxide, a surface carbide is formed, is in agreement with the work of Joyner and Roberts (8). Our results would suggest also that the mechanism of this carbide formation is one of dissociation rather than disproportionation, since, as well as Ni_2C^+, we observe peaks due to Ni_2O^+ but there were no peaks which could be assigned to a structure containing carbon dioxide.

A further set of studies were then embarked on to investigate the adsorption of CO on copper and to compare these with those obtained for Ni. It had already been suggested from infra-red studies that only one form of Cu-CO bonding occured, namely the linear form(10), so it was interesting to see whether species analogous to Ni_2CO^+ were produced from the system.

As with the experiments on nickel, the copper was cleaned and cooled down to 77 K and CO admitted in measured doses. Saturation occurred at about 80 L and the spectrum is shown in Fig. 12 . It will be noted that the only other species besides Cu^+ occurring in the spectrum is $CuCO^+$. This should be contrasted with nickel where under these conditions Ni_2CO^+ was clearly visible.

Upon the admission of 1 000 L of carbon monoxide to clean Cu at room temperature, the intensities of the Cu^+ peaks increased markedly and new peaks appeared at 154, 156 and 158 corresponding to Cu_2CO^+. The spectrum is shown in Fig. 13 . Saturation occurred at about 5000 L.

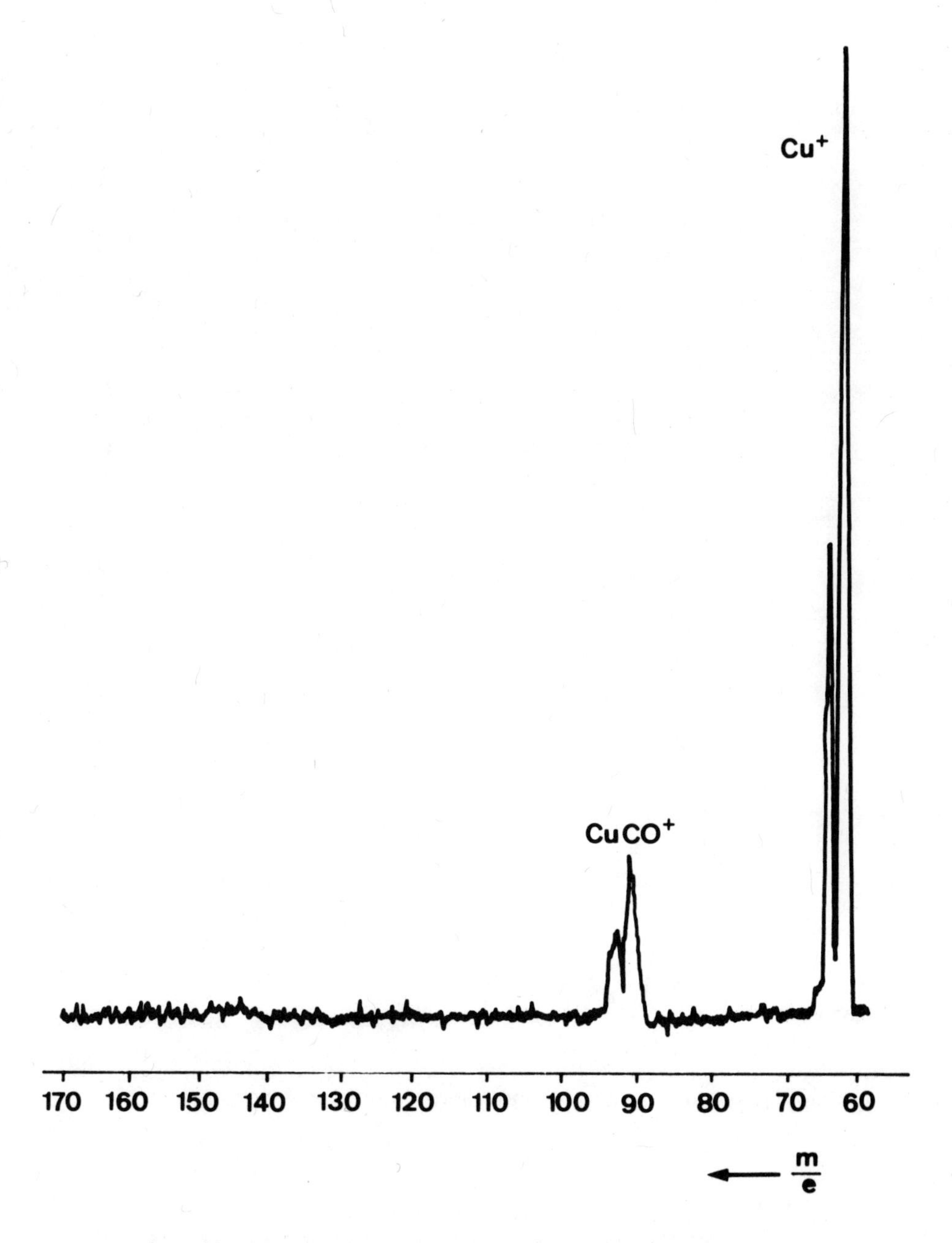

Fig. 12 : SIMS spectrum of copper foil at 77K after a saturation dose of carbon monoxide

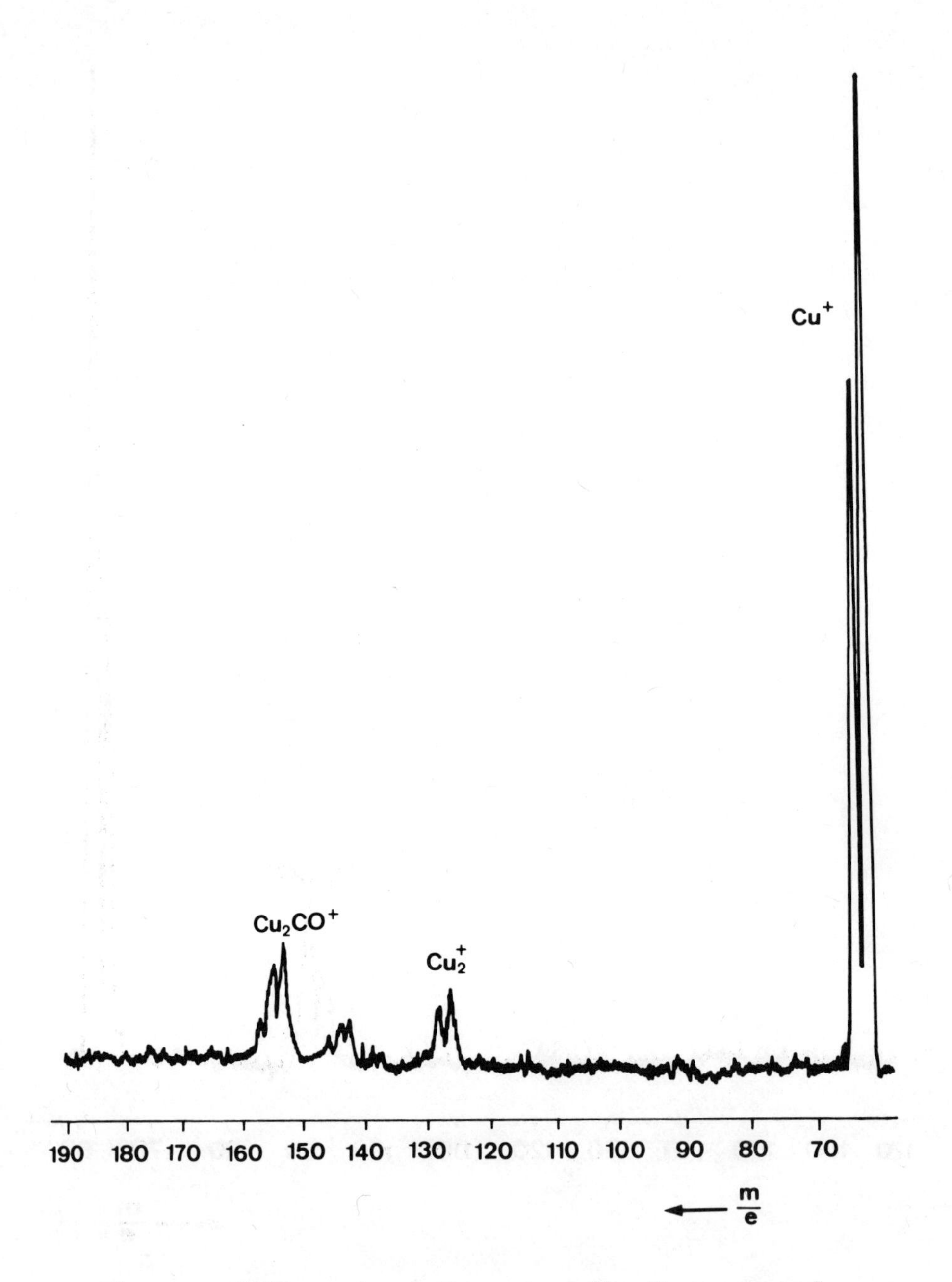

Fig. 13 : SIMS spectrum of copper foil after saturation with carbon monoxide at room temperature

Allowing carbon monoxide to flow over the sample at 390 K and 10^{-6} torr caused the appearance of various carbide species, see Fig. 14. These were assigned to CuC_3^+, CuC_4^+, $Cu_2C_2^+$, $Cu_2C_2^+$, $Cu_2C_3^+$ and $Cu_3C_2^+$. There were also very small peaks which could be assigned to $Cu_2CO_2^+$, suggesting perhaps, that the carbon is formed by disproportionation of the carbon monoxide followed by desorption of carbon dioxide.

Thus we see that in contrast with the results obtained from the adsorption of carbon monoxide on nickel at 77 K, adsorption on copper at this temperature produces only the $CuCO^+$ species and no Cu_2CO^+. Using a similar argument to that used for $NiCO^+$ it is believed that this is derived from a surface structure analogous to that in Fig. (a). This is in direct agreement with the results obtained from the IR studies[10] mentioned above. Furthermore our adsorption studies at room temperature reveal the secondary ion species Cu_2CO^+ but there is no evidence for the presence of $CuCO^+$. Arguing as before, it is probable that this is derived from a surface structure analogous to that shown in Fig.11 (d).

We have been able, therefore, in this particular study, to obtain considerable information concerning the mode of chemisorption of CO on both Ni and Cu. Fragmentation of the adsorbed species need not be invoked to explain the results which themselves seem to be in agreement with other work done by IR and PES. However, to extend the technique to a system for which extensive fragmentation may occur under sputtering conditions, we have investigated the adsorption of C_2H_4 on Ni by S.I.M.S.

As before, the nickel was cleaned and then cooled to 77 K and dosed with C_2H_4. Saturation occurred after about 40 L had been admitted. Fig. 15 shows the spectra which was obtained. As can be seen, fairly intense peaks occurred at mass numbers 86 and 88 which were assigned to $NiC_2H_4^+$. In addition, small peaks were observed at mass numbers 84 and 90 assigned to $^{58}NiC_2H_2^+$ and $^{60}NiC_2H_6^+$ (the other isotope of each of these coincides with one of the peaks due to $NiC_2H_4^+$). After correcting for the isotopic abundance of ^{58}Ni and ^{60}Ni, the intensities of the $NiC_2H_2^+$ and $NiC_2H_6^+$ signals were approximately equal.

The remaining, less intense, peaks which appeared in the spectrum, were assigned to $Ni_2C_4H_4^+$ (mass numbers 144, 146 and 148) $C_2H_x^+$ where x = 1-5 (mass numbers 25-29) and CH_y^+ where y = 1-3 (mass numbers 13-15). A number of interesting points arise as to the behaviour of various peaks during the dosing. The $NiC_2H_4^+$ species increased in intensity at a greater rate then $Ni_2C_2H_4^+$ and the intensities of the peaks due to $NiC_2H_2^+$ and $NiC_2H_6^+$ remained approximately equal and reached a constant value of about 1/10 of the intensity of $NiC_2H_4^+$ after about 50 L. On warming the saturated surface to room temperature (over a period of about 30 min) it was found

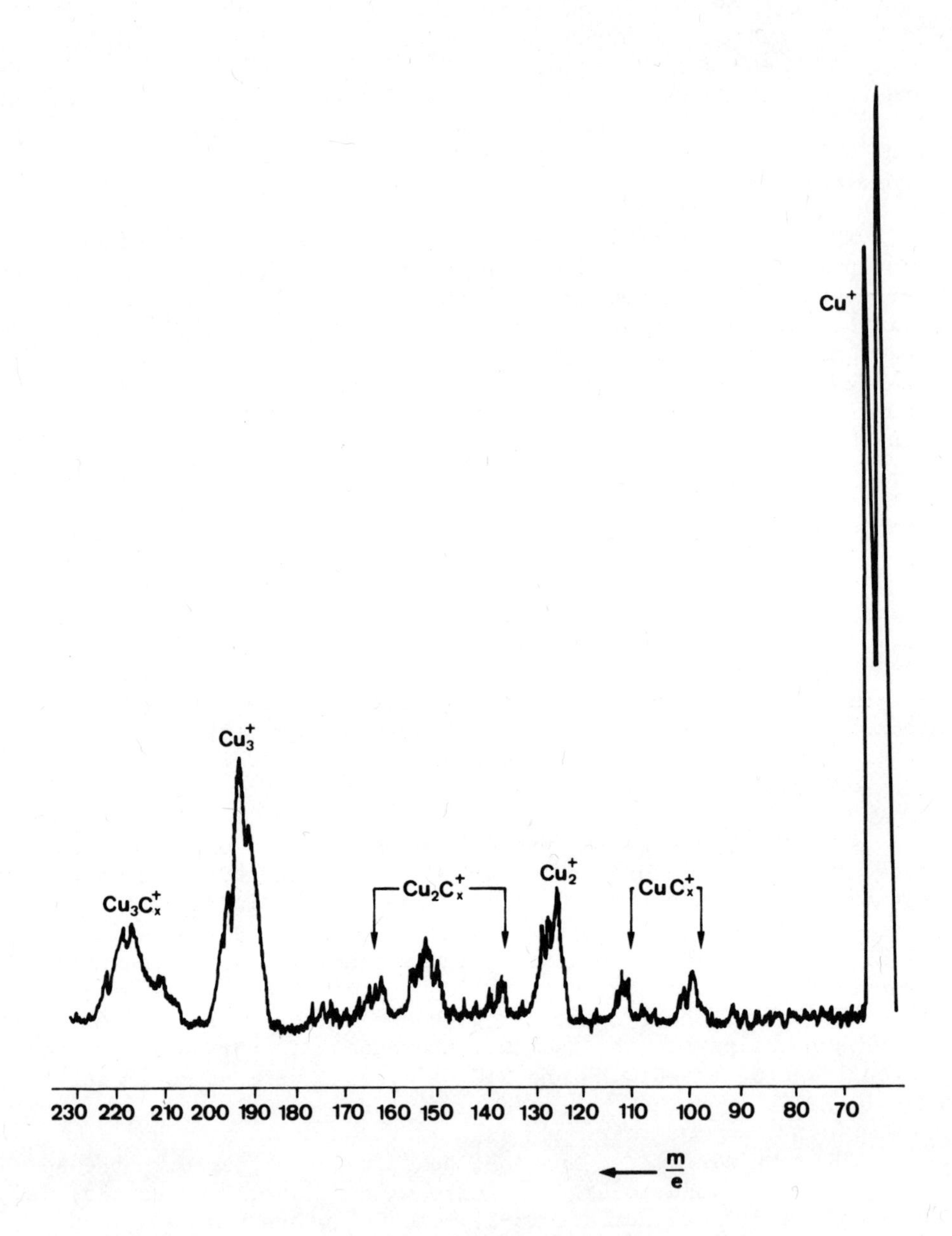

Fig. 14 : The appearance of carbide species in the SIMS spectrum after flowing carbon monoxide over the surface at 10^{-6} torr and 390 K

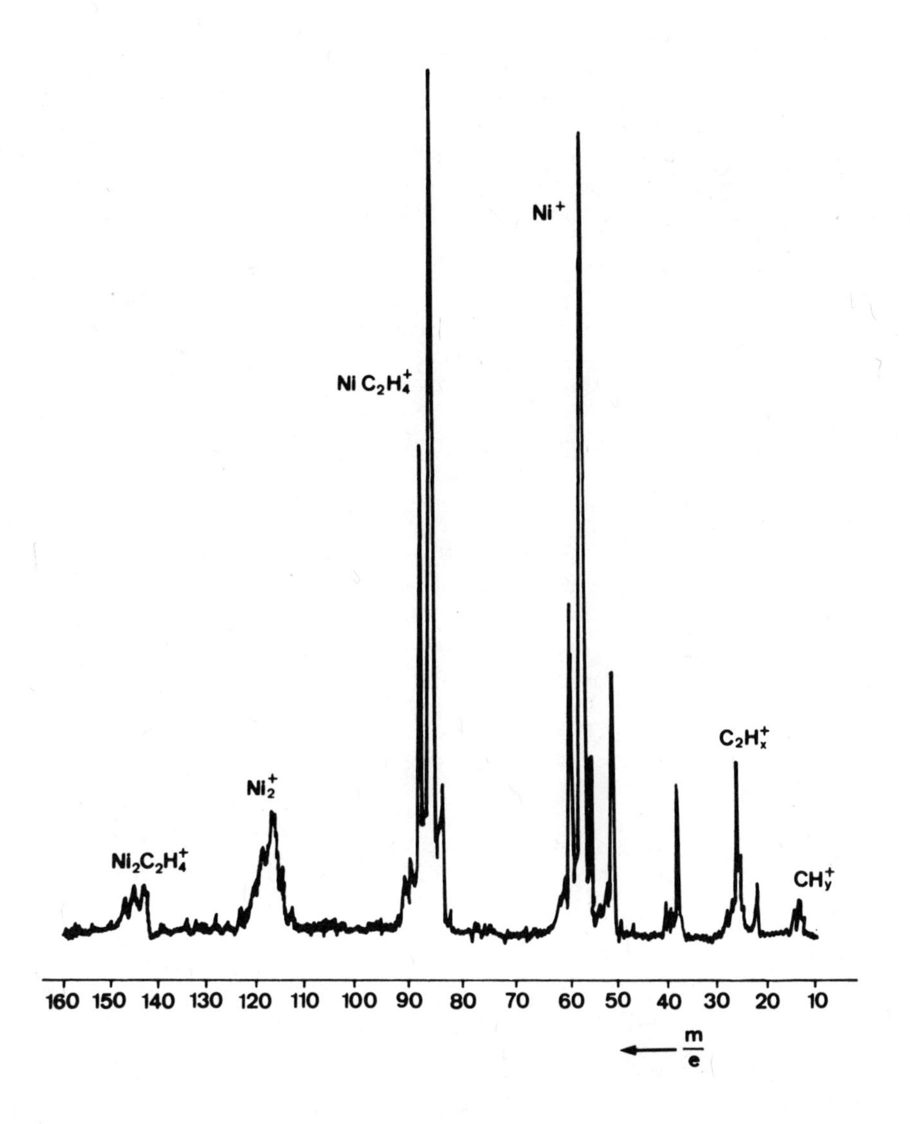

Fig. 15 : SIMS spectrum of the nickel foil after admission of a saturation dose of ethylene at 77 K

that the peak intensity of Ni^+ increased by about 30 % while that of $NiC_2H_4^+$ decreased to about 10 % of its value at 77 K. There was a 30 % decrease in the heights of each of the Ni_2^+ and $Ni_2C_2H_4^+$ peaks. The peaks due to $C_2H_x^+$ and CH_y^+ disappeared from the spectrum along with those due to $NiC_2H_2^+$ and $NiC_2H_6^+$.

Admitting ethylene to the sample at 295 K caused peaks to appear in the spectrum at mass numbers 86 and 88 which, again, were assigned to $NiC_2H_4^+$. There were no peaks which could be assigned to $Ni_2C_2H_4^+$ even after flowing ethylene over the surface at 10^{-6} torr for 1 1/2 hours.

However, there were peaks which could be assigned to hydrocarbon species and to hydrocarbon species associated with nickel. The diagramatic spectrum showing the mass numbers and relative intensities of the peaks due to these species is shown in Fig.16 . Similar experiments were carried out at 350 K, but the appearance of the spectra obtained was very similar.

Considering the above results we have once again the problem of deciding whether or not the two species $NiC_2H_4^+$ and $Ni_2C_2H_4^+$ come from a common precursor or whether they are separate surface entities. Using a similar argument to that used previously these must be distinct species. Further weight is added to this argument by the variation of the relative intensities of these two peaks with exposure. It was possible to assess the relative coverages of $NiC_2H_4^+$ and $Ni_2C_2H_4^+$ and to plot these against ethylene exposure in Fig. 17. Since the relative coverage of these species varies, the smaller one can not be a fragment of the larger one and they must be distinct entities.

The type of surface structures formed must now be considered. We would suggest that the species $NiC_2H_4^+$ is derived from a structure in which ethylene is bonded to a surface nickel atom by means of its π electrons, as suggested by photoemission(11). The $Ni_2C_2H_4^+$ species is derived, we suggest, from a structure in which each of the carbon atoms is bonded to a different nickel atom by means of σ bonds, the structure suggested by IR experiments(12).

From our spectra it is evident that the adsorbed ethylene undergoes various surface reactions to form other hydrocarbon species.

Even at 77 K there is a small amount of self hydrogenation and breakage of the C-C bond. This is evident from the peaks at mass numbers 84 and 90 and those at 13, 14 and 15. At higher temperatures it appears from our spectra that both C-C and C-H bond breaking and making can occur.

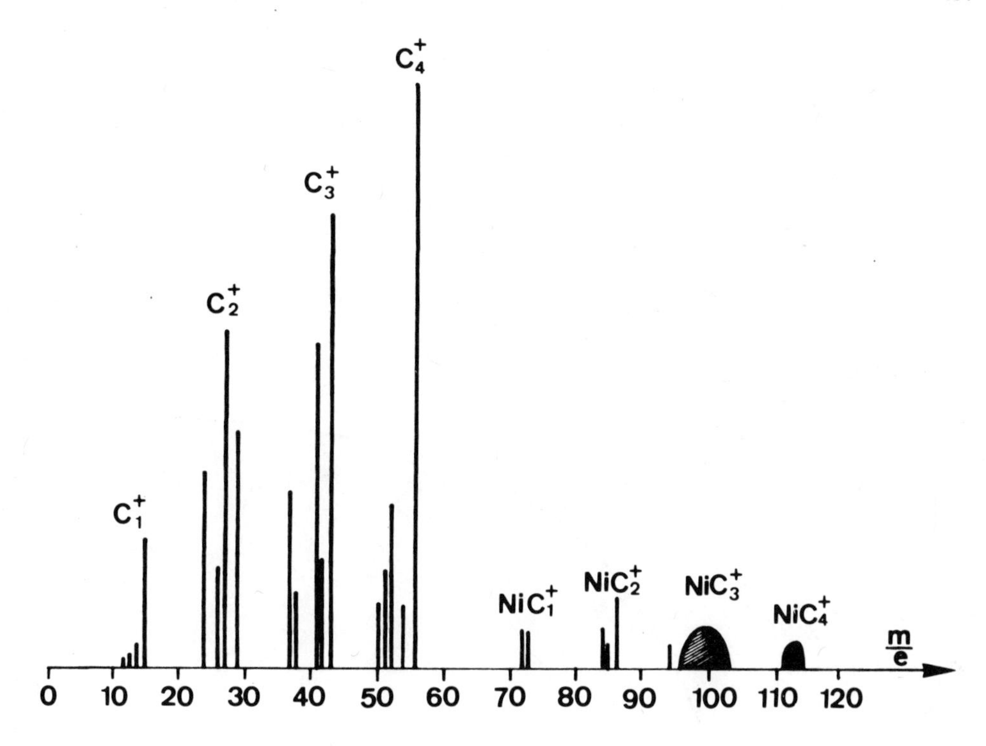

Fig. 16 : Diagramatic SIMS spectrum showing only the hydrocarbon species which appeared in the spectrum after exposing the nickel sample to ethylene at 10^{-6} torr and 295K for 1.5 hours.

The C_1 and C_2H to C_2H_3 species could have been formed by fragmentation of the ethylene by the argon ion beam. Although the results contained here do not provide incontroversible evidence that there is no induced fragmentation, two facts indicate that the species observed are those which are present on the surface. First, the hydrocarbon peaks indicate larger as well as smaller species than the parent, ethylene-containing, species and they form a continuous series. If fragmentation were occurring only smaller species would be observed. Second, although not strictly comparable, adsorption of ethylene on silver(13) produces a SIMS spectrum in which only C_2 species are observed. If ion bombardment caused the appearance of C_1, C_3 and C_4 species in the spectrum obtained from nickel then

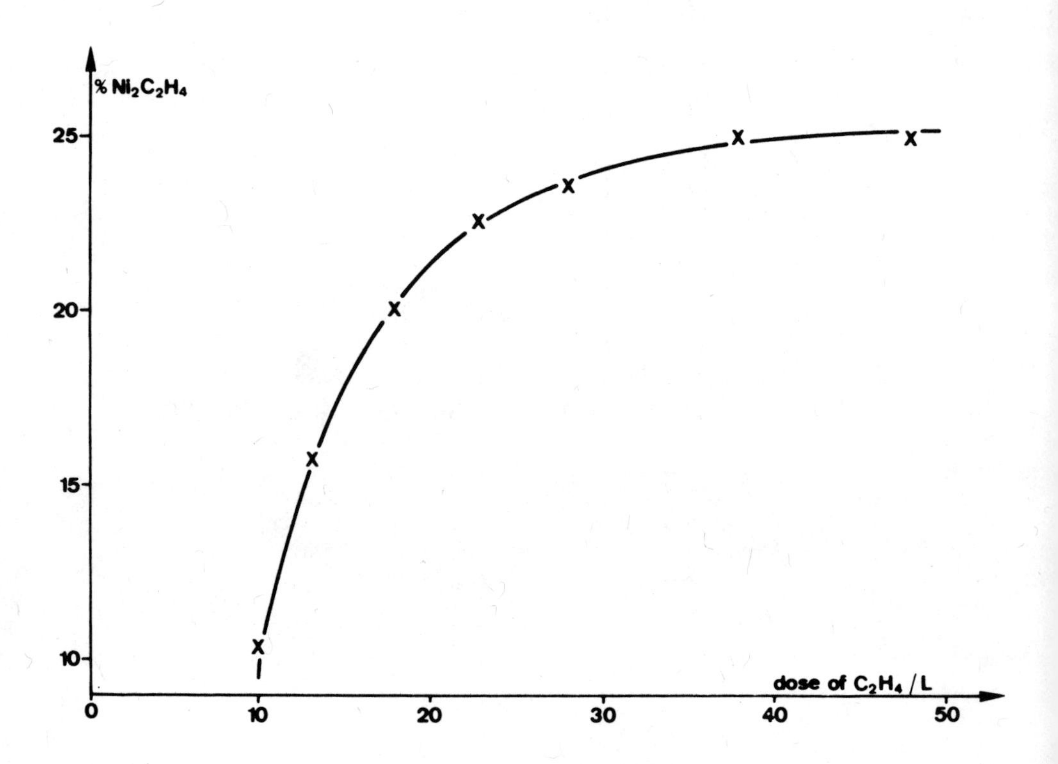

Fig. 17 : Variation of the proportion of $Ni_2C_2H_4$ on the surface with dose of ethylene at 77 K

it is likely that it would cause the appearance of these species on silver as well.

Further evidence to suggest that C_1 and C_4 species occur on the surface comes from hydrogenation experiments[12] over nickel in which butane is a major component in the gas phase at 20°C and methane is observed at 100°C. Evidence for self-hydrogenation comes from deuterium exchange work[14].

Once again we have shown that S.I.M.S. when applied to chemisorption of molecules on metals has given information which correlates well with evidence from other techniques, when available. There are still a number of problems to overcome, one being the quantitative aspects of the method, and secondly amassing enough data to show that if fragmentation or reaction due to the primary ion beam is present, then it is negligible or can be accounted for. The evidence is beginning to be obtained, and from the surface chemists' point of view the future of the technique seems to be most exciting.

REFERENCES

1. Ya. M. Fogel, Soviety Physics Uspekhi 10, (1967) 1
2. A. Benninghoven, Surface Science, 35,(1973) 427-457
3. G. Carter, J.S. Colligon, "Ion Bombardment of Solids", Heinemann Ed. Books Ltd., London, 1968
4. J.M. Schroer, T.N. Rhodin and R.D. Bradley, Surface Science, 34, (1973),571
5. Z. Sroubek, Surface Science, 44,(1974) 47-59
6. A. Benninghoven and A. Müller, Surface Science, 39,(1973) 416-426
7. A. Müller and A. Benninghoven, Surface Science, 39,(1973) 427-436
8. R.W. Joyner and M.W. Roberts, J.S.C. Faraday I, 70,(1974) 1819
9. A.M. Bradshaw and J. Pritchard, Surface Science, 17,(1969) 372
10. M.A. Chesters, J. Pritchard and M.L. Sims, Chem. Comm., (1970) 1454
11. J.E. Demuth and D.E. Eastman, Phys. Rev. Letts., 32, (1974) 1123
12. B.A. Morrow and N. Sheppard, Prox. Roy. Soc., A311, (1969) 391
13. M. Barber and J.C. Vickerman, unpublished data.
14. J. Turkevich, D.O. Schissler and P. Irsa, J. Phys. Colloid Chem., 55, (1951) 1078

FIELD-ION-MASS-SPECTROMETRY INVESTIGATING ELECTRONIC STRUCTURE AND REACTIVITY OF SURFACES

Jochen H. BLOCK

Fritz-Haber-Institut der Max-Planck-Gesellschaft
Berlin - Dahlem

INTRODUCTION

In FIMS an atoms or molecule is ionized in front of a surface by the interaction of a high electric field. The electron enters the solid. The positive ion is analyzed mass spectrometrically. Originally, the analytical aspect of a mass analysis of field ions with low fragmentation was the main emphasis of development. Principles of FIMS have been summarized (1,2,3). Now, by using mass and energy analysis of field ions simultaneously, details about the mechanism of field ionization can be studied. It will be demonstrated in this contribution that the electronic structure of a metal surface interferes with the ionization process as well as the reactivity of a metal surface. With the great advantage of extreme surface specificity on single crystal planes and with the possibility to identify individual surface molecules the energy distribution of field ions now can give additional information on electronic states of a surface and on the energetics of surface processes.

THE SIMPLE ENERGY POTENTIAL DIAGRAM

For those molecules which have no interaction with the surface, the minimum field strength F_o which is required for ionization is given by the ionization potential I_p of the molecule and the work function Φ of the emitter surface : $eF_o x_c = I_p - \Phi$. As to be seen in fig.1, electrons can reach empty electronic states of the solid only at or above the Fermi-level, which implies a minimum distance x_c from the surface. The higher the work function the smaller the distance x_c. If the distribution function of possible x_c-values could be measured, this function should indicate the density of unoccupied electronic states at or above the Fermi-level of a metal.

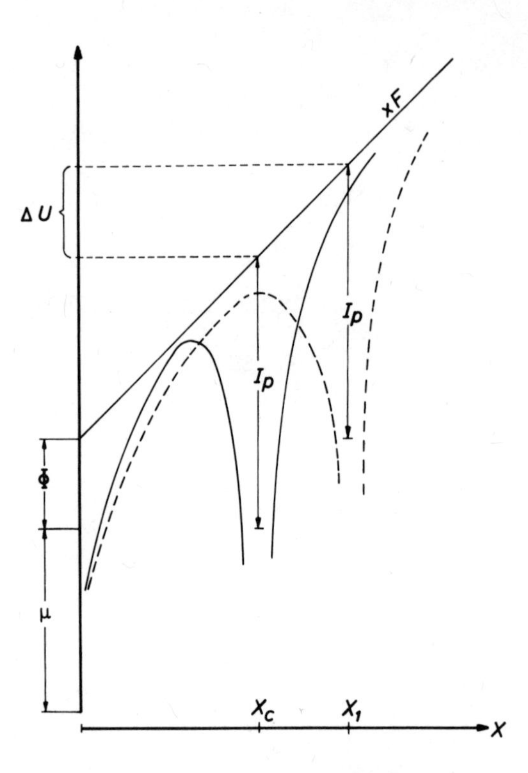

Fig.1 : Potential for electrons tunneling out of the atom. Different distances x_c and x_1 result in a potential difference ΔU,
Φ = work function,
μ = Fermi energy

The x_c-value of the ionization process can be measured by the energy loss of the field ion. The potential difference at two possible places x_o and x_1 of ionization is given by $\Delta U = eF\ (x_1 - x_c)$. This value can be measured by the energy distribution of field ions.

We can define an appearance potential, AP, which corresponds to the minimum energy which must be supplied by the electric field in order to ionize a molecule. The appearance potential depends on the effective ionization energy I_p of the molecule or, more generally, it reflects the energetics of the ionization process. The AP-values represent the low energy onset part of the energy distribution of field ions. The potential at the emitter surface (fig.2) may be U_E and at the retarding electrode U_R when the first ions reach the detector. The absolute value of the appearance potential is then

$$AP = U_E - U_R + \Phi_R$$

AP depends on the work function of the retarder electrode Φ_R and not on the work function of the emitter. This can be proven in an energetic circle. It turns out that the emitter- and retarder-Fermi potentials are linked in such a way that only the retarder work function Φ_R appears in the determination of AP. For experimental conditions this is advantageous, since those influences which change emitter work functions may be neglected.

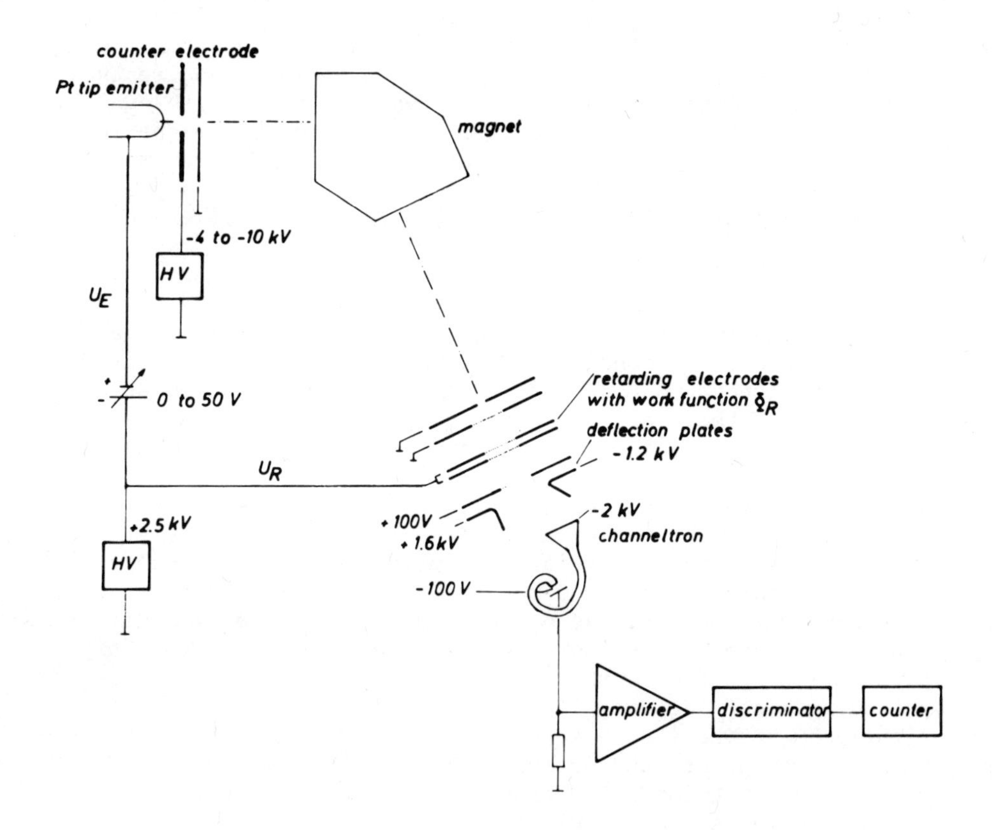

Fig. 2 : Magnetic sector type mass spectrometer and field ion source.

For an ion the AP value determines the energy uptake from the field. The AP value is related to the ionization potential I_p, to the polarization energies E_p of the neutral and E'_p of the ion in the following way

$$AP = I_p + E_p - E'_p$$

Polarization energies of ions are usually smaller than those of their neutral molecules and have frequently been neglected.

THE MEASUREMENT OF ENERGY DISTRIBUTION

Experimentally the energy distribution of ions, i.e., the kinetic energy which the particles acquire during field ionization from the electric field has been measured in three different kinds of mass spectrometers :

a) In the magnetic sector type mass spectrometer, retarding field electrodes have been applied for this purpose. Since this is an integral measurement, high stabilities in field ion currents are required. The energy analysis performed by Jason (4) (5) and by Heinen (6) used retarding electrodes in back of the magnetic mass separator as shown in fig.2. Goldenfeld, et al. (7) mounted the retarding grid between the ion source and magnetic analyzer. In this case,a spherical grid has to be used for the divergent ion beam, otherwise the distribution of momenta would be measured instead of the energy distribution.

In the design and operation of the energy analyzer in fig.2 the retarding electrode consists of two gold meshes, with about 20 strips per linear mm and 70 % transmission. The two meshes are 2 mm apart. With this construction, a minimum potential penetration is achieved, since the work function of the retarding electrode is part of the measured potential difference ($U_E-U_R-\Phi_R$), gold is a convenient material which is least affected by adsorption processes. The potential difference between the emitter electrode and the retarding grid is supplied by an additional variable low voltage source of 0 to 100 V. The ion intensity at a given mass number is measured as a function of this potential difference. Ions, which are created several Å in front of the emitter tip at a potential $U_E - \Delta U$, arrive with an energy loss and can pass the retarding electrode only if the potential difference is sufficiently high. The ion detector, which for instance is a channeltron, is placed off the optical axis of the instrument. Deflection plates focus the ion beams onto the entrance aperture of the detector whereas secondary electrons and Bremsstrahlung are not detected.

b) In the time-of-flight mass spectrometer field pulses cause uncertainties in energy distributions. Müller and Krishnaswamy (8) combined the atom probe with a Möllenstedt energy analyzer (fig.3). This is an electrostatic saddle field lens which uses the strong chromatic aberration of an out-of-axis beam. An energy resolution of 5×10^{-5} has been achieved. Another experimental possibility was verified by Müller and Sakurai (9) in the combination of a magnetic sector field with the atom probe. An energy resolution of the order of $<10^{-4}$ can be obtained.

c) The combination of a quadrupole mass analyzer with an energy filter was first used by Utsumi and Nishikawa (10). The analyzer has to be installed in front of the quadrupole field which alters the ion energies. The instrument is demonstrated in fig.4 as well as the potentials of ions in this system.

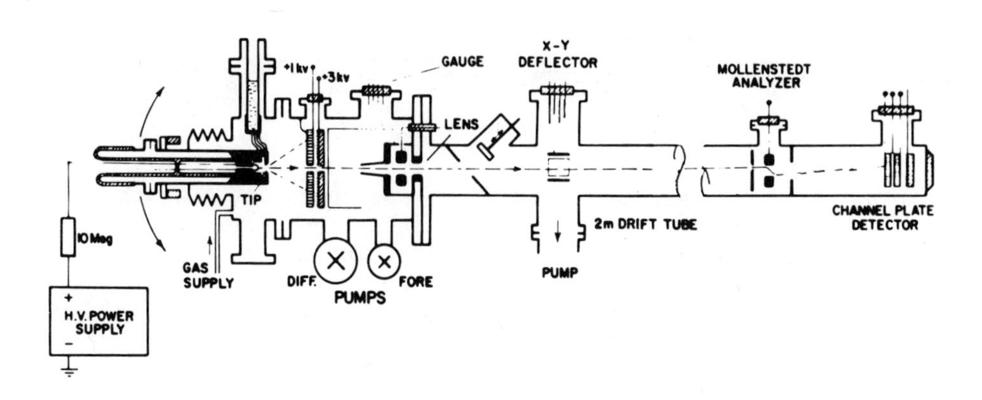

Fig.3 : The atom-probe FIM equipped with Möllenstedt energy analyzer according to Müller and Krishnaswamy (8).

RESULTS ON ELECTRONIC SURFACE STATES

The first energy distribution measurement (11) in an FIM revealed an unexpected narrow energy value, which corresponded to an ionization zone $x_c - x_1$ of 0.2 Å. Rather unexpected was the later discovery by Jason (12) who found a periodic structure in the energy distribution of field ions. At sufficiently high electric fields the intensities of parent molecular ions H_2^+, D_2^+, CO^+ or Ne^+ which were ionized at single crystal planes of tungsten exhibit a periodic struc-

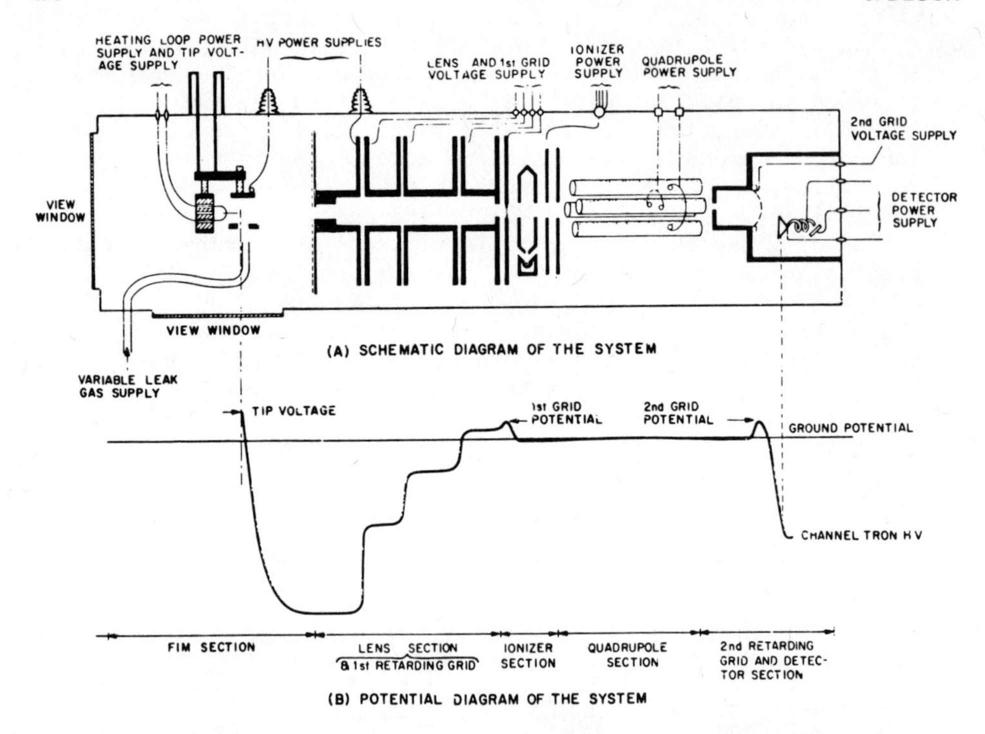

Fig.4 : A quadrupole field ion mass spectrometer with retarding field energy analyzer according to Utsumi and Nishikawa (10). The lower part shows the potential of the ions.

ture with several peaks in intensity. As one of the typical examples the H_2^+ energy distribution for different electric fields is show, for a (011) W surface in fig.5. The Jason peaks have the following properties.

a) The energies for maximum ionization are separated by several eV.
b) The energy separation of successive peaks decreases with increasing energy deficit.
c) The distance between succesive peaks increases with increased external field.
d) The peak hight decreases at higher masses of ionized molecules.
e) Near the minimum field at which field ionization becomes apparent the periodic structure is not observed.

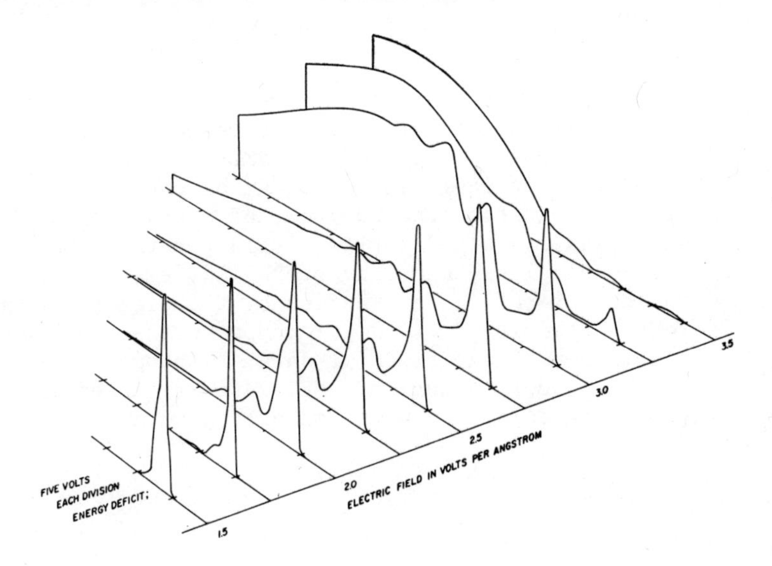

Fig.5 : The energy distribution of H_2^+ ions for different electric fields. Relative differential intensities are plotted vertically, the zero point of energy deficit is unspecified. Values according to Jason (12).

Jason (12) gave a convincing explanation for this phenomenon. He suggested a resonance tunneling to be responsible for the periodicity of the energy distribution. Resonant states are formed by the combination of the externally applied electric field and the surface potential. Under the influence of the field the space between the surface and the molecule (corresponding to x in fig.1) has to be treated wave mechanically. As to be discussed later on, this leads to resonance tunneling at certain preferred energy values. Another original idea for explaining the Jason effect was published by Lucas (13), who considered the excitation of surface plasmons during the acceleration of field ions. This mechanism was however disproved (8) (13).

In a careful investigation and with higher precision Müller and Krishnaswamy (8) repeated Jasons experiment. They found that Jasons values for the field strength should be corrected by + 50 %. Otherwise there was complete agreement for experimental data of H_2 and Ne on tungsten. Furthermore, the effect could be confirmed on other surfaces like gold, iridium, silicon or carbon. The energy distribution of Ne^+ on (011) W is demonstrated together with new theoretical calculations in fig.7.

Appelbaum and McRae (14) recently refined the Jason model. Calculations by Jason and also by Alferieff and Duke (15) for the resonance tunneling were corrected for the reflectivity of the metal surface . Using empirical data of LEED measurements, Appelbaum and McRae could account for influences of the distribution of the density of states of the surface. With a potential energy model which was three-dimensional at the ionized atom and one-dimensional at the surface and which was corrected for the field penetration into the metal, they could calculate tunneling probabilities. The model, describing the ionization rate as a function of the distance from the surface, is represented in fig.6. Fig.7 compares the results of numerical calculations with experimental data. Although the complex amplitude reflection coefficients of LEED data were used in a semi-empirical way, the agreement is astonishingly good.

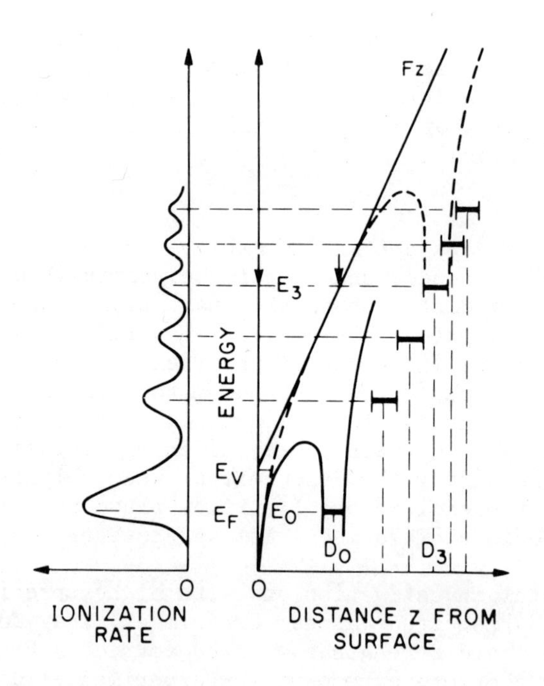

Fig.6 : Schematic representation of the energy distribution (left) and a one-dimensional energy diagram to explain the origins of the peak (right). E_F Fermi-level, Ev=vacuum-level. The orbital energy levels at the right refer to ions at various distances z from the surface. Each level characterized by a vertical dashed line corresponds to a peak.

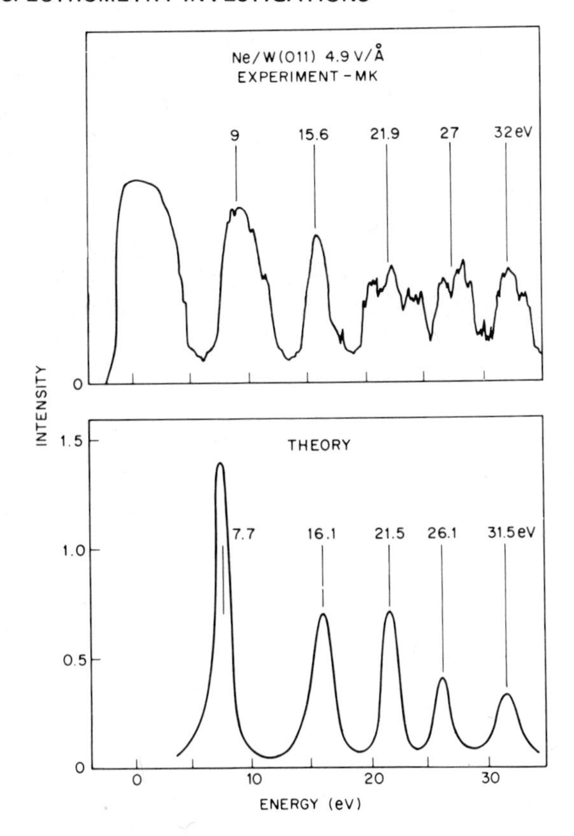

Fig.7 : Energy deficit during the field ionization of Ne on W (011), F = 4.9 V/Å. Top : Experimental results as found in (8). Bottom : Theoretical calculations by Appelbaum and McRae, according to (14).

In fig.7 the energy deficit reaches values up to 30 eV. This is due to the behavior of electrons near the potential step. Influences of the distribution of the density of states have to be expected at much smaller energy losses.

Recent investigations by Utsumi and Smith (16) claim to give this correlation. Measurements of the field ion energy distribution are made in a FIM (without mass discrimination) and with a precision of < 0.2 eV. A clean tungsten surface yields within the first Jason peak a structure which is shown in fig. 8a. The open circles, which are measured values of He^+ intensities on a (011) W surface, show additional humps on the high-energy side of this peak. These peaks are 1.3, 2.0 and 2.7 eV above the Fermi-level. They are compared with theoretical calculation by Christensen and Feuerbacher (17). Bulk density of states and surface density of states data, as obtained by calculations are given for comparison with experiments on fig.8. After the adsorption of N_2 at the (001) W surface the energy distribution of He^+ is different (fig.8b). This difference also depends whether N_2 is adsorbed at 300°K or at 21°K. These data are not understood in detail. This publication represents the first case for measurements on the density of unoccupied states by field ionization.

For the understanding of electronic properties of surfaces this would be an important tool especially since this energy region is inaccessible for photoemission spectroscopy as well as for field emission spectroscopy.

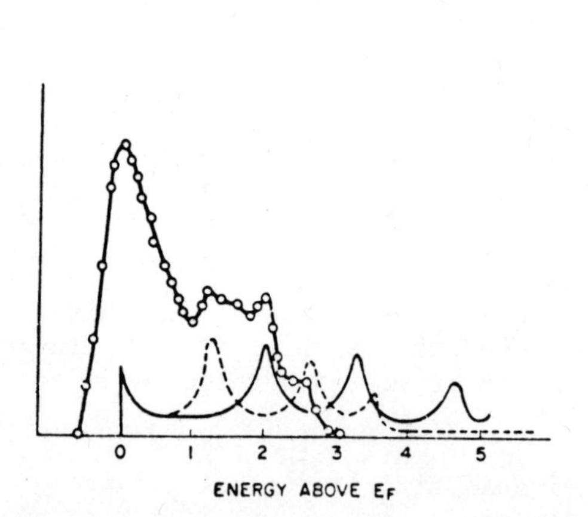

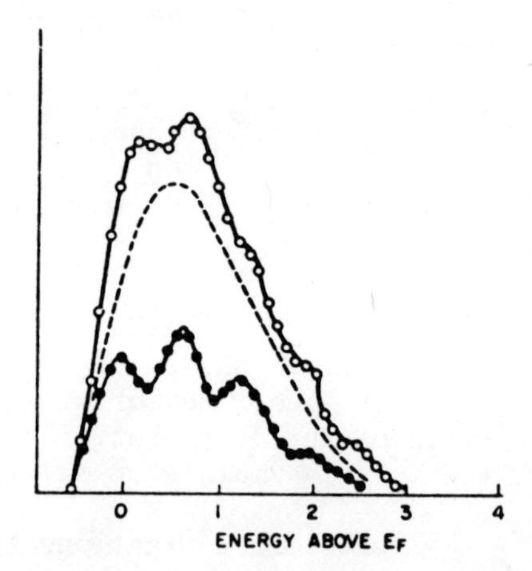

Fig.8a + 8b : The energy distribution of He^+ on clean (100) W (open circles) in the left part (a). Comparison with calculations of Christensen and Feuerbacher. After adsorption of N_2 at 300°K (filled circles in the right picture (b)), open circles adsorption of N_2 at 21°K. Absolute intensities are arbitrary.

OBSERVATION OF MOLECULAR EFFECTS

The field ionization of hydrogen gave already one exception in the occurance of Jason peaks. This was the H_3^+ ion, which was thought to be formed by a surface reaction. With the isotopes H_2, HD and D_2 another rather unexpected effect was observed (18). The energy distribution of H^+ ions showed peak maxima which the authors correlated with the molecular behavior in front of the emitter. Fig.9 shows the energy distribution of H^+ ions from H_2 and HD at different fields. The energy deficit is thought to be correlated with pendulum motions of the molecule.

REACTIVITY AT SOLID SURFACES INVESTIGATED BY FIMS

The Gibbs free energy ΔG of a chemical reaction is correlated with the potential ΔE of an electrochemical cell by the relation

$$\Delta G = n F_f \Delta E$$

where n is the number of electrons which are transfered for each molecule and F_f is the Faraday constant. If a reaction is far from equilibrium the activation energy ΔQ_a given by this potential is

$$\Delta Q_a = n F_f \Delta E$$

The value ΔE of an electrochemical cell is comparable with the value ΔU of fig.1. This means, the formation of field ions at a surface may be regarded as an electrochemical process where the necessary energy is achieved by $n F_f \Delta U$. Field ionization is an electrochemical process where one electrode is removed. The relative potential which this electrode would have to have to make the reaction energetically possible is given by ΔU. These considerations are applicable independent of the mechanism of ion formation.

The idealized case of field ionization (fig.1) applies only to rare gases on clean surfaces. If any kind of chemistry affects the interaction between the gas molecule and the surface, the mechanism of ion formation is much more complicated and reflects all kinds of chemical interactions, even chemical ionization at the surface (19).

We can classify these mechanisms as follows :

1. The field ionization of a rare gas atom will be performed at distance x_c solely by the external field F like for helium on the metal M_e^c :

$$M_e \xrightarrow{F} He^+ + e^-_{(Me)}$$

   The ionization probability can then be calculated by wave mechanics.

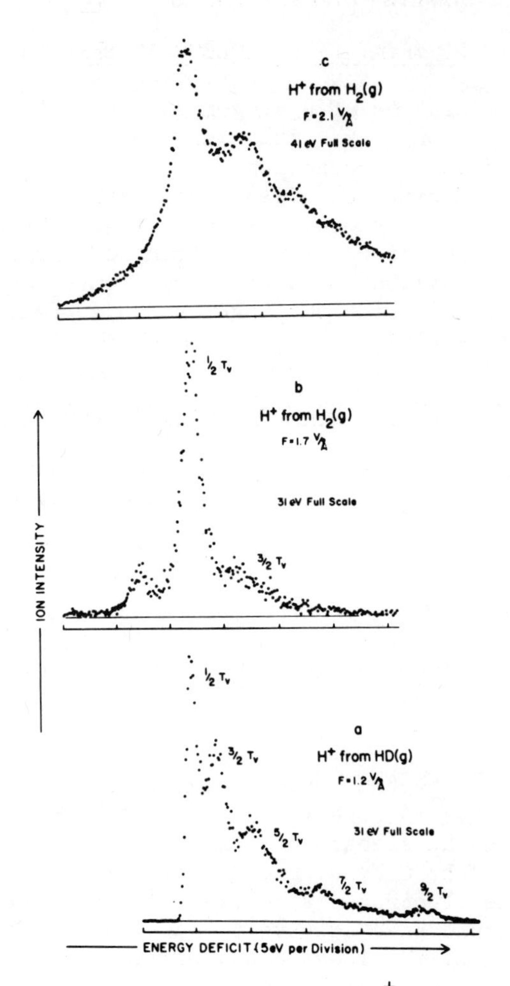

Fig.9 : The energy distribution of H^+ from H_2 and HD at different fields, according to (18). Different vibrational states τ_v are indicated which will influence ionization and energy deficit.

2. In a condensed layer, in a multilayer of gas molecule or during molecular collisions on an adsorbed layer, proton transfer processes are observed very frequently. With water molecules, for instance, molecular associates are normally observed

$$H_2O + (H_2O)_n \xrightarrow{F} H_3O^+ + (H_2O)_{n-1} + OH + e^-_{(Me)}$$

Proton transfer reactions are stimulated by acidic hydrogen atoms and are very common for alcohols, amides, etc. Molecular or dissociatively chemisorbed hydrogen is less qualified for this reaction type. With methane, for instance, CH_5^+ ions are easily observed if traces of water ($p_{H_2O} < 10^{-9}$ torr) are present, while molecular hydrogen does not yield any measurable proton attachment.

3. The ionization process at the surface of a field emitter is also greatly influenced if intermolecular interactions create charge-transfer complexes. Chemical properties of these complexes are explained by a partial electron transfer from a donor to an acceptor molecule, without actually forming separate ions. Onset fields for field ionization are considerably diminished for donor molecules. Benzene, for instance, (20) can be ionized at less than half the usual field strength if an acceptor A, like chloranil, is present at the surface.

$$\bigcirc + A \rightarrow \bigcirc^{\delta(+)} \ldots\ldots A^{\delta(-)} \xrightarrow{F} \bigcirc^{+} + A + e^{-}_{(Me)}$$

Here an intermediate with partial charges $\delta(+)$ and $\delta(-)$ is formed which facilitates the field ionization process. This kind of chemical interaction is very frequently the reason for the well-known "promotion" of field ionization, i.e. intensification of ion intensities at reduced field strengths. Taking polarization forces into account, charge transfer will also be of importance under field ionization conditions for systems which are not designed for it in the absence of an electric field.

4. The field induced cleavage of surface bonds

$$Me - R \xrightarrow{F} Me + e^{-}_{(Me)} + R^{+}$$

leads to field desorption of surface molecules or atoms. If R is identical with a bulk metal atom Me, this process is denoted as field evaporation. It is of particular importance for surface analysis and will be discussed later.

The ionization mechanisms at surface under extreme electrical fields involve a variety of intricate processes. Consequently, fundamental molecular data, like the ionization potential of an isolated molecule, have much less importance in defining onset fields in field ionization than in comparing the threshold in photo or electron impact ionization. The energetic parameters of these reactions can be measured by the energy distribution or appearance potentials of parent and fragment molecular ions.

EXAMPLES OF APPEARANCE POTENTIAL MEASUREMENTS

a) Noble gases

Fig 10 shows the retardation curves of xenon and krypton during field ionization at a Pt surface as measured by Heinen et al.(6) The shapes of curves are similar and the energy deficit $\Delta U=U_E-U_R$ (in fig.2) is 7.2 ± 0.1 eV for xenon and 9.1 ± 0.1 eV for krypton. The difference of the ΔU values corresponds within the accuracy of the measurement to the difference of ionization potentials (12.1 eV and 14 eV) as to be expected from fig.1. The work function Φ_R of the retarder electrode can be determinded Φ_R = 4.9 ± 0.1 eV. The presence of hydrocarbons which alter the work function of the emitter have no influence on the retardation.

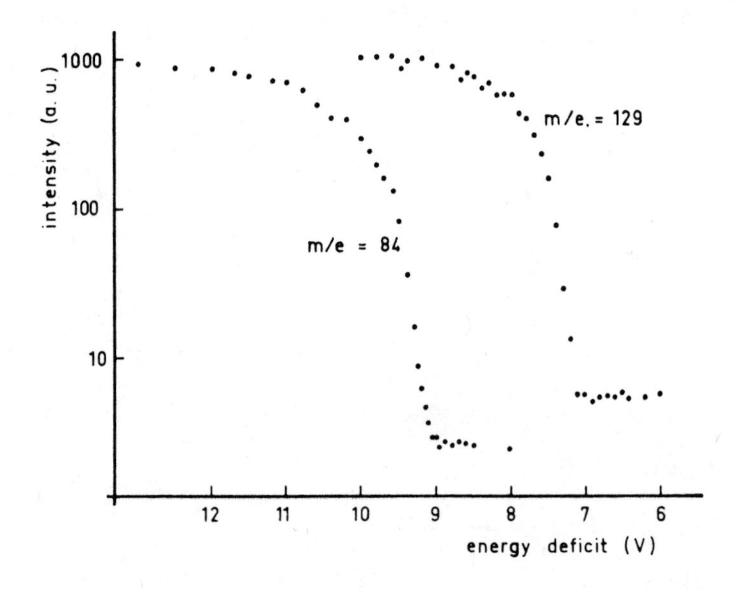

Fig.10 : Retardation curves of field ionized xenon (m/e = 129) and krypton (m/e = 84) for determination of AP values according to (6).

b) Hydrocarbons

On Pt, AP = values for hydrocarbons involve surface reactions. This will be demonstrated for n-heptane according to experimental results (6) given in fig.11. The retardation curve for the parent ion of n-heptane (M100) appears at a different position and has a different shape than that of the fragments (M43 and M29). The intensity threshold for the $C_7H_{16}^+$ ion is 5.4 ± 0.1 eV. When this value is corrected for the work function of the retarding electrode (Au, Φ_R = 4.9 eV), an AP of 10.3 eV is obtained, which is slightly above the I_p of 9.9 eV for n-heptane. In a similar way, the AP-value for $C_3H_7^+$ is 7.9 ± 0.4 eV ; for $C_2H_5^+$, two values are obtained at 9.1 ± 0.5 eV and 12.1 ± 0.3 eV.

The AP-values of the fragments of n-heptane are all below the values obtained for the molecular ion. For a molecular decomposition, AP values for the fragments should be higher than for the parent ion, $C_7H_{16}^+$. Thus, $C_3H_7^+$ and $C_2H_5^+$ must have gained energy due to a surface process. For $C_2H_5^+$, where two different AP values are found, the greater energy loss of 12.1 eV exceeds the value of 10.3 eV for $C_7H_{16}^+$ and must be attribued to a monomolecular decay process $C_7H_{16}^+ \longrightarrow C_2H_5^+ + C_5H_{11}$, which occurs in times $<10^{-12}$ sec and is field dependent. Most of the intensity of the $C_2H_5^+$ ion in fig.11 results form this homogeneous decomposition process.

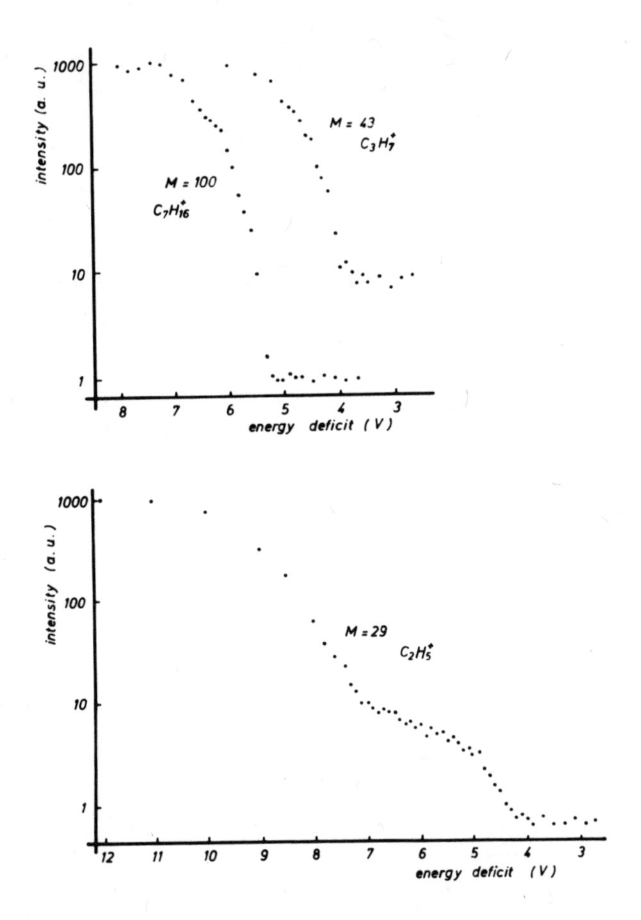

Fig.11 : Retardation curves for parent molecular ions and fragment of n-heptane according to (6)

It is still not completely understood why the AP of 10.3 eV of $C_7H_{16}^+$ exceeds its adiabatic ionization potential of 9.9 eV, although an internal excitation by Franck-Condon transitions seems likely. These transitions are not possible with rare gas atoms and appearance potentials in agreement with adiabatic ionization potentials have always been measured for noble gases.

c) Reactions of water

Water usually forms ions like H_2O^+, H_3O^+ and $(H_2O)_nH^+$, where n varies from 1 to 10. Details of the mechanisms involved in these reactions have been studied by Beckey (21), Schmidt (22), Goldenfeld (23), Anway (24) and Röllgen (25).

The formation of associated ions occurs in a multilayer of water molecules. The field induced force, which is acting on each gas molecule, is the gradient of the polarization energy $\delta E_p/\delta r = (\mu + aF)\,\delta F/\delta r$. For H_2O μ is $1.84 \cdot 10^{-18}$ esu and a is $1.57 \cdot 10^{-24}$ cm^3. At a field strength of $3 \cdot 10^7$ V/cm, the field induced force acting on H_2O molecules produces a field enhanced pressure of 16 atm (compared with a field free pressure of 10^{-5} torr). This value was not corrected for reduction of the field in the dielectric layer. This rough calculation shows that it is necessary to consider multilayer formation of H_2O molecules. Anway (24) actually measured the layer thickness form the energy deficit and reached the conclusion that up to 100 monolayers of water will be condensed on the tip surface at low fields. In general, condensed layer will form easier when the values of μ, a and I_p are high and the condensation pressures are low.
Several mechanisms have been formulated to explain the formation of H_3O^+ ions in the liquid-like surface layer. According to Onsager's theory (26), the self dissociation of water has to be field dependent. At a field strength of $3 \cdot 10^7$ V/cm, the conductivity of water does not exhibit ohmic behavior but has a parabolic dependence according to the second Wien-effect. The dissociation equilibrium, $2\,H_2O \rightleftharpoons H_3O^+ + OH^-$, is field dependent. The immediate neutralization of OH^- ions is considered as the source of secondary reactions.

The field ionization of one H_2O molecule, which subsequently reacts to form H_3O^+ by $2\,H_2O \xrightarrow[F]{-e} H_2O^+ + H_2O \longrightarrow H_3O^+ + OH$, is considered to be another possible mechanism. A third mechanism involves surface sites of the emitter Me, e.g.,

$$Me + n\,H_2O \xrightarrow[F]{-e} Me\,OH + (H_2O)_{m-1} \cdot H^+.$$

It is necessary to consider this mechanism because the threshold fields for ion formation depend on the tip material ($2 \cdot 10^7$ V/cm for W ; $5.1 \cdot 10^7$ V/cm for Ir and $6.1 \cdot 10^7$ V/cm for Pt). The formation of different reaction products at different metal surfaces also favors this mechanism. With tungsten the intensity ratio $[H_2O^+] / [H_3O^+]$ is 10^{-3} to 10^{-4} but with Pt and Ir it is about 0.5. Tungsten oxide ions were found but with noble metals no compounds were observed. However, under certain conditions, O_2^+ ions could be detected with Pt and Ir, which presumably form according to the reaction, $6H_2O_{ads} \xrightarrow{F} 4H_3O^+ + O_2^+ + 5e^-$ (Me). Furthermore, the recombination of OH-radicals leads to H_2O_2 and O by $2\ OH \rightarrow H_2O_2 \rightarrow H_2O + O$. Besides OH all particles could be measured in the form of a positive ion.

Some definite conclusions have been reached about the different mechanisms from measurements of the energy deficits and appearance potentials of field ions. Anway (24) measured the energy distributions on a tungsten field emitter, as shown in fig.12 for $H_3O^+ \cdot 3H_2O$. The broad signals with a large energy deficit originate from ionization processes occurring at the surface of the condenses layer. At higher field strength, the maximum in the distribution shifts to lower values since the thickness of the layer is decreased. This type of behavior is also found for other associated ions of water. In addition, there is a "fixed peak" at a constant low energy loss, which is produced from ionization at a minimum distance and is independent of the field strength. The appearance potential, which is different for various ionic species, is obtained from the low energy threshold of this peak. The appearance potentials of the associated ions of $H_3O^+ \cdot nH_2O$ on W, are as follows :

H_3O^+, 8.83 ± 0.05 eV ; $H_3O^+ \cdot H_2O$, 10.03 ± 0.07 eV ; $H_3O^+ \cdot 2H_2O$, 11.6, ≤ 11.5, ≤ 11.2 eV ; $H_3O^+ \cdot 3H_2O$, 11.7, ≤ 11.6, ≤ 11.1 eV. The values obtained for $H_3O^+ \cdot 2H_2O$ and $H_3O^+ \cdot 3H_2O$ are considered to be the upper limit. A comparison of these experimental values with data for different reactions given in the literature can be used to exclude several of the proposed mechanisms. The heats of formation, ΔH_r, of H_3O^+ in the following reactions are :

$$2H_2O_{liq} \longrightarrow H_3O^+ + OH^- \qquad \Delta H_r = 10.6 \text{ eV} \qquad \text{(a)}$$

$$H_2O\,(v) + H_2O_{liq} \rightarrow H_3O^+ + OH + e^- \qquad \Delta H_r = 12.0 \text{ eV} \qquad \text{(b)}$$

$$6H_2O_{liq} \longrightarrow 4H_3O^+ + O_2 + 4e^- \qquad \Delta H_r = 13.6 \text{ eV}/H_3O^+ \qquad \text{(c)}$$

$$2OH^- + 2H_2O_{liq} \rightarrow 2H_3O^+ + 2O_2 + 4e^- \qquad \Delta H_r = 8.7 \text{ eV}/H_2O^+ \qquad \text{(d)}$$

The ΔH_r values given by Anway are based on the value of 6.17 eV for the proton affinity for H_2O. His value differs slightly from those given by Heinen (6), who used the more reliable value of 7.1 eV for the reaction $H_2O(g) + H^+ \rightarrow H_3O^+(g)$.

In any event, the formation of H_3O^+ by reaction (a) is unlikely since the experimentally measured appearance potential (8.83 eV) is considerably lower than for this reaction. The reactions given by (b) and (c) are also excluded because the energy values are even higher ; reaction (d) occurs only as a side reaction of the OH radicals from (a) and (b). Based on the energy of

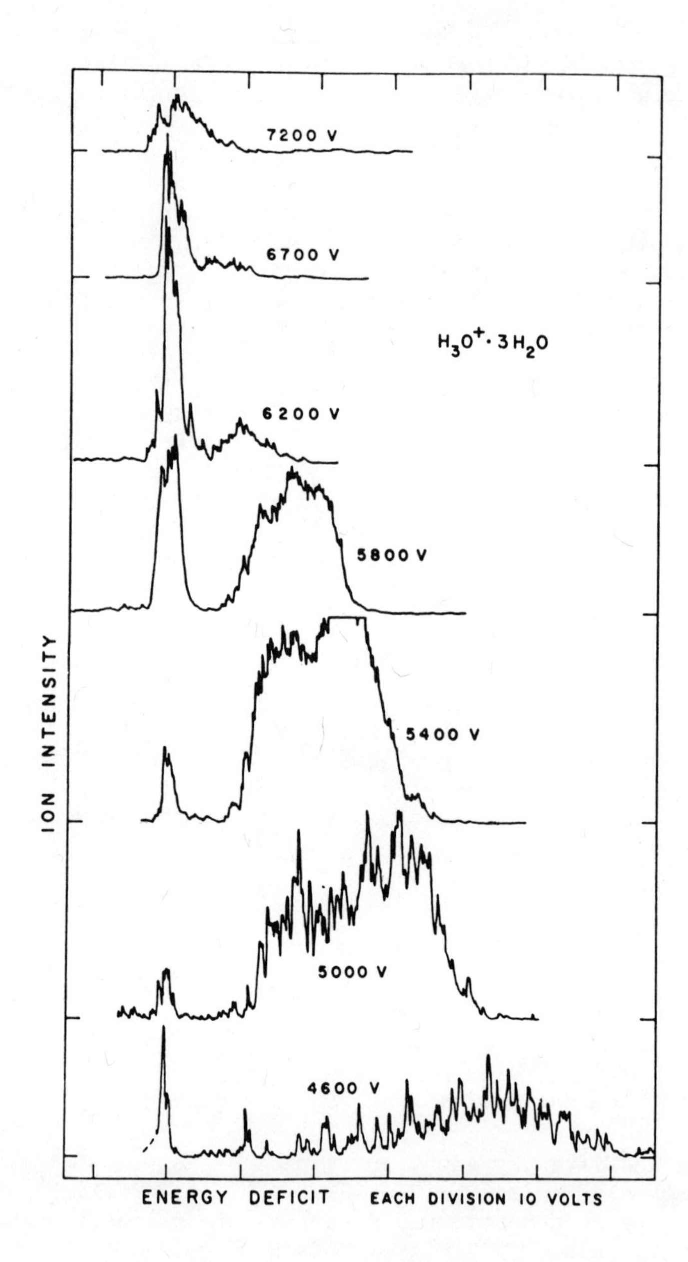

Fig. 12 : The differential energy analysis spectrum of $H_3O^+ \cdot 3H_2O$ as a function of total applied voltages (field strength) at the tip.

the reactions considered, it has to be concluded that surface sites are involved in the formation of H_3O^+ ions. This conclusion was also drawn by Heinen (6), who found an appearance potential of 10.5 eV for forming H_3O^+ on Pt-surfaces. The most likely reaction is $2H_2O \xrightarrow[F]{-e} H_3O^+ + OH$ where $\Delta H_r = 11.7$ eV. This reaction is energetically possible if the binding energy of OH radicals to the Pt surface is assumed to be 1.2 eV. The assumed binding energy of OH to the surface is not necessarily the bond energy of Pt-OH ; it is more likely that the OH radicals are attached to free radical sites on a contaminated Pt surface. Furthermore, it is suggested that the proton affinities are not too different for H_2O and $(H_2O)_n$ and that the mechanisms for the formation of $H_3O^+ \cdot (H_2O)_{n-1}$ will be comparable. A quantitative explanation for the experimental values given by Anway is still not available.

There are other observations in the field ionization of water which are still not completely understood. It is not likely that the ion $H_2O^+ \cdot H_2O$, which can be obtained on noble metals, is an ionized dimer. The appearance potential of about 11.7 eV is consistent with the suggested reaction (27) $2H_2O \xrightarrow{-e} (OH \cdot H_3O)^+ + e^-$, where the dimer, a product of a proton transfer reaction, gains the additional bonding energy of an OH radical.

CONCLUSION

The measurements of energy distribution or appearance potentials of fields ions have revealed information on the mechanism of electron tunneling into surface states as well as data on the energetics of surface reaction which form field ions. In addition to the mass spectrometric information, the mechanism of ion formation can now be analyzed on account of energetic arguments.

REFERENCES

(1) E.W. Müller and T.T. Tsong, Field Ion Microscopy Principles and Applications, American Elsevier 1969

(2) R. Gomer, Field Emission and Field Ionization, Harvard University Press 1961

(3) H.D. Beckey, Field Ionization Mass Spectrometry, Pergamon Press 1971

(4) A.J. Jason, R.P. Burns and M.G. Inghram, J. Chem. Phys. 43 (1965) 3762

(5) A.J. Jason, R.P. Burns, A.C. Parr and M.G. Inghram, J. Chem. Phys. 44 (1966) 4351

(6) H.J. Heinen, F.W. Röllgen and H.D. Beckey, Z. Naturf. 29a (1974) 773

(7) I.V. Goldenfeld, E.N. Korol and V.A. Pokrovsky, Int. J. Mass Spectr. Ion Phys. 5 (1970) 337
(8) E.W. Müller and S.V. Krishnaswamy, Surface Sci. 36 (1973) 29
(9) E.W. Müller and T. Sakurai, J. Vac. Sci. Technol. 11 (1974) 878
(10) T. Utsumi and O. Nishikawa, J. Vac. Sci. Technol. 9 (1972) 477
(11) T.T. Tsong and E.W. Müller, J. Chem. Phys. 41 (1964) 3279
(12) A.J. Jason, Phys. Rev. 156 (1967) 266
(13) A.A. Lucas and M. Sunzic, J. Vac. Sci. Technol. 9 (1972) 725
(14) J.A. Appelbaum and E.G. McRae, Surface Sci. 47 (1975) 445
(15) M.E. Alferieff and C.B. Duke, J. Chem. Phys. 46 (1967) 438
(16) T. Utsumi and N.V. Smith, Phys. Rev. Letters 33 (1974) 1294
(17) N.E. Christensen and B. Feuerbacher, Phys. Rev. B 10 (1974) 2349
(18) G.R. Hanson, G.E. Humes and M.G. Inghram, Chem. Phys. Letters 27 (1974) 479
(19) J.H. Block, Advances in Mass Spectrometry Vol.6 (1974) 109
(20) J.H. Block, Z. Phys. Chem. N.F. 64 (1969) 199
(21) H.D. Beckey, Z. Naturf. 14A (1959) 712, 15A (1960) 822
(22) W.A. Schmidt, Z. Naturf. 19a (1964) 318
(23) I.V. Goldenfeld, V.A. Nazarenko and V.A. Prokowsky, Dokl. Akad. Nank SSSR 161 (1965) 276
(24) A.R. Anway, J. Chem. Phys. 50 (1969) 2012
(25) F.W. Röllgen and H.D. Beckey, Surface Sci. 27 (1971) 321
(26) L. Onsager, J. Chem. Phys. 2 (1934) 599
(27) F.W. Röllgen and H.D. Beckey, Surface Sci. 23 (1970) 69 ; 24 (1971) 100

SPECTROSCOPIC STUDIES OF SUPPORTED METAL CATALYSTS :

Electron and Ferromagnetic Resonance and Infrared Spectroscopy

C. NACCACHE

Institut de Recherches sur la Catalyse, C.N.R.S.
Villeurbanne, France

It has long been recognized in heterogeneous catalysis that the efficiency of a catalyst can be enhanced by supporting it on a solid having large surface area. This is particularly true in the case of metal catalysts. The formation of a highly dispersed metal catalyst on the surface of a carrier comprises thermal decomposition and reduction of metal compounds followed by crystallization and aggregation of the metal thus formed. It appeared that the dispersion, the adsorption properties as well the catalytic properties of the metal depended strongly on the operating condition followed in the preparation procedure. Several physical methods have been used to study the physical and chemical properties of supported metal catalysts. In this paper some aspect of the results obtained by electron spin resonance, ferromagnetic resonance and infrared spectroscopy will be reviewed.

I. ELECTRON SPIN RESONANCE

1.1. General aspect of the esr. principle

Let us consider one unpaired electron. This electron moves around the nucleus in an orbital, this movement is associated to an orbital magnetic moment. The spinning electron is also characterized by a spin magnetic moment the value of which is given by the relation $\mu_e = - g\beta_e S$, where g is a constant called the g factor, β_e is the electronic Bohr magneton equal to $e\hbar/2mc$ where $-e$ and m denote the charge and mass of the electron and $S = \pm 1/2$. In the absence of a magnetic field the spin angular moment gives rise, for the unpaired electron, to a doubly degenerate spin energy levels. When the system containing the unpaired electron is placed in a constant magnetic field the degeneracy of the spin states is removed and there are two allowed orientations of the spin : parallel or antiparallel to the direction of the magnetic field. The energy of these two states is given by : $E = \pm 1/2\ g\beta_e H_z$. Hence the energy difference is $E = g\beta_e H_z$. The relative population of these two states in equilibrium is given by the Maxwell-Boltzmann Law $n_1/n_2 = \exp(- g\beta_e H/kT)$, where n_1 is the number of electrons in the higher energy state and n_2 the number of electrons in the lower energy state, T the temperature of the system.

If, to this system of oriented spins, a high frequency alternating field is applied the frequency of which satisfies the resonance condition $h\nu = g\beta_e H_z$, transitions between these two levels are induced and correspond to the esr absorption. Due to this absorption the populations of the levels n_1 and n_2 tend to become equal and this leads to a disruption of the Maxwell-Boltzmann equilibrium. Equilibrium is reestablished through non induced transitions from the upper to the lower level. When this disrupted equilibrium is not easily reestablished, then the populations n_1 and n_2 are rapidly equalized and the esr signal disappears. This is known as the saturation phenomenon. The line width and shape of an esr signal are determined by two main types of relaxation,namely, the spin-lattice and spin-spin which are characterized by the relaxation time T_1 and T_2. As a first approximation the line width is given by the relation

$$\Delta\nu = (1/T_1 + 1/T_2)$$

The spin-lattice relaxation time is temperature dependent and increases by decreasing the sample temperature. Hence if the main contribution to the width of the esr line is due to the spin-lattice relaxation, narrowing will be obtained by decreasing the sample temperature. In contrast T_2 is not so temperature sensitive, but strongly depends on the concentration of paramagnetic species. These differences enable one to determine whether one relaxation mode or the other predominates in a given spin system.

Until now we have examined the esr phenomenon for a free unpaired electron. However the unpaired electron is seldom free, it is rather moving in an orbital around a nucleus which may have a magnetic moment. This results in additional interactions which affect the electronic energy levels, and the resonance conditions as well. The most important information about the structure of the paramagnetic species, its electronic configuration and its crystal structure are obtained through the "hyperfine structure" and the "g-factor".

The hyperfine structure

The hyperfine structure of esr spectra is produced by the interaction of the magnetic moment of the unpaired electron with the magnetic moments of the nuclei. There are two main contributions to the hyperfine interaction, the Fermi or contact interaction which depends on the presence of a finite unpaired spin density at the nucleus (isotropic hyperfine interaction) and the dipolar or anisotropic interaction which is due to the interaction between two dipoles μ_e and μ_I separated by a distance r. Generally in solution the dipolar hyperfine term is averaged to zero and does not contribute to the esr spectrum.

The hyperfine interaction produces a splitting of the esr line into 2I + 1 hyperfine lines (I is the nuclear spin value). If the unpaired electron interacts with more than one nucleus then this will give rise to (2I + 1) (2I + 1)... hyperfine lines. Therefore when the electron equally interacts with n identical nuclei one will obtain 2nI + 1 hyperfine lines.

The g-factor

The position of the esr spectra is determined by their g-factors. In most case the g-factor of organic radicals are close in value to the g-factor of the free electron $g_e = 2.0023$ while this is not true for most inorganic radicals or paramagnetic ions. The g-deviation strongly depends upon the spin-orbit coupling which alters the effective magnetic moment of the electron μ_e and since $\mu_e = - g\beta_e H$, the g value is not constant. The extent of the deviation from g_e depends upon the spin-orbit coupling constant, the presence of a low-lying excited state and on the symmetry and magnitude of the crystal field experienced by the paramagnetic species. The g-factors have the same symmetry as the crystal field. If the crystal field symmetry is spherical the g-factor will be isotropic. For an axial field symmetry, the values of $g_\parallel$ and $g_\perp$ are found to correspond to the magnetic field being parallel or perpendicular to the distorsion axis. For a complete anisotropic crystal field the g-factor will have three components g_{xx}, g_{yy} and g_{zz}. Random tumbling of the species with average out the g-factor values and the observed value will be

$$g_{obs} = 1/3\ (g_{xx} + g_{yy} + g_{zz}) \text{ or } g_{obs} = 1/3\ (g_\parallel + 2g_\perp)$$

Conduction electron resonance

The previous theory concerns unpaired electrons in paramagnetic insulators. The features of resonance phenomena in conductors are largely determined by the fact that in these substances we are dealing with mobile electrons within the crystal structure. The electrical conductivity of metals is due to the electrons free to move throughout the interstitial space of the lattice. Those electrons in the conduction band can contribute to the magnetic moment for the metal and should exhibit in principle paramagnetic resonance. One has to recall that in a metal the Pauli susceptibility is temperature-independent. In contrast with paramagnetic transition metal-ions the orbital moments of the electrons in a metal make a negligible contribution to their total magnetic moments. Hence the g-value is close to 2.0023. In observing resonance for metals complications due to the small skin penetration of oscillating magnetic field change line shape resonance and lead to asymmetric lines. The restriction of the microwave field to a thin layer decreases the sensitivity of the method which means that small metal particles are required. Apart the so-called skin effect the spin lattice relaxation contributes also to the line width. It has been shown that despite the very weak spin-orbit coupling present in light metal atoms the coupling of the motion of the electrons with the vibration of the lattice contribute to the relaxation. For heavy metal-atoms strong spin-orbit coupling produces very short spin-lattice relaxation time and the lines are too broad which prevents observation of esr spectra. Resonances have been observed for Li, Na, Be metals. It follows from above that esr is not very appropriate for studying metal catalysts. However valuable informations can be obtained about species adsorbed on the surface, electron transfer between the metal and the carrier and finally when one is dealing with metal clusters containing very few metal atoms.

1.2. Application

Identification of paramagnetic oxygen species

During the past decade increasing attention has been paid on the catalytic oxidation of hydrocarbons with the emphasis on the intermediate oxygen species which might form on the surface of the catalyst. Based on kinetic data and on semiconductivity measurements the mechanisms which have been suggested involve such intermediates as O_2^-, O^-, O^{2-} ions. Thus it may be expected that a direct experimental identification of oxygen species formed on these catalysts would improve the understanding of the phenomena occurring during the oxidation process. In the recent years esr technique has been successfully applied in studying the nature and the electronic structure of adsorbed oxygen species.

It is well known that silver catalysts are used for selective oxidation of ethylene. In this reaction ethylene oxide and carbon dioxide are formed. Different forms of oxygen species were proposed to explain the selectivity of silver catalysts. The direct reaction of oxygen with the catalyst surface leads to the formation of Ag_2O_2 and Ag_2O superficial coumpounds respectively active for the partial and total oxidation of ethylene. Recently the esr observation of O_2^- chemisorbed on metallic silver was reported (1,2).

Molecular oxygen contains two unpaired electrons in its ground state. In the gas phase or in a condensed phase molecular oxygen gives complex esr spectra. However, owing to its large electron affinity O_2 may trap one electron and hence be converted to the O_2^- radical. The theoretical g-value expressions for the O_2^- ion in a field of orthorhombic symmetry have been derived by Känzig and Cohen (3). In a simplified form the g-value expressions are given by :

$$g_{xx} = g_e - a^2 + ab$$

$$g_{yy} = g_e - a^2 - ab + 2b$$

$$g_{zz} = g_e + 2a$$

where a is the ratio λ/Δ and b the ratio λ/E,λ represents the spin-orbit coupling constant, Δ and E the energy differences between the ground state and excited states. Therefore, while g_{xx} appears to be slightly lower than g_e, g_{yy} slightly larger but all close to g_e, g_{zz} is larger than g_e and its deviation is closely related to the magnitude of Δ.

As it has been mentioned, the O_2^- radical has one unpaired electron localized in a degenerate $\pi\oplus$ antibonding orbital when the radical experiences a spherical crystal field as in the gas phase. It results in a very short spin-lattice relaxation time which prevents the observation of an esr spectrum. However if the symmetry is partially removed, and this fact occurs when the molecule is adsorbed, the orbital degeneracy in the $2p\pi\oplus$ level will be removed. Hence the odd electron will occupy a non degenerate $2p\pi_x\oplus$ orbital. Since the g_{zz} value depends on the $2p\pi\oplus$ level splitting, from the experimental g_{zz} value, taking $\lambda = 0.014$ eV, it is possible to estimate quantitatively the electrostatic fields existing on the catalyst surface. When oxygen is adsorbed on silver dispersed on silica gel it exhibits a three g esr spectrum, $g_1 = 2.040$ $g_2 = 2.010$ $g_3=2.002$. These g-values essentially the same as those reported for O_2^- adsorbed on several oxides indicate clearly that O_2^- was formed on the surface of the silver catalyst. Additional esr spectrum was formed during the adsorption of oxygen. The intensity of this signal was enhanced after gamma irradiation of the $Ag-SiO_2$ sample at room temperature in the presence of oxygen. The esr parameters for this signal were $g_\perp = 2.036$ $g_\parallel = 2.249$. A splitting of each g component appeared which is due to the interaction of the unpaired electron with the silver nucleus having a nuclear spin of 1/2. This signal was identified to Ag^{2+} formed by the interaction of silver atom with the electron acceptor oxygen molecule. The reactivity of the oxygen species with ethylene was examined. The results obtained showed that the O_2^- species disappeared completly while the $Ag^{2+}O^{2-}$ decreased gradually. From these results it was concluded that O_2^- and oxidized silver atoms participate to the oxidation of ethylene. More recently the formation and the reactivity of O_2^- on a supported silver catalyst have been investigated by esr technique (4). The experimental data showed that the activity of the silver for O_2^- formation decreased after several cycles : outgassing, oxygen adsorption, outgassing reduction, at 160°C. After this initial decrease in activity the catalyst regains some activity. The phenomena appeared when analysis by electron diffraction revealed the formation of the stoichiometric compound Ag_2O. It was suggested that Ag_2O behaves like a semiconductor in facilitating electron transport from the bulk silver to the adsorbed oxygen.

Esr study of adsorbed carbon monoxide

Quantitative measurements of CO adsorbed on supported ruthenium catalysts have shown that at 150°C a large amount of gas was adsorbed by the metal. The measured CO/Ru ratios predicted the formation of $Ru(CO)_3$ and $Ru(CO)_4$ surface complexes. It was concluded from this work that CO chemisorption could not be used for the calculation of Ru metal surface area (5). The formation of ruthenium carbonyl compounds was confirmed by esr study. Upon CO adsorption at 25°C on the bare metal a paramagnetic species was formed showing an esr spectrum with g = 1.99. The formation of this paramagnetic species appeared to be strongly dependent on the metal particle size. When the particle size decreases it seems obvious that the proportion of surface atoms with low coordination number increases these atoms being responsable for the formation of $M(CO)_x$ complexes. In this work (6) it was postulated that $Ru(CO)_3$ complexes were formed and that an equilibrium $Ru(CO)_3 \rightarrow Ru^{3+}(CO)_3^{3-}$ exists. Ru^{3+} would have a C_{3v} symmetry and in the strong field approximation may be regarded as a one hole state i.e. has one unpaired electron. Theoretical calculation indicated that one g-value must be lower than the free electron g-value as it was found experimentally. When the CO adsorption was carried at 150°C change in the esr spectrum occured and the new signal was characterized by a three g-value $g_1 = 2.37$ $g_2 = 2.19$ $g_3 = 1.99$. The formation of $Ru(CO)_4$ complexes with a C_{4v} symmetry would account for the observed spectrum. In conclusion it seems certain that $Ru(CO)_3$ and $Ru(CO)_4$ species were formed during high temperature CO adsorption.

Carbon monoxide adsorption on zeolite supported palladium was also investigated by esr technique. Palladium metal was obtained by reduction at low temperature of Pd^{2+} exchanged Y zeolite (7). CO adsorption at room temperature resulted in the formation of a species giving rise to an esr triplet signal with $g_1 = 2.192$ $g_2 = 2.061$ $g_3 = 2.038$. This signal disappeared by pumping off carbon monoxide. Theoretical considerations have suggested that this signal must be attributed to a $Pd(CO)_X$ species the unpaired electron being highly localized in the 4d orbital of the palladium atom. For better identification of the complex formed during CO adsorption ^{13}C labeled carbon monoxide was used. The experimental results showed that each g component was split into three hyperfine lines due to the interaction of the unpaired electron with two ^{13}C nuclei (I = 1/2 for ^{13}C). Hence it was believed that CO adsorption on finely dispersed palladium atoms resulted in the formation of $Pd(CO)_2$ complexes.

Metal hydride formation

The electron spin resonance could be also used to study metal hydride formation.Palladium exchangedY zeolites were reduced at room temperature and at 150°C with hydrogen. After these treatments strong esr signal was recorded with $g_{\parallel}$ = 2.33 and $g_{\perp}$ = 2.10. If one supposesthat Pd(0) formed after the hydrogen reduction reacts with hydrogen atom to form Pd^+H^- it would result for palladium a $4d^9$ electronic configuration. Consequently this ion behaves in the same manner as Cu^{2+} ion in zeolite. The g-values for Cu^{2+} in zeolite were found $g_{\parallel}$ = 2.32 and $g_{\perp}$ = 2.06 close to those measured for hydrogen reduced palladium-zeolite. The fact that the signal disappeared upon outgassing the sample at high temperature and reappeared after a second addition of hydrogen confirmed this interpretation of hydride formation. Similar results were found in the case of Ni-zeolites.

Conclusion

From these few examples it is clear that although experimental results is difficult to obtain by application of esr technique to metal catalysts some valuable information could be obtained on the nature of chemical species formed on the surface of the solid. Furthermore one could expect that the development of experimental procedure for obtaining very small metal particles which contains a small number of atoms would permit to study more efficiently these catalysts by esr method.

2. FERROMAGNETIC RESONANCE

Ferromagnetic resonance is very similar to paramagnetic resonance but concerns system with unpaired electrons on which act very strong exchange forces. In a paramagnetic material the magnetic moments of the atoms, ions or molecules are pointed in all direction in space and the total magnetic moment is zero. When the material is placed in an external magnetic field H_o the magnetic moments allign with the direction of H_o and the material becomes magnetized. In ferromagnetic substance, because of the very strong exchange interaction between atoms, the magnetic moments of neighbouring atoms are oriented to each other even in the absence of external magnetic field. Application of magnetic field produces a magnetization which is 10^4-10^5 time higher than in paramagnetic material. This internal magnetic field changes the resonance condition and the g-value can no longer be determined by the relation $h\nu = g\beta H_o$. In ferromagnetic materials the formula for resonance has the form : $h\nu = g\beta(H_o + H_m)$ where H_m is some additional field whose value depends on the crystallographic magnetic anisotropy, the particle shape anisotropy and the relaxation processes.

The magnetic anisotropy can be understood by considering the effect of crystal field on the removal of orbital degeneracy. The crystal field leadsto an anisotropic distribution of orbital magnetic moment and hence through the spin-orbit coupling to a magnetic anisotropy. The magnetic anisotropy constants (K_1) decrease with increase of the temperature and are very small near the Curie point. For nickel the magnetic anisotropy disappears near 100°C. For single crystal the conditions of resonance along the principal directions are given by the relations

$$\text{difficult magnetization} : h\nu = g\beta(H_o + 2K_1/M_o)$$

$$\text{easy magnetization} : h\nu = g\beta(H_o + 4K_1/3M_o)$$

where M_o is the saturation magnetization.

For randomly oriented systems all orientations of the magnetization axis are equally probable with respect to the direction of the external field. This causes line broadening, asymmetric lines and shifts the g-factor. To avoid these perturbations one has to make K_1 as small as possible and hence to work at high temperature where $K_1 \cong 0$.

Shape anisotropy

When a ferromagnetic material experiences an external magnetic field H_o, a formal magnetic field H_e in opposition with the direction of the external magnetic field is generated. This field called demagnetization field is given by the relation : $H_e = -N.M_o$. The demagnetization factors N along the principal axis x, y, z are such that we have $N_x + N_y + N_z = 4\pi$. Taking into account this demagnetization factor Kittel derived the following formula for the resonance frequency :

$$h\nu = g\beta\ H_o + [(N_x - N_z)M_o]^{1/2}[H_o + (N_y - N_z)M_o]^{1/2}$$

which shows that the resonance frequency depends on the shape of the sample. For a sphere $N_x = N_y = N_z$ and resonance occurs for $h\nu = g\beta H_o$. For a disk one has $N_x = N_y = 0$ and $N_z = 4\pi$. The shape of the sample will have an important effect on the value of the external field needed for the resonance condition. For an iron specimen with M_o = 1700 gauss and at a microwave frequency of 10^4 Mc/s the values of the resonance are : 3750 gauss for a sphere, 26800 gauss for a disk in the perpendicular direction and 12500 gauss for a cylinder. When the sample corresponding to supported ferromagnetic metal catalyst contains particles of widely different shape considerable broadening occurs due to the increase in the number of possible resonance. However in contrast with crystal magnetic anisotropy, the shape anisotropy broadening is temperature insensitive provided no change in the particle shape occurs. Hence it is possible to discriminate between these two effects by analyzing the change in the line shape with the temperature.

Application

The ferromagnetic resonance until now has not been extensively used in the field of catalysis. This method can in principle give qualitatively information on the properties of ferromagnetic catalysts. However the fact that several parameters can affect simultaneously the line shape of ferromagnetic resonance of the catalyst renders difficult the interpretation. The significance of the experimental results rests yet on careful correlation of results from other sources and techniques. Progress in this field will be obtained by measurements at high temperatures to avoid line broadening through magnetic crystallographic anisotropy and when more experimental work will be done. We must therefore confine the discussion to a limited interpretation of few systems in which ferromagnetic resonance studies have been helpful.

Study of hydrogen chemisorption over nickel catalysts

Nickel metal has been reported to be a catalyst for several types of reactions. The relation which might exist between the structure of the nickel catalysts and their reactivity has lead several authors to characterize the materials by static magnetic technique. However it has been shown that ferromagnetic resonance may be employed to investigate surface phenomena occuring during the chemisorption of hydrogen on nickel. Silica supported nickel metal was prepared by reducing at 350°C with hydrogen SiO_2 sample impregnated with $Ni(NO_3)_2$ (8). The ferromagnetic resonance spectrum was recorded before and after H_2 chemisorption. It was observed that the spectrum intensity decreases by 15-25 % and that the decrease was a function of the Ni particle size. According to Selwood (8) hydrogen chemisorbed on Ni would pair off d electrons of the metal which decreases the ferromagnetic properties. It has been argued that complete demagnetization would occur for nickel particles with an average diameter of about 6.5 Å. The interaction between hydrogen and nickel was also investigated for a calcium nickel phosphate catalyst (9). The ferromagnetic spectrum was monitored during the hydrogen reduction treatment. The signal intensity as a function of the temperature of hydrogen treatment increased up to 670°C. Above this temperature there was a decrease in the relative intensity of the ferromagnetism. These results indicated that complete demagnetization occured when nickel interacts with hydrogen at high temperature due to the formation of nickel hydride in the bulk and in the surface. The ferromagnetic resonance spectrum for a sample treated at 1100°C with hydrogen and then outgassed reappeared which suggests that at this temperature diamagnetic nickel hydride was formed which decomposed into ferromagnetic nickel in vacuo.

Plot of ferromagnetic signal intensity as a function of the temperature of measurement may serve to identify the ferromagnetic species. At the Curie temperature the relative signal intensity decreases sharply which allows to determine precisely the Curie temperature. This could be very useful for unknown materials. Furthermore determination of the Curie point might bring valuable information about the ferromagnetic behaviour of very small metal clusters. The value of the Curie temperature would reflect the interactions within nickel particles and between particles. In small particles the inter-domain interaction is distorted by thermal energy leading to abnormally low Curie point.

3. INFRARED SPECTROSCOPY

The infrared spectroscopy technique has been extensively used to study the nature of chemical species which exist on various solid surfaces. The vibrational spectra in the infrared region have proved to be a rich and convenient source of information concerning both structure and bonding. A number of different molecules adsorbed on oxide or metal catalysts have been identified by this technique and several reviews have been written on this subject (10,11,12). The present article will be confined on the results obtained on supported metal catalysts and limited to the experimental works which provide information on the structure of metal particles. A large amount of work has been carried out on the identification of organic molecules adsorbed on metal catalysts but this has been neglected in the present text.

It is well known in coordination chemistry that the d-group transition metals form complexes with neutral molecules such as carbon monoxide or nitric oxide. The properties of these molecules are due to the presence of vacant π orbital in addition to lone pair. In the presence of a metal in low oxidation state CO or NO molecules form complexes through σ-π bonds. The extent of bonding of the ligand with the transition metal will be reflected by the shift of the CO or NO stretching frequencies and this shift could be usefully used to obtain information about the ability of the transition metal to act as a π-donor. Before reviewing some of the experimental works presented in the literature an elementary background in certain aspects of vibrational spectrum of diatomic molecules will be given.

A. Theoretical aspect

For diatomic molecules the number of fundamental vibrations is one. Absorption of infrared radiation by the vibrating molecule will only take place when this vibration is coupled with a changing dipole moment. According to this selection rule the vibration will give rise to infrared absorption when the atoms are different as it is the case for CO, NO molecules. The symmetric molecules such as H_2, N_2, O_2 will not absorb. The frequency at which a diatomic molecules AB has its fundamental vibration is given by the relation

$$\nu(\mathrm{Cm}^{-1}) = \frac{1}{2\,C}\sqrt{\frac{f}{\mu}}$$

where f is the force constant and μ is the reduced mass found from $\frac{1}{\mu} = \frac{1}{m_1} + \frac{1}{m_2}$, m_1 and m_2 are respectively the atomic mass of A and B. Furthermore the intensity of vibrational bands is proportional to the square of the change of the dipole moment for the corresponding fundamental vibration near the equilibrium position. Hence diatomic molecules with high values of dipole moment absorb strongly in the infrared. Strictly speaking the relation giving the frequency of the fundamental vibration is only valid for an isolated molecule. In the condensed phase molecular interaction between molecules produces a frequency shift.

The relationship between the vibration frequency ν and the force constant, associated with the fact that the force constant increases when the bond order increases, shows that the change in the bond order of the molecule will produce a shift of its fundamental vibration. By measuring the infrared stretching frequencies in the adsorbed species and comparing with that of the free molecule one can obtain important information on the bonding mode taking place in adsorption.

As stated above CO molecule may form with transition metal stable molecular compounds. The molecular orbital picture is as follow: there is first a dative overlap of the filled carbon σ orbital with the empty orbital of the metal and second a dative overlap of the filled d metal orbital of the metal with the empty π antibonding orbital of the CO. The electron back donation from the metal to the CO molecule decreases the bond order in the CO molecule and shifts the vibration frequency ν to the lower wave numbers. The magnitude of the shift will depend to the extent to which the drift of metal electrons to the CO antibonding π orbitals tends to make CO negative. Furthermore when the CO molecule is bonded to two metal atoms in the so-called "bridge form" the frequency shift will be higher compared with the CO bonded in the "linear form". In general linear M-CO groups have CO stretching frequencies in the region 2150-1900

cm^{-1} while bridging CO groups absorb in the range 1900-1700 cm^{-1}. Free CO gives an ir band at 2143 cm^{-1}.

Further informations may be obtained from the number of infrared bands appearing upon CO adsorption on metals. To illustrate the procedure consider two CO molecules adsorbed on one metal atom. In this model each CO oscillator is treated as a dipole vector and the total dipole vector for this group is the vector sum of these individual vectors.

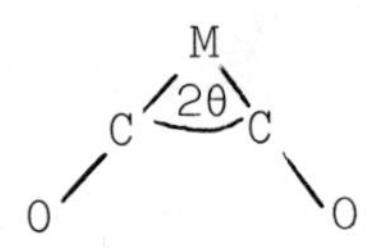

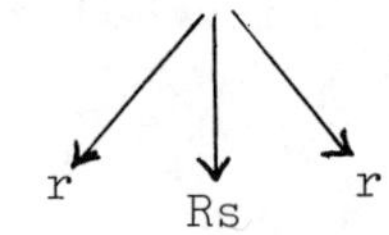

Symmetric vibration

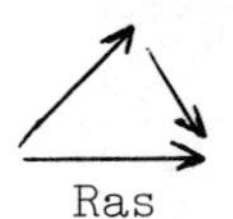

Antisymmetric vibration

It results for this group two dipole vectors giving rise to two infrared bands corresponding to the symmetric and antisymmetric vibration mode. From the ratio of the intensities of the symmetric and antisymmetric bands one can calculate the bond angle 2θ using the relation

$$\cot^2\theta = \frac{I\ sym}{I\ antisym}$$

B. Application

Although there is no general rules which can be given for the assignment of CO adsorbed on metals in linear or bridged forms it appears now almost established that, for linear form, ν_{CO} is above 2000 cm^{-1} while the ν_{CO} for the bridged form is below 2000 cm^{-1}. These assignments were first given by Eishens and co-workers (13) and latter confirmed by Sachtler and co-workers (14). However Blyholder (15) has recently expressed the view that all bands corresponded to linear form and that it is the different electronic configurations of the metal which is responsible for the ν_{CO} shifts. A large body of work devoted more particularly to this aspect was carried out by Eischens and co-workers. They showed that for supported metal the nature of carrier could affect the carbon monoxide adsorption. Comparison between CO adsorbed on Pt-SiO_2 and Pt-Al_2O_3 indicated different effects between these two carriers : CO adsorbed on Pt-Al_2O_3 is preferentially adsorbed in the bridged form while only 15 % of the total adsorbed CO is found in the bridged form on Pt-SiO_2. These authors also showed that shifts in the ν_{CO} occured when molecules such as H_2, NH_3, HCl were preadsorbed on the metal and it was concluded that these shifts were due to the effect of the added gas on the electronic nature of the metal rather than to a change in the structure of the chemisorbed CO. Since these pioneer works the spectrum of adsorbed carbon monoxide has been used as a probe to characterize the effects of carrier , the metal particle sizes, the effects of adsorbing gas on the electronic properties of the metal, the effect of alloying. Recent results will be described to show how it is possible to determine these different factors which could affect the properties of the metal.

1) Titration of supported metal

Metal finely dispersed on high surface area oxides are widely used as catalysts for many reactions. Information concerning the degree of dispersion of the metal is important for interpretating of the activity of these materials. The degree of dispersion of the metal component is generally determined by the selective chemisorption of gases such as H_2, O_2, CO. In order to be applicable for this purpose one has to know the stoichiometry of the reaction, H/Pt, O/Pt, CO/Pt. In general caution must be used when CO adsorption is employed. The discrepancy often observed between metal surface area measured by means hydrogen adsorption or CO adsorption may well be attribuable to the formation of M-CO, M_2CO or $M(CO)_2$. The formation of these species renders inapplicable the CO adsorption for surface area measurements. Hence it is always necessary to determine by infrared spectroscopy to which extend each form is present. The infrared spectra of CO adsorbed on supported palladium or platinum nickel shows distinct adsorption bands due to the linear M-CO and bridged $M_2(CO)$ forms (10). From these spectra it is somewhat hard to define the relative amounts of CO in linear and bridged forms and hence it is very difficult to determine the number of surface metal atoms from the amount of CO adsorbed. Furthermore when one is dealing with rhodium or rhenium more intricate is the interpretation of quantitative CO adsorption as it has been shown that these systems give beside the M-CO and M_2CO forms a third form ascribed to $M\text{-}(CO)_2$. The $M(CO)_2$ form is characterized by an ir doublet corresponding to the symmetric and antisymmetric vibration mode. For rhodium catalysts the bands appearing in the 2040 and 2108 cm^{-1} region were assigned to $Rh(CO)_2$ (16) and the band near 2045-2062 cm^{-1} to Rh-CO. CO adsorbed on rhenium catalysts gave ir bands at 2050-2030 cm^{-1} and 2010-1992 cm^{-1} due to $Re(CO_2)$ species and at 1950 cm^{-1} attributed to Re_2CO species (17)

2) Metal-adsorbate interaction

The effect of adsorbed molecules on the electron density of the metal particles may be studied by observation of the ir spectra of adsorbed CO. The principle is that for metal carbonyl complexes such as $M(CO)_xL$ by varying the σ-donor properties of the ligand L one modifies the electron density of the central metal atom and hence the π back donation to the CO molecule. Hence the change in the electron density of the metal will be reflected by a shift of the ν_{CO} stretching vibration (18). The σ-donor properties of the ligands are estimated by the value of the ionization potential (I.P.) which is the energy required to remove an electron from a given orbital, lower is the ionization potential stronger will be the σ-donor properties. It has been shown (19) that $Ni(CO)_3L$ complexes have ν_{CO} bands at 2111 cm^{-1} for L = PF_3 (I.P. = 13 eV), 2085 cm^{-1} for L = $P(OC_6H_5)_3$ (I.P.= 8.40) and 2069 for L = $P(C_6H_5)_3$ (I.P. = 8.2) which indicated that ν_{CO} shifts to higher frequency when the ionization potential of the ligand increases. On the basis of analogues to transition metal carbonyl complexes the shift of ν_{CO} when Lewis bases were adsorbed on metal films covered with CO was explained by the effect of electron donation to the metal by the Lewis base (20). Additional verification of the effect of adsorbed σ-donor molecules on the electron density of the metal is provided by a study of the effect of Lewis bases on the ν_{CO} stretching vibration (21). To test this viewpoint CO was adsorbed on $Pt-Al_2O_3$, the surface coverage was equal to 0.2. Ir bands at 2065 cm^{-1} and 1850 cm^{-1} corresponding respectively to the linear and the bridged forms were observed. The samples were then allowed to adsorb water (I.P.= 12.6 eV) ammonia (I.P. = 10.5 eV) and pyridine (I.P. = 9.2 eV). Infrared results indicated that the adsorption of H_2O shifts $\nu_{(CO)}$ from 2055 to 2050 cm^{-1}, NH_3 shifts $\nu_{(CO)}$ from 2065 to 2040 and pyridine shifts $\nu_{(CO)}$ from 2065 to 1990. Therefore adsorption of molecules with strong σ-donor properties would increase the electron density of the metal. Furthermore it was observed that the ν_{CO} shift towards lower wave numbers when pyridine was adsorbed depended on the amount of adsorbed pyridine. This result indicates that the collective properties of the metal are involved.

3) Interaction of metal with electron acceptor atoms or molecules

Infrared spectra of CO adsorbed by alumina-supported platinum have been also employed with the goal of establishing the nature of the species formed during the CO + O_2 interaction. The ir spectrum obtained when O_2 reacted with $Pt-Al_2O_3$ surface covered with CO showed a band at 2120 cm^{-1}. This band was also observed when CO reacted with adsorbed oxygen. It was concluded that the band at 2120 cm^{-1} was due to $Pt{<}^{O}_{CO}$ species (21). Similar species were reported when CO and O_2 reacted with rhodium catalysts (16).

The use of chlorinated supported metal catalysts for several commercial processes has led several workers to study the structure of surface species formed during the chlorination. The mechanism of the reactions between the metal catalysts and the chlorine compounds has been clarified by ir studies (22). When HCl is adsorbed at room temperature on Pt-Alumina covered with CO, the ν_{CO} shifts from 2065 cm^{-1} to 2075 cm^{-1}. HCl is dissociatively adsorbed and chlorine atoms bonded to platinum atoms produce the observed shift. Phosgene has been used as a chlorination agent. The reaction of $COCl_2$ with surface platinum gives an ir band at 2075 cm^{-1} attributed to a Pt-CO species interacting with a chlorine atom (21). At 200°C, $COCl_2$ produces $Pt(CO)Cl_2$ species identified by ν_{CO} at 2135 cm^{-1} by analogy with the ir spectrum of $(Pt\ COCl_2)$ complex which shows a band ν_{CO} at 2146 cm^{-1}. Chlorination of $Pt-Al_2O_3$ by carbon tetrachloride has also been followed by infrared. At 200°C CCl_4 reacts with $Pt-Al_2O_3$ and forms $(Pt\ COCl_2)$ surface complexes. At higher temperature chlorine is progressively desorbed. At 400°C no band which could be attributed to adsorbed CO is present and it was concluded that a compound like $PtAl_2Cl_8$ was formed which is evaporated from the carrier.

4) Alloying effects

The catalytic selectivity of metals is strongly modified by alloying. Two major effects are generally proposed as the result of alloying the active metal with the inactive one :

- the active metal might be dispersed by the alloying agent such that the number of active sites formed by the association of several metal atoms could decrease. This refers to the "geometric factor".
- the presence of the second metal could affect the electronic properties of the active metal and changes the bonding of the reactants with the surface. This was considered as the "electronic factor".

There are now several evidences which indicate that the two factors may be present in alloys. Hence any experimental method which could afford information on the surface composition of alloys will contribute to a better understanding of alloying effects on the catalytic properties of the material. Again ir spectroscopy has proved to be a powerful tool for characterizing the alloy surface.

Infrared spectra of carbon monoxide adsorbed on supported palladium and palladium-silver alloys were investigated (14). It was reported that CO adsorbed on supported palladium gave the linear and bridged forms with characteristic absorption frequencies of 2060 cm^{-1} and 1960 cm^{-1} respectively. Furthermore the studies of the intensities of the infrared bands corresponding to the linear and bridged CO decreased when palladium dispersion increased and in the highly dispersed palladium samples only the linear form was present. Palladium-silver alloys behaved in the same manner,that is when the concentration of silver increased CO adsorbed gave almost the same ir bands at the same position but the intensity of the ir band corresponding to the bridged form decreased. The rapid decay in bridged form indicated that the geometric effect was the most important. Nickel-copper alloys were also investigated by this technique (23). On pure nickel CO adsorption gave the high frequency (2020 cm^{-1}) and the low frequency bands.As for Pd-Ag alloys the intensity of the low frequency band decreased as the concentration of copper in Ni-Cu. In conclusion by these studies it was clearly showed that in the whole composition range, the electronic properties of palladium or nickel or more precisely the "intrinsic character" of the Pd or Ni are not changed markedly by alloying while pairs of adjacent Pd or

Ni atoms disappeared progressively.

5) Metal-support interaction

A fairly large amount of isolated experimental facts and observations providing evidence of the effect of the support on the properties of supported metal catalysts has now accumulated. The experimental data relating to different systems are scattered. In fact the problem concerning the possible chemical interaction between the metal and the support is not yet firmly proved. However recently spectroscopic evidence of electron transfer between small metal particles and oxide supports has been found (24). Y zeolites loaded with palladium cations by ion exchange were calcined in oxygen and subsequently reduced in hydrogen. In order to obtain information about the reduction of palladium cations experiments were made in which CO was adsorbed and infrared spectra recorded. The unreduced samples gave ir bands at 2135 and 2110 cm^{-1} which were assigned to carbon monoxide adsorbed on Pd^{2+} ions. Hydrogen reduction at 25° resulted in the decrease of the intensity of these two bands and the appearance of new ir bands at 2100, 1935 and 1895 cm^{-1}, attributed to CO adsorbed on reduced palladium. When the samples were H_2-reduced at 200°C, the ir bands related to CO adsorbed on palladium ions were not observed, while the intensities of the bands at 2100, 1935 and 1895 cm^{-1} increased. Finally H_2-reduced samples at temperature higher than 250°C showed a weak band at 2100 cm^{-1} while a new band at 2070 cm^{-1} appeared. The frequency of this band is similar to that found for CO adsorbed on Pd films (2080 cm^{-1}) and on supported palladium (2060 cm^{-1}). The various results suggest that palladium ions are reduced by hydrogen at room temperature. The increase in frequency for CO adsorbed on Pd(0) formed during the reduction treatment at relatively low temperature is the result of the decrease of the electron backdonation from the d orbitals of the metal to the antibonding orbitals of CO. Hence one could conclude that these metal particles are electron deficient. It is well known that Y zeolites have strong Lewis acid sites with strong electron acceptor properties. Electron deficient palladium particles would be due to the interaction of these metal particles with Lewis acid sites. This interaction disappears when the metal particles are sintered at high temperature.

Evidence of electron transfer between palladium metal particles and various supports was also given in (25). The frequency of CO adsorbed on supported palladium shifted to higher values as the electron acceptor properties of the support increased. The positions of infrared bands of adsorbed CO were

Pd-MgO (2065 cm^{-1}), Pd-Al_2O_3 (2075 cm^{-1}), Pd-NaY (2075 cm^{-1}),

Pd-HY (2100 cm^{-1}), Pd-MgY (2100 cm^{-1}), Pd-LaY (2105 cm^{-1}).

The ν_{CO} shift is an indication that the electron density of the palladium metal particles changed with the carrier.

In conclusion these few examples have shown that valuable informations are obtained from infrared studies of carbon monoxide adsorbed on metal catalysts. Further effort should be made by workers in order ot explain several unusual behaviours of very small metal particles, and infrared spectroscopy offers definite help in this area.

REFERENCES

(1) R.B. Clarkson and A.C. Cilliro Jr, J. Vac. Sci. Techn., 9 ,1073, (1972)
(2) N. Shimizu, K. Shimokoshi and I. Yasumori, Bull. Chem. Soc. Japan 46, 2932 (1973)
(3) W. Kanzig and M.H. Cohen, Phys. Rev. Letters, 3, 509 (1959)
(4) R.B. Clarkson and A.C. Cirillo Jr., J. Catalysis, 33, 392 (1974)
(5) M. Kobayashi and T. Shirasaki, J. Catalysis, 28, 289 (1973)
(6) M. Kobayashi and T. Shirasaki, J. Catalysis, 32, 254 (1974)
(7) C. Naccache, M. Primet and M.V. Mathieu, Adv. Chem. Ser., 121, 266 (1973)
(8) D.P. Hollis and P.W. Selwood, J. Chem. Phys., 35, 378 (1961)
(9) B.R. Loy and C.R. Noddings, J. Catalysis, 3, 1, (1964)
(10) R.P. Eischens and W.A. Pliskin, Adv. Catalysis, X, 1 (1958)
(11) L.H. Little, Infrared Spectra of Adsorbed Species, Academic Press, London (1966)
(12) G. Blyholder, Experimental Methods in Catalysis Research, ed. R.B. Anderson, Academic Press, p. 323 (1968)
(13) R.P. Eischen, S.A. Francis and W.A. Pliskin, J. Phys. Chem., 60, 194 (1956)
(14) Y. Soma-Noto and W.M.H. Sachtler, J. Catal., 32, 315 (1974)
(15) G. Blyholder, J. Phys. Chem., 68, 2772 (1964)
(16) C.W. Garland, J. Phys. Chem., 69, 1188 (1965)
J. Phys. Chem., 69, 1195 (1965)
(17) C.R. Guerra and J.H. Schulman, Surface Science, 7, 229 (1967)
(18) W.A.G. Graham, Inorg. Chem., 7, 315 (1968)
(19) C.A. Tolman, J. Am. Chem. Soc., 92, 2953 (1970)
(20) R. Queau and R. Poilblanc, J. Catalysis, 27, 200 (1972)
(21) M. Primet, J.M. Basset, M.V. Mathieu and M. Prettre, J. Catalysis, 29, 213 (1973)
(22) M. Primet, M. Dufaux and M.V. Mathieu, C.R. Acad. Sc. Paris, 280c, 419 (1975)
(23) Y. Soma-Noto and W.M.H. Sachtler, J. Catalysis, 34, 162 (1974)
(24) C. Naccache, M. Primet and M.V. Mathieu, Molecular Sieves, ed. W.M. Meier and J.B. Uytterhoeven, Adv.Chem.Ser.,121,266 (1973)
(25) F. Figueras, R. Gomez and M. Primet, Molecular Sieves, ed. W.M. Meier and J.B. Uytterhoeven, Adv. Chem. Ser., 121, 480 (1973)

SMALL METAL PARTICLES : A BRIDGE BETWEEN THEORY AND PRACTICE

Geoffrey C. BOND

School of Chemistry, Brunel University, Uxbridge, Middlesex
England

INTRODUCTION

When using metals as catalysts, common sense dictates that wherever possible they should be in the form of very small particles, having a high surface to volume ratio, and hence exposing the maximum possible fraction of the total atoms present to the gas phase. Fortunately it is relatively easy to prepare metal particles less than 100 Å in size, by methods to be described below : they exist in many catalysts in industrial use, and are easily studied in the laboratory.

Small metal particles are characterised either (i) by their average size (diameter) and by their size distribution; or (ii) by their surface area per g of metal; or (iii) by their degree of dispersion, defined as the fraction of atoms present on the surface. It is useful first to obtain a semiquantitative notion of their purely numerical properties : this may be done by using the Uniform Sphere Model, which takes 1 g of metal and subdivides it into spherical particles of uniform size. Figs. 1 and 2 show how for Pd and Pt the surface area, the degree of dispersion, the number of atoms per particle and the number of particles per g vary with particle diameter. The only assumption made in the calculations illustrated here is that the particles have the normal density of bulk metal. The calculations lose significance for sizes less than 10 Å, for no real particle can then be approximated adequately to a sphere. It is interesting to note how the almost two-fold difference in atomic mass between Pd and Pt influences the results.

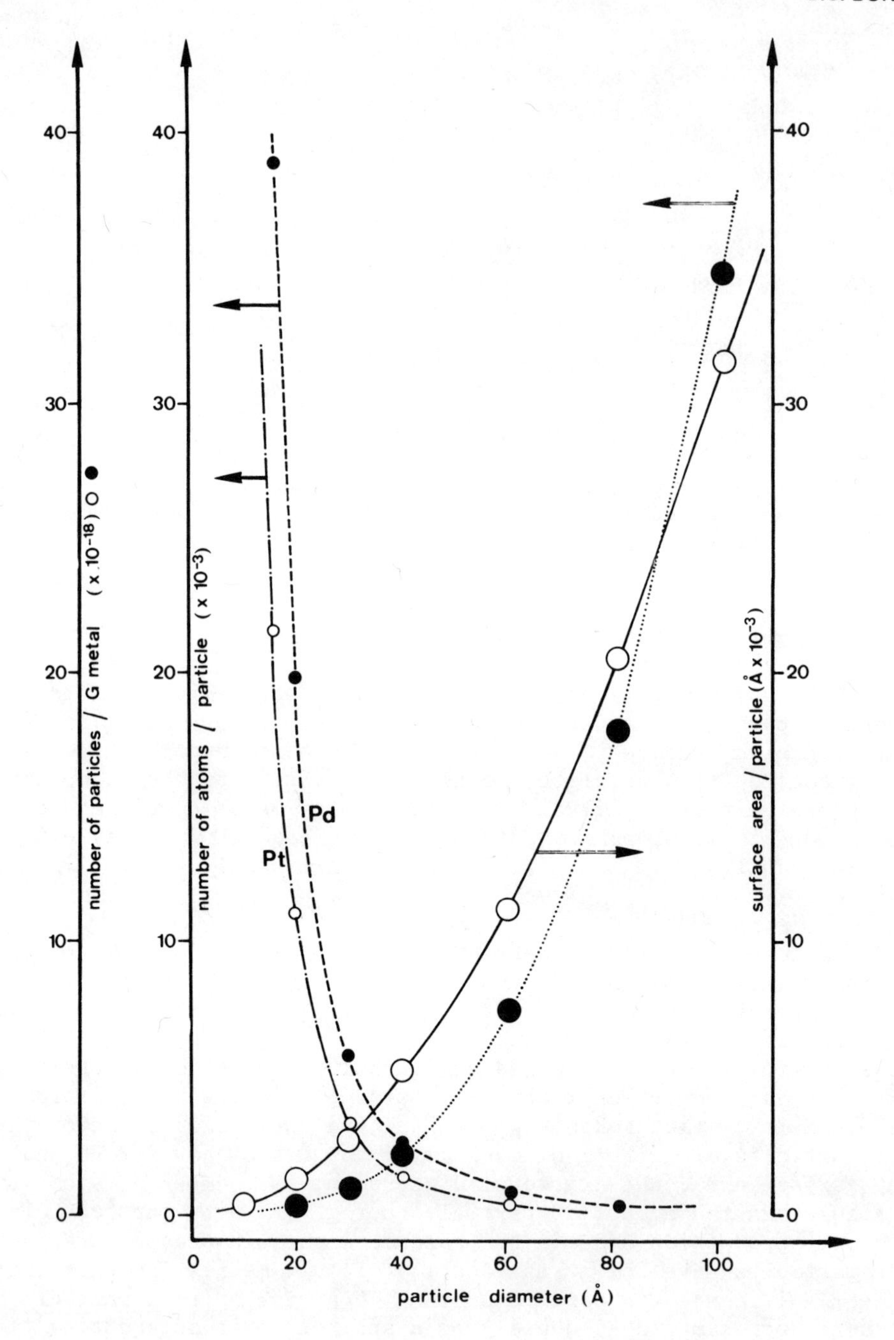

Fig.1 : The number of particles per g metal, the number of atoms per particle and the surface area per particle as a function of particle diameter according to the Uniform Sphere Model.

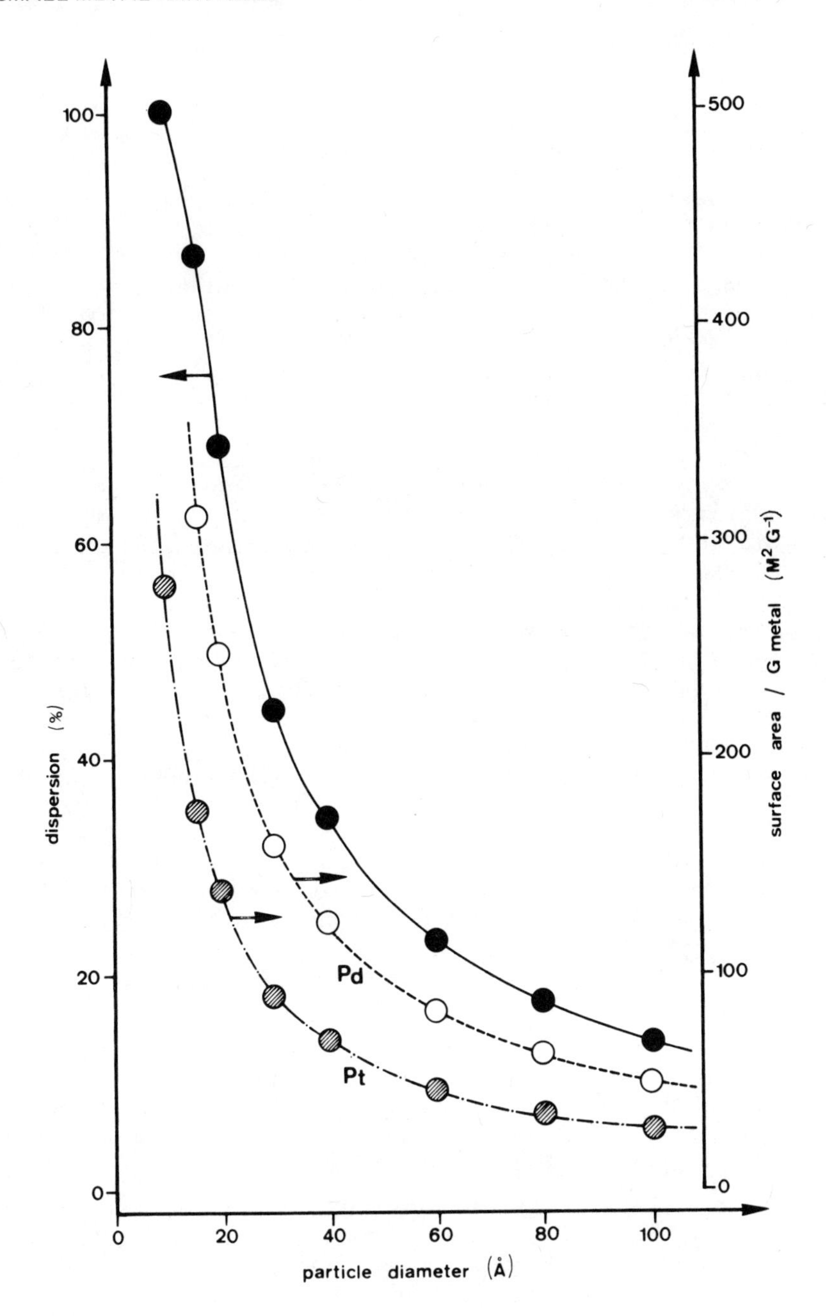

Fig. 2 : The surface area per g metal and the degree of dispersion as a function of particle diameter according to the Uniform Sphere Model.

THE PREPARATION OF SMALL METAL PARTICLES

Numerous methods are available for producing metal particles of a size such that their specific surface area is usefully large, but only three are of current interest. These are

(i) through the evaporation of metal atoms from a heated wire onto a cold substrate to produce an incomplete film;
(ii) through the techniques of colloid science;
(iii) and through the preparation of supported metal catalysts.

The first method does not produce a catalyst of practical utility, but particles so formed have clean surfaces and so are suitable for fundamental investigations (1).

Of the many techniques usable for making colloidal metals, chemical reduction of dilute solutions of salts was long ago shown to be the most satisfactory (2). Since this early and careful work on colloidal gold was performed, there have been further studies of this system (3) and the basic procedure has also been applied to platinum (4) and palladium (5), in the latter case the colloidal metal being deposited on boehmite so that its properties could be examined in the gas-solid mode. Colloidal platinum having a particle size of about 20 Å is readily made by the reduction of H_2PtCl_6 solution by sodium citrate (4). The well-known methods of colloid chemistry should permit easy control over average particle size and size distribution, although thus far only colloidal gold has been studied in detail (2) (3).

Procedures for preparing supported catalysts have long been known (6) and have been well studied : particle sizes between 10 and 100 Å are easily obtained. Four sub-divisions need to be recognised.

(i) Coprecipitation : a method not generally favoured since much of the metal is encapsulated by the support.

(ii) Impregnation : a term rightly applied to the situation where there is no chemical interaction between the support and the solution of the salt (e.g. H_2PtCl_6 + SiO_2). This method produces a wide range of particle size, and an average size which increases with metal loading (7).

(iii) Adsorption : where there is chemical affinity between the support and the salt solution (e.g. $PdCl_4^=$ + γ-Al_2O_3; $PdCl_4^=$ + activated carbon), the salt will adsorb from solution onto the support. The chemical reaction motivating the process is probably hydrolysis, but the system although widely used has never been deeply studied.

(iv) Ion exchange : a procedure applicable to both amorphous inorganic (7) and organic solids (8), (and of course to zeolites) producing smaller particles than (ii), having a narrower size distribution and an average size independent of metal loading. Ion exchange of $Pt(NH_3)_4^{2+}$ onto SiO_2 gives on reduction Pt particles of about 15 Å in size (7).

CHARACTERISATION OF SMALL METAL PARTICLES

Characterisation methods are either physical or chemical in nature. Pre-eminent among the physical methods is transmission electron microscopy, since it permits direct visual observation of the particles, and of their shape, and their size distribution (7). A critique of the technique has recently been published (9). The usefulness of X-ray line-broadening is limited to particles greater than 40-50 Å in size, and probably to pure metals, since a variation in composition among alloy particles would add its own line-broadening effect. Low-angle X-ray scattering is a most promising method (10), but the apparatus is costly and not yet readily available. ESCA may also prove to have its value (11).

While the sensitivities of the physical methods decrease as particle size decreases, the chemical methods gain is sensitivity. This is clearly because the specific surface area increases. Great use has been made of selective chemisorption to estimate the surface area and particle size of supported metals (12) (13), and very good agreement can be obtained between the results from this method and those from the physical methods (10). H_2 and CO are the most frequently used molecules, and while the H/M_s (M_s = surface metal atom) ratio of unity is well-established there is always doubt concerning the CO/M_s ratio unless determined by reference to H_2. CO cannot be used with Ni at higher temperatures, but H_2 can be used with Pd if proper precautions are taken (14). Relatively little systematic work has been done with metals other than Ni, Pd and Pt (15).

More recently surface titration techniques have been introduced, but have led to much discussion concerning the stoichiometry of the reaction under study. The hydrogen-oxygen titration was the first to be used ; if the reaction monitored is

$$\underset{*}{O} + H_2 \rightarrow H_2O + \underset{*}{H}$$

then there is a "magnification factor" of 3 compared with H_2 chemisorption on a clean surface (16). However the O/M_s ratio seems to be a function of particle size. Very recently adsorbed O has been titrated with CO (17), but the validity of the method has also been questioned (18). Chemisorbed H has been titrated with an olefin by a procedure which reveals the time-dependence of the process (19).

There is a continuing need for a rapid, cheap and reliable procedure for estimating the average size of small metal particles.

SPECIFIC PARTICLE SIZE EFFECTS

A most important question to pose is whether specific catalytic activity (i.e. activity per unit surface area) is or is not a function of particle size. The question has both theoretical and practical interest, for it may be imagined that essential catalytic properties may be lost with particles so small as not to have the normal electronic structure of the metal. It is convenient now to introduce Boudart's terminology (20) devised to describe these factors : a structure-sensitive reaction is one whose specific rate alters with particle size, while for a structure-intensitive reaction it is independent of size. Fig.3 illustrates possible modes of variation of specific activity with particle size, and notes additionally that there may be changes from positive to negative intrinsic effects (21) as the latter varies (curves 4 and 5).

A great deal of experimental evidence may be quickly summarised in the statement that a surprising number and variety of systems (defining a system as reaction + catalyst) exhibit structure-insensivity, at least to a first approximation. Some studies may not however have been pursued to sufficiently small particle sizes to reveal any effect : thus for example benzene hydrogenation, normally regarded as an "insensitive" reaction, has been shown to become "sensitive" over nickel at a particle size of about 5 Å (22). There are only three well established categories of structure-sensitive reactions : there are

(i) reactions involving C-C bond breaking (hydrogenolysis and skeletal isomerisation) and formation (dehydrocyclisation) ;
(ii) exchange of C_6H_6 with D_2 ; and
(iii) certain oxidation reactions.

Discussing these further, but in the reverse sequence, NH_3 oxidation over Pt/Al_2O_3 shows (23) a negative intrinsic effect (curve 3 of Fig.3), as does the NO-O_2-NH_3 system with the same catalyst(24). Smaller Pt particles become inactivated more rapidly than larger ones, perhaps due to stronger O_2 chemisorption, thus showing what has been termed (23) a secondary particle size effect. Other oxidation reactions manifesting similar characteristics are referred to in these papers.

The specific rate of C_6H_6 exchange with D_2 over Ni catalysts shows (25) a very marked negative intrinsic effect, due it is thought to sites of unusual geometry which are manifested in twinned crystallites which only occur in larger particles. A similar but very much smaller effect is seen with Ir catalysts (25).

The hydrogenolysis of C_2H_6 to CH_4 typifies a C-C bond-breaking reaction : this exhibits a significant positive intrinsic effect for Ni particles between 30 and 90 Å in size (26), although the procedure for altering size, namely, progressive sintering, may have induced other changes in the catalysts' surfaces. Over Rh catalysts there is a maximum in specific activity for particles of about 12 Å in size (27) (c.f. curve 5 of Fig.3). Skeletal isomerisation reactions (28) and C_7H_{16} dehydrocyclisation (29) also show structure sensitivity.

The adsorptions of N_2 and of CO also display characteristics of structure sensitivity (25).

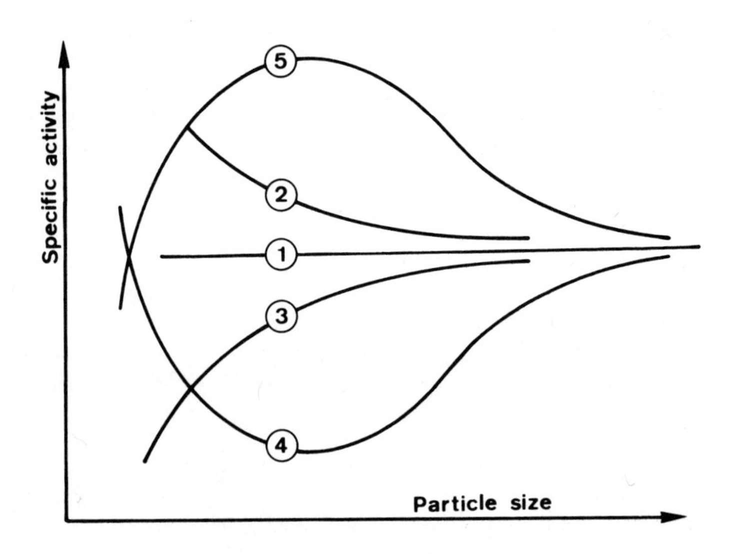

Fig.3 : Possible forms of specific particle size effect.

PHYSICAL BASES FOR SPECIFIC PARTICLE SIZE EFFECTS

The physical basis for the specific particle size effects outlined above was first sought in the topographical characteristics of small particles. Model calculations have been performed on a number of regular geometrical forms appropriate to the structure in which a particular metal crystallises ; forms considered include octahedra (30) (31), cubo-octahedra (21,30,32,33), tetrahedra and many others (30). The procedure may be briefly described for regular octahedra. Fig.4 shows that the surface of an octahedron has atoms of coordination numbers 4, 7 and 9 ; Fig.5 shows how the fractions of surface atoms having these coordination numbers vary with increasing particle size, this being derived from the number of atoms in each particle by reference to Fig.1. Similar curves have been computed for other crystal shapes (30).

It has of course been appreciated that a particle having just the requisite number of atoms to form any regular body is a quite improbable occurrence, and calculations have been extended to crystals having an incomplete outer layer (30,31). Poltorak and Boronin (31) obtained the interesting result that the average coordination number of surface atoms on octahedra having an incomplete outer layer is a reasonably smooth function of the total number of atoms comprising the particle. They recognised that for particles larger than about 40 Å there is little further change in topological characteristics, and designated the range 8-40 Å as the mitohedrical region, in which these are rapidly changing with size. van Hardeveld and Hartog (30) have extended the computations to cover sites, defined as groups of 2 to 5 atoms in specified geometrical arrays, and have drawn particular attention to two kinds of B_5 site, one of which (that occurring at steps on (110) fcc surfaces) is illustrated in Fig.6. The surface density of B_5 sites has been correlated with the quantity of strongly physically adsorbed N_2 on nickel, as well as with changes in the IR spectrum of chemisorbed CO (34) : however these calculations have only been carried out for particles having the maximum possible number of B_5 sites, i.e. those having an almost complete outer layer. The surface density of B_5 sites is then found to have a maximum value for particles of 20-25 Å in size (35). Sites of this kind are expected to exist on the stepped surfaces of single crystals shown by Somerjai (36) to have high activity in chemisorption and catalysis.

While there is thus good evidence for a topographical basis for specific particle size effects, there are other considerations which undermine confidence in the validity of this approach. There is growing evidence to suggest that microcrystals, especially those grown under conditions of supersaturation, may not have the symmetry expected from the structure of the bulk metal : in particular calculations on relative stabilities of various crystal forms have indicated that forms of 5-fold symmetry (e.g. icosahedra) should be the most stable (37,38), and indeed small crystals of this form have been observed (37,39). It will be difficult in practice to verify the symmetry of particles as small as those of use in catalysis.

It is also possible that specific particle size effects have an energetic rather than a geometric basis. Surface atoms of low coordination number cannot be expected to have the same energetic properties as those of bulk atoms. Indeed there is both theoretical and experimental evidence to show that the melting point falls as particle size decreases, and a 10 Å particle may have a melting point only half that of the bulk metal (37). Surface pre-melting may also occur, and a molten surface skin may on quenching afford an amporphous outer layer such as recently been observed with supported Pt particles using the technique of radial electron distribution (40). Surface atoms are calculated to have a larger vibrational amplitude than bulk atoms, while interatomic distances are slightly shorter (up to about 3 % for a 10 Å particle)(37). It is thus by no means certain that real metal particles have the geometrically ideal surfaces that the use of models would lead us to expect.

One other possibility deserves to be mentioned. Extremely small Pt particles, estimated to be clusters of about 6 atoms, are said to have catalytic properties more akin to those of Ir (41). This suggests that they are electron-deficient, and that specific particle size effects may have an electronic basis. Quantum-mechanical calculations might give evidence on this point.

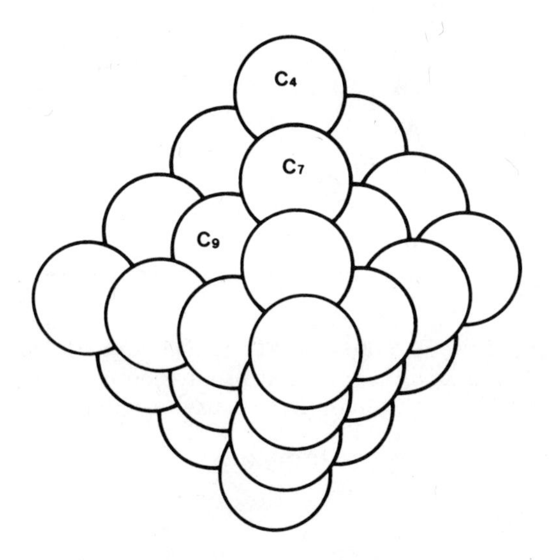

Fig.4 : A f.c.c. octahedron containing 44 atoms : the coordination numbers of apical, edge and plane atoms are indicated

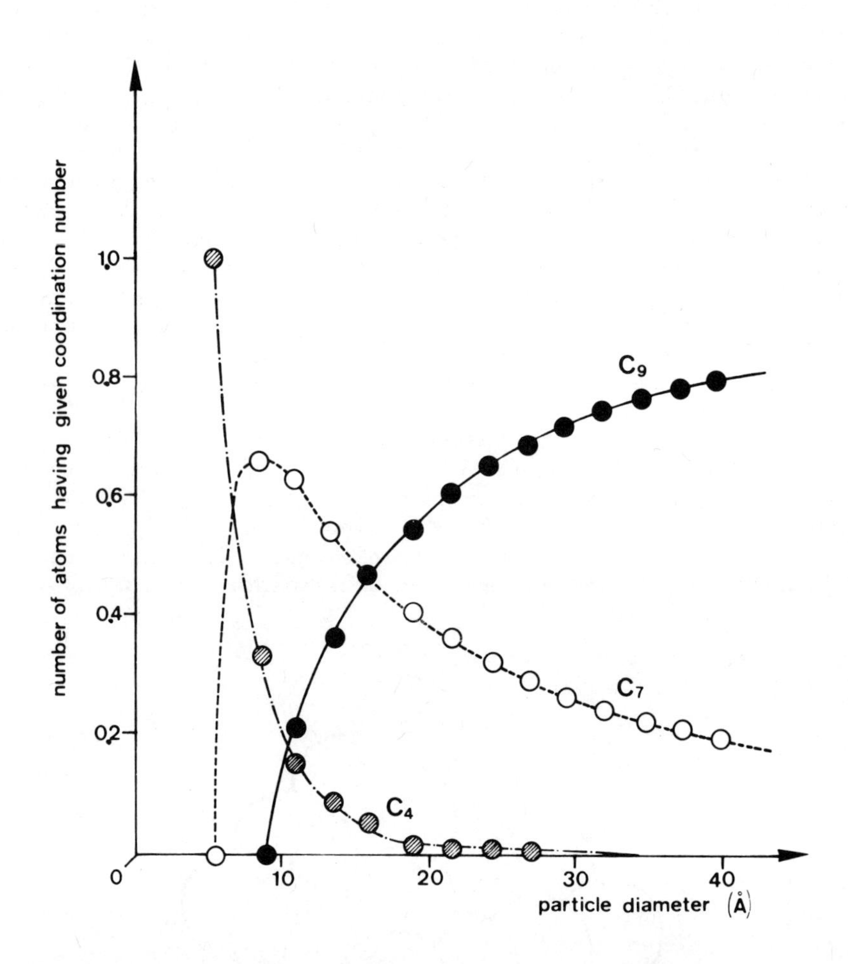

Fig.5 : Fraction of surface atoms in regular f.c.c. octahedra having coordination numbers of 4,7 and 9 as a function of particle size.

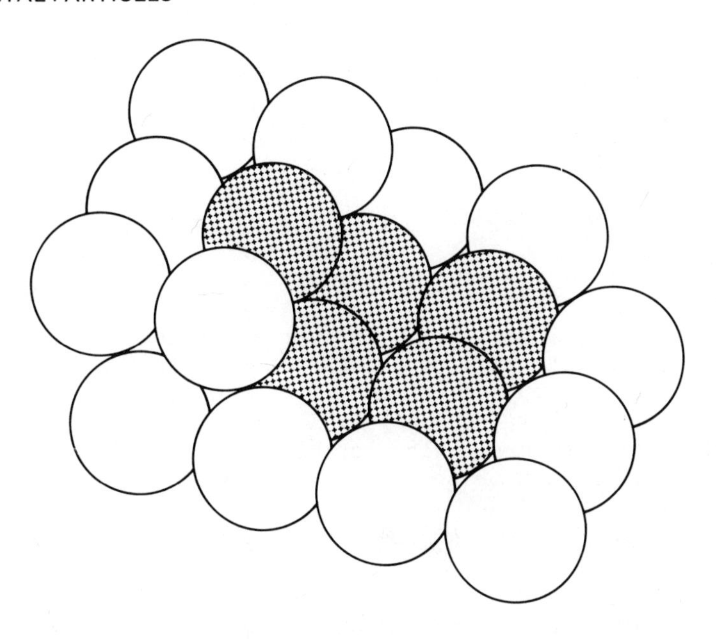

Fig.6 : Diagrammatic representation of a B_5 site occurring at a step on a (110) f.c.c. surface.

ALLOYING IN SMALL METAL PARTICLES

The introduction of a second metal produces additional complications which are currently attracting much attention. In the equilibrium state of a macroscopic alloy, basic considerations of thermodynamics dictate that the surface be enriched in whichever component has the lower surface energy (41-43) : in the case of the Ni-Cu system, quite small Cu contents lead to a surface predominantly covered with Cu (43). In very small alloy particles, which can be made on inert supports (44,45), the component of lower surface energy is likely to occupy positions of minimum coordination number preferentially (43). In the case of Pt_3Sn, surface enrichment seems to occur at the expense of the immediate sub-surface layer (41).

The discovery (46) of the apparent ability of pairs of metals not normally regarded as miscible to any significant extent to form alloys or "bimetallic clusters" when in the form of small particles has also excited much interest : Ru-Cu and Os-Cu were the pairs of metals used. Theoretical work has been carried out on the phase stability of small particles (47,48), and this is an area where new and useful discoveries are quite likely to be made.

The final complication deserving mention is that of facile diffusion normal to the surface under the influence of a strongly chemisorbing molecule. The effect was first established (49) for CO chemisorption of Pt-Ag alloys, where it was shown by work function measurements that CO drew Pt to the surface because of the stronger interaction, and that on desorption of the CO the surface relaxed to its normal composition, enriched in Ag. This effect may be quite a general one with alloys of Group VIII_3 - IB alloys, and will vastly complicate attempts to correlate chemisorption and catalytic behaviour with either bulk or normal surface composition.

QUANTUM MECHANICAL CALCULATIONS ON SMALL METAL PARTICLES

It is now proving possible to carry out quantum-mechanical calculations on metal particles of the size which can be made and studied in the laboratory : this presents an opportunity to make an exciting breakthrough in constructing a bridge between theory and practice in this branch of solid state science, to the benefit of both. Much work is thought to be in progress, and two recent papers illustrate what is being done. Extended Hückel calculations have been reported (50) for clusters of up to 55 atoms of Pd, Ag, Cu, Au, Pd-Ag and Ni-Cu : it is possible to estimate the width and position of conduction-band states and the variation of band width with size. For Pd and Pd-Ag, the number of d-band holes per atom is smaller for surface than for bulk atoms ; for Ni this number is smaller for low coordination number atoms than for highly coordinated atoms.

The SCF-Xα scattered wave method has been applied (51) to 8-atom clusters of Ni and Cu, and leads to a picture of the direction and symmetry of emergent orbitals. Work to describe the interaction of such clusters with C_2H_4 and with H atoms is in hand (51).

Theoretical work of this nature will be aided by the ability which now exists to establish by electron spectroscopy the band structure of small metal particles (52) and electronic effects due to chemisorption as sensed by changes in X-ray emission spectra (53).

REFERENCES

(1) J.R. Anderson and Y. Shimoyama, Proc. 5th Internat. Congr. Catalysis, ed. J.W. Hightower, North Holland, 1 (1973) 695
(2) J. Turkevich, P.C. Stevenson and J. Hiller, Discuss. Faraday Soc., 11 (1951) 55

(3) G. Frens, Nature, Physical Science, 241 (1973) 20
(4) G.C. Bond, Trans. Faraday Soc., 52 (1956) 1235
(5) J. Turkevich and Gwan Kim, Science, 169 (1970) 873
(6) J.W. Dobereiner, Ann. Chim. Phys., 24 (1823) 91 : see A.J.B. Robertson, Platinum Metals Rev., 19 (1975) 64
(7) B.W.J.Lynch and R.L. Moss, J. Catalysis, 20 (1971) 190
(8) D.L. Hanson, J.R. Katzer, B.C. Gates, G.C.A. Schuit and H.F. Harnsberger, J. Catalysis, 32 (1974) 204
(9) P.C. Flynn, S.E. Wanke and P.S. Turner, J. Catalysis, 33 (1974) 233
(10) A. Renouprez, C. Hoang-Van and P.A. Compagnon, J. Catalysis, 34 (1974) 411
(11) L.H. Scharpen, Chem. Abs., 82 (1975) 090498
(12) A.D. O Cinneide and J.K.A. Clarke, Catalysis Rev., 7 (1973) 214
(13) T.E. Whyte, Catalysis Rev., 8 (1974) 117
(14) P.A. Sermon, J. Catalysis, 24 (1972) 460
(15) R.A. Dalla Betta, J. Catalysis, 34 (1974) 57
(16) J.E. Benson and M. Boudart, J. Catalysis, 4 (1965) 704 ; J.E. Benson, H.S. Hwang and M. Boudart, J. Catalysis, 30 (1973) 146 ; J.M. Basset, A. Theolier, M. Primet and M. Prettre, Proc. 5th Intern. Congr. Catalysis, ed. J.W. Hightower, North Holland, 2 (1973) 915 : D.E. Meares and R.C. Hansford, J. Catalysis, 9 (1967) 125 ; G.R. Wilson and W.K. Hall, J. Catalysis, 17 (1970) 190 ; 24 (1972) 306
(17) P. Wentrcek, K. Kimoto and H. Wise, J. Catalysis, 34 (1974) 132
(18) P.C. Flynn and S.E. Wanke, J. Catalysis, 36 (1975) 244 : see also the following paper
(19) G.C.Bond and P.A. Sermon, Reaction Kinetics and Catalysis Letters, 1 (1974) 3
(20) M. Boudart, A. Aldag, J.E. Benson, N.A. Dougharty and C.G. Hawkins, J. Catalysis, 6 (1966) 92
(21) G.C. Bond, Proc. 4th Internat. Congr. Catalysis, Akadémiai Kiadó, Budapest, 2 (1971) 266
(22) J.W.E. Coenen, R.Z.C. van Meerten and H.Th. Rijnten, Proc. 5th Internat. Congr. Catalysis, ed. J.W. Hightower, North Holland (1973) 671
(23) J.J. Ostermaier, J.R. Katzer and W.H. Manogue, J. Catalysis, 33 (1974) 457
(24) R.J. Pusateri, J.R. Katzer and W.H. Manogue, AICHE Journal, 20 (1974) 219
(25) R. van Hardeveld and F. Hartog, Adv. Catalysis, 22 (1972) 75
(26) J.L. Carter, J.A. Cusumano and J.H. Sinfelt, J. Phys. Chem., 70 (1966) 2257
(27) D.J.C. Yates and J.H. Sinfelt, J. Catalysis, 8 (1967) 348
(28) J.E. Benson and M. Boudart, J. Catalysis, 4 (1965) 704
(29) M. Kraft and H. Spindler, Proc. 4th Internat. Congr. Catalysis, Akadémiai Kiadó, Budapest, 2 (1971) 286

(30) R. van Hardeveld and F. Hartog, Surf. Sci., 15 (1969) 189
(31) O.M. Poltorak and V.S. Boronin, Russ. J. Phys. Chem., 40 (1966) 1436
(32) E.G. Schlosser, Proc. 4th Internat. Congr. Catalysis, Akadémiai Kiadó, Budapest, 2 (1971) 312
(33) E.G. Schlosser, Ber. Bunsenges.Phys. Chem., 73 (1969) 358
(34) R. van Hardeveld and F. Hartog, Proc. 4th Internat. Congr. Catalysis, Akadémiai Kiadó, Budapest, 2 (1971) 295
(35) R. van Hardeveld and A. van Montfoort, Surf. Sci., 4 (1966) 396
(36) G.A. Somorjai, Catalysis Rev., 7 (1973) 87
(37) J.J. Burton, Catalysis Rev., 4 (1974) 209
(38) Y. Fukano and C.M. Wayman, J. Appl. Phys., 40 (1969) 1656
(39) E.B. Prestridge and D.J.C. Yates, Nature, 234 (1971) 345
(40) P. Ratnasamy, A.J. Leonard, L. Rodrique and J.J. Fripiat, J. Catalysis, 29 (1973) 374
(41) R.A. Dalla Betta and M. Boudart, Proc. 5th Internat. Congr. Catalysis, ed. J.W. Hightower, North Holland, 2 (1973) 1329
(42) R.A. van Santen and W.M.H. Sachtler, J. Catalysis, 33 (1974)202
R.A. van Santen and M.A.M. Boersma, J. Catalysis, 34 (1974) 13
(43) J.J. Burton, E. Hyman and D.G. Fedak, J. Catalysis, 37 (1975) 106
(44) D. Cormack, D.H. Thomas and R.L. Moss, J. Catalysis, 32 (1974) 492
(45) E.G. Allison and G.C. Bond, unpublished work
(46) J.H. Sinfelt, J. Catalysis, 29 (1973) 308
(47) D.F. Ollis, J. Catalysis, 23 (1971) 131
(48) E. Ruckenstein, J. Catalysis, 35 (1974) 441
(49) R. Bouwman, G.J.M. Lippits and W.M.H. Sachtler, J. Catalysis, 25 (1972) 350
(50) R.C. Baetzold and R.E. Mack, J. Chem. Phys., 62 (1975) 1513
(51) J.C. Slater and K.H. Johnson, Physics Today, 27 (1974) 34
(52) P.N. Ross, K. Kinoshita and P. Stonehart, J. Catalysis, 32 (1974) 163
(53) F. Freund, J. Catalysis, 32 (1974) 159

CATALYSIS BY SUPPORTED AND UNSUPPORTED METALS AND ALLOYS

V. PONEC

Gorlaeus Laboratoria
Rijksuniversiteit, Leiden
The Netherland

1. CATALYTIC ACTIVITY AND SELECTIVITY, DEFINITIONS

In this introductory lecture some basic terms used in catalytic literature will be defined and some examples of the experimental techniques (conventional and modern) will be presented.

The fundamental quantity in catalysis is the rate of the chemical reaction. It is best defined by differential quotients which are, for the homogeneous reactions, independent on the volume of the system. For a reaction like the following one

$$\nu_A A + \nu_B B \rightleftarrows \nu_C C + \nu_D D \qquad (1)$$

rate r is defined as a positive quantity by

$$r = -\frac{1}{\nu_A}\frac{dC_A}{dt} = -\frac{1}{\nu_B}\frac{dC_B}{dt} = \frac{1}{\nu_C}\frac{dC_C}{dt} \ldots \qquad (2)$$

A catalyst is a substance which changes the rate but it takes no part in the stechiometric reaction as in (1). If the catalyst and the reaction mixture form one phase, the catalyst is called homogeneous, if the catalyst forms a separate phase, it is heterogeneous. Due to its absence in (1), the catalyst cannot change the standard free energy ΔG^o of the reaction and the equilibrium constant defined as :

$$K = \frac{a_c^{\nu_c} a_d^{\nu_d}}{a_a^{\nu_a} a_b^{\nu_b}} \quad , \quad -\Delta G^o = RT \ln K \qquad (3)$$

There is no uniform and ideal way as to how to characterize the <u>activity</u> of the heterogeneous catalyst. Usually one of the following quantities is adapted to this end.

1. The rate of the catalysed reaction per unit surface area at a given temperature. With a heterogeneous catalyst the volume of the system is unimportant and the reaction rate must be related to the unit surface area of the catalyst :

$$r_S = -\frac{1}{\nu_A}\frac{dC_A}{dt}\cdot\frac{V}{S} \qquad (4)$$

$$= -\frac{1}{\nu_A}\frac{dn_A}{dt}\cdot\frac{1}{S}$$

   where n_A stands for the number of molecules (or moles, grams, etc.) and S for the surface area of the catalyst.

2. The temperature necessary to reach a certain rate r_S.

3. The rate can be written as a product of two parts :

$$r_S = k(T) \cdot f(C_A, C_B \ldots, T) \qquad (5)$$

   and the temperature dependence of <u>f</u> can usually be neglected. Then, because

$$k(T) = k_o \exp\left(-\frac{E}{RT}\right) \qquad (6)$$

   plotting of log k (or log r_S) versus $\frac{1}{T}$ leads to the determination of <u>k_o and E</u>, two factors which <u>together define the activity</u> of the catalyst. While <u>E</u> is clearly related to the energetic changes accompanying formation of active complexes (intermediates of the reaction), k_o comprises a.o. the total number of active sites, entropy changes and universal constants.

If more than one product is thermodynamically possible, e.g. the reaction follows one of the following schemes :

$$A + B \begin{cases} \xrightarrow{r_1} C \\ \xrightarrow{r_2} D \end{cases}$$

$$A + B \xrightarrow{r_1} C \xrightarrow{r_2} D, \text{ etc.} \qquad (7)$$

then, the catalyst usually determines the contribution of various reactions (1 or 2) to the total reaction. The parameter selectivity is defined ; a useful definition for cases like the one above is

$$S_1 = \frac{r_1}{r_1 + r_2} = \frac{A}{A + Be^{\frac{\Delta E}{RT}}} \qquad (8)$$

where ΔE is the difference in the activation energie of the reactions. The S(T) function is schematically shown in Fig.1 for the reforming of hexane.

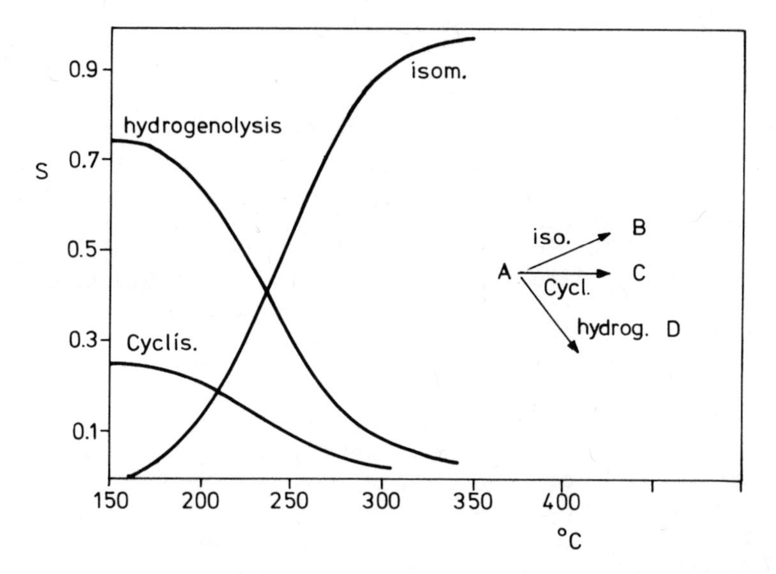

Fig.1 : Selectivities in parallel reactions; $(\text{rate})_i = r_o \exp[-E_i/RT]$ $S_j = \frac{(\text{rate})_j}{\sum_i (\text{rate})_i}$ e.g. for isomerisation $S_{iso} = [1 + A_1 \exp(\Delta E_1/RT) + A_2 \exp(\Delta E_2/RT)]^{-1}$, where $A_1 = r_o(\text{cycl})/r^o_{iso} = 10^{-9}$; $A_2 = r_o(\text{hydrogenol})/r^o_{iso} = 3.10^{-5}$; $\Delta E_1 = E_{iso} - E_{cycl} = 20$ kcal/mol ; $\Delta E_2 = E_{iso} - E_{hydr} = 20$ kcal/mol.

The catalytic effect is, as already stated by Faraday, connected with the adsorption of the reaction components on the surface, i.e., with formation of adsorption complexes. We have several useful methods to study these complexes, such as various spectroscopic methods (NMR, ESR, IR, UV - all of these were already used for it) or the study of certain reactions (dynamic methods), like exchange of isotopes, labelling of expected or detected intermediates, etc.

It is evident that the experimental basis of the whole catalytic research is the determination of the chemical rates. Therefore, let us dedicate a few words to this problem.

2. RATE MEASUREMENTS

In order to obtain reliable rate data we must make sure that the values measured are not influenced by a slow mass transfer or heat transfer, to or from the surface of the catalyst. We are not going to discuss this, as a review on these problems can be found among the materials of another A.S.I.[1] or somewhere else in the literature[2-4]. However, let me stress here that we always have to check by using one of the criteria, whether or not our data are free from these complications. A good deal of the catalytic literature before - say - 1960 should be suspected in this respect.

2.1. Conventional techniques

The simplest way to measure the rate of a chemical reaction is to follow analytically the concentration (partial pressure) of a chosen reaction component, in a closed system, as a function of time, i.e., $C_A = C_A(t)$. The rate is obtained after differentiating (numerical, graphical or directly by experimental technique applied) the $C_A(t)$ functions. If the number of molecules undergoes changes during the reaction (i.e. if $\sum_i \nu_i \neq 0$, for equations like (1)) and if only one reaction takes place, the reaction can be followed by measuring the total pressure, using one or another pressure gauge[5,6]. In a closed system the concentration of all components must be equal everywhere. A rule of thumb says that always when the total pressure is higher than 0.1 - 1 Torr[4], the gas must be stirred by a piston or propeller pump[8], etc.

Figs.2,2a and 2b show an example of such an apparatus[9]. The catalysts which can be studied are evaporated films, wires (mono- or polycrystalline) or powders. As an analytic tool, a mass spectrometer[9,10], a gas chromatograph[10] or its combination[11] may be used. Analysis can be performed continuously or by sampling. In special cases like H_2/D_2 exchange and equilibration, ortho-para H_2 conversion, even more simple gauges are sufficient for analysis[5].

Sometimes it is desirable to know the rate under the steady-state conditions. This cannot be achieved in a closed system for which it is typical that

$$C_i = f(\text{time}) \; ; \; C_i \neq F(x, y, z) \; (x \ldots \text{coordinates})$$

Then, we better use an open system with gases (or liquids) flowing through a catalytic bed with a layer of catalyst (a tubular reactor). It is typical for this system that at the steady state

$$C_i \neq f(\text{time}) \; ; \; C_i = F(x, y, z)$$

The rate r_s cannot be evaluated here by measuring $C_i(t)$! So the following procedure has to be used.

Let us consider a differential reactor of length dL where the catalyst of the total surface area dS is placed. An obvious mass-balance holds

$$\begin{aligned} r_S \cdot dS &= - \vec{V} \cdot dC_A \\ &= \vec{F}_A dC_A \cdot (C_A^o)^{-1} \\ &= + \vec{F}_A \cdot d\alpha \end{aligned} \qquad (9)$$

Here : $\vec{V}$ is the volume of gas (ml,l) passing the catalyst; $\vec{F}_A$ the flow rate (number of molecules, moles, grams - dimensions must correspond to the C_A - and r_S-dimensions); C_A^o the concentration of A for the reactor; $(C_A^o - dC_A)$ behind the reactor; α is conversion $-dC_A/C_A^o$. To determine r_S conversion α of concentration C_A is measured behind the integral reactor of length L, as a function of $\vec{V}$ or $\vec{F}_A$. If $\vec{F}$ does not depend on α (this is true for diluted reaction mixtures, low conversion, etc.) which is often the case, rate r_S is found by differentiating the $\alpha(\vec{F}_A^{-1})$ or $\alpha(\vec{V}^{-1})$ functions.

$$r_S = \frac{-dC_A}{d(\frac{S}{\vec{V}})} = \frac{d\alpha}{d(\frac{S}{\vec{F}})} \qquad (10)$$

(For a more exact analysis, see Ref.2)

Figures 3 and 4 show two typical examples of the apparatus used for measuring rates in a tubular reactor. Gas chromatography is the most frequently used analytic tool.

Gas chromatography is a well known method but to make sure that also people from theoretical laboratories will understand all details of this talk, a brief description of this method is presented. In its most common arrangement gas chromatography works as follows.

A sample of gas or a liquid (which is evaporated fast, inside

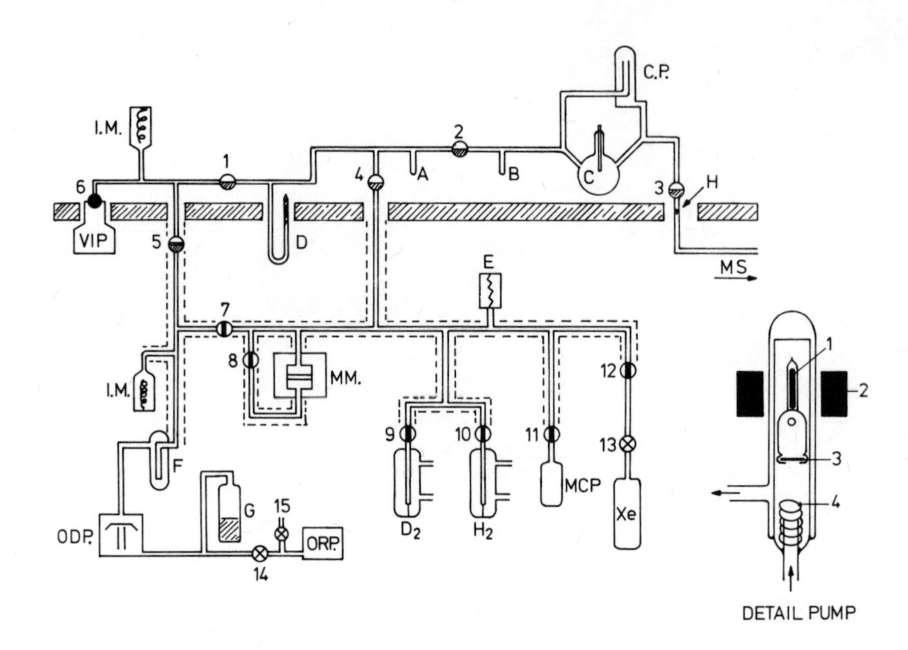

Fig.2 : Static apparatus for catalytic reactions on evaporated metal films or metal wires(9)

1 - 5 Granville-Phillips bakable valves
6 vacuum generator bakable valve
7 - 12 Hook vacuum valve (bakable to 150°C)
----- heating tapes; above the table, ovens are used
ORP - oil rotation pump; ODP-oil diffusion pump; VIP - vac ion pump ;
MM - membrane manometer (ATLAS); D - pirani gauge; E - pirani gauge ;
H_2,D_2 - palladium-silver diffusion cells for surface area measurements; A and B - freeze-out point for cleaning and bringing over condensable organic gases; C - reaction vessel with a filament (for reaction or evaporation); CP - circulation pump; IM - ionisation gauge; F - trap; G - trap with zeolites; H - leak to mass-spectrometer (MS12 AEI).

Detail : simple circulation pump; 1 - soft iron rod; 2 - coil, fed by a pulse current; 3 - light glass; 4 - spiral.

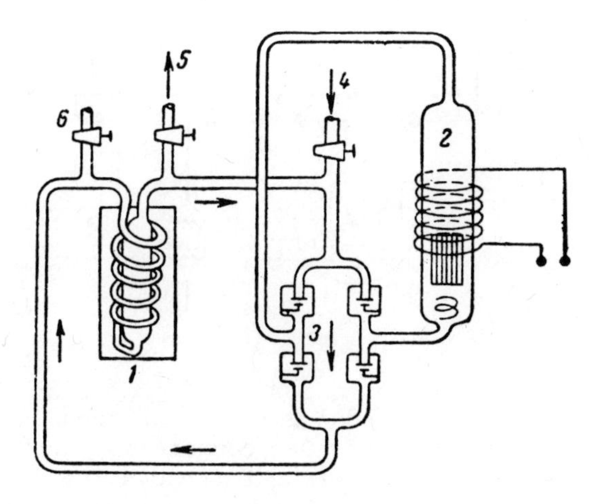

Fig.2a : Static catalytic apparatus[7].
Reactor (1) with a piston pump (2) ;
4,5,6 in- and out-lets; 3 - valves made from light glass

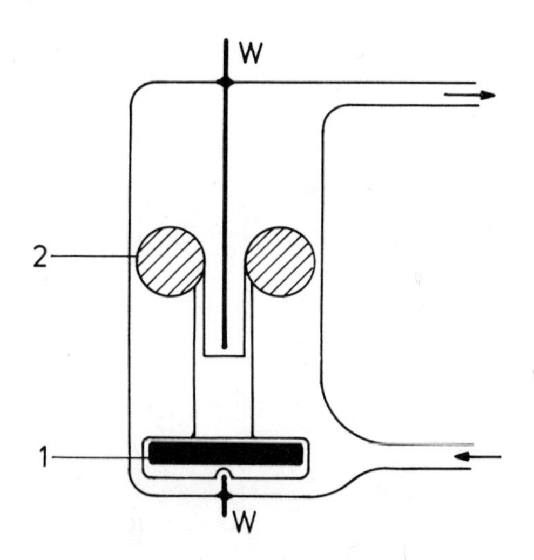

Fig.2b : Propeller pump[8] which can be used (when p > 10 Torr) in static reactors. W - tungsten rods; 1 - iron rod ; 2 - glass fans, rotated by an external magnet, placed under 1.

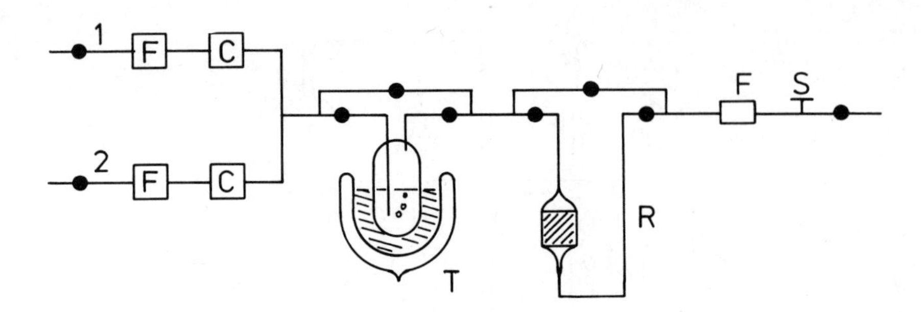

Fig.3 : Flow reactor as used for reactions when one of the initial components is a fluid. Gas components (or inert gas, diluting the mixture) are brought through 1 and 2, to a flowmeter (see e.g. Refs 5,6) and cleaning line (cooling traps, ab-, ad-sorptive and catalytic cleaning elements; Refs.5,6). Then, they enter the saturator T kept at a constant temperature and the reactor R. At point S samples are taken away for analysis, e.g. by injection syringe through a septum or by a sampling valve. F - flowmeter, C - cleaning line

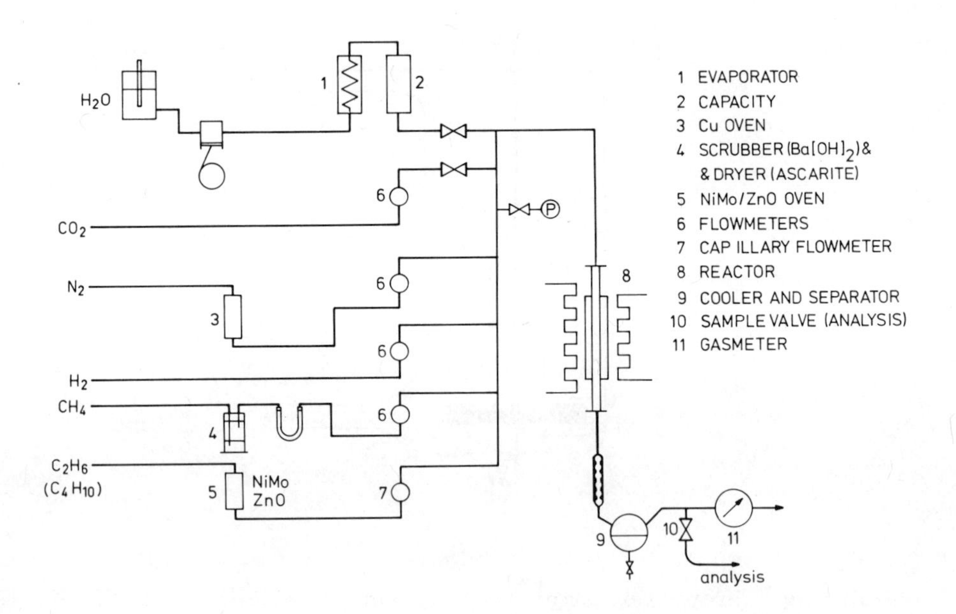

Figure 4

Captions of figure 4 : Flow reactor as used for reactions of several gases(41).
1 - evaporator ; 2 - flow stabilisator ; 3 - Cu oven ;
4 - absorber ($Ba(OH)_2$) ; 5 - oven, Ni, Mo/ZnO ;
6 - flowmeters ; 7 - capillary flowmeter ; 8 - reactor;
9 - cooler, separator ; 10 - sampling valve ;
11 - flowmeter.

the chromatograph) is injected into a stream of a carrier gas, for the chromatographic column. The last element is chosen in such a way that the various compounds travel through it with different velocities. Then, a detector sensitive to the composition of the gas detects the components of the injected mixture at the end of the column at different time periods (See Fig.5). The column may be a long capillary with walls covered by a layer dissolving - to a different extent - the components of the mixture analyzed, or it is a tube filled with solid adsorbents, etc. When a catalytic reactor is placed before the column, behind the injection place we get a so-called pulse microreactor. A pulse reactor is not well suited for gathering kinetic data (i.e., determination of the $r_S = f(C_A, C_D \ldots)$ function) but it gives usually reasonable values of the activation energy E and information on the selectivity of the catalyst. The great advantage is the small amount of compounds we used for the measurement. Sometimes, a pulse reactor is combined, to advantage with a radioactive, or in general-isotopic-labelling of reactants[14].

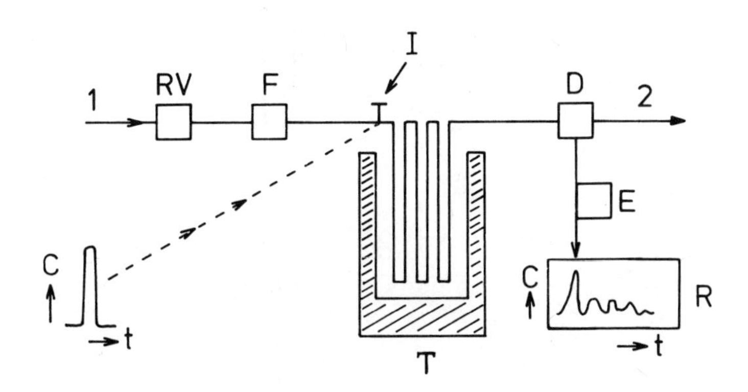

Fig.5 : Gas chromatograph, schematically.
1 - inlet; 2 - outlet; RV - flow regulating and controlling valve; F - flowmeter; I - point of injection (or sampling valve); T - thermostated oven; D - detector; E - electronic signal handling system; R - recording.

2.2. Advanced techniques

The flow systems mentioned above are usually easy to be applied in the pressure range 0.1 to several atmospheres. A steady-flow apparatus for low pressures working under 0.1 Torr was successfully developed and exploited by Frennet and his group(15).

A special and rather complicated task is to obtain reliable data for monocrystals. The problem here is to limit or exclude the catalytic background reaction on the walls of the - in most instances - metal apparatus, on the material removed upon cleaning from the crystal studied, etc. Reactions must be followed on the clean and crystallographically well-defined plane only. However, if the plane is cleaned(16) in situ, e.g. by ion bombardment or reactive cathode sputtering, a reactive material of a much higher surface

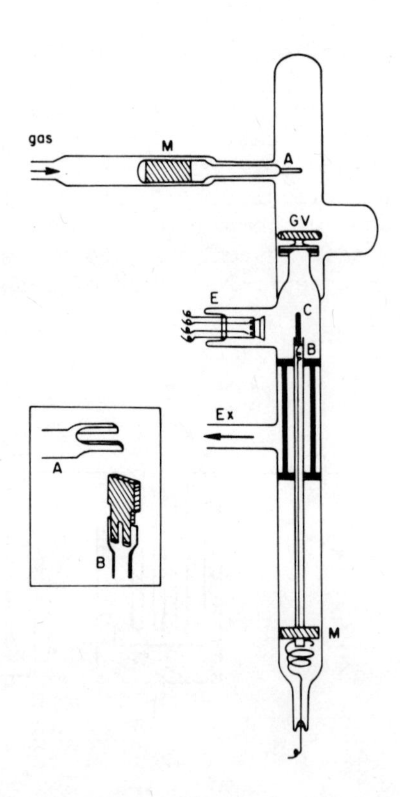

Fig.6 : Reactor for low-pressure catalytic measurements on Ge-monocrystal, cleaned by ion bombardment (E). Crystal C is cleaned at B and then moved through the ground-planed valve into the position A. There the crystal is fixed in A and the holder is driven back in B.M - magnet.

area is created inside the apparatus. This is not a serious obstacle for physical measurements which were the subject of the foregoing lectures. However, catalytic measurements are only possible when certain arrangements are performed.

In an earlier paper on this problem (see Ponec[16]) Farnsworth[17] moved a cleaned plane between two separated compartments of the ultra-high vacuum apparatus, as schematically shown in Fig.6.

In a paper published later the same author[18] used another arrangement. In order to follow the thermal desorption from only one crystallographic plane, the author used monocrystals which could be heated with the plane to be studied placed very closely to the mouth of a tube leading the desorbing gases to the fast-recording mass spectrometer (quadrupole mass spectrometer). Similar to it is an arrangement used by Somorjai et al.[19] for monitoring the reforming reactions of hexane at low pressures on monocrystal planes (Fig.7). Somorjai et al[20] also designed and used an apparatus which allowed them to study the reactions on monocrystal planes in a flow of gases at atmospheric pressure using gas chromatographic analysis. (Fig.8)

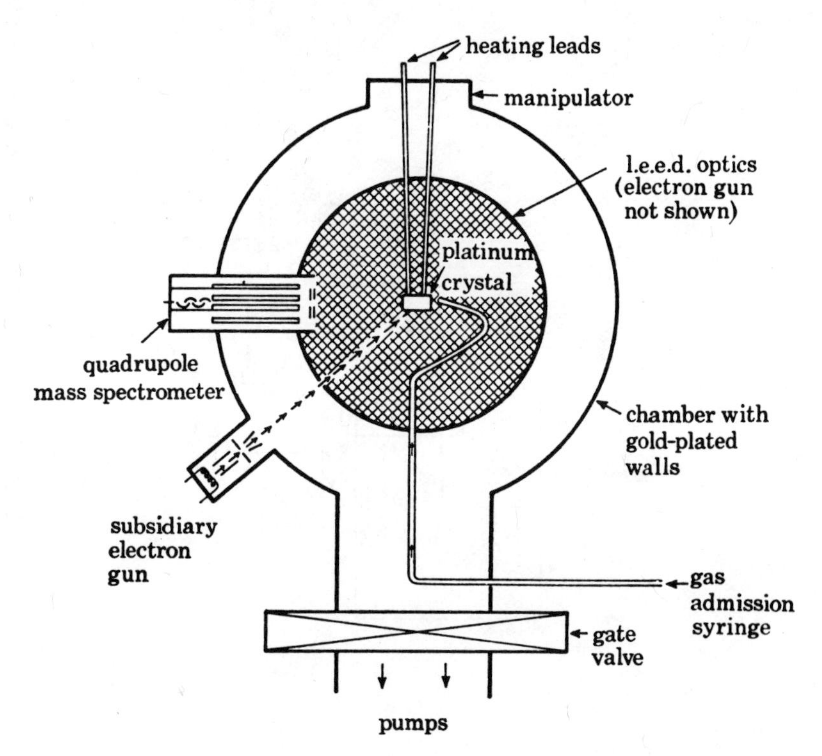

Fig. 7 : Reaction chamber used by Somorjai et al. for reactions at low pressures on monocrystals[19].

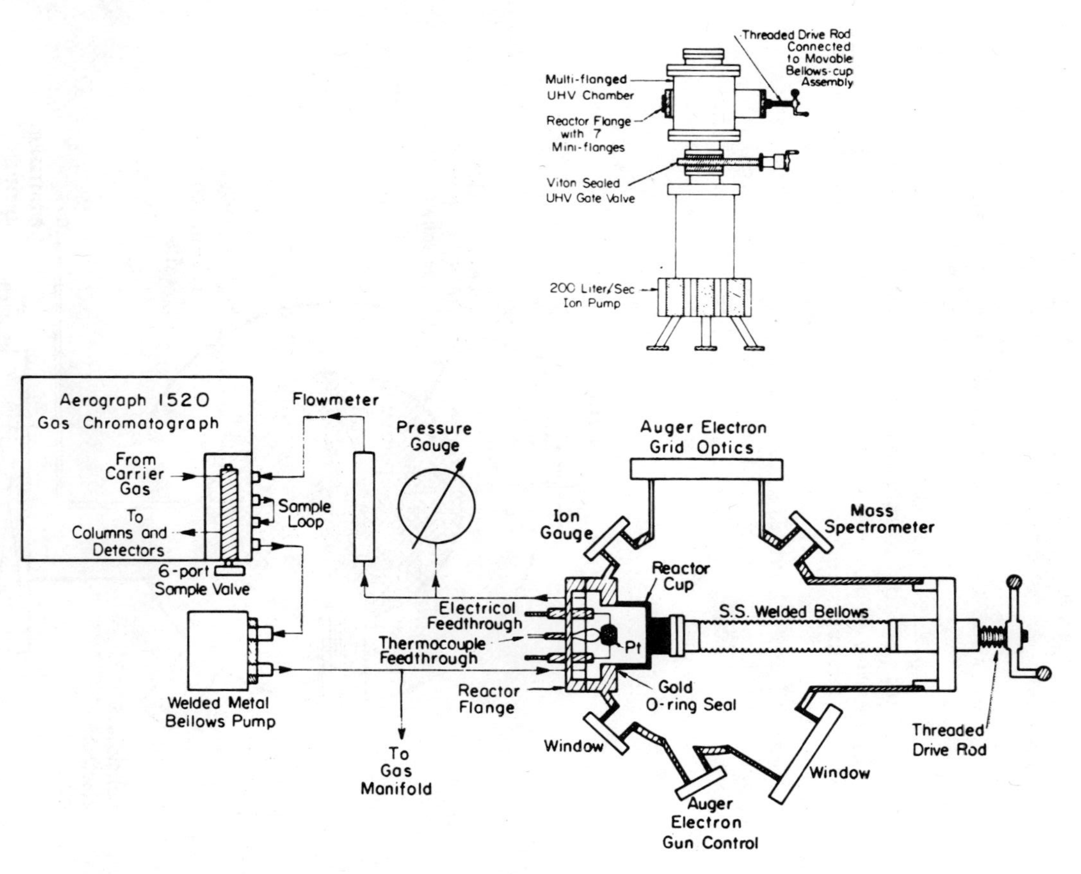

Fig.8 : Scheme of an ultra-high vacuum apparatus suitable for the work on monocrystals with a possibility to analyze gases by G.C. at 1 atm pressure[20].

The arrangements mentioned above have already much in common with the apparatus used for the studies on molecular beams, which are directed to and reflected from the monocrystal plane. Then, the beam is usually collimated by a system of thin capillaries and the energitecally homogeneous molecules are filtered off by using a velocity selector. The detector or the target is movable (or both are) so that the angular distribution of elastically and inelastically reflected (released, in the latter case) molecules can be monitored(21). It is possible to study the influence of the temperature of the beam and of the catalyst separately and owing to the geometric restriction on the beams the influence of the "background"-reaction is strongly diminished. However, extremely high pumping rates are necessary as outside the beam the pressure should be less than $\sim 10^{-4}$ Torr and the sensitivity of the detector must be very high. The chemical reaction followed must be sufficiently fast. These are rather severe restrictions, so this method is rather complementary to the classical ones covering the range of rates not covered by simple arrangements but the method, at the present stage, does not offer an equivalent alternative to them. As an example of the reactions in beams, see Fig.9, according to Ref. 22.

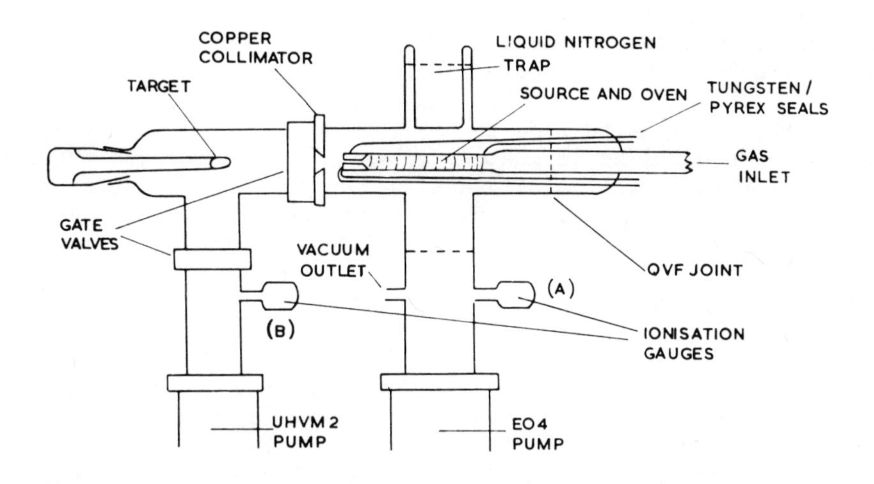

Fig.9 : Molecular beam reactions on a target, Ref. 22.

2.3. Special techniques

Sometimes it is useful to follow the behaviour of the adsorbed complexes and their composition during a running catalytic reaction. This technique may reveal which complexes are the intermediates of the reaction, or sideproducts blocking (poisoning) the active surface, etc. Examples of essentially volumetric (mass-balance) analysis of adsorbed complexes under the reactions are papers by Tamaru et al.[23] and Frennet et al.[15]. The complexes may also be analyzed by means of the IR spectra as has been shown recently[24].

For testing industrial catalysts, a so-called Well Stirred Tank Reactor is also used. However, its importance lies more in checking rate under conditions like interphase and mass transfer, heat transfer at high conversions, etc. and, therefore, we only quote the examples from the literature[25].

3. CHARACTERIZATION OF THE CATALYST SURFACES

Metals and alloys are used for catalytic experiments or in industrial practice in most varied forms, reaching from monocrystals with well-defined planes up to extremely dispersed metals on carriers which considerably differ in their properties from bulk catalysts. Characterization of alloy surfaces will be discussed in another chapter; below a few words will follow on some general questions of metal (alloy) surface characterization.

If the catalyst is used in the form of a pure metal (evaporated film, wire, powder) it is most essential to determine its total surface area. This is most easily achieved by the adsorption method and using the BET theory[26]. Otherwise, the Electron Microscopy[27] and the statistics of particles or the Low Angle Scattering of X-rays[28] may be used for the same purpose. Also the broadening of X-ray diffraction peaks can give us an idea of the particle size and the surface area[28].

At high temperatures (above 0.2 -0.3 of the melting point of the given metal) fine, pure metal powders are unstable and when sintered giving large particles. In order to prevent sintering and also to economize on the catalytic active metal (sometimes very expensive, such as Pt, Pd, Re or Ir) metal catalysts are usually prepared in a dispersed form by mounting them on a surface of a carrier[29]. Several methods were suggested to determine the metal surface area of the catalysts in the form of metal-on-carriers. Electron microscopy, X-ray diffraction and scattering methods are again very useful. At the highest resolution power, particles of about 10 - 20 Å may be counted by E.M.[30]. Physical adsorption is less helpful with these aprticles as it measures non-specifically the whole surface area of the metal and carrier together. In order to determine the metal surface area only, a specific chemisorption of a gas not being adsorbed by a carrier must be used. The most frequently used gases in this respect are H_2, CO and O_2. After the pioneering work of Boudart c.s. and Boreskov c.s. (see Reviews[31]) these chemisorption methods were developed and successfully applied by many authors and several excellent reviews on this subject are now available. I am now closing this point therefore by bringing these papers to your attention - see references under 31.

The total metal surface area is a useful parameter to know but it is clearly insufficient to characterize the surface fully. We are always interested in questions like what is the geometry (on atomic scale) of the active sites, which crystallographic planes constitute the surface of the catalyst ?

The surface of well-equilibrated particles consists of planes with the highest density of atoms like (111) and (100) planes for f.c.c. structures, etc. This is an old rule[32] which has been confirmed by experiments on macroscopic objects[33-34]. Several authors analyzed the problem of micro-particles theoretically and they suggested models of well-equilibrated, small, isolated particles[35]. The contribution of various planes in the surface[35] and the presence of some special sites (B5-sites) is then predicted[36](see e.g. Table 1). By diminishing the size of the particles the number of corner and edge atoms increases[35-37], relatively and absolutely. The corner and edge atoms are coordinatively unsaturated (as well as e.g. the B_5-sites and similar ones) and it is expected that their catalytic and adsorptive behaviour differ from that of atoms in the planes (adlineation theory of Schwab and others[38]). For more details on this problem, see other lectures of this course and recent papers by Boudart c.s.

There is no universal direct method as to how to count the special positions and, therefore, most of the conclusions are based on the theoretical analysis and theoretical predictions. There is perhaps one exception : the use of N_2 adsorption to count the B_5(or similar) sites[39], at least, if it is really so that B_5-sites are those adsorbing N_2 in a form active in IR adsorption.

The situation is most complicated in the case of the smallest particles hardly visible by E.M. Here, we have to rely entirely on the theory. Just these particles deviate most strongly from bulk crystals. They often reveal (see Fig.10) a fivefold symmetry for f.c.c. small particles, the surface of these particles has no (100) planes but only (111) planes, the lattice constant is contracted, etc. A good review on these problems is available[40].

Only with monocrystals, the surface of which can be controlled by LEED in situ[19-22] we can be sure about the geometry of sites.

This advantage is counterweighed by difficulties mentioned in Chapter 2.

Alloys will be discussed later. The reader of this paper will notice that the question of promotors (small amount of additives of other compounds improving the catalytic activity and stability of the catalysts) has been neglected. The reason is simple. It is a problem too complicated for the limited space we have for catalysis in this school.

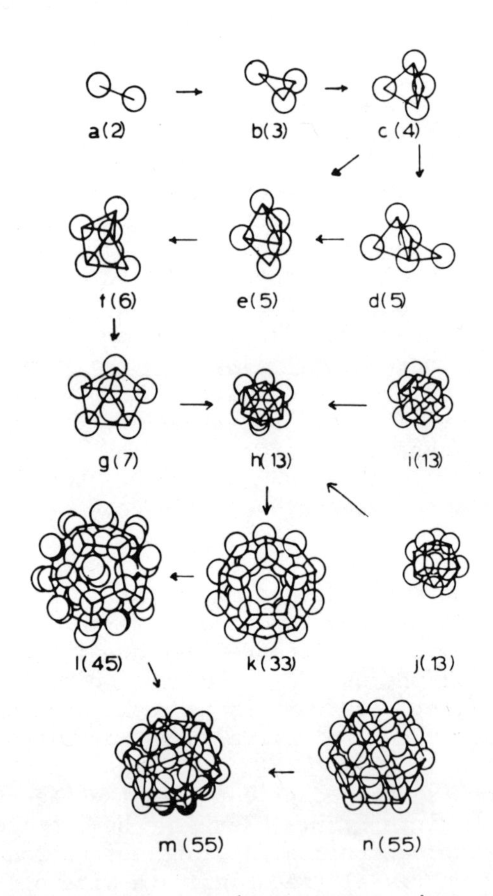

Fig.10 : Growth sequence in the formation of a 55 atom icosahedron particle (m). The f.c.c. structures d, i, j and n are unstable. According to Ref.40.

TABLE 1 : Statistics of various sites on a regular cubooctahedron particle of f.c.c. metals (Ref. 36)

	number of atoms	m		
		2	3	>3
1	N_T	38	201	$16m^3-33m^2+24m-6$
2	N_B	6	79	$16m^3-63m^2+84m-38$
3	N_S	32	122	$30m^2-60m+32$
4	on corners	24	24	24
5	in the (111) faces	8	56	$8(3m^2-9m+7)$
6	in the (100) faces	0	6	$6(m-2)^2$
7	on edges of (100) faces	0	24	$24(m-2)$
8	on edges between (111) faces	0	12	$12(m-2)$

m = number of atoms in one edge

N_T= total number of atoms constituting crystals, from which atoms :

N_B= atoms in the bulk of the cubooctahedron

N_S= atoms in the surface the cubooctahedron

B_5-sites = sites with five coordinated neighbours, e.g. see 7 in the table

TABLE 2 : Small, octahedron Pt crystals (Ref. 37)

length of edge atoms	in Å	N	N^m	fraction of atoms edges	on planes	M
4	11	44	0.87	0.63	0.21	6.94
5	13.75	85	0.78	0.55	0.36	7.46
6	16.50	146	0.7	0.47	0.47	7.76
7	19.25	231	0.63	0.41	0.55	7.97
10	27.50	670	0.49	0.29	0.69	8.31
15	41.25	2255	0.35	0.20	0.79	8.56
30	82.50	18010	0.19	0.10	0.90	8.78
∞	∞	∞	0	0	1	9

N = total number of atoms in the crystal

N^m= fraction of atoms in the surface (H/Pt ratio in H_2 chemisorptive titrations)

M = average coordinative number of atoms in the surface

4. PIECES OF INFORMATION ON THE CATALYTIC ACTIVITY AND SELECTIVITY OF PURE METALS

4.1. Activity of metals

It was a rather long and busy period of research between - say - 1948 (one of the first papers of Schwab[42] on the electronic theory of catalysis appeared) and 1968 (IVth Int. Congr. Catal., Moscow, entirely devoted to the theory of the catalysts[43]) which is characterized by attemps to find a correlation of the catalytic activity of metals to one or another physical parameter characterizing the electronic structure of metals. Various parameters were used : work function (Kemball[44]), density of states at the Fermi-level reflected e.g. by the magnetic susceptibility (Dowden[45]), the mean or the maximum kinetic energy of the electrons in the conductivity band (Dowden[46]) and the physically ill-defined Pauling d-character of metals (Beeck[47]). Prof. E. Derouane recently reviewed these attempts in his Summer-School lecture[48]and in the paper[49] published shortly afterwards. However, I would find it a rather embarrassing task if I had to state now - anno 1975 - which of these correlations should still be recommended as working.

A more empirical approach appeared to be more fruitful. Sabatier[50] stated in one of his papers that catalysis is essentially a formation and conversion of surface compounds which must be neither too strongly nor too weakly bound to the surface. This "principle of Sabatier" was also postulated in another form and on basis of new material in the papers by Balandin c.s.[31], by Schuit, Sachtler and Van Reijen[52], by Golodetz and Roiter[53] and others.

According to the Sabatier-principle the correlation, the catalytic activity-chemisorption bond strength should always reveal a maximum at the optimal bond strength. The difficulty is that the bond strength is not known and even the calorimetric data are not a sufficient substitute for it, because the extent of dissociation of the bonds in molecules and in the surface is not known. It does not supply us with the necessary information. However, an empirical analysis of data available showed that the bond strength in various series of bulk compounds as carbonates, formiates, oxalates, chlorides, oxides, etc. are intercorrelated, so the bond strength of chemisorption (surface complexes) should be as well. Therefore, Tanaka and Tamaru[54] suggested to plot the catalytic activity against the heat of formation of the highest oxides of a given metal per one metal atom, ΔH_f^o. The heat of adsorption at zero coverage is for several gases, indeed, a linear function of this parameter, as shown in Fig.11. The catalytic activity in various reactions plotted against the correlation parameter ΔH_f^o reveals the expected maxima[14], as can be seen in Fig.12. The explanation of the maximum, in Fig.12, is as follows[55]. The synbatic part of the correlation (left) means that at too low bond strengths the surface coverage is too low or the activation of the adsorbed molecules is insufficient (the activation energy of the catalytic reaction is then too high). The antibatic part of the correlation (right) means that the surface at high bond strengths is blocked (desactivated) to a too high extent by strongly adsorbed side products, or the activation energy of the steps following the adsorption step in the catalytic process is too high because upon adsorption the potential energy of the essential intermediates drops to a too low value. Such correlations can be rationalized by a simple theory, as e.g. Tanaka and Tamaru[55] showed (quantitative theory of Balandin and of other authors may be criticized on several important points[56]).

The correlation of Tanaka and Tamaru[54] for the heat of adsorption (the chemisorption bond strength is reflected by it) shows that a metal which adsorbs a certain gas more strongly than another metal, adsorbs all gases more strongly. If there were no complicating factors, if the catalytic reaction proceeded as an isolated process whithout any side-processes, the order in the catalytic activity of various metals should be the same for all reactions. To some extent this is true for a number of relatively simple reactions (see Fig.12) but generally it is not true. Let us mention some examples.

In a number of hydrogenation reactions, Pd belongs to the most active metals[57], however, for the hydrogenation of $>C = O$ bonds or for dehydrogenation of hydrocarbons (this reaction always proceeds for thermodynamic reasons at higher temperature than hydrogenation) Pd is of a surprisingly low activity. Another example are the reactions of the Fischer-Tropsch type[57]. For many simple hydrogenations Co, Ni, Pd and Pt have a comparable activity and usually the order of activity is Pd > Pt > Ni, Co ; and Fe is usually much less active than these metals. However, for reactions in $CO + H_2$ mixtures Pd and Pt are practically inactive, Ni yields CH_4 and Fe and Co give linear hydrocarbons or their derivatives. Evidently, in complicated systems of reactions (even the most simple reactions are complicated by such side-reactions as self-poisoning and modification of the catalyst by the reaction mixture), the selectivity of the catalyst is not only determining the product patterns but also the activity in the reactions monitored. We will see in the next paragraph.

4.2. Selectivity of metals

4.2.1. Virtually isolated reactions

Even the most simple reaction like hydrogenation of ethylene which gives only one product appeared to be a complicated net of several reactions. At low temperatures there is always some conversion of ethylene into higher hydrocarbons (this process is visible by IR spectra[58]) which remain anchored on the surface and desorb only at higher temperatures and hydrogen pressures. Further, at any temperature, there is always a parallel running dehydrogenation to acetylene and carbon residues which both remain bound to the metal surface desactivating it by blocking the active sites. Such dehydrogenation on the surface proceeds even in the presence of a surplus of hydrogen and at temperatures when in the gas phase the saturated hydrocarbons are thermodynamically more stable than olefines and acetylenes. It was Beeck[47] who stated that those metals which form strong metal / hydrogen and metal / carbon bonds dehydrogenated ethylene at its strongest and were to a higher extent desactivated than the metals with weak adsorption bonds. Beeck found that hydrogenation was governed by the same kinetic equation for all metals and all metals (with a possible exception of W) revealed the same activation energy. Therefore, his conclusion was that the activity of a given metal is actually determined by the area of the working non-desactivated surface. The antibatic relation activity-chemisorption bond strength found experimentally, complies well with these ideas.

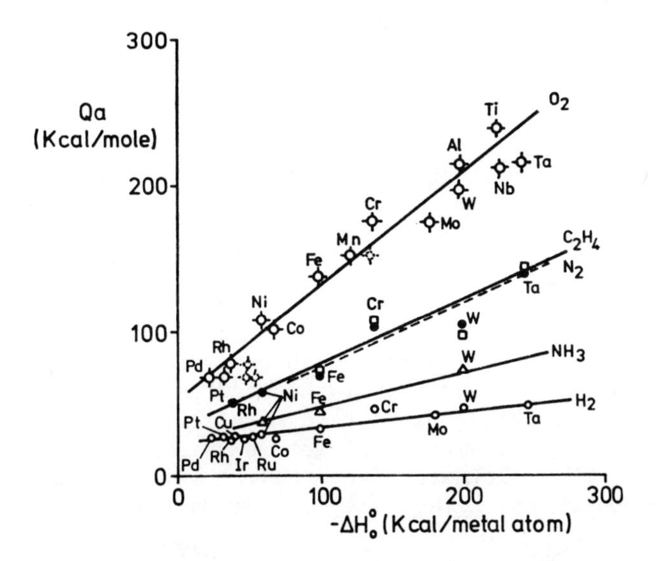

Fig.11 : Correlation of the heat of adsorption at zero coverage with parameter ΔH_f^o (heat of formation of the highest oxide of a given metal, per metal atom). According to Tanaka and Tamaru[54].

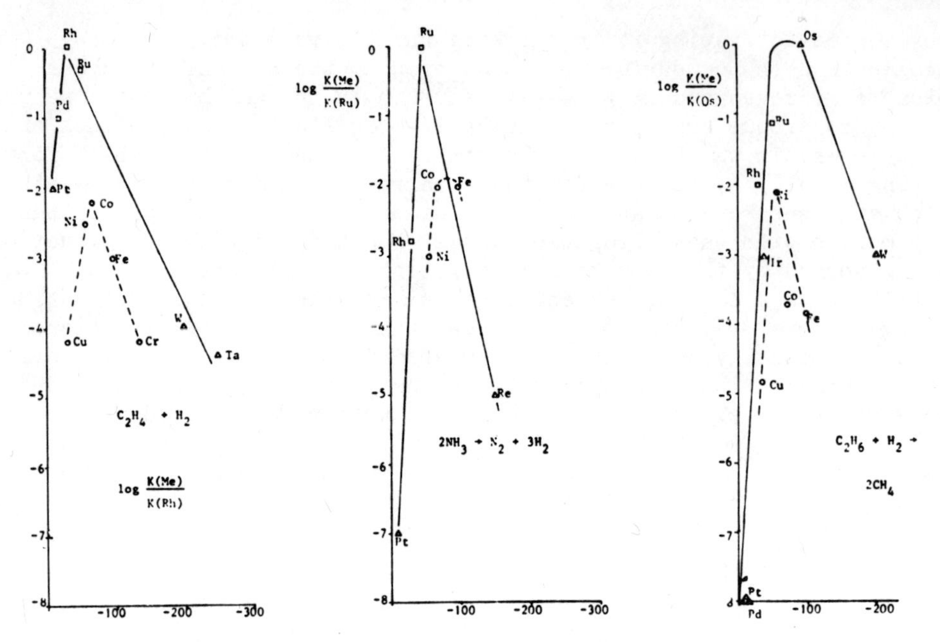

Fig.12 : Catalytic activity in three reactions as a function of the correlation parameter ΔH_f^o (Tanaka, Tamaru[54]). Activity and ΔH_f^o data from various sources (see Ref. 55).

By monitoring the electric resistance of metal films during the ethylene preadsorption, the ethylene plus hydrogen reaction and during the reaction of the unreacted hydrogen from the gas mixture with the adsorbed residues, the authors of reference 59 were able to establish that the surface of the most active metals (Rh, Pt) was desactivated to a lesser extent and more reversibly by firmly adsorbed residues than the surface of metals less active in hydrogenation (as e.g. Fe or Mo). It means that the resulting <u>activity</u> of a metal <u>was determined by</u> its <u>selectivity</u> in the following two processes :

$$(C_2H_4)_{gas} \rightarrow (C_2H_3)_{ads} \xrightarrow[1]{+H_2} (C_2H_6)_{gas} \quad \text{(hydrogenation)}$$

$$(C_2H_3)_{ads} \xrightarrow{2} (C_2H_2)_{ads} \quad \text{(surface dehydrogenation)}$$

Such a selectivity effect hidden at first sight is no exception, we will see one more example when talking about alloys.

4.2.2. Selectivity in two consecutive or parallel hydrogenation reactions

A typical example for this group of reactions is hydrogenation of acetylene(s) or di-enes, as butadiene.

$$C_2H_2 + H_2 \xrightarrow{1} C_2H_4$$

$$C_2H_4 + H_2 \xrightarrow{2} C_2H_6$$

$$H_2C = CH - CH = CH_2 + H_2 \xrightarrow{1} CH_2 = CH - CH_2 - CH_3 \xrightarrow{\text{isomers}}$$

$$CH_3 - CH = CHCH_3 \quad \text{(cis, trans)}$$

$$\text{(butene 1 or butene 2)} + H_2 \xrightarrow{2} C_4H_{10}$$

The high selectivity S defined as

$$S = \frac{(\text{rate})_1}{(\text{rate})_1 + (\text{rate})_2} \qquad (11)$$

is often required and empirically also achieved. The reactions just mentioned are always accompanied by processes mentioned already with ethylene : dehydrogenation on the surface (desactivation) and oligomerization to higher hydrocarbons.

The above-mentioned reactions were intensively studied on English schools (Sheridan, Bond, Wells and others) and several excellent reviews on this subject are available[57,60,61]. The authors came to the conclusion that two factors are determining the overall selectivity of metals : the ratio of the heats of adsorption of ethylene and acetylene (the so-called thermodynamic selectivity) and the ratio of reactivities of both molecules (mechanistic selectivity). Wells summarizes the results as follows : the high selectivity of Ni, Co, Fe and Cu is due to the low reactivity of ethylene on these metals. With these metals indeed the curve conversion (α) v.s. time has the form of 1 in Fig.13c, i.e. after acetylene has reacted, ethylene reacts further, but more slowly than acetylene . On Pd or Pt the situation is different. Here, the $\alpha(t)$ curve is like curve 2 in Fig.13c, because ethylene reacts faster than acetylene . However, on these metals, evidently, the ratio of adsorption heats is favourite for a high selectivity ; the heat of acetylene adsorption is higher so that ethylene is being continuously displaced from the adsorbed layer and it cannot react further to ethane. Therefore, the high selectivity of Pd is due to a high thermodynamic selectivity. Wells has made the following prediction : where the high thermodynamic factor is operating, the selectivity must be - in a wide range - independent of the conversion of acetylene (even the smallest amount of acetylene present is enough to prevent ethylene adsorption and to displace from the surface the ethylene already formed). When the reactivities are important, the increasing conversion leads to a lower selectivity. Fig.13b illustrates this point, the two groups of metals can be distinguished, indeed.

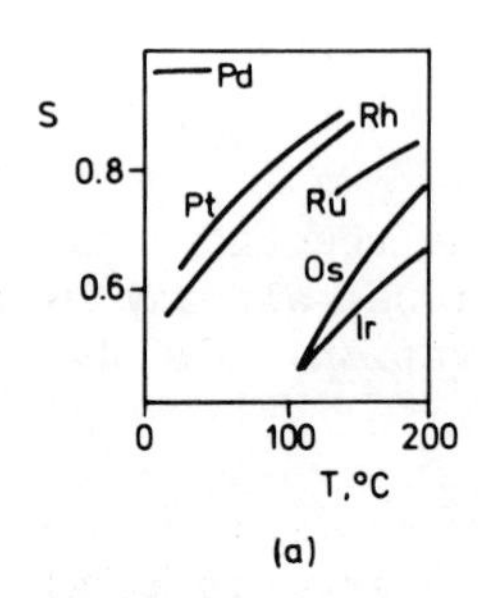

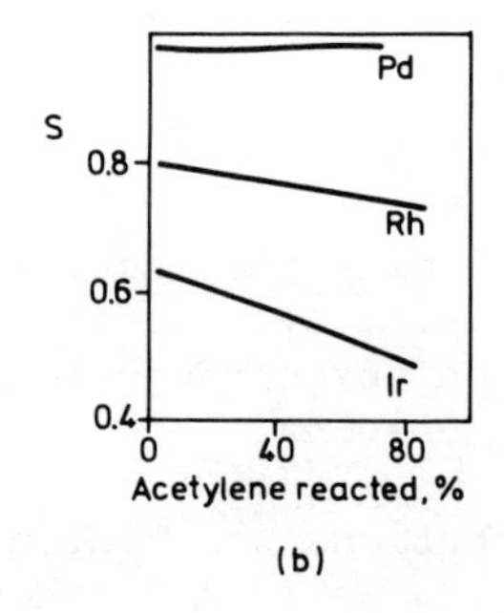

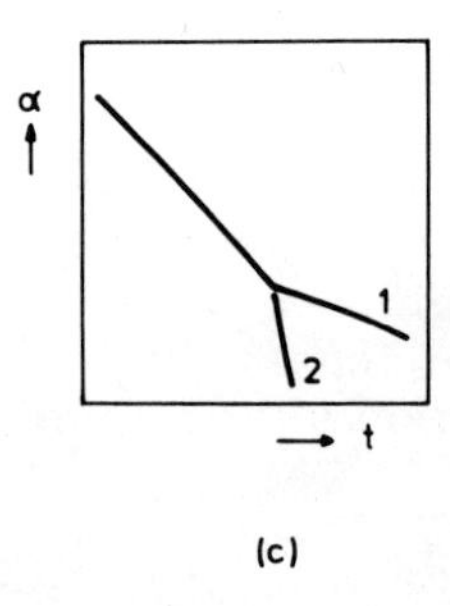

Fig.13 : Selectivity for ethylene formation upon acetylene hydrogenation[60,61].

a) as a function of temperature, for various metals; static measurements in a closed apparatus, initial pressure p (acet) = 50 Torr, $p(H_2)$ = 200 Torr, conversion $\alpha = 0.1$

b) the dependence of selectivity upon the percentage of acetylene removed by reaction; reaction conditions as in a)

c) conversion α as a function of time t, schematically. At the breaking-point most (or all) of the acetylene has reacted and hydrogenation to ethane starts to prevail.

Notice : two types (1,2) of the relation of the states (slopes) for acetylene and ethylene are possible.

The next step should be to explain why the selectivity is controlled by the reactivity on Fe, Co, Ni and perhaps also on Ir and Os. A definite answer is not yet available. Let us keep in mind the fact that the activity in ethylene hydrogenation is low with Fe, Co and Ni because a too large part of the surface is poisoned here by carbon residues. It is known from other experiments with poisoning (Hg, CO) that acetylene is less sensitive for it than ethylene. This poisoning by carbon residues can function sterically by preventing ethane, being slightly bulkier than ethylene, to be formed. It all shows that also the geometry and size of the active sites (lattice constants) might be important for selectivity, in this type of reactions.

Figure 14 shows that similar regularities are also found for the hydrogenation of 1,3-butadiene and, of course, the suggested explanation is also the same.

Interesting and unsolved, in general, selectivity problems do exist in hydrogenation reactions of molecules which have two different double-bonds, like unsaturated or aromatic aldehydes and ketones. It is known that on the most active hydrogenation catalysts like Rh and Ru, hydrogen attacks both groups of unsaturated molecules, that is, the C = C and C = O groups, with a comparable rate ; Ni is usually much more active for hydrogenation of the C = C group. The same holds for Pt and Pd but the properties of these catalysts can be modified to an important extent by adding promotors like Fe, etc.
This research work has been done mostly for preparative reasons and next to nothing is known about the physical background of the selectivity : is the selectivity dictated by preferential adsorption of either C = C or C = O groups or different reactivities of various adsorbed groups, etc. ? No definite answer has been given yet.

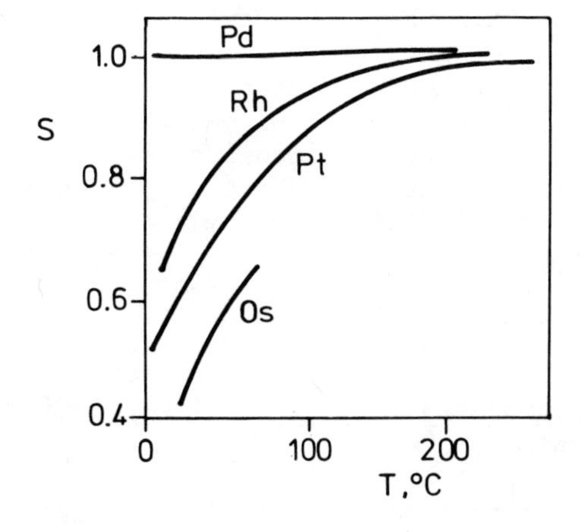

Figure 14

Selectivity for butadiene-1,3 hydrogenation to butenes as a function of temperature, for various metals. Initial pressures : butadiene 50 Torr, hydrogen 100 Torr; conversion α = 0.1(60,61)

4.2.3. Selectivity in Exchange reactions

When a mixture of hydrocarbon and a surplus of deuterium is brought into contact with a metal, exchange of protium in hydrocarbons occurs for deuterium. If we look at the distribution of various products with one, two or more deuterium atoms (D_1,D_2,D_3..... products) as it comes from the reaction at very low conversion, we will get a picture of the adsorption complexes present in the system. This approach pioneered by Kemball[62], Burwell[63] and Bond[64] led to the identification of various types of such hydrocarbon complexes and it appeared that metals are selective in their formation.

The situation is not always perfectly clear but some regularities have already emerged from the empirical data[62-65]. The most strongly binding metals like W and Mo prefer to stop the exchange during one sojourn of the molecule on the surface at an early stage, at the D_1-product. In contrast to it, metals like Pd, Pt or Rh, which are weakly binding and which are able to form special weak bonds, such as π-bonds promote the multiple exchange of most hydrocarbons up to the highest possible extent. Metals like Ni have a position between these two extremes. Experiments with properly chosen molecules disclosed that e.g. Ni favours species like (2) and (3) from Fig.15, while Pd is very poor in formation of (2) but it forms readily one or all species sub (3) - (6). As has been mentioned, W and Mo stop the exchange at (1).

Interesting regularities were observed upon exchange with molecules like toluene[66]

$$C_6H_5-CH_3$$

It has been found that almost all metals prefer the exchange on the side chain group but some ring exchange always occurs as well, when catalysts with clean surfaces are used. However, when the surface is poisoned (by CO, O or Hg) or when D_2O is used instead of D_2 (D_2O oxidizes metals like Ni or Fe) the exchange remains restricted to the side chain only. This indicated that the selectivity in the side chain/ring exchange is, at least partially, determined by sterical conditions on the surface : the small CH_2- or CH_3- groups of the side chains of alkyl-benzenes[68] are less influenced by blocking the surface than the bulky aromatic ring.

4.2.4. Selectivity in destructive and non-destructive reactions of saturated hydrocarbons

It was believed for a long time that metals can only catalyze the destructive reactions like hydrogenolysis while the non-destructive rearrangements like isomerisation or dehydrocyclisation were ascribed to the carrier (acid surface) used for these metals. However, it appeared that also pure metals like Pt, Pd or Ir and Au can isomerize without any carrier. The problem of selectivity in these reactions has been studied by various authors[69-74]. As an example, the following tables show the results of three gases : pentane, hexane and heptane.

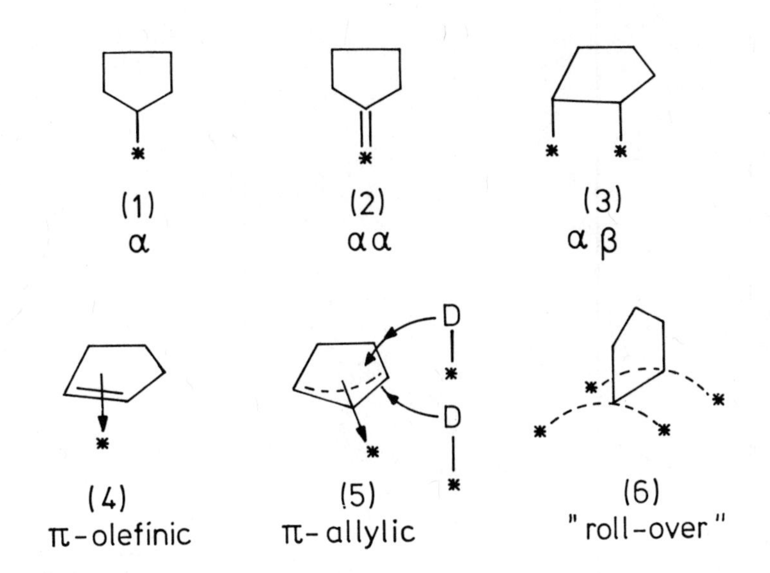

Fig.15 : Various species, suggested in the literature to explain the data in the exchange reaction with D_2 of cyclopentane (see text).

The first of these Tables (Table 3) shows already many of the important features of the respective metals. Pt differs in that it reveals also non-destructive reactions like dehydrocyclisation and isomerisation and the cracking proceeds on it also differently. While Ni breaks (cracks) the hydrocarbons from the end of the molecule (terminal splitting) Pt breaks most readily the bands in the middle of the molecule. The same trend can also be seen with hexane and heptane : Pt (and to a smaller extent Pd) is the metal which always catalyses isomerisation and dehydrocyclisation (see for details Tables 4 and 5 where S_{cycl} stands for dehydrocyclisation selectivity,i.e. the formation of all cyclic products like benzene,

TABLE 3 : Product distribution in n-pentane reactions with H_2

catalyst	T°C	p_{H_2} atm.	p_{pent}	% , molar C_1	C_2	C_3	C_4	iso-C_5	c-C_5	% S_{iso}	S_{cycl}
Pt/SiO_2 (69)	312	0.9	0.1	5	17	15	3	52	6	67	8
(16 wt%)	346	0.9	0.1	6	20	18	4	43	7	59	9
Ni/SiO_2 (70)	350	2.5	0.5	85.9	6.6	4.7	8.1	0	0	0	0
(9 wt%)	350	5.0	0.5	77.0	8.5	8.3	5.6	0	0	0	0
	350	5.0	2.0	52.0	0.5	3.0	4.4	0	0	0	0

iso-C_5 = isopentane

c-C_5 = cyclopentane

$S_{iso} + S_{cycl} + S_{cracking} = 1$ (cracking = hydrogenolysis)

Experiments performed in an open flow apparatus under the conditions indicated, by respective authors.

TABLE 4 : Selectivity in the n-hexane reactions with H_2 (Ref. 71)

catalyst	T°C	% S_{iso}	S_{cycl}	S_{crack}	cracking
Pd	405	17.5	42.7	39.8	terminal
Co	233	0	0	100	multiple
Fe	250	0	0	100	multiple
Pt	295	39.1	45.0	15.9	in the middle
Ni	250	0	0	100	multiple (terminal)

catalysts : supported metals, 0.8 - 5%

apparatus : pulse-reactor, at atm. pressure of H_2

multiple splitting : hydrocarbon is split off into C_1-pieces (CH_4) before it leaves the catalyst surface

terminal splitting : always one C_1-fragment is split off during one adsorption-sojourn on the surface

TABLE 5 : Selectivity in n-heptane reactions with H_2 (Ref.72)

catalyst	T^o_C	S_{crack}	S_{iso}	S_{cycl}
Pd	300	90.5	6.2	3.1
Pt	275	37.4	46.8	15.8
Rh	113	93.0	7.0	0
Ru	88	92.5	7.5	0
Ir	125	87	13	0

All catalysts : pure metal powders
Experiments performed in a flow apparatus under comparable conditions for all metals; hydrogen/heptane mole ratio 5:1

TABLE 6 : Selectivity in n-hexane reactions with H_2 on evaporated metal films (Ref. 73)

catalyst	T^oC	%	
		S_{iso}	S_{cycl}
Pt ultra-thin film 0.32 μg/cm^{-2}	273	13.5	76.5
Pt thick film 0.1 mg/cm^{-2}	273	37.2	27.3
Ni ultra-thin film 0.2 μg/cm^{-2}	273	4.1	6.6
Ni thick film 0.1 mg/cm^2	273	0.1	0.4

Experiments performed in a static apparatus at about 60 Torr; hydrogen/hexane ratio 10 : 1

$S_{iso} + S_{cycl} + S_{crack} = 1$

TABLE 7 : Selectivity to partial oxidation of ethylene (Ref.81)

catalyst	acetaldehyde CH_3CHO % products	acetic acid CH_3COOH % products
Ru	< 0.1	< 0.5
Rh	$\leqslant 1$	$\leqslant 1$
Pd	$\leqslant 3$	15 - 25
Ir	< 0.1	8 - 13
Pt	0.2	0.8

Catalysts : supported 5 wt% metal/silica
Apparatus : open flow, tubular reactor

toluene, methylcyclopentane, etc.).

The reaction of saturated hydrocarbons with H_2 are typical structure-sensitive, demanding reactions. The change in the metal particle size often changes the product distribution markedly. There are already numerous examples of it in the literature. Table 6 shows one of them as an illustration. Diminishing the crystal size favours the non-destructive reactions, mainly the dehydrocyclisation. According to other authors(74) it also favours the isomerisation via cyclic intermediates formed by dehydrocyclisation on the surface.

The question arises which factors determine the selectivity of particular metals in alkane reactions. Several authors have shown by making the analysis of the adsorbed layer, by measuring the adsorption(75,76) or following catalytic reactions(62) that those metals which form stable carbides and which adsorb various gases strongly like W, Mo, Fe and to a great extent Ni as well, dissociate more of the C-H bonds of the adsorbed hydrocarbons than the weakly binding metals (for weakly and strongly binding metals, see the correlation of Tanaka and Tamaru, under 4.1 of this paper) like Pd, Pt and Ir. The last mentioned metals are those which can also isomerize and dehydrocyclize. The conditions just mentioned are probably interconnected. In order to crack a hydrocarbon molecule, the catalyst must first dehydrogenate it and only then it can break the C-C bonds which became accessible after dehydrogenation, and only if the metal-carbon bond is sufficiently strong the C-C bond will be broken, indeed. Strong metal-carbon bonds imply a strong metal-hydrogen bond, i.e. also deeper dissociation. Carter et al.(72) pointed out that those metals which can isomerize are those which can easily change their valency to form various multiple bonds, etc. This could be another factor determining the selectivity in these reactions.

4.2.5. Selectivity in oxydative reactions on metals

Scarce data available on this subject are collected by several review papers(77-80). The most relevant data are, however, from the recent paper by Cant and Hall(81) who continued the work of Kemball and Patterson(82).

According to references 81 and 82, it is mainly Pd which is able to oxidize ethylene or propylene with some selectivity for other products than products of deep oxidation (CO, CO_2, H_2O) (see Tables 7 and 8). It should be noticed that most of the other transition metals (Ni, Co, Fe, Mo, W, etc.) are oxidized to respective oxides under conditions of the reaction and from the metals in Tables 7 and 8 probably Ru is the most strongly oxidized.

Experiments have shown,that selective and deep oxidation are two separate reaction routes :

$$C_2H_4 + O_2 \xrightarrow{(i)} CH_3CHO,\ CH_3COOH \qquad (i)$$

$$C_2H_4 + O_2 \xrightarrow{(ii)} CO,\ CO_2,\ H_2O \qquad (ii)$$

and not a consecutive oxidation of products of (i). There are several indications (the role of the oxygen pressure, the effect of alloying - see below) that the decisive moment is the dissociative adsorption of hydrocarbons, when the molecule is deeply dehydrogenated it is then deeply oxidized to CO, CO_2 and H_2O.

Quite specific is the position of Ag. Silver does not adsorb hydrocarbon without preadsorption of oxygen. When a molecule like ethylene is then adsorbed in the oxygen layer the molecular oxygen species leads to the formation of ethylene oxide (epoxidation) while the atomic oxygen acts non-selectively :

$CH_2{=}CH_2$ / O / ----+---- / O / Ag   or   O---CH_2 / O ‖ CH_2 / Ag   ⟶ $CH_2 - CH_2$ / O

This has been proved by using isotopes and infrared spectroscopy[83]. When the molecule contains a very weakly bound hydrogen like allylic hydrogen of propylene, dehydrogenative adsorption proceeds readily and propylene is oxidized by Ag up to CO and CO_2.

TABLE 8 : Selectivity to partial oxidation of propylene on metals (Ref. 81)

catalyst	CH_3CHO	CH_3COOH	CH_3CH_2CHO	$(CH_3)_2CO$	C_3 acids
Ru	3 - 9	6 - 10	5 - 14	2 - 7	0.5
Rh	0.5 - 2.0	2 - 5	10 - 25	6 - 9	0.5
Pd	< 0.2	1.5 - 3.0	1.0 - 3.0	2.5 - 4.0	0.5
Ir	< 0.2	28 - 32	0.5 - 1.0	3 - 4	2 - 3
Pt	< 0.1	<0.4 - 0.9	0.1	0.2 - 0.7	<0.5

catalysts : supported 5 wt% metal/silica
apparatus : open flow, tubular reactor

4.2.6. Concluding remarks

Only several examples were chosen to illustrate the selectivity of metals, some more data are available in the literature and still more has to be done in future. However, some interesting points are clear already now. The selectivity of a metal can be governed by one or more of the following factors :

1) The ratio of chemisorption bond strengths of various adsorption complexes. This is probably the most important factor with consecutive reactions.

2) The ratio of reactivities of various complexes, reactivities being probably correlated with the bond strengths.

3) The sterical or geometrical conditions on the surface. These conditions can limit the extent of dehydrogenation (dissociation, in general) of adsorption species or vice-versa - hydrogenation of intermediates, etc.

4) The state (e.g. the dehydrogenation degree) of intermediates which is influenced by both the geometrical conditions on the surface and the bond strengths.

Finally, let me say that many aspects of the selective working of metals became understandable only after the respective reactions on alloys had been studied.

5. SURFACE COMPOSITION OF ALLOYS; RECENT ADVANCES IN THEORY AND EXPERIMENTAL METHODS

5.1. Theory of the surface composition

5.1.1. Random solutions of alloys

The growing amount of experimental data on the adsorptive, catalytic, corrosive and mechanical behaviour (segregation on the crystal boundaries influences the properties of materials) of alloys has revived the interest in the theoretical predictions of the surface composition. Actually, for many years already is the thermodynamic[84], statistic mechanical[85] or kinetic derivation[86] of the necessary equations available for liquid solutions. Because there is no obstacle in applying it to the solid solutions, let us consider here the solid solutions A and B in the framework of a kinetic quasichemical model[86].

We assume that the change in surface composition, the enrichment by one component is the result of a quasichemical equilibration process of exchanges between the surface (index S) and the bulk. When there are ν_A moles of component A and ν_B moles of component B per unit volume in the surface layers (volume of one or several upmost layers which behave as a well mixed liquid) a stoichiometric equation of an exchange process reads[86] :

$$A_S + \frac{\nu_A}{\nu_B} B \rightleftarrows \frac{\nu_A}{\nu_B} B_S + A \qquad (12)$$

We introduce activities "a", molar fractions X_A, X_B in the bulk and Y_A, Y_B in the surface layer(s) ; activity coefficient f and equilibrium constants "K" and we write :

$$K_A = \frac{\left(a_S^{(B)}\right)^{\nu_A/\nu_B} \cdot a^{(A)}}{\left(a^{(B)}\right)^{\nu_A/\nu_B} \cdot a_S^{(A)}} =$$

$$= \left(\frac{Y_B}{X_B}\right)^{\frac{\nu_A}{\nu_B}} \left(\frac{X_A}{Y_A}\right) \cdot \frac{\dfrac{f^{(A)}}{f^{(B)\,\nu_A/\nu_B}}}{\dfrac{f_S^{(A)}}{f_S^{(B)\,\nu_A/\nu_B}}} \qquad (13)$$

$$K_A = \left(\frac{Y_B}{X_B}\right)^{\alpha} \left(\frac{X_A}{Y_A}\right) \frac{K_{vol}}{K_S} \qquad (14)$$

with obvious conditions $X_A + X_B = 1$ and $Y_A + Y_B = 1$.

Assuming $\alpha = 1$ and substituting $K = \frac{K_{vol}}{K_S . K_A}$,

$$Y_A = \frac{K . X_A}{1 + (K' - 1)X_A} \qquad (15)$$

The function of $Y_A = f(X_A)$ is schematically shown in Fig.16.

A further evaluation of equation (15) is simple for the so-called regular solutions. For these solutions we assume that the exchange and the non-ideality do not disturb the ideal mixing, i.e. do not change the entropy of the system. Then $\Delta G_{exchange} = \Delta H_{exchange}$ and $\Delta G_{excess} = \Delta H_{excess}$, so that

$$K_A = \exp(-\Delta H_{exchange}/kT) \qquad (16)$$

$$K' = \exp(-\Delta H_{excess}/kT) = K_{vol}/K_S \qquad (16')$$

In order to estimate the ΔH's in (16) and (16') we have to work with models and make new assumptions. The assumption usually made is that the total energy of atomization of a solid is the sum of dissociation energies of bonds of various pairs between the nearest neighbours : ε_{AA}, ε_{BB} and ε_{AB} (for some details, see also the simple model in 5.1.2, below).

If $\Delta H_{excess} = 0$ (this is equivalent to $\varepsilon_{AB} = \frac{1}{2}(\varepsilon_{AA} + \varepsilon_{BB})$), only the outmost monolayer of crystals differs from the bulk, and $\Delta H_{exchange}$ is a fraction of the difference between the atomization heats of the metals A and B. This fraction is given by the ratio of the vertical nearest neighbours missing above the surface to the total number of neighbours in the bulk. This fraction is, of course, different for various crystallographic planes.

If $\Delta H_{excess} \neq 0$, a detailed model is necessary to calculate this parameter. Parameter Ω is defined as

$$\Omega = \varepsilon_{AB} - \frac{\varepsilon_{AA} + \varepsilon_{BB}}{2} \qquad (17)$$

and $\Delta H_{excess} = g(X_A, Y_A)\Omega$ is derived by the theory.

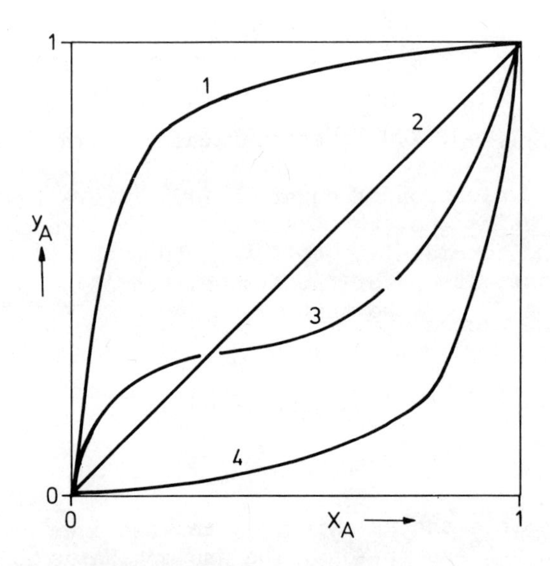

Fig.16 : Random alloys AB.
The molecular ratio Y_a of A in the surface layer as a function of molar ratio X_A in the bulk.
Several limiting cases are shown schematically.
1. $K_A \ll 1$, $K \gg 1$
2. $K = 1$
3. $K_A \sim 1$, $K = f(X_A)$
4. $K_A \gg 1$, $K \ll 1$

For the non-ideal case, the analysis cannot be restricted to the one highest layer but several layers have to be analysed theoretically. Several examples of detailed calculation are available in the literature[87-89]. Although only the most simple formulae were presented above*, let us go about general important points.

The enrichment in A occurs if $\Delta H_{exch} > 0$, i.e. when the heat of atomization of metal A is lower than that of metal B. The enrichment is dependent on the crystallographic plane (highest for high index planes) and it decreases with increasing temperature (K_A decreases with increasing T). Further, the enrichment can be either supported or counteracted by the difference in the atomic volumes (factor ν_A/ν_B).

The exact theory (to be formulated in future) must abandon the pair-wise bonding approximation and treat the problem quantum-mechanically. Then, even for the ideal solution the expression for ΔH_{exch} will be more complicated than the one used above.

5.1.2. Ordered alloys

When the temperature is lower than the critical one T_C, ordening occurs. The theory of the surface composition can be extended to this case as well. Let us analyse the surface composition of a stoichiometric compound A_kB_l, where k and l are small whole numbers.

We can again consider exchanges between the surface layer and one of the layers in the bulk and the exchanges leads to a pseudo-chemical equilibrium at which the concentration at the given place deviates by Δ from the concentration given by the stoichiometry (k/l) of the ordered compound. Formally written, the equilibrium is achieved at[87] :

$$\frac{m + \Delta}{a_1 - \Delta} = a_2 e^{E/kT} = a_2' \tag{18}$$

$$\text{or} \quad \Delta = \frac{a_2' a_1 - m}{1 + a_2'} \tag{19}$$

Parameter m is zero for an ideally ordered (long and short order, both present) compound ; $m \neq 0$ for short order. The parameters a_1 and a_2 are small numbers given by the stechiometry and crystallography of the compound and E is the energy loss or gain, related to one exchange. Δ depends on the composition (stoichiometry si expressed by the value of a_1 and a_2') in a similar way like the analogous parameters with random alloys.

---

* *Because another lecture of this School is dealing with the surface composition of alloys thermodynamically, a kinetic approach is preferred here.*

One point which is very important can be demonstrated very easily with a hypothetical linear chain of an ideally ordered alloy AB (k = l = 1), without performing any calculation[87] :

Exchange :

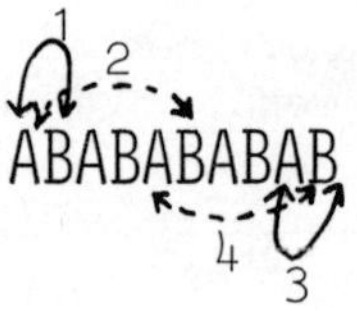

Result :

BA-ABABABAB (process 1)
ε_{AB}

BB ABA-A-ABAB (process 2)
ε_{BB} ε_{AA}

ABABABAB-BA (process 3)

ABAB-B-BABAA (process 4)

Processes 1 and 2 are favoured when ε_{AA} (dissociation energy of the AA pair bond) is higher than ε_{BB}, processes 3 and 4 occur when $\varepsilon_{AA} < \varepsilon_{BB}$. (Considering this simple scheme also helps to understand why in the random solutions $|\Delta H_{exch}| \sim |\varepsilon_{AA} - \varepsilon_{BB}|$). Further, we can easily see that when alloys are exotherm $2\varepsilon_{AB} > (\varepsilon_{AA} + \varepsilon_{BB})$, processes 1 and 3 are preferred. The endothermal alloys favour formation of bigger clusters of the same component (processes 2 and 4) but a difference in atomic volumes may cause deviation from this expected behaviour.

5.2. Experimental analysis of the surface composition of alloys

5.2.1. Electron and X-ray spectroscopy

In the literature attempts can be found to analyse the surface of alloys (or we can use data[2] therein to this end) by various spectroscopic techniques : Auger Electron exited Spectroscopy (AES), X-ray Photon exited electron emission Spectroscopy (XPS), Ultraviolet Photon exited electron emission Spectroscopy (UPS) and Micro-Probe Analysis (MPA) (electron exited X-ray emission spectroscopy, usually in a scanning performance). Various methods are, to a various extent, sensitive for the surface resp. bulk composition and they are to a various level understood theoretically. A generally oriented discussion on these methods and all problems around it would easily cover a program of a Summer School. Therefore, let us confine our discussion to one example which has been most thoroughly studied and which is also interesting from the catalytic point of view : the surface composition on Ni-Cu alloys.

First, something has to be mentioned about the bulk and its phase composition. This alloy system belongs to those reasonably well described by the regular solution approximation. The formation of alloys is endothermal and the theory predicts that at low temperatures, below the critical temperature T_C, two phases are present in equilibrium for all compositions in a certain range of concentration. Figure 17 shows the range (below the curves) of temperature and concentration where two phases, α and β coexist. The curves are drawn according to the predictions of various authors.[90-92] Our own experimental work (see below) has shown that the prediction in reference 92 is closer to the reality than other predictions.

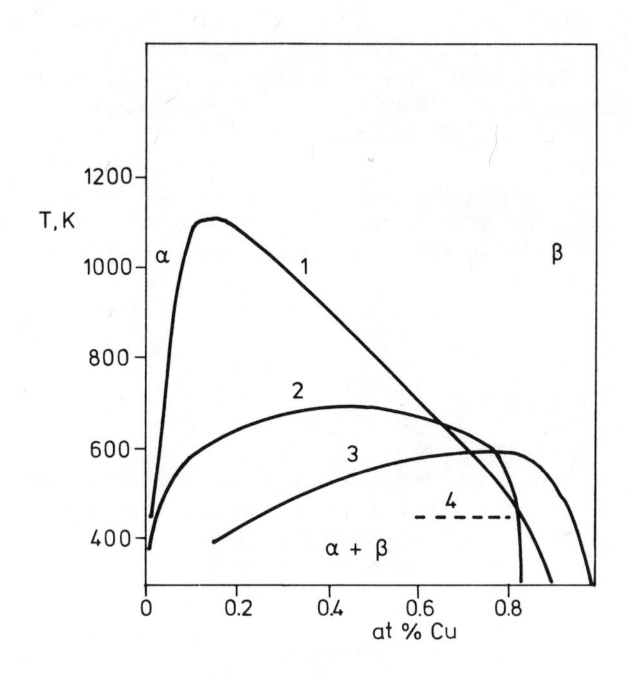

Fig.17 : Coexistence diagram for α and β phases in Ni-Cu systems. 1,2 - according to Ref.90 ; 3 - Ref.91; 4 - according to Ref.92.

What can be expected on the ground of this knowledge ? Only in systems which are brought to equilibrium at T < 180°C, two phases can be present. Probably, only with very thin films and after prolonged equilibration, there is a chance to achieve a real equilibrium at such low temperatures. (Another possibility is to induce the segregation by hydride formation and decomposition[93].) Then, two-phase alloys reveal a specific orientation of phases; the β-phase (Cu-rich) forms the surface and the α-phase the kernels of the cherry-like crystallites(90-94). (The model holds for Ni-Cu(90), Ni-Au(95) and Pt-Au(96).)

In equilibrated systems ($T > T_C$), only one phase is present in the bulk and the deviations in surface composition from the bulk composition are due to the surface (Gibbs) enrichment. Of course, the surface of the β-phase shell in two-phase systems can also deviate from the bulk composition of the β-phase or the alloy - as a whole. This point has led already to some misunderstandings in the literature. Some authors were supposing that Sachtler c.s. predicted a two-phase model for all Ni-Cu alloys without regard to the temperature, which is, of course, not true.

The Ni-Cu system has been mostly studied by AES. One point should be reminded, first : the escape depth of electrons is the parameter which determines how many layers are "seen" by AES. For Auger electrons in Cu it is about 4 Å (1-2 monolayers) at the electron energy of 90 - 100 eV and about 10 - 15 Å (5-8 monolayers) at the electron energy of 700 - 900 eV. With this in mind, many - at the first glance - contradictory results can be easily understood in a framework of one consistent picture. Namely, not the results themselves but the statements of the authors are contradictory. An important contribution to a better understanding of the apparent controversy has been supplied by Helms(97), of the laboratory of Prof. Boudart.

If the composition of the Ni-Cu alloys is tested by using the 920 eV Cu peak and the 714 eV Ni peak, the results are as follows. The composition detected by AES is equal to the bulk composition, if the alloys are cleaned mechanically or by sputtering(97-99). Only vigorous sputtering causes some depletion in Cu(100). When the sputtered samples showing a composition not very different from the bulk, were annealed at temperatures of 300 - 400°C, they showed a reversed phenomenon(97): their surface was being enriched by Cu! Similar to this, samples cleaned by an oxidation-reduction process revealed the same kind of deviations(101) of the surface composition with regard to the bulk. However, as has been mentioned, the AES peaks at the energy of about 800 eV reflect the composition of the upmost 5 - 8 layers so that only very pronounced and deep into the crystal reaching changes can be detected by monitoring these peaks. Indeed, the peaks at the energy of about 100 eV offer quite a different picture of the surface composition! Sputtered samples still show the presence of both components (although some Cu enrichment is again observable(97,102)), but samples which were annealed (500 - 600°C) after sputtering show surfaces which were very strongly enriched in Cu so that the concentration of Ni (50% in bulk) decreased under the limit of detection by AES. Also the oxidation-reduction process which can be regarded as another way of equilibration (next to sputtering plus annealing) at 300°C revealed a similar trend.

Summarizing, the samples of Ni-Cu show (at least at Ni ≈ 40 - 50 %) that after annealing (equilibration) at 300 - 600°C i.e. at a temperature not too much higher than T_C, a clear enrichment occurs by Cu, the alloy component with a lower sublimation heat.

The surface of alloys equilibrated above T_C has been also examined(106) by the electron microprobe analysis. This technique did not reveal any surface enrichment which is not very surprising because MPA analyzes a rather thick surface layer, because both high energy electrons (excitation) as well as X-rays (relaxation) have a longer penetration and escape depths. So that we can conclude : the surface of one-phase alloys (above T_C) is enriched in Cu but the deviations from the average bulk composition are limited to the upmost 2 - 3 layers of the alloy.

A related system Ni-Au reveals a similar enrichment as well, as detected by AES. On the other hand, other systems showed less tendency to segregate components at the surface, as e.g. Pd-Ni (104,122). It is for sure that much work has to be done on these problems in future.

Let us mention one more point which is relevant with regard to paragraph 5.2.2. It has been found that the work function of the sputtered Ni-Cu alloys reveals a linear relationship between the work function of the alloys and the surface composition as measured by AES high energy peaks(105). It means that in view of the results of reference 97 the work function can be put proportional to the surface composition and can be taken as a reliable measure of composition.

5.2.2. Work function measurements

First, here we will mention the definition of the work function Φ and its relation to other parameters of the metals.

The Fermi-Dirac distribution function $f(E) = [\exp(E-\eta)/kT+1]^{-1}$ which is the probability of occupation of a given level of energy E, has as a parameter of distribution the electrochemical potential $\eta_i = (\delta G/\delta n_i)_{p\,T\,n_j \neq n_i}$, with "i" for electrons. At the same time, the parameter η is the energy of the highest occupied level at T=0 K and the energy level which is occupied to 1/2 at all T > 0 K; that is the Fermi energy of the metal (this can be easily seen when the f(E) limit for T > 0 is calculated). Electrochemical potential η is also a function of e.g. the electrostatical potential (external potential) of the metal, etc. However, we want to characterize various metals and the position of their Fermi energies by a parameter which is independent of such experimental conditions as the potential applied externally to the metal. Therefore, we define the work function Φ as "the energy which is necessary to remove one electron from the crystal to the charge free vacuum at a distance of 10^{-4} cm". The distance is chosen as a distance at which we can neglect the image forces but still a distance from which the surface looks as an indefinite disk. The experience shows that this is what really corresponds to the experimental conditions at which Φ is in most cases measured. In symbols : (according to Fig.18 where the level of the zero energy and of the standard free energy are on vacuum level, corresponding with the indefinite distance of a resting electron and the metal)

$$\Phi = (-)[\eta + e\psi] = (-)[\mu - e\chi] \quad (20)$$

$$\text{where } \eta = \mu - e(\psi + \chi) \quad [\eta, \mu < 0!] \quad (21)$$

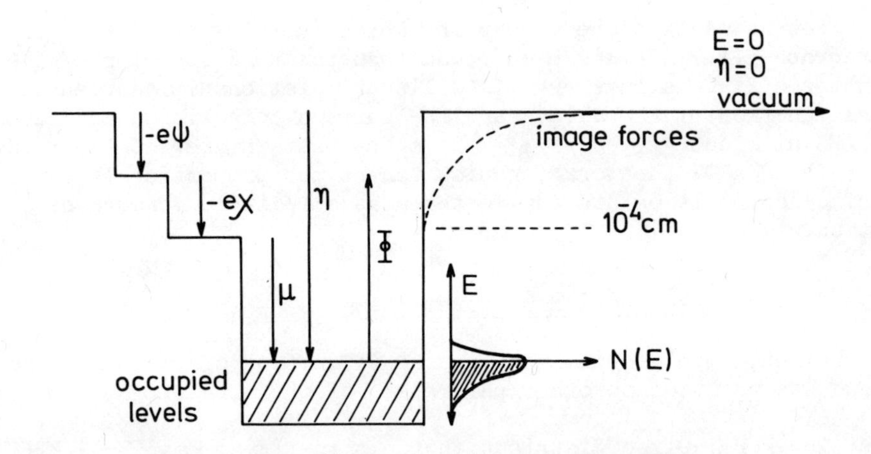

Fig.18 : Energy diagram for electrons in a metal with an electrochemical potential η (Fermi energy of the metal) $(\eta = (\frac{\delta G}{\delta n_i})_{p,T,n_j \neq n_i})$ which defines the position of the Fermi level with respect to the vacuum; with work function Φ, surface potential (surface double layer potential) χ and a positive external potential ψ. The charge of electron is (-e). N(E) is density of states.

In the equation and Fig.18 potential χ has a positive value, i.e. the surface layer is oriented in such a way that the positive charge is on the metal and the negative pole points to vacuum. This is the usual case, electrons tend to "escape" from the metal.

While per definition the parameters η and μ(μ is η for all electrical potentials zero) are the same for the metal as a whole or for several metals in equilibrium (in electrical contact), the parameter Φ may be still different for various metals in contact or for various crystallographic planes of the same metal, etc.

Work function depends, according to equations (20) and (21), on the position of the highest occupied level and on the value of the surface dipoles, which are very characteristic for a given metal or its crystallographic planes. This specificity makes it possible to use the work function measurements in order to check the surface composition of alloys. Particularly, if both components are of the same crystallographic structure (e.g., both f.c.c. metals) this is a "safe" method. However, a quantitative determination of the composition is not possible because no theory is available yet which would relate Φ to the surface composition. It is, therefore, very important that a group of Japanese authors[102] established that for the sputterd alloys (clean surfaces) Φ is a linear function of composition of Ni-Cu alloys, between two values for both pure components (note : the composition was measured by Auger spectra at high energy, but according to other data, it can be assumed that the surface layer composition did not deviate too much from the bulk one.)

In more complicated cases, like e.g. Ni-Al, when several well-defined stoeichiometric compounds are formed, the surface dipole is almost certainly no simple function of composition[108-112]. For deduction of information on the surface composition from the Φ determination the data on the crystallographic structure and considerations as in which plane of the ordered structure is the one with the lowest possible surface energy must be also employed. The measurements of Φ can be combined to advantage the measurements of $\Delta\Phi$, the change in Φ upon chemisorption of a gas which is specifically adsorbed by only one component of the alloy. An example of such an attempt to estimate the surface composition in a complicated case, is demonstrated by Fig.19. Conclusions made in this case were : at about 40 % Ni and less of the bulk Ni content, the surface is almost entirely formed by Al atoms. Indeed, the Ni_2Al_3 and $NiAl_3$ compounds have low index-planes which are from Al only. In Ni (α-phase) and Ni_3Al (β-phase) a high enrichment can be easily achieved when the surface is formed by (100) planes, with some Ni atoms missing. It is known that the Ni-Al structure can easily do without some Ni atoms and it then easily transforms into the $Ni_{1-1/3}Al$ (Ni_2Al_3) structure which has already low index-planes with Al only.

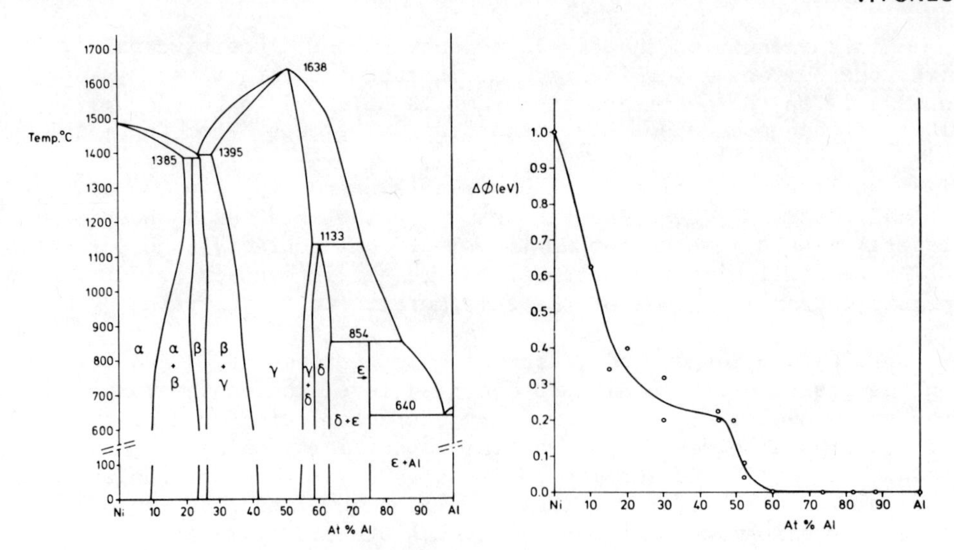

Fig.19 : Coexistence diagram for various phases in Ni-Al system. Left : ($\alpha - \varepsilon$). Right : changes of work function upon co-adsorption $\Delta\Phi$, as a function of the Al content. Alloy films ; photoemission measurements. CO is adsorbed only by Ni ; $\Delta\Phi \neq 0$ witnesses the presence of Ni in the surface. According to Ref. 107.

So far for an introduction to the problem. More about the definition of Φ, its relation to other parameters, etc. can be found in one of the reviews on this problem[110-112]. However, let us now return to the problem of the Ni-Cu alloys.

The Ni-Cu alloy was the first system with which Sachtler started his pioneering work on alloys. He attracted the attention to the crucial problem of this work : the surface composition of alloys which may - as he demonstrated by his work function measurements - deviate substantially from the bulk composition. Our recent work (P.E.C. Franken, thesis in preparation) has shown the following interesting points.

First, at temperatures above 180 - 200°C, the films studied (film thickness ± 200 Å) form one phase. At temperatures below this limit, one phase was not formed even after 40 hours of sintering. A short sintering at T = ± 200°C of a film which is originally composed from an underlayer of Cu and an upperlayer of Ni leads to a non-equilibrium state where two phases can be detected, Ni-rich and Cu-rich. Now, the interesting point is, that all three systems (one-phase films sintered at 200 - 400°C, two-phase steady-state films (T < 200°C, long sintering) and the two-phase non-homogeneous, non-equilibrium films (T ∿ 200°C, short sintering)) reveal the same Φ (surface dipole layer) as a function of composition.

The conclusion made is as follows:

i) At the moment, the measurements of Φ are more sensitive for the composition of the upmost layer than the Auger spectra ;
ii) The one-phase alloys which are equilibrated at T which is not too much higher than the critical temperature T_c, reveal a surface enrichment in Cu and their surface composition is almost the same as of the Cu-rich β-phase being an equilibrium phase at $T < T_c$;
iii) In the range 180 - 400°C the composition of the surface is almost independent of the temperature.

5.2.3. Chemisorption measurements

In the times of general believe in the rigid-band theory, it seemed virtually impossible to use a gas which would be selectively adsorbed by only one alloy component to determine the content of that particular component in the surface. According to the rigid-band theory and its original application to catalysis, we would expect that e.g. when the bulk concentration of Pd in Pd/Ag, Pt in Pt-Au or Ni in Ni-Cu drops below, say, 40 - 60 % Pd, Pt or Ni, there should be no adsorption at all of gases like CO or H_2 (Pt,Ni) even if there were Ni, Pt or Pd atoms in the surface. If, however, the electronic properties of Ni, Pd, Pt in the alloys with, say, Cu, Ag or Au were not changed substantially by alloying, selective adsorption could be a suitable method ! It were Sachtler and Van der Plank[115] who tested this idea first and the result was very satisfying. Hydrogen adsorption on Ni-Cu alloys indicated the same composition as the work function measurements !

The measurements performed, first, with evaporated (UHV) films were later extended to Ni-Cu powders[116-117] and Pt-Au films[118]. The results always showed a consistent picture : Φ measurements, ΔΦ measurements (upon adsorption of CO or another gas) and the selective chemisorption led to the same conclusions on the surface composition.

The development of the theory of the electronic structure of alloys was very fast after 1967, when Soven suggested a successful approximation (see below). It was interesting to see that both the theory as well as the experiments confirmed a posteriori the method of selective chemisorption (surface titration), originally introduced on more intuitive grounds. More about this matter in paragraph 7.

The thermal desorption as an analytic tool will be discussed in the 6th paragraph.

5.2.4. Other methods

Ion scattering

If a noble gas ion with impact energy in the order of 1 - 10 keV collides with the solid surface, then with a probability higher than 99.99 % for e.g. He^+ ions its neutralization occurs in the surface. A very small fraction of ions is rescattered by the surface as ions. If we consider an ion with an incident energy E_i and a mass M_{ion} which is scattered over an angle Θ (measured between the direction of the primary and scattered beam) by a solid state atom of mass M_{at}, the final energy E_f of a scattered ion will be :

$$E_f = \left(\frac{\cos\Theta + \{(M_{at}/M_{ion})^2 - \sin^2\Theta\}^{1/2}}{(M_{at}/M_{ion}) + 1} \right) E_i$$

for $(M_{at}/M_{ion}) \geq 1$.
Monoenergetic ions are directed on the surface and the current of scatteredions is measured as a function of energy and of the scattering angle Θ.

This very promising method, almost exclusively sensitive to the upmost layer has been applied to the alloys but recently. It is interesting to note that very recent results, mentioned at the Thin Film Conference 1975 in Warwick, showed a surface enrichment in Cu in the Ni-Cu alloys equilibrated at $T > T_C$ (120).

SIMS

The secondary ion mass spectrometry SIMS is not a method which can alone supply us with an analysis of a surface composition. However, this is - potentially - the only one method which can give us a picture of a possible clustering of components at the moment.

Closing remarks

Because the principle of Auger, XPS (ESCA) and SIMS spectroscopies are going to be dealt with by another lecture of this course, no detailed discussion on the experimental technique (including the problem of quantitative analysis) was included here.

6. ADSORPTION OF GASES BY ALLOYS AND THE THERMAL DESORPTION METHOD

6.1. Adsorption

6.1.1. Non-selective, physical adsorption

Physical adsorption due to Van der Waals forces sometimes reveals a certain sensitivity to the kind of adsorption sites, to the dipole existing on the surface of pure metals or to alloys, but by wide and large, for the purpose of analysis of the surfaces of alloy catalysts all these effects are too small, they can be actually neglected without making any large errors. Physical adsorption (its total extent) is thus better suited to determine the whole surface of alloys. Only when a gas is used which is adsorbed in a strongly polarized form and the polarization due to bonding (not only electrostatic) is sensitive to the alloy composition, some information on the surface composition can be obtained when, for example, the work function changes are monitored upon volumetric measurements of physisorption[121]. Another possibility could be a temperature programmed desorption of physisorbed gases (see below).

6.1.2. Selective chemisorption

Two conditions must be fulfilled when a firm chemisorption bond has to be formed :

(i) the orbitals of the metal on the metal atoms which should participate in the formation of the chemisorption bond must have an energy which is not very different from that of the MO's available for chemisorption on the molecules being adsorbed;
(ii) the orbital of both participants must overlap sufficiently and positively. Let us consider the case in which these conditions are met on an A-metal but not on the metal B. The question arises : (i) and (ii), change by alloying ?

The coherent potential theory and the calculation performed by it for e.g. Ni-Cu, the UPS and XPS spectra of several alloys (Ni-Cu, Pd-Ag, Cu-Ag) both showed that the shift of the energy levels of Pd or Ni in these alloys is limited. More effect should be expected for Pd than for Ni. In any case much of the original "Pd" or "Ni" character of the Pd and Ni atoms is also preserved in alloys. It is, therefore, not surprising that when a molecule is being adsorbed by the surface of such an alloy, it can always recognize its proper partner for chemisorption (from the available alloy components) quite reliably. So the chosen atoms are selectively counted by chemisorption - Ni in Ni-Cu (H_2, CO adsorption), Pd in Pd-Au, Pd-Ag (CO adsorption), Pt in Pt-Au, Pt-Cu (CO, H_2) etc. The total amount of a titrating gas adsorbed per unit surface area of alloy is a good measure for the content of the active metal in the surface of an alloy.

The method is not free from problems. If a biatomic molecule (H_2) is adsorbed and it dissociates on - say - the pure metal A using a pair of atoms A-A, how does the adsorption on a diluted alloy proceed, or does it proceed at all, where mostly -A- isolated atoms exist? If a molecule used for determination can bind more than one A atom, is it also adsorbed by diluted alloys in the same way or does the adsorption mode change?
If the chemisorption is very aggressive (corrosive), does it only count the surface atoms leaving the sublayers unchanged or does it extract atoms from the depth of the alloy to the surface?

According to the author's opinion these problems can be solved if the determination of the surface composition is performed parallel by various gases and combined with another independent method, like work function measurements, infrared spectra, etc. (Reinalda from our laboratory is, e.g., measuring the surface composition of Pt-Cu alloys, previously studied by work function measurements and hydrogen adsorption, by taking infrared spectra upon the CO adsorption.)

Still another more serious problem is, how to use the adsorption of gases in order to determine the surface composition of alloys consisting of two active components, like Pd-Ni, Pt-Pd, Pt-Ru and others. Sometimes, a suitable catalytic gas-phase reaction or a surface reaction (a reaction of adsorbed species with one gas reaction component) can be used to this end. McKee[121a] found decomposition of CH_3OH suitable to make a rough estimate of the surface composition for Pt-Ru alloys. Vlasveld[123] from our laboratory was able to detect (not to determine the concentration) both components in Ni-Pd alloys, Ni and Pd, in both limiting dilutions by using cyclopentane and deuterium exchange and its products characteristic for both components. It can be expected that also other reactions, such as reduction, oxidation in the adsorbed layer, hydrogen-deuterium exchange, can be possibly used to estimate the surface composition in some cases. There is a chance that infrared spectra of properly chosen molecules may help in this respect as well.

A method which seems to be predestinated to discriminate between the two active but in their adsorption behaviour different metals of one alloy is the thermal desorption. The following paragraph will discuss some aspects of this method.

6.2. Desorption

6.2.1. Thermal desorption - fundamentals

a - in a closed system ;
b - in a flow system (gas desorbs into a stream of an inert carrier gas like He) ;
c - in a continuously pumped system.

b and c result in the same p(t) function. Various p(t) functions are presented schematically in Fig.(20).

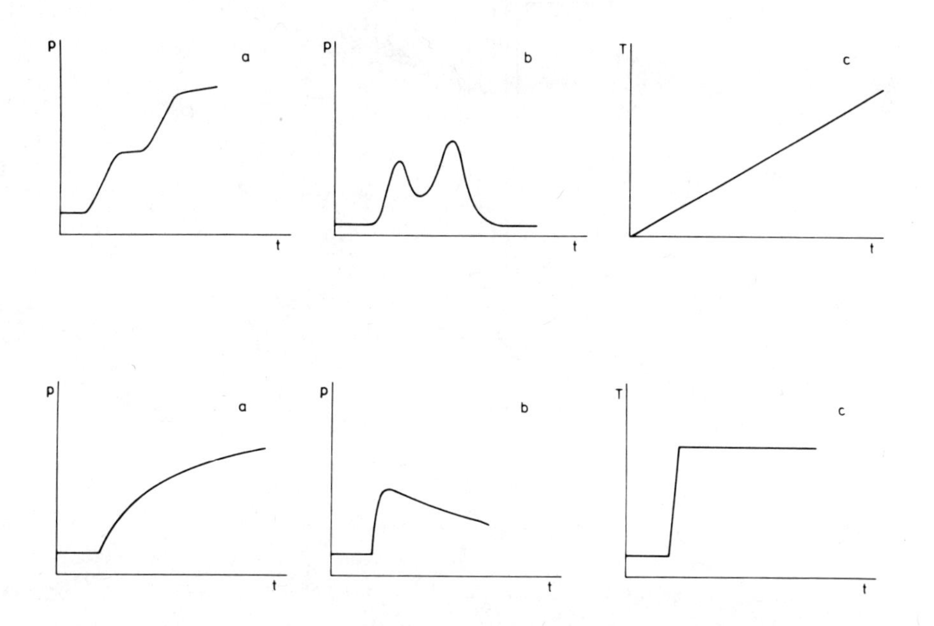

Fig.20 : above : Non-isothermal (temperature programmed) desorption (T ~ t).
below : Isothermal desorption (T = constant)
a - p(t) in a closed system
b - p(t) in a pumped system
c - temperature as a function of time t.

Thermal desorption may be performed with powders[123], films or filaments[124] (flash filament technique).
A cell in which desorption from an alloy film can be studied is shown in Fig.21. In this cell measurements were made with adsorption at pressures of 10^{-5} - 10^{-3} Torr and the thermal desorption was studied in the range of 78 - 600 K (working pressure upon desorption is 10^{-5} - 10^{-6} Torr and it is monitored via a leak by a mass spectrometer).

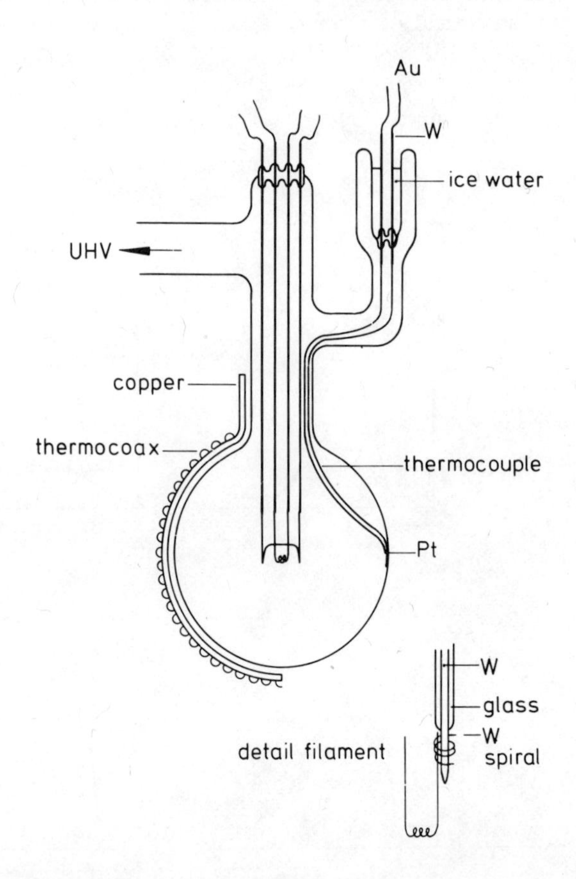

Fig.21 : Experimental vessel for adsorption/desorption experiments on evaporated metal and alloy films.

6.2.2. Basic equations for thermal desorption

Let us consider[123,124] a desorption of undissociated molecules in the model of Langmuir (homogeneous surface, no interaction of adsorbed particles, monolayer adsorption, while the molecule can be only adsorbed if it collides directly with an uncovered surface). Then, when Θ is the relative surface coverage of the surface where at maximum ${}^{m}N$ molecules can reside, the following equations holds:[123]

$$(\frac{\delta\Theta}{\delta t})_T = k_d\Theta - k_a C(1 - \Theta) = r_{des} - r_{ads} \qquad (23)$$

Here C stands for partial pressure (concentration) in the gas phase and k_d and k_a are desorption respectively adsorption rate constants. When the desorbing gas is pumped off at a rate FC which is much higher than $V\frac{dC}{dt}$ (V = volume gas phase) the following mass balance holds:

$$FC = {}^{m}Nk_d\Theta - {}^{m}Nk_a(1 - \Theta) \qquad (24)$$

When the temperature increases linearly $T = T_o + \beta t$ and when $(\frac{\delta\Theta}{\delta t})\beta^{-1} >> (\frac{\delta\Theta}{\delta T})$, we can write $(\frac{\delta\Theta}{\delta t})_T = \beta\frac{d\Theta}{dT}$. The last quotient is actually measured during the thermal desorption experiments because C (measured) $\sim \frac{d\Theta}{dT}$. By elimination in the equations above, we can see it immediately :

$$C = -{}^{m}N\frac{\beta}{F}\frac{d\Theta}{dT} \sim \frac{d\Theta}{dT} \qquad (25)$$

$$C = \frac{{}^{m}Nk_d\Theta}{F + k_d{}^{m}N(1 - \Theta)} \qquad (26)$$

Two limiting cases are obvious :

i) $F >> k_d{}^{m}N(1 - \Theta)$ and

ii) $F << k_a{}^{m}N(1 - \Theta)$

In the case i) : $C = {}^{m}N k_d\Theta F^{-1}$ (27)

and the condition for a maximum of the desorption peak is

$$\frac{dC}{dT} = 0 = [\frac{d\Theta}{dT}k_d + \frac{dk_d}{dT}\Theta] \qquad (28)$$

Because in the same limit $\frac{d\Theta}{dT} = -k_d\Theta\beta^{-1}$ which leads to the equation:

$$[-k_d\beta^{-1} + \frac{d\ln k_d}{dT}] = 0 \qquad (29)$$

In view of $k_d = k_d^o \exp(-\frac{E_d}{kT})$, the condition for a maximum of the desorption peak reads :

$$k_d = \frac{\beta E_d}{RT_m^2} \qquad (30)$$

respectively :

$$2 \log T_m - \log \beta = \frac{-E_d}{2{,}3RT_m} + \underbrace{\log \frac{E_d}{k_d^o R}}_{\alpha} \qquad (31)$$

In the other limiting case[123] ii) assuming that $E_d = -\Delta H_{ads}$, ($E_{ads} = 0$), we obtain in a similar way :

$$2 \log T_m - \log \beta = \frac{-E_d}{2{,}3RT_m} + \underbrace{\log \frac{(1-\Theta)^2\, N_m^m E_d\, k_{ads}^o}{F\, k_{des}^o\, R}}_{\alpha'} \qquad (32)$$

Interpretation is straightforward ; irrespective of the regime of desorption measurements (case i) or ii)), the temperature T_m at which the maximum occurs is simply related to the heat of adsorption respectively activation energy of desorption.

The difference in terms α and α' is not important for comparative experiments when all measurements to be compared are performed under the same regime. This is e.g. the case discussed below as an example to illustrate the power of this technique.

6.3. Some results of the adsorption/desorption measurements on alloys

Fig.22 shows several results which are relevant for the analysis of the surface composition of Pt-Au alloys. Phase diagram a shows that under the conditions used in this work two alloys α and β are coexisting in equilibrium. Work function Φ measurements revealed that the content of Pt is low for all alloys and varying only marginally (b). This is further confirmed by hydrogen adsorption c. Φ-measurements indicated that there could be some difference between the Pt-rich films (Pt > 60%) and Au-rich films (Pt < 60%). Thermal desorption confirms this conclusion (see Ref.125). The most important result is the following : the position of peak maxima is the same for Pt and for all alloys (shown for a diluted alloy). We would not expect such result on the ground of the old electronic theory of catalysis and the rigid band theory of these alloys! However, the explanation can be found in geometrical terms : Pt can bind hydrogen by ensembles of various size (1, 2, 3 or 4 atoms). By alloying, the distribution of ensembles changes according to their size and as a consequence the distribution of numbers of H atoms according to their binding energy varies in a corresponding way.

Similar results as mentioned above were obtained in our laboratory for some other system as well, e.g. CO - Ni-Cu, Pd-Ag, Pt-Au[126].

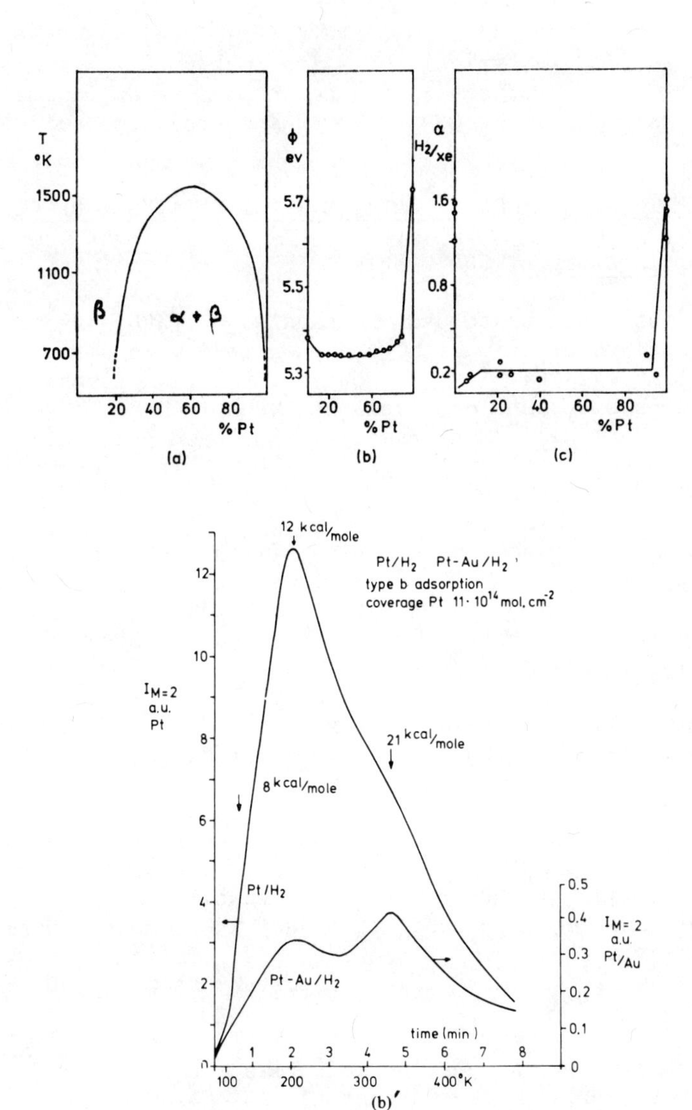

Fig. 22

Captions of Fig.22: Various pieces of information on the surface and phase composition of the Pt-Au films.

above : a - coexistence diagram of α and β phases
b - work function Φ as a function of composition[96]
c - ratio of H_2/Xe adsorption[164]

below : desorption spectra of H_2.
H_2 for pure Pt (upper curve) and Pt-Au alloys (typical for all alloys with > 40% Au); heats of adsorption corresponding approximately with peak maxima (or shoulders) are indicated.

7. THE ELECTRONIC THEORY OF CATALYSIS AND CATALYSIS BY ALLOYS

7.1. A short history of the problem

The first attempts to use the knowledge of the electronic structure of solids to explain catalytic phenomena were already done a very long time ago[42,127,128]. However, the real invasion of the solid state physics into catalysis had been launched after the first papers by Dowden[45,114] and Beeck[57] had been published. These[45] were probably the most stimulating papers published on catalysis in the whole post-war period.

In catalysis by semiconductors a similar role had been played by papers by Hauffe and Wolkenstein (see Ref. 129 for review). The whole system of ideas based on papers by Dowden, Eley, Boudart, Schwab, Beeck and others[57] is often called an "Electronic Theory of Catalysis" (ETC).

The electronic theory of catalysis was build up on the knowledge of the solid state physics and its level of development in the years 1940 - 1950. At that time it was known that transition metals deviate in their electrical resistance (after corrections for the varying lattice vibrations of various metals, the resistance of transition metals is particularly high) and the magnetic behaviour (either ferro or para-magnetic) from other non-transition metals. Catalytic people know, on the other hand, that just these metals were the most active catalysts for a number of reactions, particularly hydro/dehydrogenations. It was a small but important step to put these phenomena into relation.

For example, the saturation magnetization M_s of Ni-Cu alloys is a linear function of the Cu-content. At 60% of Cu, it decreases to zero. Mott and Jones[113] explained it by the following model for the energy band structure. Both Ni and Cu have a narrow 3d-band and an uncompletely filled broad 4s-band. The 3d-band of Ni is cut by the Fermi level so that it is uncompletely occupied (it contains d-holes) while the 3d-band of Cu is lying much lower and therefore is filled completely. When we add Cu to Ni, then according to these ideas 4s electrons of Cu fill the holes of the 3d-band of Ni without changing it (rigid bands) and the alloy so looses its ferromagnetism. Later, these ideas were applied to the ETC.

According to the ETC the reacting molecules are activated by the electron transfer, shown in the following hypothetic reaction :

$$A + B \rightarrow C$$

$$A_g \rightarrow A^+_{ads} + e\,(cat) \qquad\qquad A_g \rightarrow A^+_{ads} + e\,(cat)$$

$$A^+_{ads} + B_g \rightarrow AB^+_{ads} \qquad\qquad B_g + e\,(cat) \rightarrow B^-_{ads}$$

$$AB^+_{ads} + e\,(cat) \rightarrow C_g \qquad\qquad A^+_{ads} + B^-_{ads} \rightarrow C_g$$

Two possibilities for the mechanism are shown here, many others - in principle similar to these - may still be suggested next to it. In these schemes the essential point is : a catalyst makes the transfer of electrons easier. According to the ETC, this is the role of a catalyst.

Following these ideas simple rules for the catalytic activity of metals can be formulated :

i) an active metal has an unfilled d-band (d-holes), unpaired electrons combine most easily with electrons of the adsorbate.

ii) an active metal has to have a high density of state at the Fermi level $N(E_F)$; then, the transfer of electrons with $E \doteq E_F$ costs little energy.

Both facts can make the transfer of electrons easier during the catalytic (or electrochemical) reactions. These ideas have been first applied to reactions of hydrogen, later, other authors applied them to all reactions without difference.

Later, attempts have been made (see Ref. 57 for review) to apply in catalysis the theory of Pauling and particularly the so-called d-character was a most favourite parameter to characterize the metal in comparison with metals in catalytic reactions. However, the reasons why such correlations of the d-character with the activity should operate, were never clearly (or without errors) formulated and the worst thing was that the d-character as such was an ill-defined parameter.

However, other parameters which can really characterize the band structure of metals did not bring more success either. Various authors tried to correlate the activity with a work function[44], the mean or maximum kinetic energy of electrons in the valence band (46), the number of d-electrons (holes)[45], etc., but all this did not lead to a real progress in developing the theory.

It was expected that alloys should bring the final prove for the ETC. Bond reviewed all attempts made until 1962[57] and Boreskov expressed his views at the end of the IVth Congress on Catalysis in 1968, stating that if authors working with alloys could not even agree with the sign of the effect of alloying - a decrease or an increase in activity - alloys are evidently nothing else but "a dangerous temptation" for the scientists. However, as has been shown later, these contradictions were only apparent and neither the authors of ETC[114,129] nor their opponents[130] were right in all respects.

Before we make a step further, let us summarize the most essential conclusions and facts which were accumulated in the period since 1950.

1) An interaction gas/solid is generally not confined to a "clean" electron transfer. In the majority of cases the bands between metals and adsorbed species are covalent-like and have a highly localized character[131-134]. Some gases cause a small perturbation of surfaces (probably H_2 adsorption is an example for it) and these make use of what used to be called "free" valences on the surface. Other gases (O_2, hydrocarbons, HCOOH) extract a metal surface atom partially from the lattice (demetallization[135]). It is obvious that the sensitivity of the activity for small changes and nuances in the electronic structure of metals and alloys is high only for the first group of gases. For the aggressive gases the atomic structure of metal atoms is more important.
2) Even the simplest reactions are accompanied by a side-process ; monitoring the activity of a metal (alloy) from the main reaction, we are testing it simultaneously for the selectivity in the side-reactions.
3) We shall see below that the electronic structure of alloys is theoretically better described by the so-called Coherent Potential Approximation (CPA) than the Rigid Band Theory (RBT).

It will appear that when we take all these aspects into consideration, much of the older controversy disappears.

7.2. Some new facts and ideas on the electronic structure of alloys and the catalytic effects of alloying

7.2.1. Electronic structure of alloys

The most important and recent results are those of UPS and XPS investigations on the band structure of alloys. The physics of the photoemission of electrons is such that the distribution of photoelectrons according to their energy reflects (with only little deformation) the density of states N(E) in the metal or alloy. Fig. 23 compares schematically the predictions made by RBT (Fig.23a) and by CPA (Fig.23b).

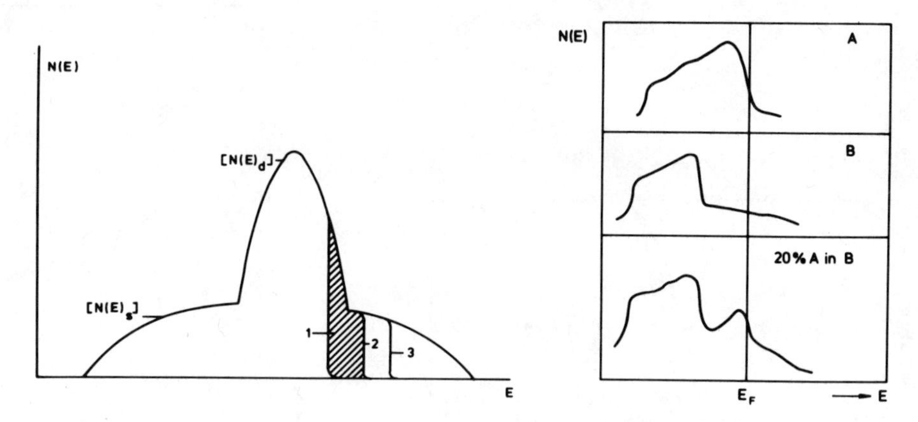

Fig. 23 : Predictions for the N(E) of alloys AB.

Left : according to rigid band theory (RBT) the form stays rigid and E_F limit shifts upon alloying. Contribution of densities with prevailing s or d character is indicated. (a)

Right: According to the minimum polarization model or coherent potential approximation (CPA), the N(E) for alloys is almost a weighted average of N(E)'s for pure components. (b)

Fig.24 shows the experimental results[136]. It is evident that RBT cannot actually be applied but CPA predicts the results well. We have already mentioned its consequence for chemisorption (see Chapter 6) and for catalysis, the conclusions are very similar : in many respects important from the point of view on catalysis ; a transition metal atom keeps its chemical character, also preserved in alloys.

The results of XPS[136] also revealed a substantial difference between the Ni-Cu and Pd-Ag alloys. In both cases the position of the d-band centre is almost unchanged by alloying. However, the width of the d-band (indicating the strength of interactions of a given atom with its environment, or - simplified said - the overlap of d-orbitals is indicated in this way) is much larger with Pd than with Ni and with the Pd-content it varies more strongly upon alloying than with Ni. As a consequence, even in diluted Ni alloys the E_F level crosses the high density of states $N(E_F)$ (originating from Ni atoms) while in diluted Pd alloys the levels coming from Pd are low-lying and they are filled completely when the d-band is narrowed by alloying (diluting Pd in Ag).

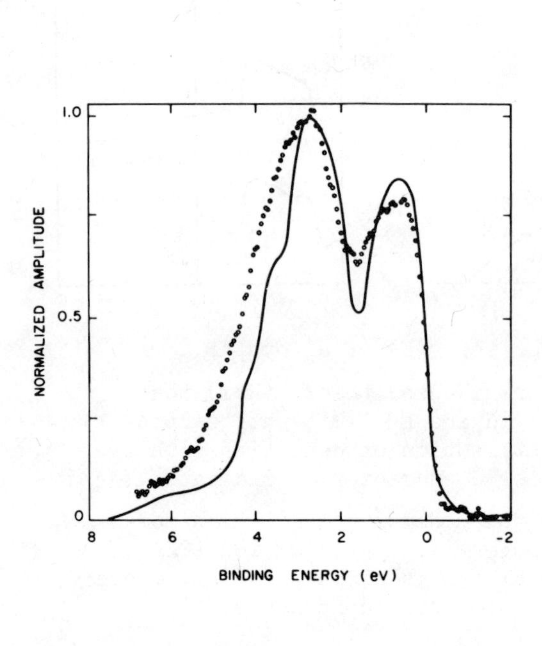

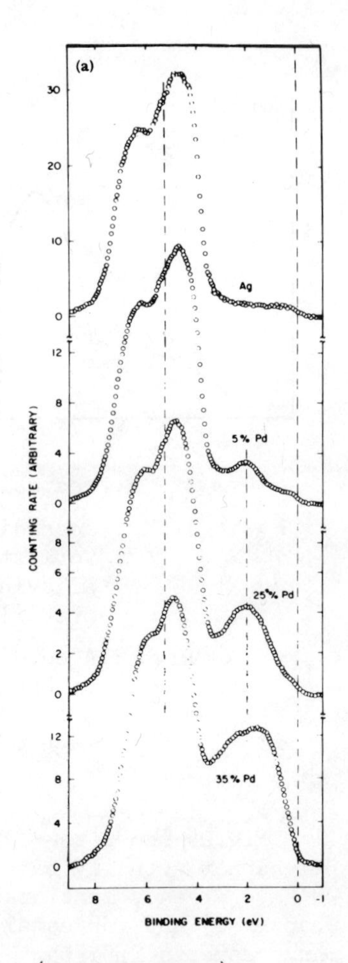

Fig.24 : Left : comparison of XPS spectra (valence band) of Cu(0.6)Ni(0.4) alloy and spectra predicted by CPA
Right: XPS spectra of various Pd-Ag alloys
Notice: at Pd concentrations of <35%, the d-band does not cross E_F (no d-holes). Data according to Ref. 136.

Because in diluted alloys Ni-N(E_F) >> Pd-N(E_F), Ni is <u>para</u>magnetic even in very diluted alloys, while Pd-Ag alloys become <u>dia</u>magnetic upon alloying! In other words, the number of d-holes on Ni atoms varies much less (the estimate is that it varies within the limits of 0.5 ± 0.1 per atom)(137) than on Pd where it is expected that the holes disappear completely or that they decrease drastically to about 0.15 per atom (from 0.4 per atom in Pd)(138). Not only in Ag but also in Pd dissolved hydrogen atoms can force Pd to redistribute its electrons between the s and d-orbitals(139).

Another important source of information were the soft X-ray emission spectra or better still, their integrated intensities. Wenger and Steineman(40) used the integrated intensities of certain special transitions as a measure of the number of electrons of a given orbital symmetry. They used for Ni the Lα band (4s,3d → 2p) to estimate the number of 3d-electrons and the L_1 band (3s → 2p) to estimate the number of Ni atoms. Similarly, for Al they used Kα (2p → 1s) and Kβ (3p → 1s) lines to estimate the number of atoms resp. valence-p-electrons.So the authors were able to establish that the number of Ni-3d-electrons in Ni-Cu alloys remains unchanged within the limits of error (±0.1 per atom). Further they found that e.g. in Ni-Al alloys both the numbers of Ni-3d- and Al-3p-electrons increase simultaneously upon alloying (shematically see Fig. 25).

Ni-Al alloys were also investigated by X-ray diffraction. An exact analysis of the structural factor revealed that in comparison with the lattice of pure metals a higher density of electrons is found in alloys in positions between atoms where the 3dNi- and 3pAl-orbitals overlap. The results mentioned above mean together that the changes upon alloying are qualitatively alike upon formation of covalent bonds. Of course, also the Ni-Al alloy has a metallic character (= delocalized bonding) but the changes mentioned before, the changes in mechanical properties, etc. nevertheless justify a simplified statement made.

The dependence of the saturation moment M_s of Ni as a function of the Cu-content in the alloy has got a new explanation : when by increasing the Ni concentration, a Ni atom gets more Ni-atoms as nearest neighbour than a certain limit (say, 8 - 9 from 12), it becomes ferromagnetic(143). The M_s (% Cu) curve can be derived on grounds of these ideas in good agreement with the experiments. It is important for catalysis that in alloys like Ni-Cu (Ni-Au, Pt-Au, etc.) clustering may occur and further, it is no more necessary to assume a d-band filling in order to explain the magnetic data. The other way round, the magnetic data do not bring any prove for the d-band filling in Ni-Cu.

The systems Ni-Cu, Pd-Au, Ni-Al are representatives for the principal kinds of effects caused by alloying. Summarizing, by alloying various situations are created :

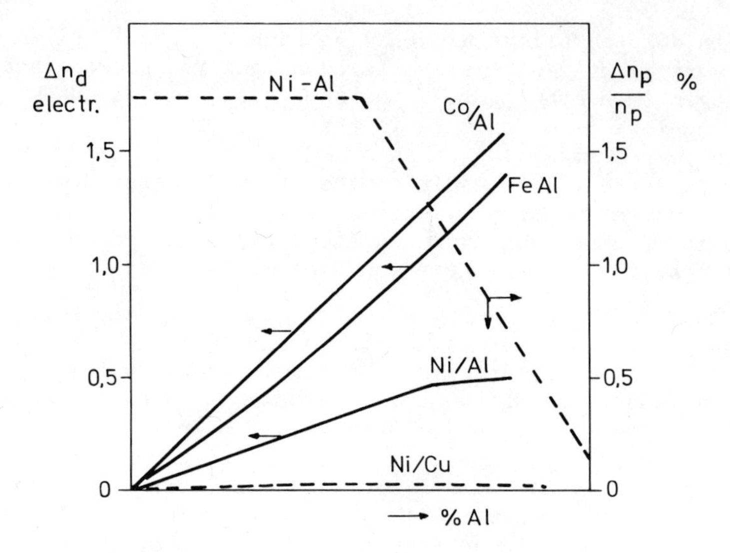

Fig. 25 : The changes in the number of electrons with d-character (Δnd) or p-character upon alloying. Various alloys. According to Ref. 137

1) The mutual perturbation of components is weak, the heat of formation $-\Delta H_f$ is negative and low ($\Delta H_f > 0$). This is a favourable situation for the formation of clusters. The electronic structure of metal atoms in such alloys differs only marginally from that in pure metals. Alloying functions mostly like diluting ; the electron transfer among components is almost negligible.

2) The mutual perturbation is weak, $\Delta H_f \leqslant 0$, no or very limited redistribution of electrons takes place. High dispersion of components (less clustering) may be expected, some transfer of electrons (see : Ag, Au) or redistribution among s and d orbitals (Pd-Ag) occurs.
Effect of diluting is here possibly combined with a change in the bonding character of chemisorption bonds ("ligand" effects, in the terminology used by Sachtler[94]).

3) The mutual perturbation is strong, $-\Delta H_f \gg 0$. Covalent-like changes in the electron structure occur in the alloys (Pt-Sn, Pt-Zr, Ni-Al and similar alloys). Dispersion of components is high, ordening is probably occurring. In the extreme case, this is a si-

tuation between that under 2) and - say - the formation of metal-chalcogenides and similar compounds. (For example, some sulphides preserve the metallic character of binding in some crystallographic directions.) At high temperatures the surface of such an alloy always contains a certain number of uncompletely coordinated atoms of both components (Schottky-defects) which can operate - essentially - as very diluted components.

7.2.2. Catalytic effects of alloying

The classical ETC tried to explain all catalytic effects of alloying by one effect in the electronic structure of alloys, viz. the change in population of the d-band. However, as we have learned above (7.2.1.) alloys display a variety of changes in the electronic structure and each of them can have a different effect on the catalytic properties of metals. It is obvious that also the catalytic effects of alloying should be discussed with more differentiation than previously. Some possible catalytic effects of alloying follow.

1) The energy of the high-lying d-levels decreases by alloying, d-levels become fully occupied, d-electrons paired. An expected effect : an increase in the activation energy of the catalytic reaction (breaking of paired spins is necessary, the energy level of electrons is lower).

2) The energy of the high-lying d-levels decreases, the chemisorption bond strength varies and the reactivity of adsorbed species varies accordingly. If the decrease (or increase) in the bond strength is not directly proportional for all reaction components and intermediates, variations in selectivity may be expected as well (population of d-orbitals does not vary).

3) Due to the change in energy and the population of the d-levels, a certain type of bonds is no more possible, such as π-olefin and π-allyl bonds, multiple bonding with the metal surface (>C=M), etc.

4) When two active metals are combined (the activity may even differ so strongly as for hydrocarbons on Ni and Cu), a multi-site adsorption of reaction components can take place partially on metal A, partially on metal B. In this way active ensembles may be formed which are present neither in A or B alone.

5) Let us now consider two effects of alloying of an "active" metal A and a relatively inactive metal B.

   a - In the alloy pairs of triplets or A are still present but the interatomic distance (the lattice constant varies upon alloying) is different now. This can lead to a change in activation energy.

   b - Atoms or ensembles of A atoms vary their mutual distance upon alloying ; if a spill-over is important this may have consequences for catalysis.

c - The most important "geometric" effect of alloying can be illustrated by the following example. Imagine two reactions running parallel on pure metal A ; one of the reactions requires only single atoms for the adsorption of reaction components (or intermediates) while the other needs doublets or triplets of atoms. When A is diluted in B which is inactive (or much less active) the first reaction is much less affected than the second.

The classical ETC neglected this effect which has been discovered and proven quite recently[144-147]. We believe that an example for such effects are a parallel running hydrogenation/hydrogenolysis reaction or deuterium/C-H exchange and hydrogenolysis, etc.

These were all direct effects of alloying on the given catalytic reaction. However, alloying may influence the reactions by affecting the side-reaction which itself can be virtually invisible because it is limited to the adsorbed layer and does not lead to a particular gas product. If any of the I - V effects affects the side-reactions importantly, activity and selectivity changes may occur with reactions which are being followed ("visible" in the gas phase). We believe that hydrogenation of benzene a.o. is a good example for this indirect effect.

In the past, a matter of much controversy was the question whether alloying of Ni with Cu does lead to a <u>decrease</u> or an <u>inc</u>rease in the activity of the catalyst. Van Barneveld[148] from our laboratory performed, therefore, the following experiments. He prepared three Ni-Cu catalysts with low concentrations of Cu (0, 5 and 10 %) and at temperatures when full miscibility can be expected (400°C). With these catalysts he observed that at low temperatures ($T < 100°C$) the order of activity per unit area was : cat (0 %) > cat (5 %) > cat (10 %). At low temperatures pure Ni revealed more signs of reversible poisoning than alloys. When the temperature was increased traces of products of side-reactions (less than 0.1 % conversion) were detected in the gas phase. At temperatures higher than the temperature of appearance of side-products, the activity was of the reverse order : cat (10 %) > cat (5 %) > cat (0 %) !! With this in mind, the authors[148] reconsidered all the older data (the necessary references are quoted in Ref. 148) in the literature and it appeared that those authors working at low temperatures, low benzene pressure, under non-steady state of the catalyst surface (i.e. with "fresh" surfaces), found a <u>dec</u>rease of activity of Ni after an admixture of Cu. On the other hand, authors who worked at higher T, higher benzene pressure and their technique (usually a tubular flow reactor) allowed them to reach a steady state, they found an <u>inc</u>rease of activity after alloying with Cu. Actually, there was no real controversy in the literature, only different authors used different conditions without realizing how important it was for the activity (actually, selectivity) of the catalysts.

With this case we learned again that there are no really simple reactions ; even when apparently there is only one gas product, the reaction may be complicated. Analyzing the effects of alloying we will always have to consider a possible role of the side-reactions, including the self-poisoning of the surface by the reaction itself.[149]

7.3. Is there any room for the role of the d-holes in catalysis ?

We have discussed (7.2.2.) how the electronic effects of alloying (1 - 3) can operate. Here, we shall concentrate our attention on one particular question : the role of the d-holes in catalysis.

The material to check this role are the alloys Ni-Cu on one hand and Pd-Au (or Ag ; the equivalency of Au and Ag is anticipated throughout the discussion) on the other. With Ni-Cu the number of d-holes varies marginally (maybe does not vary at all) ; with Pd-Au the d-holes disappear by alloying (or this number decreases drastically), which catalytic property does reflect this difference ?

The literature[57,150,151] offers some material, some additional experiments were performed in our laboratory[152]. The conclusions for the comparison of various reactions are as follows[152].

Hydrogenation of acetylene on Pd-Au, Ni-Cu (at higher T's), hydrogenation of methylacetylene on Ni-Cu, hydrogenation of cyclopropane on Pd-Au and Ni-Cu, recombination of H atoms on Pd-Au and Ni-Cu, all show the same picture : activity as a function of composition passes through a maximum. For both systems Ni-Cu and Pd-Au as well. It means that we can immediately exclude from considerations effects like "an optimum d-hole concentration" as reasons for such curves with maxima! Hydrogenation of benzene and ethylene is slightly more complicated. According to the reaction conditions the activity passes through a maximum or it decreases monotonously (see 7.2.2.). This equally holds again for Ni-Cu and Pd-Au alloys.

Hydrogenolysis of hydrocarbons is in both cases strongly inhibited by alloying the transition metal with Au resp.Cu. Again, there is no room for the role of d-holes in this effect!

With Ni-Cu as well as with Pd-Au, we observe the same sharp contrast of the influence of the Ib metal on hydrogenation reactions on one side and on hydrogenolysis on the other.

We conclude that the various phenomena mentioned above cannot be put into relation with the number and role of d-holes. However, there is one difference in the behaviour of Pd-Au and Ni-Cu which could be brought back to the effect of the varying number of d-holes. Let us turn our attention to the course of the decrease of activity with a composition at high concentrations of a Ib metal.

With Ni-Cu alloys the decrease of activity in hydrogenation/ dehydrogenation reactions sets in at 70 - 80 % Cu in Ni (see e.g. Fig. 27, below) while with Pd-Au the decrease begins already at 40 - 50 % Au. With Pd-Au the decrease is usually much sharper and continues up to virtually zero activity. This has been observed for the hydrogenation of olefines, aromatics, carbonyl groups, for recombination of H atoms, o - p H_2 conversions, etc. The difference of Pd-Au/Ni-Cu in this respect is more pronounced when reaction components are relatively weakly adsorbed gases (H_2) and the reaction temperatures are low. What could be the reason for it ?

i) In both cases the transition metal concentration in the surface decreases. The decrease is not necessarily linear with the bulk concentration and can be sharper for Pd-Au than for Ni-Cu. Eventually, Pd atoms can even completely disappear from a surface with more than - say - 50 % Au.

ii) The decrease in activity of Pd-Au alloys due to the decrease of the physical presence of Pd in the surface is strengthened by the fact that the Pd atoms have no d-holes at a concentration of Au higher than - say - 50 %.

It is thus of great importance to check the presence of Pd in the surface of diluted alloys by other than chemisorption methods. If Pd is found to be present, - this seems to be a more likely alternative - then the low activity of Pd atoms in simple hydrogenation/ dehydrogenation reactions at low temperatures must be related to the presence of d-holes. Hydrogen is evidently one of the most sensitive reaction components for this factor (Note : this has been actually foreseen in the earliest papers on ETC).

If the reaction proceeds at high temperatures, complications occur. First, d-holes can be produced by thermal excitations (this would imply 60 - 70 kcal/mol activation energy for Ib metals, but only several kcal/mol for Pd) d $\rightarrow$ s band. Further, a strong adsorption of one component may "extract" Pd from the surface so that it gets back its essentially free atom structure, which allows e.g. H_2 to interact with it, etc.

Finally, hydrogen adsorption can be assisted by the strongly adsorbed species, the so-called Twigg's reactive chemisorption may occur. In one of these ways, both components can be eventually active or Pd "recovered" from the effects of alloying.

It must not be forgotten that catalytic reactions of aggressive gases like O_2, HCOOH and similar, may always occur on both components, but with a different mechanism. This may sometimes obscure other effects.

We have seen that with regard to a possible role of d-holes, many restrictions appeared in comparison with the original ideas of the classical ETC.

8. SELECTIVITY CHANGES CAUSED BY ALLOYING, SOME EXAMPLES

8.1. Hydrogenolytic and hydrogenation reactions of hydrocarbons and some related reactions.

While hydrogenation and hydrocarbon/deuterium exchange are both related to the reactivity of the C-H bonds and proceed also at rather low temperatures (- 100°C) and with isolated catalytically active sites, such as mononuclear complexes in homogeneous reactions, hydrogenolysis usually proceeds at much higher temperatures ($T > 100°C$). To my knowledge, there is no hydrocracking (hydrogenolysis) of hydrocarbons known which has been <u>proven</u> to proceed definitely on isolated sites.

It is not difficult to suggest numerous variants of hydrogenation or exchange mechanisms on ensembles of several sites. Therefore, we will mention only the suggestion (several examples) for a possible reaction on isolated sites :

insertion reaction of olefines

$$H_2C \overset{\downarrow}{=} CH_2 \; (\text{M-H}) \longrightarrow \underset{\text{M-}}{\overset{|}{CH_2}} - CH_3 \; + \; \xrightarrow{H} (C_2H_6)_{gas}$$

$$\left[\begin{array}{c} CH_2 \\ / \quad \backslash \\ CH_2 - CH_2 \end{array} \right] \text{phys.ads gaseous} \quad + \underset{M}{\overset{H}{|}} \longrightarrow \underset{M}{|} CH_2 - CH_2 - CH_3 \; + \xrightarrow{H} (C_3H_8)_{gas}$$

$$(c\text{-}C_3H_6)_{gaseous} \longrightarrow \underset{M}{\overset{CH_2}{CH_2 \cdots CH_2}} \xrightarrow[ads]{H} \underset{M}{\overset{}{CH_2}} - CH_2 - CH_3 + \xrightarrow{H} (C_3H_8)_{gas}$$

exchange reactions

[H]gas + M   M — H   + D ...   M-D   [D]gas

M ⟶ M-H ⟶ M

However, when a C-C bond has to be broken, then at least 2H atoms should be split off (most probably, more than two) and simultaneously two sites next to each other have to be available to bind the fragments of molecules. We shall see that if we make the assumption that the reaction on C-H bonds requires less active sites (i.e; smaller ensembles) than the reactions on C-C bonds, many results mentioned below will be rationalized easily.

Recently a review on this subject has been published[153] so that we shall confine ourselves to the presentation of several examples and references to the original papers.

The influence of alloying on the parallel running exchange and hydrogenolysis or addition (hydrogenation) and hydrogenolysis has been studied in our laboratory. Fig. 26 shows the results of Roberti et al.[154] for methyl-cyclopentane/deuterium exchange and the side-reaction of the hydrogenolysis of cyclopentane. Fig. 27 presents results of Beelen et al.[155] for parallel running addition of hydrogen and hydrogenolysis (to methane and ethane) of cyclopropane on the same series of catalysts. We can immediately see the main effect of alloying : Cu depresses the activity in C-C bond breaking. Fig. 27, left-side, has a key position : it shows that the number of active sites in the surface of various alloys is almost the same for alloys in a broad composition range and it is lower than on pure Ni. A decreasing number of sites (dilution of sites) thus, leads to slowing down or elimination of hydrogenolytic reactions (see Fig.26-28) (on the Cu-rich side the catalysts are unstable and their behaviour reminds of that of pure Cu). On the other hand, the so-called multiplicity of exchange reactions and activity in C-H reactions are much less (and partially positively) influenced by addition of Cu to Ni. Similar results like those in Fig.27 (here, for cyclopropane) have been obtained by Sinfelt and Yates[117] for two separate reactions performed with the same series of catalysts, viz. hydrogenolysis of ethane (C-C) and dehydrogenation (C-H) of cyclohexane.

Sinfelt who contributed very much to the recent revival of interest in catalysis by alloys also performed other experiments[156]

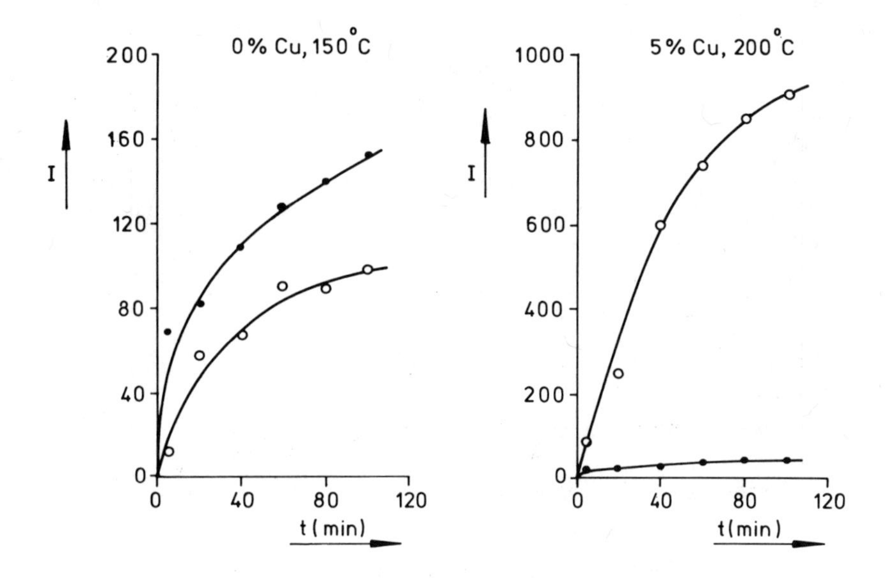

Fig. 26 : Reactions of methylcyclopentane with deuterium

(O) - C_6D_{12} is the main product of exchange reactions

(●) - CD_4 is the main product of hydrogenolysis

Concentrations of products in arbitrary but direct comparable units.

Data for Ni and 5 % Cu alloy.

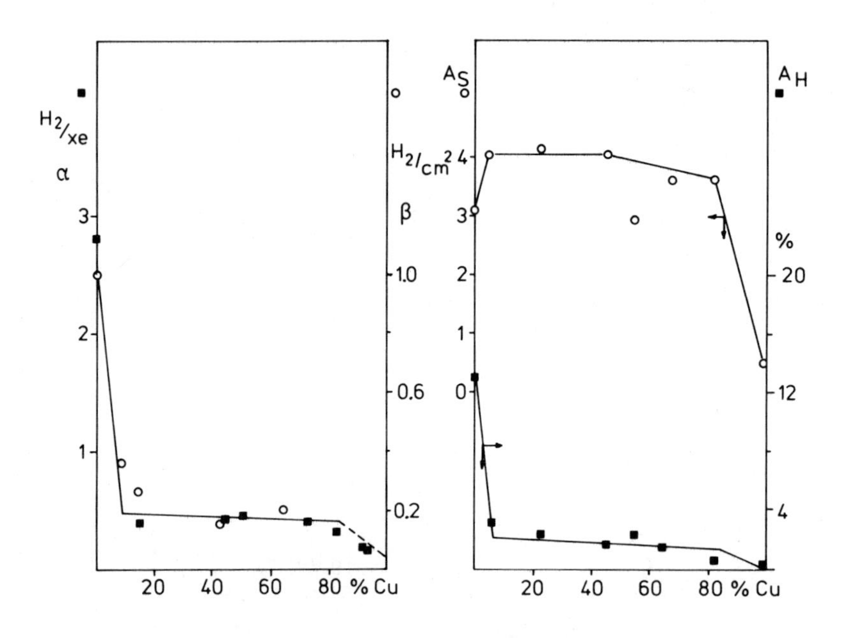

Fig. 27 : Ni-Cu alloy, H_2 adsorption (left) and cyclopropane (right).

α - hydrogen atoms (293 K)/xenon atoms (78 K) ratio for evaporated films[145]

β - hydrogen adsorption irreversible at 293 K/1 cm^2 surface area ; in relative units β = 1 for pure Ni[117]

A_S - activity (arbitrary units) for total conversion of cyclopropane

A_H - activity (arbitrary units) for hydrogenolysis of cyclopropane

α and β parameters indicate the Ni content in the surface of metal films resp. powders.

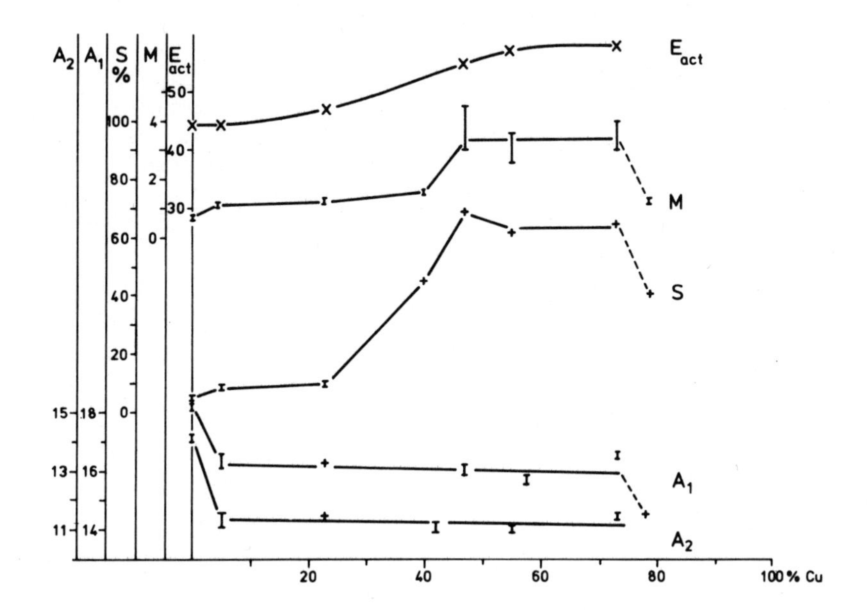

Fig. 28 : Several parameters characterizing activity and selectivity in hexane reforming reactions(146) on Ni-Cu alloys.

E_{act} - activation energy;

M - parameter characterizing the way of splitting of molecules(146);

S - total selectivity for all non-destructive reactions (all isomerisations and dehydrocyclisations);

A_1 - activity per gram catalyst;

A_2 - activity per square meter of catalyst.

which showed similar effects : when a Ib metal was added to a transition metal, the activity in C-C bond breaking was always slowed down much more than in the reactions on C-H bonds. It was also true in the case where a Ib metal and the transition metal did not form a solution. This fact also stresses the geometric (blocking) character of the changes evoked by the addition of a Ib metal (see effect V in 7.2.2.).

The phenomenon of selective blocking and the role of the number of available sites (ensemble size) in the selectivity for "C-C" and "C-H" reactions is not confined to Ni-Cu alloys. Gray et al.[157] showed that blocking of the Pt surface by carbon deposits has the same effect, it also depresses hydrogenolysis. Van Schaik and Dessing from our laboratory discovered the same influence of Au admixtures in Pt (for powders, see Ref. 158; for evaporated films, see Ref. 159). Other authors report similar results for Pd-Au alloys[152]. The phenomenon observed is, thus, of a quite general character.

8.2. Oxidation reactions

Not many reactions lead to interesting products when organic molecules are oxidized on metals (Notice : most of the metals are oxidized upon oxidation reactions). One example of an interesting reaction is the oxidation of ethylene. A convenient starting point for a discussion of this example is the mechanism suggested by Kemball and Patterson[160]. These authors showed that the reaction proceeds via two routes, I and II (the mechanism has been confirmed by a later paper[161]) :

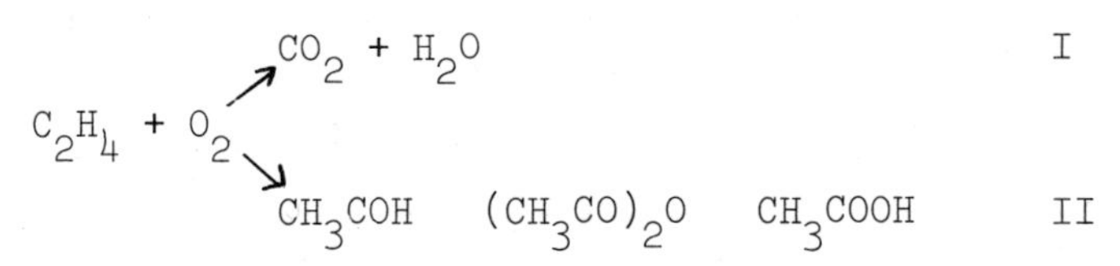

When the molecule becomes too deeply dehydrogenated, route I (selectivity S-I) is opened to a complete combustion.

The authors[161] studied this reaction on Pd-Au alloys and found that the S-II selectivity revealed a maximum as a function of the Pd concentration in the alloy (see Fig. 29). The addition of the first portion of Au to Pd probably causes a decrease in deep dehydrogenation which leads to an increase in S-II.

As in the previous example of hydrogenolysis the geometric effects (V in 7.2.2.) are able to explain the results mentioned here. From all metals it is actually only Ag which is important by its <u>epoxidation</u> activity (route II, below) :

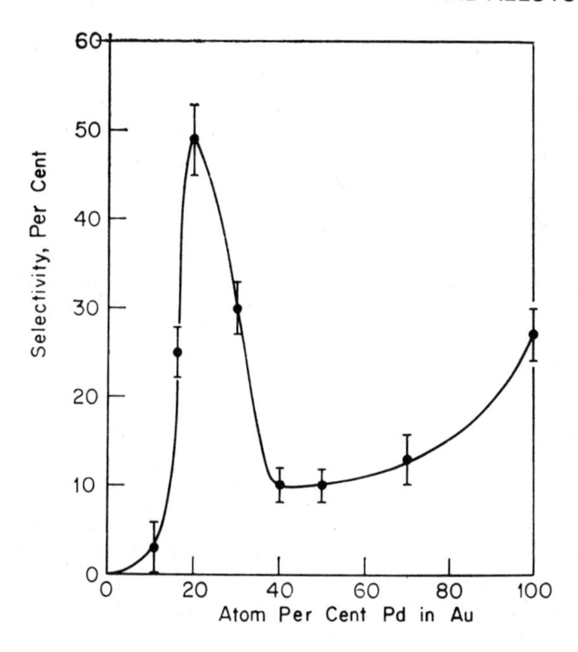

Fig. 29 : Selectivity S-II (see route II in the text) in the oxidation of C_2H_4 on Pd-Au alloys[161].

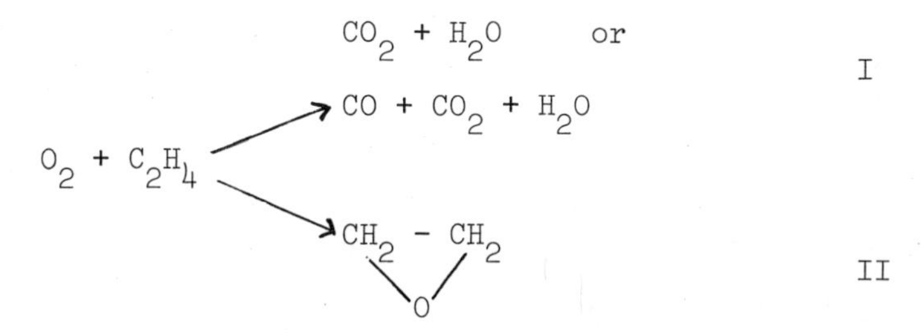

If we add Pd to Ag we increase the activity of the catalyst considerably but we decrease S-II strongly. Pd added to Ag increases evidently the dissociation degree of adsorbed ethylene. The existence of non-dissociated ethylene is a necessary but not sufficient condition for epoxidation to occur. It has been shown[162] that oxygen for epoxidation must be in a molecular non-dissociated form. Its concentration can be increased by adding Au (or metalloids) to

Ag. Here Au probably also functions through essential geometric (blocking) effects. They key-role of dehydrogenation in the determination of the selectivity can also be derived from the fact that it is more difficult not to burn the reaction molecule completely (S-I route) in the presence of metals when the molecule contains a reactive hydrogen atom like allylic hydrogen in propylene, etc.

The oxidation reactions are thus another example of the role of geometric factors (ensemble size) in the determination of the selectivity of a given catalyst. Moreover, a role of a specific adsorption form (molecular O_2) and the influence of alloying on its presence or absence have been demonstrated.

8.3. Other reactions

Some other reactions of potential interest in selectivity studies are e.g. dehydrogenation of alcohols (running parallel with the formation of ethers and dehydration to olefines), hydrogenation of C=O bonds (accompanied by various reactions of products), reactions of molecules with two reactive centres in one molecule (C=C next to C=O and similar), reactions of amines, nitriles, etc. Material on the selectivity effects in these reactions is rather fragmentary and at this moment does not allow a systematic explanation as with the reactions mentioned in the two foregoing paragraphs.

The experimental difficulties are also more considerable. While e.g. for hydrogenolysis we can easily find pairs of metals which differ very strongly in their activity, the rather aggressive adsorption of molecules with C=O, COH or C≡N, $C-NH_2$ groups takes place on all metals and this leads to a catalytic activity of almost all solids.

Snel from our laboratory[163] studied the reactions of propylamine on Au-Cu alloys. At higher temperatures the main reaction on Cu is dehydrogenation to acrylonitrile, on Au it is the metathesis into di- and tri-propyl and propylidene amines. The selectivity for -C≡N formation varies smoothly and linearly between 0 and 75 % Au in the alloy, probably indicating a smooth variation of the concentration of particular alloy components in the surface (Fig.30).

8.4. Concluding remarks

The investigation on alloys would be stimulated if a theoretical analysis and answers were available for some important questions like the following. First, a question dealing with special chemisorption bonds. Is it a realistic picture of atoms (such as C or N) double-bound to the surface or eventually with one metal atom bearing more than one of such groups? If so, how is the formation of these bonds influenced by alloying?

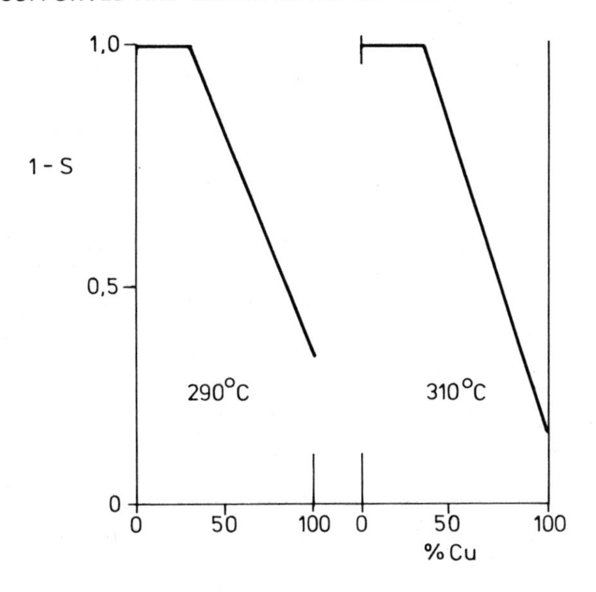

Fig.30 : S - selectivity for nitrile formation (parallel running are various metathesis reactions) in prim. propylamine reactions as a function of composition for Au-Cu alloys[163] (cut melts).

Another question is that of the adsorption site which can be formulated as follows : if a certain molecule prefers the position in the surface hole (valley among several atoms) for its adsorption on a pure metal, does the same also hold for an alloy ? Or, which position in the surface (site) is necessary to activate the molecule for dissociation of the C-H, C-C or C≡O bonds, etc. and what does alloying mean for it ? It has been suggested above that in many cases the role of a Ib metal in the alloys with the group VIII metals is to prevent dissociation of C-H and C-C bonds. Shall spectroscopic measurements of adsorbed species e.g. confirm this point (NMR, IR) ?

A point of particular importance is the question of the surface composition of alloys, prior to the reaction (in vacuum) or upon the reaction under the influence of corrosive chemisorption of reaction components, etc. Also the question of detection and analysis of the size of clusters of alloy components deserves more attention.

Many of the reactions mentioned above are structure sensitive (demanding) reactions. It would be very interesting to see the effect of alloying on extremely diluted metal carriers. Alloying in combination with variations in the crystal size of a catalyst may open new ways to governing the mechanism of catalytic reactions.

REFERENCES

(1) Fundamental Principles in Heterogeneous Catalysis, NATO A.S.I., Venice, Italy 1971 - lecture by L. Carra

(2) Satterfield C.N., Mass Transfer in Heterogeneous Catalysis, M.I.T. Press, Cambridge, U.S.A. (1970)

(3) Hougen O.A., Ind. Eng. Chem. 53 (1961) 509

(4) Weisz P.B. and Prater C.D., Adv. Catal. 6 (1954) 144

(5) Melville H. and Gowenlock B.G., Experimental Methods in Gas Reactions, McMillan, London (1964)

(6) Comprehensive Chemical Kinetics, ed. C.H. Bamford and C.F.H. Tipper, Vol.1, Elsevier, Amsterdam (1969)

(7) Temkin M.A., Kinetika i Kataliz 3 (1962) 509

(8) Campbell K.C. and Thomas S.J., Trans. Faraday Soc. 55 (1959) 985

(9) Roberti A., Ponec V. and Sachtler W.M.H, J. Catal. 28 (1973) 381. Van der Plank P. and Sachtler W.M.H., J. Catal. 12 (1968) 35.

(10) Anderson J.R. and Avery N.R., J. Catal. 8 (1967) 48

(11) Kemball C., Nisbet J.D., Robertson P.J. and Scurell M.S., Proc. Roy. Soc. London A338 (1974) 299

(12) Kokes R.J., Tobin H. and Emmett P.H., J. Am. Chem. Soc. 77 (1955) 5860

(13) Blanton W.A., Byers C.H. and Merrill R.P., Ind. Eng. Chem. 7 (1968) 611

(14) Galeski J.B. and Hightower J.W., Canad. J. Chem. Eng. 48 (1970) 151

(15) Frennet A. and Liénard G., Proc. IVth Int. Congr. Catal., Moscow, 1968 ; paper 51, p.65 - Akademiai Kiado, Budapest, 1971.
Frennet A., Catal. Rev. Vol.10, in press

(16) Ponec V., Preparation of Ultra-clean Surfaces, in Treatese on Adhesion, ed. R.C. Patrick, 1969, M. Dekker, N.Y., p. 485

(17) Shooter D. and Farnsworth H.E., J. Phys. Chem. 66 (1962) 169

(18) Onchi M. and Farnsworth H.E., Surface Sci. 11 (1960) 203

(19) Somorjai G.A., Joyner R.W. and Lang B., Proc. Roy. Soc. London A334 (1972) 335

(20) Kahn D.R., Petersen E.E. and Somorjai G.A., J. Catal. 34(1974) 294

(21) Merrill R.P., Catal. Rev. 4(1970) 115

(22) Mc Carroll J.J. and Thomson S.J., J. Catal. 19(1970) 144

(23) Tamaru K., Trans. Faraday Soc. 59(1959) 824
Fukuda K., Onishi T. and Tamaru K., Bull. Soc. Chem. Japan 42

(1969) 1192.
(24) Noto Y., Fuduka K., Onishi T. and Tamaru K., Trans. Faraday Soc. 63(1967) 3081
Fukuda K., Nagashima S., Noto Y., Onishi T. and Tamaru K., Trans. Faraday Soc. 64 (1968) 522
(25) Tajbl D.G., Simons J.B. and Carberry J.J., I. & E.C. Fundamentals 5(1966) 171
Mahoney J.A., J. Catal. 32 (1974) 247
(26) Emmett P.H., Adv. Catal. 1 (1948) 65
(27) Watson J.H.L., Anal. Chem. 20 (1948) 576
Moss R.L., Platinum Metal Rev. 11 (1967) 141
Renouprez A., Hoang-Van C. and Compagnon P.A., J. Catal. 34 (1974) 411
Flynn P.C., Wanke S.E. and Turner P.S., J. Catal. 33 (1974) 237
Adams C.R., Benesi H.A., Curtis R.M. and Meisenheimer R.G., J. Catal. 1 (1962) 336
(28) Jellinek M.H. and Fankuchen I., Adv. Catal. 1 (1948) 257
Scherer P., Nachr. Ges. Wiss. Göttingen, Math. Phys. k12 (1918) 98
Jones F.W., Proc. Roy. Soc. London A166 (1938) 16
Splindler H., Z. Chemie 13 (1973) 1
Smith W.L., J. Appl. Crystal. 5 (1972) 127
Renouprez A., Hoang-Van C. and Compagnon P.A., J. Catal. 34 (1919) 411
Guinier A., Ann. Phys. 12 (1939) 162 ; Guinier A. and Fournet G., Small Angle Scattering, ed. J. Wiley & Son, N.Y., 1955
(29) Beecroft T. and Miller A.W., Preparation and Characterization of Solid Catalysts, in Repts on Progress Appl. Chem. 55 (1970) 385.
Ciapetta F.G. and Plank C.J. in Catalysis, ed. P.H. Emmett, Vol.1, Reinhold, N.Y. (1974)
(30) Prestridge E.B. and Yates D.J.C., Nature 234 (1971) 345
(31) Müller J., Rev. Pure and Appl. Chem. (Austr.) 19 (1969) 151
Sermon P.A. and Bond G.C., Catal. Rev. 8 (1973) 211
Kral H., Z. physik. Chem. NF 48 (1966) 129
Schlosser E.G., Chem. Ing. Technik 39 (1967) 409
Spindler H., Z. Chem. 13 (1973) 1
(32) Stranski I.N., Forms of Equilibrium of Crystals, in Disc. Faraday Soc. 5 (1949) 13
Poltorak O.M., Zhur. Foz. Khim. 34 (1960) 1
(33) Sachtler W.M.H., Dorgelo G. and Vand der Knapp W., J. Chim. Phys. 5 (1954) 494
(34) Anderson J.R. and Avery N.R., J. Catal. 5 (1966) 446
Anderson J.R., Baker B.G. and Sanders J.V., J. Catal. 1 (1962) 443
(35) Schlosser E.G., Ber. Bunsenges. Physik. Chem. 73 (1969) 358
(36) Van Hardeveld R. and Hartog F., Surface Sci. 15 (1969) 189
(37) Poltorak O.M. and Boronin V.S., Zhur. Fiz. Kchim. 40 (1966) 1436
(38) Boudart M., Vannice M.A. and Benson J.E., Z. Physik. Chem. NF

64 (1969) 171
Schwab G.M. and Pietch E., Z. Physik. Chem. Abt. B 1 (1928) 385
Taylor H.S., Proc. Roy. Soc. A 108 (1925) 105
(39) Van Hardeveld R. and Hartog F., IVth Int. Congr. Catal., Moscow 1968, paper 70 ; Akademiai Kiado Budapest, 1971, Vol.2. p.295
(40) Burton J.J., Catal. Rev. - Sci.Eng. 9 (1974) 209
Allpress J.G. and Sanders J.V., Austr. J. Phys. 23 (1970) 23
(41) Rostrup-Nielsen J.B., J. Catal. 31 (1973) 173
(42) Schwab G.M. and Holz G., Z. Anorg. Allg. Chem. 252 (1944) 305
Schwab G.M., 67 (1955) 433
Schwab G.M., Disc. Faraday Soc. 8 (1950) 166
(43) Proceedings of the IVth International Congress on Catalysis, Moscow, 1968 ; ed. Akademiai Kiado, Budapest, Hungary, 1971
(44) Kemball C., Proc. Roy. Soc. London A 214 (1952) 413
(45) Dowden D.A., J. Chem. Soc. 1950, 242 ; Ind. Eng. Chem. 44 (1952) 997
(46) Dowden D.A., Chemisorption, ed. W.E. Garner, Butterworth, London, 1957, p.3
(47) Beeck O., Disc. Faraday Soc. 8 (1950) 118
Beeck O., Rev. Mod. Phys. 17 (1945) 61
(48) Derouane E.G., Mémoires Soc. Roy. Sci. Liège, 6ième série, tome 1, Fasc.4, p. 297, 1971
(49) Derouane E.G., Ind. Chim. Belg. 36 (1971) 359
(50) Sabatier P., Berichte 44 (1911) 2001
(51) Balandin A.A., Adv. Catal. 19 (1969) 1
(52) Fahrenfort J.J., Van Reijen L.L. and Sachtler W.M.H., Z. Elektrochem. 64 (1960) 216
Schuit G.C.A., Van Reijen L.L. and Sachtler W.M.H., Act. 2ième Congr. Int. Catal., ed. Technip, Paris, 1961, p. 893
(53) Makishima S., Yoneda Y. and Satto Y., Act.2ième Congr. Int. Catal., ed. Technip, Paris, 1961, p. 671
Golodetz G.J. and Roiter V.A., Ukr. Khim. Zhur. 29 (1963) 667
(54) Tanaka K. and Tamaru K., J. Catal. 2 (1963) 366
(55) Tanaka K. and Tamaru K., Kin. i Kat. 7 (1966) 242
Ponec V. and Schuit G.C.A., Chem. Wkbl. 1974
(56) Ponec V. and Cerny S., Disc. Czechosl. Acad. Sci. 75 (1965) No. 5.
Ponec V., Proc. IIIrd Int. Congr. Catal., Amsterdam ; ed. Sachtler W.M.H. et al., North-Holland Publ. Comp., 1965, p.544 (Discussion)
Knor Z., ibid. p. 545 (Discussion)
(57) Bond G.C., Catalysis by Metals, Acad. Press, London, 1962
(58) Perri B., Disc. Faraday Soc. 41 (1966) 121
(59) Kouskova A., Adamek J. and Ponec V., Coll. Czech. Chem. Commun. 35 (1970) 2538
(60) Bond G.C. and Wells P.B., Adv. Catal. 15 (1964) 91
(61) Wells P.B., Chemistry & Industry (Britain) 1964, 1742
(62) Kemball C., Catal. Rev. 5 (1971) 33
(63) Burnell R.L. Jr, Accounts Chem. Res. 2 (1969) 289 ; Catal. Rev. 7 (1) (1972) 25.

(64) see e.g. Bond G.C. and Turkevich T.J., Trans. Faraday Soc. 50 (1954) 1335
Addy J. and Bond G.C., Trans. Faraday Soc. 33 (1957) 386, 383, 388
(65) Rooney J.J., Chemistry in Britain 242 (1966)
(66) Horrex C., Moyes R.B. and Squire R.C., Proc. IVth. Int. Congr. Catal., Moscow, 1968 ; ed. Akademiai Kiado, Budapest, (1971) p. 333
(67) Hirota K. and Ueda T., Proc.IIIrd Int. Congr. Catal., Amsterdam, 1964 ; ed. Sachtler W.M.H. et al., North-Holland Publ. Comp. (1965) p. 1238
(68) Harper R.J. and Kemball C., Proc.IIIrd Int. Congr. Catal., Amsterdam, 1964 ; ed. Sachtler W.M.H. et al., North-Holland Publ. Comp. (1965), p. 1145
Crawford E. and Kemball C., Trans. Faraday Soc. 58 (1962)2452
Phillips M.J., Crawford E. and Kemball C., Nature 197 (1963) 487
(69) Van Schaik J.R.H., Dessing R.P. and Ponec V., J. Catal., in press
(70) Kikuchi E. and Morita Y., J. Catal. 15 (1969) 217
(71) Matsumoto H., Saito Y. and Yoneda Y., J. Catal. 22 (1971) 182
(72) Carter J.L., Cusumano J.A. and Sinfelt J.H., J. Catal. 20 (1971) 223
(73) Anderson J.R., Mac Donald R.J. and Shimoyama Y., J. Catal. 20 (1971) 147
(74) Corrolleur C., Tomanova D. and Gault F.G., J. Catal. 24 (1972) 401
(75) Merta R. and Ponec V., Proc. IVth Int. Congr. Catal., Moscow, 1968 ; ed. Akademiai Kiado, Budapest, Vol.2 (1971) p. 53
(76) Ross J.R.H., Roberts M.W. and Kemball C., J. Chem. Soc., Faraday Trans. I, 68 (1972) 914
(77) Margolis L.Y., Adv. Catal. 14 (1963) 420
(78) Voge H.H. and Adams C.R., Adv. Catal. 17 (1967) 151
(79) Margolis L.Y., Catal. Rev. 8 (1) (1974) 241
(80) Sachtler W.M.H., Catal. Rev. 4 (1970) 277
(81) Cant N.W. and Hall W.K., J. Catal. 16 (1970) 226
(82) Patterson W.R. and Kemball C., J. Catal. 2 (1967) 465
(83) Kilty P.A., Rol N.C. and Sachtler W.M.H., Proc. Vth Congr. Catal., Palm Beach, 1972 ; ed. Hightower H.J., North-Holland Publ. Comp. Vol. 2, (1973) 929
(84) Siskova M. and Erdos E., Coll. Czech. Chem. Commun. 25 (1960) 1729
Defay R., Prigogine I., Belleman A. and Everett D.H., Surface Tension and Adsorption, Langman, Green, N.Y., 1966
(85) Gugenheim E.A., Trans. Faraday Soc. 41 (1945) 150
(86) Elovich S.Yu and Larionov C.G., Izv. Akad. Nauk SSSR, Otd. Khim. Nauk 1962, 531
(87) Van Santen R.A. and Boersma M.A.M., J. Catal. 34 (1974) 13
Van Santen R.A. and Sachtler W.M.H., J. Catal. 33 (1974) 202

(88) Williams F. and Nason D., Surface Sci. 45 (1974) 377
(89) Meijering J.L., Acta Metall. 14 (1966) 251
(90) Sachtler W.M.H., Dorgelo G.J.H. and Jongepier R., Basis Problems in Thin Film Physics, Göttingen, VanderHoek & Ruprecht, 1966, p. 218
Van der Plank P. and Sachtler W.M.H., J. Catal. 12 (1968) 35
(91) Elford L., Müller F. and Kubashewski O., Ber. Bunsenges. 73 (1969) 601
(92) Meijering J.L., Acta Met. 8 (1957) 257
(93) Palczewska W. and Majchrzak S., Bull. Acad. Polon. Sci. Série Chim. 17 (1969) 681
(94) Sachtler W.M.H., Le Vide, No. 164, 1973, p.67
(95) Results of P.E.C. Franken, this laboratory, will be published soon
(96) Bouwman R. and Sachtler W.M.H., J. Catal. 19 (1970) 127
(97) Helms C.R., J. Catal. 36 (1975) 114
(98) Ertl G. and Küppers J., J. Vac. Sci. Technol. 9 (1972) 829
(99) Quinto D.T., Sundaram V.S. and Robertson W.D., Surface Sci. 28 (1971) 504
(100) Nakayama K., Ono M. and Shimitzu H., J. Vac. Sci. Technol. 9 (1972) 749
(101) Takahassa Y. and Shimizu H., J. Catal. 29 (1973) 479
(102) Ono M., Takasu Y., Nakayama K. and Yamashina T., Surface Sci. 26 (1971) 313
(103) Williams F.L. and Boudart M., J. Catal. 30 (1973) 438
(104) Mathieu H.J. and Landolt D., paper 14 ; Stoddart G.T.H., Moss R.L. and Pope P., paper 15 ; Surface Sci. Conference, Warwick, 1975. Will be publised in Surface Sci.
(105) Takasu Y., Konno H. and Yamashina T., Surface Sci. 45 (1974) 321
(106) Hardy W.A. and Linnett J.W., Trans. Faraday Soc. 66 (1970) 487
(107) Franken P.E.C. and Ponec V., J. Catal. 35 (1974) 417
(108) Sachtler W.M.H., Ber. Bunsenges. Phys. Chem. 59, No.2, (1955) 119
(109) Albrecht H.E., Phys. Stat.Sol. (a) 6 (1971) 135
(109) Lang N.D. and Kohn W., Phys. Rev. B3 (1971) 1215
Lang N.D., Phys. Rev. B4 (1971) 4234
Smith J.R., Phys. Rev. 181 (1969) 522
Mahan G.D. and Schaich W.L., Phys. Rev. B10 (1974) 2647
(110) Rivière J.C., Work Function, Measurements and Results, in Solid State Surface Sci. ; ed. Green, M. Dekker, N.Y., 1969, Vol.1, p.179
(111) Culver R.V. and Tompkins F.C., Adv. Catal. 11 (1959) 68
(112) Tompkins F.C., The Solid-Gas Interface ; ed. Flood E.A., M. Dekker, N.Y. 1967
(113) Mott N.F. and Jones H., Theory and Properties of Metals and Alloys, Oxford, 1936

(114) Dowden D.A., Proc. IVth Int. Congr. Catal., Moscow, 1968 ; ed. Akademiai Kiado, Budapest, 1971
(115) Van der Plank P. and Sachtler W.M.H., J. Catal. 7 (1967) 300
(116) Cadenhead D.A. and Wagner N.J., J. Phys. Chem. 72 (1968) 2775
(117) Sinfelt J.H., Carter J.L. and Yates D.J.C., J. Catal. 24 (1972) 283
(118) Stephan J.J. and Ponec V., J. Catal. 37 (1975) 81
(119) Brongersma H.H., J. Vac. Sci. Technol. 11 (1974) 231
(120) Brongersma H.H., Private Commun. (also an invited paper at the Surface Sci. Conf., Warwick, 1975 ; will be published in Surface Sci.)
(121) Klemperer D., Private Commun.
(121a) McKee D.W., Trans. Faraday Soc. 64 (1968) 2200
McKee D.W. and Norton F.J., J. Phys. Chem. 68 (1964) 481
(122) Vlasveld J.L., M. Sc. Thesis, Leiden University, 1975
(123) Cvetanovic R. and Amenomiya J., Catal. Rev. 6 (1972) 12
(124) Ehrlich G., Adv. Catal. 14 (1963) 256
Redhead P.A., Vacuum 12 (1962) 203
Carter G., Vacuum 12 (1962) 245
Smith A.W., J. Coll. Interfaces Sci. 34 (1970) 491
(125) Stephan J.J., Ponec V. and Sachtler W.M.H., Surface Sci. 47 (1975) 403
(126) Stephan J.J., Thesis, Leiden University, 1975
(127) Nyrop J.E., The Catalytic Action of Surfaces, Levin & Munksgaard, Copenhagen, 1937
(128) Schwab G.M., Disc. Faraday Soc. 8 (1950) 166
Couper A. and Eley D.D., ibid. 8 (1950) 172
Dowden D.A. and Reynolds P.W., ibid. 8 (1950) 184
(129) Wolkenstein Th.Th., The Electronic Theory of Catalysis on Semiconductors, Pergamon Press, Oxford, 1963
(130) Boreskov G.K., Kin. i Kat. 10 (1969) 5 (in English)
(131) Koutecky J., Trans. Faraday Soc. 54 (1958) 1038 ; Z. Elektrochem. 60 (1956) 835
(132) Grimley T.B., J. Vac. Sci. Technol. 8 (1971) 31
Knor Z., J. Vac. Sci. Technol. 8 (1971) 57
(133) Thorpe B.I., Surface Sci. 33 (1972) 306
Newns D.M., Phys. Rev. 178 (1969) 1123
Penn D.R., Surface Sci. 39 (1973) 333
(134) Kelly M., Surface Sci. 43 (1974) 587
(135) Sachtler W.M.H. and Dorgelo G.J.H., Z. Physik. Chem. NF 25 (1960) 69
(136) Sachtler W.M.H., Proc. IIIrd Int. Vac. Congr., Vol.1, p.41 (1966)
(136) Hüfner S., Wertheim G.K. and Wernick J.H., Phys. Rev. B 8 (1973) 1973
(137) Lang N.D. and Ehrenreich H., Phys. Rev. 168 (1968) 605
Kouvel J.S. and Comly J.B., Phys. Rev. Letters 24 (1970) 598
Seib D.H. and Spicer W.E., Phys. Rev. Letters 20 (1968) 1441
(138) Stocks G.M., Williams R.W. and Faulkner J.S., J. Phys. F. Metal Physics 3 (1973) 1688

(139) Gibb T.R.P. Jr, McMillan J. and Roy R.J., J. Phys. Chem. 70 (1966) 3024
(140) Wenger A. and Steinemann S., Helv. Phys. Acta 47 (1974) 321
(141) Cooper M.J., Phil. J. Mag. 8 (1963) 811
(142) Mozer B., Keating D.T. and Moss S.C., Phys. Rev. 175 (1968) 868
Hicks T.J., Rainford B., Kouvel J.S., Louw G.G. and Comly J.B., Phys. Rev. Letters 22 (1969) 531
Kidron A., Phys. Letters 30A (1969) 304
(143) Robbins C.G., Claus H. and Beck P.A., Phys. Rev. Letters 22 (1969) 1307 ; Vogt E., phys. stat. sol. (b) 50 (1972) 653
Perrier J.P., Tissier B. and Tournier R., Phys. Rev. Letters 24 (1970) 313
(144) Soma-Noto Y. and Sachtler W.M.H., J. Catal. 32 (1974) 315
(145) Ponec V. and Sachtler W.M.H., J. Catal. 24 (1972) 250
(146) Ponec V. and Sachtler W.M.H., Catalysis, Proc. Vth Int. Congr. Catal., Miami Beach, 1972 ; ed. Hightower, North-Holland Publ. Comp., 1973, p. 645, paper 43
(147) Rushford H.G. and Whan D.A., Trans. Faraday Soc. 67 (1971) 3577
(148) Van Barneveld W.A.A. and Ponec V., Rec. Trav. Chim. 93 (1974) 243
(149) Jongepier R. and Sachtler W.M.H., J. Res. Inst. Catal., Hokkaido Univ. 16 (1968) 69
(150) The literature not quoted in Ref. 57, relevant for the discussion on Ni-Cu alloys : Ref. 145 - 149 of this paper.
Van der Plank P. and Sachtler W.M.H., Surface Sci 18 (1969) 62
Ljubarskii G.D., Evzerikhin E.I. and Slinkin A.A., Kin. i Kat. 5 (1964) 311
Bond G.C. and Mann R.S., J. Chem. Soc. 3566 (1959)
Mann R.S. and Khulbe C., Canad. J. Chem. 48 (1970) 2075
Mann R.S. and Khulbe K.C., ibid. 45 (1967) 2755 ; 47 (1969) 215
Cadenhead D.A. and Masse N.G., J. Phys. Chem. 70 (1966) 3553
Sinfelt J.H., Carter J.L. and Yates D.J.C., J. Catal. 24 (1972) 283
Cadenhead D.A., Wagner N.J. and Thorp R.L., Proc. IVth Int. Congr. Catal., Moscow, 1968 ; ed. Akademiai Kiado, Budapest, 1971, Vol.1, p. 341, paper 26
(151) The literature not quoted in Ref. 57, relevant for the dicussion on Pd-Au or Pd-Ag alloys :
Hardy W.A. and Linnett J.W., Trans. Faraday Soc. 66 (1970) 447
Inami S.H. and Wise H., J. Catal. 26 (1972) 92
Eley D.D., J. Res. Inst. Catal. Hokkaido Univ. 16 (1968) 101
Joice B.J., Rooney J.J., Wells P.B. and Wilson G.R., Disc. Faraday Soc. 41 (1966) 223
McKee D.W., J. Phys. Chem. 70 (1966) 525
Cinneide A.O. and Clark J.K.A., J. Catal. 26 (1972) 233
(152) Visser C., Zuidwijk J.G.P. and Ponec V., J. Catal. 35(1974)457

(153) Ponec V., Catal. Rev. - Sci. Eng. 11 (1) (1975) 41
(154) Roberti A., Ponec V. and Sachtler W.M.H., J. Catal. 28 (1973) 381
(155) Beelen J.M., Ponec V. and Sachtler W.M.H., J. Catal. 28 (1973) 376
(156) Sinfelt J.H., J. Catal. 29 (1973) 308
(157) Gray T.J., Masse N.G. and Oswin H.G., Act.2 ième Congr. Int. Catal., Paris, 1960 ; ed. Technip Paris, 1961, Vol.2, p.1697, paper 83
(158) Van Schaik J.R.H., M. Sc. Thesis ; see also Ref. 69
(159) Dessing R.P., Thesis, Leiden University, 1974
(160) Kemball, C. and Patterson W.R., Proc. Roy. Soc. London A 270 (1962) 219
(161) Gerberich, H.R., Cant, N.W. and Hall, W.K., J. Catal. 16 (1970) 204
(162) for review see : Kilty, R. and Sachtler, W.M.H., Catal. Rev. 10 (1974)
(163) Snel, R., M.Sc.Thesis, Leiden University, 1974
(164) Kuijers, F.J., M.Sc. Thesis, Leiden University, 1973

Notes added in proof

1) Hydrogenation of acetylene and methylacetylene

When the work on the manuscript of this paper had been finished, a new and very interesting paper appeared on this subject (N. Yoshida and K. Hirota, Bull.Chem.Soc. Japan,Vol.46,No.1. - 184 (1975)). The Japanese authors have shown that for methylacetylene:

i) in all cases they studied (Pt, Ni, Rh, Ru, Ir) the reaction of ethylene is faster than that of acetylene.
ii) the selectivity was independent on the conversion in the whole range of reactions.
iii) a clear correlation was found between the presence of higher deuterated propylene (product of the first addition) and a low selectivity. This could indicate that the original methylacetylene (acetylene?) can be dehydrogenated to a different extent, the higher dissociated species reacting through up to propane (ethane?).

2) δ% character of metal bonds

Discussions during the courses have shown that some words should be added to make the statements in this paper more convicing with regard to parameter δ%. Therefore, let us analyse two points:

i) how is this parameter determined?
ii) how is it related to catalysis?

Re i)

Pauling starts his procedure by stating that the relation between the multiplicity and the bond length (atom radii R) is expressed for C-C, C=C and C ≡ C by an equation

$$R(n) - R(1) = - 0.31 \log n$$

R(1) stands for a "single"bond, R(n) for a multiple bond (order n). For metals with more nearest neighbours than valence electrons $n < 1$ and $R(n) < R(1)$. Pauling assumes, however, the same equation to be valid for metals as well as for C-C bonds (Notice: as Hume Rothery already mentioned, the above relation does not hold for multiple bonds in bi-phosphorus, bi-nitrogen or other similar bonds.)
For simple metals, Pauling finds empirically the following relation for R(1) calculated in a described way (1st period metals) :

$$R_1^o = 1.825 - 0.043\ Z$$

The bonds in transition metals are shorter and this is according to Pauling a consequence of the contribution of the d-orbitals to the metal-metal binding. Pauling suggests that for transition metals it holds

$$R(1)_{trans} = R^o(1) - (a - bZ)\delta\%$$

where a and b are constants to be determined and δ% the percentage to which the metal-metal binding is due to the binding (valence) d-orbitals.
The constants a and b are determined by using data of the following three metals : Ni, Co and Fe. For each of these metals Pauling suggests two arbitrary resonance structures mixed in such ratio that a measured magnetic moment results from calculations with these structures.
Pauling divides the orbitals of transition metals into three groups: atomic orbitals bearing unpaired (responsible for the magnetic moment) or paired d-electrons not taking part in the metal binding, valence orbitals of d, s or p character with electrons paired by metallic binding and unoccupied metallic orbitals. Various resonance structures (chosen according to the principle of minimum of hypothesis!) differ in the division of electrons among these three groups of orbitals.
For example, Pauling sugges for cobalt :

Structure A, having 3 unpaired d-electrons (on atomic non-bonding orbitals),
2-valence d-electrons, 1-4s-valence electrons and 3-valence p-electrons.

Structure B, having 2 paired electrons, d-electrons (resulting moment is zero), one unpaired atomic d-electron, 3-valence d-electrons, 1-4s-valence electrons and 2-valence p-electrons. One p-orbital is free.

Mixing the structures in a ratio 35/65, Pauling obtains a valence 6 for Co and a resulting atomic moment of 1.7 Bohr magnetons.

For the mentioned metals - Fe, Co and Ni - all values which result are near 40% (40% ± 1%).
From the structures of the three metals where δ% is known, constants a and b are calculated and then the δ% character is determined for all other metals by using R(1) radii and the last equation.

The author of this paper finds the procedure too speculative and the numerical values of the δ% character of little physical content, particularly when the δ% values found by extrapolation reach from 20% to 50%.

Re ii)

Beeck suggested : with increasing δ% character metals have less orbitals available for bonding with the adsorbate. However, the δ% does not vary parallel (see the resonance structures in the original literature) to any type orbitals.
Other authors assumed that the use of δ% character to correlate the catalytic data is justified because it is somehow a measure of the metal-metal bond strength. However, this bond strength is more directly characterized by the heat of sublimation, melting points or coefficients of elastic deformation, etc. However, none of these parameters varies among metals exactly parallel with the δ% character. It is the author's opinion (which may be questioned, of course) that it would be better not to use the parameter for correlating the catalytic data as activity and selectivity of various metals and alloys.

Relevant literature

Pauling, L. Proc. Roy. Soc. London, A 196, 343 (1949)

Criticism on it :
Baker and Jenkins, Adv. Catal. 7, 1 (1955)
Coles and Hume Rothery, Adv. Physics 3, 149 (1954)

SHORT COMMUNICATIONS

LIST OF PARTICIPANTS

I. ABBATI, Istituto di Fisica, Politecnico di Milano, Piazza L. di Vinci, 32 - 20133 Milano - Italy

H. ALBERS, Van't Hoff Laboratory, Padualaan, 8, De Uithof - Utrecht Netherlands

G. ALLAN, I.S.E.N., Laboratoire Physique des Solides, Rue Fr.Baes, 3 - 59046 Lille - France

G.P. ALLDREDGE, University of Texas, Physic Department RLM 5.208, Austin - Texas 78712 - U.S.A.

M. ALNOT, Centre de Cinétique Physique et Chimique du C.N.R.S., Route d' Andoeuvre - Villers-Nancy - France

S. ANDERSSON, Chalmers University of Technology, Department of Physics, Fack - 40220 Göteborg 5 - Sweden

J.-M. ANDRE, Facultés Universitaires de Namur, Groupe Chimie-Physique, Rue de Bruxelles, 61 - 5000 Namur - Belgium

C. BACKX, Potgieterlaan, 67 - Hazerswonde - Netherlands

M. BARBER, UMIST, P.O. Box 88, Manchester - M60 1DQ - England

I. BATRA, IBM Research Laboratory, K33/281, San Jose - California 95193 - U.S.A.

P. BILOEN, Koninklijk/Shell-Laboratory, Badhuisweg, 3 - Amsterdam-Noord - Netherlands

J.M. BLOCK, Fritz-Haber Institüt der Max Planck Gesellschaft, Faradayweg 4-6 - 1 Berlin 33 / Dahlem - Germany

J. B.NAGY, Facultés Universitaires de Namur, Groupe de Chimie-Physique, Rue de Bruxelles, 61 - 5000 Namur - Belgium

G.C. BOND, Brunel University, Department of Chemical engineering, Kingston Lane, Uxbridge - Middlesex UB8 3PH - England

M. BOUDART, Stanford University, Stauffer III, Department of Chemical Engineering, Stanford - California 94305 - U.S.A.

R. BRAKO, Institute "Ruder Boskovic", P.O.B. 1016 - 41001 Zagreb-Yugoslavia

C.R. BRUNDLE, I.B.M. Research Laboratory, San Jose - California 95114 - U.S.A.

A. CAMPERO, University Autonoma Metropolitana, Department of Chemistry, Avenue San Pablo, Azcapotzalco - Mexico 16, D.F. - Mexico

T.W. CAPEHART, Cornell University, Clark Hall, Physics Department, Ithaca - New York 14850 - U.S.A.

G. CASALONE, Istituto di Chimica Fisica, University di Milano - 20133 Milano - Italy

D. COCKE - Fritz-Haber Institüt der Max-Planck Gesellschaft, Faradayweg 4-6 - 1 Berlin 33 / Dahlem - Germany

I. DALLA LANA, University of Alberta, Department of Chemical engineering, Edmonton - Alberta - Canada

A. DATTA, Imp. College of Science and Technology, Department of Mathematics, Exhibition Road - London SW7 2RH - England

C. DEFOSSE, Place Croix du Sud, 1 - 1348 Louvain-la-Neuve - Belgium

K. DENISSEN, N.V. Philips, Department Elcoma H.O.C., Building BL - Eindhoven - Netherlands

E. DEROUANE, Facultés Universitaires de Namur, Groupe de Chimie-Physique, Rue de Bruxelles, 61 - 5000 Namur - Belgium
B. DJAFARI ROUHANI, C.E.N. Saclay, Section Etudes interactions gaz-solides, B.P. 2 - Gif-sur-Yvette 91190 - France
M. DOMKE, Fritz-Haber Institüt der Max-Planck Gesellschaft, Faradayweg 4-6 - 1 Berlin 33 - Germany
P. ECHENIQUE, Cambridge University, Cavendish Laboratory, Madingley-road - Cambridge CB3-OHE - England
T. ENGEL, Physikalisch-Chemisches Institüt der Universität, Sophienstraass 11 - 8 Müchen 2 - Germany
H. ENGELHARDT, Technische Universität München, Physik Department E20 - 8046 Garching - Germany
W. ERLEY, Institüt für Grenzflächenforschung und Vakuumphysik der Kernforschungsanlage Jülich GmbH - 517 Jülich 1 - Germany
G. FERRARIS, Montedison, Istituto Ricerche G.Donegani, Via del Lavaro 4 - 28100 Novara - Italy
L. FIERMANS, University of Gent, Laboratorium voor Kristallografic Krijgslaan, 271 - 9000 Gent - Belgium
J. FIGAR, Battelle, Route de Drize, 7 - 1227 Carouge/Geneve - Switzerland
A. FRENNET, Ecole Royale Militaire, Laboratoire de Catalyse FNRS, 30 Avenue Renaissance - 1040 Bruxelles - Belgium
J.J. FRIPIAT, C.R.S.O.C.I. - CNRS, Rue de la Ferrolerie - 45045 Orléans - France
J. FRIPIAT, Facultés Universitaires de Namur, Groupe de Chimie-Physique, rue de Bruxelles, 61 - 5000 Namur - Belgium
J.W. GADZUK, N.B.S., Surface and Electron Physic Division, Department of Physic Division, I.B.S., Washington - D.C. 20234 - U.S.A.
P.O. GARTLAND, Norwegian Institute of Technology, Department of Physics - 7034 Trondheim - Norway
J.P. GASPART, I.L.L., B.P. 156 - Grenoble - France
J.W. GEUS, University of Utrecht, Division of Inorganic Chemistry, Croesestraat, 77A - Utrecht - Netherlands
J. GOODWIN, University of Michigan, Department of Chemical Engineering, Ann Arbor - Michigan 48104 - U.S.A.
T.B. GRIMLEY, University of Liverpool, Donnan Laboratories, P.O. Box 147 - Liverpool L69 3BX - England
M. GRUNZE, Universität München, Physikalisch-Chemisches Institüt Sophienstraass 11 - 8 München 2 - Germany
O. GUNNARSSON, Institute of Theoretical Studies, Fack - 402 20 Göteborg - Sweden
K. HANSEL, Technische Universität, Lehrstuhl für Theoretische Chemie, Arcisstraass 21 - 8 München 2 - Germany
N. HERBST, National Bureau of Standards, B-214 Metrology Building - Washington, D.C. 20234 - U.S.A.
P. HERTEL, Institüt für Theoretische Physik, Jungiusstraass, 9 - 2 Hamburg 36 - Germany
B. HOLLAND, University of Warwick, Department of Physics - Coventry CV4 7AL - England

K. HORN, Queen Mary College, Department of Chemistry, Mile end Road - London E1 4N5 - England
R. HUBIN, Rue du Chêne, 8 - 4051 Plainevaux - Belgium
P.G. JAMES, BP Research Centre, Chertsey Road, Sunburry on Thames - Middlesex - England
D. JEWSBURY, University of Leicester, University Road - Leicester LE1 7RH - England
M. JONSON, Institute of Theoretical Studies, Fack - 40220 Göteborg 5 Sweden
Y. JUGNET, Laboratoire de Chimie Nucléaire, Boulevard du 11 Novemre 1918 - 69621 Villeurbanne - France
J. KATZER, University of Delaware, Department of Chemical Engeneering Newaek - Delaware 19711 - U.S.A.
N. KAUFHERR, Institut de Recherche sur la Catalyse, Boulevard du 11 Novembre 1918, 39 - 69621 Villeurbanne - France
P. KINET, Université Libre de Bruxelles, Faculté des Sciences Appliquées, 50 , Avenue F.D. Roosevelt - 1050 Bruxelles - Belgium
D. KOLB, Fritz-Haber Institüt der Max-Planck Gesellschaft, Faradayweg 4-6 - 1 Berlin 33 - Germnay
N. LANG, I.B.M., T.J. Watson Research Centre, P.O. 218, Yorktown Heights - New York 10598 - U.S.A.
A. LECLOUX, C/O Solvay & Cie., 310 Rue de Ransbeek - 1120 Bruxelles-Belgium
G. L'HOMME, Institut de Chimie Industrielle, Laboratoire de Génie Chimique, Rue A. Stewart 2 - 4000 Liège - Belgium
A. LUCAS, Facultés Universitaires de Namur, Groupe de Chimie Physique, Rue de Bruxelles, 61 - 5000 Namur - Belgium
S. MALO, Instituto Mexicano del Petroleo, Av. Cien Metros 500 Apdo. Pos. 14-805 - Mexico 14 - Mexico
P. MATSCHKE, Max-Planck Institüt Plasmaphysik und Forschungsgelande 8046 Garching Munich - Germany
M. MESTDAGH, Université Catholique de Louvain, Laboratoire Chimie-Physique Minérale, Place Croix du Sud 1, 1348 Louvain-la-Neuve - Belgium
J.P. MUSCAT, I.S.E.N., Laboratoire de Physique des Solides, Rue F. Baes, 3 - 59046 Lille - France
C. NACCACHE, Institut de Recherche sur la Catalyse, 39, Boulevard du 11 Novembre 1918 - 69626 Villeurbanne - France
C. NIJBERG, Chalmers University of Technology, Department of Physics, Fack - 402 20 Göteborg - Sweden
R. NIEMINEN, Helsinki University of Technology, Department of Technical Physics - 02150 Otaniemi - Finland
K. PEDERSEN, Haldor Topsoe A/S, P.O. Box 49 - 2860 Søborg - Denmark
J.P. PIRARD, Institut de Chimie Industrielle, Laboratoire du Génie Chimique, Rue A. Stevart, 2 - 4000 - Liège - Belgium
V. PONEC, Rijkuniversiteit Leiden, Gorlaeus Laboratories, P.B. 75 - Leiden - Netherlands
D. REINALDA, Gorlaevs Laboratory, Postbox 75 - Leiden - Netherlands

G. RHEAD, Ecole Nationale Supérieure de Chimie de Paris, Laboratoire de Métallurgie et Physico-Chimie des Surfaces, 11, Rue Paul et Marie Curie - 75231 - Paris Cédex 05 - France
T.N. RHODIN, Cornell University, School of Applied and Engeneering Physics, College of Engeneering, Ithaca - New York 14850 - U.S.A.
M.A. ROSALES MEDINA, University of Liverpool, Donnan Laboratories, P.O. Box 147 - Liverpool L69 3BX - England
L. SALEM, Université de Paris Sud, Laboratoire de Chimie Théorique, 91405 Orsay - France
K. SCHWAHA, Physikalisch-Chemisches Institüt, Innrain 52A - A-6020 Innsbruck - Austria
K. SCHWARZ, Technische und Elektrochemische Institüt, Technische Hochschule Wien, Getreidemarkt, 9 - 1060 Wien - Austria
J.M. STREYDIO, PCES, 1, Place Croix du Sud - 1348 Louvain-la-Neuve - Belgium
O. TAPIA, Quantum Chemistry Group, Box 518 - 75120 Uppsala - Sweden
M. TESCARI, Universita di Milano, Istituto di Chimica Fisica - 20 133 Milano - Italy
TRAN MINH DUC, Institut de Physique Nucléaire, Laboratoire de Chimie Nucléaire, 43, Boulevard du 11 Novembre 1918 - 69621 Villeurbanne - France
VAN DER AVOIRD, University of Nijmegen, Institute of Theoretical Chemistry, Toernooiveld - Nijmegen - Netherlands
J. VAN DER MEIJDEN, University of Utrecht, Division of Inorganic Chemistry, Croesestraat, 77A - Utrecht - Netherlands
J.H.C. VAN HOOF, Eindhoven University of Technology, Department of Inorganic Chemistry - Eindhoven - Netherlands
W. VAN RIEL, Universiteit Instellingen Antwerpen, Department of Natuurkunde, Universiteit Plein 1 - 2610 Wilrijk - Belgium
R. VAN SANTEN, Koninklijk Schell Laboratorium, Badhuisweg, 3 - Amsterdam - Netherlands
J. VEDRINE, CNRS, Institut de Recherche sur la Catalyse, 39, Boulevard du 11 Novembre 1918, 69100 Villeurbanne - France
B. VU THIEN, Département de Physique des Matériaux, 43, Boulevard du 11 Novembre 1918 - 69621 Villeurbanne - France
D. VERCAUTEREN, Facultés Universitaires de Namur, Groupe de Chimie-Physique, 61, Rue de Bruxelles - 5000 Namur - Belgium
J. WABER, Northwestern University, Department of Materials Science - Evanston, IL 60201 - U.S.A.
C.A. WARD, University of Missouri, Materials Research, Rolla - Missouri 65401 - U.S.A.
J. WEAVER, University of Wisconsin-Madison, Physics Science Laboratory, Stoughton - Wisconsin 53589 - U.S.A.
S.P. WEEKS, University of Pennsylvania, Department of Physics, Philadelphia - Pennsylvania 19175 - U.S.A.
B. WOOD, 333 Ravenswood Avenue, Menlo Park - California - U.S.A.
P. WOOD, University of Bradford, Department of Chemistry - Bradford Yorkshire BD7 1DP - England

SUBJECT INDEX